OXFORI ... ENCE SERIES

OXFORD STATISTICAL SCIENCE SERIES

1. A. C. Atkinson: *Plots, transformations, and regression*
2. M. Stone: *Coordinate-free multivariable statistics*
3. W. J. Krzanowski: *Principles of multivariate analysis: a user's perspective*
4. M. Aitkin, D. Anderson, B. Francis, and J. Hinde: *Statistical modelling in GLIM*
5. Peter J. Diggle: *Time series: a biostatistical introduction*
6. Howell Tong: *Non-linear time series: a dynamical system approach*
7. V. P. Godambe: *Estimating functions*
8. A. C. Atkinson and A. N. Donev: *Optimum experimental designs*
9. U. N. Bhat and I. V. Basawa: *Queuing and related models*
10. J. K. Lindsey: *Models for repeated measurements*
11. N. T. Longford: *Random coefficient models*
12. P. J. Brown: *Measurement, regression, and calibration*
13. Peter J. Diggle, Kung-Yee Liang, and Scott L. Zeger: *Analysis of longitudinal data*
14. J. I. Ansell and M. J. Phillips: *Practical methods for reliability data analysis*
15. J. K. Lindsey: *Modelling frequency and count data*
16. J. L. Jensen: *Saddlepoint approximations*
17. Steffen L. Lauritzen: *Graphical models*
18. A. W. Bowman and A. Azzalini: *Applied smoothing methods for data analysis*
19. J. K. Lindsey: *Models for repeated measurements* Second edition
20. Michael Evans and Tim Swartz: *Approximating integrals via Monte Carlo and deterministic methods*
21. D. F. Andrews and J. E. Stafford: *Symbolic computation for statistical inference*
22. W. J. Krzanowski: *Principles of multivariate analysis: a user's perspective* Revised edition

Principles of Multivariate Analysis

A User's Perspective

Revised Edition

W. J. Krzanowski

School of Mathematical Sciences
Exeter University

OXFORD
UNIVERSITY PRESS

Great Clarendon Street, Oxford OX2 6DP

Oxford University Press is a department of the University of Oxford.
It furthers the University's objective of excellence in research, scholarship,
and education by publishing worldwide in

Oxford New York

Athens Auckland Bangkok Bogotá Buenos Aires Calcutta
Cape Town Chennai Dar es Salaam Delhi Florence Hong Kong Istanbul
Karachi Kuala Lumpur Madrid Melbourne Mexico City Mumbai
Nairobi Paris São Paulo Shanghai Singapore Taipei Tokyo Toronto Warsaw
with associated companies in Berlin Ibadan

Oxford is a registered trade mark of Oxford University Press
in the UK and in certain other countries

Published in the United States
by Oxford University Press Inc., New York

First published 1988
Revised edition 2000

A catalogue record for this book is available from the British Library

Library of Congress Cataloging in Publication Data
(Data available)

ISBN 0 19 850708 9

Typeset by Newgen Imaging Systems (P) Ltd, Chennai, India
Printed in Great Britain
on acid-free paper by
Bookcraft Ltd., Midsomer Norton, Avon

Preface to revised edition

This edition includes a new appendix, which traces developments that have taken place in the years since publication of the first edition and which clarifies some issues raised by readers of the text. I will gratefully receive any further comments or suggestions for improvement.

School of Mathematical Sciences W. J. K
University of Exeter

Preface

This book is intended as a relatively gentle but reasonably comprehensive introduction to the principles and techniques of practical multivariate analysis. It is aimed at the user or potential user of multivariate methods, whether this be the researcher who has collected multivariate data which need analysis, the statistician who is not well versed in multivariate methodology but who is faced with the need to analyse multivariate data, or the student who is encountering multivariate ideas for the first time. To cater best for such a broad and predominantly applied intended readership, the book has what I believe to be a number of distinctive features: it is organized according to a problem-orientated format, in contrast to the traditional technique-orientated format of most other multivariate texts; geometrical reasoning is presented where possible in preference to algebraic manipulation; and mathematical passages that are not essential for practical understanding are highlighted, so that they may be omitted without interrupting the thread of ideas. My hope is that these devices will help to make this exposition of multivariate methodology more readily accessible to those users who at present have to struggle with the techniques, either blindly applying existing computer programs to their data or attempting to understand technical accounts pitched at a very mathematical level.

My approach to multivariate analysis has inevitably been moulded by the environment in which I have learned and practised it. After a traditional mathematical education, I was indeed fortunate to have had my interest in multivariate methods kindled by working with John Gower at Rothamsted Experimental Station, where there existed a line of multivariate work dating back to J. Wishart in the 1920s. To this foundation has been added the firm applied approach of my colleagues at the University of Reading, which has ensured that practical objectives always take precedence over mathematical niceties. The experience of advising researchers and teaching students that I have gained at these institutions has greatly influenced the style and content of this book.

On a professional level, I much enjoyed involvement in a series of one-week courses on multivariate analysis organized by the Institute of Statisticians between 1979 and 1983, and learned a lot from the lectures and practical sessions conducted by my colleagues on these courses. I am grateful to them all in general, and in particular to Colin Chalmers and Dave Collett who will recognize some of their ideas in the following pages. I would also like to thank Robert Curnow for the Cystic Fibrosis data analysed in Chapters 7 and 12, Tank Waddington for the preference scaling example in Chapter 4, John Kent for providing comments on the manuscript, and all authors who willingly gave permission for me to reproduce tables and figures from their publications. Relevant details and acknowledgements to publishers are included in the text as appropriate.

On a personal level, I would like to acknowledge warmly the help of a special quartette of ladies, without whom this book would never have seen the light of day. The germ of the idea for it was planted by Agnes Herzberg, who by supplying me with a constant stream of books for review in the *ISI Short Book Reviews* made me aware of a potential need for a book aimed at the multivariate 'user'. If the book would never have been started without this impetus, it would certainly never have been produced without the expert assistance of Rosemary Stern, who did all the artwork beautifully, and Lynne Rogers who not only typed a fiendish manuscript superbly but also managed to preserve her equanimity at all times. Above all, the book could not have been finished without the constant support of my wife Wendy, who never failed to bolster my flagging morale in spite of much disruption to domestic routine. To them all I am most grateful, as I will indeed be to any reader who either draws my attention to errors in the book or suggests any improvements that can be made to it.

Department of Applied Statistics W. J. K
University of Reading
February 1987

Preface to paperback edition

For this new edition, I have taken the opportunity to correct a number of typographical and other small errors. I would like to acknowledge gratefully the help of Norman Geary in spotting many of these lapses.

Department of Applied Statistics W. J. K.
University of Reading
March 1990

To Wendy, Adam, and Helen

Contents

Part III: Analysing ungrouped data

Part IV: Analysing grouped data

Introduction

At one time, the idea of measurement was associated almost entirely with scientific pursuit, by which was meant a preoccupation with the physical sciences. These admit very strict and exactly reproducible laws, the determination of which requires very precise measuring instruments. Much effort was therefore expended in refining instrumentation to achieve the required degree of precision, to enable such laws to be detected. Scientists were nevertheless aware that, no matter how precise the instruments that were used, some error of measurement must inevitably ensue, and hence there was a need to allow for such error in scientific calculation. From these beginnings, via the work of such pioneers as Gauss and Laplace on theories of errors, developed the subject of Statistics. With greater knowledge about the likely behaviour of errors, scientists became more confident in their ability to separate 'signal' from 'noise' (to use present-day terminology). The process of measurement and quantification thus spread into less exact sciences, gradually permeating through the Biological and then the Social Sciences. Hand in hand with this developing confidence went a greater thirst for numerical information of all kinds, together with the increasing technology to cope with processing of this information. We now live in an age obsessed by numbers, and are bombarded with numerical and statistical information from all sides. Statistical methods are thus required in all manner of disciplines, in some of which (such as History, Languages and Linguistics, Classics, etc.) the idea of quantitative study would have been unthinkable not many years ago.

A consequence of this development is that the research worker in nearly any discipline finds himself or herself having to collect numerical data on the individuals of some population that is under study. The objectives of the research may prescribe that certain specified measurements have to be made on each of these individuals. More frequently, the objectives are less precise or there exist various possible offshoots of the project, and the research worker will want to measure as many aspects of each individual as is possible within the time and budget available for the work. In all such cases, the end product of the data collection is a *multivariate* set of data, in which a number of individuals (n, say) each carries the value of more than one measurement. In general, we can envisage p such measurements. The researcher then has to make sense of this mass of numbers, and to try and meet the objectives of the research by selecting an appropriate method of analysis. It is with such problems that the subject of multivariate analysis is concerned.

The origins of multivariate analysis are to be found in work of mathematicians of the last century, who developed the linear algebra and multidimensional geometry that form the basis of many multivariate methods. It is an unfortunate feature of the subject, however, that for the first fifty years or so its development was almost entirely mathematical. The heavy computational demands that these techniques make, rendered their application to any but the smallest data sets virtually impossible with hand calculation only. Thus the techniques were inaccessible to those workers who lacked a strong mathematical background. The last twenty years, however, have seen an explosion in computer development. Virtually unlimited computing power is now, if not actual reality, a fairly immediate prospect. Sophisticated computer software is readily available, and the research worker in any discipline has immediate access to many multivariate techniques via such readily available general statistical package programs as BMDP, GENSTAT, SAS, and SPSS, not to mention special-purpose suites such as CLUSTAN or MDS(X). Computation is now no object, but to use such packages correctly and efficiently, the worker must have a good *understanding* of the various techniques that are available. It is only with such understanding that the researcher will be able to select the appropriate technique for the data and problem in hand, and will then be able to understand and interpret correctly the output from the computer.

While there exist currently many fine textbooks on multivariate analysis, it is my belief that the needs of the average research worker, as described above, are not adequately catered for. Most textbooks are still written with the mathematician or theoretical statistician in mind, and demand a high level of mathematical ability from the reader. Those books that do attempt to avoid such mathematical dependence, moreover, swing to the other extreme and present a rather shallow primer of computing techniques. More seriously, perhaps, nearly all books written on the subject are *technique-orientated*. This means that they tend to be divided up according to multivariate techniques, and within each chapter or section a technique is presented, its fundamental mathematics and properties are worked through, and some examples of its use are given. The research worker, however, is faced with a *problem*, and would like an outline of which techniques may be used to solve this problem together with sufficient detail about each technique to enable him/her to decide which one is most appropriate and how it may be used. This suggests that a *problem-orientated* approach would be the most useful one for the general user of multivariate methods.

The aim of this book, therefore, is to present a problem-orientated discussion of multivariate analysis that has the user firmly in mind. While this suggests that it is the non-mathematician who is the primary target, it is also hoped that this text will provide a good introduction to the subject for the student of Statistics or for the practising Statistician. Once the basic ideas have been firmly grasped, then reference can be made to any of the more mathematical texts to fill in the details. Clearly, however, the inclusion of a certain amount of mathematics is inescapable, as many of the multivariate methods to be discussed are merely restatements in a specific context of results from linear algebra. Two devices have been adopted to try and make this side of things more palatable. As a general philosophy, algebraic manipulation has been replaced wherever possible by geometric reasoning. It is often forgotten that there is a close correspondence between many aspects of linear algebra and those of coordinate geometry, and those scientific workers who would be totally dismayed by a page of algebraic manipulation may be entirely at ease with a few geometric concepts which achieve the same results. The first chapter of the book therefore develops some elementary geometric ideas and their algebraic equivalents, as a prelude to this pictorial approach which will thread its way through the remainder of the book. Of course algebra cannot be eliminated entirely, but will be kept to a minimum. The second device for trying to ensure readability of the text as a whole is that those passages that present interesting but non-essential pieces of mathematics will be displayed in smaller print, and may be omitted either at first reading or even entirely by less mathematical readers. Some elementary algebraic ideas and manipulations are also given in Appendix A to enable any gaps to be filled by such readers.

Having established the aims of a text, it is customary to attempt a brief summary of its scope. In keeping with the proposed problem-orientated approach, however, it may be more appropriate at this stage simply to list some of the more common problems encountered in a range of disciplines by workers who have obtained a multivariate data set, thereby demonstrating the possibilities afforded by the techniques that will be described in the following pages. In *Agriculture*, we may wish to classify soil types, to develop new crop varieties, or to determine those combinations of crop characteristics that show largest differences in response to various fertilizer treatments. In *Anthropology* it may be of interest to distinguish between different hominid populations, or to identify to which population some skull remains belong. In *Education*, we may wish to explain the variation among students' academic abilities, or to relate pupils' domestic situation to their academic performance. In *Industry*, effort may be channelled into relating production costs to costs of materials, or company performance to the economic environment. Alternatively we may be interested in determining the factors that lead to optimum performance of some manufacturing process. In *Marketing*, common requirements are to classify consumers, to relate new products to established products, or to explain the differences between various products on the market. *Medicine* has increasingly looked to statistical support, particularly in diagnosis problems. Here interest may focus on methods of distinguishing diseased from healthy patients, or on methods of identifying the complaint of a patient. *Psychology* is an area in which multivariate analysis has a long tradition, and typical problems might be the measurement of consistency in judges' sensory assessments, or the determination of factors affecting children's cognitive ability. In *Sociology*, we might wish to classify respondents to a questionnaire, to display similarities between tribal languages, or to determine the factors affecting the behaviour of delinquents. As a final example, *Taxonomy* is an area which encompasses most of the Biological Sciences. Here we may be interested in distinguishing different populations of organisms, or in classifying species of flora.

The final task of an introduction is to draw together the common strands linking the problems in such diverse areas, to describe the underlying structure of data that may be collected, and to establish some fundamental terminology for use throughout the rest of the book. The common strand is, of course, that in each of the above problems the investigator has made more than one measurement on, or noted more than one attribute of, each of a number of experimental units. These units may be soil samples (Agriculture); skulls (Anthropology); students (Education); industrial plants (Industry); products (Marketing); patients (Medicine); judges (Psychology); tribal languages (Sociology); or nematodes (Taxonomy). Typical measurements or observations that might be made on each of these units are as follows: pH of soil, colour of soil, texture of soil, number of stones in sample (Agriculture); lengths of various bones or parts of skull (Anthropology); marks obtained by each student in a range of examinations, age, presence or absence of various socio-economic indicators (Education); amounts of various materials used at each plant, temperature of production process, number of staff employed, rates or rents charged for premises (Industry); retail price of product, percentage share of market, number of direct competitors, average rating on ten-point scale of quality as obtained from sample questionnaire (Marketing); laboratory measurements such as blood sugar content made on each patient, physiological responses such as heart rate and blood pressure, presence or absence of various symptoms as judged by the physician (Medicine); and so on.

Various terms are used more or less synonymously in the literature for the entities in a multivariate data set, and it is as well to establish such terminology right from the outset. The experimental (or observational) *units* (examples of which were given in the last paragraph) may also be referred to quite generally as *individuals*, or more specifically as *subjects* (particularly in Psychological, Sociological, or experimental contexts) or *plots* (in Agriculture and allied areas). The measurements or observations made on each unit are collectively known as *variates* or *variables*, while if a distinction is to be made between measurement and observation then *responses* usually denote the former and *attributes* the latter.

The common structure in all the problems outlined above, therefore, is that more than one variable is observed on each individual. There is usually no connection between these individuals, and the statistical principle of random sampling will ensure that different units are at least uncorrelated if not totally independent. Since each variable is observed on every unit, however, there will consequently be *correlations* existing between the variables and it is the existence of these correlations that is central to multivariate analysis. Whether such correlations must be allowed for in the interpretation of a data set, or exploited in achieving the objectives of the analysis, *some* action is necessary and it is with this action that multivariate methods are primarily concerned.

Given this common structure in the problems that were outlined earlier, it is clear that every multivariate data set has a common form. This form can be exhibited most simply as a *data matrix*, whose rows refer to the units in the sample and whose columns refer to the variables measured on each unit. Any multivariate data set can therefore be denoted symbolically by the following data matrix:

$$\mathbf{X} = \begin{pmatrix} x_{11} & x_{12} & \dots & x_{1p} \\ x_{21} & x_{22} & \dots & x_{2p} \\ \vdots & \vdots & & \vdots \\ x_{n1} & x_{n2} & \dots & x_{np} \end{pmatrix}$$

Here x_{ij} denotes the observation corresponding to the jth variable as measured on the ith individual. If p variables are observed on each of n individuals then the matrix will have n rows and p columns. It may happen that for some reason a particular single observation is not available. Thus, for example, one of the skulls in the Anthropological example may be damaged, and some of the bones may be missing and hence unmeasurable. Alternatively, one of the respondents in the Marketing questionnaire may have elected not to answer some of the questions. In such cases there are *missing values* in the sample, and hence there will be gaps in the corresponding data matrix. Special methods for handling such incomplete data matrices will be discussed where appropriate.

While the data matrix forms a unique entity common to all the problems detailed earlier, it can be recognized that a grouping of these problems is possible, and that each group of problems corresponds to a further imposition of structure on the data matrix. In some of the problems there is no a priori categorization of either units or variables in the sample, and all rows of the data matrix are treated on an equal footing as are all the columns. Such circumstances obtain, for example, in the Educational problem of explaining the variation among students' academic abilities, or the Psychological problem of determining the factors affecting children's cognitive abilities. In both cases we wish to explore *all* variables and *all* individuals to uncover some underlying structure, without imposing any external constraints. Consider, however, either the Anthropological example of distinguishing between different hominid populations, or the Medical problem of distinguishing between diseased and healthy patients. Here we have the additional information that the individuals come from each of a number of *groups*, and this information can be incorporated by a corresponding grouping of the rows of the data matrix. The objectives of the analysis will therefore consist of trying to find some feature in the variables, i.e. the columns of the matrix, which will show up this grouping of the rows as advantageously as possible. An alternative structure underlies, for example, the Educational problem of relating pupils' domestic situation to their academic performance, or the Industrial problem of relating production costs to costs of materials. Here it is the variables that are put into groups (e.g. production costs in one group, costs of materials in the other, where both groups are present in each Industrial plant) so that the columns of the data matrix may be partitioned correspondingly. The objectives of the analysis now consist of trying to find some feature in the rows of the matrix which shows up this grouping of the columns as advantageously possible.

The structure of this book reflects these possible a priori patterns of the data matrix. In Part I, consisting of Chapters 1 to 5, we first examine possible methods of multivariate data display. The ability to *look* at the data often goes a long way towards answering many of the questions that have been posed, so methods of displaying complex sets of data form a vital first stage in any systematic analysis. Part II is a link between simple data display and formal statistical analysis, encompassing a discussion of the ideas of a model and an outline of the chief probabilistic tools for modelling multivariate data. The remaining three parts then treat techniques of multivariate analysis appropriate to each of the three data structures described above: Part III is concerned with the case where no a priori structure is imposed on the data matrix, Part IV with the case where the rows of the data matrix are grouped, and Part V with the case where the columns are grouped. This last part also deals much more generally with techniques aimed at exploring relationships among the columns of the data matrix. In these five parts it is hoped to raise most of the common problems that multivariate data sets pose for the research worker, and to describe appropriate methods for dealing with these problems.

Part I: Looking at multivariate data

1
Motivation and fundamental concepts

An experimenter, or statistician, faced with a multivariate data set will often feel overwhelmed by the sheer bulk of numbers it contains. An archaeologist will not think it excessive to measure twenty different variables on each skeleton that he or she inspects, neither will a physician think it unusual to obtain twenty different indicants (e.g. signs, symptoms, laboratory measurements) from each patient presented for diagnosis. Yet with only fifty such skeletons, or fifty such patients, both scientists will be confronted with 1000 numbers for inspection. At this stage, a second problem becomes apparent, namely that these are not 1000 separate (unrelated) numbers but that in examining them some allowance must be made for the fact that relationships exist between the twenty variables or indicants. These relationships (possibly complex ones) arise as a consequence of *all* variables being measured on *each* individual observed by the scientist, and this feature is indeed emphasized by writing the 1000 numbers in the form of a 50×20 data matrix. A scientist's natural inclination when presented with a set of numbers is to scan them in the hope of detecting some interesting features or patterns in the data. The above discussion implies that any single number in the data matrix must be judged in relation to all the other numbers in the same row as well as the same column of the matrix, so that a simple visual inspection of the data matrix is unlikely to show up immediately any patterns that may exist in the numbers. The problems are exacerbated the more variables or individuals that are measured. Consequently, formal procedures are needed to help the scientist or statistician in this hunt for pattern, and it is with such procedures that we shall concern ourselves in this first part of the book.

In elementary statistical work, it is generally true that a graphical presentation of data is more likely to provide an immediate impression of pattern than a tabular presentation of the same data. Indeed, with even quite small sets of data, a graph may show up quite striking features which remain hidden when the numbers themselves are presented. Consider the hypothetical data, representing measurements made on two variables x and y for eight individuals, that are presented in Table 1.1.

Table 1.1 Measurements made on two variables for each of eight units

x	y
1.1	1.9
1.2	0.7
1.5	1.8
1.7	1.2
0.8	1.6
1.9	1.0
1.3	2.2
1.3	1.0

No pattern is directly evident in such a set of numbers, yet when these values are plotted in a scatter diagram (Fig. 1.1), it is immediately clear that the points representing the even rows of Table 1.1 (the circles) are distinctly separated from the points representing the odd rows (the crosses).

Detection of such pattern may often be an end in itself, while in other situations it may provide an impetus for further research on the part of the scientist or for further analysis on the part of the statistician. Whatever the consequences, however, it is clear that the pattern will be evinced from the data only if an adequate method of display is employed.

Many different methods exist for displaying multivariate data, and the best method to employ in a particular case will obviously depend upon a variety of factors and circumstances relating to the data in hand. However, perhaps the main factor which distinguishes between classes of available techniques is the type of data to be represented.

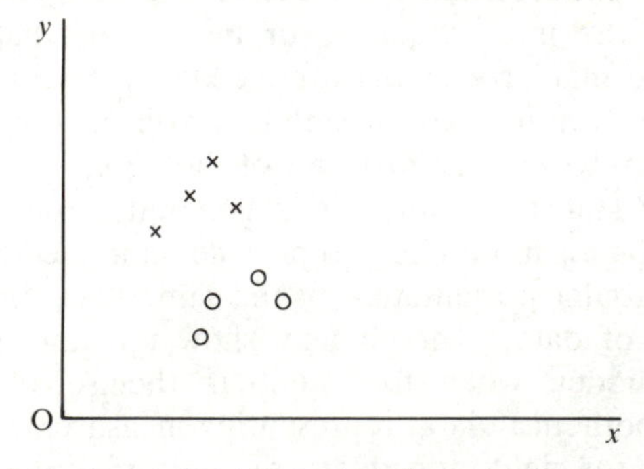

Fig. 1.1 Data of Table 1 plotted as a scatter diagram

1.1 Types of data

Broadly speaking, there are two main categories of data encountered in practice. Variates such as length, weight, number of leaves, or pressure can be measured quantitatively on an interval scale. They assign a unique numerical value to each individual observed, and so are called *quantitative* variates. Variates such as colour, shape, texture, taste or smell, on the other hand, provide a classification of objects into categories which describe the qualities possessed by each object. They are therefore called *categorical* or *qualitative* variates. Each type of variate can be further refined, to provide a fuller description of the amount of information that it contains. Thus quantitative variates can be either *discrete* or *continuous*. In the former case the set of values that the variate can assume is either finite or countably infinite and the values themselves are usually integers. Examples are the number of leaves on each of a set of trees, the number of insects succumbing to a given dose of insecticide, and the number of days that elapse before a patient recovers from a given disease after a particular treatment. A continuous variable, on the other hand, is one with an uncountably infinite set of possible values, these usually being all values contained in a closed interval of the real line. The only restriction imposed is by the accuracy of the measuring instrument. Thus the height of an adult person can take any value between (say) 1 metre and 2.5 metres, the weight of precipitate in a chemical experiment can take any value between (say) 0 and 10 milligrams, and so on. Turning next to qualitative variables, ones that provide a simple categorization of the objects are said to be measured on a *nominal* scale. This would be the case, for example, with the categories: round, square, rectangular, or oval for the variate, 'shape'. If there is an implied ranking or ordering, however, then the variate is said to be on an *ordinal* scale. An example of this would be the categorization: absent, mild, moderate, severe for the variate, 'level of pain', associated with a given medical condition. Qualitative variates are often recorded numerically by some coding system, to facilitate subsequent computer processing. Such numerical coding is usually purely arbitrary in the case of a nominal variate, but may be more meaningful when applied to an ordinal variate. If a qualitative variate has just two possible states, e.g. yes/no, present/absent, off/on, then it is referred to as a *dichotomous* or *binary* variate. A common coding in this case is zero/one.

The data presented in Table 1.2 provide an illustration of the various classifications described above. The table is taken from a larger set of data obtained during an investigation into psycho-social influences on breast cancer in women. The variables measured, and their categorizations, are as follows: x_1 is a binary variable giving the menopausal status

Table 1.2 Psycho-social influences on incidence of breast cancer; example of a data set

x_1	x_2	x_3	x_4	x_5	x_6	x_7	x_8	x_9	x_{10}	x_{11}	x_{12}	x_{13}
1	49	0	2	3	7	0	6	2	15	1	1	52.46
0	59	0	2	2	6	3	7	2	17	1	1	60.83
0	49	2	0	6	8	4	9	0	11	1	0	48.70
0	58	0	2	1	1	1	7	3	14	1	1	57.33
1	49	0	2	1	0	0	3	0	13	0	1	64.26
1	43	0	2	0	2	0	4	0	11	1	0	53.90
0	46	1	1	4	7	0	8	1	16	1	1	52.45
0	59	0	2	7	6	3	4	0	14	1	1	66.18
0	53	0	1	4	8	0	1	2	14	1	0	49.54
0	51	1	0	4	3	0	7	1	11	1	1	53.88
0	50	2	0	10	0	0	3	1	21	0	0	58.46
1	50	1	1	3	8	4	2	0	16	1	1	50.22

of the subject (0 = pre-menopause, 1 = post-menopause); x_2 is the age of the subject in years, a discrete quantitative variable; x_3 and x_4 are both three-state nominal variables—arbitrarily coded 0, 1, or 2—giving the subject's classification for the attributes labelled 'temper' and 'feelings'; x_5 to x_9 are all ordinal qualitative variables, being assessors' ratings for the subject on a 0–10 scale of the psychosocial factors 'acting out hostility', 'criticism of others', 'paranoid hostility', 'self-criticism', and 'guilt'; x_{10} is the age in years at menarche of the subject, another discrete quantitative variable; x_{11} and x_{12} are binary variables, denoting presence or absence of allergy and thyroid respectively; while x_{13} is the subject's weight in kilograms, a continuous quantitative variable.

1.2 Towards a pictorial representation

It has been suggested above that a graphical or pictorial representation will help the human eye to pick out any inherent pattern in a multivariate data set, smoothing out the 'noise' in the system in the process, and that different types of data may need different types of representation to achieve this. How might we set about producing such a pictorial representation? In this section we introduce some general principles in fairly abstract fashion; particular methodology and concrete examples will follow in Chapters 2 to 5.

Any multivariate data matrix contains information on two distinct entities, the rows and the columns. These correspond, respectively, to the units sampled or observed, and to the variables measured on these units. Before attempting to represent the matrix pictorially, it should therefore be asked whether the patterns to be investigated are those between units, those between variables, or those encompassing both units and variables simultaneously. This last case can be viewed most naturally as some type of combination of the previous two, and its discussion should logically follow theirs. For the present, therefore, let us assume that we wish to concentrate on either the units or on the variables as the focus for the representation.

If we wish to highlight any pattern existing between the units, then each unit must be represented individually by some geometrical or pictorial construction. This might be a point, a curve, or a connected shape (such as a circle or rectangle) totally enclosing a part of space; these constructions correspond in geometry to bodies in zero, one, or two dimensions respectively. Patterns between the units will then be readily highlighted by the similarities or dissimilarities between the corresponding constructions, as given either by visual similarity between solid bodies or by closeness of points or lines to each other. Exactly the same principles apply to the representation of variables in a system, except that it will now be each variable rather than each unit that will be represented by the point, curve, or body.

Any pictorial representation is constrained by the fact that it has to be made on a sheet of paper, and hence for all practical purposes it has to be two-dimensional. There are some exceptions to this, of course. It is perfectly possible to extend the concepts of drawing geometrical constructions on a sheet of paper to the making of three-dimensional geometrical models. Also, modern advances in computer graphics have led to the possibility of three-dimensional screen displays. However, the overwhelming demand for multivariate pictorial methods is still for the traditional two-dimensional representation. Of the various methods currently available for achieving such representation, three main types can be distinguished. One may be called the *direct* two-dimensional approach. This approach is a simple generalization of such familiar univariate display techniques as the pie chart, bar chart and graph. It involves merely choosing an appropriate curve or solid to represent each unit or variable, and then plotting or drawing these constructions directly in two dimensions. Any axes used in such a graph often have no direct relevance to the problem at hand, and such displays are most often appropriate when the units or variables are represented by (one-dimensional) lines or curves, or (two-dimensional) solids. If the representation is of (zero-dimensional) points, then one of the two remaining approaches is usually more appropriate.

The first of these approaches may be called the *projection* approach,

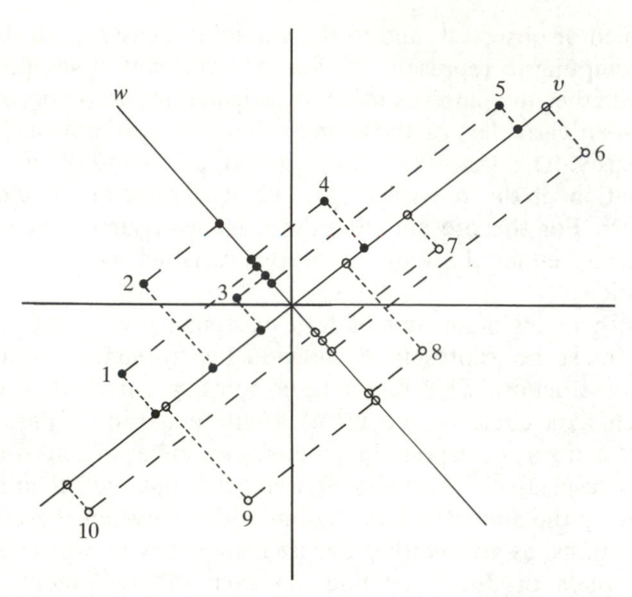

Fig. 1.2 Points labelled 1 to 10 projected on to two lines labelled v and w

and involves rather more geometric reasoning than the direct approach. Here it is tacitly assumed that there exists a 'true' representation of the multivariate data, but this 'true' representation is too complicated to be observed directly. Instead we can obtain an approximation to it by 'projecting' the true configuration into two suitably chosen dimensions. In this case, interpretation of the axes in these two dimensions is often also of direct relevance to the problem. The idea of projection is illustrated in Fig. 1.2.

Suppose that the ten points plotted in Fig. 1.2 as a scatter diagram represent the true configuration of a multivariate data set in two dimensions, but that we wish to obtain an approximation to this configuration in one dimension (for plotting, say, along a single line). Different projections will achieve different approximations to the same configuration. If we wish to obtain the best approximation for displaying the total spread of the points, then we might project them on to direction v. This simply means that we place each point at the foot of the perpendicular from that point to the line v. On the other hand, if we wish to find the approximation which best exhibits the difference between the open circles and the solid circles, then projecting on to the line w will be more appropriate. The projections of the ten points on to both lines are shown at the feet of the respective perpendiculars in Fig. 1.2. Any approximation is a good one if the points all undergo small displacement when projected, and a poor one if some or all points undergo a large displacement. Thus projection on to v gives a much better approximation to the overall configuration than projection on to w in Fig. 1.2.

The reason why, in most cases, the 'true' representation of a multivariate data set cannot be observed directly is that it contains too many dimensions for assimilation by the human eye. Living in a three-dimensional world, we are perfectly accustomed to the ideas of points, distances, and angles in this world, and we readily handle a variety of geometrical ideas in everyday life. All these concepts can be formulated, via coordinate geometry, into algebraic equations for mathematical treatment. The fundamental rules governing the handling of these equations can then be applied to systems other than the ones representable simply as three-dimensional physical models, to provide a consistent geometry of such systems. In particular, there is no reason why one should be restricted to considering only systems containing just three dimensions. It is a perfectly natural mathematical extension to conceive of a system containing p dimensions (where p can take *any* integral value), and then to apply all the fundamental geometric rules to this system. Such a p-dimensional geometric model will no longer be directly observable by the human eye and will have no counterpart in the physical world, but will nevertheless be entirely consistent and easily handled mathematically. We shall see later that a perfectly reasonable model of either the units or the variables in a data matrix can be formulated by imagining these entities as points in a space usually comprising more than three dimensions. This 'true' picture can not be visualized directly, so an approximation to it has to be sought in two or three dimensions. This resultant approximation can then either be plotted on graph paper, or represented as a three-dimensional solid model, and hence inspected visually. The way in which this approximation is obtained is by choosing an appropriate two-dimensional 'slice' through the true p-dimensional space, and then projecting all the points in the space on to this chosen slice. Different objectives for representation will lead to different possible slices, as illustrated previously in Fig. 1.2. A variety of such choices may together give a good overall impression of the whole space, but more usually just one choice would be selected for any given problem.

In the projection approach, an objective must be specified before the appropriate slice can be determined. This objective is quantified by expressing it as a function of the points in the geometrical representation. The slice for projection of the points that is finally selected will, in general, be the one among all possible slices that optimizes this objective function. For example, if the aim is to find the two-dimensional slice for which the spread of all points projected on it is a maximum, we must first define some suitable function V that is a measure of the spread of all points in two dimensions. Such a function might, for example, be given by

$$V=\sum_{i=1}^{n}(x_{i1}-\bar{x}_1)^2+\sum_{i=1}^{n}(x_{i2}-\bar{x}_2)^2,$$

where (x_{i1}, x_{i2}) are the coordinates of the ith point in the two dimensions, and $\bar{x}_1=\frac{1}{n}\sum_{i=1}^{n}x_{i1}$, $\bar{x}_2=\frac{1}{n}\sum_{i=1}^{n}x_{i2}$ are the means of the two sets of coordinates. Given any particular slice, we can project all points on to it and calculate V. The best slice to choose is thus the one which yields maximum V.

The third general approach for obtaining a pictorial representation extends the projection approach in one of two ways. We either assume there exists a 'true' representation in a large number of dimensions to which we require an approximation (but in looking for this approximation we do not wish to be restricted by only considering orthogonal projection methods), or we assume that there is no single 'true' underlying representation but we wish to construct a representation which best approximates specified objectives. In either case we can specify some objective function V, and then attempt by trial and error (with the help of a computer) to construct a representation which optimizes V. Techniques based on this approach bear certain resemblances to, and in some cases lead to the same results as, projection techniques, but it will be convenient to treat them separately. Accordingly we will designate such an approach as the *optimization* approach.

Our aim in this first part of the book is to provide a description and discussion of the most useful methods of multivariate data display; for this purpose the categorization into the three general approaches *direct, projection,* and *optimization* will be adopted. Geometric ideas will clearly play a dominant role in many of the techniques to be discussed. It is thus worthwhile to establish first some fundamental results concerned with p-dimensional Euclidean spaces, and to demonstrate how these results are just natural extensions of familiar geometric ideas in two- or three-dimensional Euclidean space. Some of the concepts introduced here will be needed in many of the techniques described throughout the book, so that a fundamental understanding of them is valuable right from the outset. Often, also, less mathematical readers of multivariate texts are intimidated by seemingly complex algebraic expressions and manipulations, and do not realize that the ideas underlying these operations often carry a very simple and intuitive geometric basis. Such readers may thus miss valuable insights into the associated multivariate techniques, and hence fail to realize the true range of possible uses of these techniques. The following section will thus also highlight some of this correspondence between algebraic operations and geometric ideas, as an aid to understanding multivariate methods.

1.3 Some geometrical concepts

1.3.1 Two-dimensional basics

Consider the simple situation depicted in Fig. 1.3 where two points P and Q are plotted in a two-dimensional space. In order to be able to make such a plot, we need to specify two rectangular *axes,* and the *coordinates* of the two points referred to these axes. The points are then placed on the diagram at the intersection of lines constructed at each coordinate value for each axis and parallel to the other axis, as demonstrated in Fig. 1.3. In order to describe this situation mathematically we need to label the axes, and a convenient method of labelling is to denote the first (horizontal) axis by X_1 and the second (vertical) axis by X_2. Denoting the origin of the axes by O, we thus designate the lines OX_1 and OX_2 on the diagram as the axes. If we notionally think of P as the first point, then we can let its coordinates on the axes OX_1 and OX_2 be x_{11} and x_{12} respectively. The first suffix attached to x in each case thus gives the ordinal number of the *point* referred to, while the second suffix gives the ordinal number of the *axis* referred to. The points P and Q can then be referred to by the pairs (x_{11}, x_{12}) and (x_{21}, x_{22}), which give their coordinates relative to the chosen axes. The point $(0, 0)$ denotes the origin O. The *distance d* between P and Q is found by simple application of Pythagoras' theorem. By this theorem, $\mathrm{PQ}^2 = \mathrm{PA}^2 + \mathrm{AQ}^2$. Now since

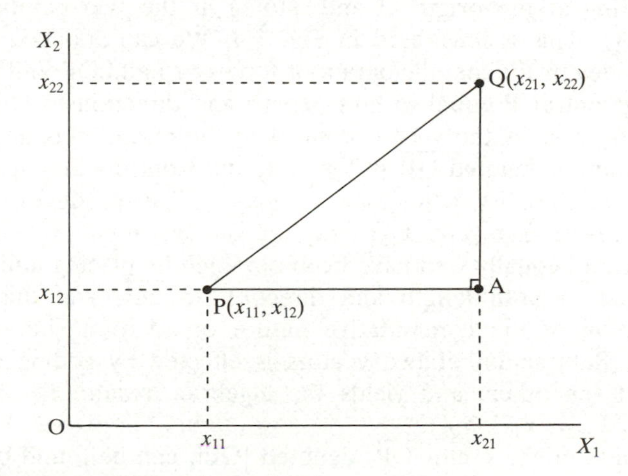

Fig. 1.3 Distance between two points

the X_2 coordinate values of P and A are both the same (i.e. x_{12}), then PA must be the difference in X_1 coordinate values of P and A, i.e. $x_{21}-x_{11}$. Similarly, since the X_1 coordinate values of A and Q are both x_{21}, then AQ must equal $x_{22}-x_{12}$.

Hence,

$$PQ^2=(x_{21}-x_{11})^2+(x_{22}-x_{12})^2.$$

Thus, using summation notation, we may write

$$d^2=\sum_{s=1}^{2}(x_{2s}-x_{1s})^2$$

or

$$d=\left\{\sum_{s=1}^{2}(x_{2s}-x_{1s})^2\right\}^{\frac{1}{2}}=\sqrt{\left\{\sum_{s=1}^{2}(x_{2s}-x_{1s})^2\right\}}.$$

By similar reasoning, if there were n points on the diagram then the distance between any two of them, say the ith and jth points, would be given by

$$d_{ij}=\left\{\sum_{s=1}^{2}(x_{is}-x_{js})^2\right\}^{\frac{1}{2}}.$$

There is a further consequence of the above formulation, relating to the definition of P and Q as (x_{11}, x_{12}) and (x_{21}, x_{22}). Mathematically, these two couples are known as two-dimensional *vectors* having *elements* x_{11}, x_{12} and x_{21}, x_{22} respectively. Vectors are quantities which possess both length and direction. Thus an alternative representation of these two couples is as two directed line segments $\overrightarrow{OP}$ and $\overrightarrow{OQ}$, starting at the origin O and ending at the two points P and Q respectively. This is illustrated in Fig. 1.4. We can add two vectors by using the idea of the 'parallelogram of forces': to add $\overrightarrow{OP}$ and $\overrightarrow{OQ}$, place a line segment at P equal in both length and direction to $\overrightarrow{OQ}$, and the result of the sum is the vector from O to the end of this line segment. This resultant is denoted $\overrightarrow{OR}$ in Fig. 1.4, and from the diagram it is clear that $\overrightarrow{OR}$ is given by $(x_{11}+x_{21},\ x_{12}+x_{22})$. This produces the simple algebraic result that $(x_{11},\ x_{12})+(x_{21},\ x_{22})=(x_{11}+x_{21},\ x_{12}+x_{22})$. Note that $\overrightarrow{OR}$ could equally well have been obtained by placing a line segment at Q equal in both length and direction to $\overrightarrow{OP}$, and this is just a manifestation of the commutative nature of addition (i.e. $x_{11}+x_{21}=x_{21}+x_{11}$). Subtraction of two vectors is effected by adding one to the reverse of the other, and yields the algebraic result $(x_{11},\ x_{12})-(x_{21},\ x_{22})=(x_{11}-x_{21},\ x_{12}-x_{22})$.

The *length* of the vector $\overrightarrow{OP}$, denoted |OP|, can be found by applying Pythagoras' theorem to the distance between O and P, and in the case of

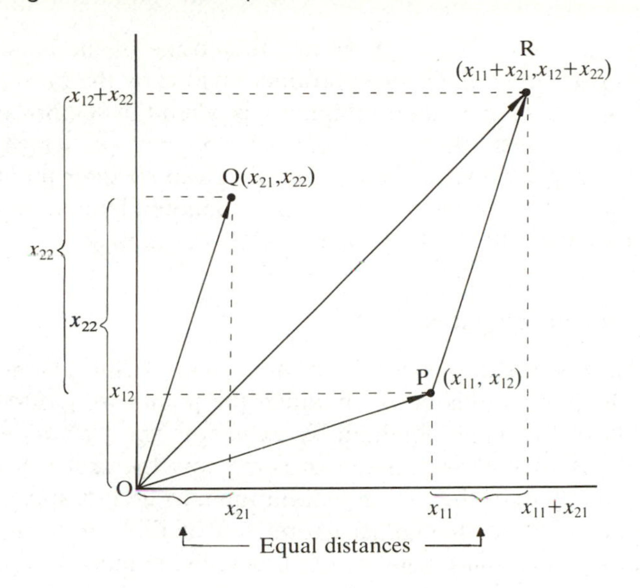

Fig. 1.4 Addition of vectors

$\overrightarrow{OP}$ defined above we find

$$|OP| = (x_{11}^2 + x_{12}^2)^{\frac{1}{2}} = \left(\sum_{s=1}^{2} x_{1s}^2\right)^{\frac{1}{2}}.$$

To find the *angle* between two vectors, we need to apply the cosine rule of elementary geometry. Consider the vectors $\overrightarrow{OP}$ and $\overrightarrow{OQ}$ in Fig. 1.4 and let α be the angle between them. By the cosine rule,

$$|QP|^2 = |OQ|^2 + |OP|^2 - 2\,|OQ| \cdot |OP| \cdot \cos \alpha.$$

Use of the foregoing expressions for distance between two points, and length of a vector, shows that

$$(x_{11} - x_{21})^2 + (x_{12} - x_{22})^2 = x_{11}^2 + x_{12}^2 + x_{21}^2 + x_{22}^2 - 2\,|OQ| \cdot |OP| \cdot \cos \alpha.$$

Expanding brackets and cancelling,

$$-2x_{11}x_{21} - 2x_{12}x_{22} = -2\,|OQ| \cdot |OP| \cdot \cos \alpha$$

so that

$$\cos \alpha = \frac{x_{11}x_{21} + x_{12}x_{22}}{|OQ|\,|OP|} = \frac{\sum_{s=1}^{2} x_{1s}x_{2s}}{|OQ|\,|OP|}.$$

The numerator of this expression is known as the *inner product* of the two vectors (x_{11}, x_{12}), (x_{21}, x_{22}). Hence the cosine of the angle between two vectors is given by their inner product divided by the product of their lengths. A special case of some simplicity is when the vectors are each of unit length, i.e. such that $x_{11}^2 + x_{12}^2 = x_{21}^2 + x_{22}^2 = 1$, in which case the cosine of the angle between them is simply given by their inner product. For simplicity, vectors will from now on be denoted by a lower-case bold letter, namely $\boldsymbol{x}_1 = \overrightarrow{OP} = (x_{11}, x_{12})$, $\boldsymbol{x}_2 = \overrightarrow{OQ} = (x_{21}, x_{22})$.

1.3.2 Lines and subspaces

The two-dimensional space we have been considering above is represented by the page of this book on which the points are plotted, and the continuation of this page obtained by extending the axes indefinitely in both directions. If now we restrict in some way the space in which the points may be plotted, then we constrain them to lie in a *subspace* of the original space. The mathematical expression of this constraint is to say that while the whole space consists of all possible points (x_1, x_2), i.e. such that $-\infty < x_1 < \infty$ and $-\infty < x_2 < \infty$, a subspace consists of those points (x_1, x_2) for which the ranges of x_1 and/or x_2 are restricted in some way. An example might be the set of points (x_1, x_2) for which $0 \leqslant x_1 < \infty$ and $0 \leqslant x_2 < \infty$, in which case we are restricted to the positive quadrant of the whole space, and the subspace can loosely be thought of as occupying a quarter of the original space. One noteworthy feature of this subspace is the fact that, despite the restriction on x_1 and x_2, it is still a two-dimensional space in its own right, and all geometric constructions in this subspace themselves obey the rules of two-dimensional geometry. This is because x_1 and x_2 can vary their values *independently* of each other in this subspace. A more fundamental restriction is imposed by introducing a *dependence* between x_1 and x_2. Thus if we create a subspace by allowing one of x_1, x_2 to take any value in the range $(-\infty, \infty)$, but then constrain the value of the other coordinate to be fixed, we have in fact created a subspace *of smaller dimensionality* than the original space. The dimensionality here is one, since only one coordinate is free to be varied at will. One possible constraint is given by the equation $ax_1 + bx_2 = c$, where a, b, and c are specified constants. One of x_1, x_2 may be chosen at will, but once it has been chosen the other coordinate is determined by the equation. This equation corresponds to a *line* in two dimensions, so the subspace consists of all points on this line. The most convenient form of equation for this line results when the above expression is divided by b, and terms are arranged to give the equation $x_2 = mx_1 + k$.

Lines prove to be important more generally in defining subspaces of larger dimension spaces, so are worth examining in a little more detail.

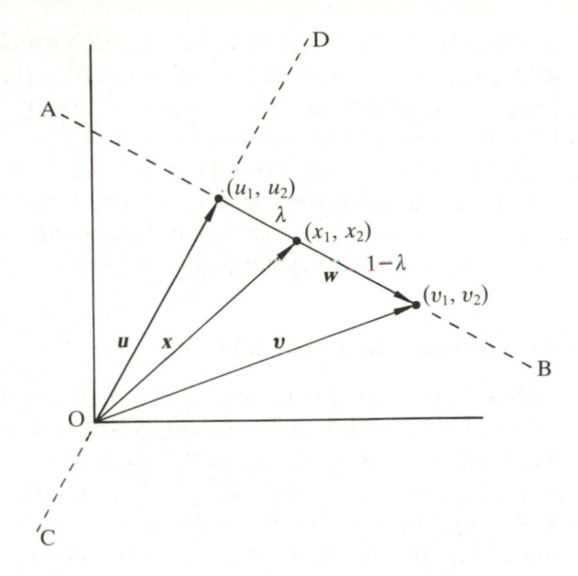

Fig. 1.5 Vector form for the equation of a straight line

To determine the equation of any line, we must find two points through which that line passes. Suppose (u_1, u_2) and (v_1, v_2) are two such points. Then the equation is obtained simply by replacing x_1, x_2 in $x_2 = mx_1 + k$ successively by u_1, u_2 and v_1, v_2, and then solving the resulting pair of simultaneous equations in m and k. However, it is more instructive to consider the vector form of equation. Figure 1.5 shows the line AB whose equation is to be found, passing through the points (u_1, u_2) and (v_1, v_2). These two points form the ends of two vectors emanating from O, denoted by $\boldsymbol{u}$ and $\boldsymbol{v}$ respectively. Let (x_1, x_2) be any point on AB; this forms the end point of a third vector $\boldsymbol{x}$. Now suppose that $\boldsymbol{w}$ denotes a vector starting at (u_1, u_2), lying along AB and finishing at (v_1, v_2). Then if (x_1, x_2) divides this vector in the ratio $\lambda : 1-\lambda$ for some value λ, it follows that the vector starting at (u_1, u_2) and ending at (x_1, x_2) will be $\lambda\boldsymbol{w}$, while the one starting at (x_1, x_2) and ending at (v_1, v_2) will be $(1-\lambda)\boldsymbol{w}$. From the addition of vectors, we thus have

$$\boldsymbol{x} = \boldsymbol{u} + \lambda\boldsymbol{w}$$

and

$$\boldsymbol{v} = \boldsymbol{x} + (1-\lambda)\boldsymbol{w}.$$

Eliminating $\boldsymbol{w}$ between the two equations yields $\boldsymbol{x} = \lambda\boldsymbol{v} + (1-\lambda)\boldsymbol{u}$.

It can be shown that the same expression results whatever the position of $\boldsymbol{x}$ along AB, and hence allowing λ to take any real value yields the equation of the line AB. Each different value of λ determines a unique vector $\boldsymbol{x}$ and vice versa; we therefore call λ a *parameter* in this equation.

A useful special case concerns lines passing through the origin. Suppose we want the equation of the dotted line CD in Fig. 1.5. This line passes through $(0, 0)$ and (u_1, u_2), so taking $\boldsymbol{v} = (0, 0)$ in the above general form we find the equation to be $\boldsymbol{x} = \lambda \boldsymbol{u}$. Hence a useful additional property of the vector form of a point is that it defines the equation of a line passing through the origin and that point.

1.3.3 Extensions to higher dimensionality

The above concepts have been presented in terms of two-dimensional pictures, in which the various rules can be readily verified. If we now wish to construct a three-dimensional representation, we simply need to add to the existing representation a third axis OX_3 which is perpendicular to both OX_1 and OX_2, and then to attach coordinate values along this third dimension to any points in the representation. Thus points P and Q of Fig. 1.3 would now have coordinates (x_{11}, x_{12}, x_{13}) and (x_{21}, x_{22}, x_{23}), and these triples could equally well be used to denote vectors (or lines) emanating from the origin O and ending at P and Q respectively. By simple extension of the earlier rules we can deduce that the distance d between P and Q is given by $d^2 = (x_{21} - x_{11})^2 + (x_{22} - x_{12})^2 + (x_{23} - x_{13})^2$, *i.e.*

$$d = \left\{ \sum_{s=1}^{3} (x_{2s} - x_{1s})^2 \right\}^{\frac{1}{2}};$$

the angle α between $\overrightarrow{OP}$ and $\overrightarrow{OQ}$ is given by

$$\cos \alpha = \frac{x_{11}x_{21} + x_{12}x_{22} + x_{13}x_{23}}{|OP|\,|OQ|} = \frac{\sum_{s=1}^{3} x_{1s}x_{2s}}{|OP|\,|OQ|}$$

where $|OP|^2 = \sum_{s=1}^{3} x_{1s}^2$ etc.; and the resultant $\overrightarrow{OR} = \overrightarrow{OP} + \overrightarrow{OQ}$ is given by $(x_{11} + x_{21},\ x_{12} + x_{22},\ x_{13} + x_{23})$. In a three-dimensional space, the possibility exists for considering one-, two-, or three-dimensional subspaces. One-dimensional subspaces are lines, as before, and the vector form of the equation of a straight line holds as good in three dimensions as in two. Thus the equation of the line passing through $\boldsymbol{u}$ and $\boldsymbol{v}$ is still $\boldsymbol{x} = \lambda \boldsymbol{v} + (1 - \lambda)\boldsymbol{u}$, but with each vector now having three coordinates. Similarly, the line passing through O and $\boldsymbol{u}$ has equation $\boldsymbol{x} = \lambda \boldsymbol{u}$. Two-dimensional subspaces are planes. The scalar equation of a plane is $x_1 + ax_2 + bx_3 = c$, but it is perhaps most conveniently defined by a pair of lines at right angles to each other. If these lines (and hence the plane itself) include the origin O, then we can verify that they are at right angles by choosing one vector on each and making sure that the cosine of the angle between these vectors is zero. Thus if $\overrightarrow{OP}$ and $\overrightarrow{OQ}$ above lie on the two lines respectively, a sufficient condition for them to be perpendicular to each other is that $\sum_{s=1}^{3} x_{1s}x_{2s} = 0$.

We can now extend all these concepts and rules to the case where each point has an *arbitrary* number of coordinate values, associating each coordinate with one of an arbitrary number of rectangular axes. If we specify p such axes and coordinates, and use a generalization of all the above rules in two or three dimensions, then we are led to what mathematicians call *p-dimensional Euclidean space*. This space is entirely conceptual for p greater than 3, and the various results cannot be physically verified. Nevertheless it is still a space in which all our familiar ideas of points, lines, planes, distances, etc., hold good. Thus points P and Q would have coordinates $(x_{11}, x_{12}, \ldots, x_{1p})$ and $(x_{21}, x_{22}, \ldots, x_{2p})$ respectively, and these p-tuples equally define two vectors $\boldsymbol{x}_1$, $\boldsymbol{x}_2$ emanating from the origin O and terminating at P and Q respectively. The lengths $|\boldsymbol{x}_1|$ and $|\boldsymbol{x}_2|$ of these two vectors are $\left(\sum_{s=1}^{p} x_{1s}^2\right)^{\frac{1}{2}}$ and $\left(\sum_{s=1}^{p} x_{2s}^2\right)^{\frac{1}{2}}$ respectively, and the angle α between them is given by

$$\cos\alpha = \frac{\sum_{s=1}^{p} x_{1s}x_{2s}}{\left(\sum_{s=1}^{p} x_{1s}^2 \cdot \sum_{s=1}^{p} x_{2s}^2\right)^{\frac{1}{2}}}.$$

The distance between P and Q is

$$d = \left\{\sum_{s=1}^{p} (x_{2s} - x_{1s})^2\right\}^{\frac{1}{2}},$$

while the resultant $\boldsymbol{x} = \boldsymbol{x}_1 + \boldsymbol{x}_2 = (x_{11} + x_{21}, x_{12} + x_{22}, \ldots, x_{1p} + x_{2p})$. Subspaces of all dimensions up to $p - 1$ can be defined for this space, but again the most important practical cases will be one- and two-dimensional subspaces, which correspond to ordinary lines and planes. Subspaces of dimensionality greater than two are often referred to as hyperplanes. Equations of lines are exactly as before, and a q-dimensional hyperplane is defined by q mutually orthogonal lines for any $q = 2, 3, \ldots, p - 1$.

1.3.4 Rotation of axes

It has been assumed so far that the axes of a geometrical representation have been fixed, and will remain unchanged in any representation. When

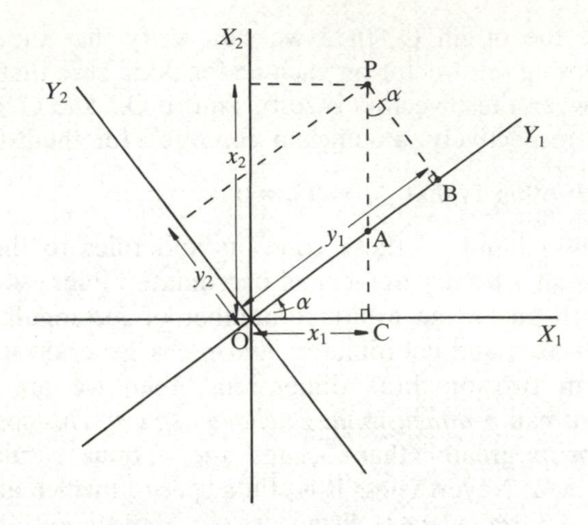

Fig. 1.6 Effect of a rotation of axes on the coordinates of a point

the relationships between sets of points are of prime concern, however, it is the *configuration* of the points that is important while the frame of reference is only of secondary interest. Thus it is sometimes helpful to change the frame of reference, particularly if this change leads to some subsequent simplification. One change that is frequently employed is a *rotation* of the axes. Consider the point P of Fig. 1.6. Suppose that it has coordinates (x_1, x_2) with reference to the original axes OX_1, OX_2, and now we rotate the axes anticlockwise through an angle α to position OY_1, OY_2, as marked. If the coordinates of P referred to these new axes are y_1, y_2 then using the distances and angles marked on Fig. 1.6 we obtain:

$$
\begin{aligned}
y_1 &= \mathrm{OA} + \mathrm{AB} \\
&= \frac{x_1}{\cos\alpha} + y_2 \tan\alpha
\end{aligned}
$$

so that

$$x_1 = y_1 \cos\alpha - y_2 \sin\alpha. \tag{1.1}$$

Also

$$
\begin{aligned}
x_2 &= \mathrm{PA} + \mathrm{AC} \\
&= \frac{y_2}{\cos\alpha} + x_1 \tan\alpha
\end{aligned}
$$

and substituting from (1.1) this yields

$$x_2 = \frac{y_2}{\cos\alpha} + \frac{\sin\alpha}{\cos\alpha}(y_1 \cos\alpha - y_2 \sin\alpha)$$

i.e.

$$x_2 = y_1 \sin \alpha + \frac{y_2}{\cos \alpha}(1 - \sin^2 \alpha)$$

$$= y_1 \sin \alpha + y_2 \frac{\cos^2 \alpha}{\cos \alpha}, \qquad \text{since } \sin^2 \alpha + \cos^2 \alpha = 1,$$

or

$$x_2 = y_1 \sin \alpha + y_2 \cos \alpha. \tag{1.2}$$

Now if (1.1) is multiplied by $\cos \alpha$, (1.2) is multiplied by $\sin \alpha$, and the two resultant equations are added, we find that

$$x_1 \cos \alpha + x_2 \sin \alpha = y_1 \cos^2 \alpha + y_1 \sin^2 \alpha,$$

or, in other words,

$$y_1 = x_1 \cos \alpha + x_2 \sin \alpha. \tag{1.3}$$

Similarly, on multiplying (1.2) by $\cos \alpha$, (1.1) by $\sin \alpha$, and subtracting the two, we obtain

$$y_2 = -x_1 \sin \alpha + x_2 \cos \alpha. \tag{1.4}$$

Thus if we know the coordinates (x_1, x_2) of a point P referred to a pair of orthogonal axes, and we wish to deduce its new coordinates (y_1, y_2) when the axes are rotated (anticlockwise) through an angle α, we can use eqns (1.3) and (1.4) to do so. In vector and matrix notation, these equations are written

$$\begin{pmatrix} y_1 \\ y_2 \end{pmatrix} = \begin{pmatrix} \cos \alpha & \sin \alpha \\ -\sin \alpha & \cos \alpha \end{pmatrix} \begin{pmatrix} x_1 \\ x_2 \end{pmatrix}$$

or

$$\boldsymbol{y} = \mathbf{A}\boldsymbol{x}$$

where

$$\boldsymbol{y} = \begin{pmatrix} y_1 \\ y_2 \end{pmatrix}, \qquad \boldsymbol{x} = \begin{pmatrix} x_1 \\ x_2 \end{pmatrix},$$

and

$$\mathbf{A} = \begin{pmatrix} \cos \alpha & \sin \alpha \\ -\sin \alpha & \cos \alpha \end{pmatrix}. \tag{1.5}$$

(Readers unfamiliar with matrices, and elementary operations on them, should refer at this juncture to Appendix A.)

The reverse operation to an anticlockwise rotation of axes through an angle α is a clockwise rotation through an angle α. Selecting the former rotation (arbitrarily) as being in the positive sense, the latter rotation can then be described as one through an angle $-\alpha$. If we know that the coordinates of a point P are (y_1, y_2), and we then rotate the axes through an angle $-\alpha$, we can obviously use eqns (1.1) and (1.2) to find its coordinates (x_1, x_2) relative to its new axes. In matrix notation these equations can be written

$$\boldsymbol{x} = \mathbf{B}\boldsymbol{y},$$

where

$$\boldsymbol{x} = \begin{pmatrix} x_1 \\ x_2 \end{pmatrix}, \qquad \boldsymbol{y} = \begin{pmatrix} y_1 \\ y_2 \end{pmatrix}$$

as before, and

$$\mathbf{B} = \begin{pmatrix} \cos\alpha & -\sin\alpha \\ \sin\alpha & \cos\alpha \end{pmatrix}.$$

Matrices **A** and **B** exhibit some interesting properties. First, consider their product (cf. Section A2 of Appendix A):

$$\begin{aligned} \mathbf{AB} &= \begin{pmatrix} \cos\alpha & \sin\alpha \\ -\sin\alpha & \cos\alpha \end{pmatrix}\begin{pmatrix} \cos\alpha & -\sin\alpha \\ \sin\alpha & \cos\alpha \end{pmatrix} \\ &= \begin{pmatrix} \cos^2\alpha + \sin^2\alpha & -\cos\alpha\sin\alpha + \sin\alpha\cos\alpha \\ -\sin\alpha\cos\alpha + \cos\alpha\sin\alpha & \sin^2\alpha + \cos^2\alpha \end{pmatrix} \\ &= \begin{pmatrix} 1 & 0 \\ 0 & 1 \end{pmatrix}. \end{aligned}$$

Similarly, we find that $\mathbf{BA} = \begin{pmatrix} 1 & 0 \\ 0 & 1 \end{pmatrix}$.

The matrix $\begin{pmatrix} 1 & 0 \\ 0 & 1 \end{pmatrix}$ is known as the *identity* (2×2) matrix, denoted by **I**. It can be verified that the effect of multiplying any vector $\boldsymbol{a} = \begin{pmatrix} a_1 \\ a_2 \end{pmatrix}$ by **I** is to leave it unchanged, i.e. $\mathbf{I}\boldsymbol{a} = \boldsymbol{a}$. If we consider the interpretation of **A** and **B** as rotations of axes, we see that this is a perfectly reasonable result. Consider the coordinates of the point P. With respect to the original axes they are given by $\boldsymbol{x}$. Rotating axes through an angle α gives coordinates $\boldsymbol{y} = \mathbf{A}\boldsymbol{x}$. Now rotating axes back through an angle $-\alpha$ yields coordinates $\mathbf{B}\boldsymbol{y} = \mathbf{BA}\boldsymbol{x}$. But this is the *inverse* rotation, which brings the axes back into their original positions, so these coordinates must be $\boldsymbol{x}$. Thus $\mathbf{BA}\boldsymbol{x} = \boldsymbol{x}$, and so **BA** must be the identity matrix. Starting from the axes in the position $\mathrm{O}Y_1$, $\mathrm{O}Y_2$, and performing a rotation through an angle $-\alpha$ followed by one through an angle α will verify that $\mathbf{AB}\boldsymbol{y} = \boldsymbol{y}$, so that **AB** must also be the identity matrix. These relationships are expressed by saying that **B** is the *inverse* of **A** (and thus **A** is the inverse of **B**; cf. Section A3 of the Appendix).

A second interesting aspect of these matrices is the relationship between their elements. If we write the rows of $\mathbf{A}$ as columns of a (2×2) matrix, we simply obtain the matrix $\mathbf{B}$. This relationship is expressed by saying that $\mathbf{B}$ is the *transpose* of $\mathbf{A}$ and writing symbolically $\mathbf{B} = \mathbf{A}'$. Thus we find from the earlier equations that $\mathbf{A}'\mathbf{A} = \mathbf{I}$ and $\mathbf{A}\mathbf{A}' = \mathbf{I}$. These two equations, which show that the inverse of $\mathbf{A}$ is its transpose $\mathbf{A}'$, characterize $\mathbf{A}$ as an *orthogonal* matrix (see Section A3 of Appendix A).

A final point to notice about the transformations defining the changes in coordinates of P on rotating the axes through an angle α is that eqns (1.1) to (1.4) can be written as

$$x_1 = b_{11}y_1 + b_{12}y_2 \tag{1.6}$$

$$x_2 = b_{21}y_1 + b_{22}y_2 \tag{1.7}$$

$$y_1 = a_{11}x_1 + a_{12}x_2 \tag{1.8}$$

$$y_2 = a_{21}x_1 + a_{22}x_2 \tag{1.9}$$

where a_{11}, b_{11}, etc. are constants (given by $a_{11} = \cos\alpha$, $b_{11} = \cos\alpha$, etc.). Thus a rotation of axes induces a *linear transformation* in the coordinate values of any point P. Such a linear transformation is written most concisely in vector and matrix form, as $\boldsymbol{y} = \mathbf{A}\boldsymbol{x}$ and $\boldsymbol{x} = \mathbf{B}\boldsymbol{y}$, given earlier.

The ideas given above for two dimensions generalize immediately to the abstract case of p-dimensional Euclidean space. If the coordinates of a point P referred to p mutually orthogonal axes $OX_1, \ldots, OX_p$ are $(x_1, \ldots, x_p)$, and if these axes undergo a rigid rotation to some new position $OY_1, \ldots, OY_p$, then the coordinates $(y_1, \ldots, y_p)$ of P referred to the new axes are given by a linear transformation

$$\begin{aligned} y_1 &= a_{11}x_1 + \ldots + a_{1p}x_p \\ y_2 &= a_{21}x_1 + \ldots + a_{2p}x_p \\ &\vdots \\ y_p &= a_{p1}x_1 + \ldots + a_{pp}x_p. \end{aligned}$$

The a_{ij} here are constants, determined by the precise position of rotation. This linear transformation is written most compactly in the matrix form

$$\boldsymbol{y} = \mathbf{A}\boldsymbol{x}$$

where $\boldsymbol{x}$ and $\boldsymbol{y}$ are the p-component vectors $\begin{pmatrix} x_1 \\ \vdots \\ x_p \end{pmatrix}$, $\begin{pmatrix} y_1 \\ \vdots \\ y_p \end{pmatrix}$, and $\mathbf{A}$ is the $(p \times p)$ matrix $\begin{pmatrix} a_{11} & \ldots & a_{1p} \\ a_{21} & \ldots & a_{2p} \\ \vdots & & \vdots \\ a_{p1} & \ldots & a_{pp} \end{pmatrix}$.

This matrix is orthogonal, so that $\mathbf{A}'\mathbf{A} = \mathbf{A}\mathbf{A}' = \mathbf{I}_p$. (The suffix p is attached to $\mathbf{I}$ to emphasize that it is the identity matrix having p rows and columns.) It is worth noting in passing that vectors will always be considered to be *column* vectors, unless explicitly stated otherwise. Thus they are equivalent to $(p \times 1)$ matrices. To save space it sometimes helps to write them as rows (i.e. $(1 \times p)$ matrices), and if this is done then the column nature of the vector $\boldsymbol{x}$ will be emphasized by denoting the row $(x_1, \ldots, x_p)$ as the transpose $\boldsymbol{x}'$.

A final point to note is that *any* linear transformation $\boldsymbol{y} = \mathbf{A}\boldsymbol{x}$ can be viewed as a change of axes of the coordinate system. However, if $\mathbf{A}$ is not an orthogonal matrix then this change of axes will not be a rigid rotation, but will involve some deformation of the original system. On the other hand, any orthogonal matrix defines a rigid rotation and vice versa.

1.3.5 Latent roots and vectors

Now consider the effect of a transformation matrix $\mathbf{A}$ on some vector $\boldsymbol{x}$. In principle, $\mathbf{A}$ could be any square matrix. In practice, however, we will find that in nearly all subsequent applications $\mathbf{A}$ is a *symmetric* matrix (cf. Section A1 of Appendix A), so we restrict attention here to just such matrices. The effect of $\mathbf{A}$ is to transfer $\boldsymbol{x}$ from its original position in space to some new position $\boldsymbol{y}$. This is illustrated in Fig. 1.7.

Now consider the dotted line through vector $\boldsymbol{x}$. Any point on this line has coordinates $k\boldsymbol{x}$, for some k. In other words, any vector $\boldsymbol{z}$ (other than $\boldsymbol{x}$) which lies on this line is either a stretching $(k > 1)$ or a shrinking $(k < 1)$ of $\boldsymbol{x}$. What does the transformation $\mathbf{A}$ do to $\boldsymbol{z}$? By applying the elementary rules given in Section A2 of Appendix A, it can be verified easily that $\mathbf{A}\boldsymbol{z} = \mathbf{A}(k\boldsymbol{x}) = k\boldsymbol{y}$. Thus transforming a stretched or shrunken $\boldsymbol{x}$ produces the correspondingly stretched or shrunken $\boldsymbol{y}$, so the transformation $\mathbf{A}$ sends the dotted line through $\boldsymbol{x}$ to the dotted line through $\boldsymbol{y}$.

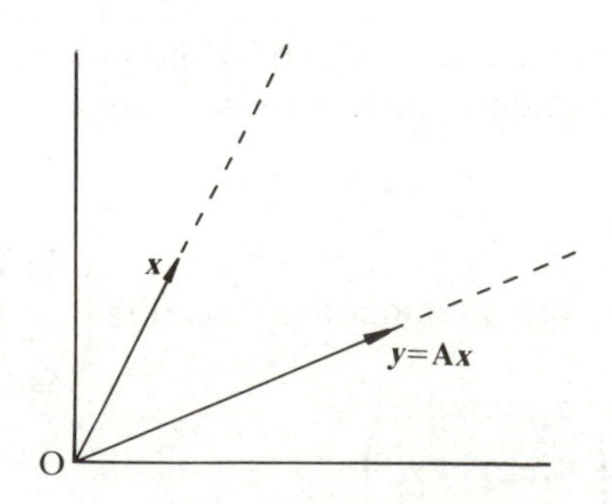

Fig. 1.7 Linear transformation of a vector

Every line in this space (whether 2- or p-dimensional) will be transformed to some *other* line in the space, in general. However, it can be proved that there are just p exceptions to this rule in p-dimensional space. There exist p directions in this space which, under the transformation $\mathbf{A}$, transform back on to themselves. Because of this invariance, such directions are often called *fundamental* or *canonical* directions. Every transformation $\mathbf{A}$ has its own such set of directions. If $\boldsymbol{x}$ is a vector along one of these directions, then the effect of applying the transformation $\mathbf{A}$ to $\boldsymbol{x}$ is either to stretch $\boldsymbol{x}$, shrink $\boldsymbol{x}$, or leave $\boldsymbol{x}$ unchanged. Thus

$$\boldsymbol{x} \text{ must satisfy } \mathbf{A}\boldsymbol{x} = \lambda\boldsymbol{x} \text{ for some scalar } \lambda. \qquad (1.10)$$

Moreover, all possible vectors along any one such direction will be affected in the same manner. Thus to describe the p canonical directions for any given $\mathbf{A}$ we need to specify each direction, and quote the amount of stretching or shrinking associated with each. These quantities are obtained by solving eqn (1.10). The directions $\boldsymbol{x}$ are called (interchangeably) the *eigenvectors, characteristic vectors* or *latent vectors* of $\mathbf{A}$, while the corresponding terms for the associated stretch/shrink factors λ are *eigenvalues, characteristic roots* or *latent roots* of $\mathbf{A}$ respectively. Since directions are usually specified by unit vectors, it is generally assumed that a latent vector is standardized to have unit length before being quoted or used, so that the sum of squares of the components of $\boldsymbol{x}$ is unity, i.e. $\boldsymbol{x}'\boldsymbol{x} = 1$. (For further algebraic details, see Section A5 of Appendix A.)

Many multivariate techniques require the calculation of the eigenvalues and eigenvectors of particular (symmetric) matrices. Solution of eqn (1.10) is not a trivial matter, but can now be done quickly, accurately, and efficiently using a variety of computer algorithms; a procedure for so doing is included in most statistical packages and suites of programs. A final point to notice is that the angle between any (fundamental) direction $\boldsymbol{x}$ and each of the coordinate axes can be determined readily. Let $\boldsymbol{y}$ be a unit vector along any coordinate axis. If the axis to be considered is the ith, then $\boldsymbol{y}$ has the ith component unity, and the remaining components zero. If α is the angle between $\boldsymbol{x}$ and $\boldsymbol{y}$, then earlier results show that $\cos\alpha = \boldsymbol{x}'\boldsymbol{y} = x_i$. Thus the elements of $\boldsymbol{x}$ give the *direction cosines* between $\boldsymbol{x}$ and each of the coordinate axes.

This concludes our preliminary survey of useful geometrical concepts. The last section in this chapter provides a bridge to the next chapter, by introducing the idea of similarity between physical entities, and linking this to the distance between points representing these entities in a geometric configuration.

1.4 Similarity, dissimilarity, and distance

The 'projection' and 'optimization' approaches to the construction of a graphical representation of a multivariate sample, as described in Section 1.2, both rely on the idea that there exists some 'true' configuration of points which faithfully represents the patterns existing among the entities being compared. The two approaches then endeavour to produce either this true configuration or an acceptable approximation to it. A basic requirement for such a configuration to represent faithfully the patterns existing among the entities is that similar entities should be represented by points that are close together, and the more dissimilar the entities are, the more distant should be the points representing them. There is thus a direct connection between the dissimilarity of two entities and the distance between them in the geometrical representation. While the actual construction of such configurations will be considered in the next few chapters, knowing how to *measure* the dissimilarity between entities is evidently of more fundamental interest and hence this preliminary discussion of the concept. Also, dissimilarity between entities forms the starting point of various techniques, so it is worth gathering together the basic ideas in one section. Since dissimilarity is so closely linked to the idea of distance, one natural way of measuring it is by the use of a familiar metric such as Euclidean distance. However, this is not the only possibility, and over the years a plethora of dissimilarity measures have been proposed. Without any guidance, an inexperienced researcher may thus be thoroughly confused when required to choose a measure for some investigation, so a further purpose of the brief survey that follows is to try and establish some guidelines for such a choice. The choice of measure is usually closely tied to the nature and purpose of the individual study, but gross mistakes will usually be avoided by paying careful attention to the type of data that have been collected, and to the nature of the entities between which the dissimilarity is to be computed.

The description that follows of the various dissimilarity measures is therefore structured according to these two criteria. Before setting out these measures, however, a few general comments may be made. First, it has been emphasized that the focus of interest is usually on the dissimilarity between two entities. Frequently, however, a *similarity* measure is quoted in the literature or calculated in a computer package. If s is the similarity between two entities (usually in the range $0 \leqslant s \leqslant 1$), then the dissimilarity d is the direct opposite of s and hence may be obtained by using any monotonically *decreasing* transformation of s. The most common such transformation is $d = 1 - s$. Secondly, all dissimilarity measure formulae naturally assume that a set of data is available on which to apply them. *Choosing* the appropriate data base is a subjective aspect but a very important first stage in any analysis. There are two questions which often cause problems (when dissimilarity between individuals is to be calculated). Should the variables be scaled in any way? Should a variable that is highly correlated with another one be rejected or not? Clearly, if some variables exhibit much larger values or a much greater range of values than the others, then if left unscaled they will dominate any dissimilarity calculation. This may, on the other hand, be valid in the context of some studies. Similarly, highly correlated variables may distort a dissimilarity calculation by giving undue weight to outlying individuals, but this may in turn be justified if they are measuring differences that are relevant to the problem in hand. Such questions must always be resolved *before* dissimilarities are calculated and the problem is tackled.

We now turn to a consideration of the various measures of dissimilarity that have been proposed over the years. Following the above comments, we will consider separately each possible type of data, and within each category we will distinguish the case where the entities to be compared are individuals, from the case where they are variables. We suppose that there exists a sample of n individuals on each of which p variables have been measured; we shall use subscripts i, j to indicate individuals, and subscripts k, l to indicate variables in the following formulae.

(a) Quantitative data

The numerical value x_{ik} is observed for the kth variable on the ith individual in the sample.

Dissimilarity d_{ij} between individuals i and j

(i) Euclidean distance: $d_{ij} = \left\{ \sum_{k=1}^{p} (x_{ik} - x_{jk})^2 \right\}^{\frac{1}{2}}$.

(ii) Scaled Euclidean distance: $d_{ij} = \left\{ \sum_{k=1}^{p} w_k^2 (x_{ik} - x_{jk})^2 \right\}^{\frac{1}{2}}$ for some suitable weights w_k. Typical choices would be either $w_k =$ (standard deviation of kth variable)$^{-1}$ or $w_k =$ (range of kth variable)$^{-1}$, the effect of both of these choices being to equalize the importance of each variable in the sample.

(iii) Minkowski metric: $d_{ij} = \left\{ \sum_{k=1}^{p} |x_{ik} - x_{jk}|^{\lambda} \right\}^{1/\lambda}$ for some integer λ. The case $\lambda = 1$ is known as the 'city block' metric, while the case $\lambda = 2$ gives the Euclidean metric in (i) above. The consequence of increasing λ is increasingly to exaggerate the more dissimilar units relative to the similar ones.

(iv) Canberra metric: $d_{ij} = \sum_{k=1}^{p} \frac{|x_{ik} - x_{jk}|}{(x_{ik} + x_{jk})}$.

(v) Czekanowski coefficient: $d_{ij} = 1 - \frac{2 \sum_{k=1}^{p} \min(x_{ik}, x_{jk})}{\sum_{k=1}^{p} (x_{ik} + x_{jk})}$.

Dissimilarity d_{kl} between variables k and l

When the entities to be compared are variables, the correlation coefficient (or some variant of it) provides a sensible measure of *similarity*. Thus we need to convert this to a dissimilarity, as discussed above. However, before we can do this we must decide whether a high but negative correlation coefficient represents close agreement between the variables or marked disagreement between them. Thus:

(vi) $d_{kl} = 1 - s_{kl}$, where $s_{kl} = \frac{\sum_{i=1}^{n} (x_{ik} - \bar{x}_k)(x_{il} - \bar{x}_l)}{\left\{ \sum_{i=1}^{n} (x_{ik} - \bar{x}_k)^2 \sum_{i=1}^{n} (x_{il} - \bar{x}_l)^2 \right\}^{\frac{1}{2}}}$, is appropriate if a correlation coefficient of -1 represents the maximum *disagreement* between two variables; while

(vii) $d_{kl} = 1 - s_{kl}^2$ with s_{kl} as in (vi) above is appropriate if correlation coefficients of -1 and $+1$ are treated equivalently as showing maximum *agreement* between variables.

(viii) $d_{kl} = 1 - \frac{\sum_{i=1}^{n} x_{ik} x_{il}}{\left(\sum_{i=1}^{n} x_{ik}^2 \sum_{i=1}^{n} x_{il}^2 \right)^{\frac{1}{2}}}$ is an alternative possibility.

(b) Dichotomous data

x_{ik} can take just two values, which can be arbitrarily coded + and −, say. To compute a dissimilarity between two entities, therefore, the relevant data can be reduced to the 2×2 table:

		Entity 2 +	Entity 2 −
Entity 1	+	a	b
	−	c	d

This table shows that, out of all the possible pairwise comparisons between the two entities, a show + in both positions, d show − in both positions, b show the disagreement + −, while c show the reverse disagreement − +.

Dissimilarities between individuals

In this case we compare the p variable values for the two individuals, so $a+b+c+d=p$. The three most common measures are as follows:

(ix) $d_{ij} = 1 - (\text{simple matching coefficient}) = 1 - \frac{a+d}{p} = \frac{b+c}{p}$. Thus dissimilarity between two units is measured as the proportion of variables that show disagreement in their recorded values between the two units, and this is an acceptable measure in most studies.

(x) Jaccard coefficient: $d_{ij} = \frac{b+c}{a+b+c}$. This coefficient is particularly appropriate when the two 'values' + and − of the dichotomy represent respectively 'presence' and 'absence' of an attribute. In this case, one can frequently argue that absence of a trait in both units should not contribute to their degree of 'likeness' (e.g. absence of wings in both mice and elephants is irrelevant when assessing their similarity). In such cases, the divisor $a+b+c$ is more appropriate than the divisor $p=a+b+c+d$ given in (ix) above.

(xi) Czekanowski coefficient: $d_{ij} = \frac{b+c}{2a+b+c}$. This is in the same spirit as (x), but each + + match is given extra weight to compensate for the neglect of all − − matches.

Dissimilarities between variables

In this case, $a+b+c+d=n$ (the number of units in the sample) so that a gives the number of units showing + for both variables compared, etc. The analogue of a correlation coefficient for dichotomous data is a measure of association arising from the resultant (2×2) contingency table. The most common such measure is the familiar χ^2 measure given by

$$\chi^2 = \frac{(ad-bc)^2(a+b+c+d)}{(a+b)(a+c)(c+d)(b+d)}.$$

However, the size of χ^2 is influenced by the size of n. Since $\chi^2 \leqslant n$, this dependency is readily removed on using the measure

(xii) $d_{kl} = 1 - \sqrt{\left(\frac{\chi^2}{a+b+c+d}\right)}$.

(c) Qualitative data

If each response is categorical with more than two categories, a simple similarity measure is given by Sneath's matching coefficient c_{ij}. This takes as its value the number of attributes on which the two units have the same category, divided by p. Then

(xiii) $d_{ij} = 1 - c_{ij}$. This coefficient can be used whether dissimilarities between units or between variables are required, there being no specially suitable measures in the latter case.

The above formulae (i)–(xiii) summarize the most common measures of dissimilarity that are in widespread use in multivariate analysis. There is nothing to prevent the reader devising his or her own measure of association for use in a particular circumstance, particularly if this personal measure seems more appropriate than the standard ones given above. In most common applications such extra work should not be necessary, but nonetheless there is one major problem with the direct application of the above formulae to the calculation of dissimilarity between units, namely that for any one of them to be applicable all variables in the data set must be of the same type. In any realistic application, it is more than likely that different types of data will be measured on the same individual. Typically, one's measurements may consist of a mixture of numeric values, counts, rankings, binary attributes and categorical variables. One possible approach is to compute a separate dissimilarity value between any two units from the observations within each type of response, and then to derive a final dissimilarity value between the units as an average (possibly weighted) of these individual dissimilarity values. If a weighted average is contemplated, however, the choice of weights is highly subjective and may, of course, be quite crucial to the resulting analysis. These ideas have all been formalized by Gower (1971a), who has defined a general coefficient of similarity suitable for all types of data. This is the similarity coefficient used in the GENSTAT statistical computing package, and it is in essence an average of similarity values for each permitted variable in the computation. If two individuals i and j can be compared on a variable k, they are assigned a score s_{ijk}. This score is zero when i and j are considered different, and some positive fraction or unity otherwise. The individuals may not be comparable on a given variate either because information is missing or in the case of dichotomous variables because a character is non-existent in both i and j (see comments following (x) above). The possibility of making comparisons can be represented by a quantity δ_{ijk}, equal to 1 when i and j can be compared on variable k, and 0 otherwise. When $\delta_{ijk} = 0$, s_{ijk} is unknown but conventionally is set to zero. Then the similarity c_{ij} and dissimilarity d_{ij} between i and j are defined by

(xiv) $$c_{ij} = \sum_{k=1}^{p} s_{ijk} \Big/ \sum_{k=1}^{p} \delta_{ijk}; \qquad d_{ij} = 1 - c_{ij}$$

where for *quantitative* variables, $s_{ijk} = 1 - \dfrac{|x_{ik} - x_{jk}|}{(\text{range of variable } k)}$, for *qualitative* variables, $s_{ijk} = \begin{cases} 1 & \text{if } i \text{ and } j \text{ agree on variable } k, \\ 0 & \text{otherwise} \end{cases}$ and for *dichotomous* variables s_{ijk} and δ_{ijk} have the settings shown below.

		Values of variable k			
Individual	i	+	+	−	−
	j	+	−	+	−
	s_{ijk}	1	0	0	0
	δ_{ijk}	1	1	1	0

Furthermore, Gower allows the possibility of modifying this to a weighted coefficient and discusses the problems involved in making a rational choice of weights. As implemented in the GENSTAT package, however, the most common form of the coefficient is the unweighted version.

A final point worthy of comment is the treatment of missing values in the computation of a dissimilarity coefficient. A common method of handling missing values in many general multivariate problems is to employ *imputation*, that is the insertion of an estimate for each missing value, thereby completing the data set. At the crudest level this estimate is just the mean of the particular variable over the values that *are* present in the sample, while at the most sophisticated level this estimate is obtained by an iterative least-squares or maximum likelihood procedure (see, for example, Beale and Little 1975). Such procedures are not recommended when calculating similarity or dissimilarity coefficients, particularly when there are many missing values in a data set. The reason for this is that either very similar or identical estimates may be imputed for different individuals and hence the similarity between these individuals may be grossly exaggerated. It is far better to ignore any variables that have any values missing when calculating the dissimilarity between two units, and only to work with those variables that have all values present for both the units concerned. If this means working with r variables instead of p, the resulting dissimilarity or similarity can be inflated by a factor (p/r) subsequently, to make it comparable with those values that have been computed from p variables.

We conclude this section with a simple numerical example to illustrate the calculation of these dissimilarity measures. Consider evaluation of all the above coefficients that yield a distance between two individuals, on the first two rows of Table 1.2. First, consider the quantitative variables x_2, x_{10}, and x_{13}.

(i) Euclidean distance:

$$d_{12} = \{(59-49)^2 + (17-15)^2 + (60.83-52.46)^2\}^{\frac{1}{2}}$$
$$= (100 + 4 + 70.06)^{\frac{1}{2}} = \sqrt{174.06} = 13.19.$$

(ii) Euclidean distance standardized by range. The range of x_2 is $59-43=16$, that of x_{10} is $21-11=10$, and that of x_{13} is $66.18-48.70=17.48$. Thus

$$d_{12} = \left\{\frac{(59-49)^2}{16^2} + \frac{(17-15)^2}{10^2} + \frac{(60.83-52.46)^2}{17.48^2}\right\}^{\frac{1}{2}}$$
$$= \sqrt{0.6599} = 0.812.$$

(iii) Minkowski metric with $\lambda = 1$:

$$d_{12} = |59-49| + |17-15| - |60.83-52.46|$$
$$= 10 + 2 + 8.37 = 20.37.$$

(iv) Canberra metric:

$$d_{12} = \frac{|59-49|}{(59+49)} + \frac{|17-15|}{(17+15)} + \frac{|60.83-52.46|}{(60.83+52.46)}$$
$$= 0.0926 + 0.0625 + 0.0739 = 0.229.$$

(v) Czekanowski coefficient:

$$d_{12} = 1 - \frac{2(49+15+52.46)}{59+49+17+15+60.83+52.46}$$
$$= 1 - \frac{2\times 116.46}{253.29} = 1 - 0.920 = 0.08.$$

Next, the binary variables x_1, x_{11}, and x_{12}. For the first two rows we obtain the (2×2) table:

		Row 1	
		1	0
Row 2	1	2	0
	0	1	0

Thus

(ix) $$d_{12} = \frac{1+0}{3} = \frac{1}{3};$$

(x) $$d_{12} = \frac{1+0}{2+1+0} = \frac{1}{3};$$

(xi) $$d_{12} = \frac{1+0}{4+1+0} = \frac{1}{5}.$$

Next the qualitative variables x_3, x_4, x_5, x_6, x_7, x_8, and x_9. There are three matches for the first two rows (x_3, x_4, and x_9) out of seven variables, so

(xiii) $$d_{12} = 1 - \frac{3}{7} = \frac{4}{7}.$$

Note that, since x_4 to x_8 were all stated to be ordinal variables, they could also have been treated as quantitative. Such treatment would imply, however, that all values of each variable were 'equally spaced'.

To obtain a single dissimilarity value between the first two rows, we would thus need an average of three selected values from above, one

Table 1.3 Computation of Gower's Coefficient

k	s_{12k}	δ_{12k}
1 (i.e. x_1)	0	1
2	$1 - \frac{10}{16} = 0.375$	1
3	1	1
4	1	1
5	0	1
6	0	1
7	0	1
8	0	1
9	1	1
10	$1 - \frac{2}{10} = 0.8$	1
11	1	1
12	1	1
13	$1 - \frac{8.37}{17.48} = 0.521$	1
Total	6.696	13

from each type of variable. We might thus choose (iv), (x), and (xiii) to give a value $\frac{1}{3}(0.229 + 0.333 + 0.571) = 0.378$.

Alternatively, we could compute Gower's coefficient. For this calculation we have the values shown in Table 1.3 (treating x_4 to x_9 again as qualitative variables).

Thus $c_{12} = \frac{6.696}{13} = 0.515$, so $d_{12} = 1 - 0.515 = 0.485$.

The above calculations illustrate the differences that exist between the various possible dissimilarity measures when applied to any particular set of data, and hence highlight the necessity for careful choice of dissimilarity measure before any analysis of data is attempted. Techniques that require the calculation of dissimilarities will be encountered in various places throughout the remainder of the book.

2
One-way graphical representation of data matrices

Having completed a preliminary look at some underlying geometrical concepts, we are now in a position to discuss the various techniques that are available for graphical representation of multivariate data. In keeping with the problem-orientated approach, the discussion will be structured according to the form in which the multivariate data are presented. Since all multivariate data can be presented in the form of a matrix, the most convenient classification will consider the entities that define the rows and columns of this matrix. It should be remembered that virtually *any* matrix of numbers can be input into *any* particular multivariate routine, and the computer will nearly always produce an answer. Whether or not this answer is *sensible,* however, will depend on whether or not the right type of matrix was input to the routine (mindful of the old computing adage: garbage in, garbage out). Care must therefore be exercised in deciding what sort of matrix one is faced with in any given application, what the objectives of the analysis are, and hence which set of techniques is appropriate for the problem in hand.

Techniques to be described in this chapter are primarily aimed at graphically representing *either* the n units *or* the p variables in an $(n \times p)$ data matrix. In such a matrix, the rows and columns define *essentially different entities* which are not linked in any way other than by the fact that the columns refer to the characteristics measured on the rows. The techniques in this chapter will thus be equally appropriate for any other such matrix for which a *one-way* representation (of either rows or columns) is required. Some matrices, however, do not match this specification. In multivariate statistics, many instances arise of matrices in which the rows and columns have the *same* classification. Thus a correlation matrix has the same variables labelling both rows and columns, while similarity matrices giving the similarities between every pair of individuals in a sample have the same individuals labelling both rows and columns. Such matrices require the methods of the next chapter for their representation, rather than the methods of this chapter.

In other situations, it is possible to be misled into thinking that a matrix falls into the classification of this chapter when this is not the case. For example, we might have information on the correlations between every variable in one set of m variables and every variable in another set of n variables, but not on the $m \times m$ and $n \times n$ within-set correlations. Hence we have a rectangular $m \times n$ matrix in which the row labels are not the same as the column labels, and it might be thought that methods of this chapter are applicable. Note however that, although the actual labels are different, the *entities* in both rows and columns are the same (i.e. both sets relate to *variables*). Hence, while this $(m \times n)$ matrix can be formally subjected to the methods of this chapter, the results may not be very illuminating. An exactly analogous situation applied to units is when we have the similarities or dissimilarities between every individual in one set of m units and every individual in another set of n units, but not the $(m \times m)$ and $(n \times n)$ within-set similarities or dissimilarities. These situations will be discussed in Chapters 3 and 4.

A slightly more contentious case arises when the rows and columns do relate to two genuinely different entities, but the data may be viewed more as dissimilarities than as observations. An example might be where each of n individuals in a sample has been asked to rate each of p consumer products (such as soap powders, say) on a scale from 0 to 10. The data matrix thus has individuals in the rows, products in the columns, and responses from 0 to 10 as its entries. It thus seems to be quite suitable for the techniques in this chapter. On the other hand, the columns are really also units (albeit of different kind to the rows) rather than variables. Also, the entries can be thought of as providing dissimilarity information between the products, with associated information on the similarities and dissimilarities between individuals. This is therefore another case that falls within the scope of Chapter 4, and the techniques discussed there may provide a better graphical representation of the data than the techniques of the present chapter.

From the above introduction, it should be clear that some thought must be put into deciding which techniques will be both appropriate and useful for representing a given set of data. Also, more than one technique may prove appropriate, with different techniques shedding light on different aspects of the data. With this in mind, let us now consider the various techniques appropriate for a straightforward data matrix. We shall use the categorization into direct two-dimensional, subspace projection, and optimization introduced in Section 1.2, and distinguish where necessary the case of representing units graphically from the one of representing variables graphically. Within each such categorization, we shall also pay due regard to the type of data for which the technique is appropriate.

2.1 Direct two-dimensional representations

2.1.1 Extensions of simple graphs

Right from the outset of a statistician's training, he or she is taught that a simple plot of the data is an essential prelude to *any* statistical analysis. This will help to highlight any interesting patterns in the data, isolate obviously outlying observations, and give some idea as to whether any necessary assumptions underlying the prospective analysis are satisfied or not. If just one variable has been measured on each individual, then everyone is familiar with such representations as the bar chart or pie chart for qualitative data, and the frequency polygon or histogram for quantitative data. If the variable values form a time sequence, then a two-dimensional plot of variable value against time is the most natural one (and indeed is probably the most commonly used graph). When two quantitative variables x_1 and x_2 have been measured on each individual, then a simple scatterplot of x_1 against x_2 is almost as well-known, this being a traditional prelude to a simple regression analysis. When more than two variables have been measured, then one obvious possibility for graphical representation is to provide a scatter plot for every possible pair of variables in the sample; with p variables there will be $\frac{1}{2}p(p-1)$ such plots. This may be unduly confusing, however, partly because of the number of separate plots that may arise and partly because each variable appears in a number $(p-1)$ of these plots. Various simple extensions to these familiar one- and two-variable plots have therefore been proposed, to try and circumvent such problems of interpretation.

When the data are qualitative, then the technique which Anderson (1960) has called a *glyph* may be appropriate. In this technique, each individual in the sample is represented by a circle and each variable by a ray emanating from this circle. The position of the ray indicates which variable is being represented, and the length of the ray indicates the category occupied by the given individual for that variable. Sometimes the lengths of the rays are obviously determined by the categories, as for example in the case of the categories absent, low, medium, and high for the variable 'level of pain' in a medical study, when the lengths of the corresponding ray would be zero, short, medium, and long. An example taken from Anderson (1960) (and reproduced with permission from the American Statistical Association) illustrates this type of usage. Table 2.1 gives the values observed for four individuals on each of five variables, where each variable has the three possible categories 'low', 'medium', and 'high'. The resulting glyph is presented in Fig. 2.1; the identification of variables from rays is given at the top of the figure; a long ray signifies the category 'high', a short ray the category 'medium', and absence of a

Table 2.1 Example of a data matrix with qualitative data

Units	Variables *A*	*B*	*C*	*D*	*E*
1	Low	Medium	High	Low	Low
2	High	High	High	High	Medium
3	Medium	Medium	High	Medium	Low
4	Low	Low	Medium	Medium	Low

ray the category 'low'. If the categories do not have any implied ranking (e.g. sandy, stony, peaty as classifications of a soil) then the correspondence between category and length of ray must be made arbitrarily.

At the other extreme, if the variables are quantitative then the technique may still be used, with length of ray corresponding *exactly* to the value observed. In this case, however, allowance must also be made for negative values. This can be done by representing negative values as rays emanating from the opposite side of the circle to their positive counterparts. Thus negative values would appear as rays *below* the circles in Fig. 2.1. An alternative treatment of quantitative variables is to divide the range of each variable into a number of categories, and then treat the data as being qualitative. This clearly reduces the information content of the data, but the result may still be accurate enough as regards pictorial representation (which in any case can never hope to reproduce all the details of the original data).

A further possibility is to remove two of the rays that correspond to

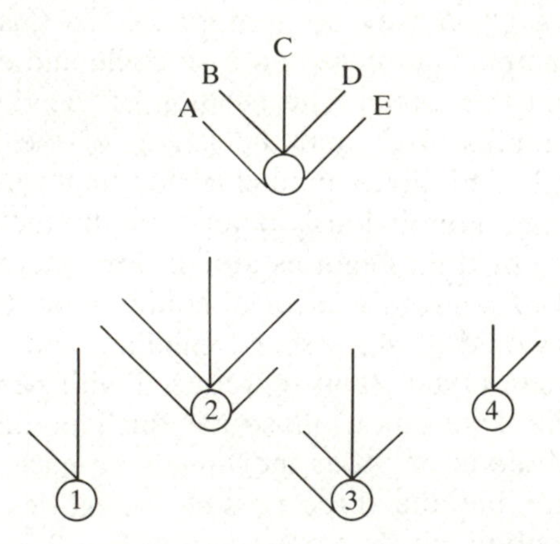

Fig. 2.1 Glyph of data in Table 2.1; example taken from Anderson (1960) and reproduced with permission from the American Statistical Association

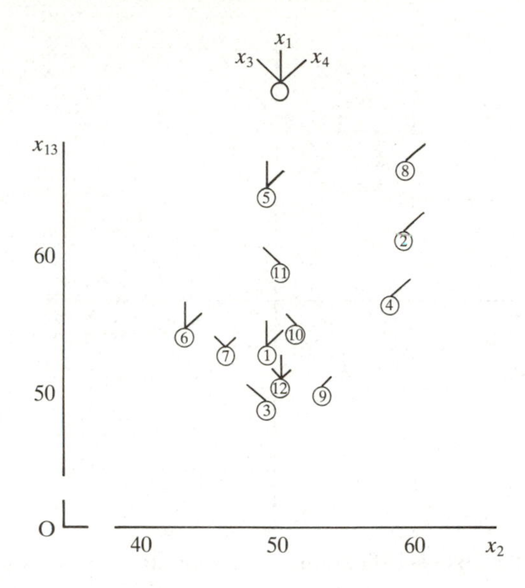

Fig. 2.2 Glyph of data in Table 1.2

quantitative variables, and to use these two variables as axes in a scatter diagram on which the circles are plotted. This is particularly useful when a pictorial representation is required for data comprising both qualitative and quantitative variables. To illustrate this type of representation, Fig. 2.2 depicts variables x_1, x_2, x_3, x_4, and x_{13} of Table 1.2. x_2 and x_{13} are used as the axes of a scatter plot on which the twelve individuals are plotted as circles. x_1, x_3, and x_4 are plotted as rays emanating from each circle in the positions shown at the top of Fig. 2.2. x_1 is binary, so a ray is present if $x_1 = 1$ and absent if $x_1 = 0$. x_3 and x_4 are both three-state with values 0, 1, 2: a ray is absent if the value is 0, is short if the value is 1, and is long if the value is 2. Examining Fig. 2.2, an interesting pattern is immediately evident: for all individuals to the right of number 10 (i.e. older than 51 years), the rays corresponding to x_1 and x_3 are both absent. Such a pattern would merit further investigation.

If the data are entirely quantitative, then various possibilities exist for a simple two-dimensional representation. The basis of all these is the scatter plot, and two variables must therefore be chosen, perhaps arbitrarily, to form the initial axes. Further variables may then be incorporated in various ways. One obvious possibility is to use the same idea as for the glyph, with the amendments mentioned already for dealing with quantitative variables. A systematic approach would be to represent the third variable as a line originating from each point on the

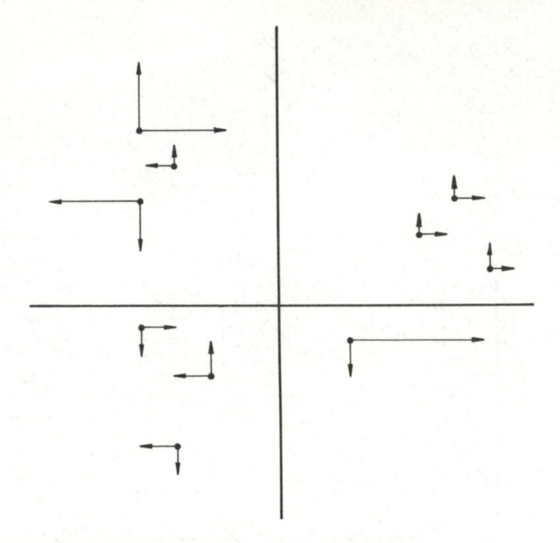

Fig. 2.3 Four-dimensional (fictitious) data plotted on a two-dimensional graph

scatter diagram, with length proportional to the value observed on the third variable and directed eastwards if this value is positive, westwards if negative. A fourth variable could use the north–south directions, a fifth the NE–SW, and a sixth the NW–SE directions. Figure 2.3 shows a typical such plot for a four-variable set of data. The small cluster of three points in the NW quadrant thus breaks up when the third and fourth variables are added, but the cluster in the NE quadrant remains together in the four dimensions.

To emphasize the displacement from the original two-dimensional configuration of the scatter plot induced by the third and fourth variables, the ends of the east–west and north–south lines could be joined to form triangles. Small triangles thus denote observations close to the original plane while large triangles denote observations far away from the original plane. Figure 2.4 is an example of such a representation, reproduced from Brothwell and Krzanowski (1974) with permission from Academic Press.

Other variations on this theme readily suggest themselves. For instance, the third variable may be grouped into a number of classes and these may be represented by different symbols (e.g. large, medium, small circles) or by different colours or by different shadings. Some further examples may be found in Marriott (1974, Chapter 1) or Everitt (1978, Chapter 2). Both of these authors make the point that such representations are clearly limited to a maximum of about five or six variables, and become confusing if there are many points plotted on the diagram.

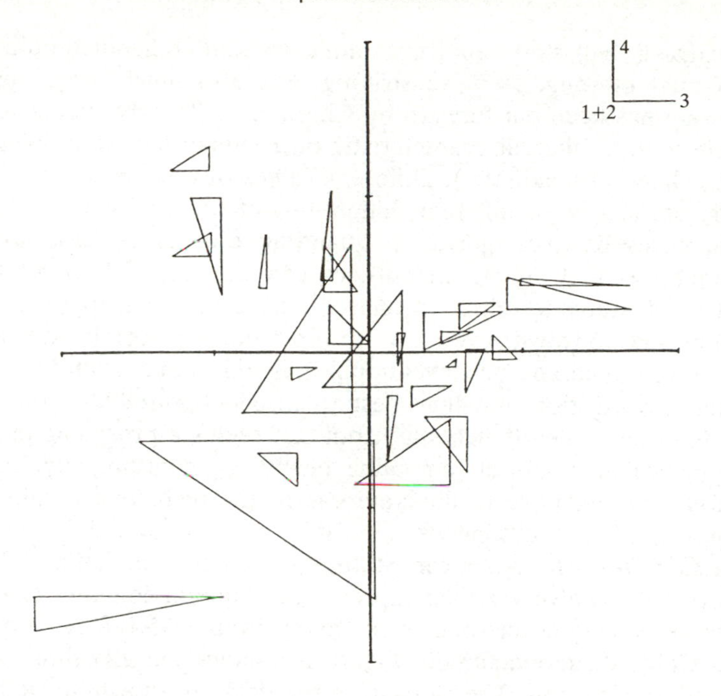

Fig. 2.4 Four-dimensional (archaeological) data plotted on a two-dimensional graph. Reproduced from Brothwell and Krzanowski (1974) with permission from Academic Press

They may, on the other hand, be very useful representations for plotting the results of other multivariate techniques discussed later in this book, and will be referred to in this context as appropriate. When used in this capacity, there is often a natural ordering of the variables in the sample. This resolves the unsatisfactory feature of arbitrariness in the choice of the two variables to be used as axes in the above examples.

2.1.2 Pictorial representations

As an alternative to representing data by lines and circles, one possibility is to associate the value of each variable with the measurement of one characteristic of a familiar object. A simple version of this approach is discussed by Hartigan (1975), who associates the values of one or two variables with the dimension of one side of a box. Thus each individual in the sample is represented by a box, and similar-sized and -shaped boxes indicate similar values along all variables for the corresponding individuals.

Hartigan's shapes are relatively simple and can be handled quite easily for manual plotting. A more striking, but also much more complex, suggestion has been put forward by Chernoff (1973) who associates each variable with a different characteristic of a human face (e.g. eyes, ears, mouth, shape of head, etc.). Different values of the variable lead either to different shapes or different dimensions of this characteristic. It may thus be necessary to categorize the quantitative variables before using the technique, so that simple stereotypes can be obtained. If we choose correlated features to reflect correlated attributes (e.g. happy/sad mouth positions and no frown/frown lines to represent two correlated variables) then conventional and unconventional individuals may soon be isolated. Further investigation into how best to associate variables with specific facial features is clearly necessary, but this seems a promising technique as sophisticated graphical computing facilities become readily available. The following example to illustrate use of the technique is taken from Chernoff (1973), reproduction of Table 2.2 and Fig. 2.5 being with permission from the American Statistical Association. Table 2.2 shows the raw data, twelve variables representing mineral contents assayed on fifty-three equally spaced specimens taken from a 4500-foot core drilled into a Colorado mountainside. Figure 2.5 shows the fifty-three faces as obtained by Chernoff. The patterns in the data are shown up well by the

Table 2.2 Data on twelve variables representing mineral contents from a 4500-foot core drilled from a Colorado mountainside. Reproduced from Chernoff (1973) with permission from the American Statistical Association

Specimen identity	X_1	X_2	X_3	X_4	X_5	X_6	X_7	X_8	X_9	X_{10}	X_{11}	X_{12}
200	320	105	057	050	001	001	001	060	020	250	210	370
201	280	150	040	050	001	001	001	060	040	210	130	420
202	260	165	033	050	001	001	001	060	010	250	090	440
203	305	110	044	040	001	001	001	050	050	260	140	250
204	290	160	035	035	001	001	001	050	020	210	060	510
205	275	130	047	035	001	001	001	050	020	230	090	570
206	280	155	035	035	001	001	001	080	020	270	170	400
207	300	115	050	060	001	001	001	120	010	280	190	300
208	250	130	041	030	005	001	001	070	030	250	110	330
209	285	120	047	040	001	001	001	070	010	240	170	280
210	280	105	047	070	001	001	001	060	020	370	070	300
211	300	135	050	040	001	001	001	120	060	250	160	200
212	280	110	056	050	001	001	001	150	010	280	270	280
213	305	080	065	080	005	001	001	130	010	300	260	260
214	230	175	029	035	001	001	001	270	030	250	140	240

Table 2.2 (Continued)

Specimen identity	X_1	X_2	X_3	X_4	X_5	X_6	X_7	X_8	X_9	X_{10}	X_{11}	X_{12}
215	325	060	052	090	001	001	001	160	010	280	260	170
216	270	170	025	040	001	001	001	160	010	290	070	330
217	250	185	031	025	001	001	001	120	001	260	080	330
218	260	185	030	015	001	001	001	270	080	480	010	330
219	270	185	032	010	005	001	001	180	040	450	020	220
220	325	045	053	005	020	001	001	600	080	660	020	250
221	315	090	047	005	020	001	001	410	200	600	060	260
222	335	100	047	010	040	001	001	360	080	590	110	170
223	310	010	049	005	080	018	001	640	240	630	060	190
224	410	001	049	001	075	032	001	760	440	800	001	001
225	360	001	048	001	080	055	001	770	260	770	010	010
226	310	015	051	001	105	036	001	660	380	640	001	010
227	420	005	049	001	095	056	001	620	520	680	001	001
228	415	020	049	005	025	036	001	370	220	340	001	001
229	420	005	041	001	070	060	001	630	510	580	001	001
230	450	005	040	001	090	070	001	690	570	630	001	001
231	395	001	025	015	100	071	001	580	530	560	001	010
232	380	010	027	025	035	039	001	350	320	400	001	270
233	430	010	025	030	030	025	001	340	340	360	001	200
234	410	075	022	010	005	015	001	170	170	170	001	060
235	520	055	024	040	005	001	001	210	190	190	001	180
236	385	135	018	010	005	008	001	140	200	260	001	020
237	535	065	010	020	001	001	001	110	230	270	001	070
238	550	095	001	010	001	001	001	050	230	270	001	030
239	510	100	001	001	001	001	001	190	150	230	001	110
240	510	095	001	040	001	001	001	140	100	150	001	040
241	385	180	010	001	001	001	001	050	050	300	001	050
242	505	125	001	001	001	001	001	001	200	130	001	030
243	470	090	001	020	001	001	001	160	300	380	001	060
244	465	110	001	035	001	001	001	260	440	500	001	060
245	400	140	001	015	001	023	001	330	400	390	001	040
246	415	105	015	025	040	032	001	220	190	270	001	010
247	435	075	010	015	001	069	001	370	360	500	001	010
248	370	145	010	010	005	012	040	130	080	330	001	030
249	380	210	001	001	001	001	020	070	001	050	001	030
250	430	065	001	005	020	001	075	130	070	300	001	020
251	420	080	030	001	005	026	001	050	100	350	001	050
252	425	060	035	005	001	001	030	100	010	340	001	010
Min.	250	001	001	001	001	001	001	001	001	050	001	001
Max.	520	210	065	090	105	071	075	770	570	800	270	570

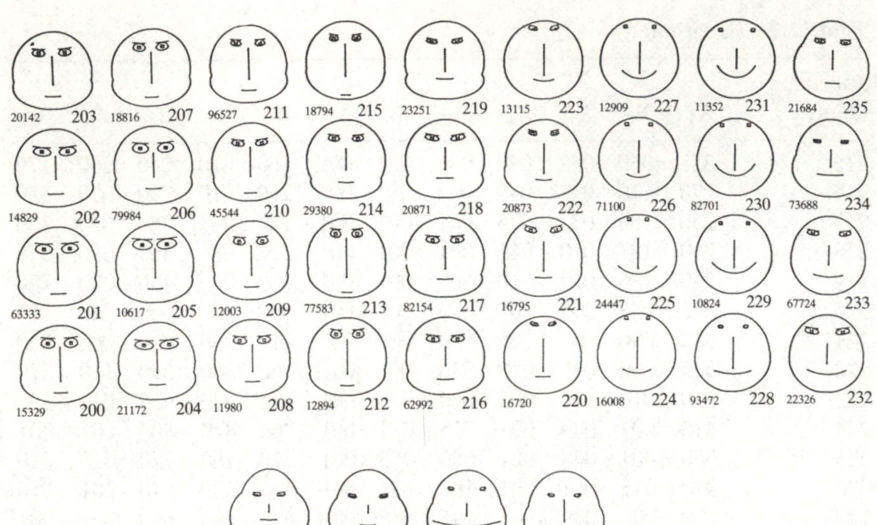

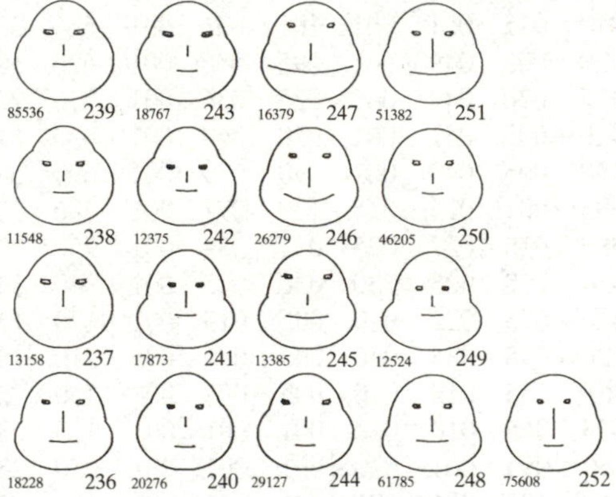

Fig. 2.5 The faces obtained by Chernoff (1973). Reproduced with permission from the American Statistical Association

contrasts in these faces. For example, note the three distinct groups of head shapes (specimens 200–219, 220–233, and 234–252 roughly), with prominent smiles for the middle group.

As with the earlier glyphs, a different association of variables and features would have led to a very different visual impression of the faces. This aspect has been investigated by Chernoff and Rizvi (1975), who estimate that the effect of a random permutation in the assignment of parameters may affect the error rate in a classification task using these faces by a factor of about 25 per cent. Further investigation of the assignment of variables to features is reported by Jacob (1983).

As an intermediate picture between the simple boxes and the more complex faces, Kleiner and Hartigan (1981) have suggested trees and castles as suitable representations for multivariate observations. Their technique involves some prior analysis before production of the representation, and so is outside the scope of the present section. However, in addition to the new ideas that are put forward, this reference also provides an excellent summary of the other methods of multivariate display discussed in the present chapter.

2.1.3 Function representations

In the pictorial representations described above, each multivariate observation which has values $x_1, x_2, \ldots, x_p$ on p variables is represented by an object familiar in everyday life, such as a box or a face. As an alternative to an everyday object, the observation may be represented by some *mathematical function.* The most familiar such function in present use is the Fourier curve suggested by Andrews (1972). The *Andrews curve* corresponding to a p-variable observation $\boldsymbol{x}' = (x_1, \ldots, x_p)$ is obtained by calculating the function

$$f_{\boldsymbol{x}}(t) = x_1/\sqrt{2} + x_2 \sin t + x_3 \cos t + x_4 \sin 2t + x_5 \cos 2t + \ldots$$

and plotting this function over the range $-\pi < t < \pi$. A set of observations will thus appear as a set of curves drawn across the plot. The programming effort required for this plot is small, but an output device with relatively high precision is required. The function $f_{\boldsymbol{x}}(t)$ possesses a number of interesting and useful properties. These are discussed by Andrews (1972), but may conveniently be summarized as follows.

1. The function representation preserves the means, i.e. if $\bar{\boldsymbol{x}}$ is the mean of a set of n multivariate observations $\boldsymbol{x}_i$, then at each point t in $-\pi < t < \pi$ the function corresponding to $\bar{\boldsymbol{x}}$ is the mean of the n functions corresponding to the individual observations.

2. The function representation preserves the Euclidean distances, i.e. the Euclidean distance d between two observations $\boldsymbol{x}' = (x_1, \ldots, x_p)$ and $\boldsymbol{y}' = (y_1, \ldots, y_p)$ is directly proportional to the Euclidean distance D between the two corresponding functions. We have already encountered the Euclidean distance between $\boldsymbol{x}'$ and $\boldsymbol{y}'$ (dissimilarity coefficient (i) in Section 1.4) as $d^2 = \sum_{i=1}^{p} (x_i - y_i)^2$, but now need to define the distance D between the two corresponding functions $f_{\boldsymbol{x}}(t)$ and $f_{\boldsymbol{y}}(t)$. To do this, treat D^2 as being composed of the sum of squares of all possible differences $(x_i - y_i)$, and make the usual association between summation for discrete variables and integration for continuous ones. Thus at the value t_0 the difference between the two functions is $f_{\boldsymbol{x}}(t_0) - f_{\boldsymbol{y}}(t_0)$, and t_0 can take any value between $-\pi$ and π. Hence we can define $D^2 = \int_{-\pi}^{+\pi} \{f_{\boldsymbol{x}}(t) -$

$f_y(t)\}^2\, dt$. With these definitions Andrews establishes that $D^2 = \pi d^2$, so that one distance can be obtained from knowledge of the other.

3. The representation preserves the variances. If the variables in the data matrix are uncorrelated with common variance σ^2 then the function value at t has variance $\sigma_f^2 = \sigma^2(\frac{1}{2} + \sin^2 t + \cos^2 t + \sin^2 2t + \cos^2 2t + \ldots)$. If p is odd this variance reduces to the constant $\frac{1}{2}\sigma^2 p$, while if p is even then the variance lies between $\frac{1}{2}\sigma^2(p-1)$ and $\frac{1}{2}\sigma^2(p+1)$. In either case the relative dependence of σ_f^2 on t is either very slight or non-existent, so that the variability of the plotted function is almost constant across the graph.

These properties suggest various uses that can be made of Andrews curves. We shall encounter some more uses later in this book, but the present emphasis is on methods of exploratory data representation, and the salient property in this context is (2) above. This property tells us that two observations with very similar sets of variable values will be represented by curves that remain close together for all values of t. Two observations whose variable values differ markedly, on the other hand, will be represented by curves which also differ markedly, at least for some values of t. Hence we can use this property for the possible identification of clusters, outliers, or other peculiarities of the data.

However, two major problems may arise with this use of Andrews plots, both of which are common also to the other techniques of data display that have already been discussed. The first is that only a fairly limited number of observations may be plotted on the same diagram, since a multitude of disparate curves is more likely to lead to confusion than to clarity. In general, the most useful procedure may be to supplement an overall plot of the data with separate plots of subsets of the data. The first plot will give a general impression of the whole sample while the subsequent plots will provide the necessary detail that may be missed in the larger plot. The second main problem is one that has already been mentioned in connection with both glyphs and Chernoff faces, namely that quite a different picture can be obtained if the variables are permuted and the functions $f_{x_i}(t)$ are recalculated. In some cases, such as the use of Andrews plots for representing canonical variates (see Chapter 11), the variables have a natural ordering and there is no question as to which variable to choose as x_1, which as x_2, and so on. If this is not the case, however, then care must be exercised in choosing the order of variables for plotting. In general, the low frequencies (i.e. x_1, x_2, x_3) are distinguished more readily on the plots than the high frequencies (i.e. x_{p-2}, x_{p-1}, x_p). For this reason, it is best to associate those variables that are thought to be the most important for the purpose at hand with the low frequencies (although this may not always be easy to determine in practice). Andrews recommends that a principal component analysis (see next section) be done *before* the functions are calculated, and x_1 in the functions should be associated with the first principal component, x_2 with the second principal component, and so on.

To illustrate some of the above points, consider the famous set of iris data first discussed by Fisher (1936). These data are reproduced in Table 2.3 with permission from Cambridge University Press, and consist of the four variables sepal length, sepal width, petal length, and petal width measured on fifty specimens of each of three types of iris, namely *Iris setosa, Iris versicolor,* and *Iris virginica.* Figure 2.6 shows the Andrews curves obtained when the first ten individuals from each of the three groups are plotted using x_1 as sepal length, x_2 as sepal width, x_3 as petal length, and x_4 as petal width. The ten *setosa* individuals form a fairly compact group of curves, having a smaller mean level and smaller amplitudes than the remaining twenty curves. Distinguishing between the *versicolor* and *virginica* groups is much more difficult, the overall curves not appearing to fall into two distinct groups. A possible distinguishing feature is obtained at a particular value of t near the positive end of the range. Taking $t = 2.8$ we see that at this value there appear to be three tight groups of curves and one outlier. This illustrates a further property mentioned by Andrews, namely that the function values associated with a particular value t_0 of t present the projections of the original data points on to the vector $(1/\sqrt{2}, \sin t_0, \cos t_0, \sin 2t_0, \cos 2t_0, \ldots)$. The projection on to this one-dimensional space may reveal patterns obscured by other dimensions of the data. Hence it is always worthwhile examining Andrews plots to see if particular values of t exist for which striking patterns (e.g. clusters and/or outliers as in Fig. 2.6) are evident.

Now consider a permutation of the variables. Figure 2.7 shows the Andrews plots obtained with x_1 as petal length, x_2 as petal width, x_3 as sepal length, and x_4 as sepal width. Examination of the data in Table 2.3 suggests that, with this arrangement of variables, the *setosa* individuals will have the smallest mean level, the *virginica* individuals will have the largest mean level, and the *versicolor* individuals will come between the two (but nearer to *virginica* than to *setosa*). This is in fact the case, with the *setosa* individuals forming the bunch of curves at the bottom of the set in Fig. 2.7, and the *versicolor* and *virginica* individuals mixed above them. Note, however, that because of the relative similarity of variables x_3 and x_4 across all groups, the frequencies of all curves are very similar. Hence Fig. 2.7 gives a poorer means of distinguishing the three groups than Fig. 2.6. The outlier is evident in both figures, but is also more pronounced in Fig. 2.6 than in Fig. 2.7.

In addition to the use of such plots for simple data representation,

Table 2.3 Fisher's *Iris* data. Reproduced from Fisher (1936) with permission from Cambridge University Press

Iris setosa				*Iris versicolor*				*Iris virginica*			
Sepal length	Sepal width	Petal length	Petal width	Sepal length	Sepal width	Petal length	Petal width	Sepal length	Sepal width	Petal length	Petal width
5.1	3.5	1.4	0.2	7.0	3.2	4.7	1.4	6.3	3.3	6.0	2.5
4.9	3.0	1.4	0.2	6.4	3.2	4.5	1.5	5.8	2.7	5.1	1.9
4.7	3.2	1.3	0.2	6.9	3.1	4.9	1.5	7.1	3.0	5.9	2.1
4.6	3.1	1.5	0.2	5.5	2.3	4.0	1.3	6.3	2.9	5.6	1.8
5.0	3.6	1.4	0.2	6.5	2.8	4.6	1.5	6.5	3.0	5.8	2.2
5.4	3.9	1.7	0.4	5.7	2.8	4.5	1.3	7.6	3.0	6.6	2.1
4.6	3.4	1.4	0.3	6.3	3.3	4.7	1.6	4.9	2.5	4.5	1.7
5.0	3.4	1.5	0.2	4.9	2.4	3.3	1.0	7.3	2.9	6.3	1.8
4.4	2.9	1.4	0.2	6.6	2.9	4.6	1.3	6.7	2.5	5.8	1.8
4.9	3.1	1.5	0.1	5.2	2.7	3.9	1.4	7.2	3.6	6.1	2.5
5.4	3.7	1.5	0.2	5.0	2.0	3.5	1.0	6.5	3.2	5.1	2.0
4.8	3.4	1.6	0.2	5.9	3.0	4.2	1.5	6.4	2.7	5.3	1.9
4.8	3.0	1.4	0.1	6.0	2.2	4.0	1.0	6.8	3.0	5.5	2.1
4.3	3.0	1.1	0.1	6.1	2.9	4.7	1.4	5.7	2.5	5.0	2.0
5.8	4.0	1.2	0.2	5.6	2.9	3.6	1.3	5.8	2.8	5.1	2.4
5.7	4.4	1.5	0.4	6.7	3.1	4.4	1.4	6.4	3.2	5.3	2.3
5.4	3.9	1.3	0.4	5.6	3.0	4.5	1.5	6.5	3.0	5.5	1.8
5.1	3.5	1.4	0.3	5.8	2.7	4.1	1.0	7.7	3.8	6.7	2.2
5.7	3.8	1.7	0.3	6.2	2.2	4.5	1.5	7.7	2.6	6.9	2.3
5.1	3.8	1.5	0.3	5.6	2.5	3.9	1.1	6.0	2.2	5.0	1.5
5.4	3.4	1.7	0.2	5.9	3.2	4.8	1.8	6.9	3.2	5.7	2.3
5.1	3.7	1.5	0.4	6.1	2.8	4.0	1.3	5.6	2.8	4.9	2.0
4.6	3.6	1.0	0.2	6.3	2.5	4.9	1.5	7.7	2.8	6.7	2.0
5.1	3.3	1.7	0.5	6.1	2.8	4.7	1.2	6.3	2.7	4.9	1.8
4.8	3.4	1.9	0.2	6.4	2.9	4.3	1.3	6.7	3.3	5.7	2.1
5.0	3.0	1.6	0.2	6.6	3.0	4.4	1.4	7.2	3.2	6.0	1.8
5.0	3.4	1.6	0.4	6.8	2.8	4.8	1.4	6.2	2.8	4.8	1.8
5.2	3.5	1.5	0.2	6.7	3.0	5.0	1.7	6.1	3.0	4.9	1.8
5.2	3.4	1.4	0.2	6.0	2.9	4.5	1.5	6.4	2.8	5.6	2.1
4.7	3.2	1.6	0.2	5.7	2.6	3.5	1.0	7.2	3.0	5.8	1.6
4.8	3.1	1.6	0.2	5.5	2.4	3.8	1.1	7.4	2.8	6.1	1.9
5.4	3.4	1.5	0.4	5.5	2.4	3.7	1.0	7.9	3.8	6.4	2.0
5.2	4.1	1.5	0.1	5.8	2.7	3.9	1.2	6.4	2.8	5.6	2.2
5.5	4.2	1.4	0.2	6.0	2.7	5.1	1.6	6.3	2.8	5.1	1.5
4.9	3.1	1.5	0.2	5.4	3.0	4.5	1.5	6.1	2.6	5.6	1.4
5.0	3.2	1.2	0.2	6.0	3.4	4.5	1.6	7.7	3.0	6.1	2.3
5.5	3.5	1.3	0.2	6.7	3.1	4.7	1.5	6.3	3.4	5.6	2.4
4.9	3.6	1.4	0.1	6.3	2.3	4.4	1.3	6.4	3.1	5.5	1.8
4.4	3.0	1.3	0.2	5.6	3.0	4.1	1.3	6.0	3.0	4.8	1.8

Table 2.3 (Continued)

Iris setosa				*Iris versicolor*				*Iris virginica*			
Sepal length	Sepal width	Petal length	Petal width	Sepal length	Sepal width	Petal length	Petal width	Sepal length	Sepal width	Petal length	Petal width
5.1	3.4	1.5	0.2	5.5	2.5	4.0	1.3	6.9	3.1	5.4	2.1
5.0	3.5	1.3	0.3	5.5	2.6	4.4	1.2	6.7	3.1	5.6	2.4
4.5	2.3	1.3	0.3	6.1	3.0	4.6	1.4	6.9	3.1	5.1	2.3
4.4	3.2	1.3	0.2	5.8	2.6	4.0	1.2	5.8	2.7	5.1	1.9
5.0	3.5	1.6	0.6	5.0	2.3	3.3	1.0	6.8	3.2	5.9	2.3
5.1	3.8	1.9	0.4	5.6	2.7	4.2	1.3	6.7	3.3	5.7	2.5
4.8	3.0	1.4	0.3	5.7	3.0	4.2	1.2	6.7	3.0	5.2	2.3
5.1	3.8	1.6	0.2	5.7	2.9	4.2	1.3	6.3	2.5	5.0	1.9
4.6	3.2	1.4	0.2	6.2	2.9	4.3	1.3	6.5	3.0	5.2	2.0
5.3	3.7	1.5	0.2	5.1	2.5	3.0	1.1	6.2	3.4	5.4	2.3
5.0	3.3	1.4	0.2	5.7	2.8	4.1	1.3	5.9	3.0	5.1	1.8

Andrews has also considered questions of significance and confidence limits for these plots assuming that the p variables in the sample are uncorrelated. Property (3) listed at the beginning of this section is used in this context. Also, Goodchild and Vijayan (1974) have extended Andrews' results to the case of correlated variables. These applications will be discussed in Chapter 9.

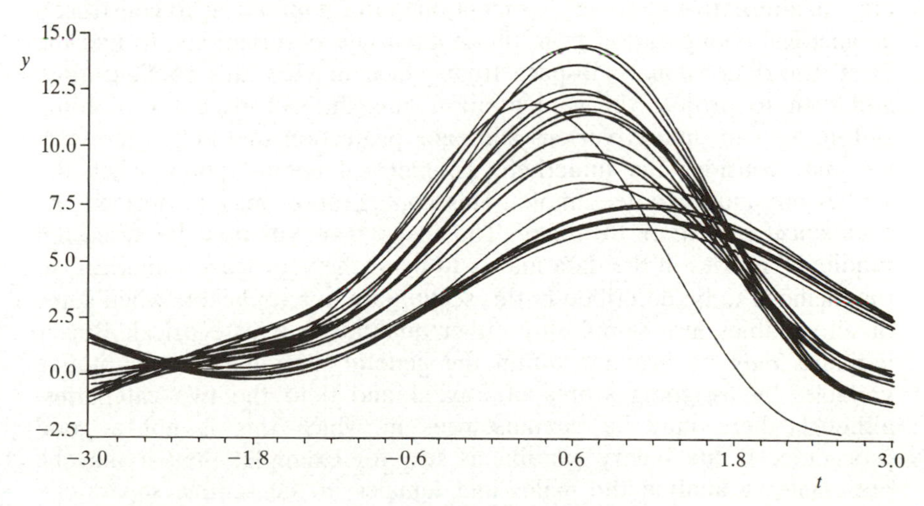

Fig. 2.6 Andrews curves representation of thirty *Iris* specimens

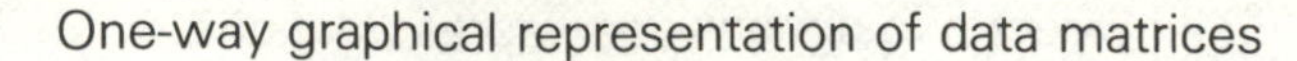

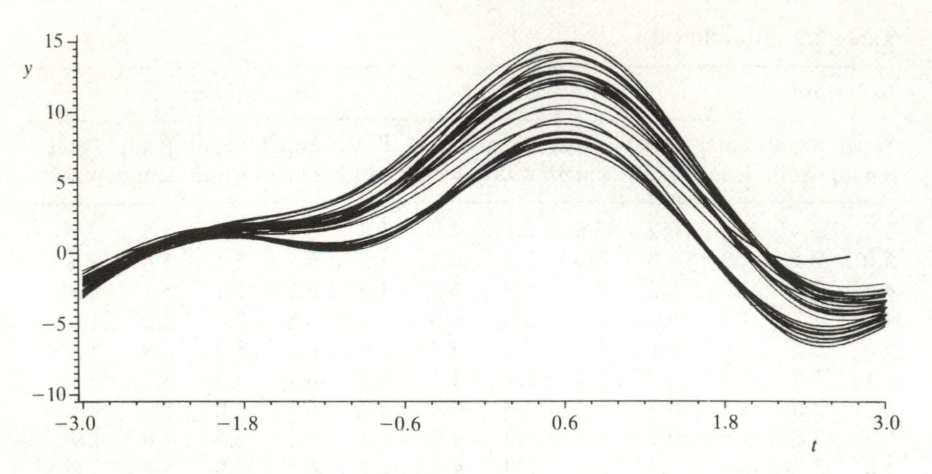

Fig. 2.7 Andrews curves representation of the same thirty *Iris* specimens as in Fig. 2.6, but with variables permuted

2.2 Representations based on subspace projection

2.2.1 Geometrical models of data

The methods described in the previous section all, in essence, provide a transformation of each multivariate observation in such a way as to enable that observation to be depicted graphically in direct form. Thus the observations can be transformed into boxes, faces, castles, curves, etc. An alternative approach, as mentioned in Chapter 1, is to construct a geometrical configuration from the multivariate observations, to find the 'best' two-dimensional subspace from which to view this configuration, and then to project the configuration into this subspace for plotting. Before we can discuss various subspace projection methods, therefore, we must consider the underlying geometrical configuration which describes our multivariate sample. This configuration may be termed the *geometrical model* of the data. The ideas that will now be discussed require all entries in the data matrix to be at the very least numerical, so the methods to be described in this section are not applicable when some or all variables in a sample are either qualitative or categorical. Binary variables *may* be brought within the general umbrella of quantitative variables by assigning scores of, say, 1 and 0 to the two categories, although there may be circumstances in which this is not a good procedure. If the binary variable is sex, for example, then it may be preferable to analyse the males and females in the sample separately, rather than to conduct an overall analysis which includes a variable on which all males are scored as 1, and all females as 0 (or vice versa). Similar considerations can be applied to any categorical variable, of course. However, if such a variable has many categories then there may be too few observations in each category to enable separate analyses to be conducted and an overall analysis incorporating the categorical variable is necessary. Extensions of the current techniques to incorporate such cases will be described in the next chapter. For the present we assume that all variables in the sample are quantitative.

Two separate, but complementary, geometrical models may be arrived at, depending on whether it is the individuals or the variables in the sample that are the focus of attention. Consider first the individuals. In this case the sample consists of n entities to be represented, and each of these entities has p responses recorded for it. If p is equal to 2 then we have a situation familiar to many readers, since this is precisely the starting point of simple regression or correlation analysis in which the relationship is sought between the two variables that have been measured on each of n individuals. A familiar tool in such an analysis is the scatter plot, in which the two values observed on a given individual are taken as the coordinates of a point representing that individual, referred to two perpendicular axes. The whole sample is thus represented by n points in two dimensions, these dimensions being determined by the usual horizontal and vertical axes of a piece of graph paper. Bearing in mind the discussion about abstract p-dimensional Euclidean spaces of Chapter 1, the extension of the scatter plot to the general multivariate case is immediate. We now have p values observed on each individual in the sample. If we associate each variable with an axis in p-dimensional space, these axes being orthogonal to each other, and take the p responses measured on a given individual as the coordinates of a point representing this individual along each of these axes, then we have a geometrical model of the sample as n points in p-dimensional space. If p is greater than 3 then we cannot hope ever to construct this model in its entirety, and the objective of the techniques in this section is to find the 'best' approximation to this model that we *can* construct. However, it is clear there must also be a dual representation of the sample, in which the individuals and the variables interchange their roles. If the focus of attention now passes to the variables, then we must consider p entities, each of which has n observations attached to it. Using a similar reasoning to that above, we can associate each of the n sample members with an (orthogonal) axis in n-dimensional space and treat the n values attached to a given variable as the coordinates of a point representing that variable referred to these axes. The sample is now represented by a configuration of p points in n dimensions.

In the former representation (n points in p dimensions), the points relate to the objects or individuals in the sample, so where necessary this representation will be referred to as the *object space*. The latter representation will therefore be designated the *variable space,* since the p points in n dimensions represent the variables in the sample. Most common multivariate techniques use the former representation, and this representation will generally be assumed in most of the following unless the contrary is stated. However, appreciation of the latter is often of implicit benefit in understanding a technique, even if it is not used explicitly in that technique.

2.2.2 Properties of the models

Before moving on to consider various subspace projection techniques, it is worth pausing to examine in a little more detail the two geometrical models introduced above, in particular establishing some of their fundamental properties. Let us consider first the object space. Fig. 2.8 shows an example of such a space in the case $p = 3$. It is common to find that such a set of points is referred to in textbooks as a data 'swarm' in p dimensions; but we should not forget that as p increases so does the amount of space available for the points to occupy, and so the swarm becomes less and less dense. The points are customarily regarded as lying within a hyperellipsoid as shown in Fig. 2.8 (for reasons that will be explained in Chapter 6). The point A has coordinates given by the means of the p variate values in the sample, and hence may be regarded as the centre of gravity, or just the centre, of this hyperellipsoid. The data are said to be *mean-centered* if the mean of the corresponding variate is

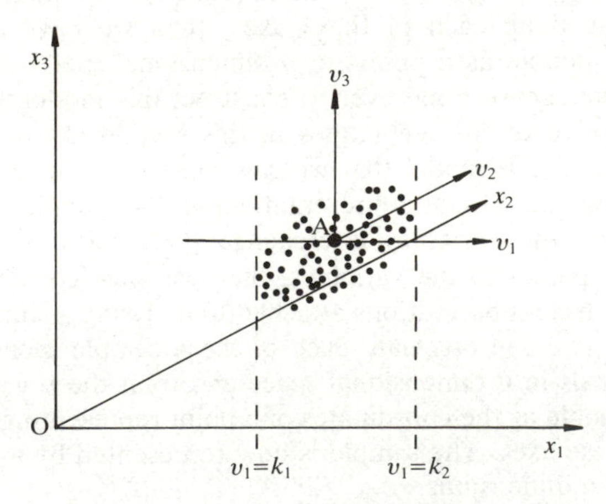

Fig. 2.8 Swarm of points in three-dimensional 'object' space

subtracted from each data value. Mean-centering corresponds to a rigid translation of the axes. Thus if the data represented in Fig. 2.8 are mean-centered then the points all remain as they are, but the axes x_1, x_2, and x_3 are moved so that they are parallel to the original ones with the origin at A instead of O. The new axes are denoted by v_1, v_2, and v_3 in Fig. 2.8.

The data swarm may be sandwiched between two hyperplanes for any single variate x_i, and these hyperplanes are given by $x_i = k_1$ and $x_i = k_2$ (say) where k_1 and k_2 are constants. An example is shown in Fig. 2.8 for the axis v_1 (after mean-centering). The distance between the hyperplanes defines the 'spread' for that variate, and will be related to the standard deviation of the data values for that variate. If the swarm is a normal data swarm (see Chapter 7), then the distance between the hyperplanes will be approximately six times the standard deviation for that variate. The data are said to be *standardized* if each data value has the mean of the corresponding variate subtracted from it, and if the result is then divided by the standard deviation of the same variate. In the geometrical representation this standardization induces the rigid translation of axes described above followed by a scaling of the units on each axis so that the distance between the bounding hyperplanes is approximately the same (about 6 units) for each axis. Clearly the effect of this is to confine the swarm to a box with side 6, but the swarm will still be elliptical and not spherical. This is illustrated in Fig. 2.9.

The *correlation* between any two variates is a measure of the departure from circularity of the projection of the swarm (after standardization) on to the plane formed by the two variate coordinate axes. In other words, it is a measure of the eccentricity of the ellipse so formed. We may also, in this space, consider the relationship between individuals, and this is done most naturally by associating the dissimilarity between two individuals with the Euclidean distance between their corresponding points in the space (dissimilarity coefficient (i) of Section 1.4). For this reason,

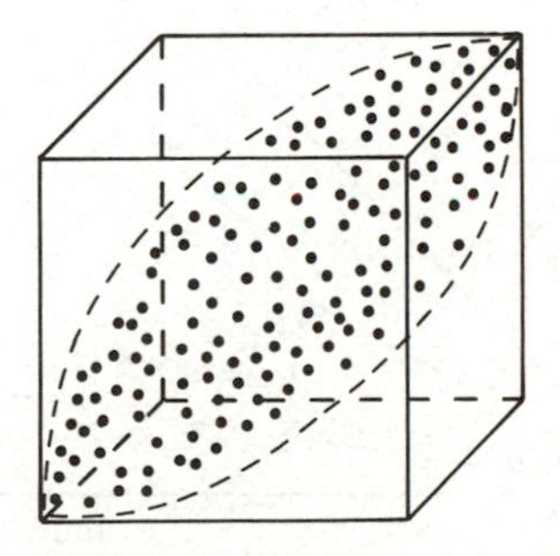

Fig. 2.9 Ellipsoidal data swarm enclosed by cubic box

coefficient (i) is the most common measure of dissimilarity between individuals when quantitative variables are involved.

Let us now turn to the variable space, in which the p variables are represented by points in a space of n dimensions. In fact we usually consider the data to have been mean-centered, so this space is actually of $n-1$ dimensions (the hyperplane perpendicular to the unit vector), but this loss of a dimension is not of any material significance. More pertinently, although it is correct to say that each variable is represented by a point in this space, it is most fruitful to think of it as represented by the *vector* from the origin to that point. Properties of the variables are then reflected by the properties of these vectors. Clearly such a representation will be conceptual for all but those trivial samples where $n \leqslant 3$, but the following properties can be related to the simple illustration in Fig. 2.10.

The length $|\boldsymbol{x}_j|$ of $\boldsymbol{x}_j$ is the corrected sum of squares $c_{jj} = \sum_{i=1}^{n} (x_{ij} - \bar{x}_j)^2$ for the jth variate, while the inner product of $\boldsymbol{x}_i$ and $\boldsymbol{x}_j$ is the corrected sum of products $c_{ij} = \sum_{s=1}^{n} (x_{si} - \bar{x}_i)(x_{sj} - \bar{x}_j)$ between the ith and jth variates. The angle between two vectors (e.g. θ, shown between variates 1 and 2 in Fig. 2.10) is a measure of how much the variates deviate from describing the same data apart from scaling, i.e. the converse of the correlation between the variates. In fact, the cosine of this angle is the correlation coefficient between the corresponding variates.

Simple regression has a direct representation in this space also. The simple regression of variable i on variable j is the projection of $\boldsymbol{x}_i$ on to $\boldsymbol{x}_j$ (see also Fig. 2.10). The resolution of $\boldsymbol{x}_i$ into components along and perpendicular to $\boldsymbol{x}_j$ is the division into model (i.e. regression of $\boldsymbol{x}_i$ on $\boldsymbol{x}_j$)

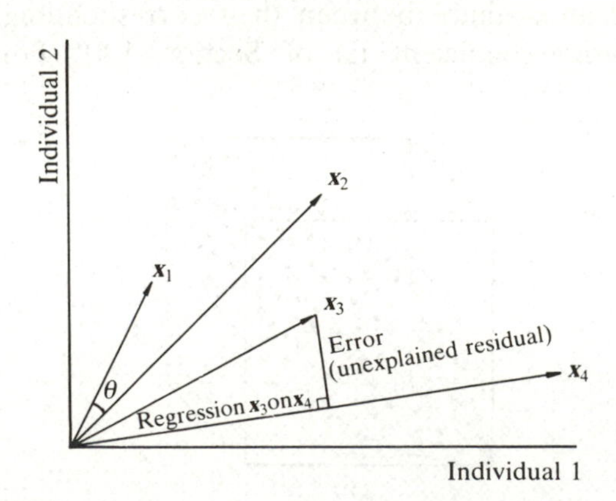

Fig. 2.10 Representation of variables by vectors in the 'variable' space

and error. The length of the projection is the model sum of squares, and that of the perpendicular the residual sum of squares. This resolution can be readily generalized to multiple regression by projecting the dependent variable on to the hyperplane of the regressor variables.

The object and variable spaces are clearly not independent entities. Since both are derived from the same data matrix then there must be some connection between them. There is, however, a certain lack of symmetry in the manner in which relationships are calculated in the two spaces, namely as a distance in the object space and as an angle in the variable space. Multivariate techniques that have as their starting point an $(n \times n)$ matrix of distances are often referred to as *Q-techniques,* while those that have as their starting point a $(p \times p)$ matrix of correlations, covariances, or angles are often referred to as *R-techniques.* We shall be encountering similarities and differences between these classes of techniques in succeeding chapters.

2.2.3 Principal component analysis: geometrical concepts

Now let us turn to the object space, and consider the representation of a sample of n individuals in this space. In order to allow a natural labelling of axes, and to avoid confusion in some of the mathematical details later, we shall denote variates by upper-case letters and values observed on them by lower-case letters in this chapter. (This distinction will be dropped in later chapters.) As a simple example, consider an imaginary data set in which the heights and weights of n university students have been recorded. This two-dimensional sample might then be represented (after mean-centering) by a scatter plot such as given in Fig. 2.11.

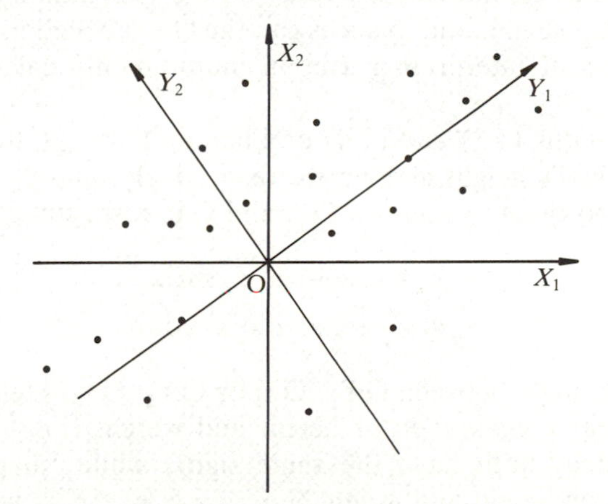

Fig. 2.11 Imaginary data on heights (x_1) and weights (x_2) of n individuals

The axes OX_1 and OX_2 of this representation have been determined by the variables height and weight respectively, in accordance with our postulates for such a representation. However, it is the *configuration of points* that is of intrinsic interest to the data analyst, and the frame of reference with respect to which these points are plotted is to some extent irrelevant. For example, we could rotate the axes to the new positions OY_1 and OY_2 without altering the configuration of points, and relate the points to these new axes for any future analysis without changing the outcome of that analysis.

It may also occur that such new axes may actually carry some useful meaning to the investigator; indeed they may even be more meaningful than the original measurements that were taken. Consider again the new axes OY_1 and OY_2 of Fig. 2.11, and imagine all the data points successively compressed on to each of these axes. Points at the extreme right-hand end of OY_1 will correspond to individuals that have large values (in general) of both height and weight, i.e. to 'large' individuals; while points at the extreme left-hand end of OY_1 will correspond to individuals that have small values (in general) of both height and weight, i.e. to 'small' individuals. Thus an individual's value on the OY_1 axis is a reflection of that individual's size, and hence this axis can be labelled as a 'size' axis. Now, consider axis OY_2. Points at the top of this axis will tend to correspond to individuals whose weight (X_2) is large in relation to their height (X_1); while points at the bottom of this axis will tend to correspond to individuals whose height is large in relation to their weight. The former are short fat individuals, while the latter are tall thin individuals. A point in the middle of this axis will tend to come from an individual whose height and weight are approximately in the correct proportion. Thus an individual's value on the OY_2 axis is a reflection of that individual's shape, and this axis can thus be labelled as a 'shape' axis. Such a process of interpreting axes in multidimensional space is called *reification*.

Now, from eqns (1.3) and (1.4) of Chapter 1 we see that if x_1 and x_2 are an individual's height and weight respectively, and y_1 and y_2 are that individual's coordinate values on OY_1 and OY_2 respectively, then

$$y_1 = x_1 \cos \alpha + x_2 \sin \alpha$$

$$y_2 = -x_1 \sin \alpha + x_2 \cos \alpha$$

where α is the angle between OX_1, OY_1 or OX_2, OY_2. Hence we see that 'size' is a linear *combination* of height and weight (i.e. $y_1 = a_1x_1 + a_2x_2$ where a_1 and a_2 both have the same sign), while 'shape' is a linear *contrast* between height and weight (i.e. $y_2 = b_1x_1 + b_2x_2$ where b_1 and b_2 are of opposite sign).

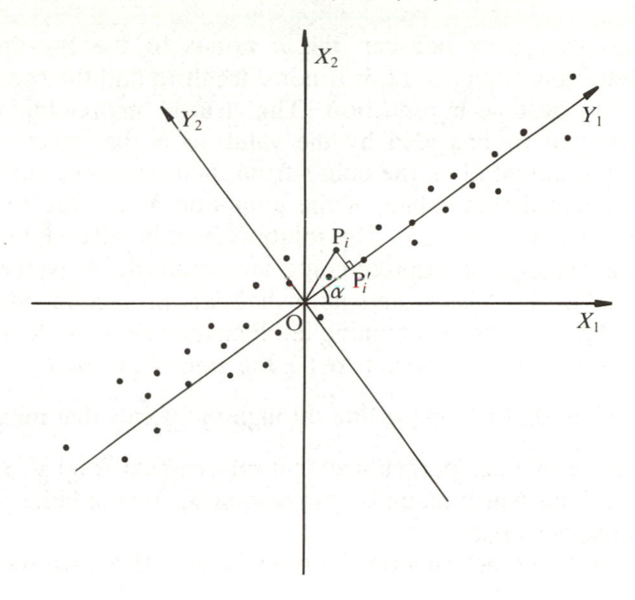

Fig. 2.12 Second sample of heights (x_1) and weights (x_2) of n individuals

Next, suppose that we have a further sample of n individuals whose heights (X_1) and weights (X_2) have been measured, and this sample is represented in Fig. 2.12. Once again the 'size' (Y_1) and 'shape' (Y_2) axes have been drawn in. This time, however, we see that while there is a wide spread of sample values (i.e. projection of the points) on the Y_1 axis, the spread of values on the Y_2 axis is relatively small. This means that the individuals in the second sample have widely differing sizes but very similar shapes. Following the discussion in Section 1.2, we would conclude that the one-dimensional approximation to the two-dimensional picture of Fig. 2.12 that is obtained by projecting the points on to the Y_1 axis is a reasonably good approximation, since the displacements of all the points in this projection are quite small. In other words, we could characterize the differences between the n individuals sufficiently well if, instead of quoting the height x_1 and weight x_2 for each, we were simply to quote its index of size, $y_1 = x_1 \cos \alpha + x_2 \sin \alpha$. Replacing the two original variables X_1 and X_2 by a single derived variable Y_1 in this way effects a reduction in dimensionality from 2 to 1, because we could now represent the sample data by plotting the individuals according to their Y_1 values. Different values of α give different derived variables Y_1 and hence different plots. Among all these plots there will be one that is deemed to be 'best', in the sense that it provides the truest impression of the relationships that exist between the n points in the two-dimensional picture. Returning to Fig. 2.12, it is not difficult to find the condition that will yield the best such reduction. The 'truest' impression of all the relationships will be provided by the value of α that gives rise to the smallest displacement of all the points from their original positions. Since the eventual coordinate values of the points on Y_1 are their orthogonal projections on to the line OY_1, the solution must be given by the line for which these orthogonal displacements are smallest. A typical point is indicated in Fig. 2.12 by P_i and its orthogonal projection on to OY_1 is denoted by P_i'. The above reasoning led Pearson (1901) to define the line OY_1 of closest fit to the points to be the one obtained by minimizing $\sum_{i=1}^{n} (P_iP_i')^2$. Note that this is the line through the points that minimizes the sum of squares of their perpendicular displacements from it, in contrast to regression lines which minimize the sum of squares of either horizontal or vertical displacements.

Applying Pythagoras' Theorem to the triangle OP_iP_i', we see that

$$(OP_i)^2 = (OP_i')^2 + (P_iP_i')^2.$$

Summing over all the points P_i, it thus follows that

$$\sum_{i=1}^{n} (OP_i)^2 = \sum_{i=1}^{n} (OP_i')^2 + \sum_{i=1}^{n} (P_iP_i')^2.$$

Hence

$$\frac{1}{n-1}\sum_{i=1}^{n} (OP_i)^2 = \frac{1}{n-1}\sum_{i=1}^{n} (OP_i')^2 + \frac{1}{n-1}\sum_{i=1}^{n} (P_iP_i')^2$$

on dividing by $(n-1)$.

Now the left-hand side of this equation, $\frac{1}{n-1}\sum_{i=1}^{n} (OP_i)^2$, is fixed for any given sample irrespective of the coordinate system that is employed. Hence choosing OY_1 to minimize $\frac{1}{n-1}\sum_{i=1}^{n} (P_iP_i')^2$ must be equivalent to choosing OY_1 to maximize $\frac{1}{n-1}\sum_{i=1}^{n} (OP_i')^2$. Since O is the centroid of the points, $\frac{1}{n-1}\sum_{i=1}^{n} (OP_i')^2$ is just the sample variance when the individuals have values given by their Y_1 coordinate. Thus finding the line OY_1 that minimizes the sum of squared perpendicular deviations of the points from it is exactly equivalent to finding the line OY_1 such that the projections of the points on it have maximum variance. This is an operationally more effective approach, and was the starting point for Hotelling's (1933) derivation of *principal components*. This approach will be discussed below in more detail. In the meantime, let us note that for the simple two-dimensional case illustrated in Fig. 2.12, the link between these approaches is intuitively sensible. Choosing OY_1 to ensure the smallest possible perpendicular deviation of all the points is equivalent to the choice of rectangular axes that gives the smallest spread of projections on OY_2 and hence the largest spread of projections on OY_1.

If we now consider a p-dimensional set of data, with an associated $(n \times p)$ data matrix, we can envisage a sequence of steps of the above form. The data are modelled as usual by a swarm of n points in p dimensions, each axis corresponding to a measured variable. We can thus look for a line OY_1 in this space such that the spread of the n points when projected on to this line is a maximum. This operation defines a derived variable of the form $Y_1 = a_1X_1 + a_2X_2 + \ldots + a_pX_p$, with coefficients a_i satisfying $\sum_{i=1}^{p} a_i^2 = 1$ (see Chapter 1) and determined by the requirement that the variance of Y_1 be maximized. Having obtained OY_1, consider the $(p-1)$ dimensional subspace *orthogonal to* OY_1, and look for the line OY_2 in this subspace such that the spread of points when projected on to this line is a maximum. This is equivalent to seeking a line OY_2 at right angles to OY_1, such that the spread of points when they are projected on to OY_2 is as large as possible (although, clearly, this spread must be no greater than the spread along OY_1). Having obtained OY_2, we then consider the $(p-2)$ dimensional subspace orthogonal to *both* OY_1 and OY_2. Thus we look for a line OY_3 which is at right angles to both OY_1 and OY_2 such that the spread of points when projected along OY_3 is as large as possible after the spreads on OY_1 and OY_2 have been taken into account. This process can be continued until we have obtained p mutually orthogonal lines OY_i $(i = 1, \ldots, p)$. Each of these lines defines a derived variable $Y_i = a_{i1}X_1 + a_{i2}X_2 + \ldots + a_{ip}X_p$ $(i = 1, \ldots, p)$, where the constants a_{ij} are determined by the requirement that the variance of Y_i is a maximum but subject to the constraint of orthogonality with each Y_j $(j < i)$, i.e. subject to $\sum_{k=1}^{p} a_{ik}a_{jk} = 0$ for $j < i$ (see Section 1.3.3) as well as $\sum_{k=1}^{p} a_{ik}^2 = 1$ for each i. The Y_i thus obtained are called the (sample) *principal components* of the system, and the process of obtaining them is called *principal component analysis*.

The usefulness of principal component analysis in multivariate data display should now become evident. The p-dimensional geometrical model formed from the sample can be considered to be the 'true' picture of the data. If we wish to obtain the 'best' r $(<p)$ dimensional representation of this p-dimensional 'true' picture, then we simply need

to project the points into the r-dimensional subspace defined by the first r principal components $Y_1, Y_2, \ldots, Y_r$. This is done very easily, as follows. The original data matrix contains the values x_{ij} observed on the jth variable of the ith individual in the sample ($i = 1, \ldots, n; j = 1, \ldots, p$). Principal component analysis has yielded the r sets of coefficients in the definitions of the principal components $Y_i = a_{i1}X_1 + a_{i2}X_2 + \ldots + a_{ip}X_p$ ($i = 1, \ldots, r$). The r principal component values for the sth sample member are thus given by

$$y_{s1} = a_{11}x_{s1} + a_{12}x_{s2} + \ldots + a_{1p}x_{sp}$$
$$y_{s2} = a_{21}x_{s1} + a_{22}x_{s2} + \ldots + a_{2p}x_{sp}$$
$$\vdots$$
$$y_{sr} = a_{r1}x_{s1} + a_{r2}x_{s2} + \ldots + a_{rp}x_{sp}.$$

These r values $y_{s1}, y_{s2}, \ldots, y_{sr}$ are usually called the r *principal component scores* for the sth individual. Repeating this calculation for all sample members converts the original data matrix having n rows and p columns into a matrix of principal component scores having n rows and r columns. These values can then be (notionally) plotted against r orthogonal axes to obtain the required projection of the sample into the r-dimensional subspace defined by the first r principal components. Of course, by far the most useful case is when $r = 2$, as we can then actually plot the (y_{i1}, y_{i2}) pairs as a scatter diagram. This plot gives us a two-dimensional approximation to the true configuration of points in p dimensions.

That we can reasonably describe this as the 'best' two-dimensional representation for viewing the data is justified by the variance property of principal components, and its connection with orthogonal projection. Since the first two principal components are the two linear combinations of the original variables that have the largest sample variance associated with them, the plane defined by these two components is the two-dimensional subspace of the original p-dimensional space which has associated with it the greatest scatter of all n points. Hence it is the plane in which there is the greatest possibility of showing up all the essential features of the original swarm of n points. The link between scatter and projection of points discussed earlier in the context of Fig. 2.12 also extends easily to this case. The plane defined by the first two principal components is the best-fitting plane to the n points in Pearson's (1901) sense of giving rise to the smallest perpendicular displacement of the points.

The beauty of this method is that if we now wish to find the best-fitting three-dimensional representation of points, we need only add the scores on the third principal component Y_3 to the scores that already exist on the first two components Y_1 and Y_2, and use the triples (y_{i1}, y_{i2}, y_{i3}) as the coordinates of the points in three dimensions. Thus if we have already plotted our sample in the plane of the first two components, and we now wish to see how the configuration changes when we pass to a three-dimensional representation, we need only attach an arrow to each point in the two-dimensional plot pointing due east if that point's third component score is positive, and due west if the score is negative. If the length of the arrow is made proportional to the absolute value of the score, then all these arrows show how far the points would be displaced vertically from their current two-dimensional positions when passing to the third dimension. A fourth dimension can be added in similar fashion by representing a point's score on the fourth component by an arrow running in the north (positive)–south (negative) direction. Such a diagram has already been illustrated in Fig. 2.3.

This procedure thus provides us with a method of displaying high-dimensional multivariate data in a few dimensions. We must never forget, however, that the resulting display is merely an *approximation* to the true configuration. Before drawing conclusions about any pattern in such a display, we must be on the look-out for any distortion that may have been caused by projecting the points into the selected subspace. If we have chosen to look at the first r principal components, such distortion will be detected either by a large score y_{st} for an individual s on a component t greater than r, or by a large 'residual sum of squares' $y_{s,r+1}^2 + y_{s,r+2}^2 + \ldots + y_{s,p}^2$ for an individual s. Such large values indicate that this individual has undergone a large displacement when being projected into the space of the first r components, and will almost certainly be misrepresented in the r-dimensional picture.

Principal component analysis is catered for within any general-purpose statistical computing package, and the user need only supply the data matrix together with a few simple instructions to obtain all the quantities described above (principal component coefficients a_{ij}, principal component scores y_{st}) as well as, perhaps, some automatic computer plots of the scores. For the benefit of more mathematically minded readers, we now augment the above description of principal component analysis by giving some mathematical and technical details. Some general notation and terminology is first established, and this should be fully assimilated as it will be used throughout the remainder of the book. Section 2.2.4 may be omitted without loss of continuity, however, and less mathematical readers may pass directly to the practical details of Section 2.2.5 if they wish.

In general, we suppose that a $(p \times 1)$ vector $\boldsymbol{X}' = (X_1, X_1, \ldots, X_p)$ of variate values is observed on each of n independent units, giving rise to the usual $(n \times p)$ data matrix whose (i, j)th element is x_{ij}. Denote the p

values observed on the ith unit by the vector $\boldsymbol{x}_i$. The mean of the jth variable in the sample is given by $\bar{x}_j = \frac{1}{n}\sum_{i=1}^{n} x_{ij}$. Let all these means be collected together in the *sample mean vector* $\bar{\boldsymbol{x}}' = (\bar{x}_1, \bar{x}_2, \ldots, \bar{x}_p)$. The variance of the jth variable is given by $s_{jj} = \frac{1}{n-1}\sum_{i=1}^{n} (x_{ij} - \bar{x}_j)^2$, and the covariance between the jth and kth variables is given by

$$s_{jk} = \frac{1}{n-1}\sum_{i=1}^{n} (x_{ij} - \bar{x}_j)(x_{ik} - \bar{x}_k).$$

Let all these variances and covariances be collected together in the *sample covariance matrix* $\mathbf{S}$ which has (j, k)th element s_{jk}. By expanding the right-hand side as a matrix product (Appendix A, Section A2), it can be verified easily that $\mathbf{S} = \frac{1}{n-1}\sum_{i=1}^{n} (\boldsymbol{x}_i - \bar{\boldsymbol{x}})(\boldsymbol{x}_i - \bar{\boldsymbol{x}})'$. The matrix

$$\mathbf{C} = (n-1)\mathbf{S} = \sum_{i=1}^{n} (\boldsymbol{x}_i - \bar{\boldsymbol{x}})(\boldsymbol{x}_i - \bar{\boldsymbol{x}})',$$

with (j,k)th element $c_{jk} = \sum_{i=1}^{n} (x_{ij} - \bar{x}_j)(x_{ik} - \bar{x}_k)$ is often referred to as the *corrected sum of squares and products* (*SSP*) *matrix*.

2.2.4 Principal component analysis: mathematical details

Consider forming a scalar variate Y as a linear combination $a_1X_1 + a_2X_2 + \ldots + a_pX_p$ of the original variates X_i. Thus we can more concisely write $Y = \boldsymbol{a}'\boldsymbol{X}$, where $\boldsymbol{a}' = (a_1, a_2, \ldots, a_p)$. Then the value of Y corresponding to the ith sample member will be given by $y_i = a_1x_{i1} + a_2x_{i2} + \ldots + a_px_{ip} = \boldsymbol{a}'\boldsymbol{x}_i$, while the mean of Y over the n sample members will clearly be equal to $\bar{y} = a_1\bar{x}_1 + a_2\bar{x}_2 + \ldots + a_p\bar{x}_p = \boldsymbol{a}'\bar{\boldsymbol{x}}$. Next, consider the sample variance of Y. This is given by $S_Y^2 = \frac{1}{n-1}\sum_{i=1}^{n} (y_i - \bar{y})^2$. But $y_i - \bar{y} = \boldsymbol{a}'\boldsymbol{x}_i - \boldsymbol{a}'\bar{\boldsymbol{x}} = \boldsymbol{a}'(\boldsymbol{x}_i - \bar{\boldsymbol{x}})$ from above. Also, since the transpose of a scalar is equal to that scalar, we can equivalently write $(y_i - \bar{y}) = (\boldsymbol{x}_i - \bar{\boldsymbol{x}})'\boldsymbol{a}$. Thus, on multiplying these two expressions together, we find

$$(y_i - \bar{y})^2 = \boldsymbol{a}'(\boldsymbol{x}_i - \bar{\boldsymbol{x}})(\boldsymbol{x}_i - \bar{\boldsymbol{x}})'\boldsymbol{a}$$

so that

$$\begin{aligned}\sum_{i=1}^{n} (y_i - \bar{y})^2 &= \sum_{i=1}^{n} \boldsymbol{a}'(\boldsymbol{x}_i - \bar{\boldsymbol{x}})(\boldsymbol{x}_i - \bar{\boldsymbol{x}})'\boldsymbol{a} \\ &= \boldsymbol{a}'\left\{\sum_{i=1}^{n} (\boldsymbol{x}_i - \bar{\boldsymbol{x}})(\boldsymbol{x}_i - \bar{\boldsymbol{x}})'\right\}\boldsymbol{a}\end{aligned}$$

since $\boldsymbol{a}$ is constant over i, and hence

$$S_Y^2 = \frac{1}{n-1}\sum_{i=1}^{n}(y_i - \bar{y})^2 = \boldsymbol{a}'\left\{\frac{1}{n-1}\sum_{i=1}^{n}(\boldsymbol{x}_i - \bar{\boldsymbol{x}})(\boldsymbol{x}_i - \bar{\boldsymbol{x}})'\right\}\boldsymbol{a}$$

$$= \boldsymbol{a}'\mathbf{S}\boldsymbol{a} \quad \text{from the definition of } \mathbf{S} \text{ above.}$$

It is easy to verify, by multiplying out the vectors, that the *quadratic form* $\boldsymbol{a}'\mathbf{S}\boldsymbol{a}$ is expressible in terms of the elements of $\boldsymbol{a}$ and $\mathbf{S}$ as $\sum_{i=1}^{p}\sum_{j=1}^{p} a_i a_j s_{ij}$.

We can now turn to the derivation of the principal components. In the course of the earlier geometrical discussion of this chapter, we showed that the first principal component could be viewed as the line through the p-dimensional space containing our n points with respect to which the projections of the n points have maximum variance. The fundamental geometry of Chapter 1 established that a line could be simply represented by a vector $\boldsymbol{a}$ of unit length. The condition for such a vector is that the sum of squares of its elements is unity, i.e. that $\boldsymbol{a}'\boldsymbol{a} = \sum_{i=1}^{p} a_i^2 = 1$. Hence, algebraically, we can *define* the first principal component to be the linear combination $Y_1 = \boldsymbol{a}_1'\boldsymbol{X}$ of the original variables that gives rise to the maximum value of $\boldsymbol{a}_1'\mathbf{S}\boldsymbol{a}_1$ subject to the constraint that $\boldsymbol{a}_1'\boldsymbol{a}_1 = 1$. We therefore require to find the vector $\boldsymbol{a}_1$ that satisfies these conditions. It can be verified from any standard text on differential calculus (e.g. Courant (1936), pp. 190–9) that the conditions for vector $\boldsymbol{a}_1$ to maximize $\boldsymbol{a}_1'\mathbf{S}\boldsymbol{a}_1$ subject to the constraint $\boldsymbol{a}_1'\boldsymbol{a}_1 = k$ are precisely the same as those for the vector $\boldsymbol{a}_1$ to maximize $\boldsymbol{a}_1'\mathbf{S}\boldsymbol{a}_1 - l_1(\boldsymbol{a}_1'\boldsymbol{a}_1 - k)$, where l_1 is a constant known as the *Lagrange* (*undetermined*) *multiplier*. This latter maximization can be effected in the usual way by differential calculus. In the present case we thus proceed as follows.

Let

$$V_1 = \boldsymbol{a}_1'\mathbf{S}\boldsymbol{a}_1 - l_1(\boldsymbol{a}_1'\boldsymbol{a}_1 - 1)$$

$$= \sum_{i=1}^{p}\sum_{j=1}^{p} a_{1i}a_{1j}s_{ij} - l_1\left(\sum_{i=1}^{p} a_{1i}^2 - 1\right),$$

where $\boldsymbol{a}_1' = (a_{11}, a_{12}, \ldots, a_{1p})$. Then

$$\frac{\partial V_1}{\partial a_{1k}} = 2\sum_{j=1}^{p} s_{kj}a_{1j} - 2l_1 a_{1k} \qquad (k = 1, \ldots, p).$$

To find the vector $\boldsymbol{a}_1' = (a_{11}, \ldots, a_{1p})$ maximizing V_1, we thus set $\frac{\partial V_1}{\partial a_{1k}} = 0$ for all k and solve the resulting set of simultaneous equations.

Now

$$\frac{\partial V_1}{\partial a_{1k}} = 0 \rightarrow \sum_{j=1}^{p} s_{kj}a_{1j} = l_1 a_{1k}.$$

The left-hand side of this equation is the kth element of $\mathbf{S}\boldsymbol{a}_1$, while the right-hand side is the kth element of $l_1\boldsymbol{a}_1$. Thus when all k equations are treated simultaneously, it follows that the maximizing value of $\boldsymbol{a}_1$ must satisfy

$$\mathbf{S}\boldsymbol{a}_1 = l_1\boldsymbol{a}_1 \tag{2.1}$$

i.e.

$$(\mathbf{S} - l_1\mathbf{I})\boldsymbol{a}_1 = \boldsymbol{0}. \tag{2.2}$$

This is a homogeneous set of p equations in p unknowns, and the theory of equations tells us that for a non-trivial ($\boldsymbol{a} \neq \boldsymbol{0}$) solution we require

$$|\mathbf{S} - l_1\mathbf{I}| = 0. \tag{2.3}$$

Hence l_1 is an eigenvalue of $\mathbf{S}$, and the solution $\boldsymbol{a}_1$ is its corresponding eigenvector normalized so that $\boldsymbol{a}_1'\boldsymbol{a}_1 = 1$. However, there are p eigenvalues of $\mathbf{S}$, so we must still determine which of these is the required one. If we pre-multiply eqn (2.1) by $\boldsymbol{a}_1'$, we obtain $\boldsymbol{a}_1'\mathbf{S}\boldsymbol{a}_1 = l_1\boldsymbol{a}_1'\boldsymbol{a}_1$, and since we must have $\boldsymbol{a}_1'\boldsymbol{a}_1 = 1$ this yields $l_1 = \boldsymbol{a}_1'\mathbf{S}\boldsymbol{a}_1$. Thus l_1 is the sample variance of Y_1, and since we are attempting to maximize this variance then l_1 must be chosen to be the *largest* eigenvalue of $\mathbf{S}$. It therefore follows that the coefficients $\boldsymbol{a}_1$ in the first principal component $Y_1 = \boldsymbol{a}_1'\boldsymbol{X}$ are given by the elements of the eigenvector $\boldsymbol{a}_1$ that corresponds to the largest eigenvalue l_1 of $\mathbf{S}$.

Now consider the second principal component. The earlier geometrical discussion established this to be the line, in p-dimensional space, that is orthogonal to the line defining the first principal component, and that is such that the projections of the n points on this line give rise to the largest variance possible among all such lines. Thus we look for a second linear combination $Y_2 = \boldsymbol{a}_2'\boldsymbol{X}$ of the original variables. Conditions for this combination to define a line are as before, namely $\boldsymbol{a}_2'\boldsymbol{a}_2 = \sum_{i=1}^{p} a_{2i}^2 = 1$, where $\boldsymbol{a}_2' = (a_{21}, a_{22}, \ldots, a_{2p})$. In addition, this line must be orthogonal to the one defining the first principal component, the condition for which is that $\boldsymbol{a}_2'\boldsymbol{a}_1 = \boldsymbol{a}_1'\boldsymbol{a}_2 = \sum_{i=1}^{p} a_{1i}a_{2i} = 0$. The variance of Y_2 is clearly $\boldsymbol{a}_2'\mathbf{S}\boldsymbol{a}_2$, so maximization of this variance subject to the above constraints will again involve Lagrange multipliers. With two constraints we need two such multipliers, l_2 and m say, and we thus require to maximize

$$V_2 = \boldsymbol{a}_2'\mathbf{S}\boldsymbol{a}_2 - l_2(\boldsymbol{a}_2'\boldsymbol{a}_2 - 1) - m(\boldsymbol{a}_2'\boldsymbol{a}_1)$$
$$= \sum_{i=1}^{p}\sum_{j=1}^{p} a_{2i}a_{2j}s_{ij} - l_2\left(\sum_{i=1}^{p} a_{2i}^2 - 1\right) - m\left(\sum_{i=1}^{p} a_{1i}a_{2i}\right).$$

Thus

$$\frac{\partial V_2}{\partial a_{2k}} = 2\sum_{j=1}^{p} s_{kj}a_{2j} - 2l_2a_{2k} - ma_{1k} \qquad (k = 1, \ldots, p).$$

Identifying the terms on the right-hand side as the kth elements of $2\mathbf{S}\boldsymbol{a}_2$, $2l_2\boldsymbol{a}_2$ and $m\boldsymbol{a}_1$ respectively, we see that setting $\frac{\partial V_2}{\partial a_{2k}} = 0$ for all k leads to the equation

$$(\mathbf{S} - l_2\mathbf{I})\boldsymbol{a}_2 = \tfrac{1}{2}m\boldsymbol{a}_1. \tag{2.4}$$

On pre-multiplying (2.4) by $\boldsymbol{a}_1'$, and remembering that $\boldsymbol{a}_1'\boldsymbol{a}_1 = 1$ while $\boldsymbol{a}_1'\boldsymbol{a}_2 = 0$, we see that

$$\boldsymbol{a}_1'\mathbf{S}\boldsymbol{a}_2 = \tfrac{1}{2}m. \tag{2.5}$$

Pre-multiplying (2.1) by $\boldsymbol{a}_2'$ and remembering that $\boldsymbol{a}_2'\boldsymbol{a}_1 = 0$, however, yields $\boldsymbol{a}_2'\mathbf{S}\boldsymbol{a}_1 = 0$. Since $\boldsymbol{a}_1'\mathbf{S}\boldsymbol{a}_2$ is a scalar quantity and $\mathbf{S}$ is a symmetric matrix, then $\boldsymbol{a}_1'\mathbf{S}\boldsymbol{a}_2 = \boldsymbol{a}_2'\mathbf{S}\boldsymbol{a}_1 = 0$. Substitution in (2.5) thus yields $m = 0$, and hence from (2.4) we see that the coefficients $\boldsymbol{a}_2$ of the second principal component also satisfy $(\mathbf{S} - l_2\mathbf{I})\boldsymbol{a}_2 = \boldsymbol{0}$. Entirely analogous reasoning to that used for the first component, together with the fact that the variance of the second component must be maximum *after the first component has been accounted for,* shows that the coefficients of the second principal component $Y_2 = \boldsymbol{a}_2'\boldsymbol{X}$ are given by the elements of the eigenvector $\boldsymbol{a}_2$ corresponding to the second largest eigenvalue l_2 of $\mathbf{S}$.

The above process can be continued for all principal components. Thus in the case of the jth component, we look for the linear combination $Y_j = \boldsymbol{a}_j'\boldsymbol{X}$ of the original variables which is orthogonal to all previous combinations, i.e. $\boldsymbol{a}_j'\boldsymbol{a}_i = 0$ for all $i < j$, and which has maximum variance. Maximization of the variance subject to these constraints leads to the maximization of an expression involving j Lagrange multipliers. The formal algebra is lengthy, but the essential results are obvious generalizations of the ones above. They are summarized at the start of the next section.

2.2.5 Principal component analysis: properties and practical considerations

To summarize the essential definitions and algebraic results, the *jth principal component* is the linear combination $Y_j = \boldsymbol{a}_j'\boldsymbol{X}$ which has greatest sample variance for all $\boldsymbol{a}_j$ satisfying $\boldsymbol{a}_j'\boldsymbol{a}_j = 1$ and $\boldsymbol{a}_j'\boldsymbol{a}_i = 0$ $(i < j)$. The *coefficients* $\boldsymbol{a}_j'$ are given by the elements of the eigenvector corresponding to the jth largest eigenvalue l_j of $\mathbf{S}$, and $\text{var}(Y_j) = l_j$. If $\boldsymbol{a}_j' = (a_{j1}, a_{j2}, \ldots, a_{jp})$, then $a_{jk}\sqrt{(l_j)}$ is sometimes called the *loading* of the kth original variable on the jth component. Finally, if $\boldsymbol{x}_i' = (x_{i1}, x_{i2}, \ldots, x_{ip})$ are the values of the original p variables on the ith individual, then that individual's value on the jth principal component is computed as $y_{ij} = \boldsymbol{a}_j'\boldsymbol{x}_i = a_{j1}x_{i1} + a_{j2}x_{i2} + \ldots + a_{jp}x_{ip}$ $(j = 1, \ldots, p)$. This is known as the ith individual's *score* on the jth component.

Principal component analysis has been motivated within this chapter as a suitable projection method for viewing a high-dimensional set of (quantitative) data in a few dimensions. For this purpose, we simply need to calculate and plot the scores of the n sample members using the first few principal components as axes. Thus calculation of the first two principal components and evaluation of the $2n$ resultant scores would be sufficient for this purpose, and we do not require any more detail of the technique than has been given above. Such usage, however, not only ignores the considerable extra interpretative ability regarding a set of data that is afforded by a principal component analysis, but also skates over some awkward problems associated with the technique. We therefore consider these aspects further in this section. First it is worth clearing up some terminological confusion that can arise. As has been mentioned above, the principal components in any application are defined to be the new variables Y_i $(i = 1, \ldots, p)$ that are derived as linear combinations $\boldsymbol{a}_i'\boldsymbol{X}$ of the original variables, while the elements of $\boldsymbol{a}_i$ are called the coefficients of the ith principal component. Popular usage, however, often ascribes the word 'component' to the vector of coefficients, so that $\boldsymbol{a}_i$ is then itself called the ith principal component. This ties in with the geometric interpretation discussed earlier, wherein a vector represents a line in space, and the principal components are thought of as lines in the p-dimensional sample space defining the hyperplane on to which the sample members are projected. Providing that this is kept in mind, there should usually be no confusion arising from such duality of definition.

The first problem encountered with principal component analysis is in deciding precisely on what data the analysis should be performed. Should we take the raw data as they are presented to us, or should we first transform them in some way? It might be thought that if the aim is simply to 'look' at the data in a suitably chosen set of dimensions, then we should not tamper with the data in any way before the analysis. Various objections may be raised to this, however, particularly if we want to go beyond this simple usage of the technique and we want to use the derived principal components either in some interpretative fashion (to be discussed later) or as input to some subsequent analyses. Recollect that the first principal component is the linear combination $Y_1 = \boldsymbol{a}_1'\boldsymbol{X}$ of the original variables which has maximum variance among all such linear combinations. There are two potentially unsatisfactory features of this definition as regards application to some data sets. Suppose, for example, all the measured variables represent vastly different entities (e.g. X_1 is the height of a plant in millimetres, X_2 is the dry weight of the plant in milligrams, X_3 is the number of leaves on the plant, X_4 is the presence or absence of a certain fertilizer in the soil in which the plant was grown). What does a linear combination of such different entities possibly represent, and will it be at all useful for interpretative purposes? Alternatively, suppose that all the measured variables represent the same or similar entities, but some are much more variable than others. Such differences in variance might arise because of different scales of measurement, different magnitudes of measurement, or some combination of the two factors (e.g. X_1 is the height of a plant in centimetres, X_2 is the average length of its leaves in millimetres, X_3 is the average width of its leaves in millimetres). The variables that exhibit a lot of variance will correspond to axes in the geometrical representation along which the sample members have wide scatter, while the other variables correspond to axes that by comparison have very small scatter. Following the earlier reasoning about the derivation of principal components, therefore, the new axes along which the sample members have maximum scatter will correspond almost directly with these dominant original axes, and the corresponding principal components will be simply equal to the X_i with largest variances. We will not learn anything new about the sample by performing the technique.

Both of the above possibilities therefore suggest that principal component analysis will be of maximum use, and help in understanding a sample, when all the measured variables are at least comparable in magnitude of variance as well as in terms of their units of measurement. When this is not the case, many statisticians argue that the data should be *standardized* prior to principal component analysis. Standardization is simply effected by considering each variate separately, subtracting the variate mean from each sample observation, and dividing the result by that variate's standard deviation. The outcome is to convert the sample covariance matrix **S** to the sample correlation matrix **R**, and hence the principal components are obtained as the eigenvectors of **R**. These components are, of course, different from the ones that derive from the covariance matrix **S**; what is more, as the following mathematics demonstrates, knowledge of one of these sets of components (i.e. from **R** or from **S**) does *not* enable the other set to be derived.

Let **D** denote the diagonal matrix whose diagonal elements are the same as those of **S** and whose off-diagonal elements are all zero. For non-trivial standardization we further require that the diagonal elements of **D** are not all equal.

Then $\mathbf{R} = \mathbf{D}^{-\frac{1}{2}}\mathbf{S}\mathbf{D}^{-\frac{1}{2}}$, where $\mathbf{D}^{-\frac{1}{2}}$ is the diagonal matrix whose diagonal entries are the inverses of the square roots of those of **D** (see Chapter 14). Hence if $\boldsymbol{a}$ is a vector of coefficients for a principal component derived from **R**, then

$$\mathbf{R}\boldsymbol{a} = l\boldsymbol{a} \tag{2.6}$$

where l is the variance of this principal component.

Suppose also that $\boldsymbol{b}$ is a vector of coefficients for a principal component derived from **S**, and m is the variance of this principal component, then

$$\mathbf{S}\boldsymbol{b} = m\boldsymbol{b}. \tag{2.7}$$

But since $\mathbf{R} = \mathbf{D}^{-\frac{1}{2}}\mathbf{S}\mathbf{D}^{-\frac{1}{2}}$ then $\mathbf{S} = \mathbf{D}^{\frac{1}{2}}\mathbf{R}\mathbf{D}^{\frac{1}{2}}$ so that substitution in (2.7) yields $\mathbf{D}^{\frac{1}{2}}\mathbf{R}\mathbf{D}^{\frac{1}{2}}\boldsymbol{b} = m\boldsymbol{b}$,

i.e.

$$\mathbf{R}\mathbf{D}^{\frac{1}{2}}\boldsymbol{b} = m\mathbf{D}^{-\frac{1}{2}}\boldsymbol{b}. \tag{2.8}$$

If we set $\boldsymbol{c} = \mathbf{D}^{\frac{1}{2}}\boldsymbol{b}$, then (2.8) shows that

$$\begin{aligned} \mathbf{R}\boldsymbol{c} &= m\mathbf{D}^{-1}\boldsymbol{c} \qquad (2.9) \\ &\neq l\boldsymbol{c} \quad \text{for any scalar } l \\ &\text{(since elements of } \mathbf{D} \text{ are not all equal).} \end{aligned}$$

Comparison of (2.6) and (2.9) thus establishes that the principal components derived from **R** are not equal to those derived from **S** and that knowledge of one set does not allow simple transformations to the other set.

This shows that the results of a principal component analysis will be greatly affected by whether or not the data are first standardized. A careful decision must therefore be made *before a principal component analysis is attempted* as to whether or not standardization is desirable. This decision will clearly be influenced by any peculiar circumstances pertaining to a given application of the technique. As a general guideline, however, it would seem sensible to standardize first whenever the measured variables show marked differences in variances, or whenever they concern very different measured entities or units. Unstandardized data are preferable whenever the measured variables are comparable both with respect to units and variances, or when a simple and straightforward data plot is the sole purpose of the analysis.

Irrespective of the choice of matrix (**R** or **S**) for extraction of the components the next major problem is encountered when we attempt to quantify how 'good' any given number of components are as an approximation to the 'true' p-dimensional representation of the sample. Turning this question round to extend the use of principal component analysis, we might prefer to ask in fact whether just two components are adequate to represent all the essential features of the sample. If they *are* adequate, then a two-dimensional plot will be a faithful reproduction of the sample features. If they are *not* adequate, then just how many components should we take? In fact, what is the 'essential dimensionality' of the data representation? Even if we go beyond the immediate aim of finding the best two-dimensional representation of the sample, a primary use of principal component analysis is still that of data reduction and simplification: can we reduce the number of variables from p to some smaller number m, and if we can then what is the smallest value of m that ensures sufficient detail in data pattern is retained? This objective is usually termed *dimensionality reduction*. But variance formed the basis of the criterion that was used to determine the principal components. Hence a measure of the adequacy with which the projection of the n sample points into the subspace defined by the first r principal components approximates the true scatter of n points might be based on some function of the variance of these projected points. Furthermore, we know that each eigenvalue l_i of the covariance matrix **S** gives the variance of the points with respect to the corresponding component. If the centroid of the points is at the origin of the coordinate axes, then use of Pythagoras' theorem in the same manner as that in Fig. 2.12 shows that the total variance of the points S_r^2 in the subspace defined by the first r principal components is given by the sum of the variances of the points when they are projected on to each component as axis.

Thus

$$S_r^2 = l_1 + l_2 + \ldots + l_r = \sum_{i=1}^{r} l_i.$$

Now the total variance S_p^2 of the original system is just the sum of the variances of each original variate X_i, and this is the sum of the diagonal elements of the matrix **S**. This sum is mathematically denoted as trace(**S**), and mathematical theory establishes that trace$(\mathbf{S}) = l_1 + l_2 + \ldots + l_p$. This equation is just a reflection of the fact that no change is being made to the system, the principal component transformation being simply a rotation of the frame of reference of the points, and the total variance of the X_i must equal the total variance of the Y_i. Thus $S_p^2 = \sum_{i=1}^{p} l_i$, and the proportion of total variance of the system that is 'explained' by the first r principal components is given by

$$P_r = \frac{S_r^2}{S_p^2} = \sum_{i=1}^{r} l_i \Big/ \sum_{i=1}^{p} l_i = \sum_{i=1}^{r} l_i \Big/ \text{trace}(\mathbf{S}).$$

The term P_r thus seems to be a perfectly reasonable measure of adequacy, or 'goodness-of-fit', of the projected configuration in the subspace defined by the first r principal components to the true configuration in p dimensions, and it has been used as such for many years. At a descriptive level, the analyst would hope that the chosen number of components explained at least 75–80 per cent of the available variation before the subspace representation was deemed to be an adequate approximation to the true configuration. This line of reasoning is then easily turned round when one is faced with the problem of *choosing* an appropriate number of components to represent one's data, and a common conclusion is to decide to choose the minimum r for which $P_r > 0.75$ (say). More stringently, one might require $P_r > 0.80$ or even $P_r > 0.85$. It is clear, therefore, that basing one's choice of r on this measure involves much subjectivity and arbitrariness.

Various principles have been put forward over the years in an attempt to introduce some objectivity into the process. One popular approach is to plot the l_i against i $(i = 1, \ldots, p)$ in a so-called *scree diagram.* A typical pattern is illustrated in Fig. 2.13, where the first few l_i show a

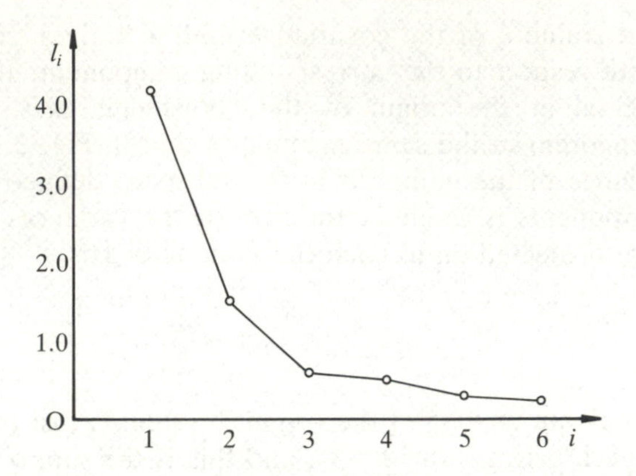

Fig. 2.13 Example of a scree diagram

sharp decline, followed by a much more gradual slope. If one argues that those dimensions corresponding to the flat portion of this graph represent undifferentiable 'noise' components of the system, then one should logically choose r as being at the foot of the initial steep decline. This leads to the popularly held belief that r should be chosen at the value of i at which the 'elbow' of the scree diagram occurs; in Fig. 2.13 this would probably be at l_3. An alternative procedure in popular use for choice of r requires computation of the average of the l_i, i.e. $\bar{l} = \frac{1}{p} \sum_{i=1}^{p} l_i$. Here it is reasoned that if the data were roughly spherical, in other words if there were no directions in the multivariate space that gave better projectional representation than the original axes, then each of the principal components would have variance approximately equal to $\bar{l}$. Thus in any principal component analysis, those components with $l_i > \bar{l}$ should be 'important', while those with $l_i < \bar{l}$ should be ignorable. Hence the suggestion is to choose r as the largest value of i for which $l_i \geqslant \bar{l}$. Whenever components are extracted from a correlation matrix then $\bar{l} = 1$, hence the common belief that the best value of r when dealing with standarized data is the number of latent roots greater than unity.

It must be stressed that all the above methods represent *ad hoc* suggestions, albeit ones that have proved popular and useful over the years by a variety of users of principal component analysis. There is no formal statistical justification for any of these methods. The ultimate test of any method must be whether or not it gives a sensible result in a particular application, and whether or not the user understands the conclusions and finds them useful. However, it is worth pointing out to interested readers that various other approaches to the problem of choice of number of components have been suggested in recent years, based on more statistically motivated principles. One such principle is error of prediction, and cross-validation is a method that uses this principle in choosing a statistical model from a range of alternatives. The use of cross-validation in this fashion has been described fully by Stone (1974), but roughly speaking it works as follows. Suppose there exists a set of predictive (statistical) models C from which a choice has to be made, and a data matrix $\mathbf{X}$ containing n individuals can be used to make this choice. If we delete the ith individual $\boldsymbol{x}_i$ from the data set, let us denote the data matrix containing the remaining $(n-1)$ individuals by $\mathbf{X}_{-i}$. For any particular model m among the set C, involving a set of parameters $\boldsymbol{\beta}^{(m)}$, we can use $\mathbf{X}_{-i}$ to obtain estimates $\hat{\boldsymbol{\beta}}^{(m)}$ of the parameters and then use the model together with estimated parameters to predict the value of $\boldsymbol{x}_i$ (or of some part of $\boldsymbol{x}_i$). If we denote this predicted value by $\hat{\boldsymbol{x}}_i^{(m)}$, we can then quantify in some suitable fashion the error between observed and predicted values. For example, if the observation to be predicted is a single scalar element x_i then a suitable measure of discrepancy might be $(x_i - \hat{x}_i^{(m)})^2$. In general, let us denote the discrepancy by $f(\hat{\boldsymbol{x}}_i^{(m)}, \boldsymbol{x}_i)$. We can repeat this process, deleting each observation $\boldsymbol{x}_i$ $(i = 1, \ldots, n)$ in turn from $\mathbf{X}$ and thus obtaining an overall measure of discrepancy $E(m) = k \sum_{i=1}^{n} f(\hat{\boldsymbol{x}}_i^{(m)}, \boldsymbol{x}_i)$ for model m. Repeating the whole process for all models $m \in C$ gives us a set of values of the discrepancy measure, and the model that should be chosen is the one yielding the smallest value of this measure.

We have so far encountered geometrical models but not statistical models. The latter will be introduced in Chapter 6. However, in the context of principal component analysis, an appropriate statistical model m has 'signal' given by the number of components to be retained, and random 'error' is superimposed on this. One possible discrepancy measure $E(m)$ is the sum of squared differences between observed data values and those predicted from an m-component model, the sum being taken over all np values in the data matrix. Such a procedure has been discussed by Wold (1978) and by Eastment and Krzanowski (1982); see also the article by Golub *et al.* (1979). An investigation, by Monte Carlo simulation, of the performance of such a method has been given by Krzanowski (1983a), where a comparison is also made with the other methods (based on the l_i) that are described above.

This concludes the discussion of the two main problems that arise either before or during a principal component analysis, namely deciding whether or not to standardize the data, and deciding on the number of components to extract in the analysis. Once these issues have been settled, and the mathematical analysis has been completed, however, there is one further stage possible at which subjectivity is heavily involved. Since the derived components are linear combinations of the original variates, it is possible that these combinations may carry some meaningful physical interpretations. If such interpretations can be attached to the components, then not only will comprehension of the results of the analysis be aided, but valuable insight may also be gained into the mechanism or structure of the situation that is being studied.

Interpretation of components is, of course, the same as interpretation of new axes in multidimensional space, which we have already discussed briefly in connection with Fig. 2.11 where the term reification was applied to the process. We must first decide which original variates X_j are 'important' in defining each principal component $Y_i = \boldsymbol{a}_i'\boldsymbol{X}$. Adopting a process of elimination, we can ignore those X_j whose coefficients a_{ij} for the ith component are near zero, as such variables do not contribute to that component. However, there is no objective, or hard and fast, rule that tells us how small a given a_{ij} must be before its corresponding variate can be ignored. Such judgement is inevitably highly subjective, and furthermore is usually made in relation to all the other coefficients in that component. Often, a coefficient may be deemed to be 'sufficiently small' in one component (relative to all the other coefficients in that component), whereas one of similar size in another component might *not* lead to discarding the corresponding X_j. This will be demonstrated in the example below. Having settled on a set of X_j which are considered to be 'important' in a given component, we must then pay due regard to the sizes and signs of their coefficients in attempting a physical interpretation of that component. The process is best illustrated by considering a detailed example.

Table 2.4 presents the correlation matrix of six bone measurements made on each of 275 white leghorn fowl, while Table 2.5 gives the results of a principal component analysis of the standardized data (Wright, 1954).

Table 2.4 Correlation matrix for white leghorn fowl

	X_1	X_2	X_3	X_4	X_5	X_6
X_1 (skull length)	1.000	0.584	0.615	0.601	0.570	0.600
X_2 (skull breadth)		1.000	0.576	0.530	0.526	0.555
X_3 (humerus)			1.000	0.940	0.875	0.878
X_4 (ulna)				1.000	0.877	0.886
X_5 (femur)					1.000	0.924
X_6 (tibia)						1.000

Table 2.5 Principal component analysis of white leghorn fowl data (after standardization)

Original variable	Component coefficients a_{ij}					
	1	2	3	4	5	6
Skull: length (X_1)	0.35	0.53	0.76	−0.04	0.02	0.00
breadth (X_2)	0.33	0.70	−0.64	0.00	0.00	0.03
Wing: humerus (X_3)	0.44	−0.19	−0.05	0.53	0.18	0.67
ulna (X_4)	0.44	−0.25	0.02	0.48	−0.15	−0.71
Leg: femur (X_5)	0.44	−0.28	−0.06	−0.50	0.65	−0.13
tibia (X_6)	0.44	−0.22	−0.05	−0.48	−0.69	0.17
Variances l_i	4.58	0.71	0.41	0.17	0.08	0.05
Percentage of total	76%	12%	7%	3%	1%	1%

The six original measurements break down naturally into three sets of two measurements each: skull measurements (length, breadth), wing measurements (humerus, ulna), and leg measurements (femur, tibia). Owing to the different sizes of these various bones, it makes sense to standardize the data at the outset and thus use the correlation matrix as the basis for the analysis. Let us now consider the interpretation of the six components in turn.

The first component by itself explains about three-quarters of the total variation in the (standardized) sample, and hence is overwhelmingly the main source of variation among the fowl. Coefficients of each X_i in this component are given in the first column of Table 2.5, and it can be seen that these coefficients are all approximately equal (at about $a_{1j} = 0.4$). While the coefficients pertaining to X_1 and X_2 are slightly smaller than the rest, we must recognize there exists a certain amount of sampling fluctuation and instability of coefficients (Krzanowski, 1984a), and we are thus justified in pronouncing such coefficients as 'equal' to the rest. Hence the first principal component can be approximately represented as $Y_1 = 0.4\ (X_1 + X_2 + \ldots + X_6)$. This combination can be seen to separate out the fowl according to their size, as large fowl will tend to have large values for each of the X_i and hence a large Y_1, while small fowl will tend to have small values for each of the X_i and hence a small Y_1. Thus we designate Y_1 as a measure of 'size', and conclude that the main distinguishing feature among the birds is simply their overall size. The second component Y_2 explains about 12 per cent of the total variation. Thus, while coming a long way behind Y_1 as a source of variation, it is nevertheless still of some significance (in the non-technical sense). Since the coefficients of Y_1 were all positive, and Y_1 is orthogonal to all succeeding components, then all components after Y_1 are bound to be contrasts, i.e. will have a mixture of positive and negative signs among their coefficients. In the case of Y_2, we note two positive, and not very different, coefficients attached to X_1 and X_2 together with four negative, and approximately equal, coefficients attached to the other variates. Thus Y_2 is approximately of the form $Y_2 = a(X_1 + X_2) - b(X_3 + X_4 + X_5 + X_6)$. The most extreme positive value on this component will be taken by a bird that has large values for both X_1 and X_2 (i.e. a large head) but small values for X_3, X_4, X_5, X_6 (i.e. small wings and legs). At the other extreme, a large negative value will be taken for Y_2 by a bird that has a small head but large wings and legs (i.e. interchanging the size of values on X_1, X_2 and X_3, X_4, X_5, X_6). Thus we see that Y_2 is a measure of 'shape' of fowl, specifically the shape obtained by contrasting head size with wing plus leg (i.e. body) size. This contrast is, moreover, the feature which best distinguishes between fowl once allowance has been made for overall size.

When we turn to Y_3, we see that the coefficients of X_3, X_4, X_5, and X_6 are all approximately zero and can be thus ignored. Coefficients of X_1 and X_2 are similar in size but opposite in sign, so that Y_3 is approximately of the form $Y_3 = a(X_1 - X_2)$. A large positive value will thus be obtained for Y_3 by a bird with large X_1 but small X_2 (i.e. one that has a long, narrow head), while a large negative value by a bird with small X_1 but large X_2 (i.e. one that has a short, broad head). 'Skull shape' is thus the third most important source of variability among fowl, and this feature explains about 7 per cent of the total observed variation. The remaining three components between them explain only a further 5 per cent of variation, and might therefore be dismissed as unimportant in a formal analysis. They each provide a well-defined interpretation, however, and so are considered further here. In the case of Y_4, we see that coefficients of X_1 and X_2 are effectively zero, while all the remainder are approximately 0.5 in magnitude. Thus Y_4 is approximately given by $Y_4 = 0.5(X_3 + X_4 - X_5 - X_6)$. Birds at the positive end of this scale are those with large wings (X_3, X_4) but small legs (X_5, X_6), while those at the negative end will have small wings but large legs. Y_4 is thus a measure of 'body shape', contrasting specifically the wings with the legs.

Y_5 and Y_6 then both provide an interesting example of the arbitrariness and subjectivity inherent in the interpretation of components. Each of these components contains one pair of coefficient values about ± 0.7, and one pair about ± 0.2. It could be argued that the latter are so heavily dominated by the former that they should be ignored (in addition to those coefficients much closer to zero). As a counter argument, it could be pointed out that the coefficients lying between -0.2 and -0.3 in Y_2

were *not* ignored. These latter, however, were four in number and their cumulative contribution was therefore at least twice as much as the ones in Y_5 and Y_6. The balance of all these arguments would suggest approximating the last two components as $Y_5 = 0.7(X_5 - X_6)$ and $Y_6 = 0.7(X_3 - X_4)$, which are readily seen to represent 'leg shape' and 'wing shape' respectively, using the same sort of reasoning as for Y_3. This completes the interpretation of the six principal components, and demonstrates that the overall variation between the fowl can be broken down into six orthogonal, physically meaningful sources. Since these sources are also ranked in order of importance by the analysis, and a quantitative measure is made as to the extent of this importance for each source, we have managed to build up a fairly comprehensive picture of the differences that exist between such fowl.

Several points that arise from this analysis should either be stressed or clarified. The first is the arbitrary and subjective nature of the emphasis to be placed on each coefficient. Such arbitrariness is most readily highlighted in the discussion of components Y_5 and Y_6. A reasonable case could perhaps be made by an analyst for *not* ignoring the coefficients ± 0.2 in these components; that analyst's interpretation of these components would then be different from the interpretations given above. Despite this minor uncertainty, we see that all six components given in Table 2.5 have a fairly clear-cut physical interpretation. The main point to stress in this discussion, however, is that this is a *very unusual occurrence* in practical applications. Much more commonly in practice, very few (if any) of the principal components will have a clear-cut interpretation, and those that do will be open to much more argument than was the case with the leghorn fowl. A typical principal component analysis will not yield coefficients that are incontrovertibly either 'ignorable' or 'retainable', and most analyses will not be so kind as to provide coefficients so comparable in magnitude as were the ones above, in Y_1, Y_3, and Y_4 in particular. In this connection, it is worth quoting the following from Marriott (1974).

> It must be emphasised that no mathematical method is, or could be, designed to give physically meaningful results. If a mathematical expression of this sort has an obvious physical meaning, it must be attributed to a lucky chance, or to the fact that the data have a strongly marked structure that shows up in the analysis. Even in the latter case, quite small sampling fluctuations can upset the interpretation; for example the first two principal components may appear in reverse order, or may become confused altogether. Reification, then, requires considerable skill and experience if it is to give a true picture of the physical meaning of the data.

The final point concerning interpretation is slightly more technical, and relates to the signs of the coefficients. Since these coefficients are elements of the eigenvectors of either **R** or **S**, and the eigenvectors are normalized to have unit sum of squared elements, there is no effect on the solution if the signs of all coefficients in a given component are reversed. (This is equivalent to multiplication of the appropriate eigenvector by -1, which may occur arbitrarily within a computer package anyway.) Such reversal of signs simply changes the polarity of the given component. For example, changing the signs in Y_3 above would yield $Y_3 = a(X_2 - X_1)$; the positive end of this component would now correspond to birds with short, broad skulls rather than long, narrow ones, but the interpretation of the component as one measuring 'skull shape' would not be altered. In some circumstances, however, such reversal of polarity may be advantageous, and in certain cases it may even aid the interpretation. It may be essential, in any case, if we want to compare an analysis with some previous results.

2.2.6 Principal component analysis as method of data display: illustrative example

Having digressed to the extent of providing details about the technique of principal component analysis, we now return to the main theme of this chapter and illustrate the use of this technique in plotting multivariate data. The following example is taken from Jeffers (1967), and is concerned with the study of variation in forty individual *alate adelges* (winged aphids). These individuals were caught in a light trap, and the nineteen variables given in Table 2.6 were measured on each. The object of the study was to determine how many distinct taxa were present in the habitat from which the trap was taking samples, and these variables were selected as being possibly of diagnostic value. *Adelges* are difficult to identify with certainty by the conventional taxonomic keys, so inspection of the data is necessary to see whether distinct grouping of the insects is evident. The sample appears to be of nineteen dimensions, so principal component analysis should be useful in providing a good projection into a few dimensions. Before doing the analysis we must first decide whether to transform or standardize the data. The variables form a mixture of lengths and counts, so standardization would seem to be a reasonable preliminary. The starting point for the analysis is thus the correlation matrix, given in Table 2.7.

There is a high degree of correlation between many of the variables, which suggests that the data actually lie in far fewer than nineteeen dimensions. In fact, Table 2.8 shows that the first two principal components explain 85 per cent of the total variability in the sample after standardization, so that we should obtain a good representation by projecting the sample into the space of the first two components. To do this we calculate the scores on the first two components for each sample

Table 2.6 Variables measured on each *Adelges* specimen

1. LENGTH	body length
2. WIDTH	body width
3. FORWING	fore-wing length
4. HINWING	hind-wing length
5. SPIRAC	number of spiracles
6. ANTSEG 1	length of antennal segment I
7. ANTSEG 2	length of antennal segment II
8. ANTSEG 3	length of antennal segment III
9. ANTSEG 4	length of antennal segment IV
10. ANTSEG 5	length of antennal segment V
11. ANTSPIN	number of antennal spines
12. TARSUS 3	leg length, tarsus III
13. TIBIA 3	leg length, tibia III
14. FEMUR 3	leg length, femur III
15. ROSTRUM	rostrum
16. OVIPOS	ovipositor
17. OVSPIN	number of ovipositor spines
18. FOLD	anal fold
19. HOOKS	number of hind-wing hooks

member, and plot these as a scatter diagram. In doing this we must remember that we are dealing with the standardized data, so the values of the orginal variables that are applied in each linear combination must be the sample values *after standardization*. We obtain the plot given in Fig. 2.14.

The plotted data suggest that there were four major groups of aphids caught in the trap. The principal component coefficients (not reproduced here) indicated that the first component Y_1 was a measure of the general size of the organism, while the second component Y_2 was essentially the number of ovipositor spines possessed by the organism. Differences in these two measures should therefore distinguish between the four groups (probably corresponding to species). The recognition of these four groups is facilitated by the identification and interpretation of the components. Furthermore, these components can be used to isolate the most important variables for distinguishing between the aphids, and in future work the number of measured variables might be reduced from nineteen to about four or five.

2.2.7 Other subspace projection methods

We have treated principal component analysis at some length, because it is a much used (and much abused) statistical technique and hence

Table 2.7 Coefficients of correlation between *Adelges* variables

1																	
+0.934	2																
+0.927	+0.941	3															
+0.909	+0.944	+0.933	4														
+0.524	+0.487	+0.543	+0.499	5													
+0.799	+0.821	+0.856	+0.833	+0.703	6												
+0.854	+0.865	+0.886	+0.889	+0.719	+0.923	7											
+0.789	+0.834	+0.846	+0.885	+0.253	+0.699	+0.751	8										
+0.835	+0.863	+0.862	+0.850	+0.462	+0.752	+0.793	+0.745	9									
+0.845	+0.878	+0.863	+0.881	+0.567	+0.836	+0.913	+0.787	+0.805	10								
−0.458	−0.496	−0.522	−0.488	−0.174	−0.317	−0.383	−0.497	−0.356	−0.371	11							
+0.917	+0.942	+0.940	+0.945	+0.516	+0.846	+0.907	+0.861	+0.848	+0.902	−0.465	12						
+0.939	+0.961	+0.956	+0.952	+0.494	+0.849	+0.914	+0.876	+0.877	+0.901	−0.447	+0.981	13					
+0.953	+0.954	+0.946	+0.949	+0.452	+0.823	+0.886	+0.878	+0.883	+0.891	−0.439	+0.971	+0.991	14				
+0.895	+0.899	+0.882	+0.908	+0.551	+0.831	+0.891	+0.794	+0.818	+0.848	−0.405	+0.908	+0.920	+0.921	15			
+0.691	+0.652	+0.694	+0.623	+0.815	+0.812	+0.855	+0.410	+0.620	+0.712	−0.198	+0.725	+0.714	+0.676	+0.720	16		
+0.327	+0.305	+0.356	+0.272	+0.746	+0.553	+0.567	+0.067	+0.300	+0.384	−0.032	+0.396	+0.360	+0.298	+0.378	+0.781	17	
−0.676	−0.712	−0.667	−0.736	−0.233	−0.504	−0.502	−0.758	−0.666	−0.629	+0.492	−0.657	−0.655	−0.687	−0.633	−0.186	+0.169	18
+0.702	+0.729	+0.746	+0.777	+0.285	+0.499	+0.592	+0.793	+0.671	+0.668	−0.425	+0.696	+0.724	+0.731	+0.694	+0.287	−0.026	−0.775

Table 2.8 Eigenvalues of the correlation matrix of *Adelges* variables

		Percentage of variability	
Component	Eigenvalue	Component	Cumulative
1	13.86	73.0	73.0
2	2.37	12.5	85.5
3	0.75	3.9	89.4
4	0.50	2.6	92.0

requires clear understanding. From the viewpoint of multivariate data display, it provides a convenient way of projecting the p-dimensional data swarm into a two-dimensional subspace so that a graphical display may be constructed easily. Its popularity stems partly from the ease of the associated computations, partly from its simple intuitive explanation, and partly from the attractive properties possessed by the components (which enable considerably more information to be extracted from the analysis than just data display). It is often thought, however, that it is the *only* approach to displaying multivariate quantitative data, which is by no means the case. Various other possibilities exist within the general categorization of 'projection methods', and these will now be considered briefly. The main problem attaching to most of these methods, however, is computational. Principal component analysis admits an analytical solution, and hence the associated computations take only fractions of a

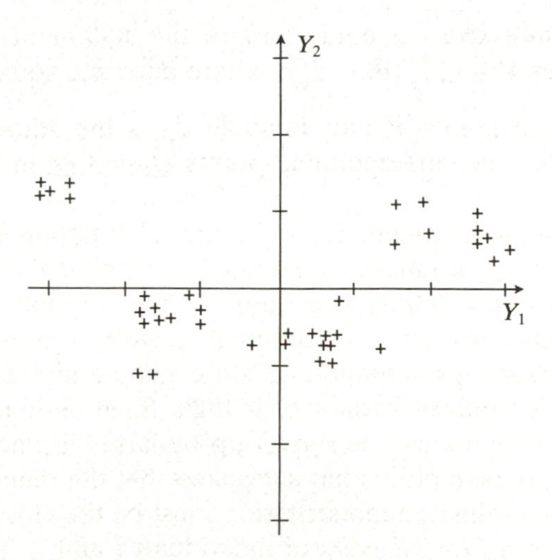

Fig. 2.14 Plotted values of the first two components for individual insects

second on modern computers. Most other projection methods are not so fortunate, and rely on iterative numerical algorithms for their solution. At best this means that considerably more computer time is required, even on very modern machines, while at worst it can mean that a solution is not obtainable because of non-convergence of the iterations. Also, excessive detail of these techniques is not warranted in a text such as this, because once the objective function defining the required projection has been specified, the remainder of the work is left in the hands of the computer. Major aspects of these methods will thus be highlighted here, and sufficient references will be provided to enable the interested reader to follow up the computational details.

It will be most convenient to contrast the following techniques directly with principal component analysis. Also, although all of these techniques are applicable for finding a representation of the p-dimensional data in any dimensionality q of subspace, we shall refer explicitly to the case $q = 2$ which is most useful for direct data display. If $P_1, P_2, \ldots, P_n$ denote the points representing the sample members in the original p dimensions, and $Q_1, Q_2, \ldots, Q_n$ denote the corresponding points in a given two-dimensional approximation, then the following two properties characterize the two dimensions found by the principal component approach:

1. The points $P_1, P_2, \ldots, P_n$ are projected *perpendicularly* into $Q_1, \ldots, Q_n$.
2. The orientation of the two-dimensional planar approximation is such that it accounts for the greatest possible *variation* of points.

Property (2), moreover, is equivalent to the statement that the chosen plane minimizes $V = \sum_i \sum_j (d_{ij}^2 - \hat{d}_{ij}^2)$, where d_{ij}^2 is the squared (Euclidean) distance between points P_i and P_j while $\hat{d}_{ij}^2$ is the squared (Euclidean) distance between the corresponding points Q_i and Q_j in the approximating plane.

Now the simple mathematics of principal component analysis is a consequence of requirement (1), namely orthogonal projection. This, however, places an arbitrary constraint on the solution. A disadvantage attaching to this constraint, in conjunction with requirement (2), is the possibility of poor representation of some sample members even though the overall proportion of variability is high. Such poor representation of some sample members will be shown up by large displacement in higher dimensions. Anderson (1971) has suggested that the main criterion of the adequacy of a coordinate representation must be the closeness with which $\hat{d}_{ij}$ approximates d_{ij} for *all pairs* of individuals i and j. This requirement leads us to seek the representation that minimizes the quantity $L =$

$\sum_i \sum_j (d_{ij} - \hat{d}_{ij})^2$, or more generally, $L^* = \sum_i \sum_j w_{ij}(d_{ij} - \hat{d}_{ij})^2$, where the w_{ij} provide differential weights for the comparisons. For example, if more importance attaches to the accurate representation of large distances, we might use $w_{ij} = d_{ij}$. The coordinates of $Q_1, \ldots, Q_n$ that minimize L or L^* can be found only by iterative approximation from an initial configuration. Anderson (1971) suggests using a steepest descent method, with the following strategy.

1. Start with the principal components solution in $t = 4$ or 5 dimensions.
2. Find the best solution in t dimensions.
3. Find the best of the t solutions in $(t-1)$ dimensions derived from starting configurations obtained by dropping each axis in turn. Reduce t to $t-1$.
4. repeat (2) and (3) until the best solution at $t = 2$ is obtained.

A detailed computational scheme is provided by Sammon (1969) for the special case of minimizing L^* with $w_{ij} = 1/d_{ij}$ (i.e. more weight attaches to the accurate representation of *small* distances). He terms this case *non-linear mapping*.

Anderson (1971) provides an example, contrasting the results obtained on minimizing L with those obtained by principal component analysis (i.e. on minimizing V). The data concern measurements taken on twenty-three Glamorganshire soils and were described by Rayner (1966). Figure 2.15 reproduces the two-dimensional representation obtained by Rayner with principal component analysis. By looking at the residuals from this analysis, Anderson suggests that sample members 8, 18 and 20 are placed misleadingly, since they all have large displacements along components higher than two; the internal evidence of Fig. 2.15 confirms the indication that sample member 8 is badly placed and the brown earths are not compact. Fig. 2.16 gives the two-dimensional configuration obtained by minimizing L. Sample member 8 has now been placed with the gley soils, and the brown earths form a more compact set, but the two lessivated soils are now separated owing to the improved position of sample 20. (Figs. 2.15 and 2.16 are reproduced from Anderson (1971) with permission from Plenum Publishing Corporation.) Anderson concluded that minimization of L has provided the best representation.

The two-dimensional projection obtained by minimizing V (i.e. principal component analysis) is an attempt to obtain the best plottable approximation to the overall p-dimensional space, in the sense of reproducing as faithfully as is possible in two dimensions the *overall* features of the sample. The modifications produced by minimizing L or L^* still have more or less this objective in view, as the resulting configurations aim to reproduce all inter-point distances as well as

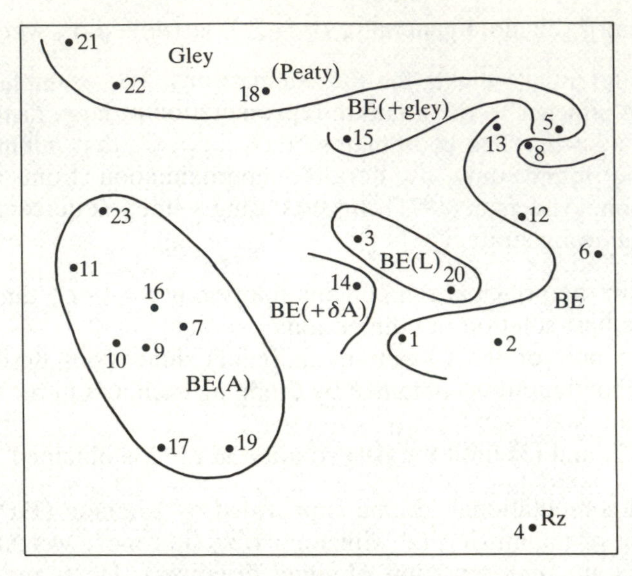

Fig. 2.15 Principal component representation of Glamorganshire soils: BE, brown earth; BE(A), acid brown earth; BE(L), lessivated brown earth; Rz, rendzina. Reproduced from Anderson (1971) with permission from Plenum Publishing Corporation

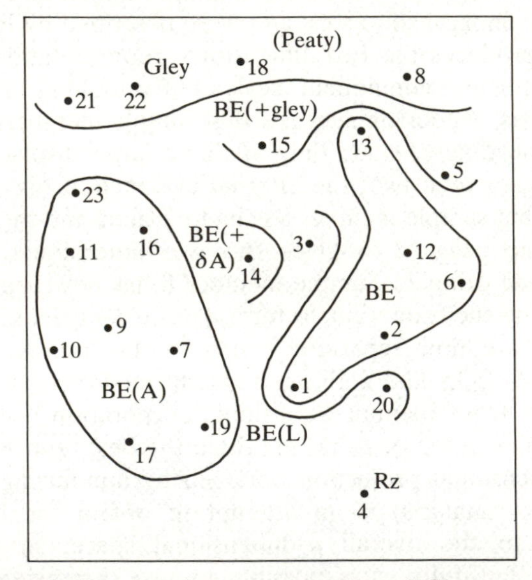

Fig. 2.16 Quadratic loss function representation of Glamorganshire soils: BE, brown earth; BE(A), acid brown earth; BE(L), lessivated brown earth; Rz, rendzina. Reproduced from Anderson (1971) with permission from Plenum Publishing Corporation

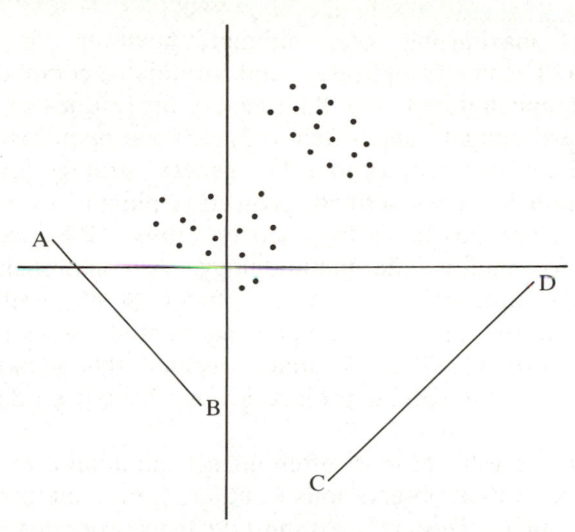

Fig. 2.17 Different perspectives on a set of data: points projected on to AB give a single group, but when they are projected on to CD the split into two groups is evident

possible. With an initial p-dimensional configuration, however, there must be many other possible two-dimensional approximations. These can be thought of as windows through which the space is viewed. Pursuing this analogy, it is evident that viewing the space through different windows will inevitably highlight different features of the sample. For example, if we consider the two-dimensional set of points in Fig. 2.17, we could envisage setting up two different (one-dimensional) windows positioned respectively at AB and CD in the diagram. If we viewed the points through the window AB (i.e. projected on the line AB) we would see a single mass of points, but if we viewed through window CD we would see two separate groups of points.

If we have some definite feature of the sample that we wish to investigate, there will exist a window through which this feature will be displayed to its best possible advantage. There has been increasing interest in recent years, therefore, in seeking low-dimensional projections of a multivariate data set in such a way as to display specific aspects of, or structure in, the data. All such methods rely on setting up a quantitative measure of the degree to which a given projection reflects the aspect of interest, and then choosing a projection (by numerical iteration) to optimize this measure. Principal component analysis and non-linear mapping are, of course, both examples of such a process, with quantitative measures V and L^* respectively. Once we go beyond the simple objectives of maximizing total variance, however, the quantitative measures may become complicated, and formidable computational problems may be encountered. For this reason, projections of at most two dimensions are sought, and often only a one-dimensional projection proves to be a viable proposition. The general strategy for this type of data description has been termed 'projection pursuit' by Friedman and Tukey (1974), and has been discussed by Sibson (1984) as providing a way ahead for multivariate methodology. A comprehensive general review of the area is provided by Huber (1985), with additional interesting material in the accompanying invited discussion (see also Jones and Sibson (1987)). We shall conclude this section by briefly reviewing some of the specific projections that have been documented to date.

Aspects of interest that most often attract the analyst of a set of data are extreme individual observations ('outliers'), or some general subdivision of the sample ('clusters'). Sibson (1984) incorporates both of these aspects when he suggests that the most useful projections to seek are those which display extreme non-normality. His proposed quantitative measure is the entropy of the projected sample, on the grounds that since (for fixed variance) entropy is a maximum for the normal distribution, then minimizing entropy will produce the most interesting projection. His proposed strategy is first to subject the observed variables $\boldsymbol{X}$ to the transformation $\boldsymbol{Y} = \mathbf{T}^{-1}(\boldsymbol{X} - \bar{\boldsymbol{x}})$, where $\bar{\boldsymbol{x}}$ is the sample mean vector, $\mathbf{S}$ is the sample covariance matrix, and $\mathbf{T}$ is obtained from the Cholesky decomposition of $\mathbf{S}$ as $\mathbf{S} = \mathbf{TT}'$. This transformation ensures that the new variates $\boldsymbol{Y}$ have zero sample mean and identity covariance matrix. Any given projection is defined by linear combinations of the Y_i, so for a given projection the entropy can be estimated by first making a density estimate of the projected points. The projection that maximizes the negative of the entropy is then sought by numerical means. For full technical details see Jones and Sibson (1987). Other methods have been proposed that exploit, either directly or indirectly, properties of the normal distribution. Andrews, Gnanadesikan, and Warner (1971) also first transform $\boldsymbol{X}$ to $\boldsymbol{Y}$. They then seek a single direction for projecting the points as a weighted sum of the transformed observation vectors, where the weighting for each observation is a power of its distance from the sample mean. When the power is positive, this direction will tend to point towards groups of extreme observations; when the power is negative the direction will tend to point towards groupings of near observations. Malkovich and Afifi (1973) look for single directions which optimize various quantitative measures. Those directions of most interest seem to be the ones which either maximize or minimize the sample skewness or kurtosis of the projected points. Belyavin, in an unpublished

Ph.D. dissertation of the University of Reading (1981), generalizes this idea to a search for q-dimensional subspaces ($q > 1$) in which the multivariate skewness or kurtosis of the projected points is either maximized or minimized. He provides an interesting example in which extreme observations are identified by a two-dimensional projection of this nature, whereas a corresponding principal component projection would not uncover these outliers. Cox and Small (1978) exploit some second-order properties of the multivariate normal distribution (see Chapter 7) in choosing a pair of variates such that the partial F-ratio for the regression of the first variate on the square of the second variate, in the presence of the first power of the second variate, should be maximized. This technique was advanced as a method for detecting sample non-normality, but should be equally sensitive to the presence of boomerang-shaped arrangements of observations. More general non-linear structural relationships among the variates (and hence certain non-linear types of data structure) may be detected by including powers and products of variates in a principal component analysis, following Gnanadesikan and Wilk (1969). A good review of techniques for detecting multivariate non-normality, which includes some discussion of many of the methods mentioned above, is given by Gnanadesikan (1977).

2.3 Non-numerical data and missing values

The methods discussed in the preceding sections have assumed tacitly that all variates observed in the sample yield numerical data. This assumption is necessary in order to construct the original p-dimensional geometrical configuration that models the sample, from which various projections into small-dimensional subspaces may be contemplated. What happens when some, or all, of the variates are non-numerical attributes (e.g. colour of eyes, socio-economic class of parents, sex), and the initial geometrical model cannot be constructed or correlation matrices calculated? It may be possible to convert such categorical variables into numerical ones, either by some arbitrary scoring or by replacing a k-category discrete variable by $(k-1)$ dummy binary variables scored 0 or 1. If this is done then previously described methods may be formally invoked. Often, however, the data matrix is expanded disproportionately by this device, and the resulting analyses are hard to interpret. Scoring nominal variates may also be hazardous. A better approach is to use the idea of a dissimilarity measure between two sample members, as introduced in Chapter 1. Applying one of the measures given there to every pair of individuals will convert the $(n \times p)$ data matrix into an

$(n \times n)$ matrix giving the dissimilarity δ_{ij} between every pair of individuals i and j.

If we now equate the concept of dissimilarity between a pair of individuals with the distance between a pair of points representing these individuals, then we can see an alternative basis for constructing a geometrical configuration. Instead of arriving at a two-dimensional representation by projection of the points existing in p dimensions in such a way as to optimize a criterion like V, L or L^* defined in Section 2.2.7, we might try *directly to construct* such a two- (or higher) dimensional representation by optimizing V, L or L^* but with the d_{ij} replaced by the dissimilarities δ_{ij}. This was termed the *optimization* approach in Chapter 1. Note that, whereas the projection approaches generally require information on pairs of variates (such as $(p \times p)$ covariances or correlations) and are thus R-techniques, optimization approaches usually start from information on pairs of units (such as $(n \times n)$ dissimilarity matrices) and are thus Q-techniques. We consider such techniques in the next chapter.

Missing observations can often occur in sets of multivariate data. If the experimental units are animals or plants they may well die before all the variates have been measured on them. In anthropology or archaeology, often a specimen such as a skull may be damaged and certain variates cannot be measured. In surveys, a respondent may have omitted to answer certain questions. In all these cases, the data matrix will have some rows in which not all columns contain values, and the question arises as to whether this partial information can still be used in the techniques that have been discussed in this chapter.

Most of these techniques represent each row of the data matrix graphically by applying some correspondence or function to the entities x_i in each column of that row. Thus Andrews curves are formed by computing the linear function of the x_i whose coefficients are the sine or cosine terms of Section 2.1.3; Hartigan's boxes are constructed by associating the x_i with the dimensions of a box; principal component scores are the linear functions of the x_i whose coefficients are the elements of certain eigenvectors of the sample correlation or covariance matrix, and so on. If any of the x_i are absent, the appropriate function cannot be evaluated or the correspondence made. The only way in which the units in question can be represented graphically, therefore, is for the missing values to be replaced by some suitable numerical entries. This process is termed *imputation*.

The simplest (but crudest) form of imputation is to replace each missing value by the mean of all values that are present on the corresponding variate (for a given group if necessary). A more sophisticated approach was proposed by Buck (1960), who suggested the use of such imputation of means as the first step of a two-step process. The second step consists of setting up regression equations in which each variate containing any missing values is treated as the dependent variate, and all other variates are the explanatory variables. Parameters in these regression equations are estimated from the data, and hence each missing value can be predicted from the appropriate equation. This predicted value is the one imputed. Beale and Little (1975) then showed that even better estimates can be obtained by an iterated version of this procedure. In their scheme, each set of imputed values is used in estimating the regression parameters in the various regression equations, and a new set of imputed values is derived as predictions from the updated regressions. The process converges quickly to give predicted values which change minimally between iterations, and these predicted values are the final imputed ones. The Beale and Little procedure is a special case of the so-called E–M algorithm (Dempster *et al.*, 1977).

If the user does not wish to employ any form of imputation, then the only satisfactory general alternative is to exclude any units containing missing values entirely from the analysis. In a few of the techniques, however, a limited set of objectives may be achieved by using just those (and all those) values present. In principal component analysis, for example, the variance–covariance matrix may be computed in this way. Thus sample means for each variate are computed from all those values present on that variate, and similarly for the variances of each variate, while covariances between pairs of variates are computed from all units that have an observed value for that pair of variates. These variances and covariances are put together into the covariance matrix, and eigenvalues and eigenvectors of the latter can be extracted. Note, however, that such an approach can only be used with the limited objective of calculating and interpreting eigenvalues/eigenvectors, and imputed values are still needed for obtaining principal component scores. If this latter is desired, then the best procedure is to impute the values *before* the sample covariance matrix is calculated. Calculating the matrix element by element, as described above, destroys some of the normal features of the matrix and leads occasionally to odd results. For example, correlations outside the range -1 to $+1$ can be obtained from such covariance matrices. Devlin *et al.* (1981) suggest some ways of overcoming these problems, but extreme caution should be exercised, and this method is not recommended if many data entries are missing.

3
Graphical methods for association or proximity matrices

In Chapter 2 we considered graphical methods of representing either rows or columns of a matrix, where these rows and columns are linked to two sets of entities that are not formally related. These methods could be applied to any rectangular matrix satisfying this requirement, but were principally aimed at graphical representations of $(n \times p)$ data matrices. By contrast, in the present chapter we shall be concerned with matrices whose rows and columns are both linked to the *same* entities. Such matrices must also by definition be square, since the same number of entities indexes both rows and columns. Furthermore, since it is the same set of entities indexing rows and columns, then the elements of the matrix must represent *relationships* between pairs of entities in this set.

The two main types of matrices encountered in multivariate analysis that fall within the scope of this chapter, therefore, are *correlation* or *covariance* matrices between variables, and *similarity* or *dissimilarity* matrices between units. The former can be said to contain information on *associations* between variables, the latter to contain information on *proximities* between individuals.

An example of a correlation matrix has already been given in Table 2.7; the variables indexing rows and columns are detailed in Table 2.6.

As an example of a similarity matrix, consider the data described by Rayner (1966) and already referred to in Chapter 2. The data set comprised twenty-three soil profiles from Glamorganshire, and each profile typically contained about four 'horizons' or layers. Fifty variables were measured on each horizon of each soil, these variables being a mixture of binary, nominal, ordinal, and numeric types. A general measure of similarity, such as described in Section 1.4, was then used to obtain the similarity between every pair of horizons in the sample. A similarity matrix for the twenty-three Glamorganshire soils was finally obtained, the similarity between any pair of soils being given by the average of the similarities between their matched pairs of horizons. This similarity matrix is reproduced from Rayner (1966) in Table 3.1 (with

Table 3.1 Matrix of similarities between twenty-three Glamorganshire soils (reproduced from Rayner (1966) with permission from Blackwell Scientific Publications)

Soil	1	2	3	4	5	6	7	8	9	10	11	12	13	14	15	16	17	18	19	20	21	22
2	0.858																					
3	0.800	0.819																				
4	0.679	0.730	0.649																			
5	0.767	0.794	0.758	0.702																		
6	0.793	0.855	0.757	0.783	0.865																	
7	0.790	0.758	0.766	0.649	0.659	0.696																
8	0.755	0.713	0.711	0.653	0.829	0.771	0.681															
9	0.761	0.742	0.790	0.644	0.687	0.685	0.796	0.653														
10	0.729	0.679	0.768	0.608	0.648	0.639	0.749	0.646	0.864													
11	0.734	0.723	0.794	0.600	0.636	0.646	0.838	0.595	0.841	0.800												
12	0.779	0.843	0.823	0.738	0.784	0.854	0.706	0.725	0.708	0.661	0.708											
13	0.732	0.762	0.770	0.697	0.806	0.825	0.682	0.802	0.681	0.671	0.684	0.855										
14	0.807	0.805	0.793	0.686	0.790	0.759	0.801	0.758	0.783	0.752	0.752	0.775	0.774									
15	0.796	0.782	0.814	0.620	0.774	0.760	0.758	0.746	0.775	0.715	0.753	0.816	0.801	0.812								
16	0.735	0.738	0.751	0.643	0.650	0.675	0.807	0.630	0.803	0.759	0.850	0.748	0.713	0.800	0.798							
17	0.751	0.696	0.696	0.688	0.624	0.649	0.797	0.632	0.792	0.778	0.767	0.653	0.668	0.783	0.712	0.840						
18	0.725	0.702	0.740	0.563	0.751	0.706	0.769	0.716	0.741	0.699	0.727	0.703	0.725	0.817	0.799	0.745	0.740					
19	0.819	0.762	0.758	0.719	0.695	0.696	0.801	0.704	0.861	0.844	0.800	0.684	0.711	0.796	0.757	0.768	0.815	0.722				
20	0.799	0.827	0.814	0.681	0.742	0.767	0.706	0.698	0.712	0.680	0.713	0.822	0.771	0.770	0.783	0.723	0.654	0.671	0.778			
21	0.663	0.606	0.700	0.486	0.634	0.597	0.731	0.616	0.763	0.752	0.800	0.642	0.670	0.677	0.736	0.751	0.686	0.769	0.694	0.644		
22	0.694	0.672	0.705	0.556	0.701	0.633	0.763	0.628	0.759	0.743	0.842	0.683	0.714	0.697	0.739	0.753	0.711	0.742	0.715	0.667	0.817	
23	0.713	0.694	0.745	0.584	0.656	0.680	0.791	0.622	0.804	0.805	0.854	0.712	0.722	0.739	0.759	0.836	0.749	0.694	0.785	0.710	0.800	0.844

permission from Blackwell Scientific Publications); the units indexing the rows and columns are the twenty-three soils. It can be converted into a dissimilarity matrix by using any suitable monotonic transformation. We shall see later that a particularly appropriate choice is to convert the similarity s_{ij} between the ith and jth soils into the dissimilarity $d_{ij} = \{2(1 - s_{ij})\}^{\frac{1}{2}}$. (If it is not necessary to distinguish explicitly between similarities and dissimilarities, the word *proximities* can be used as a substitute for either.)

It has already been seen in Section 1.4 that correlations between variables can be thought of as similarities between them, and various suitable transformations have been suggested for converting correlations into dissimilarities. We therefore lose no generality if we focus our attention in this chapter exclusively on graphical methods for $(n \times n)$ matrices of dissimilarities between every pair of n units. In most cases encountered in practice, the matrix of dissimilarities will be *symmetric* as well as square, as the dissimilarity between units i and j is obviously the same as that between units j and i. This is reflected in Tables 2.7 and 3.1 by the fact that only the lower half of each matrix is printed—it is tacitly assumed that the upper half is the transpose of the lower half. There are a few important exceptions, however, in which an $(n \times n)$ matrix whose rows and columns index the same entities is not symmetric, i.e. is such that $d_{ij} \neq d_{ji}$. Examples of such matrices can arise in psychology (d_{ij} is the number of times in a series of trials that item i is said to be the same as item j, taken in that order); genetics (d_{ij} is the number of progeny when a male of line i is crossed with a female of line j); sociology (d_{ij} is the number of people who commute daily from city i to city j), and many other fields. Such matrices will be considered at the end of this chapter and in the next chapter. For the present, however, we assume that we are dealing only with symmetric matrices.

Before proceeding to discuss graphical methods, however, we must first consider the question: why bother with methods for $(n \times n)$ dissimilarity matrices at all? Could we not step back one stage to the original $(n \times p)$ data matrix and use methods of the last chapter on this matrix, rather than first computing the $(n \times n)$ dissimilarity matrix? The answer to this is that in certain cases we could, but there are very many situations in which focusing on the dissimilarity matrix is either more fruitful or indeed essential. In the first place, the methods of the last chapter imply particular measures of dissimilarity. In the main, these methods require Euclidean distance as dissimilarity between units and Pearson product moment correlation as similarity between variables. If a different measure is thought to be more appropriate in a certain situation, then the methods of the previous chapter will not be applicable. Even more crucially, with some types of data the methods of the previous chapter can not be implemented at all. In particular, we have already seen that the only way of treating some nominal variables numerically is to convert the $(n \times p)$ data matrix into a $(n \times n)$ dissimilarity matrix. Hence no progress would be possible if we did not do this. Finally, it happens in some experiments that the dissimilarities themselves are the observations. For example, in psychological experiments subjects may be presented with pairs of stimuli such as Morse code signals (Rothkopf, 1957) or samples of different soft drinks (Chauhan *et al.*, 1983), and asked to assess their similarity directly. Thus the 'data matrix' is actually the $(n \times n)$ dissimilarity matrix itself.

For all these reasons, it can be seen that graphical methods for examining $(n \times n)$ dissimilarity matrices are highly desirable (particularly when n is large, as overall relationships existing within the sample are then very difficult to detect from the matrix by eye). As in the previous chapter, there are two distinct lines of approach. We may attempt to produce a direct, two-dimensional, pictorial representation. Alternatively, we may try to construct a configuration of n points (representing the n units) in which the distance between any two points is equal to the dissimilarity between the corresponding units. Such a configuration may need to be high-dimensional for accurate representation of dissimilarities, so a projection into a suitable two-dimensional subspace may then be necessary. These two approaches will be treated separately below, in Sections 3.1 and 3.2, for the case when all recorded dissimilarities are numeric. It may happen, however, that either we only have ordinal information on the dissimilarities, or we only wish to use their rank order. Section 3.3 deals with this situation. Finally, Section 3.4 considers the special case of asymmetric matrices.

3.1 Direct two-dimensional representation of units, derived from numerical dissimilarities

3.1.1 The dendrogram

All the direct two-dimensional representations discussed in Chapter 2 rely on the depiction of each sample unit as a shape such as a box, a face, a curve, or a harmonic function. This is possible because each unit has associated with it a number of variate values, and these values can be mapped on to the various dimensions of the shape used to represent the unit. We are now assuming that our input data comprise numerical values giving dissimilarities between every pair of units in the sample, so a radically different approach to two-dimensional representation is necessary. Since the data give information on the degree of 'closeness' or proximity of each unit to every other unit, a reasonable way of proceeding might be to look for *groups* of units such that all units in a group are relatively 'similar' to each other but relatively 'different' from all units in other groups. Any direct two-dimensional representation needs therefore to give a pictorial impression of this grouping.

An alternative way of expressing the fact that the units in the sample can be grouped in some way is to say that the units 'cluster'; the process of finding the appropriate grouping can therefore be termed *cluster analysis*. This is a very wide topic, which has its roots not only in Statistics but also in such diverse fields as Numerical Taxonomy, Systematics, and Pattern Recognition, and it is outside our present brief to pursue a full investigation of all its aspects here. The interested reader is directed to such clear expository texts as those by Sneath and Sokal (1973), Everitt (1980) and Gordon (1981). Our primary interest here is to focus on those aspects that lead to a useful pictorial representation of the data, so only a brief outline will be given of the background ideas.

Our aim, as stated in the first paragraph of this section, is to find a grouping of the units such that all units in a group are relatively similar to each other. For this aim to have any operational meaning, we must quantify the term 'relatively similar'. To fix ideas, let us suppose that the measure of dissimilarity between units that is used results in values that lie in the range 0 to C. If we feel that two units are 'relatively similar' whenever the dissimilarity between them is less than $\frac{1}{2}C$, then we might demand of our procedure that it produces groups of units such that any two units in the same group have dissimilarity less than $\frac{1}{2}C$, and that any two units in different groups have as large a dissimilarity as possible. Suppose now that we retain this overall approach, but make more stringent our definition of two units being 'relatively similar' by requiring the dissimilarity between them to be less than $\frac{1}{4}C$. We will now obtain *more* groups by our clustering procedure, as in general there will be fewer units in any given group that mutually satisfy this more stringent criterion. In keeping with natural human ideas of classification and evolutionary theories, an attractive property that might be demanded of a clustering procedure is that the larger number of groups obtained under the more stringent criterion be produced by splitting into subgroups those groups obtained under the less stringent criterion. An alternative way of viewing this is to think of the fewer number of groups obtained under the less stringent criterion as being produced by amalgamation of some of the groups obtained under the more stringent criterion. Any method of cluster analysis that possesses this property is said to lead to *hierarchic* clustering; if the groups are produced by successive splitting then the method is said to be *divisive,* but if the groups are produced by successive amalgamation then the method is said to be *agglomerative*.

Whichever method of clustering is envisaged, it is easy to see what happens when our definition of 'relatively similar' takes one of its two possible extreme values. If two units are deemed to be relatively similar only when the dissimilarity between them is zero, then no units will combine to form groups (unless there happen to be two or more units in the sample that are identical in all respects). If, on the other hand, two units are deemed to be relatively similar when the dissimilarity between them is less than or equal to C, then *all* the units satisfy this and can be thought of as forming a single group. If we use the term *clustering coefficient* to denote the value of dissimilarity at which two groups coalesce, then we see that the clustering coefficient α can take on values between 0 and C. When $\alpha = 0$ there are in general n groups in the data, each comprising a single unit; when $\alpha = C$ there is just one group present, consisting of all n units; when α takes an intermediate value between 0 and C, there can be any number of groups between 1 and n. There exist many ways of defining a clustering coefficient, just as there exist many ways of defining a similarity coefficient (cf. Section 1.4). Each different definition gives rise to a different (hierarchical) *cluster method*, and some of the more popular options will be reviewed below. Whichever cluster method is chosen, however, the whole clustering process can be summarized diagrammatically in tree form. Such a diagram is called a *dendrogram*; a simple example is given in Fig. 3.1.

A dendrogram shows the sequence of successive fusions or divisions that occur among the existing groups as the clustering coefficient is systematically varied between its extreme values. At the bottom of the dendrogram, all units form separate groups. At the top of the dendrogram, all units fall into a single group. A divisive method thus starts at

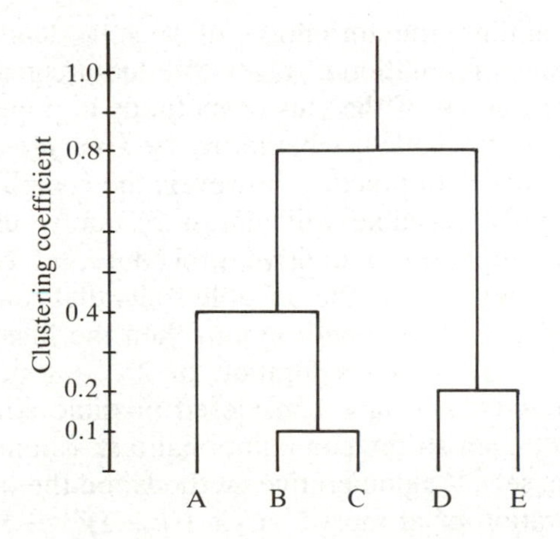

Fig. 3.1 Simple example of a dendrogram

the top of the dendrogram and works downwards, while an agglomerative method starts at the base of the dendrogram and works upwards. The clustering coefficient scale may be placed at the side of the dendrogram, to provide a graduation. Note that this scale can be presented either in terms of similarity or in terms of dissimilarity; in Fig. 3.1 it is in terms of dissimilarity. Fig. 3.1 thus shows that the sample contains five members, A, B, C, D, and E. The two most similar ones are B and C, which join into a group at a value 0.1 of the clustering coefficient. The two next most similar are D and E, which join to form a group at the value 0.2 of the clustering coefficient; at this stage there are three groups in the data, namely the single unit A and the pairs (B, C) and (D, E). Unit A joins the pair (B, C) to form the group (A, B, C) at the value 0.4 of the clustering coefficient, and finally at the value 0.8 the two existing groups join to form a single group containing all five members. The diagram thus provides a good overall impression of the information contained in a matrix of dissimilarities. One can see immediately which groups of individuals are relatively similar to each other, and by reference to the scale of the clustering coefficient one can judge which groups are most, and which ones are least, homogeneous. Furthermore, the dendrogram can be 'cut' at any chosen position to see which members fall into which groups, either at a chosen level of clustering coefficient or at a chosen number of groups. For example, if we think that there might be three groups in the data represented by Fig. 3.1, and we want to determine the best split into three groups, then we need only cut the dendrogram by a horizontal line at any point at which there are three vertical lines (i.e. between values 0.2 and 0.4 of the clustering coefficient). We then readily see that the three groups are given by the partition A, (B, C), (D, E) of the sample.

In order to outline the formation of such a dendrogram, we now describe the main computational stages of a hierarchical cluster analysis, and briefly review some of the cluster methods in popular use. The first decision that has to be made is whether to use an agglomerative approach or a divisive approach. In practice, however, the computational resources at present generally available will almost certainly rule out the latter. This is because, to apply a divisive procedure, at each stage of the partition we must consider all the possible splits that *could be made*. Thus if we start at the top of the dendrogram, then the first split of n objects into two groups requires consideration of $2^{n-1}-1$ possible partitions. Scott and Symons (1971) have considered possible strategies to reduce computation, but even so the computational task is mammoth for all but the smallest data sets. Agglomerative methods, on the other hand, simply require consideration of at most ${}^{k-1}C_2 = \frac{1}{2}(k-1)(k-2)$ members at the stage at which k groups have been formed. Computation is thus usually very fast, and nearly all available algorithms for hierarchical cluster analysis are agglomerative ones.

The actual production of a hierarchy from a dissimilarity matrix using an agglomerative method can be achieved by the following very simple algorithm. Let d_{ij} denote the dissimilarity between the ith and jth individuals, and D_{ij} denote the dissimilarity between the ith and jth groups formed.

Step 1: Define each individual as a group (i.e. $D_{ij} = d_{ij}$ initially), and store the $(n \times n)$ dissimilarity matrix of D_{ij} values.

Step 2: Find the smallest element of this matrix (making a random choice if necessary in the case of equality). If this is D_{km}, fuse existing groups k and m and note the coefficient value D_{km}.

Step 3: Calculate the dissimilarity between the new group and each of the existing groups. Replace the kth and mth rows and columns of the dissimilarity matrix by the single row/column of new dissimilarity values, thereby reducing the order of the dissimilarity matrix by one.

Step 4: Recycle through steps 2 and 3 until left with just one group.

Such an algorithm is ideal for implementation on a computer. However, to proceed we must make step 3 explicit, i.e. we must give a rule for calculating the dissimilarity between two groups each of which may contain more than one sample member. Various rules have been proposed over the years, and each rule is associated with a different cluster method. The following are the methods in most frequent use. In all the following, we suppose that the two groups are denoted i and j, and that there are n_i and n_j units in these groups respectively.

The *nearest neighbour* (or *single-link*) method: D_{ij} is the *smallest* of the $n_i n_j$ dissimilarities between each element of i and each element of j.

The furthest neighbour (or *complete-link*) method: D_{ij} is the *largest* of the $n_i n_j$ dissimilarities between each element of i and each element of j.

The *group average* method: D_{ij} is the *average* of the $n_i n_j$ dissimilarities between each element of i and each element of j.

The *centroid* method: D_{ij} is the squared Euclidean distance between the *centroids* of group i and group j, where the elements of these groups are represented by points in space.

The *median* method: D_{ij} is the distance between the *medians* of group i and group j, in this geometric representation.

The *minimum variance* method: D_{ij} is the 'between-group sum-of-squares' for groups i and j.

The first three methods are thus straightforward 'arithmetic' methods, while the remaining three are more complicated 'geometric' or 'algebraic' ones. For all methods, however, there exists a common algorithmic expression for calculating the distance $D_{k.ij}$ between any group k and the *union* of groups i and j. This is the crucial calculation that needs to be done at step 3 of the hierarchical algorithm, and it takes the following forms for the different clustering methods.

Nearest neighbour

$$D_{k.ij} = \tfrac{1}{2}(D_{ki} + D_{kj} - |D_{ki} - D_{kj}|) \qquad (3.1)$$

Furthest neighbour

$$D_{k.ij} = \tfrac{1}{2}(D_{ki} + D_{kj} + |D_{ki} - D_{kj}|) \qquad (3.2)$$

Group average

$$D_{k.ij} = \frac{n_i}{n_i + n_j} D_{ki} + \frac{n_j}{n_i + n_j} D_{kj} \qquad (3.3)$$

Centroid

$$D_{k.ij} = \frac{n_i}{n_i + n_j} D_{ki} + \frac{n_j}{n_i + n_j} D_{kj} - \frac{n_i n_j}{(n_i + n_j)^2} D_{ij} \qquad (3.4)$$

Median

$$D_{k.ij} = \tfrac{1}{2}D_{ki} + \tfrac{1}{2}D_{kj} - \tfrac{1}{4}D_{ij} \qquad (3.5)$$

Minimum variance

$$D_{k.ij} = \{(n_i + n_k)D_{ki} + (n_j + n_k)D_{kj} - n_k D_{ij}\}/(n_i + n_j + n_k) \qquad (3.6)$$

Note that each expression above is a special case of the general form $D_{k.ij} = \alpha_i D_{ki} + \alpha_j D_{kj} + \beta D_{ij} + \gamma\,|D_{ki} - D_{kj}|$ (Lance and Williams, 1966), and hence all the above methods can be implemented easily in a single computer program for hierarchical dendrogram production. Starting with $D_{ij} = d_{ij}$ for all i, j triggers the procedure, and then the hierarchic clustering proceeds automatically for any chosen method.

This last fact is both a strong attraction and a grave disadvantage. The attraction comes from the extreme simplicity of the numerical algorithm, making the automatic processing of even quite large data sets very easy. The disadvantage arises from the multiplicity of methods available and the fact that each clustering method will lead to a different partition of the sample and hence to a different dendrogram. Considerable thought must therefore be put into the choice of clustering method, *before the analysis is attempted.* Too often the ease of computing is made an excuse for trying as many methods as possible, and then selecting the results that 'look best'. Not only does this approach produce exceptionally wasteful amounts of computer output (far more than the original data often warrant), but there is a grave danger that by adopting such an approach one will choose the solution that best reflects one's original preconceptions about the problem. Hence, the person who sets out with the intention of finding an entirely objective solution to the problem often ends up by producing a highly subjective one to which false credence has been attached. A more satisfactory approach is to consider carefully the desired objective of the analysis, and to select from the outset the clustering method that best meets this objective. Each method has some distinctive features and properties that may assist in making this choice, so we conclude our brief discussion of clustering methods with a summary of these features.

One important consideration may be the nature of the clusters produced. The complete-link, group average, and centroid methods all lead to 'spherical' clusters exhibiting high internal affinity; the single link method, however, often leads to elongated clusters in which pairs of very dissimilar units may occur. This feature is often referred to as 'chaining', and arises because in this method a unit can join a group on account of its similarity with just *one* existing member of that group. It should be noted that such features are not unreasonable in subjects like taxonomy where evolutionary chain mechanisms may be at work (and where the single-link method was first popularized). On the mathematical side, only the single-link and complete-link methods are invariant under monotonic transformation of the d_{ij} values. Thus, for example, if we were to transform all elements in the dissimilarity matrix by taking their logarithms, then only these two methods would give the same dendrogram before and after transformation. This property was held to be very important by Jardine and Sibson (1971). Other points in favour of the single-link method are that it shows up well the presence of any 'outliers' in the data, and it is extremely cheap on both computer time and computer space. Good algorithms allow clustering of thousands of units relatively easily (see, e.g. Roger and Carpenter (1971) and Section 3.1.2 below). As regards general shape, the minimum-variance method tends to force the data into equal-sized clusters, the complete-link method tends to force the data into clusters of equal 'diameter', while the median method weights towards the units most recently added to the cluster. Problems may arise with the centroid and median methods, where it is possible to have $D_{k.ij} \leqslant D_{ij} \leqslant \min(D_{ik}, D_{jk})$ and an illegal dendrogram. However, these two methods together with the minimum-variance method have appealing geometric properties (not shared by the other methods), which can be tied in with the methods of data display to be discussed in the next section.

As a final cautionary note, it should be stressed that cluster analysis is *not* appropriate if we already possess a priori knowledge about the existence of groups in the data. Suppose, for example, that an experimenter has obtained a number of individuals from each of a number of species of nematode. He is experimenting with a new microscopic technique, has managed to measure on each individual a set of variables never before observed, and now wishes to find out whether these variables will suffice to characterize the different species. A common approach suggested in such circumstances by inexperienced researchers is to use cluster analysis in an attempt to see whether the analysis will recover the groups known to exist. This will almost certainly never happen, and some individuals are sure to be mixed up between groups with whatever clustering method is tried. The researcher will find himself or herself trying more and more clustering methods, producing ever-increasing sets of computer output, and ending up in total bewilderment. The correct approach is always to use all the information one possesses *at the outset,* and to select a statistical technique that takes account of this information. In the present example the experimenter knows which individuals come from which groups, so he should use this information in the analysis. The correct question to ask is thus: 'Do the measured variables show up the differences known to exist between these groups?'; appropriate techniques for doing this will be discussed in Chapter 11.

In the meantime, let us work through a simple example to demonstrate the algorithms described earlier, and then provide some illustrations of the use of dendrograms in practice. Suppose that a simple data set consists of five individuals, and that use of one of the dissimilarity coefficients (described in Chapter 1) on the measurements made on these five individuals yields the following half-matrix of dissimilarity values between individuals:

Unit	1	2	3	4	5
1	0				
2	4	0			
3	1	4	0		
4	4	2	4	0	
5	5	5	3	4	0

Consider the production of a dendrogram, using two of the clustering methods described earlier.

(i) *The nearest-neighbour method*

Step 1

The smallest dissimilarity is the value 1 in the third row and first column of the matrix. (The zero self-dissimilarities are of course ignored.) Thus units 1 and 3 are most similar, and are merged to form the first group. The clustering criterion value is 1 for this join.

We now compute the distances $D_{i.13}$ from the joined group $(1+3)$ to each of the other units i ($i = 2, 4, 5$). The nearest neighbour distance is the *smaller* of d_{i1} and d_{i3} in each case i.

Thus

$$D_{2.13} = \min(d_{21}, d_{23}) = \min(4, 4) = 4$$
$$D_{4.13} = \min(d_{41}, d_{43}) = \min(4, 4) = 4$$
$$D_{5.13} = \min(d_{51}, d_{53}) = \min(5, 3) = 3$$

Alternatively, we can use the general expression (3.1) to obtain:

$$D_{2.13} = \tfrac{1}{2}(d_{21} + d_{23} - |d_{21} - d_{23}|) = \tfrac{1}{2}(4 + 4 - |4 - 4|) = 4$$
$$D_{4.13} = \tfrac{1}{2}(d_{41} + d_{43} - |d_{41} - d_{43}|) = \tfrac{1}{2}(4 + 4 - |4 - 4|) = 4$$
$$D_{5.13} = \tfrac{1}{2}(d_{51} + d_{53} - |d_{51} - d_{53}|) = \tfrac{1}{2}(5 + 3 - |5 - 3|) = 3$$

Either way, the new dissimilarity matrix, after grouping units 1 and 3, becomes

Group	(1 + 3)	2	4	5
(1 + 3)	0			
2	4	0		
4	4	2	0	
5	3	5	4	0

Step 2

The smallest element of the above matrix is the value 2 in the row labelled 'group 4' and the column labelled 'group 2'. Hence units 2 and 4 are now merged to form the second group, at a clustering criterion value of 2. We now compute the distances $D_{(1,3).24}$ and $D_{5.24}$ from the joined group $(2, 4)$ to the existing groups $(1, 3)$ and 5 as follows:

$$D_{(1,3).24} = \min(D_{2.13}, D_{4.13}) = \min(4, 4) = 4$$
$$D_{5.24} = \min(d_{52}, d_{54}) = \min(5, 4) = 4$$

The reader can check that the same values are obtained by use of (3.1). The dissimilarity matrix thus becomes:

Group	(1 + 3)	(2 + 4)	5
(1 + 3)	0		
(2 + 4)	4	0	
5	3	4	0

Step 3
The smallest element of the above is the value 3, being the dissimilarity between unit 5 and group (1 + 3). At this stage, therefore, unit 5 is added to the group already consisting of units 1 and 3, at clustering criterion value 3. The distance between this new group (1 + 3 + 5) and the only other existing one (2 + 4) is thus $\min\{D_{(1,3).24}, D_{5.24}\} = \min(4, 4) = 4$. Hence the new matrix is

Group	(1 + 3 + 5)	(2 + 4)
(1 + 3 + 5)	0	
(2 + 4)	4	0

Step 4
The only possibility left is to merge (2 + 4) with (1 + 3 + 5) to form a single group of 5 elements, and the clustering criterion is 4 (this being the minimum remaining).

The history of all the above steps is plotted in the dendrogram in Fig. 3.2(a).

(ii) *The furthest-neighbour method*
The same overall strategy is followed, except that at each step we recompute the dissimilarity between each existing group and the new group formed at that step as the *maximum* of two existing dissimilarities (rather than the minimum used above). Alternatively, expression (3.2) is used in place of (3.1) for the computations.

The following steps occur in the process; the reader is invited to work through the details, and verify that the given matrices are indeed derived.
Step 1
Since the same starting matrix is used, the same initial merge of units 1 and 3 takes place at clustering criterion value 1. The new dissimilarity matrix now becomes:

Group	(1 + 3)	2	4	5
(1 + 3)	0			
2	4	0		
4	4	2	0	
5	5	5	4	0

Step 2
Merge units 2 and 4 at clustering criterion value 2. This gives the

following matrix:

Group	(1 + 3)	(2 + 4)	5
(1 + 3)	0		
(2 + 4)	4	0	
5	5	5	0

Step 3
Merge (1 + 3) with (2 + 4) at clustering criterion value 4 to give:

Group	(1 + 2 + 3 + 4)	5
(1 + 2 + 3 + 4)	0	
5	5	0

Step 4
All units join to form one group at clustering criterion value 5.

The dendrogram deriving from the above procedure is given in Fig. 3.2(b).

The chief difference between the two methods is thus in the treatment of unit 5. The single-link method joins this unit to the (1, 3) group at the penultimate stage, while the complete-link method leaves it as a single-element group to the very end. The reason for this is that the dissimilarity between unit 5 and each of units 1 and 2 is the highest value (5) in the original dissimilarity matrix. A unit will only join a group in the complete-link method if its dissimilarity to *all* the units in that group is below the threshold value; so, in this case, unit 5 has to remain until the end before joining any group (since units 1 and 2 are in different groups). The single-link method, however, permits a unit to join a group if that unit's dissimilarity with *any* unit is below the threshold value. The

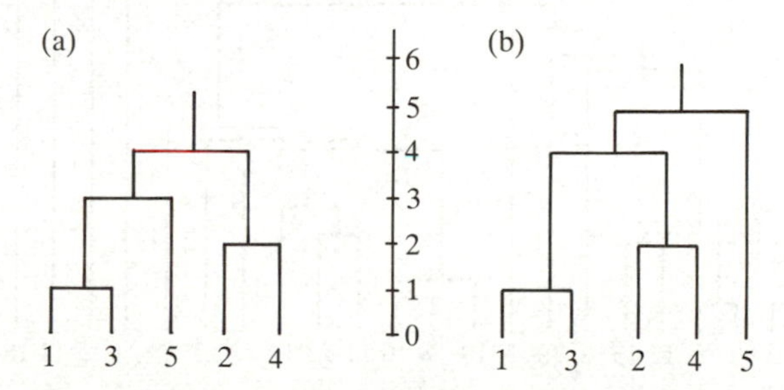

Fig. 3.2 Dendrograms derived from simple data set: (a) nearest neighbour method; (b) furthest neighbour method

dissimilarity value of 3 between units 3 and 5 thus accounts for unit 5 joining the (1, 3) group at threshold value 3. The fact that a unit does not join any existing groups until the very end of the dendrogram with the complete-link method is therefore *not* an indication that the unit is an outlier. The corresponding behaviour with the single-link method, however, *would* suggest it was an outlier.

As a practical illustration of the use of the dendrogram for graphical representation of high-dimension data, let us return to the soil profile data on Glamorganshire soils. Description of the data is given in Rayner (1966); a principal component representation of the data has already been given in Fig. 2.15, and a quadratic loss function representation in Fig. 2.16. Now consider dendrograms constructed from the data, using the similarity matrix of Table 3.1 as the intermediate stage. The single-link dendrogram is given in Fig. 3.3, and the complete-link dendrogram in Fig. 3.4. These dendrograms are reproduced from Gordon (1981) with permission from Chapman and Hall.

Numbering of soil samples is as in Figs 2.15 and 2.16, so that the dendrogram representations can be compared directly with these two figures. Consider first the single-link dendrogram of Fig. 3.3. The impression conveyed is that two reasonably compact groups exist, comprising soils 1, 2, 6, 5, 12, 13 and 19, 9, 10, 11, 23, 16, 22, 17, 7. Soils 8, 20, 3 then loosely join the first major group, soil 4 is a complete outlier, while soils 21, 15, 14, 18 only link after amalgamation of the two major groups. Reference to Fig. 2.15 shows that the first major group above (i.e. 1, 2, 6, 5, 12, 13) is in fact precisely the set of Brown Earths

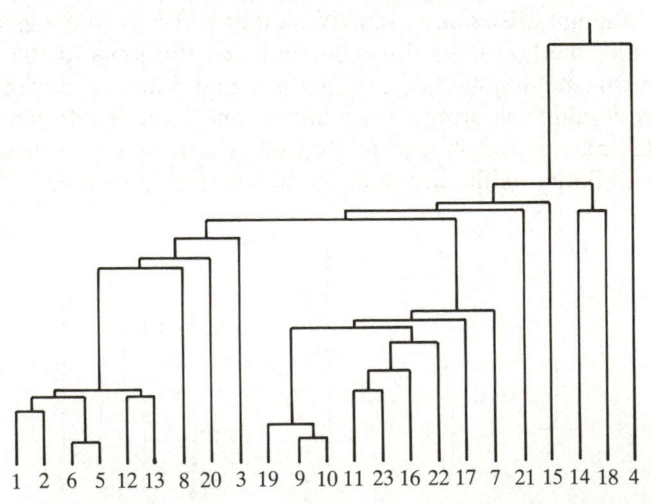

Fig. 3.3 Single-link dendrogram of Glamorganshire soils. Reproduced from Gordon (1981) with permission from Chapman and Hall

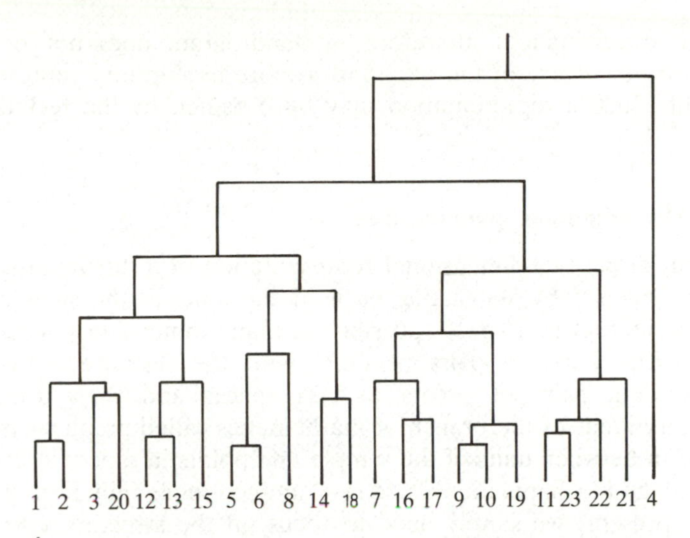

Fig. 3.4 Complete-link dendrogram of Glamorganshire soils. Reproduced from Gordon (1981) with permission from Chapman and Hall

with the addition of one Brown Earth (gley) sample, number 5. Two of the three samples that loosely join this group (20 and 3) are the two Lessivated Brown Earths. The second major group above (i.e. 19, 9, 10, 11, 23, 16, 22, 17, 7) is in fact precisely the set of Acid Brown Earths, with the inclusion of one Gley interloper (sample 22). Sample 4 is the single rendzina specimen in the set. Finally, samples 21, 15, 14, 18 comprise a mixture of the less numerous other types of soil. The overall impression conveyed by this dendrogram is thus much in accord with the opinions of a soil surveyor as regards a possible grouping or classification of these 23 soil samples.

The complete-link dendrogram produces some similar features, but also some marked differences. The Brown Earths group is not now so obviously compact, although perhaps a more satisfactory categorization of the Acid Brown Earths is achieved. Overall, however, the complete-link dendrogram is perhaps less successful as a summary of the data, if accord with the surveyor's judgement is to be the criterion of 'success'. A more precise quantification of the degree to which the two dendrograms match can be obtained via the methods discussed in Section 6.4 of Gordon (1981); see also Finden and Gordon (1985). It is evident, however, that if several different clustering methods are applied to a set of data then conflicting dendrograms are liable to result, and confusion may ensue. Also, the very nature of the techniques ensures that grouping of the units is implied, even when such grouping is in fact very tenuous.

In some circumstances, therefore, a dendrogram does not provide a satisfactory summary of the data and a more exploratory representation is sought. Such a representation may be obtained by the techniques of Section 3.2.

3.1.2 The minimum spanning tree

A second direct two-dimensional representation of a dissimilarity matrix may be obtained by associating each of the units in the sample with a point (or vertex) in a space (possibly of many dimensions), each of the dissimilarities between pairs of units with the distance between the corresponding pairs of points in this space, and then using some elementary ideas of the branch of mathematics called graph theory. This association between units of the sample and points in space, already used to advantage in Chapter 2, will be re-examined more fully in Section 3.2. For the present, we simply need to focus on the structure known as a *tree*. This is any set of straight-line segments joining pairs of points such that

(1) every point is connected to every other point by a set of lines;
(2) each point is visited by at least one line; and
(3) no closed circuits appear in the structure.

If the length of any such segment is given by the distance between the points that it connects, the length of a tree is defined to be the sum of the lengths of its segments. The *minimum spanning tree* (MST) of a set of n points is then the tree of shortest length among all trees that could be constructed from these points.

Gower and Ross (1969) review some problems posed in the literature that require the computation of a MST, and they discuss various algorithms for this computation. The most popular algorithms operate iteratively; at any stage the segments belong to one of two sets: set A containing those segments assigned to the MST, and set B, those not assigned. The most common approach is to assign iteratively to A the shortest segment in B that does not form a closed loop with any of the segments already in A. Initially A is empty and iteration stops when A contains $(n-1)$ segments. Figure 3.5 is the simple example given by Gower and Ross of a tree with integer segment lengths; thus segment BE is 5 units long. If any of the pairs of vertices AC, AG, CG, GH, etc. were joined then closed circuits would be introduced and the resulting figure would not be a tree. The length of the tree is 28 units, and it can be readily seen from the algorithm described above that it must be the MST for this set of points. The joins specified by this algorithm occur in the sequence: (EF) (BC) (FH) (AB) (BE) (EG) (DF).

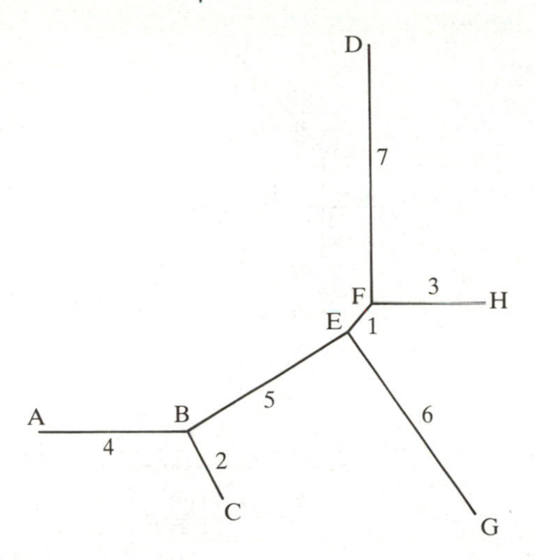

Fig. 3.5 A tree with eight vertices and integer segment lengths

Consider now the soil profile data analysed earlier. The matrix of similarities between every pair of the twenty-three soil profiles is given in Table 3.1; by subtracting these values from 1.0 we can obtain dissimilarities between the profiles and hence distances between the corresponding points. Application of the algorithm described above then yields the MST of Fig. 3.6.

The MST has a number of useful features. Primarily it is a convenient method for highlighting close neighbours in a sample, as demonstrated in Fig. 3.6. It is also closely related to the dendrogram obtained from the single-link clustering method. The form of the algorithm described above shows that the clusters at any level δ can be obtained from the MST by deleting all segments of length greater than δ, and therefore that the dendrogram can also be derived from the MST. For example, consider the threshold value 0.15 for the soil data. Deleting all segments having lengths greater than this value from the tree of Fig. 3.6 isolates, among others, the group consisting of the points 1, 2, 6, 5, 12, 13, which is precisely the same set as the first group at the left-hand end of the single-link dendrogram in Fig. 3.3. Increasing the threshold to 0.18 adds exactly the points 8, 20, and 3 to this group, as also happens in Fig. 3.3. Roger and Carpenter (1971) show how a MST can be constructed cumulatively, thus obviating the need for all the data to be stored in the computer. This unlocks the door to the construction of very large MSTs and hence to single-link cluster analyses of very large data sets.

The MST itself will also help in the interpretation of other cluster

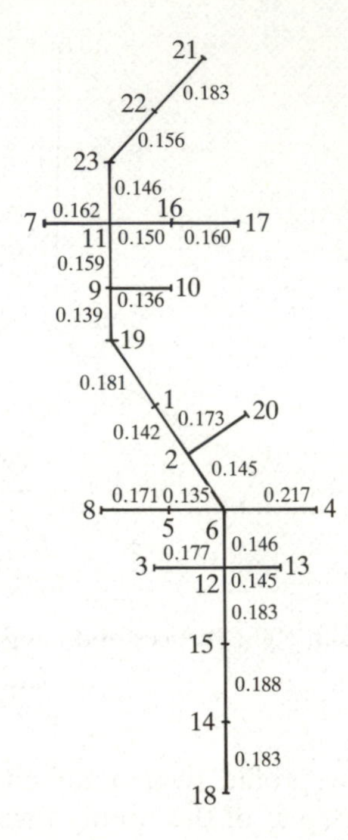

Fig. 3.6 Minimum spanning tree for Glamorganshire soil profiles

analysis methods. For example, close neighbours assigned to different clusters will be revealed and the adequacy of the clusters can be judged. The MST is also useful in highlighting any distortions produced by subspace projection methods, and this usage will be discussed further in the next section.

3.2 Subspace representation of units, derived from numerical dissimilarities

The methods of the previous section aim to describe an $(n \times n)$ matrix of dissimilarities between every pair of individuals in a sample by splitting the sample up into a (small) number of relatively homogeneous groups. However, as has been already stressed, such an approach *forces* us to split the sample. A useful complement would be a technique that enabled us to *examine* the relative positions of the n sample members to each other. This brings us back to the realm of geometric models, the underlying basis for such techniques as the principal component analysis of Chapter 2. There, the starting point was a geometric model of the $(n \times p)$

data matrix as n points in p-dimensional Euclidean space E^p. Since such a model could not in general be constructed in the physical world, approximations were sought that *could* be constructed in two or three dimensions. Principal component analysis (PCA) enables the r-dimensional subspace E^r to be isolated in which the overall variability of the points is a maximum. Furthermore, it was mentioned in Chapter 2 that if d_{ij} is the Euclidean distance between points i and j in E^p, and $\hat{d}_{ij}$ is the corresponding distance in an r-dimensional subspace, then the PCA subspace E^r is the one for which $V = \sum_i \sum_j (d_{ij}^2 - \hat{d}_{ij}^2)$ is a minimum. It was also demonstrated in Chapter 2 that other subspaces may be preferable in certain circumstances, and computational schemes for finding the subspaces that minimized $L = \sum_i \sum_j (d_{ij} - \hat{d}_{ij})^2$ or $L^* = \sum_i \sum_j w_{ij}(d_{ij} - \hat{d}_{ij})^2$ were discussed.

We are now supposing that, instead of a data matrix, we have available an $(n \times n)$ matrix of dissimilarities between all pairs of entities in a sample. Denote this matrix by $\boldsymbol{\delta}$, and the dissimilarity between the ith and jth entities by δ_{ij}. A useful geometrical model of the sample would be one in which each entity was represented by a point, and the distance between any two points was equal to the dissimilarity between the two corresponding entities. In other words we make a direct correspondence not only between entities and points, but also between dissimilarities and distances. Such a model would then give a good visual impression of the relationships existing between the members of the sample.

The first question to ask is therefore: Can such a model be constructed? (Note it is now necessary to *construct* the model from information on distances, whereas associating rows of the data matrix with sets of coordinate values immediately ensured the existence of the model in Chapter 2.) It turns out that such a model *can* be constructed, if the dissimilarities between entities satisfy certain basic properties. These conditions are specified in the following fundamental scaling theorem, for further details of which the reader is referred to the book by Torgerson (1958).

Scaling theorem

Given a matrix $\boldsymbol{\delta}$ of positive distances δ_{ij} between all pairs of n objects, a necessary condition for the n objects to be represented by n points in $E^{(n-1)}$ such that inter-point distance d_{ij} is equal to δ_{ij} for all i, j is that the elements of $\boldsymbol{\delta}$ satisfy the metric inequality $\delta_{ij} \leq \delta_{ik} + \delta_{kj}$ for all k, i, j.

Thus if the measure of dissimilarity that was used to calculate the values δ_{ij}, or if the subjective assessment used to derive them, is such that $\delta_{ij} \leqslant \delta_{ik} + \delta_{kj}$ for all k, i, j, then a geometrical model of the n entities is generally constructable. The problem here, as with the ideas considered in Chapter 2, is that this model may be in too many dimensions for comprehensibility, as existence is only guaranteed for a space of $(n-1)$ dimensions and n may be very large. We therefore need a second stage, namely the identification of a subspace of this space into which the points can be projected and in which the inter-entity dissimilarities δ_{ij} are *approximated* as well as possible by the corresponding interpoint distances $\hat{d}_{ij}$. One feasible approach to the problem is thus to set $d_{ij} = \delta_{ij}$ for all i, j, and then use the computer routines described in Chapter 2 to find the r-dimensional subspace minimizing either L or L^*. The same approach could, in principle, be used to find the subspace minimizing V. However, just as principal component analysis provides an exact analytical solution to this last minimization when starting from a data matrix, so there exists an exact analytical solution when starting from a dissimilarity matrix. This analytical solution has variously been termed classical scaling, metric scaling, or principal coordinate analysis (to distinguish it from principal component analysis with which it can be easily confused). We now give the mathematical argument underpinning this solution.

If we know the exact coordinates of n points in a p-dimensional Euclidean space, we can easily calculate the Euclidean distances between each pair of points. Let the coordinates of the points be given by the rows of the $(n \times p)$ matrix $\mathbf{X}$ and let $\mathbf{Q} = \mathbf{XX}'$. Thus if the $(n \times n)$ matrix $\mathbf{Q}$ has (i, j)th element q_{ij}, then

$$q_{rs} = \sum_{j=1}^{p} x_{rj} x_{sj}. \tag{3.7}$$

Let d_{rs}^2 be the squared Euclidean distance between points r and s, then

$$\begin{aligned} d_{rs}^2 &= \sum_{j=1}^{p} (x_{rj} - x_{sj})^2 \\ &= \sum_{j=1}^{p} x_{rj}^2 + \sum_{j=1}^{p} x_{sj}^2 - 2\sum_{j=1}^{p} x_{rj} x_{sj} \\ &= q_{rr} + q_{ss} - 2q_{rs}. \end{aligned} \tag{3.8}$$

Hence, given $\mathbf{X}$, we can readily find all the d_{rs}^2. We are interested in exactly the reverse: given all the d_{rs}^2, can we reconstruct $\mathbf{X}$? There is no unique solution unless we impose a location constraint. The most convenient one is to put the 'centre of gravity' of the points, $\bar{\mathbf{x}}$, at the origin so that $\sum_{r=1}^{n} x_{rj} = 0$ for all j. Using this constraint and (3.7), it then follows that the elements in any row or column of $\mathbf{Q}$ must sum to zero. Hence summing (3.8) over r, over s, and over both r and s, respectively, and setting $A = \sum_{r=1}^{n} q_{rr} = \text{trace}(\mathbf{Q})$, we find:

$$\sum_r d_{rs}^2 = A + nq_{ss} \tag{3.9}$$

$$\sum_s d_{rs}^2 = nq_{rr} + A \tag{3.10}$$

$$\sum_r \sum_s d_{rs}^2 = 2nA. \tag{3.11}$$

Substituting for q_{rr} and q_{ss} in (3.8) from (3.10) and (3.9), and using (3.11) to eliminate A, we then obtain:

$$q_{rs} = -\tfrac{1}{2}(d_{rs}^2 - d_{r.}^2 - d_{.s}^2 + d_{..}^2) \tag{3.12}$$

where a dot represents the average of values over the corresponding suffix, i.e. $d_{r.}^2$ is the average of the rth row of matrix $\mathbf{D} = (d_{ij}^2)$, $d_{.s}^2$ is the average of the sth column of $\mathbf{D}$, and $d_{..}^2$ is the average of all elements of $\mathbf{D}$.

Thus, given the matrix of all squared inter-point distances, we may regenerate the $\mathbf{Q}$ matrix. But $\mathbf{Q}$ is symmetric, so using the spectral decomposition of such a matrix (see Appendix A, Section A5) we can write $\mathbf{Q} = \mathbf{T\Lambda T}'$ where $\mathbf{\Lambda}$ is a diagonal matrix of eigenvalues of $\mathbf{Q}$ and $\mathbf{T}$ is the matrix whose columns are the corresponding eigenvectors of $\mathbf{Q}$. If $\mathbf{Q}$ is positive semi-definite then the entries in $\mathbf{\Lambda}$ are all non-negative, and we can obtain their square roots in the diagonal matrix $\mathbf{\Lambda}^{\frac{1}{2}}$. Thus $\mathbf{Q} = \mathbf{T\Lambda T}' = \mathbf{T\Lambda}^{\frac{1}{2}}\mathbf{\Lambda}^{\frac{1}{2}}\mathbf{T}' = \mathbf{XX}'$ where $\mathbf{X} = \mathbf{T\Lambda}^{\frac{1}{2}}$. Hence from the inter-point distances we have reconstituted the coordinates $\mathbf{X}$.

Given an $(n \times n)$ matrix of dissimilarities $\boldsymbol{\delta} = (\delta_{ij})$ between every pair of entities in a sample, we can thus find a geometrical representation of the n entities by n points in the following manner.

Step 1. Form the matrix $\mathbf{E}$ with (i, j)th element $e_{ij} = -\frac{1}{2}\delta_{ij}^2$.

Step 2. Subtract from each element of $\mathbf{E}$ the means of the row and column in which it is located, and add to it the mean of all elements of $\mathbf{E}$. Denote the resulting matrix by $\mathbf{F}$.

Step 3. Find the eigenvalues and eigenvectors of $\mathbf{F}$. Rank the eigenvalues in decreasing order; denote the ith largest eigenvalue by λ_i and its corresponding eigenvector by $\mathbf{v}_i$ (normalized so that $\mathbf{v}_i'\mathbf{v}_i = 1$).

Step 4. The coordinates of the n points on the jth axis of the Euclidean representation are given by the elements of $(\sqrt{\lambda_j})\mathbf{v}_j$.

The fact that the coordinates are obtained from the eigenvalues and eigenvectors of $\mathbf{F}$ means that the representation is referred to its principal axes, while the fact that the eigenvalues are ranked in decreasing order means that the successive eigenvectors refer to dimensions of decreasing importance. Consequently, the 'best' r dimensions in which to represent the sample will be given by the eigenvectors corresponding to the first r eigenvalues, just as with principal components. Another feature that can be carried over from principal component analysis is the measure of 'goodness of fit' of the first r dimensions to the 'true' $(n-1)$ dimensional representation, given by $G = \sum_{i=1}^{r} \lambda_i \Big/ \sum_{i=1}^{n-1} \lambda_i$. However, care must be exercised with the use of this measure for reasons outlined below.

In the crucial step 4 for actual derivation of the coordinate representation, we require to extract square roots of the eigenvalues of $\mathbf{F}$. Clearly, these square roots will only yield real quantities if the eigenvalues are non-negative, and for this to be so the matrix $\mathbf{F}$ must be at least positive semi-definite. It can be shown that this will always be the case if the input dissimilarities δ_{ij} obey the metric inequality $\delta_{ij} \leq \delta_{ik} + \delta_{kj}$ for all i, j, k. Thus if the dissimilarities obey this property, then the metric scaling algorithm will provide a good, approximate, low-dimensional representation of the relationships existing between the n entities, and G will give an accurate idea of the amount of distortion caused by reducing the dimensionality from $(n-1)$ to r. If s_{ij} is the similarity between individuals i and j calculated by any of the measures described in Section 1.4, then setting $\delta_{ij}^2 = 2(1 - s_{ij})$ should lead to a good geometric representation because Gower (1971a) establishes positive semi-definiteness of $\mathbf{F}$ for a range of these similarity measures. In particular, this property holds for the general measure (xv) (see Section 1.4), and hence it also holds for any of the more restrictive, commonly used definitions. The property also holds for the correlation coefficient as a measure of similarity between variables.

However, it may happen that either the measure of dissimilarity that is used, or the subjective method of deriving dissimilarities that is employed, leads to a set of δ_{ij} for which the metric inequality is *not* obeyed. In this case $\mathbf{F}$ will not be positive semi-definite, and will thus have one or more negative eigenvalues. The size of these negative eigenvalues reflects the amount of deviation from the metric conditions required of the dissimilarities. The immediate consequence is that some dimensions of the representation will be imaginary, requiring for the coordinates the square root of a negative number. Possible courses of action in such a case are reviewed by Mardia (1978), who concludes that the best representation is obtained by restricting attention to the positive eigenvalues and associated eigenvectors alone. Little distortion will ensue if the negative eigenvalues are small in absolute value, but grave doubts must be cast on the worth of the approximation if one or more large negative eigenvalues are involved (see Sibson (1979) for further discussion). It should also be noted that the measure G is distorted in the presence of negative eigenvalues, because $\sum_{i=1}^{n-1} \lambda_i$ is then less than the sum of all the positive λ_i. Consequently, G will be larger than it should be for small values of r, and an over-optimistic view will be presented of the worth of the chosen representation. A possible remedy is to use $\sum_{i=1}^{n-1} |\lambda_i|$ in place of $\sum_{i=1}^{n-1} \lambda_i$ in the denominator of G.

Much of the confusion between principal component and principal coordinate analyses stems from the fact that if the dissimilarity matrix $\boldsymbol{\delta}$ is computed as the matrix of Euclidean distances from the data matrix $\mathbf{X}$ (i.e. using dissimilarity coefficient (i) of Section 1.4), then metric scaling of $\boldsymbol{\delta}$ yields precisely the principal component scores obtained from a principal component analysis of $\mathbf{X}$. The two techniques are dual to each other in this sense, but it should be clearly borne in mind that principal component analysis is an R-technique while metric scaling is a Q-technique. The latter will thus be a computationally faster alternative to principal component analysis if n is smaller than p, but the two techniques are only in direct competition for use when we are dealing with Euclidean distance as dissimilarity measure. Gower (1966a) gives a clear discussion of the inter-relationships between the two techniques, together with guidance as to when each should be employed.

As an illustration of the use of metric scaling, consider the investigation into the taxonomic status of British water voles, genus *Arvicola*, reported by Corbet *et al.* (1970). The primary aim of the study was to compare populations of British water voles with European ones, and to investigate the hypothesis that more than one species of the genus might be present in Britain. Information on the presence or absence of thirteen characteristics was obtained from about 300 skulls subdivided into fourteen groups; sample sizes ranged from 11 to 50 in each group. The percentage incidence $x_1, \ldots, x_{13}$ of the thirteen characteristics in each sample is given in Table 3.2. These incidences x_i were first subjected to an angular transformation $y_i = \sin^{-1}(1 - 2x_i/100)$. For a given pair of samples (of sizes n_1, n_2 respectively), the Euclidean squared distance was calculated using the y_i values and divided by 13. A correction for bias of the form $(1/n_1 + 1/n_2)$ was then subtracted, and the square root of the result was taken to be the dissimilarity between the corresponding two vole populations. The bias correction had the slightly unsatisfactory feature of introducing two (very small) negative squared dissimilarities, but since they were so small it was felt that little distortion would be introduced by using their absolute values when taking square roots. The resulting dissimilarity matrix is given in Table 3.3. When metric scaling was applied to this dissimilarity matrix, over 80 per cent of the trace was

Table 3.2 Percentage incidence of thirteen variates in each of fourteen samples of *Arvicola* (reproduced with permission from the Zoological Society, London)

	Surrey	Shrops	Yorks	Perths	Aberd.	Eilean Gamhna	Alps	Yugosl.	Germany	Norway	Pyrenees I	Pyrenees II	North Spain	South Spain
1. Fenestra flocculi present	48.5	67.7	51.7	42.9	18.1	65.0	57.1	26.7	38.5	33.3	47.6	60.0	53.8	29.2
2. Preorbital foramen double	89.2	67.0	84.8	50.0	79.6	81.8	76.2	53.1	67.9	83.3	92.9	90.9	88.1	74.0
3. Anterior frontal foramen present	7.9	2.0	0	0	4.1	9.1	21.4	23.5	17.9	27.8	26.7	13.6	7.1	16.0
4. Posterior frontal foramen double	42.1	23.0	31.3	50.0	44.9	31.8	38.1	38.2	21.4	29.4	10.0	68.2	33.3	46.0
5. Maxillary foramen I present	92.1	93.0	88.2	77.3	79.6	81.8	66.7	44.1	82.1	86.1	36.7	40.9	88.1	86.0
6. Maxillary foramen II present	100.0	100.0	100.0	100.0	100.0	100.0	97.6	94.1	100.0	100.0	100.0	100.0	100.0	100.0
7. Inframaxillary crest medium or large	100.0	86.0	94.1	90.9	77.6	59.1	14.3	11.8	60.7	63.9	50.0	18.2	19.0	18.0
8. Foramen sphenoidale medium present	35.3	44.0	18.8	36.4	16.7	20.0	23.5	11.8	35.7	53.8	14.3	100.0	85.7	88.0
9. Processus pterygoideus present	11.4	14.0	25.0	59.1	37.1	30.0	9.5	18.2	24.0	18.8	7.4	5.0	9.8	16.3
10. Foramen ovale double	100.0	99.0	100.0	100.0	100.0	100.0	100.0	100.0	100.0	100.0	100.0	80.0	73.8	72.0
11. Foramen hypoglossi double	71.9	97.0	83.3	100.0	90.4	100.0	91.4	94.9	91.7	83.3	86.4	90.0	72.2	80.4
12. Basioccipital foramen present	31.6	31.0	33.3	38.9	9.8	5.0	11.8	12.5	37.5	8.3	90.9	50.0	73.7	69.6
13. Mental foramen double	2.8	17.0	5.9	0	0	9.1	17.5	5.9	0	34.3	3.3	0	2.4	4.0
No. of skulls	19	50	17	11	49	11	21	17	14	18	16	11	21	25

Table 3.3 Dissimilarity matrix for *Arvicola* populations (reproduced with permission from the Zoological Society, London)

	Surrey	Shrops	Yorks	Perths	Aberd.	Eilean Gamhna	Alps	Yugosl.	Germ.	Norway	Pyren. I	Pyren. II	N. Spain
Shropshire	0.099												
Yorkshire	0.033	0.022											
Perthshire	0.183	0.114	0.042										
Aberdeen	0.148	0.224	0.059	0.068									
Eilean Gamhna	0.198	0.039	0.053	0.085	0.051								
Alps	0.462	0.266	0.322	0.435	0.268	0.025							
Yugoslavia	0.628	0.442	0.444	0.406	0.240	0.129	0.014						
Germany	0.113	0.070	0.046	0.047	0.034	0.002	0.106	0.129					
Norway	0.173	0.119	0.162	0.331	0.177	0.039	0.089	0.237	0.071				
Pyrenees I	0.434	0.419	0.339	0.505	0.469	0.390	0.315	0.349	0.151	0.430			
Pyrenees II	0.762	0.633	0.781	0.700	0.758	0.625	0.469	0.618	0.440	0.538	0.607		
North Spain	0.530	0.389	0.482	0.579	0.597	0.498	0.374	0.562	0.247	0.383	0.387	0.084	
South Spain	0.586	0.435	0.550	0.530	0.552	0.509	0.369	0.471	0.234	0.346	0.456	0.090	0.038

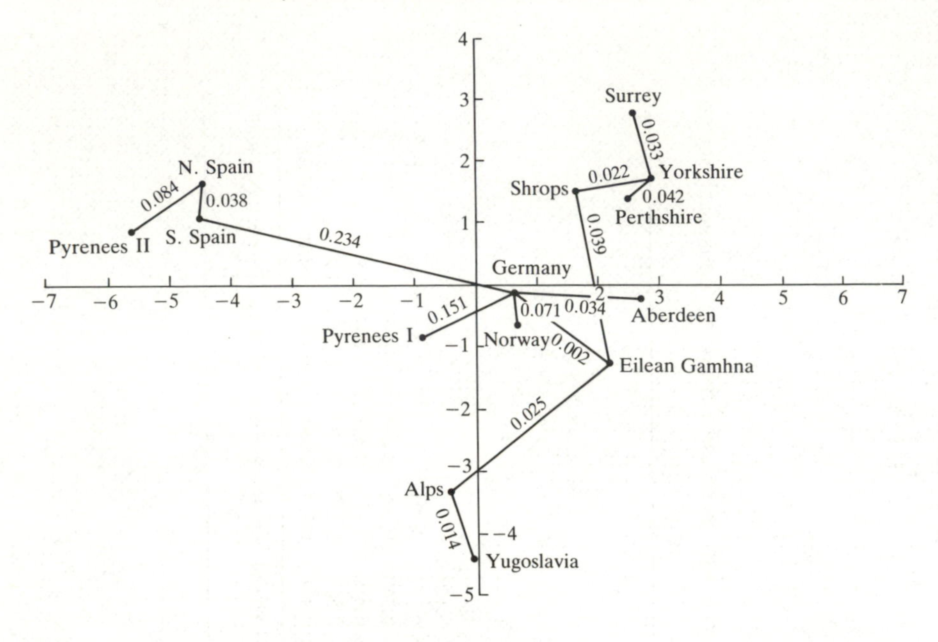

Fig. 3.7 *Arvicola* sample means plotted on the first two principal coordinate axes, with a superimposed minimum spanning tree

accounted for by the first two eigenvalues and a two-dimensional projection (of the 13-dimensional space) was thus felt to be an adequate representation of the relationship existing between the vole populations. This representation is shown in Fig. 3.7. (Tables 3.2 and 3.3, and Fig. 3.7, are reproduced with permission from the Zoological Society, London.)

It has already been suggested (in Section 3.1.2) that the minimum spanning tree may be a useful check on the adequacy of projections as pictorial representations. The minimum spanning tree obtained from Table 3.3 is therefore superimposed on the points in Fig. 3.7. Remembering that this figure is a two-dimensional projection of a 13-dimensional space, it should always be borne in mind that some distortion may be introduced by such a projection. Points far apart in the true dimensionality may after projection appear close to each other. Since the minimum spanning tree tends to connect neighbours, it will quickly point up such false impressions. Thus points that are close together on the diagram, but that are *not* connected in the MST, are likely positions of such distortion. The only such instance in Fig. 3.7 is the sample from Aberdeen. This appears from the diagram to be closest to the sample from Eilean Gamhna, but the MST connects it to the more distant-looking sample from Germany. Closer inspection of the distances attached to each branch of the MST in this region suggests that it is the point representing Eilean Gamhna that is in a false position, and this point should in fact be almost coincident with the point representing Germany. Apart from a slight distortion in scale of distances between points, no other points seem to be in obviously false positions and the two-dimensional projection thus gives a reasonably faithful impression of the true picture.

As regards the conclusions that were drawn from the study, the metric scaling indicated that there were no grounds for supposing there to be two distinct species of *Arvicola* present in the British Isles. The theory had previously been put forward that incidence of both *Arvicola terrestris* and *Arvicola sapidus* could be expected in Britain. In the present study, *terrestris* populations were the ones from Germany, Norway, Alps, Yugoslavia, and Pyrenees I, while *sapidus* populations were the ones from North Spain, South Spain, and Pyrenees II. It can be seen that *all* British populations align with the *terrestris* group, and they are all distant from the *sapidus* group. Furthermore, there is no evidence of a North–South divide within the British group. Hence a single species within the British isles may be inferred.

3.3 Subspace representation of units, derived from ordinal data

So far in this chapter we have assumed that the dissimilarities in our $(n \times n)$ starting matrix are given in numerical form, and the construction of diagrams has been described with the aim of representing individuals or objects by points in such a way that all inter-point distances match all inter-object dissimilarities as closely as possible. Situations may occur in practice, however, in which considerably more slack is permitted in the representation. The most common of these situations is where faithfulness of representation is demanded only to within *rank-order* of the dissimilarities. Thus we might be prepared to accept as satisfactory a configuration of points in which the rank-order of the inter-point distances exactly matches the rank-order of the inter-object dissimilarities, although there exist large discrepancies between some of the actual numerical values.

Why should we wish to make this modification? There are two main practical reasons, and one theoretical or statistical reason. The most obvious practical reason is that sometimes the data are actually obtained in ordinal form, so that all we have to start with are ranks in the dissimilarity matrix rather than numerical values. In an investigation into the taste of a selection of soft drinks, for example, each subject may be presented with pairs of unidentified drinks and asked to compare them according to taste. The best that the subject may be able to do is to say that drink A is more like drink B than drink C, or that drinks A and B are more alike than drinks C and D, and so on. Gathering together data in this fashion will lead to a dissimilarity matrix consisting of ranks rather than numerical values. A second common practical situation is where numerical dissimilarities *have* been obtained, but the precision of the individual values is suspect for some reason (e.g. high degree of subjectivity, or errors in measuring technique) and it is felt that only the ranks of the dissimilarities are sufficiently trustworthy for analysis. It should perhaps be mentioned here (as well as later in Chapter 9) that the effect on the metric scaling solution of small perturbations in the input dissimilarities has been the subject of some study, notably by Sibson (1979). However, it is often difficult to build realistic perturbation patterns into an analysis of data, and so, to a first approximation, the replacement of the numerical dissimilarities by their ranks is a sensible procedure. Finally, the statistical reason for use of ranks concerns the dimensionality of the constructed representation. Since relaxation of the objective to the matching of rank orders of distances and dissimilarities introduces some slack into the system, it is generally found that satisfactory configurations can be obtained in smaller dimensionalities than when attempting to match numerical distances and dissimilarities. Thus if replacement of numerical values by ranks is considered sensible, then a satisfactory representation in two or three dimensions becomes much more feasible than before.

The construction of geometrical representations of objects (or stimuli) from information about their pairwise dissimilarities has long been of interest to quantitative psychologists, who use the term *scaling* for such a procedure. Since the methods nearly always result in configurations occupying more than one dimension, such scaling is generally known as *multidimensional scaling*. The classical metric scaling theory, as described in the previous section, was the accepted methodology at the start of the 1960s; the book by Torgerson (1958) is an excellent description of the state of the art at this juncture.

With the advent of the large-scale use of electronic computers in the 1960s, there was also a rapid increase of interest in multidimensional scaling methods. This was fuelled by the pioneering work of Shepard (1962a,b), who argued that the nature of data encountered in quantitative psychology dictated that the interpoint distances of the constructed configuration should be *monotonically* related to the given dissimilarities between the stimuli, and hence that the rank orders of distances and dissimilarities are of major importance. Methods deriving from this source have been called *non-metric multidimensional scaling* methods, although latterly the term *ordinal* is thought to be more appropriate than 'non-metric'. The amount of research in this area has in fact been prodigious, and we shall restrict ourselves here to a brief discussion of the fundamental method. The reader is referred to the two volumes edited by Shepard *et al.* (1972) for a more detailed discussion of some of the theory and possible applications (including a historical review of the development of multidimensional scaling methods in the introduction to the first volume), and to Gordon (1981) for a brief description of the more important later work.

Although the idea had been first put forward by Shepard in 1962, his solution resulted in a heuristic algorithm which contained a number of unsatisfactory features. These features were discussed by Kruskal (1964a), who proceeded to formalize Shepard's approach by defining a function that measures the departure from monotonicity between dissimilarities and distances. A companion paper (Kruskal, 1964b) then described a computer algorithm for optimizing this function and hence for obtaining the geometric configuration in which inter-point distances and input dissimilarities have as nearly monotonic a relationship as possible. These two papers quickly became classics of the psychometric literature, and the method remains in popular use today in largely the form in which it was devised over twenty years ago.

In essence, the method consists of the same stages that comprise any of the 'optimization' methods discussed in Section 2.2.7. The user must:

1. Choose the desired dimensionality t of representation.
2. Select an arbitrary initial configuration of n points in t dimensions.
3. Define the function S that specifies the degree of departure from monotonicity between inter-point distances in the configuration and inter-stimuli input dissimilarities.
4. Use a standard computer function minimization procedure (e.g. steepest descent) to find the configuration which minimizes S.

If it is also required to find the 'best' dimensionality in which to represent the sample, the following stages are additionally necessary:

5. Repeat the whole process for a range of values of t. In practice $1 \leq t \leq 5$ would be a sensible range, if it was hoped to obtain a representation that could be plotted easily. By general consent the best strategy appears to be to start in the highest dimensionality t_{max}, to find the best configuration (and hence the minimum value of S in this dimensionality), and then to take the coordinates of the points in the first $t_{max} - 1$ dimensions as the initial configuration for a solution in dimensionality one less than before. Continue in this way, reducing by one dimension at each iteration.

6. By plotting the minimum value of S against the dimensionality t and looking for an 'elbow' in the plot (as with the scree plots of Section 2.2.5), a suitable value of t for the final representation can generally be deduced.

The function S was called STRESS (short for *ST*andardized *RE*sidual *S*um of *S*quares) by Kruskal, and was arrived at as follows. Suppose that (at any stage of the process) there exists a current configuration in which d_{ij} is the distance between points i and j corresponding to a dissimilarity δ_{ij} between stimuli i and j. Plotting the d_{ij} against their counterpart δ_{ij} will show up the degree to which they are monotonically related. A perfect monotonic relationship exists if the smallest d_{ij} corresponds to the smallest δ_{ij}, the next smallest d_{ij} to the next smallest δ_{ij}, and so on. A plot resulting from such a relationship is illustrated in Fig. 3.8. The feature of this plot is that as one moves from any particular d_{ij} value to the next larger one (i.e. from any circle on the diagram to the nearest one to it on the right), then the corresponding δ_{ij} value must become larger (i.e. the next circle is *higher* than the previous one). Thus the line joining the circles progresses steadily upwards as it moves from left to right.

In any typical configuration this will not be the case, with the d_{ij} and δ_{ij} *not* in perfect monotonic relationship. Thus, for example, Fig. 3.9 shows a situation in which the second lowest δ_{ij} corresponds to the fourth lowest d_{ij}, while the second largest δ_{ij} corresponds to the third largest d_{ij}. The non-monotonicity is detected by moving *downwards* when passing from the third empty circle (from the left) to the fourth and also when passing from the sixth empty circle (from the left) to the seventh. A line joining the empty circles progressively from left to right would move downwards in some places and upwards in others.

If the five empty circles marked on Fig. 3.9, however, were moved to the positions indicated by the solid circles, then monotonicity would be

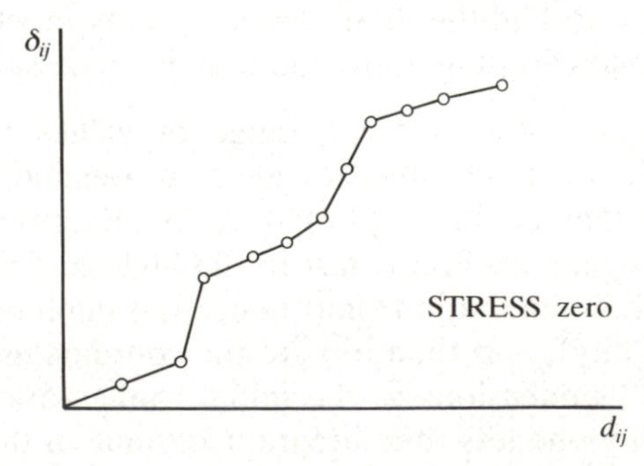

Fig. 3.8 Perfect monotonic relationship between δ_{ij} and d_{ij}

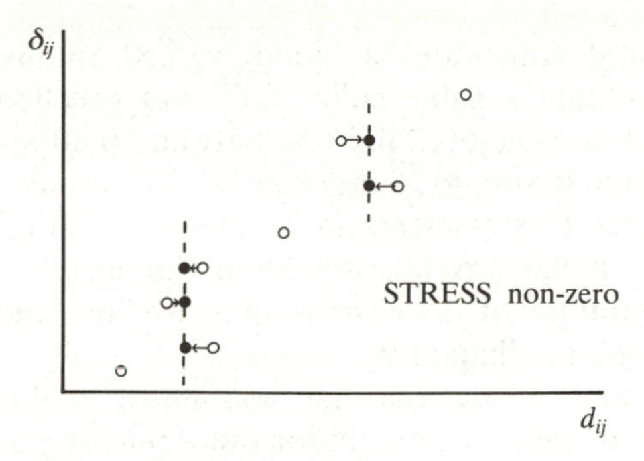

Fig. 3.9 Non-monotonic relationship between δ_{ij} and d_{ij}

restored. In making these moves none of the dissimilarity values δ_{ij} is altered, but all five distance values d_{ij} are changed to new values $\hat{d}_{ij}$. Interest focuses on the *smallest* change from d_{ij} to $\hat{d}_{ij}$ that will effect monotonicity of relationship. This is a problem in *monotone regression,* a topic discussed in general by Barlow *et al.* (1972). Kruskal (1964b), however, provides the specific solution needed in multidimensional scaling (and illustrated in fact in Fig. 3.9). All that is required is to choose $\hat{d}_{ij}$ to be the average of the d_{ij} for each block of non-monotone points (although the computer routine needed to effect this is non-trivial). With this solution we can now define the function STRESS as

$$\text{STRESS} = \sqrt{\left\{\frac{\sum\sum_{i<j} (d_{ij} - \hat{d}_{ij})^2}{\sum\sum_{i<j} d_{ij}^2}\right\}}.$$

Thus, for a given configuration of n points in t dimensions, Kruskal's monotone regression routine enables the STRESS to be computed. By viewing STRESS as a function of nt parameters (the coordinates of n points in t dimensions), it is then possible to apply a steepest descent procedure to find the configuration minimizing STRESS; Kruskal (1964b) describes the implementation of such a procedure.

The STRESS of the final configuration in t dimensions gives a rough indication of the extent to which a non-metric multidimensional scaling has provided a reasonable representation of a set of dissimilarities. Thus STRESS is zero in Fig. 3.8 because a perfect monotone relationship obtains, but non-zero in Fig. 3.9. As is to be expected, increasing t in general reduces STRESS since better representations can be obtained by raising dimensionality. It is thus usually the aim of the experimenter to find the smallest value of t for which a 'good' value of STRESS is obtained, but what constitutes a 'good' value? In his original work, Kruskal suggested that a value below 0.05 was excellent, one between 0.05 and 0.10 was satisfactory, and one between 0.10 and 0.15 probably acceptable but giving rise to some doubts. More-recent investigations suggest these blanket assessments to be over-simplistic, and one would certainly need to make any assessment in the light of the number of points and the number of dimensions that are involved. Some further comments are made in Chapter 9.

In Fig. 3.10, the two-dimensional non-metric scaling configuration derived from the dissimilarity matrix for the *Arvicola* populations (Table 3.3) is presented. The actual configuration obtained from the computer program has been rotated through 180° to facilitate comparison with the metric scaling configuration of Fig. 3.7. It can be seen that there is a high degree of resemblance between the two configurations, the only point with marked difference between the two being the point representing the Pyrenees I population. However, this point is still more of a neighbour to the Alps/Yugoslavia/Norway group than to the S. Spain/N. Spain/Pyrenees II group, so the overall conclusions would remain the same as before. Note also that the STRESS value for this configuration is 0.125, so that a three-dimensional solution would be preferable if a more accurate representation were desired. The Pyrenees I population would then probably be moved into a position more consistent with the metric solution.

It is worth noting that the more 'extreme' populations or points are liable to greater fluctuation in an ordinal solution than in a metric one. This is because there is a large amount of empty space on the edges of a

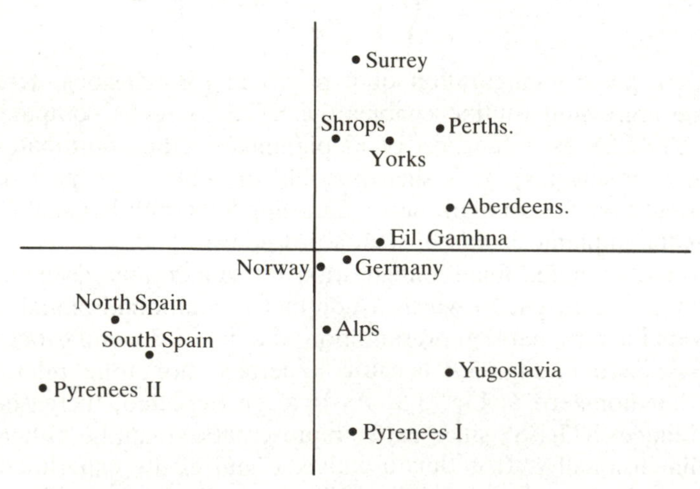

Fig. 3.10 Non-metric scaling configuration for *Arvicola* data

configuration, and a population or individual that has a large dissimilarity δ to all other populations or individuals can generally be placed anywhere in this space without disturbing the monotonicity between δ_{ij} and d_{ij}. Conversely, the more populations or individuals there are in the data, the more tightly constrained are the points by the order restrictions on the d_{ij} and hence the more closely do the metric and the non-metric configurations match. With many points in the representation, therefore, a metric solution will be acceptable even when the ordinal assumption is more strictly appropriate, and will have the benefit of requiring far less computational effort.

To summarize, three basic methods have now been described for producing a configuration of points from an input matrix of dissimilarities: the non-metric scaling method described in this section, the metric scaling method of the previous section, and the non-linear mapping (minimization of L^*) described in Section 2.2.7. Anderson (1971) has compared the three approaches on the Glamorgan soil data discussed earlier. Of these three techniques, metric scaling is the simplest computationally. Its main weakness is the possibility of poor representation of some points even though the overall variability is high. Non-linear mapping overcomes this, but involves much greater computational effort. Also, both non-linear mapping and non-metric scaling rely on iterative procedures which may only converge to local minima. It is often suggested that these iterative techniques be performed several times, with different starting configurations, to ensure that a global optimum has been reached. Note also that the STRESS function of non-metric scaling is not continuous, and experiment has shown that, for n less than about 30, points in the configuration may be subjected to appreciable displacements without any alteration in the minimum value. Nevertheless, all these techniques have proved to be very useful in representing dissimilarity matrices, and they are now firmly established as appropriate methods of pattern display.

In conclusion, we may note that the last twenty years have seen an explosion of interest in scaling methods, particularly in the Psychometry literature, and numerous variants or extensions of the above three basic methods now exist. Possibilities exist for either extension or variation to be made in any of the three key areas:

(1) measurement level of the scaling process (e.g. metric or ordinal);
(2) definition of goodness-of-fit function;
(3) method of fitting the configuration.

Of all the packages that have been proposed to date, the most comprehensive is that described by Takane *et al.* (1977). In their method, they permit all four levels of measurement (nominal, ordinal, interval, and ratio), thereby subsuming all existing types of scaling under one umbrella. As goodness-of-fit, Takane *et al.* use STRESS but applied to *squared* distances (i.e. d_{ij} and $\hat{d}_{ij}$ of STRESS are replaced by d_{ij}^2 and $\hat{d}_{ij}^2$), while the method used to fit the configuration is alternating least squares. They thus coin the acronyms SSTRESS for the function, and ALSCAL for the method. Furthermore, facility exists in their package for dealing with individual differences, but discussion of this aspect is delayed until Chapter 6. They provide full details of the necessary mathematics and computation. For a review of other available scaling packages, see Coxon (1982).

3.4 Methods for handling asymmetric matrices

In all the techniques discussed so far in this chapter, it has been assumed that graphical representation is desired using as input a *symmetric* dissimilarity or distance matrix. The majority of dissimilarity matrices will, of course, satisfy this assumption. Nevertheless, asymmetric relationships between pairs of items or populations may arise in practice. Examples are confusion matrices (giving the number of times in a series of trials that item i is identified as item j, taken in that order); diallel-cross experiments (giving the number of progeny or yield when a male of line i is crossed with a female of line j); immigration/emigration statistics (or the number of people who live in place i and carry out some activity in place j); food preference studies (giving the proportion of members of a panel who prefer soft drink i to soft drink j); and so on. All such situations yield a square matrix $\mathbf{D}$ of observations d_{ij} relating row i to column j. The rows and columns index the same entities, in common with all other square matrices considered in this chapter, but now in general $d_{ij} \neq d_{ji}$. Thus, although it may be reasonable to suppose that a distance interpretation partially underlies such dissimilarities, the lack of symmetry rules out the use of any of the methods described so far in this chapter. The appeal of the latter techniques is often so great, however, that asymmetric matrices are first 'symmetrized' to allow one of these methods to be used. The simplest device for achieving this purpose is to replace each d_{ij}, d_{ji} pair by the single value $\frac{1}{2}(d_{ij} + d_{ji})$, which is equivalent to the analysis of the matrix $\frac{1}{2}(\mathbf{D} + \mathbf{D}')$. This analysis ignores departures from symmetry that may be informative. Indeed, in the food preference example cited above, the matrix $\frac{1}{2}(\mathbf{D} + \mathbf{D}')$ will have all off-diagonal entries equal to 0.5, and it is *only* the departures from symmetry that are of interest. Constantine and Gower (1978) have addressed the problem of finding suitable graphical methods for representing such asymmetric matrices adequately and have described two appropriate techniques. One is multidimensional unfolding, which will be outlined in the next chapter, while the other is the canonical analysis of skew-symmetry. A brief discussion of this analysis now concludes the present chapter.

We can write any square matrix $\mathbf{D}$ as

$$\mathbf{D} = \tfrac{1}{2}(\mathbf{D} + \mathbf{D}') + \tfrac{1}{2}(\mathbf{D} - \mathbf{D}')$$
$$= \mathbf{M} + \mathbf{N},$$

and it is evident that $\mathbf{M}$ is a symmetric matrix (i.e. $\mathbf{M} = \mathbf{M}'$) while $\mathbf{N}$ is a skew-symmetric matrix (i.e. $\mathbf{N} = -\mathbf{N}'$). Graphical representation of $\mathbf{D}$ may therefore be accomplished in two parts: one picture, using any of the methods described already in this chapter, will describe the symmetric part $\mathbf{M}$ of the matrix, while a second picture can be produced to describe the skew-symmetric component $\mathbf{N}$. This second picture has as its basis the canonical decomposition of a skew-symmetric matrix, which states that we can write $\mathbf{N}$ in the form $\sum_{i=1}^{[p/2]} \lambda_i(\boldsymbol{u}_i\boldsymbol{v}_i' - \boldsymbol{v}_i\boldsymbol{u}_i')$. Here $\boldsymbol{u}_i$ and $\boldsymbol{v}_i$ $(i = 1, \ldots, [p/2])$ may all be taken to be the columns of an orthogonal matrix, λ_i are positive values which may be taken in decreasing order, and $[p/2]$ denotes the greatest integer less than or equal to $p/2$. The λ_i, $\boldsymbol{u}_i$, and $\boldsymbol{v}_i$ may be found by subjecting the symmetric matrix $\mathbf{NN}'$ $(= -\mathbf{N}^2)$ to an eigenvalue/eigenvector decomposition, which will result in the production of pairs of equal roots λ_i^2 with associated pairs of vectors $\boldsymbol{u}_i$, $\boldsymbol{v}_i$. Let λ denote the *largest* root found, and let $\boldsymbol{u}$ and $\boldsymbol{v}$ denote the vectors associated with this root. Then we can approximate $\mathbf{N}$ by $\lambda(\boldsymbol{u}\boldsymbol{v}' - \boldsymbol{v}\boldsymbol{u}')$, and this will be the best rank-2 approximation to $\mathbf{N}$ that can be found. If $\mathbf{N}$ is a $(p \times p)$ matrix (i.e. p entities index the rows and columns of the original matrix), then $\boldsymbol{u}' = (u_1, \ldots, u_p)$ and $\boldsymbol{v}' = (v_1, \ldots, v_p)$. By plotting the vectors $\boldsymbol{u}$, $\boldsymbol{v}$ as p points with coordinates (u_i, v_i) relative to orthogonal axes, we obtain a two-dimensional configuration representing $\mathbf{N}$.

Although the configuration that is obtained in this way resembles the sort of configurations obtained by principal component analysis, or by scaling methods, the geometrical interpretation of this configuration needs care. Constantine and Gower (1978) show that the interpretation does not centre on the distance between two points, but on the area of the triangle formed by those two points and the origin O of the coordinates. Suppose that P_1 and P_2 are two such points. Then the lack of symmetry in the relationship between the stimuli corresponding to P_1 and P_2 is proportional to the area of the triangle OP_1P_2. If the relationship is symmetrical this area should therefore be zero, so that O, P_1, and P_2 will all lie on a straight line. Identification of points collinear with the origin is

Table 3.4 Amounts of response reduction caused by cross-adaptation

	Test stimuli						
Adapting stimuli	1	2	3	4	5	6	7
1. Citral	—	12.0	4.1	10.3	16.0	20.0	11.4
2. Cyclopentanone	9.5	—	15.7	9.6	8.5	14.4	11.8
3. Benzyl acetate	2.4	11.0	—	8.3	21.1	17.4	15.4
4. Safrole	13.3	6.9	3.1	—	4.9	18.5	9.7
5. Butyl acetate	10.2	5.1	19.1	6.6	—	17.7	12.5
6. *m*-Xylene	6.6	5.6	7.2	8.6	19.1	—	19.5
7. Methyl salicylate	5.5	4.5	12.4	4.3	18.5	11.6	—

thus a prime objective. Also, since triangles on the same base that includes a vertex at the origin and with the other vertex on a line parallel to the base are equal in area, parallel lines play an important part in the interpretation.

To illustrate the use of this technique, consider Table 3.4 (taken from a 1971 University of Utrecht thesis by Köster, and reproduced by Constantine and Gower (1978)). This table gives a measure of the effect of one odour (the 'adapting stimulus') on the adaptability of a subject to detect a second odour (the 'test stimulus'). Suppose that a proportion p of subjects presented with a particular test stimulus can actually detect it. If these subjects are first presented with an adapting stimulus then the proportion who can correctly detect the test stimulus will fall to p'. A weak adapting stimulus followed by a strong test stimulus should yield a small difference $(p-p')$, while a strong adapting stimulus followed by a weak test stimulus should yield a large difference $(p-p')$. Table 3.4 gives the values of $(p-p')$ multiplied by 100 for such an experiment. Of interest are patterns of gross asymmetry, as these will indicate odours of very different strengths. Figure 3.11 (reproduced from Constantine and Gower (1978) with permission from the Royal Statistical Society) gives the canonical analysis of the skew-symmetric part of this matrix. The first two dimensions account for 77 per cent of the total sum of squares of the skew-symmetric elements of Table 3.4, so the fit appears to be quite good. It is evident that the points representing citral, cyclopentanone, benzyl acetate, safrole, and butyl acetate are nearly collinear with the origin, which correctly indicates the general symmetry of Table 3.4 for these stimuli. The gross asymmetry of *m*-xylene and methyl salicylate, both with each other and with all other stimuli, is also revealed. Clearly such a two-dimensional representation is only an approximation to the true picture, and will be prone to some inaccuracies. Most obvious here is an exaggerated asymmetry between *m*-xylene and butyl acetate, and an

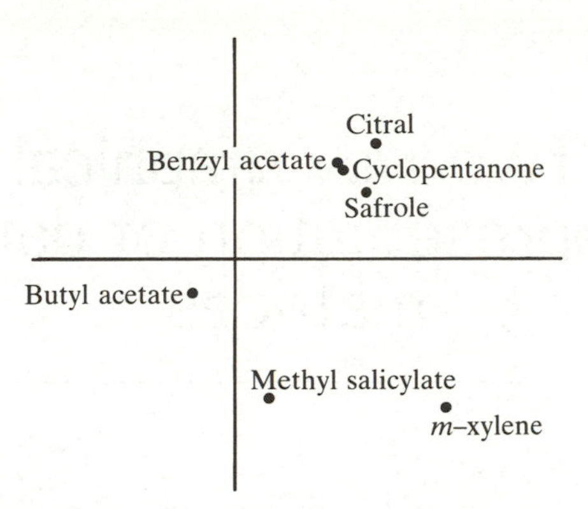

Fig. 3.11 Canonical analysis of the skew-symmetric part of Table 3.4. Reproduced from Constantine and Gower (1978) with permission from the Royal Statistical Society

exaggerated symmetry between citral and butyl acetate. On the whole, however, the representation of the asymmetry in the table is remarkably successful.

3.5 Postscript: missing values

A cautionary note should be added if the user wishes to compute dissimilarities from a raw data matrix as a prelude to using any of the techniques in this chapter, and there are missing entries in the data. Use of imputation of missing values, as described in Section 2.3, may lead to spuriously low dissimilarity values, particularly if the crude imputation of means is used. This is because the *same* values may be inserted in corresponding positions of several units, and hence these units will appear to be more similar than they actually are. In this case it is better to calculate the similarity between a pair of units from just those variates which are present on both units. If there are q such variates and p variates in total, the similarity obtained in this way should then be multiplied by a factor (p/q) to bring its numerical value into line with those similarities computed from all p variates. Note that Gower's general coefficient of similarity is set up to handle missing values automatically in this fashion.

4
Two-way graphical representation of data matrices

Various graphical methods have been described in Chapters 2 and 3 for representing the entities indexing *either* the rows *or* the columns of a matrix, where the entries in the matrix may be of various possible types. It sometimes happens, however, that valuable information may be extracted from a visual display of *both* the rows *and* the columns of the matrix, considered simultaneously.

For example, Gabriel (1971) analysed a set of data in which the percentage of houses possessing various facilities (e.g. toilet, bath, kitchen, water, electricity) was measured for each of a number of localities in Israel (e.g. old city quarters, modern city quarters, rural localities). One of the methods already described could be used to provide a graphical representation in which the localities are denoted by points, such that closeness of two points implies similarity of the two corresponding localities. By means of his biplot technique (to be described in this chapter), however, Gabriel was additionally able to superimpose on this plot a representation of the other set of entities, the facilities, in such a way as to highlight the main facilities shared by groups of similar localities. The resultant plot gave a more complete picture of the data matrix, as well as providing valuable insight into the relationships between the facilities and the localities.

Many other investigations would benefit from such simultaneous handling of both sets of entities referenced in a matrix. For example, the market researcher surveying the purchasing pattern of a particular consumer product (say a breakfast cereal) among various subgroups of the population would welcome a pictorial representation that not only showed the relationships among the subgroups, but also characterized each cluster of subgroups by one or two cereals popular among those subgroups. Such a representation would help to direct sales promotion and advertising more efficiently, and might also identify gaps in the market where a new cereal could make an impact. Alternatively, the sociologist investigating people's perception of various jobs or professions might ask a sample of individuals to assess the importance of each of a number of personal attributes (e.g. patience, perseverance, ability to communicate, ability to get on with people) in relation to each of a number of jobs or professions. In addition to a pictorial representation that shows the perceived similarities among the various professions, a simultaneous characterization of each group of professions according to the most important personal attributes required by them would be very revealing. Such methodology would also be of great interest to the industrial psychologist hoping to place applicants to a large firm or company into the most suitable jobs, so as to maximize employee satisfaction.

There is thus demand for graphical methods of displaying information on more than one set of entities, and a description of some of the more common such methods will be given in this chapter. As with the methods discussed in previous chapters, the appropriate technique to select will be determined primarily by the nature of the input data. However, an additional consideration in the present case is the desired nature of the output configuration. We shall concentrate on methods analogous to the subspace techniques of previous chapters. In such techniques, the principal mode of representing an entity is as a point in space, although an alternative device used from time to time is the vector in space. When two sets of entities are to be represented, we may thus choose to represent both sets by points, both sets by vectors, or one set by points and the other by vectors. One technique giving rise to each mode of representation will be described below. A further technique enables a simultaneous decomposition of the data to be effected according to each set of entities, but the representation of each set is then done separately. Description of this technique will conclude the chapter. Although all of these techniques aim for simple data exposition, it is probably fair to warn the reader that they share fairly sophisticated mathematical foundation, and this chapter is thus perhaps a bit more complex than the previous ones.

The mathematical bases underlying each of these techniques have much in common, although there are marked differences between the techniques both in regard to the assumed input data as well as to the complexity of numerical manipulation and final output. In particular, a fundamental decomposition of the input matrix lies at the heart of each technique and provides the driving force for the representation. This common decomposition means that the techniques are more similar than at first meets the eye. For mathematical completeness, a description of this decomposition and some of its properties is provided in the next section. Less mathematical readers may prefer to omit this section, and proceed directly to a discussion of the practical techniques.

4.1 The singular value decomposition of a (rectangular) matrix

Suppose that $\mathbf{X}$ is an arbitrary real $m \times n$ matrix of rank k. Then $\mathbf{X}$ can be expressed as the sum of k matrices of rank 1 in a variety of ways, but perhaps the most notable and useful of these ways is the 'singular value decomposition':

$$\mathbf{X} = \sigma_1 \boldsymbol{u}_1 \boldsymbol{v}_1' + \sigma_2 \boldsymbol{u}_2 \boldsymbol{v}_2' + \ldots + \sigma_k \boldsymbol{u}_k \boldsymbol{v}_k' \tag{4.1}$$

where the numbers σ_i are real and positive, the column vectors $\boldsymbol{u}_i$ are orthonormal (i.e. orthogonal and of unit length) and each has m elements, while the row vectors $\boldsymbol{v}_i'$ are orthonormal and each has n elements. The orthonormal sets $\boldsymbol{u}_1, \ldots, \boldsymbol{u}_k$ and $\boldsymbol{v}_1', \ldots, \boldsymbol{v}_k'$ can be completed to sets $\boldsymbol{u}_1, \ldots, \boldsymbol{u}_m$ and $\boldsymbol{v}_1', \ldots, \boldsymbol{v}_n'$, although there is non-uniqueness in this operation if $k+2 \leqslant \max(m, n)$. A complete singular value decomposition of $\mathbf{X}$ is then $\mathbf{X} = \sum_{j=1}^{t} \sigma_j \boldsymbol{u}_j \boldsymbol{v}_j'$, where $t = \min(m, n)$ and $\sigma_j = 0$ for $j = k+1, \ldots, t$ if $k < t$.

It is evident from the above, therefore, that if we write the $\boldsymbol{u}_j$ as columns of an $(m \times m)$ matrix $\mathbf{U}$, the $\boldsymbol{v}_j'$ as the rows of an $(n \times n)$ matrix $\mathbf{V}'$, and define the $(m \times n)$ matrix $\boldsymbol{\Sigma}$ to have elements σ_{ij} where $\begin{cases} \sigma_{jj} = \sigma_j \\ \sigma_{ij} = 0 \ (i \neq j) \end{cases}$, then the singular value decomposition has the equivalent form

$$\mathbf{X} = \mathbf{U}\boldsymbol{\Sigma}\mathbf{V}'. \tag{4.2}$$

Furthermore, orthonormality of the $\boldsymbol{u}_i$ and of the $\boldsymbol{v}_i'$ ensures that $\mathbf{U}$ and $\mathbf{V}$ are orthogonal matrices, i.e. $\mathbf{U}'\mathbf{U} = \mathbf{U}\mathbf{U}' = \mathbf{I}_m$ and $\mathbf{V}'\mathbf{V} = \mathbf{V}\mathbf{V}' = \mathbf{I}_n$. The vectors $\boldsymbol{u}_i$ are found as the eigenvectors of the $(m \times m)$ matrix $\mathbf{X}\mathbf{X}'$, the vectors $\boldsymbol{v}_i$ as the eigenvectors of the $(n \times n)$ matrix $\mathbf{X}'\mathbf{X}$, while the numbers σ_i $(i = 1, \ldots, k)$ are the square roots of the positive eigenvalues of either $\mathbf{X}'\mathbf{X}$ or $\mathbf{X}\mathbf{X}'$.

When $\mathbf{X}$ is square and symmetric, then $\mathbf{U} = \mathbf{V} =$ the matrix of eigenvectors of $\mathbf{X}$, while $\boldsymbol{\Sigma}$ is the diagonal matrix of eigenvalues of $\mathbf{X}$; the singular value decomposition is in this case just the familiar eigenvalue/eigenvector, or *spectral,* decomposition of $\mathbf{X}$. Existence of the singular value decomposition has been known for a long time, but its importance (in both Matrix Algebra and Statistics) has been much under-rated and it is mentioned in relatively few books on these subjects. Its popularization in Psychometry, and its subsequent use in the various techniques to be described in the rest of this chapter, stems from the paper by Eckart and Young (1936). A useful summary of the chief properties, and uses in Statistics, of this decomposition is given by Good (1969), from which we abstract the following as being of direct relevance in the remainder of the chapter.

The property that probably has most relevance in the present context is the least-squares property established by Eckart and Young. Suppose that $\mathbf{X}$ is an $(m \times n)$ matrix of rank k having the singular value decomposition (4.1) or (4.2), and we wish to approximate $\mathbf{X}$ as closely as possible by another $(m \times n)$ matrix $\mathbf{Y}$ of smaller rank, r, say. There are various possible definitions of what is meant by 'as closely as possible', but a commonly acceptable definition in Statistics would be in the least squares sense of minimizing the sum of squared discrepancies

$S = \sum_{i=1}^{m} \sum_{j=1}^{n} (x_{ij} - y_{ij})^2 = \text{trace}\{(\mathbf{X}-\mathbf{Y})(\mathbf{X}-\mathbf{Y})'\}$ between observed elements x_{ij} of $\mathbf{X}$ and their fitted counterparts y_{ij} of $\mathbf{Y}$. In this case, the best-fitting matrix $\mathbf{Y}$ is given by the singular value decomposition of $\mathbf{X}$, but with all σ_i for i greater than r set to zero, i.e.

$$\mathbf{Y} = \mathbf{U}\boldsymbol{\Omega}\mathbf{V}'$$

where

$$\boldsymbol{\Omega} = \left(\begin{array}{cccc|c} \sigma_1 & 0 & \dots & 0 & \\ 0 & \sigma_2 & & \vdots & \\ \vdots & & \ddots & \vdots & 0 \\ 0 & & \dots & \sigma_r & \\ \hline & & 0 & & 0 \end{array}\right).$$

The correspondence between representations (4.2) and (4.1) thus shows that $\mathbf{Y}$ is the sum of the first r terms on the right-hand side of (4.1). Furthermore, the minimum value of S achieved by this means is $S_{\min} = \sigma_{r+1}^2 + \sigma_{r+2}^2 + \dots + \sigma_k^2$. There is an immediate connection between the singular value decomposition and principal component analysis. If $\mathbf{X}$ is the *mean-centered* $(n \times p)$ data matrix $(n > p$; rank $p)$, then it is evident from above that we can account for a fraction $(\sigma_1^2 + \dots + \sigma_r^2)/(\sigma_1^2 + \dots + \sigma_p^2)$ of the total sample variance by means of a matrix of rank r. The first r columns of $\mathbf{V}$ thus yield the first r principal components of $\mathbf{X}$, while the principal component scores can be derived from $\mathbf{U}$ and $\boldsymbol{\Omega}$. There are thus practical computational advantages in obtaining a principal component analysis via the singular value decomposition of $\mathbf{X}$ rather than the spectral decompositions of $\mathbf{X}'\mathbf{X}$ and $\mathbf{X}\mathbf{X}'$.

A second important use of the singular value decomposition is in finding the generalized inverse of a matrix. If $\mathbf{X}$ is an $(m \times n)$ matrix of rank k, then any generalized inverse $\mathbf{Z}$ of $\mathbf{X}$ is an $(n \times m)$ matrix satisfying the following four properties:

(i) $\mathbf{XZX} = \mathbf{X}$,

(ii) $\mathbf{ZXZ} = \mathbf{Z}$,

(iii) $(\mathbf{ZX})' = \mathbf{ZX}$, and

(iv) $(\mathbf{XZ})' = \mathbf{XZ}$.

If $\mathbf{X}$ has the singular value decomposition given by (4.1) and (4.2), then it is easy to verify that all of these conditions are satisfied by the matrix $\mathbf{Z} = \mathbf{V}\boldsymbol{\Lambda}\mathbf{U}'$ where

$$\boldsymbol{\Lambda} = \left(\begin{array}{cccc|c} \sigma_1^{-1} & 0 & \dots & 0 & \\ 0 & \sigma_2^{-1} & & \vdots & \\ \vdots & & & 0 & 0 \\ 0 & \dots & 0 & \sigma_k^{-1} & \\ \hline & & 0 & & 0 \end{array}\right).$$

The easiest way of obtaining a generalized inverse is thus via this decomposition.

Further important applications of the singular value decomposition occur in the solution of linear equations, in the specification of interaction models in two-way analysis of variance, and in the analysis of contingency tables; these applications, however, are not of direct relevance to this book.

To calculate the singular value decomposition (4.1) of the matrix **X**, we need to find the values σ_i and the vector pairs $\boldsymbol{u}_i$, $\boldsymbol{v}_i$ for $i = 1, \ldots, k$. These quantities satisfy the following equations for each i:

$$\boldsymbol{u}_i'\mathbf{X} = \sigma_i \boldsymbol{v}_i' \tag{4.3}$$

$$\mathbf{X}\boldsymbol{v}_i = \sigma_i \boldsymbol{u}_i \tag{4.4}$$

$$\mathbf{X}\mathbf{X}'\boldsymbol{u}_i = \sigma_i^2 \boldsymbol{u}_i \tag{4.5}$$

$$\mathbf{X}'\mathbf{X}\boldsymbol{v}_i = \sigma_i^2 \boldsymbol{v}_i \tag{4.6}$$

where $\sigma_1 \geqslant \sigma_2 \geqslant \ldots \geqslant \sigma_k > 0$, and $\boldsymbol{u}_i'\boldsymbol{u}_j = \boldsymbol{v}_i'\boldsymbol{v}_j = \delta_{ij}$, the Kronecker delta (i.e. $\delta_{ij} = 1$ if $i = j$ and $\delta_{ij} = 0$ if $i \neq j$).

Any solution of a pair of equations—(4.3) and (4.4), (4.3) and (4.5), or (4.4) and (4.6)—will satisfy the remaining two equations. Numerical algorithms for computing the singular value decomposition have been given by various authors; a suitable reference is Golub and Reinsch (1970).

4.2 Representing a matrix by two sets of vectors: the biplot

The biplot is a technique devised by Gabriel (1971) for representing exactly, by means of two sets of vectors, the rows and columns of any matrix whose rank is 2. Since most matrices encountered in multivariate practice will have rank greater than 2, it follows that an exact pictorial representation simultaneously of the rows and columns of a general $(m \times n)$ matrix **X** will not be possible. An approximate representation must therefore be sought.

The approximate biplot representation that is suggested by Gabriel involves constructing a rank-2 approximation **Y** to the original matrix **X**, and then obtaining an exact biplot of **Y**. It has been shown in the previous section that the best rank-2 approximation **Y** to **X** (in the least squares sense) is provided by the singular value decomposition of **X**. If this decomposition is given by

$$\mathbf{X} = \sigma_1 \boldsymbol{u}_1 \boldsymbol{v}_1' + \sigma_2 \boldsymbol{u}_2 \boldsymbol{v}_2' + \ldots + \sigma_k \boldsymbol{u}_k \boldsymbol{v}_k' \tag{4.7}$$

then

$$\mathbf{Y} = \sigma_1 \boldsymbol{u}_1 \boldsymbol{v}_1' + \sigma_2 \boldsymbol{u}_2 \boldsymbol{v}_2' \tag{4.8}$$

$$= (\boldsymbol{u}_1, \boldsymbol{u}_2) \begin{pmatrix} \sigma_1 & 0 \\ 0 & \sigma_2 \end{pmatrix} \begin{pmatrix} \boldsymbol{v}_1' \\ \boldsymbol{v}_2' \end{pmatrix}. \tag{4.9}$$

Now if the vectors $\boldsymbol{u}_i$ are given by $\boldsymbol{u}_i' = (u_{i1}, u_{i2}, \ldots, u_{im})$ for $i = 1, 2$, and if the vectors $\boldsymbol{v}_i$ are given by $\boldsymbol{v}_i' = (v_{i1}, v_{i2}, \ldots, v_{in})$ for $i = 1, 2$, then we have

$$\mathbf{Y} = \begin{pmatrix} u_{11} & u_{21} \\ u_{12} & u_{22} \\ \vdots & \vdots \\ u_{1m} & u_{2m} \end{pmatrix} \begin{pmatrix} \sigma_1 & 0 \\ 0 & \sigma_2 \end{pmatrix} \begin{pmatrix} v_{11} & v_{12} & \ldots & v_{1n} \\ v_{21} & v_{22} & \ldots & v_{2n} \end{pmatrix}. \tag{4.10}$$

To obtain a biplot, it is first necessary to write $\mathbf{Y}$ as the product of two matrices $\mathbf{GH}'$, where $\mathbf{G}$ is an $(m \times 2)$ matrix and $\mathbf{H}$ is an $(n \times 2)$ matrix. It is evident that this can be done in a variety of ways, but (4.10) suggests three simple factorizations:

$$\mathbf{Y} = \begin{pmatrix} u_{11}\sqrt{\sigma_1} & u_{21}\sqrt{\sigma_2} \\ u_{12}\sqrt{\sigma_1} & u_{22}\sqrt{\sigma_2} \\ \vdots & \vdots \\ u_{1m}\sqrt{\sigma_1} & u_{2m}\sqrt{\sigma_2} \end{pmatrix} \begin{pmatrix} v_{11}\sqrt{\sigma_1} & v_{12}\sqrt{\sigma_1} & \ldots & v_{1n}\sqrt{\sigma_1} \\ v_{21}\sqrt{\sigma_2} & v_{22}\sqrt{\sigma_2} & \ldots & v_{2n}\sqrt{\sigma_2} \end{pmatrix}, \tag{4.11}$$

$$\mathbf{Y} = \begin{pmatrix} u_{11}\sigma_1 & u_{21}\sigma_2 \\ u_{12}\sigma_1 & u_{22}\sigma_2 \\ \vdots & \vdots \\ u_{1m}\sigma_1 & u_{2m}\sigma_2 \end{pmatrix} \begin{pmatrix} v_{11} & v_{12} & \ldots & v_{1n} \\ v_{21} & v_{22} & \ldots & v_{2n} \end{pmatrix} \tag{4.12}$$

or

$$\mathbf{Y} = \begin{pmatrix} u_{11} & u_{21} \\ u_{12} & u_{22} \\ \vdots & \vdots \\ u_{1m} & u_{2m} \end{pmatrix} \begin{pmatrix} v_{11}\sigma_1 & v_{12}\sigma_1 & \ldots & v_{1n}\sigma_1 \\ v_{21}\sigma_2 & v_{22}\sigma_2 & \ldots & v_{2n}\sigma_2 \end{pmatrix}. \tag{4.13}$$

Now denote the ith row of $\mathbf{G}$ by $\boldsymbol{g}_i'$ and the ith row of $\mathbf{H}$ (i.e. the ith column of $\mathbf{H}'$) by $\boldsymbol{h}_i'$; for the factorization (4.11), for example, we would have $\boldsymbol{g}_i' = (u_{1i}\sqrt{\sigma_1} \quad u_{2i}\sqrt{\sigma_2})$ and $\boldsymbol{h}_i' = (v_{1i}\sqrt{\sigma_1} \quad v_{2i}\sqrt{\sigma_2})$. The biplot consists of plotting the $m + n$ vectors $\boldsymbol{g}_i'$ $(i = 1, \ldots, m)$ and $\boldsymbol{h}_i'$ $(i = 1, \ldots, n)$ in a plane. Thus for the factorization (4.11), $\boldsymbol{g}_i$ would be represented by a vector emanating from the origin with end point given by the coordinates $(u_{1i}\sqrt{\sigma_1}, \quad u_{2i}\sqrt{\sigma_2})$, while $\boldsymbol{h}_i$ would be represented by a vector emanating from the origin and having end point at $(v_{1i}\sqrt{\sigma_1}, v_{2i}\sqrt{\sigma_2})$. The vectors $\boldsymbol{g}_1, \ldots, \boldsymbol{g}_m$ are termed the *row effects* or *row markers* of $\mathbf{Y}$ while the vectors $\boldsymbol{h}_1, \ldots, \boldsymbol{h}_n$ are the *column effects* or *column markers*. Each element y_{ij} of $\mathbf{Y}$ is represented as the inner product $\boldsymbol{g}_i'\boldsymbol{h}_j$ of the corresponding row effect and column effect vectors. In the biplot, an inner product of two vectors may be appraised visually by considering it as the product of the length of one of the vectors times the length of the other's projection on to it. This is useful in allowing rapid visual appraisal of the structure of the matrix. For example, one can readily see which rows or columns are proportional to which other rows or columns (same directions of the corresponding vectors), which entries are zero (right angles between row and column effects), and so on.

As has been demonstrated in (4.11), (4.12), and (4.13), $\mathbf{Y}$ has many possible factorizations into the form $\mathbf{GH}'$ and each will give a different biplot. Choice of factorization may be important, but in practice the three versions (4.11), (4.12), and (4.13) cover the most useful cases. These three versions correspond respectively to a general factorization where emphasis is placed neither on rows nor columns, and two special factorizations placing direct emphasis on rows and columns in turn. Thus, for example, if we want the relations between rows of $\mathbf{Y}$ to be represented by corresponding relations of $\boldsymbol{g}$ vectors, we require the following to be satisfied for any two rows $\boldsymbol{y}_i'$ and $\boldsymbol{y}_j'$ of $\mathbf{Y}$:

$$\boldsymbol{y}_i'\boldsymbol{y}_j = \boldsymbol{g}_i'\boldsymbol{g}_j,$$

$$|\boldsymbol{y}_i| = |\boldsymbol{g}_i|,$$

$$|\boldsymbol{y}_i - \boldsymbol{y}_j| = |\boldsymbol{g}_i - \boldsymbol{g}_j|,$$

and

$$\cos(\boldsymbol{y}_i, \boldsymbol{y}_j) = \cos(\boldsymbol{g}_i, \boldsymbol{g}_j)$$

where $|\boldsymbol{x}|$ denotes the length of the vector $\boldsymbol{x}$ and $(\boldsymbol{x}, \boldsymbol{y})$ denotes the angle between vectors $\boldsymbol{x}$ and $\boldsymbol{y}$ (see Section 1.3.3). These relationships will be satisfied if $\mathbf{YY}' = \mathbf{GG}'$, which in turn requires that $\mathbf{H}'\mathbf{H} = \mathbf{I}_2$. This latter is satisfied by (4.12). Similarly, one can ensure that relations between columns of $\mathbf{Y}$ are represented by corresponding relations of $\boldsymbol{h}$ vectors if $\mathbf{Y}'\mathbf{Y} = \mathbf{HH}'$, which requires $\mathbf{G}'\mathbf{G} = \mathbf{I}_2$ and (4.13).

To illustrate the technique, consider the example analysed by Gabriel (1971). Table 4.1 presents a matrix $\mathbf{X}$ giving percentages of households having various facilities and appliances in East Jerusalem Arab areas, by quarters of the town. The average percentages in each quarter indicate the standard of living in that area, and the average percentage of each facility or appliance indicates its overall prevalence.

If it is assumed that the main effects 'standard of living' in each area and 'overall prevalence' of each appliance combine multiplicatively, and there is no area–appliance interaction, then the entries of Table 4.1 are estimated by the first component $\mathbf{X}_{(1)} = \sigma_1 \boldsymbol{u}_1 \boldsymbol{v}_1'$ in the singular value decomposition of $\mathbf{X}$. If this component is subtracted from $\mathbf{X}$ then Table 4.2 results. This table corresponds to the residual matrix $\mathbf{X} - \mathbf{X}_{(1)}$ from a multiplicative model, and enables the differential prevalence of different facilities and appliances in the different quarters to be studied. Thus, for example, large positive values in Table 4.2 for radios and toilets in Sur-Bahar indicate that these two facilities have a higher prevalence in this area than expected from a simple multiplicative model. From Table 4.1 it is evident that values for all appliances and facilities in this area are

Table 4.1 Facilities and equipment in East Jerusalem in 1967, by subquarter (reproduced from Gabriel (1971) with permission from the Biometrika trustees)

	Old city quarters				Modern		Other		Rural
	Christian	Armenian	Jewish	Moslem	Am. Colony Sh. Jarah	Shaafat Bet-Hanina	A-Tur Isawiye	Silwan Abu-Tor	Sur-Bahar Bet-Safafa
Percentage of households possessing:									
Toilet	98.2	97.2	97.3	96.9	97.6	94.4	90.2	94.0	70.5
Kitchen	78.8	81.0	65.6	73.3	91.4	88.7	82.2	84.2	55.1
Bath	14.4	17.6	6.0	9.6	56.2	69.5	31.8	19.5	10.7
Electricity	86.2	82.1	54.5	74.7	87.2	80.4	68.6	65.5	26.1
Water*	32.9	30.3	21.1	26.9	80.1	74.3	46.3	36.2	9.8
Radio**	73.0	70.4	53.0	60.5	81.2	78.0	67.9	64.8	57.1
TV set	4.6	6.0	1.5	3.4	12.7	23.0	5.6	2.7	1.3
Refrigerator***	29.2	26.3	4.3	10.5	52.8	49.7	21.7	9.5	1.2

* In dwelling. ** Or transistor radio. *** Electric.

Table 4.2 Residual matrix after subtracting out multiplicative least squares fit (first singular component) from Table 4.1 with next two singular values, rows and columns (reproduced from Gabriel (1971) with permission from the Biometrika trustees)

	Christian	Armenian	Jewish	Moslem	American	Shaafat	A-Tur	Silwan	Sur-Bahar	$\boldsymbol{u}_2$	$\boldsymbol{u}_3$
Toilet	1.60	2.17	21.16	10.62	−17.12	−16.72	−2.44	4.81	12.64	0.394	0.185
Kitchen	−3.11	0.43	1.04	0.15	−5.87	−5.52	3.65	8.58	6.05	0.107	0.213
Bath	−15.49	−11.80	−17.56	−17.09	20.71	35.12	3.14	−8.09	−7.20	−0.575	0.371
Electricity	11.71	8.83	−4.21	8.18	−1.25	−5.28	−2.83	−3.27	−18.51	0.023	−0.768
Water	−11.66	−13.53	−14.02	−12.90	27.19	23.04	3.57	−4.94	−16.89	−0.525	0.059
Radio	2.30	0.85	−2.73	−2.65	−2.76	−3.33	0.10	−0.47	14.76	0.071	0.244
TV set	−3.27	−1.74	−4.70	−3.63	3.36	13.95	−1.94	−4.56	−3.41	−0.173	0.042
Refrigerator	2.78	0.31	−16.53	−13.10	21.42	19.31	−3.64	−14.89	−14.62	−0.437	−0.358
$\boldsymbol{v}_2'$	0.171	0.172	0.381	0.307	−0.495	−0.574	−0.027	0.195	0.297	$\sigma_2 = 88.35$	
$\boldsymbol{v}_3'$	−0.486	−0.340	0.151	−0.223	−0.070	0.209	0.152	0.207	0.679	$\sigma_3 = 33.67$	

lower than in the other areas. However, the two high positive values noted above highlight the fact that the drop in prevalence compared with the other areas is not as marked for radios and toilets as for the other facilities and appliances. Now, evidently,

$$\mathbf{X} - \mathbf{X}_{(1)} = \sum_{i=2}^{8} \sigma_i \boldsymbol{u}_i \boldsymbol{v}_i'$$

is a singular value decomposition of the residual matrix, so the first two singular values are given by the second and third singular values σ_2, σ_3 of the original matrix $\mathbf{X}$. Table 4.2 shows these values, along with the corresponding vectors $\boldsymbol{u}_2$, $\boldsymbol{u}_3$, $\boldsymbol{v}_2'$ and $\boldsymbol{v}_3'$. The goodness of fit of these components to the residual matrix $\mathbf{X} - \mathbf{X}_{(1)}$ is $(\sigma_2 + \sigma_3)/(\sigma_2 + \sigma_3 + \ldots + \sigma_8) = 0.937$, so that a matrix of rank 2 should give a very close approximation to the residuals. The biplot of Fig. 4.1 has been constructed using factorization (4.13); it is reproduced from Gabriel (1971) with permission from the Biometrika trustees, as are Tables 4.1, 4.2, and the following interpretation of this biplot.

The Old City quarters are opposite a cluster of the modern quarters. The one rural area is roughly orthogonal to both clusters, somewhat nearer to the Old City than to the modern quarters, whereas the other quarters are less prominent in a similar direction, with the poorer Silwan and Abu-Tor areas closer to the poorest of the Old City quarters, and the richer A-Tur and Isawiye slightly in the direction of the modern quarters.

The modern quarters appear to have a particularly high prevalence of baths, water inside the dwelling, and refrigerators, whereas the poorer quarters have relatively high prevalences of toilets and electricity. Evidently the last two items were pretty generally available in all urban sections and thus are not indicative of better living conditions, whereas the former three items were much more available in better-off homes.

It is interesting to note that electricity is noticeably rarer in the rural area than in all urban quarters, whereas radios, presumably battery operated, and kitchens are the items that least reflect the low general level of the rural area.

In the present example, attention was focused on residuals from a multiplicative fit so that the second and third components were biplotted. In other instances it might be more interesting to biplot the first two components and study the data matrix itself. More recently, Bradu and Gabriel (1978) have shown how the biplot can be used to assess the nature of the best linear model that can be fitted to a data matrix, while Gabriel and Odoroff (1986) have extended this work to three-dimensional biplots.

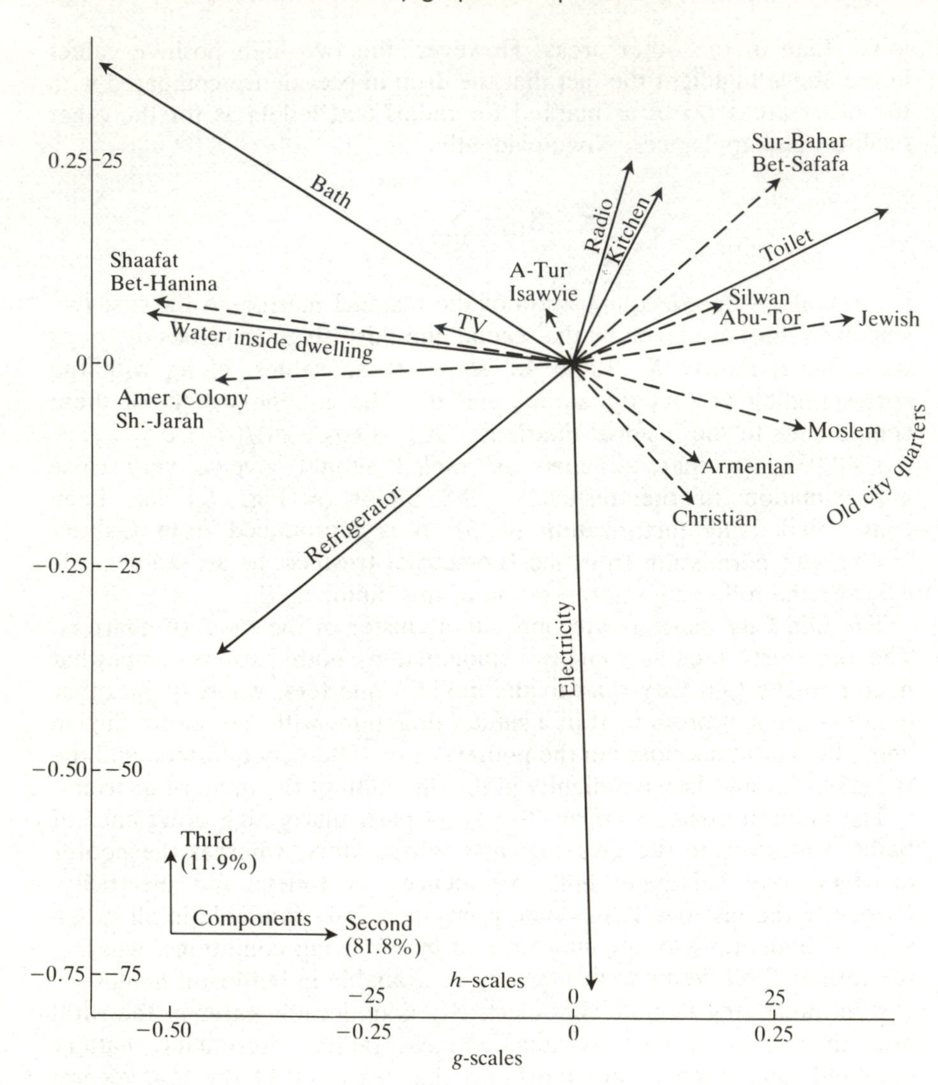

Fig. 4.1 Differential prevalence of facilities and equipment in East Jerusalem households (Arab) in 1967 (reproduced from Gabriel (1971) with permission from the Biometrika trustees)

4.3 Representing a matrix by one set of points and one set of vectors: preference scaling

The technique to be discussed in this section is mathematically exactly equivalent to the biplot technique of the previous section, but it warrants a separate description for a number of reasons. In the biplot, both the rows and the columns of the input matrix are represented by vectors. Interpretation of relationships between these vectors has been discussed in Section 4.2. While the same decomposition of the input matrix is at play in the present technique, one set of vectors is now regarded as a set of *points* in space and this leads to a particular form of interpretation of the resulting configuration. This interpretation is particularly apposite for a certain type of input matrix, and analysis of such matrices has recently become very popular in the Social Sciences. Within such subject areas as Psychology and Sociology the technique is variously known as *preference scaling* or *repertory grid analysis*, and many examples exist of its substantive applications (see, e.g. Slater (1976), Fransella (1977), Waddington (1983)). It is therefore of interest to set out the aims and fundamental mathematics of the technique, and to point out its essential similarity to the biplot technique of Section 4.2.

In the type of application envisaged here, the input data consist of the ratings or rankings of a set of p *stimuli* made by each of n *subjects*, arranged in an $(n \times p)$ matrix. Rows of the matrix thus correspond to subjects and columns to stimuli. For example, a set of n tasters may be asked to assess the sweetness of each of p soft drinks, assigning a score between 0 (very bitter) and 7 (very sweet) to each drink; or a sample of n people may be asked to rank a set of p attributes (e.g. patience, diligence, altruism) in what they consider to be the order of importance for a particular career. The 'score' assigned to the jth attribute by the ith subject in this latter case will thus be the rank order (between 1 and p) of that attribute as assessed by that subject. The representation of such an $(n \times p)$ data matrix generally found most useful is as a set of p points and n vectors in space. The points represent the stimuli and the vectors the subjects, and it is usually most convenient to standardize the vectors to a common length, typically unit length. The points and vectors are chosen so that the length of the projection of the jth point on to the ith vector (measured from the end of that vector) is as closely proportional to the entry in the ith row and jth column of the data matrix as possible. If the entries in the ith row of the matrix are the rankings of the p stimuli, therefore, the projections of the p points on to the n vectors should give as close a match as possible to the n sets of rankings in the data matrix.

To illustrate the ideal solution, consider the matrix shown in Table 4.3 which presents hypothetical rankings of four stimuli as produced by three subjects. Figure 4.2 shows a pictorial representation constructed according to the above specification.

Here the three subjects are represented by the three vectors labelled 1, 2, 3 while the four stimuli are represented by the four points A, B, C, D. The vectors are all of common length and thus their end points lie on the circumference of a (unit) circle whose centre is the origin O of the space;

Table 4.3 Hypothetical ranking of four stimuli by each of three subjects

	Stimulus			
	A	B	C	D
Subject 1	1	2	3	4
2	4	3	2	1
3	3	4	2	1

each vector is also extended by a dotted line to the diametrically opposite point on the circle. Taking the (arrowed) end of each vector as the starting point for each ranking, it can thus be seen that the projections of the points A, B, C, and D are in the orders ABCD, DCBA, and DCAB for vectors 1, 2, and 3 respectively. Since these are the rank orders of the four stimuli as presented by the three subjects, the diagram is a completely faithful representation of the data in the table.

The essential aspects of interpretation of such diagrams are also brought out by this simple example. Subjects who present similar rankings for the stimuli will be represented by vectors that cluster together. Thus subjects 2 and 3 only differ in the interchange of rankings for stimuli A and B, and their corresponding vectors are close together. Subject 1, on the other hand, has a ranking exactly the reverse of that given by subject 2, and the vectors representing these two subjects are at right angles to each other. Furthermore, stimuli whose rankings are generally in close proximity will be represented by points that are close together, as is also evident from Fig. 4.2.

Of course, perfect representation such as in this example is possible only for a fairly limited choice out of all possible rankings that could be generated among p stimuli. Moreover, if the input data are *ratings* rather than *rankings,* then the *length* of each projection from the end of each

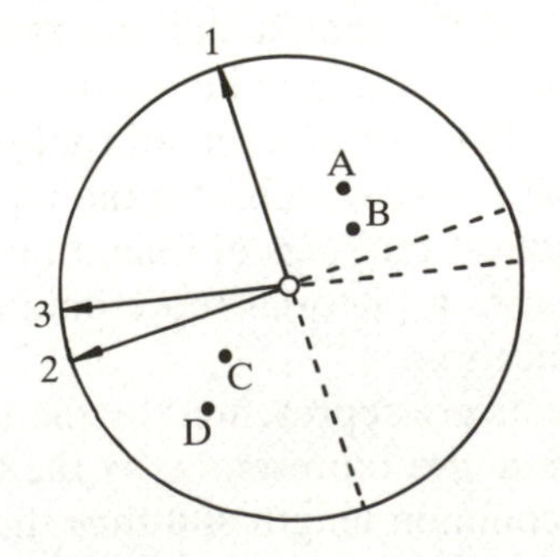

Fig. 4.2 Point/vector representation of data from Table 4.3

vector must be proportional to the corresponding entry in the data matrix, and perfect correspondence is even less likely. Consequently, in practical applications, it is almost certain than an *approximate* representation is the best that can be achieved. Brief consideration of the mathematics behind the requirements outlined above for the pictorial representation will now highlight the connection of this technique with Gabriel's biplot. To establish this connection, we must temporarily replace the points in the configuration representing stimuli by vectors emanating from the origin O and terminating at these points, and recollect from Chapter 1 that the length of the projection of one vector on to another vector of unit length is given by the inner product of the two vectors.

Assume, therefore, that we seek an approximate pictorial representation of the data in a small number of dimensions, d, say. Typically, of course, d will equal 2. Let $\mathbf{Z}$ denote the input data matrix, $\boldsymbol{x}_j$ the vector from the origin O to the point representing stimulus j in the chosen space, and $\boldsymbol{y}_i$ the unit-length vector representing subject i in this space. According to our prescription above, which requires the preference value assigned by subject i to stimulus j to be proportional to the projection of $\boldsymbol{x}_j$ on $\boldsymbol{y}_i$, the entry z_{ij} in the ith row and jth column of $\mathbf{Z}$ should be approximated by the inner product $\boldsymbol{y}_i'\boldsymbol{x}_j$. Setting the vectors $\boldsymbol{y}_i$ as the rows of an $(n \times d)$ matrix $\mathbf{Y}$, and the vectors $\boldsymbol{x}_j$ as the rows of a $(p \times d)$ matrix $\mathbf{X}$, therefore requires $\mathbf{Z}$ to be approximated by a matrix product $\mathbf{YX}'$. If $d = 2$, this is exactly the same requirement as that of finding the matrix product $\mathbf{GH}'$ for the biplot of Section 4.2.

Taking least-squares as the criterion for approximation, leads immediately to the solution via the singular value decomposition of $\mathbf{Z}$. Denote this decomposition by $\mathbf{Z} = \mathbf{U\Sigma V}'$ as in Section 4.1. Write $\mathbf{U}^{(d)}$ as the matrix containing the first d columns of $\mathbf{U}$; $\boldsymbol{\Sigma}^{(d)}$ as the $(d \times d)$ top left portion of $\boldsymbol{\Sigma}$; and $\mathbf{V}^{(d)}$ as the matrix containing the first d columns of $\mathbf{V}$. Then the required solution is given by $\mathbf{Y} = \mathbf{U}^{(d)}$ and $\mathbf{X} = \mathbf{V}^{(d)}\boldsymbol{\Sigma}^{(d)}$.

To illustrate this technique, consider the following example in which job applicants were asked to rate a number of different professions with respect to a set of attributes (P. A. J. Waddington, personal communication). The subjects who took part in the exercise were all final-year university undergraduates applying to a large industrial company for their first employment, and the applicants were sorted into groups according to the progress made in the particular job application (e.g. withdrawn after preliminary enquiry; rejected after first interview; rejected after second interview; job offered). The objective was to determine whether or not the successful applicants had a different perception of a wide range of professions from the unsuccessful applicants, but for illustrative purposes we shall consider only one such group of applicants here. Each applicant was asked to rate a number of professions according to a range of attributes. The professions were: Social Worker, Market Researcher, Prison Governor, Chartered Accountant, Journalist, Commercial Manager in Industry, Solicitor, Civil Servant, Banker, Nurse, Police Officer, Teacher. For each profession, twenty-two attributes were presented in bipolar form. The two poles for each attribute are given in Table 4.4, and by ticking one of seven boxes between the two poles the candidate assigned a rating between 1 and 7 to each attribute.

Both attributes and polarity within attributes were randomized for professions before being presented to the applicants, and an average (or 'consensus') attribute by profession matrix was computed for each group of applicants. These consensus matrices were then subjected to separate repertory grid analyses, and Fig. 4.3 shows the resulting two-dimensional representation for one such group (those who withdrew after preliminary enquiry).

Here the various professions are represented as points, while the attributes are represented as vectors with positions marked around the circumference of the unit circle (together with the appropriate polarity). We find some natural profession groupings (Accountant + Solicitor +

Table 4.4 Attributes rated by job applicants for each of a number of professions

1. Male dominated / Female dominated
2. Highly qualified / Unqualified
3. Good career prospects / Poor career prospects
4. Legal / Non-legal
5. Conventional / Unconventional
6. High social status / Low social status
7. Young person's job / Older person's job
8. Financial / Non-financial
9. Sedentary / Active
10. Have authority / Subject to authority
11. Work independently / Part of a team
12. Use initiative / Restricted by rules
13. Paperwork important / Paperwork incidental
14. Routine / Unpredictable
15. Private enterprise employee / Public sector employee
16. Have contact with public / Isolated from public
17. Profit seeking / Non-profit oriented
18. Self-interested motive / Altruistic motive
19. Deal with under-privileged / Deal with privileged people
20. Compassionate / Hard-headed
21. Help people / Use people to achieve other ends
22. A job I would like / A job I would dislike

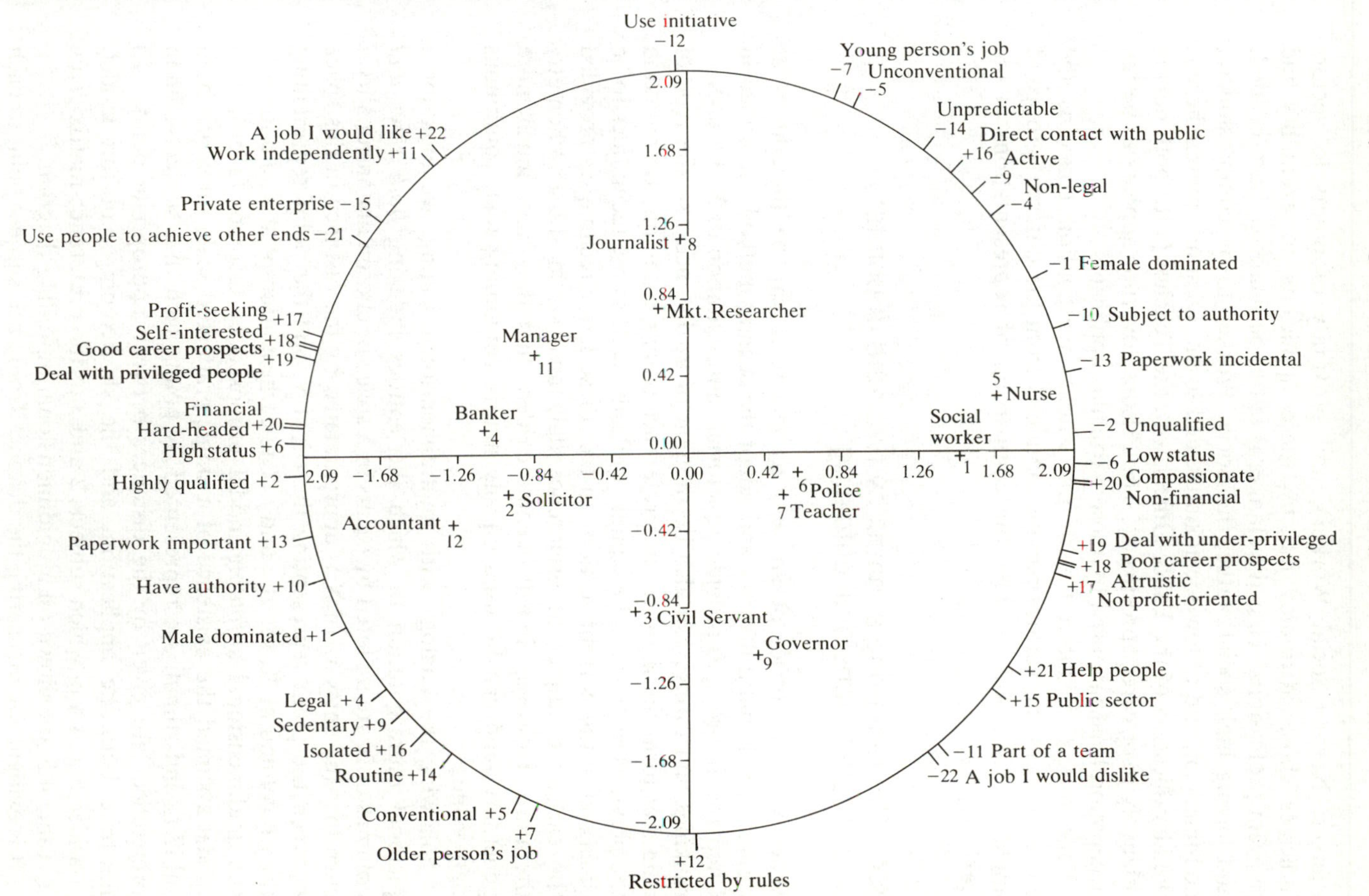

Fig. 4.3 First two dimensions in a repertory grid analysis of job applicants' data

Banker; Nurse + Social Worker), and proximity of points to vectors indicates those attributes felt by these applicants to exemplify the relevant professions. Thus the Banker is felt to be financial, hard-headed, and having high social status; the Social Worker to be compassionate, non-financial, and of low social status, and so on. Of some interest here is the identification of the Police Officer with the Teacher (with both having relatively poor career prospects). The Prison Governor appears to be the occupation least favoured by these particular applicants.

One aspect of interest with the whole investigation concerns the comparison of the various pictures obtained from the different groups of applicants. We will consider the comparison of representations in the next chapter.

4.4 Representing a matrix by two sets of points: multidimensional unfolding

Whenever the entries of an $(m \times n)$ matrix $\mathbf{X}$ can be thought of as *distances,* it is natural to try and adapt the scaling methods described in Chapter 3 in order to provide a pictorial representation of $\mathbf{X}$. Of course, rankings and ratings can also be thought of as distances, so the type of data that form the input to the preference scaling technique described in the previous section can be equally well subjected to the techniques to be discussed in the current section. In fact these latter techniques provided the earliest examples of preference data analysis in the Psychometry literature. However, it will become evident that they have much more generality, and can be used on any matrix containing real, continuous data values.

Following the analogy with multidimensional scaling, we look for a pictorial representation in which the entities indexing *both* rows *and* columns of $\mathbf{X}$ are denoted by points in a space of chosen dimensionality d such that the entry x_{ij} of $\mathbf{X}$ is approximated by the distance in this space between the point representing the ith row, and the point representing the jth column of $\mathbf{X}$, for all i and j. The earliest attempt at a solution was the unidimensional *unfolding* model proposed by Coombs (1950). This model assumed the existence of a single graduated scale on which both subjects and stimuli are represented by points such that subject–stimuli preferences are given by the distances between appropriate points. To illustrate both the application, and the shortcomings, of this model, consider Fig. 4.4 in which subjects 2 and 3 of the preference matrix given in Table 4.3 are shown on a unidimensional unfolding diagram.

It is evident, by measuring the distance from each subject point to each stimulus point, that the stimulus rankings for each subject are the ones

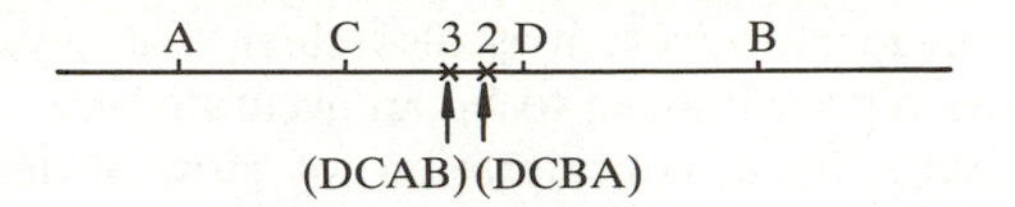

Fig. 4.4 Illustration of one-dimensional unfolding model, using data of Table 4.3. Preference order for each subject given in parentheses

presented in Table 4.3 (and indicated in parentheses on Fig. 4.4). Coombs called this the 'unfolding' model because the preference order for a given subject can be generated by folding the stimulus scale at the corresponding subject point. To recover the stimulus scale from the preference data, then, it is necessary to unfold simultaneously all these preference scales for subjects. This procedure will produce the common stimulus scale. In such a diagram, the further a stimulus point is from a subject point the less that subject will like that stimulus. Therefore, each subject point corresponds to that subject's 'ideal' stimulus (namely the hypothetical stimulus that the subject would prefer above all others).

However, it is also evident that such a unidimensional scale will be even more limited in its ability to represent exactly preference data than the method of the previous section. In fact, subject 1 cannot be accommodated without error on Fig. 4.4, whereas all three subjects were perfectly represented on Fig. 4.2. In order to allow greater accuracy of representation, we must include more dimensions in our configuration. The generalization to *multidimensional unfolding,* first proposed by Bennett and Hays (1960) is fairly straightforward. A two-dimensional representation of Table 4.3 is shown in Fig. 4.5. Note that one or two of the points on this diagram can be appreciably displaced without invalidating the solution; this is the usual consequence of attempting to represent non-metric information, and less freedom would be accorded to the configuration if the entries in Table 4.3 had been numerical rather

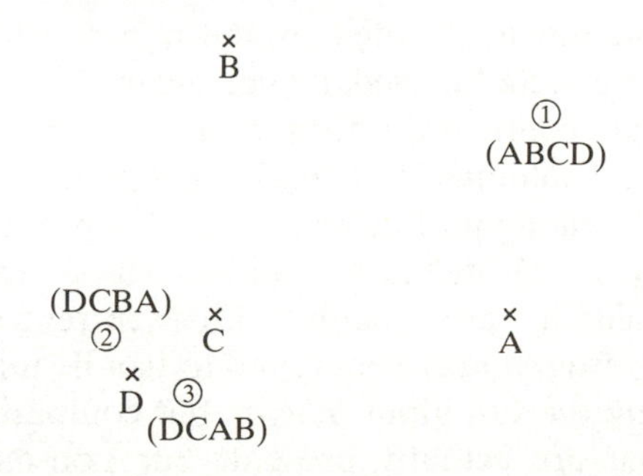

Fig. 4.5 Two-dimensional unfolding analysis of data in Table 4.3. Preference order for each subject given in parentheses

than simple rankings. However, it is also clear that several dimensions are needed for the representation to be an accurate one.

Let us now turn to a consideration of how a multidimensional unfolding configuration may be computed, given an input matrix **X**. Recollect that the objective is to obtain the coordinates in a (small-dimensional) space of two sets of points, one representing the m rows of **X** and the other representing the n columns of **X**, such that the distance between any one point of the first set and any one from the second set approximates as closely as possible the appropriate entry x_{ij} of **X**. As with multidimensional scaling, a crucial decision to be made concerns what is meant by 'approximates as closely as possible'. Assuming that the entries of **X** are all numerical (i.e. real, continuous values), two major possibilities exist. In a *metric* solution, the actual distances between points are important, and a configuration is sought in which the numerical discrepancies between these distances and the entries of **X** are as small as possible. A least-squares criterion is generally considered to be an adequate measure of fit. In a *non-metric* solution, on the other hand, we seek a configuration in which the rank order of inter-point distances matches the rank order of elements of **X** as closely as possible, using a criterion such as STRESS (defined in Section 3.3). At an early stage, Coombs looked for a metric version of his unfolding model (Coombs and Kao 1960; Coombs 1964). These methods were taken up by Ross and Cliff (1964), but it was soon discovered that even in simple cases the solution was very difficult to obtain. It was the work of Schönemann (1970) that first provided a practical solution to the problem, and this forms the basis of most computer programs in recent use (Lyons 1980). The mathematical details are somewhat intricate and will therefore not be presented here; the interested reader is referred to Schönemann's paper for a full description. It may simply be noted in passing that, once again, the singular value decomposition plays a central role in the technique.

Easier implementation is afforded by the non-metric solution, and to see how this can be effected consider once again the nature of the input data. Since these data come in the form of an $(m \times n)$ rectangular matrix in which the rows and columns in general correspond to different entities, the input data can be thought of as an $(m+n) \times (m+n)$ distance matrix, in which the $\frac{1}{2}m(m-1)$ distances 'within rows' and the $\frac{1}{2}n(n-1)$ distances 'within columns' are 'missing'. Most current non-metric multidimensional scaling programs are equipped to handle missing values. This is done by specifying the full input matrix, but computing STRESS from only the values that are actually present. For non-metric multidimensional unfolding the full matrix has size $(m+n) \times (m+n)$, but the STRESS is computed from the mn elements of **X**.

In all the above it has been assumed that $\mathbf{X}$ is rectangular, $(m \times n)$, with rows and columns indexed by different entities. It may happen, however, that $\mathbf{X}$ is a square $(m \times m)$ matrix with its rows and columns similarly classified. An example is when $\mathbf{X}$ is an asymmetric distance matrix, as discussed in Section 3.4. If such a matrix is subjected to multidimensional unfolding, then m points are obtained to represent the row items and m points to represent the column items. Interpretation of the analysis of an asymmetric matrix lies mainly in investigating the distances between the two sets, and there is only a secondary interest in within-set distances. Since the output configuration includes points for the row entities and points for the column entities, however, both row-column and column-row distances are represented. Multidimensional unfolding of an asymmetric distance matrix therefore attempts to represent the *whole* matrix, in contrast to the canonical analysis of Section 3.4 which represents only the skew-symmetric portion of the matrix. This may not necessarily lead to a better interpretation, however, and Constantine and Gower (1978) found that multidimensional unfolding analysis of their data sets gave rather disappointing results. Although the data were fitted reasonably well in two dimensions, little insight was given into the nature of the asymmetry. Figure 4.6 presents the two-dimensional non-metric unfolding configuration obtained by Constantine and Gower for the data given in Table 3.4 (for which the canonical analysis configuration has already been given in Fig. 3.11).

The figure suggests that cyclopentanone, benzyl acetate, methyl salicylate, and butyl acetate have similar effects as adapting stimuli, while the large asymmetry in response of citral, benzyl acetate, and safrole with m-xylene comes out clearly. However, as Constantine and Gower point

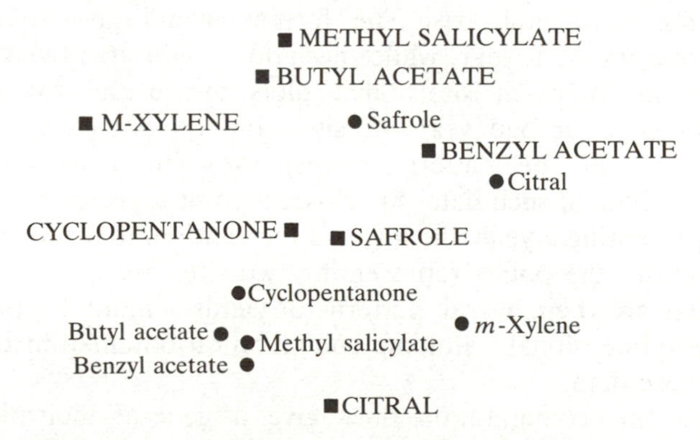

Fig. 4.6 Multidimensional unfolding configuration obtained from the data of Table 3.4. Circles and lower-case letters refer to row labels, while squares and upper-case letters refer to column labels

Table 4.5 Yields of winter wheat (kg/unit area)

Site	*Year* 1970	1971	1972	1973
1. Cambridge	46.81	39.40	55.64	32.61
2. Cockle Park	46.49	34.07	45.06	41.02
3. Harpers Adams	44.03	42.03	40.32	50.23
4. Headley Hall	52.24	36.19	47.03	34.56
5. Morley	36.55	43.06	38.07	43.17
6. Myerscough	34.88	49.72	40.86	50.08
7. Rosemaund	56.14	47.67	43.48	38.99
8. Seale-Hayne	45.67	27.30	45.48	50.32
9. Sparsholt	42.97	46.87	38.78	47.49
10. Sutton Bonington	54.44	49.34	24.48	46.94
11. Terrington	54.95	52.05	50.91	39.13
12. Wye	48.94	48.63	31.69	59.72

out, although the representation is a faithful one it gives no special insight into interpreting the data. The STRESS value of 21.2 per cent could be improved upon, but whether any improvement in interpretation would result is open to doubt.

The above example uses a non-metric approach to produce a configuration; the following simple example was presented by Lyons (in the 1980 M.Sc. dissertation cited above) as an illustration of metric multidimensional unfolding. Table 4.5 gives the yields of winter wheat as harvested over four years in twelve different sites. A typical graphical display of such data would be as a plot of yields, either using sites or years as the horizontal axis. The former would give four separate 'profiles', one for each year, while the latter would give twelve separate 'profiles', one for each site. Such plots are useful for identifying individually good or bad years or sites, but do not easily lead to an overall picture of the pattern of responses. In a multidimensional unfolding analysis of such data, the closer a point representing a site is to a point representing a year, the lower is the yield on that site in that year; while the closer the points representing two sites are to each other, the more similar are their overall patterns of yields. Figure 4.7 presents the metric multidimensional unfolding configuration obtained in two dimensions for these data.

Although this configuration does give a general indication of the relationship between sites and years, there are several problems that arise in interpretation. Firstly, a two-dimensional configuration is too much of a simplification and gives too much distortion for detailed interpretation;

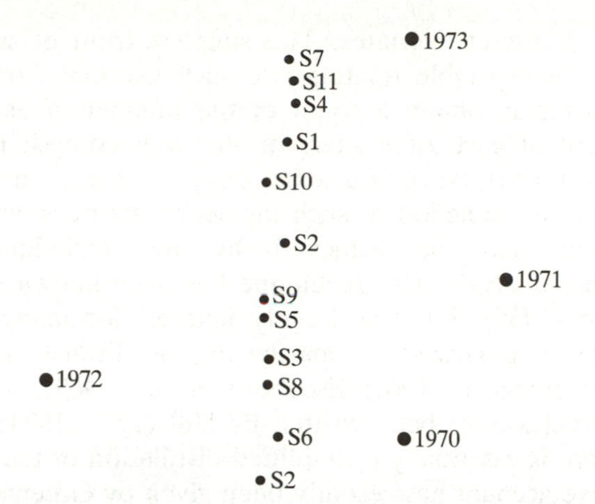

Fig. 4.7 Metric multidimensional unfolding configuration obtained from the data in Table 4.5

higher dimensionality is necessary for more accurate representation. Secondly, the presence of interaction between sites and years and, to a lesser extent, random error militates against concise summary. Such interaction is exhibited by non-parallel profiles on a plot of yield by site or yield by year, and means that any simple row-by-column model of the matrix (whether additive or multiplicative) will not be able to fit the data well. One further possibility would therefore be to remove main effects of sites and years first, before attempting a pictorial representation of the remaining interaction, but this will not be pursued further here.

Despite the various shortcomings of these two examples, they do illustrate the range of possible situations to which unfolding techniques may be applied, and they suggest that certain aspects of the data, which would not be highlighted by a standard graphical approach, may be extracted by an unfolding analysis.

4.5 Representing incidence matrices: correspondence analysis

The techniques described so far in this chapter have all aimed at the simultaneous depiction of the rows and columns of a matrix containing either ordinal or numerical data. Another situation that arises frequently in practice is where a count is made of the collection of attributes possessed by a set of individuals. This results in qualitative data, exhibited in an *incidence matrix* which gives the counts of the joint occurrences of discrete variates. The simplest form of such data is an $(m \times n)$ contingency table relating two such variates. For example, the plant ecologist may obtain a count of the number of each of m floral species present at each of n sites, or the archaeologist may count the number of each of m types of artefact discovered in each of n excavated graves. Graphical depiction of such incidence matrices would clearly be valuable, and may be achieved by the technique known as *correspondence analysis*. This technique has been known since the work of Hirschfeld (1935), but was largely ignored for many years. It has recently been popularized, primarily by the French school of data analysts (e.g. Benzecri, 1969; Escofier-Cordier, 1969), and some good expository articles have been written by Hill (1973, 1974). The description given here is essentially a simplified distillation of the latter. A fully comprehensive account has recently been given by Greenacre (1984).

Consider two discrete variables I and J having m and n categories respectively, and let k_{ij} be the number of times that category i of I and category j of J is simultaneously observed in a sample of k individuals. Then all these values k_{ij} can be exhibited in the $(m \times n)$ incidence matrix (or contingency table) $\mathbf{K} = (k_{ij})$. One possible approach for graphical depiction of this matrix is to obtain a scale of values for each discrete variable, thus enabling the variables to be plotted. This is equivalent to finding a 'score' for each category of each discrete variable, this score being used as that category's coordinate value in the graphical representation. Such an approach can thus be viewed as a natural extension of either multidimensional scaling or principal component analysis to categorical data.

Formulated in this way, it is evident that what is sought is a pair of real-valued variates to replace the original two categorical variates, the choice of these real-valued variates being determined by some imposed optimality criterion. Fisher (1940) tackled the problem by choosing the two derived variates in such a way as to maximize their correlation. Correspondence analysis is formally equivalent to this approach, but can be introduced as an iterative process in the following way. Start by assigning arbitrarily a set of scores to the categories of one of the two discrete variables. It is immaterial which one is chosen, but for definitiveness suppose that we choose the variable indexing the n columns and let $y_j^{(0)}$ be the chosen scores $(j = 1, \ldots, n)$. Then the average score for the ith row of the incidence matrix is given by

$$x_i^{(0)} = \frac{\left(\sum_{j=1}^{n} k_{ij} y_j^{(0)}\right)}{\left(\sum_{j=1}^{n} k_{ij}\right)}.$$

The $x_i^{(0)}$ are then taken to be the initial scores for the m categories of the discrete variable indexing the rows. From the two initial sets of scores, $\boldsymbol{y}^{(0)} = (y_1^{(0)}, y_2^{(0)}, \ldots, y_n^{(0)})'$ and $\boldsymbol{x}^{(0)} = (x_1^{(0)}, x_2^{(0)}, \ldots, x_m^{(0)})'$, an iterative scheme can be set up, where at each stage of the iteration the new column scores are given by the averages of the current row scores, and reciprocally the new row scores are given by the averages of the current column scores (a process termed 'reciprocal averaging' by Hill). The equations governing this process are:

$$x_i^{(l)} = \frac{\left(\sum_{j=1}^{n} k_{ij} y_j^{(l)}\right)}{\left(\sum_{j=1}^{n} k_{ij}\right)}$$

and

$$y_j^{(l+1)} = \frac{\left(\sum_{i=1}^{m} k_{ij} x_i^{(l)}\right)}{\left(\sum_{i=1}^{m} k_{ij}\right)} \quad \text{for } l = 0, 1, 2, \ldots.$$

In vector and matrix form we may write these equations as

$$\boldsymbol{x}^{(l)} = \mathbf{R}^{-1}\mathbf{K}\boldsymbol{y}^{(l)}$$

and

$$\boldsymbol{y}^{(l+1)} = \mathbf{C}^{-1}\mathbf{K}'\boldsymbol{x}^{(l)}$$

where $\mathbf{R} = \operatorname{diag}\left(\sum_{j=1}^{n} k_{ij}\right)$ and $\mathbf{C} = \operatorname{diag}\left(\sum_{i=1}^{m} k_{ij}\right)$ are the diagonal matrices of row and column totals. (If any of these totals is zero then inversion of the corresponding matrix is not possible. However, if the total is zero then no individual is observed with the given row or column category, so this category can be deleted from the corresponding discrete variable and either the row or column of the matrix removed without any detriment. We therefore assume throughout that this has already been done as necessary.)

The above iterative process will converge to give final solutions $\boldsymbol{x}$ and $\boldsymbol{y}$. In fact there will be more than one possible solution, and Hill defines the triple $(\rho, \boldsymbol{x}, \boldsymbol{y})$ as a solution of the correspondence analysis of $\mathbf{K}$ if $\rho\boldsymbol{x} = \mathbf{R}^{-1}\mathbf{K}\boldsymbol{y}$ and $\rho\boldsymbol{y} = \mathbf{C}^{-1}\mathbf{K}'\boldsymbol{x}$ (where the constant ρ is included to allow for arbitrary normalization of $\boldsymbol{x}$ and $\boldsymbol{y}$).

By writing $\mathbf{R}^{\frac{1}{2}} = \operatorname{diag}\left\{\sqrt{\left(\sum_{j=1}^{n} k_{ij}\right)}\right\}$ and $\mathbf{C}^{\frac{1}{2}} = \operatorname{diag}\left\{\sqrt{\left(\sum_{i=1}^{m} k_{ij}\right)}\right\}$, an equivalent condition for $(\rho, \boldsymbol{x}, \boldsymbol{y})$ to be a solution is that

$$\rho(\mathbf{R}^{\frac{1}{2}}\boldsymbol{x}) = (\mathbf{R}^{-\frac{1}{2}}\mathbf{K}\mathbf{C}^{-\frac{1}{2}})(\mathbf{C}^{\frac{1}{2}}\boldsymbol{y})$$

and

$$\rho(\mathbf{C}^{\frac{1}{2}}\boldsymbol{y}) = (\mathbf{R}^{-\frac{1}{2}}\mathbf{K}\mathbf{C}^{-\frac{1}{2}})'(\mathbf{R}^{\frac{1}{2}}\boldsymbol{x}). \tag{4.14}$$

Return for a moment to the singular value decomposition of a matrix $\mathbf{X}$ as defined by eqn (4.1). Orthonormality of the $\boldsymbol{u}_i$ means that

$$\boldsymbol{u}_i'\boldsymbol{u}_j = \begin{cases} 0 & \text{if } i \neq j, \text{ so that } \boldsymbol{u}_i'\mathbf{X} = \sigma_i\boldsymbol{v}_i' \text{ or } \sigma_i\boldsymbol{v}_i = \mathbf{X}'\boldsymbol{u}_i, \\ 1 & \text{if } i = j. \end{cases}$$

Similarly, orthonormality of the $\boldsymbol{v}_i$ leads to $\sigma_i\boldsymbol{u}_i = \mathbf{X}\boldsymbol{v}_i$. It can thus be seen that these two equations are exactly comparable to those in (4.14), where σ_i is identified with ρ; $\boldsymbol{u}_i$ with $\mathbf{R}^{\frac{1}{2}}\boldsymbol{x}$; and $\boldsymbol{v}_i$ with $\mathbf{C}^{\frac{1}{2}}\boldsymbol{y}$.

A singular value decomposition of $\mathbf{X} = \mathbf{R}^{-\frac{1}{2}}\mathbf{K}\mathbf{C}^{-\frac{1}{2}}$ will produce all possible triples $(\rho_i, \boldsymbol{x}_i, \boldsymbol{y}_i)$. If the decomposition is given by $\mathbf{X} = \sum_{j=1}^{t} \sigma_j\boldsymbol{u}_j\boldsymbol{v}_j'$, then

$$\rho_i = \sigma_i$$

$$\boldsymbol{x}_i = \mathbf{R}^{-\frac{1}{2}}\boldsymbol{u}_i$$

and

$$\boldsymbol{y}_i = \mathbf{C}^{-\frac{1}{2}}\boldsymbol{v}_i. \tag{4.15}$$

In the same way as with all such ordination methods (e.g. principal components, metric multidimensional scaling) the solutions are arranged in decreasing order of importance by their singular values ρ_i. A given $\boldsymbol{x}_i$, $\boldsymbol{y}_i$ pair yield the scores to be assigned to the row and column categories when producing new derived (continuous) variates X and Y to replace I and J, while ρ_i gives the correlation between these derived variates. An adequate graphical representation may need several dimensions, in which case the solutions corresponding to the largest singular values must be chosen as the axes of the configuration. However, there is one peculiarity of the present technique, which singles it out from the other ordination methods already described. This arises as a consequence of the reciprocal averaging process, which ensures that the trivial solution, $\rho = 1$, $\boldsymbol{x} = (1, 1, \ldots, 1)'$, and $\boldsymbol{y} = (1, 1, \ldots, 1)'$, is always obtained; moreover it is always the maximal solution (i.e. the one with largest ρ). Since this solution represents no more than the standard mean centering of a matrix (albeit in two 'directions' in the present instance), it must be ignored; an r-dimensional representation is thus obtained from the r solutions corresponding to the r largest ρ_i other than $\rho_1 = 1$.

When it comes to the graphical representation, it is evident that various different graphs may be obtained from the computed solutions. A simultaneous depiction of both rows and columns may be obtained by plotting the largest non-trivial $\boldsymbol{x}$ against the largest non-trivial $\boldsymbol{y}$; however, this means that only one dimension can be represented for each variable. More accuracy may be obtainable if rows and columns are plotted separately, by plotting the two largest non-trivial $\boldsymbol{x}_i$ against each other, and the two largest non-trivial $\boldsymbol{y}_i$ against each other. As an example of all these representations, Hill (1974) discussed the correspondence analysis of the Munsingen–Rain incidence matrix, which specifies the occurrences of seventy types of artefact in fifty-nine excavated tombs. This matrix is presented in Kendall (1971), who posed the problem of seriating the matrix, i.e. placing the tombs in their correct temporal order, bearing in mind that any particular type of artefact was probably only in use for a limited period. The best joint row/column representation of the incidence matrix as derived by Hill is presented in Fig. 4.8, while the best separate two-dimensional diagrams of tombs and artefacts are given in Figs. 4.9 and 4.10 respectively. (These figures are reproduced with permission from the Royal Statistical Society).

The order shown in Fig. 4.8 is almost identical to the one obtained by Kendall using a different method, and Hill discusses various additional aspects of the structure of the data that are highlighted by the correspondence analysis. The most striking aspect is that the seriation is by no means unequivocally continuous, but in fact there exist two major groups of individuals between which the connections are rather tenuous.

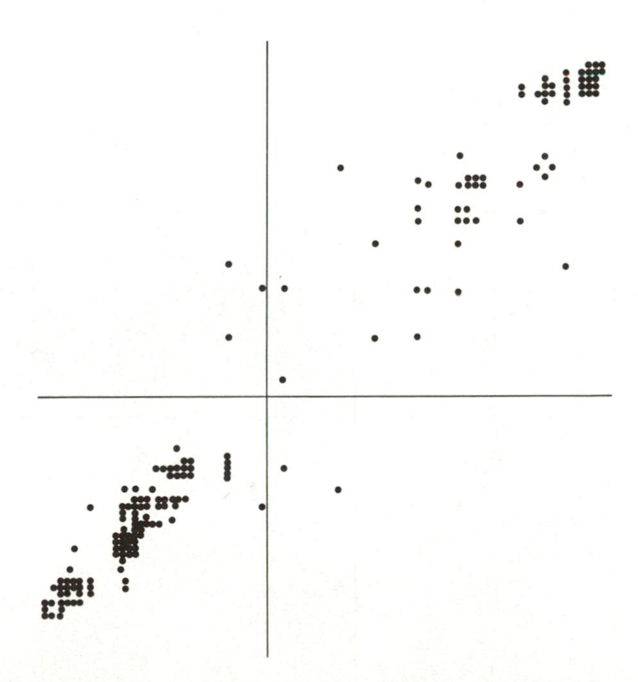

Fig. 4.8 The Munsingen–Rain incidence matrix seriated by correspondence analysis. Horizontal axis represents tombs, vertical axis artefacts. $\rho^2 = 0.96$. Reproduced with permission from the Royal Statistical Society

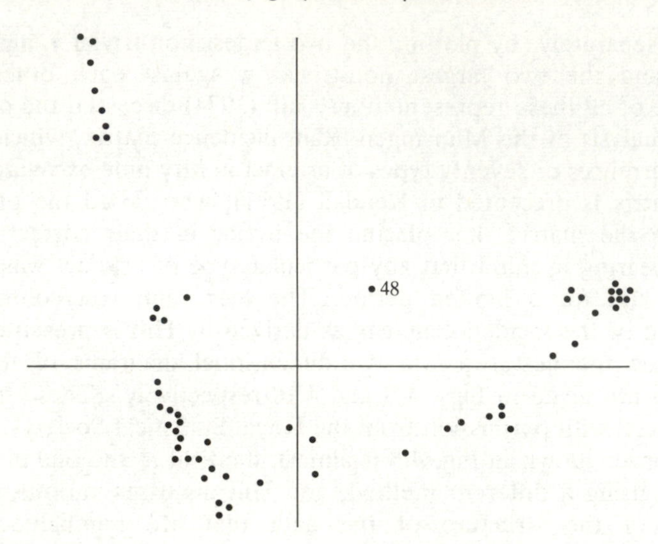

Fig. 4.9 Scatter diagram of tombs from Munsingen data, using the first two derived variates. $\rho_1^2 = 0.96$ (horizontal) and $\rho_2^2 = 0.90$ (vertical). Reproduced with permission from the Royal Statistical Society

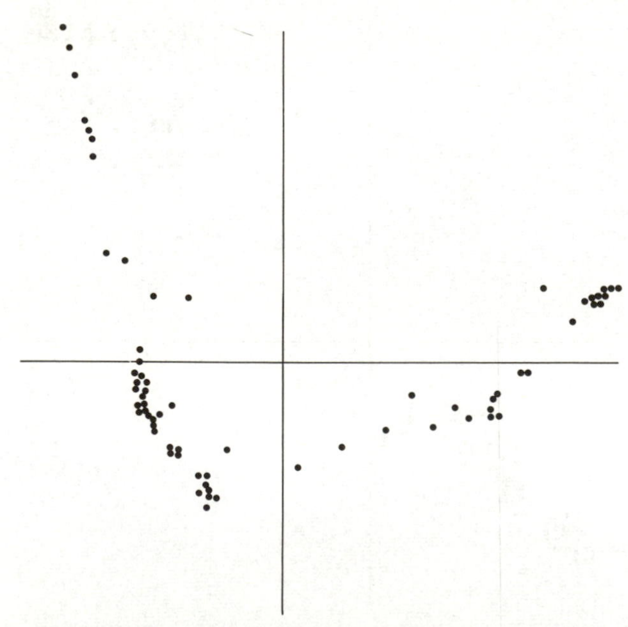

Fig. 4.10 Scatter diagram of artefacts from Munsingen data, using the first two derived variates. $\rho_1^2 = 0.96$ (horizontal) and $\rho_2^2 = 0.90$ (vertical). Reproduced with permission from the Royal Statistical Society

The addition of a second dimension in each of the scatter diagrams for artefacts and tombs (Figs 4.9 and 4.10) adds the further valuable information that the large group at the bottom of the incidence diagram in Fig. 4.8 can itself be split into two groups, which is confirmed by a closer examination of the incidence matrix. Also, the existence of outliers such as tomb number 48 is readily highlighted by the two-dimensional plots.

The basic correspondence analysis as described above operates on an $(m \times n)$ incidence matrix in which counts of simultaneous occurrence of two discrete variables for individuals in a sample are recorded. Such a matrix can be expanded to display the actual data that made up its entries. For example, suppose the incidence matrix consists of two rows and three columns with the following entries:

$$\mathbf{K} = \begin{pmatrix} 1 & 0 & 2 \\ 0 & 2 & 1 \end{pmatrix}.$$

If the rows are labelled a, b while the columns are labelled c, d, e, the information conveyed in $\mathbf{K}$ is that in a sample of five individuals, there was one observation (a, c), two observations (b, d), one observation (b, e), and two observations (a, e), where the pairs (i, j) represent the categories of the two discrete variables observed on each individual. If we now consider a five-vector $\boldsymbol{x}$ as giving the categories observed on a particular individual in the order (a, b, c, d, e), then, for example, the first individual (for which a and c were observed) would be denoted $\boldsymbol{x}_1' = (1, 0, 1, 0, 0)$. Similarly for the other five observations $\boldsymbol{x}_2$, $\boldsymbol{x}_3$, $\boldsymbol{x}_4$, $\boldsymbol{x}_5$, and $\boldsymbol{x}_6$ in the sample. These $\boldsymbol{x}_i$ can be written as columns of a (5×6) matrix $\mathbf{Z}$ as follows:

$$\mathbf{Z} = \begin{pmatrix} 1 & 0 & 0 & 0 & 1 & 1 \\ 0 & 1 & 1 & 1 & 0 & 0 \\ 1 & 0 & 0 & 0 & 0 & 0 \\ 0 & 1 & 1 & 0 & 0 & 0 \\ 0 & 0 & 0 & 1 & 1 & 1 \end{pmatrix}.$$

$\mathbf{Z}$ is thus an expanded matrix, giving the individual observations for a sample commensurate with the incidence matrix $\mathbf{X}$. In general, it has number of rows equal to the number of rows plus the number of columns of $\mathbf{X}$, i.e. the total number of categories of the original two discrete variables, and number of columns equal to the number of sample observations.

Correspondence analysis can now be applied to the matrix $\mathbf{Z}$ in analogous fashion to analysis of $\mathbf{X}$. This analysis will give a simultaneous scaling of discrete variable categories and sample members. Hill (1974) terms this a *first-order* correspondence analysis, in distinction from the correspondence analysis of **X**, which he terms a *zero-order* analysis. He discusses various properties and consequences of the analysis, which we shall not go into here beyond saying that a first-order analysis is a useful device for extending the technique to analysis of higher-order contingency tables. Consider, for example, a three-way contingency table in which the incidences of categories of three discrete variables are recorded for a sample of n individuals. If the number of categories of each variables are c_1, c_2, and c_3, then the incidence matrix **X** is written as a $c_1 \times c_2 \times c_3$ contingency table. Expanding **X** into the form **Z** above will give a $(c_1 + c_2 + c_3) \times n$ matrix, in which each column will now have three ones and the remaining entries zero. Correspondence analysis can then be applied to **Z** in the usual fashion.

A final point to note about correspondence analysis is that, while the technique involves a simultaneous decomposition of the data matrix into row and column components, valuable insight into the data structure can be provided by separate plotting of these components. Such facility was not afforded by the other techniques of this chapter.

This discussion of correspondence analysis has, of necessity, been very brief. For a comprehensive treatment of the subject, see Greenacre (1984).

5
Analytical comparison of two or more graphical representations

In this final chapter dealing with graphical representations, we shall concentrate on techniques that fall under either the *projection* or the *optimization* categories introduced in Chapter 1. In the former category, the input data are thought of as a swarm of points (representing sample individuals) in a high-dimensional space, and we obtain a pictorial representation by projecting this swarm into a low-dimensional subspace of the original space. Principal component analysis is a typical example of such a technique. In the latter category, no explicit high-dimensional geometrical model exists for the data, but instead we construct directly a low-dimensional space in which points represent the sample individuals, by optimizing some suitable criterion function. Multidimensional scaling is an example of such a technique. Both approaches yield low-dimensional spaces in which sample individuals are represented by points, but the projection approach additionally enables an interpretation of the derived subspace to be made in terms of the dimensions of the original space.

Now it frequently happens that several such configurations are obtained in the course of a practical investigation, and a comparison is required between the resulting pictures. Consider, for example, the data of Table 3.2 on percentage incidence of thirteen variates in each of fourteen samples of water voles. Using the dissimilarity measure described in Section 3.2, the dissimilarity matrix between these samples is given in Table 3.3. A metric multidimensional scaling of this matrix yields the two-dimensional configuration of Fig. 3.7, while a non-metric multidimensional scaling of this matrix yields the two-dimensional configuration of Fig. 3.10. Comparison of these two configurations indicates the similarity or otherwise in the outcomes of these two different scaling techniques on the same dissimilarity matrix. One can list a whole range of other possible comparisons that might be required. For example, suppose that the raw data of Table 3.2 had been converted into a dissimilarity matrix between samples using a different dissimilarity measure to the one described in Section 3.2. This would have yielded a matrix of the form of Table 3.3 but with different entries, and this matrix could be analysed by either metric or non-metric scaling. A comparison of the metric scaling configuration with that of Fig. 3.7 would immediately highlight any changes in inference caused by the different dissimilarity coefficients, and similarly for a comparison between the non-metric scaling configuration and that of Fig. 3.10. Another possibility is that a second zoologist might have measured a different set of variates on the same sample of water voles, thus obtaining a different data matrix and hence a different dissimilarity matrix between these samples. Once more, a comparison of the multidimensional scaling configuration obtained from this matrix with either Fig. 3.7 or Fig. 3.10 would be of interest, to examine the influence of the measured variables on the relationships among the samples. Thus at least six configurations of the fourteen samples are possible, and one might also wish to make an overall comparison of all six configurations simultaneously.

All of the above representations arise from an optimization approach to the construction of a configuration. The possibility of comparison between configurations arises because the points of all configurations relate to the same n entities, but it is only the comparison between corresponding points of two configurations that is of interest. Consider now the white leghorn fowl data analysed by means of principal component analysis in Section 2.2. The six variables measured on each of the 276 sample individuals are described in Section 2.2.5, the sample correlation matrix is given in Table 2.4, and the details of the principal component analysis are presented in Table 2.5. While this analysis can be used to plot the 276 fowl in the subspace of, say, the first two components, such a plot was not of primary interest in the original analysis. Of greater interest was the *interpretation* of the derived components. Such interpretation was conducted by inspecting the sizes and signs of the coefficients in each principal component, and was then used in attempting to discover the main sources of variability among the fowl. Suppose now that a second sample of fowl were collected from, say, a different continent, and the same six variables were measured on each individual. A principal component analysis on this second sample could be used in exactly the same way as before to identify the main sources of variability among the new individuals. A comparison of the two sets of principal component coefficients would thus indicate whether there were common sources of variability in the two samples or not. However, a simple comparison of coefficients may be misleading. Recollect that the first r principal components define the r-dimensional subspace of the original space into which the sample individuals can be projected with least displacement. In the present example, since the same six variables are measured on the two samples of fowl, the original six-dimensional spaces in which the two samples are notionally embedded coincide. Thus a comparison of the two sets of r components is equivalent to a comparison of the two corresponding r-dimensional subspaces of this common space. It is possible for two sets of vectors to look quite different from each other, but still to define the same subspace. Hence we would only be certain that the major sources of difference between the fowl of the two sets were *not* the same if the two principal component subspaces differed in some way.

In each of the two general situations described above, there is a need for an analytical technique that will provide a precise *numerical* measure of the extent to which two or more graphical representations differ (despite there being an obvious, intuitive way of providing an informal assessment). In the first case we assume that the n points of each of a set of configurations relate to the same n entities, and we wish to define a quantity (or quantities) measuring the departure from coincidence of the sets of points. In the second case we assume that the p axes of a set of configurations are the same, and we wish to define a quantity (or quantities) measuring the departure from coincidence of a set of subspaces defined in relation to these axes. Each of these cases will now be treated in turn.

5.1 Comparison of two n-point configurations: procrustes analysis

Let $\mathbf{X}$ and $\mathbf{Y}$ be, respectively, an $(n \times p)$ and an $(n \times q)$ matrix giving the coordinates of n points in each of two configurations. In each case, the coordinates of the ith point referred to orthogonal axes in Euclidean space are given by the values in the ith row of the matrix. The first configuration is thus in p-dimensional space and the ith point has coordinates $(x_{i1}, x_{i2}, \ldots, x_{ip})$, while the second configuration is in q-dimensional space and the ith point has coordinates $(y_{i1}, y_{i2}, \ldots, y_{iq})$. Suppose first that $p > q$. In this case, the second configuration is in a subspace of a p-dimensional space, and can be treated as if it were in this latter space simply by attaching $(p - q)$ coordinate values of zero to each point. This is done by adding $(p - q)$ columns of zeros to the right-hand side of $\mathbf{Y}$, thus converting it into an $(n \times p)$ matrix. Without loss of generality, therefore, we will assume throughout that $p = q$, and that this condition is achieved where necessary by adding an appropriate number of columns of zeros to the smaller of the two matrices.

We now wish to compare the two configurations, assuming that corresponding rows of the two matrices relate to the same entity. As a

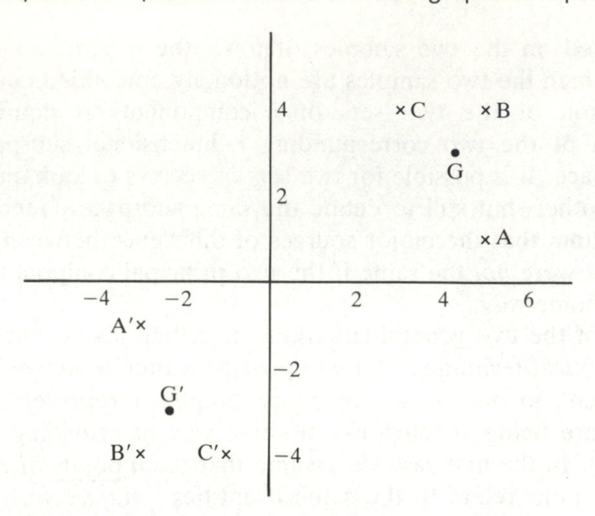

Fig. 5.1 Two simple configurations for comparison

simple illustration of the ideas involved, consider the two three-point configurations whose coordinates (in two dimensions) are given by the matrices

$$\mathbf{X} = \begin{pmatrix} 5 & 1 \\ 5 & 4 \\ 3 & 4 \end{pmatrix} \quad \text{and} \quad \mathbf{Y} = \begin{pmatrix} -3 & -1 \\ -3 & -4 \\ -1 & -4 \end{pmatrix}.$$

With the general assumption that $p = q$, we can always plot the two configurations against the same p orthogonal axes. If this is done for the above example, we obtain the diagram of Fig. 5.1, where A, B, C denote the first, second and third rows of $\mathbf{X}$ while A′, B′, C′ are the corresponding points of the second configuration $\mathbf{Y}$. The centroids of the two configurations are given by G(4.33,3) and G′(−2.33, −3).

A simple measure of the degree of coincidence exhibited by two such configurations, which would be of intuitive appeal to statisticians, is the sum of squared distances between corresponding points, $M^2 = \sum_{i=1}^{n} \left\{ \sum_{j=1}^{p} (x_{ij} - y_{ij})^2 \right\}$. For the simple example above this becomes

$$\begin{aligned} M^2 &= (\mathrm{AA}')^2 + (\mathrm{BB}')^2 + (\mathrm{CC}')^2 \\ &= \{(5+3)^2 + (1+1)^2\} + \{(5+3)^2 + (4+4)^2\} + \{(3+1)^2 + (4+4)^2\} \\ &= 276. \end{aligned}$$

Such direct computation of M^2 between two configurations is not very meaningful, however, because of the arbitrary position, orientation, and scale of one configuration relative to the other. When we talk of comparing two configurations, we implicitly mean a comparison of the *internal relationships* among the n points as exhibited by the two representations. The internal relationships among the points of a configuration are not disturbed by any of the following operations.

1. A fixed displacement of all the points through a constant distance in a common direction, i.e. a rigid *translation* of the whole configuration.
2. A fixed displacement of all the points through a constant angle, keeping the distance of each point from the centroid unchanged, i.e. a rigid *rotation* of the whole configuration.
3. A stretching or shrinking of all the points by a constant amount in a straight line from the point to the centroid of the configuration, i.e. a uniform *dilation* of the whole configuration.

There is one further rigid-body operation that preserves internal relationships among the points of a configuration, namely a *reflection* of the points about a point, line, plane, etc. However, a reflection in q dimensions can always be effected by a rotation in $(q+1)$ dimensions. For example, the one-dimensional reflection of the points S, T about the origin O into the points S′, T′ in Fig. 5.2 is achieved also by rotating the portion OST about O through 180° in the whole plane. Reflections can thus be considered as covered under 2 above, after the addition of a zero coordinate to each point of the configuration.

Before calculating a statistic such as M^2, therefore, it would make sense first to translate, rotate, and dilate the two configurations to positions of best fit relative to each other. Once this has been done M^2 measures the remaining 'lack of fit' of one configuration to the other, and hence provides a meaningful comparison of the two configurations. Now it is evident that operations 1 to 3 performed on *both* configurations will

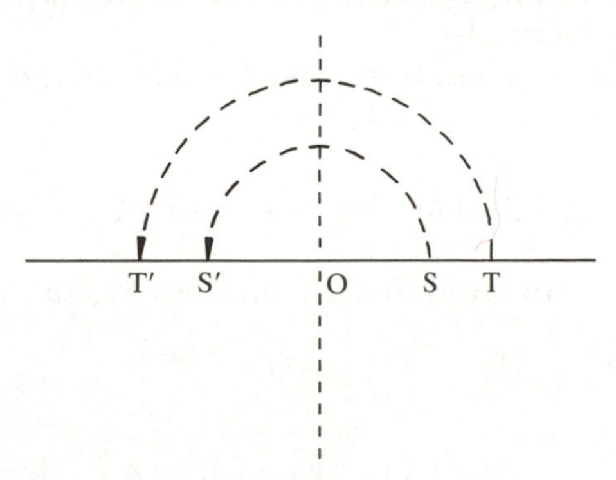

Fig. 5.2 Reflection in one dimension as a rotation in two dimensions

always yield negligible 'residual sum of squares' M^2 (e.g. by shrinking both configurations to minute size and superimposing one on top of the other). The appropriate procedure is thus to keep one configuration fixed and match the other one to it. In our simple example of Fig. 5.1, translating the configuration in the bottom left quadrant so that G′ coincides with G and then rotating this translated configuration through 180° will result in A′ coinciding with A, B′ coinciding with B, and C′ with C. Thus $M^2 = 0$, as these two configurations match exactly.

We now consider the general theory for this technique, which has been termed 'procrustes rotation' or 'procrustes analysis' in the literature. This technique has a long history in connection with factor analysis, but the approach that we describe here is in the spirit of multidimensional scaling applications as discussed by Schönemann and Carroll (1970), and by Gower (1971b), a rigorous mathematical statement for which has been given by Sibson (1978). Following the discussion of the previous paragraph, we will fix the configuration whose coordinates are given by **X** and match the configuration with coordinates **Y** to it. The matching will be accomplished by carrying out operations 1 to 3 above in sequence, in such a way as to make the final value of M^2 for the matched configurations as small as possible.

1. Matching under translation

From the definition,

$$M^2 = \sum_{i=1}^{n} \left\{ \sum_{j=1}^{p} (x_{ij} - y_{ij})^2 \right\}. \tag{5.1}$$

Let $\bar{x}_j = \frac{1}{n} \sum_{i=1}^{n} x_{ij}$ and $\bar{y}_j = \frac{1}{n} \sum_{i=1}^{n} y_{ij}$ $(j = 1, \ldots, p)$, so that the centroids G_X and G_Y of the two configurations have coordinates $(\bar{x}_1, \bar{x}_2, \ldots, \bar{x}_p)$ and $(\bar{y}_1, \bar{y}_2, \ldots, \bar{y}_p)$ respectively.

Add and subtract $\bar{x}_j$, $\bar{y}_j$ inside the round bracket of expression (5.1) and collect terms:

$$M^2 = \sum_{i=1}^{n} \left[\sum_{j=1}^{p} \{(x_{ij} - \bar{x}_j) - (y_{ij} - \bar{y}_j) + (\bar{x}_j - \bar{y}_j)\}^2 \right].$$

Now take the first two terms in curly brackets together and expand the square, noting that $\sum_{i=1}^{n} (x_{ij} - \bar{x}_j) = \sum_{i=1}^{n} (y_{ij} - \bar{y}_j) = 0$:

$$M^2 = \sum_{i=1}^{n} \sum_{j=1}^{p} \{(x_{ij} - \bar{x}_j) - (y_{ij} - \bar{y}_j)\}^2 + n \sum_{j=1}^{p} (\bar{x}_j - \bar{y}_j)^2. \tag{5.2}$$

But $(x_{ij} - \bar{x}_j)$ and $(y_{ij} - \bar{y}_j)$ are elements of the matrices **X** and **Y** *after mean centering,* and are therefore coordinates of the configurations represented by **X** and **Y** when it is arranged that their centroids both coincide at the origin of the axes (see Fig. 2.8 and Section 2.2.2). Thus

$$M^2 = M_{\mathrm{O}}^2 + n \sum_{j=1}^{p} (\bar{x}_j - \bar{y}_j)^2 \qquad (5.3)$$

where M_{O}^2 is the residual sum of squares between the two configurations after they have both been translated so that their centroids are at the origin. However, $\sum_{j=1}^{p} (\bar{x}_j - \bar{y}_j)^2$ is just the Euclidean distance between these two centroids G_X and G_Y, so that

$$M^2 = M_{\mathrm{O}}^2 + n(\mathrm{G}_X\mathrm{G}_Y)^2.$$

Clearly, therefore, the best fit under translation occurs when the two configurations have the same centroid. In this case, $\mathrm{G}_X\mathrm{G}_Y = 0$ and $M^2 = M_{\mathrm{O}}^2$. The simplest way of ensuring this is to mean-centre both **X** and **Y** at the outset.

2. Matching under rotation

Let us therefore assume that **X** and **Y** have both been mean-centered, and consider the next step of the matching process. Any rotation of **Y** relative to **X** can be expressed as an orthogonal matrix **Q** (Fig. 1.6 and Section 1.3.4), and after this rotation the coordinates of the configuration are given by the rows of **YQ**.

Now we can write $M^2 = \sum_{i=1}^{n} \left\{ \sum_{j=1}^{p} (x_{ij} - y_{ij})^2 \right\}$ as $M^2 = \mathrm{trace}\{(\mathbf{X} - \mathbf{Y})(\mathbf{X} - \mathbf{Y})'\}$. On expanding the matrix product inside the trace operator we obtain

$$M^2 = \mathrm{trace}(\mathbf{XX}' + \mathbf{YY}' - 2\mathbf{XY}'). \qquad (5.4)$$

After rotation by **Q**, therefore, **Y** becomes **YQ** and the residual sum of squares becomes

$$M^2 = \mathrm{trace}(\mathbf{XX}' + \mathbf{YQQ}'\mathbf{Y}' - 2\mathbf{XQ}'\mathbf{Y}').$$

Because **Q** is an orthogonal matrix, $\mathbf{QQ}' = \mathbf{I}$ and this expression is

$$M^2 = \mathrm{trace}(\mathbf{XX}' + \mathbf{YY}' - 2\mathbf{XQ}'\mathbf{Y}'). \qquad (5.5)$$

Thus to make M^2 as small as possible, we must choose **Q** to make trace $(2\mathbf{XQ}'\mathbf{Y}')$ as large as possible (since $\mathrm{trace}(\mathbf{A} \pm \mathbf{B}) = \mathrm{trace}(\mathbf{A}) \pm \mathrm{trace}(\mathbf{B})$

from Appendix A, Section A2). This is an exercise in constrained maximization, since the elements of $\mathbf{Q}$ are subject to the constraints of orthogonality (i.e. $\mathbf{QQ}' = \mathbf{I}$), and can be solved by standard differential calculus using a set of Lagrange multipliers. Details of this calculation are unnecessary here; suffice it to say that the required rotation $\mathbf{Q}$ is given by $\mathbf{Q} = \mathbf{VU}'$, where $\mathbf{U\Sigma V}'$ is the singular value decomposition of the matrix $\mathbf{X}'\mathbf{Y}$ (as defined in Section 4.1).

3. Matching under dilation

If the scales on which the two configurations are expressed differ, a final step of the matching procedure is to scale the coordinates in the matrix $\mathbf{Y}$ by a factor c, and estimate the value of c which minimizes M^2. Remembering that $\mathbf{Y}$ has already been rotated to $\mathbf{YQ}$, the effect of a scale change by a factor c is to convert the coordinates to $c\mathbf{YQ}$. Hence, from (5.5),

$$M^2 = \text{trace}(\mathbf{XX}' + c^2\mathbf{YY}' - 2c\mathbf{XQ}'\mathbf{Y}').$$

(A scale change will clearly not affect the optimal choice of rotation, so the results of 2 above remain the same whether or not c is included in (5.5).)

Thus

$$M^2 = c^2\,\text{trace}(\mathbf{YY}') - 2c\,\text{trace}(\mathbf{XQ}'\mathbf{Y}') + \text{trace}(\mathbf{XX}'). \tag{5.6}$$

This expression is a quadratic in c, and application of simple differential calculus shows that the minimum value of M^2 is obtained when $c = \text{trace}(\mathbf{XQ}'\mathbf{Y}')/\text{trace}(\mathbf{YY}')$. Note furthermore that

$$\begin{aligned} \text{trace}(\mathbf{XQ}'\mathbf{Y}') &= \text{trace}(\mathbf{Q}'\mathbf{Y}'\mathbf{X}) \\ &\qquad (\text{since trace}(\mathbf{AB}) = \text{trace}(\mathbf{BA}) \text{ for any } \mathbf{A}, \mathbf{B}) \\ &= \text{trace}(\mathbf{UV}'\mathbf{V\Sigma U}') \\ &\qquad (\text{using the results of 2}) \\ &= \text{trace}(\mathbf{V}'\mathbf{V\Sigma U}'\mathbf{U}) \\ &= \text{trace}(\mathbf{\Sigma}) \\ &\qquad (\text{by properties of the singular value decomposition}). \end{aligned}$$

Hence, given the singular value decomposition of 2, the optimum dilation factor c is given by

$$c = \text{trace}(\mathbf{\Sigma})/\text{trace}(\mathbf{YY}'). \tag{5.7}$$

This completes the three stages of the whole process. A nice feature of the process is that these three stages can all be accomplished independently of each other. Furthermore, having performed all three stages, the resultant value of M^2 is the smallest value achievable. Denote this value by $M^2_{\min}$; substitution from (5.7) into (5.6), and use of the other relationships in 3, shows that

$$c^2\,\mathrm{trace}(\mathbf{YY}') + M^2_{\min} = \mathrm{trace}(\mathbf{XX}'). \tag{5.8}$$

This identity may be used as the basis for an analysis of variance, interpreted as a partition of the total sum of squares among the vertices of the fixed configuration into the total sum of squares among the vertices of the fitted configuration plus a residual sum of squares.

Unfortunately, it is also evident that the process of fitting one configuration to another by this means is not symmetric, because the best system of scaling $\mathbf{Y}$ relative to $\mathbf{X}$ is not the inverse of the best system of scaling $\mathbf{X}$ relative to $\mathbf{Y}$ (as can be seen directly from (5.7)). To avoid problems of interpretation arising from this feature, it is common practice to standardize both configurations at the outset of the analysis so that they have unit total squared distance from their respective centroids. This standardization ensures that $\mathrm{trace}(\mathbf{XX}') = \mathrm{trace}(\mathbf{YY}') = 1$, and if this is done then the analysis of variance simplifies to $c^2 + M^2_{\min} = 1$ where $c = \mathrm{trace}(\boldsymbol{\Sigma})$, a value independent of whether we rotate $\mathbf{Y}$ to fit $\mathbf{X}$, or vice-versa.

To illustrate the technique, consider the data presented in Table 5.1 giving ABO gene frequencies for various human populations. These data

Table 5.1 ABO gene frequencies for various populations. Reproduced from Kurczynski (1970) with permission from the Biometric Society

Population	Allele				Sample size
	A_1	A_2	B	O	
Ainu	0.2773	0.0000	0.2042	0.5185	271
Austrians (Vienna)	0.2282	0.0595	0.1082	0.6041	10 000
Bantu	0.0969	0.0811	0.1143	0.7077	858
Chinese	0.1788	0.0000	0.1526	0.6686	250
English	0.2096	0.0696	0.0612	0.6602	3 459
Eskimos	0.4226	0.0000	0.0780	0.4994	180
Italians (Ferrara)	0.1864	0.0480	0.0574	0.7083	279
Lapps	0.1575	0.3577	0.1394	0.3454	183
Maori	0.3527	0.0000	0.0038	0.6435	267
Melanesians (Bougainville)	0.2555	0.0000	0.0656	0.6789	2 858
Navaho	0.1240	0.0000	0.0000	0.8760	361
Poles (Cracow)	0.2313	0.0539	0.1523	0.5625	5 201

Table 5.2 Distances between populations of Table 5.1 using distance measure D_K. Reproduced from Kurczynski (1970) with permission from the Biometric Society

	Austrians	Bantu	Chinese	English	Eskimos	Italians	Lapps	Maori	Melanesians	Navaho	Poles
Ainu	0.383	0.640	0.284	0.619	0.389	0.557	0.864	0.637	0.523	0.849	0.265
Austrians		0.302	0.279	0.171	0.462	0.214	1.059	0.452	0.289	0.538	0.134
Bantu			0.398	0.322	0.889	0.327	0.829	0.780	0.636	0.543	0.354
Chinese				0.427	0.512	0.401	0.882	0.618	0.349	0.656	0.260
English					0.529	0.102	0.977	0.444	0.324	0.460	0.294
Eskimos						0.539	0.843	0.445	0.363	0.863	0.467
Italians							0.807	0.503	0.651	0.526	0.307
Lapps								1.047	2.151	1.260	1.070
Maori									0.307	0.526	0.497
Melanesians										0.425	0.373
Navaho											0.608

originated in Mourant *et al.* (1958), and were presented by Kurczynski (1970) as part of an investigation comparing three measures of distance between multinomial populations. Krzanowski (1971b) listed three further measures of distance between such populations that had been proposed in the literature. There were thus six possible measures, each enabling a (12×12) distance matrix to be computed between the populations of Table 5.1. Let us denote these measures D_B, D_C, D_E, D_K, D_G, and D_X, following Krzanowski (1971b) where further details of their definition may be found. For example, the distance between every pair of populations using D_K is given in Table 5.2. This distance matrix is of the same form as those in Chapter 3, and can be subjected to a metric scaling to provide a configuration of twelve points representing the populations. Each of the other distance measures will provide a corresponding distance matrix and hence a corresponding configuration of twelve points representing these populations. A comparison of any two such configurations is thus a comparison of the performance of the corresponding distance measures on this data set. Every possible pair of these configurations was compared using procrustes rotation, and the values of $M^2_{\min}$ are given in Table 5.3 for all these comparisons.

Now these $M^2_{\min}$ values give a measure of dissimilarity between any two configurations. Sibson (1978) demonstrated that $M_{\min}$ defines a metric, so these $M^2_{\min}$ values can be treated as 'squared distances' between corresponding configurations and hence, by implication, between the distance measures that generated them. Consequently, the $M^2_{\min}$ values in Table 5.3 can themselves be subjected to multidimensional scaling, and the outcome will be a configuration of six points. Each point in this configuration will represent one of the measures D_B, D_C, D_E, D_G, D_K, and D_X, and the distance between any two such points will represent the dissimilarity (as given by $M_{\min}$) between the configurations generated by the corresponding measures from the data of Table 5.1.

Table 5.3 Values of $M^2_{\min}$ for every pair of configurations obtained from six distance measures applied to data in Table 5.1. Reproduced from Krzanowski (1971b) with permission from the Biometric Society

D_X	0.0					
D_C	0.176	0.0				
D_E	0.165	0.011	0.0			
D_K	0.343	0.355	0.384	0.0		
D_G	0.370	0.371	0.406	0.003	0.0	
D_B	0.343	0.345	0.367	0.007	0.007	0.0
	D_X	D_C	D_E	D_K	D_G	D_B

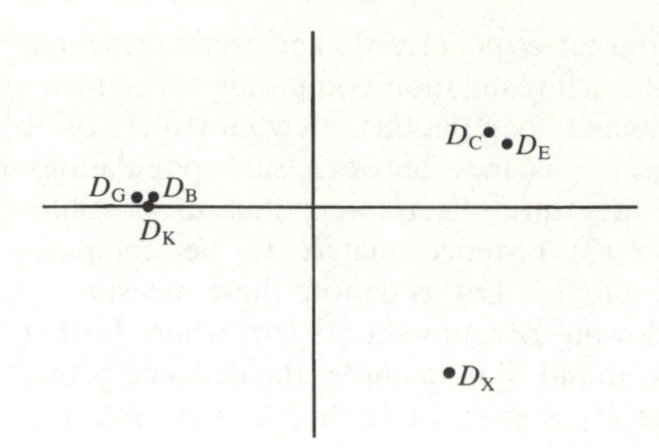

Fig. 5.3 First two principal axes for comparison of all six distance measures. Reproduced from Krzanowski (1971b) with permission from the Biometric Society

Although Sibson (1978) demonstrated by counter-example that M_{min} does not always lead to a Euclidean metric, in many data sets the Euclidean property will be satisfied, and metric scaling will give a faithful representation of the M^2_{min} data. Applying metric scaling to the matrix of Table 5.3 yields the two-dimensional diagram of Fig. 5.3. It is evident from this diagram that as a primary source of difference the measures split into two groups (D_B, D_G, D_K) and (D_C, D_E, D_X). The former is homogeneous, but the latter splits at the secondary level into the pair (D_C, D_E) and the single measure D_X. (For further details and discussion, see Krzanowski (1971b).)

5.2 Simultaneous comparison of *g* *n*-point configurations: generalized procrustes analysis

In the example just considered, the six configurations obtained by applying six different distance measures to the data of Table 5.1 were compared, but this was effected by comparing all possible pairs of configurations and then constructing a further configuration from the results of all these pairwise comparisons. Gower (1975) has generalized the idea of procrustes analysis to allow a *single* analysis, in which all *g* configurations are simultaneously translated, rotated, reflected, and scaled so that a goodness-of-fit criterion is optimized. An analogue of the two-configuration analysis of variance breakdown is also provided, to enable sources of difference between the *g* configurations to be investigated. The underlying theory and algebraic development are much more intricate than for the two-configuration case, however, so they will not be detailed here. We shall content ourselves with a brief outline of the main ideas, and leave the interested reader to refer to Gower (1975) for a full exposition of the mathematics and computing involved.

Let $\mathbf{X}^{(i)} = (x_{st}^{(i)})$ be an $(n \times p)$ matrix giving the coordinates of the n points in p dimensions for the ith configuration $(i = 1, \ldots, g;\ s = 1, \ldots, n;\ t = 1, \ldots, p)$. As before, it is assumed that the sth row of each of the $\mathbf{X}^{(i)}$ refers to the same entity $(s = 1, \ldots, n)$, so that the g configurations are compared by seeing how well corresponding points match. There is no loss of generality in assuming that the dimensionality of all configurations is the same, i.e. p. If in fact the ith configuration has dimensionality p_i, and the p_i are all different, then we can treat all the configurations as being embedded in a larger space, whose dimensionality equals that of the largest configuration. This is simply done by setting $p = \max(p_1, p_2, \ldots, p_g)$, and appending $(p - p_i)$ columns of zeros to $\mathbf{X}^{(i)}$ $(i = 1, \ldots, g)$.

Consider now a typical point s of all configurations. Its coordinates in the ith configuration are $(x_{s1}^{(i)}, x_{s2}^{(i)}, \ldots, x_{sp}^{(i)})$. Let $\bar{x}_{st} = \frac{1}{g} \sum_{i=1}^{g} x_{st}^{(i)}$ $(t = 1, \ldots, p)$. If all g configurations are plotted against the same p orthogonal axes, then there will exist a set of g points corresponding to each entity being represented. Thus we can say that there are n sets of g 'like points'. The sth such set of points will have a centroid G_s whose coordinates are given by $(\bar{x}_{s1}, \bar{x}_{s2}, \ldots, \bar{x}_{sp})$. If the g configurations match very well then the g points representing each entity will be close to each other, so each of these n sets of 'like points' will be very compact, but if the g configurations match very poorly then some (or all) of these sets of points will be very diffuse. As a simple illustration, Fig. 5.4 (reproduced from Gower's 1975 *Psychometrika* article with permission from the Psychometric Society) shows three (well-fitting) configurations of four points in two dimensions, where the sth point of the ith configuration is denoted $P_s^{(i)}$.

A simple measure of 'diffuseness' of the sth set of g points is the sum of squared distances between each point in the set and the centroid of that set, i.e. $\sum_{i=1}^{g} \sum_{t=1}^{p} (x_{st}^{(i)} - \bar{x}_{st})^2$. Consequently, a reasonable measure of the goodness of fit of the g configurations to each other is the sum of this diffuseness measure over all n sets,

$$\bar{M}^2 = \sum_{i=1}^{g} \sum_{s=1}^{n} \sum_{t=1}^{p} (x_{st}^{(i)} - \bar{x}_{st})^2. \tag{5.9}$$

However, it can be shown that the sum of squared distances of g points from their centroid is equal to the sum of squares of all $\frac{1}{2}g(g-1)$

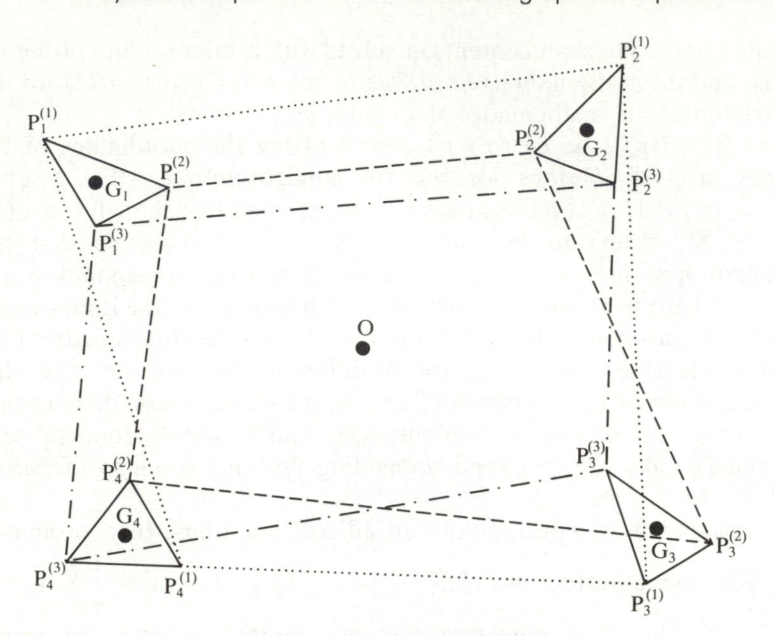

Fig. 5.4 Geometrical representation of three configurations ($g = 3$) each with two dimensions ($p_1 = p_2 = p_3 = 2$) and four vertices ($n = 4$) referring to the same four entities. Each entity therefore gives rise to a triangular cluster of vertices whose centroids G_j ($j = 1, 2, 3, 4$) are marked. The centroid of the whole system is at 0. Reproduced from Gower (1975) with permission from the Psychometric Society

inter-point distances divided by g. To avoid calculating centroids, therefore, it is more convenient to replace (5.9) by the criterion

$$\bar{M}^2 = \sum_{j=2}^{g} \sum_{i=1}^{j-1} \sum_{s=1}^{n} \sum_{t=1}^{p} (x_{st}^{(i)} - x_{st}^{(j)})^2. \tag{5.10}$$

A measure of goodness-of-fit between g configurations is thus the value of (5.10) after the g configurations have been matched as well as possible under translation, rotation/reflection, and dilation. The goodness of fit is thus the minimum value of $\bar{M}^2$ after each $\mathbf{X}^{(i)}$ has been subjected to the following operations:

(1) translation—it has had a constant matrix added to it;
(2) rotation—it has been post-multiplied by an orthogonal matrix $\mathbf{H}_i$;
(3) dilation—it has been multiplied by a scalar c_i.

To avoid the scaling problems encountered in the two-configuration case, Gower recommends subjecting *all* matrices $\mathbf{X}^{(i)}$ to these steps, imposing suitable constraints to prevent degenerate solutions such as $c_i = 0$ for all i. Translation is again taken care of by ensuring that all configurations have the same centroid, and it is most convenient to take the origin of axes at this common centroid. Mean-centering all matrices $\mathbf{X}^{(i)}$ thus accomplishes this step. With the present formulation, however, the rotation and dilation steps are no longer independent, and moreover the optimum solution to one step requires the optimum value from the other. Thus the solution can only be obtained by a numerical iterative procedure, fully described by Gower. Having obtained the solution, the total sum of squares of all g configurations can be partitioned into between- and within-configuration components for interpretation. See Gower (1975) for a detailed example.

5.3 Comparison of subspaces

We now turn to the second type of comparison discussed in the introduction to this chapter. To motivate the discussion, consider the following problem encountered during a British Council investigation into the educational achievements of Venezuelan students at colleges in the North of England. There were approximately 160 such students, who were distributed randomly among ten colleges, and each student spent a year at his/her designated college. Each student was subjected to a centrally administered examination at the start, in the middle, and at the end of the year; each examination consisted of tests in a set of eight subjects and these subjects are listed in Table 5.4. One aim of the investigation was to compare student performance between colleges, so for this comparison the data were broken up into ten subsets. Each subset consisted of the score in the final examination for each of the eight subjects as obtained by each student in a given college. Thus if there were n_i students in the ith college, then the data matrix for this college had n_i rows and eight columns. A principal component analysis was done on each of these ten data matrices in an attempt to summarize descriptively the main sources of variation between students in each college. Some way of comparing these analyses was then necessary. Since each college contained *different* students, comparisons among the colleges in the style of Sections 5.1 and 5.2 are inappropriate, particularly as the number of students varies from college to college. In this case, it is not the extent of variation from college to college in a *particular* student's result that is of interest, but whether the main sources of variation among students *in general* differ from college to college. Since the *same* eight tests were scored for each student in all colleges, a comparison of principal component coefficients between colleges seems to be indicated.

Table 5.4 Coefficients of first four principal components for each of three colleges (reproduced from Krzanowski (1979) with permission from the American Statistical Association)

	College A				College B				College C			
Subject	I	II	III	IV	I	II	III	IV	I	II	III	IV
Comprehension	0.227	0.458	0.218	0.804	0.207	0.655	0.669	0.016	0.402	0.465	0.181	0.291
Essay	0.322	−0.191	0.331	0.063	0.376	0.285	−0.277	−0.009	0.405	−0.040	0.128	−0.553
Cloze 1	0.319	0.004	−0.525	0.000	0.250	−0.098	0.169	0.110	0.402	−0.193	−0.581	0.225
Cloze 2	0.275	−0.014	−0.416	−0.040	0.184	0.091	0.137	0.023	0.321	−0.209	−0.505	0.097
Structure	0.741	−0.123	0.385	−0.312	0.400	0.070	−0.211	−0.839	0.403	0.465	0.181	0.291
Dictation	0.343	0.152	−0.455	0.116	0.684	−0.559	0.220	0.231	0.405	−0.040	0.128	−0.553
Spanish cloze 1	0.014	−0.711	−0.171	0.398	0.256	0.337	−0.382	0.413	0.269	−0.580	0.383	0.209
Spanish cloze 2	−0.019	−0.458	0.099	0.284	0.161	0.202	−0.435	0.244	0.106	−0.384	0.401	0.339

Such a comparison should highlight whether there is comparability in sources of variation (indicating that teaching emphasis was similar among the colleges) or whether variability in some subject areas is less for some colleges than for others (indicating greater teaching emphasis in those areas for the former colleges).

Interpretation of principal components was shown in Section 2.2.5 to be highly subjective, so some formal procedure is necessary for their comparison. We shall illustrate the methodology by reference to a subset of the British Council results, namely the coefficients of the first four principal components for three colleges labelled A, B, and C as given in Table 5.4. Now, since the same eight tests were administered to all students, we can imagine all students to be represented by points in a single eight-dimensional space, with separate sets of points denoting separate colleges. Each of the ten principal component analyses referred to thus defines a distinct subspace of this eight-dimensional space; for example, the three analyses reported in Table 5.4 define three four-dimensional subspaces. A comparison of sources of variation for any two colleges will thus require a comparison of the two subspaces defined by their respective principal component analyses, while a simultaneous comparison of sources of variation for all ten colleges will require a simultaneous comparison of all ten subspaces. More pertinently, we will require a measure quantifying the degree of departure from coincidence of these subspaces, since coincident subspaces imply identical sources of variation. We will consider the problem in two stages, first treating the comparison between two subspaces, and then describing a generalization to simultaneous comparison of g subspaces (in analogous fashion to the approach to procrustes analysis above).

First, therefore, consider the two-sample comparison. Let A and B denote two multivariate samples of n_1 and n_2 individuals, respectively, each individual having the same p variables measured on it. Suppose each sample has undergone a principal component analysis and that the same number, k, of components have been considered as adequate for the purposes of representing each sample. It will be shown later that this restriction is unnecessary, but as it leads to simplified discussion and interpretation it will be adopted for the present. If the original variables measured on each individual are denoted $x_1, \ldots, x_p$ and the principal components of samples A and B are denoted $y_1, \ldots, y_k$ and $z_1, \ldots, z_k$ respectively, then

$$y_i = \sum_{j=1}^{p} l_{ij} x_j$$

and $\qquad\qquad (i = 1, \ldots, k)$

$$z_i = \sum_{j=1}^{p} m_{ij} x_j$$

where the l_{ij} and m_{ij} are the coefficients of the principal components for the two samples. All the measured individuals are represented by points in a p-dimensional space having orthogonal coordinate axes $x_1, x_2, \ldots, x_p$. Of these points, the ones representing individuals in sample A are effectively located in one k-dimensional subspace of this space while the ones representing individuals in sample B are effectively located in a second k-dimensional subspace, and these two subspaces are defined by the two sets of orthogonal axes $y_1, \ldots, y_k$ and $z_1, \ldots, z_k$ respectively.

The general strategy for obtaining a measure of closeness of the two subspaces, as described by Krzanowski (1979), bears many points of resemblance with the strategy for obtaining a measure of closeness between two configurations using procrustes analysis. However, as we are now concerned with closeness of two sets of directions in space (i.e. the principal components) rather than closeness of two sets of points, a more natural measure of closeness will be based on angles between lines rather than on distances between points. One possibility is just to measure the angles between corresponding pairs of principal components for the two samples. However, it was shown in procrustes analysis that distances between corresponding points in the two raw configurations were not meaningful, and the configurations should first be matched under translation, rotation/reflection, and dilation before measuring these distances. Similarly, the angles between corresponding pairs of principal components might be quite large even though the two sets of components define the same k-dimensional subspace. Hence, before computing any angles, we must first find the 'best-matching' sets of orthogonal axes for the two subspaces to be compared. Following the procedure adopted in procrustes analysis, we consider one of the subspaces to be 'fixed' and find the best-matching set of orthogonal axes in the second subspace. The following results enable this to be done. They are stated here without any proofs, but their derivation depends only on fairly simple properties of orthogonal projection and coordinate geometry; for details see Krzanowski (1979).

We assume that the principal component coefficients in samples A and B are given, as above, by l_{ij} and m_{ij} respectively ($i = 1, \ldots, k$; $j = 1, \ldots, p$). Write $\mathbf{L}$ and $\mathbf{M}$ as the $(k \times p)$ matrices having (i, j)th elements l_{ij} and m_{ij}. (Thus the principal components of sample A come from the rows of $\mathbf{L}$, and the principal components of sample B come from the rows of $\mathbf{M}$.) Define $\mathbf{N} = \mathbf{LM'ML'}$ ($= \mathbf{TT'}$, where $\mathbf{T} = \mathbf{LM'}$). Then Krzanowski proves the following results.

Result 1. The minimum angle between an arbitrary vector in the space of the first k principal components of A and the one most nearly parallel to it in the space of the first k principal components of B is given by $\cos^{-1}\sqrt{\lambda_1}$, where λ_1 is the largest eigenvalue of $\mathbf{N}$.

Result 2. Let λ_i be the ith largest eigenvalue of $\mathbf{N}$, $\boldsymbol{a}_i$ its associated eigenvector, and $\boldsymbol{b}_i = \mathbf{L}'\boldsymbol{a}_i$ $(i = 1, \ldots, k)$. Then $\boldsymbol{b}_1, \ldots, \boldsymbol{b}_k$ form a set of mutually orthogonal vectors embedded in subspace A and $\mathbf{M}'\mathbf{M}\boldsymbol{b}_1, \ldots, \mathbf{M}'\mathbf{M}\boldsymbol{b}_k$ a corresponding set of mutually orthogonal vectors in subspace B into which the differences between the subspaces can be partitioned. The angle between the ith pair $\boldsymbol{b}_i$, $\mathbf{M}'\mathbf{M}\boldsymbol{b}_i$ is given by $\cos^{-1}\sqrt{\lambda_i}$ $(i = 1, \ldots, k)$.

It can also be shown easily that the sum of the eigenvalues of $\mathbf{N}$ equals the sum of squares of the cosines of the angles between each of the principal components of A and each one of B. This sum clearly lies between the values 0 (if the two subspaces are entirely disjoint) and k (if the two subspaces are entirely coincident), so forming a natural measure of total similarity between the two subspaces. Results 1 and 2 above then show that the similarities between A and B can be exhibited solely through the pairs $\boldsymbol{b}_i$, $\mathbf{M}'\mathbf{M}\boldsymbol{b}_i$, with λ_i measuring the contribution of the ith pair to the total similarity. Thus the line (i.e. one-dimensional portion) of subspace A that is closest to subspace B is given by $\boldsymbol{b}_1$; $\mathbf{M}'\mathbf{M}\boldsymbol{b}_1$ is the line in subspace B that it is closest to, and $\cos^{-1}\sqrt{\lambda_1}$ is the angle between them. If we now wish to define the plane (i.e. two-dimensional portion) of subspace A that is closest to subspace B, we need simply to add to $\boldsymbol{b}_1$ the line orthogonal to it that is next closest to B. This line is $\boldsymbol{b}_2$, and $\mathbf{M}'\mathbf{M}\boldsymbol{b}_2$ is the line in subspace B that it is closest to. Thus $(\boldsymbol{b}_1, \boldsymbol{b}_2)$ and $(\mathbf{M}'\mathbf{M}\boldsymbol{b}_1, \mathbf{M}'\mathbf{M}\boldsymbol{b}_2)$ define the planes in A and B that are closest to each other. The extent to which these planes are non-coincident is then measured by the two 'critical angles' $\cos^{-1}\sqrt{\lambda_1}$ and $\cos^{-1}\sqrt{\lambda_2}$. So we can continue, adding the pair $\boldsymbol{b}_3$, $\mathbf{M}'\mathbf{M}\boldsymbol{b}_3$ to define the most similar three-dimensional portions of A and B; $\mathbf{b}_4$, $\mathbf{M}'\mathbf{M}\boldsymbol{b}_4$ to define the most similar four-dimensional portions; and so on. A quantification of the extent to which two r-dimensional portions differ is provided by the r critical angles $\cos^{-1}\sqrt{\lambda_1}, \ldots, \cos^{-1}\sqrt{\lambda_r}$. Thus if the two subspaces A and B intersect in r dimensions then these r critical angles will all be zero. Hence the first r eigenvalues of $\mathbf{N}$ are all equal to 1, and $\boldsymbol{b}_1, \ldots, \boldsymbol{b}_r$ form a basis for this common r-dimensional subspace.

In order to interpret the nature of the similarities or differences between A and B, we can consider the pair of eigenvectors $\boldsymbol{b}_i$ and $\mathbf{M}'\mathbf{M}\boldsymbol{b}_i$ associated with each eigenvalue λ_i. These vectors are defined with respect to the original p axes, and hence may be interpreted in the same way as any principal component is interpreted (see Section 2.2.5). Furthermore, the vector in the original p-dimensional space that is closest to both $\boldsymbol{b}_i$ and $\mathbf{M}'\mathbf{M}\boldsymbol{b}_i$ is the bisector of the angle between them, in the plane in which they lie. This bisector is given by

$$\boldsymbol{c}_i = \{2(1+\sqrt{\lambda_i})\}^{-\frac{1}{2}}\left(\mathbf{I}+\frac{1}{\sqrt{\lambda_i}}\mathbf{M}'\mathbf{M}\right)\boldsymbol{b}_i.$$

The set $\boldsymbol{c}_1, \ldots, \boldsymbol{c}_k$ will thus define the k-dimensional subspace that is the 'average' (or 'consensus') of subspaces A and B.

Finally, we return to the point mentioned at the outset, namely that it will frequently happen that A and B have been characterized by a different number of principal components. Suppose that k_1 components define A, and k_2 components define B. Then if $k = \min(k_1, k_2)$, there will only be k non-zero eigenvalues of $\mathbf{N}$, and the comparison of A and B can proceed on the basis of these non-zero eigenvalues and their corresponding eigenvectors.

To illustrate the technique, let us compare the principal components of colleges A and B in the British Council investigation outlined earlier. The first four components for each college are given in Table 5.4. A useful description is afforded by comparing the first r components of each college sequentially for $r = 1, 2, 3$, and 4. Thus the first components of each college are compared, then the planes defined by the first two components of each college, then the three-dimensional spaces defined by the first three components of each college, and finally the four-dimensional spaces defined by all four components of each college are compared. Results of these comparisons are given in Table 5.5. Part (a) gives the angles of separation of the two principal component spaces, and the eigenvalues λ_i are quoted as well as the corresponding angles $\cos^{-1}\sqrt{\lambda_i}$.

Part (b) of Table 5.5 gives the vectors $\boldsymbol{c}_i$ that are closest to both spaces for each dimensionality of comparison.

We see from this table that the first principal component of college A is not very similar to the first principal component of college B, there being an angle of 34° between them. Similarly, when the first two components of each college are compared, the nearest the two planes come to each other is 31°; otherwise they are almost orthogonal to each other with an angle of 81°. When we proceed to component spaces of dimensions 3 and 4, however, we see that a common subspace of three dimensions is strongly indicated by angles as low as 4°, 6° and 18°. Interpreting these results, we infer that the prime sources of variation between the students in the two colleges are rather different, but most of the differences disappear if each college is characterized by three or more principal components. Table 5.4 indicates that the primary components for the two colleges load very differently on the variables 'structure' and 'dictation',

Table 5.5 Comparison of principal component spaces of colleges A and B (reproduced from Krzanowski (1979) with permission from the American Statistical Association)

(a) Eigenvalue analysis

	Eigenvalues				Corresponding critical angles (degrees)			
Component spaces of dimension 1	0.689				33.9			
Component spaces of dimension 2	0.024	0.739			81.1	30.7		
Component spaces of dimension 3	0.173	0.764	0.977		65.5	29.1	8.7	
Component spaces of dimension 4	0.521	0.910	0.989	0.996	43.9	17.5	6.0	3.6

(b) Vectors closest to both spaces

Dimension of comparison	1	2		3			4			
Vector	1	1	2	1	2	3	1	2	3	4
Comprehension	0.227	0.217	−0.076	−0.325	0.194	0.766	0.181	−0.753	0.268	−0.558
Essay	0.365	0.402	−0.217	0.382	0.254	0.288	0.471	0.075	0.151	−0.057
Cloze 1	0.297	0.280	0.143	−0.155	0.394	−0.217	0.049	−0.276	0.030	0.401
Cloze 2	0.240	0.242	−0.006	−0.085	0.308	−0.078	0.092	−0.210	0.030	0.248
Structure	0.596	0.599	0.034	0.318	0.474	0.310	0.797	0.094	−0.468	−0.016
Dictation	0.537	0.458	0.586	−0.278	0.637	−0.378	0.077	−0.449	−0.013	0.656
Spanish cloze 1	0.141	0.254	−0.644	0.557	0.130	−0.146	0.226	0.193	0.694	0.180
Spanish cloze 2	0.074	0.146	−0.410	0.480	0.016	−0.044	0.206	0.244	0.451	0.034

but the compromise ('average') subspace defined by the vectors in part (b) of Table 5.5 presents a rather different pattern. When component spaces of dimension 4 are compared, for example, the vector nearest to both spaces has a very high loading on structure (0.797), and a secondary contribution from essay (0.471). This, then, is the component with respect to which the between-student scatter is most similar in the two colleges.

Now let us turn to a simultaneous comparison of the principal components obtained in more than two groups. Suppose in fact that g groups exist, with n_i units in the ith group and the same p variables have been measured on each unit. Also, each group has been described by its first k principal components. Similar remarks apply here as applied earlier to the case where each group has been described by a different number of principal components; if k_i principal components have been extracted for the ith group, then we set $k = \min(k_1, \ldots, k_g)$; the procedure to be described will generate only k non-zero critical angles. Write $\mathbf{L}_t$ as the $(k \times p)$ matrix whose (i, j)th element gives the coefficient of the jth measured variable for the ith principal component in the tth group $(i = 1, \ldots, k;\ j = 1, \ldots, p; t = 1, \ldots, g)$. Krzanowski (1979) proves the following.

Result 3: Let $\boldsymbol{b}$ be an arbitrary vector in the original p-dimensional data space, and let δ_t be the angle between $\boldsymbol{b}$ and the nearest line in the subspace defined by the k principal components of group t $(t = 1, \ldots, g)$. Then the value of $\boldsymbol{b}$ that maximizes $V = \sum_{t=1}^{g} \cos^2 \delta_t$ is given by the eigenvector $\boldsymbol{b}_1$ corresponding to the largest eigenvalue μ_1 of $\mathbf{H} = \sum_{t=1}^{g} \mathbf{L}_t'\mathbf{L}_t$.

It should be stressed that there are many possible ways of defining the closeness of an arbitrary vector $\boldsymbol{b}$ to a set of g k-dimensional subspaces, of which V is but one example. However, if V is accepted as being a sensible measure of closeness, then the 'average' component that agrees most closely with all g sets of principal components is given by $\boldsymbol{b}_1$ as defined in result 3. A measure of discrepancy between this component and the k-dimensional subspace of the tth group is $\delta_t = \cos^{-1}\{(\boldsymbol{b}_1'\mathbf{L}_t'\mathbf{L}_t\boldsymbol{b}_1)^{\frac{1}{2}}\}$.

Completing the eigenvalue/eigenvector decomposition of $\mathbf{H}$ will lead to the subspace of dimension k that resembles *all* g subspaces as closely as possible. The eigenvector $\boldsymbol{b}_2$ corresponding to the second-largest eigenvalue μ_2 of $\mathbf{H}$ will give the vector orthogonal to $\boldsymbol{b}_1$ for which the criterion V has next largest value. The angle between $\boldsymbol{b}_2$ and the vector most nearly parallel to it in the subspace for the tth group is found as δ_t (above), but with $\boldsymbol{b}_2$ replacing $\boldsymbol{b}_1$. This procedure can be continued for all k eigenvalues and eigenvectors, and these k vectors define the required subspace. It is easily shown that this procedure reduces to the consensus solution already derived above when $g = 2$.

To illustrate this extension of the technique, consider a comparison of all three four-dimensional subspaces defined by the four principal components of each of colleges A, B, and C, as given in Table 5.4. Part (a) of Table 5.6 presents the coefficients of each of the four orthogonal directions closest to all three principal component subspaces (in relation to the original eight variables). These four directions thus correspond to $\boldsymbol{b}_1$, $\boldsymbol{b}_2$, $\boldsymbol{b}_3$, and $\boldsymbol{b}_4$ above, and define the four-dimensional 'average' subspace closest to all three principal component spaces. Part (b) of Table 5.6 gives the angular separation between each direction and each of the three groups. Values for the sum of squared cosines of these angles are given at the foot of each column; these sums are simply the eigenvalues μ_1, μ_2, μ_3, and μ_4 of $\mathbf{H}$, and they give the critical values of the criterion V (which has maximum possible value 3 in the present instance).

All three groups are close together along the first two axes, but begin

Table 5.6 Comparison of principal component spaces of colleges A, B, and C (reproduced from Krzanowski (1979) with permission from the American Statistical Association)

	Dimension 1	2	3	4
(a) Directions closest to each subspace				
Comprehension	0.586	−0.138	0.247	−0.756
Essay	0.334	0.223	0.075	0.246
Cloze 1	0.219	0.052	−0.497	−0.047
Cloze 2	0.188	0.048	−0.375	−0.029
Structure	0.587	−0.172	0.350	0.589
Dictation	0.336	0.076	−0.608	0.108
Spanish cloze 1	0.069	0.769	0.059	−0.077
Spanish cloze 2	0.016	0.547	0.223	−0.006
(b) Angles formed by each group with each direction				
College A	1.99°	9.38°	17.48°	15.01°
College B	3.06°	11.18°	27.73°	10.74°
College C	1.57°	6.33°	29.71°	73.75°
Sum of squared cosines	2.995	2.924	2.448	1.977

to diverge after this. Inspection of the vector loadings gives an indication of the component with respect to which the between-student scatter is most similar in all three colleges. The first direction is loaded most heavily on 'structure' (as in the four-dimensional comparison of colleges A and B) and on 'comprehension'. This latter item was absent in the previous comparison, and it reflects the higher weighting given to this variable in the first two principal components of college C. The fact that it comes out as important when all colleges are considered together implies that it was hidden but not negligible in colleges A and B. The second direction in the present comparison is weighted almost exclusively on the Spanish cloze tests (i.e. selection of a word from a list to complete a sentence), and corresponds to the third-most-important vector in Table 5.5. This indicates that only two dimensions are really common between the first four principal components of the three colleges. 'Structure' and 'comprehension' form the major common cause of differences between students, followed by their performance on the Spanish tests. Differences among the colleges are highlighted by the fourth component, on which C diverges markedly from A and B. The difference is attributable to the contrast between 'structure' and 'comprehension'.

(Tables 5.4, 5.5, and 5.6, as well as the summary conclusions, are reproduced from Krzanowski (1979) with permission from the American Statistical Association.)

Part II: Samples, populations, and models

6
Data inspection or data analysis?

6.1 Basic concepts

The material in Part I of this book has been concerned exclusively with *data inspection*: given a multivariate data set, are there any interesting patterns that can be observed in the *numbers themselves*? We have seen that inspection of a large data matrix is not an easy task, and the whole of Part I was therefore concerned with methods of constructing pictorial representations of such matrices, in the hope that any such interesting patterns in the data might be highlighted more easily. In order to construct these pictorial representations, an initial connection had to be made between the elements of the data matrix and points, lines, or shapes in Euclidean space. This connection enabled a geometrical model of the data to be set up, and many concepts familiar in geometry could then be used to achieve simplification of the initial data representation. The important aspect of all the techniques of Part I, however, is that in no case was any structure *imposed* on the data representation. Clearly some information always has to be lost when a high-dimensional set of data is represented in relatively few dimensions, but in no case was any of the lost information directly attributable to the imposition of external constraints acting on the elements of the data matrix. In other words, no preconceived ideas about the behaviour of the system governed the type of output observed in each technique. Rather, the data were able to speak for themselves, and the pictorial representations attempted to display the resulting message as faithfully as possible.

Now, while simple data inspection is obviously very valuable in any statistical investigation, it clearly stops short of an adequate *analysis* of the situation. This is because every data set contains many irregular and idiosyncratic fluctuations, as well as some systematic underlying features. True analysis of the data will hope to uncover these systematic features by removing or 'smoothing out' the irregularities. Data inspection, on the other hand, may all too often highlight the irregularities and disguise the systematic features, particularly if an inappropriate technique is chosen for the data display. How therefore can we penetrate the disguise and arrive at the main message conveyed by the data? The standard statistical approach presupposes that we already have some general idea, perhaps in rather vague or broadly qualitative terms, of how the system is expected to behave. We specify this idea as a *statistical* (or *stochastic*) *model*, which is a more or less precise expression, in quantitative terms, of this expected behaviour. Any such model is specified in terms of probability distributions, involving a number of (unknown) *parameters*, that permit flexibility in the situations to which the model can be applied. The model is then *fitted* to the data set under consideration using a general principle (such as maximum likelihood or least squares). The fitting process entails the estimation of the unknown parameters in such a way as to give maximum agreement between the observed data values and their corresponding predictions from the model. The systematic features of the data are then described by some or all of the estimated parameters, and these components enable the data to be analysed. For example, a single data set is analysed by interpreting the estimated coefficients, the differences between two data sets are analysed by comparing the two sets of estimated coefficients, and so on.

Note, therefore, that we now have two types of model: the geometrical model underlying data display, and the statistical model underlying data analysis. The former was of major importance throughout Part I (although the latter was briefly mentioned in Section 2.2.5, in the context of selection of number of principal components to describe a data set). However, from now on the statistical model becomes the focus of attention while the geometrical model passes more into the background. Thus when we refer to 'the model' we shall mean the statistical model, unless the contrary is explicitly stated.

What, therefore, are the essential ingredients of any model for the analysis of a data set? While many individual data sets exhibit features peculiar to themselves alone, requiring some special model formulation, it is nevertheless true that two distinct components can be identified in most models put forward for the analysis of data: the *systematic* component and the *random* component.

The former is usually the component that quantifies the analyst's belief about the expected *general* behaviour of the system. This is the part of the model most likely to involve parameters that have to be estimated from the data to hand. It may, for example, express the supposed dependencies among a set of measured variables, or the difference in average response to a certain stimulus between two distinct groups of individuals, or the effects of different combinations of treatments (such as fertilizers) on the harvested grain in different localities. All such systematic components express the behaviour of an 'ideal' subject or individual.

The individuals that are represented in any data set, however, have usually been selected from a (large) set of possible individuals. In fact, we typically record the values of the chosen attributes or responses on a *sample* of individuals from a *population,* any of whose members are available for selection to the sample. Thus if we have measured the heights and weights of a set of fifty undergraduate students, this is a sample of size 50 from the population of all undergraduate students in the university (or state, or country, etc.). If we have obtained some bone measurements on a set of two-hundred wild fowl, this is a sample of size 200 from the population of all wild fowl in the area under study, and so on. The methods available for selecting the individuals for the sample from the population are numerous, and will not be detailed here (see, e.g., Barnett (1974) for a full description); however, a simple way of ensuring that the sample chosen is representative of the population under study is to make it a *random* sample. This is a sample chosen in such a way that every individual in the population has an equal chance of selection. Selecting a random sample enables the sample individuals to be assumed to be uncorrelated, which is a necessary assumption for most methods of analysis. We will henceforth suppose that all samples are of this type; the reader is referred to Barnett (1974) for the mechanistic details.

Now, all individuals exhibit a certain amount of idiosyncratic behaviour, and the values recorded for a sample of individuals on any variable will show a certain amount of fluctuation. This fluctuation will be more or less prominent according to the variable being measured. For example, a human subject's heart rate can vary markedly from individual to individual, and individuals can also vary considerably in their heights and weights, but a person's body temperature will be remarkably consistent among (healthy) subjects, varying at most by some tenths of a degree. We have learnt to cope with such variability, as for example in popular medicine, by quoting an 'average' height, weight, heart rate, or body temperature (intended to represent the corresponding variable value for a 'typical' subject of a given category such as male, female, child), together with the range of possible variation in values for the given population. These ideas carry over to *every* population that is studied, whether animate or not, and must therefore be reflected in the common elements of a model. If the systematic component describes the 'average' (or 'ideal') response expected from an individual in the population, then the *random* component describes the variation in responses that might occur from individual to individual.

Different levels of complexity can be assumed in the definition of the random component; the greater the complexity, the more detail will usually be built into the model. At the simplest level, we might merely assume that individual realizations of the random component are independent observations of a random variable that has some unspecified distribution. At the most complex level, we would assume that the joint distribution of all realizations of the random component was described by some fully specified multivariate distribution. The former assumption would simply imply that individual observations fluctuated in some way about the systematic part of the model; the latter assumption imposes far more structure upon the form of this fluctuation.

The final ingredient, before the model is fully specified, concerns the way in which the random and systematic components are combined. Most simple models are either *additive* or *multiplicative,* i.e. the random and systematic components are either added together or multiplied together. In recent years a class of models called *generalized linear models* has been popularized, in which a more complex mode of combination is possible. In this class of models it is assumed that the random component is a random variable belonging to a certain family of distributions, the exponential family. The parameter of this family constitutes the systematic part of the model and may perhaps be written as a linear combination ('predictor') of some concomitant variables. This parameter can be expressed as some function of the mean of the random variable, and this function is known as the *link function.* Specification of the model thus requires specification of *three* components: (1) the linear predictor (the systematic part), (2) the appropriate member of the exponential family (the random part), and (3) the link function. However, we shall not need to have recourse to this more complex class of models. For most models in multivariate analysis, it is sufficient to assume simple additivity, with perhaps the extra distributional assumption of multivariate normality (see Chapter 7) for the random component. Estimation of parameters in such models is then effected by the use of either *least squares* (in the absence of specific distributional assumptions) or *maximum likelihood* (in the presence of specific distributional assumptions); detailed discussions of these general principles of estimation can be found in most books on statistical inference.

Our chief interest in the remainder of this book will be on models for multivariate data analysis, based mainly on the multivariate normal distribution, as such models underlie the majority of multivariate techniques. In any case, some specific distribution must be assumed for the data if statistical inference is contemplated, so the emphasis of multivariate model description can be expected to be directed towards the distribution-specific models. We thus need to establish first some fundamental ideas about multivariate populations; this will be done in Section 6.3 of this chapter. Chapter 7 will then be devoted to a description of the multivariate normal distribution and its properties.

Before proceeding to these distributional ideas, however, we shall build a small bridge connecting simple data display and model-based data analysis. We shall describe a technique that aims primarily at data display, but, because of the dimensional complexity of the input data, this objective can only be achieved by postulating a particular model and fitting it by least squares. Here the purpose of the model is purely to reduce the problem to a manageable size, and distributional assumptions are not invoked at all.

6.2 Three-way scaling: individual differences analysis

The model to be discussed here had its origins in psychometric work, and is thus most naturally described within this framework (although its potential applicability is much wider and it is now finding its way into such diverse areas as Food Studies, Geography, and Sociology). The technique is appropriate whenever the starting data are in the form of n separate $(k \times k)$ dissimilarity (or similarity) matrices, with each matrix giving the similarities or dissimilarities between every pair obtained from the *same* set of entities. Thus the situation for which the technique was first envisaged is one in which n different individuals are each asked to estimate the dissimilarity between every pair out of a set of k stimuli, and the *same* stimuli are presented to each individual. A typical practical example would be sensory evaluation, where n different tasters were asked to compare the same k soft drinks in a trial in which every possible pair of drinks was presented for comparison to each taster. Equivalent data could be generated if each taster were asked to rate each soft drink by assigning a score between, say, 0 and 10 on each of p characteristics of a drink (e.g. sweetness, colour, aftertaste), and then dissimilarities between all pairs of drinks were calculated for each taster using one of the dissimilarity measures described in Section 1.4. It is in fact immaterial how the raw data are treated, as long as they can be reduced to n separate $(k \times k)$ dissimilarity matrices with the same entities indexing rows and columns of all matrices. It is therefore also evident that situations treated previously under procrustes analysis in Chapter 5 come within this framework. For example, we might have n dissimilarity matrices among the same k entities obtained by subjecting a single $(k \times p)$ data matrix to n different dissimilarity measures; or the p variables observed on each of k individuals may be split into n subsets, and a single dissimilarity measure applied to each. It is then of interest not only to examine relationships between the k stimuli, but also to determine what differences, if any, exist between the n separate sets of such relationships. Reverting to the psychometric application, this latter objective is equivalent to asking what differences exist in the n individual perceptions of the k stimuli; hence the analysis of such data is usually termed *individual differences analysis*.

Now, such data can obviously be inspected using some of the techniques already described. The earliest approach to such inspection was to average the n dissimilarity matrices and then to conduct a multidimensional scaling (metric or non-metric) of this single average dissimilarity matrix. Resulting output yields a representation of the relationships that exist on average between the k stimuli, over the set of n individuals. The disadvantage of such an approach, however, is that the differences between individual dissimilarity matrices are essentially attributed to random error. In practice this would not generally be realistic, as pronounced individual differences are to be expected between subjects, and a more useful approach would be one which allowed for (and hence displayed) such systematic differences. The most complete description of such differences can be obtained by conducting a separate multidimensional scaling of each $(k \times k)$ dissimilarity matrix. This gives n pictorial representations, each one indicating one particular subject's perception of the k stimuli. Individual differences are then highlighted by differences between respective pictures, and can be quantified by procrustes analyses as described in Chapter 5. Thus one could conduct sets of pairwise comparisons of individual representations, followed by a multidimensional scaling of the $M^2_{\min}$ values (as in the example in Section 5.1) to yield a pictorial representation of the differences between the individuals. Alternatively one could conduct a single generalized procrustes analysis on the n stimulus representations, as described in Section 5.2. It could be argued, however, that just as multidimensional scaling of a single average dissimilarity matrix *loses* too much information, so n separate multidimensional scalings followed by procrustes analysis attempts to *provide* too much information. Being a descriptive approach, it will be too prone to local fluctuations in the individual perceptions. As was argued in the previous section, a useful *analysis* will be provided if a realistic model can be postulated that is sufficient to describe the individual differences but that also permits the more irregular local fluctuations to be smoothed out.

Such a model was proposed by Carroll and Chang (1970), who called the resulting analysis Individual Differences Scaling; this has become popularized under the acronym INDSCAL, used by Carroll and Chang for their original computer program implementation of the technique. This model relates structures from different sources in a strong way, but also permits large differences to exist among them. Suppose that S_i denotes the space in which the k stimuli can be represented by points in such a way that inter-point distances match as closely as possible the inter-stimulus dissimilarities as perceived by the ith individual. The basic premise of the INDSCAL model is that the n spaces $S_1, S_2, \ldots, S_n$ have a common set of dimensions but that these dimensions are differentially important for different individuals. In other words, there exists a fundamental space S (called the 'group stimulus space') with a set of rectangular axes $OX_1, OX_2, \ldots, OX_p$, and the space S_i for each individual can be obtained from S by weighting each of these axes by an appropriate amount. Let $\sqrt{w_{ij}}$ be the weight assigned by the ith individual to the jth axis OX_j; this reflects the 'importance' attached by that individual to this particular dimension in his/her perception of the k stimuli. As a special case, of course, some of the weights may be zero; if $w_{ij} = 0$ then the jth 'group' dimension does not enter into the ith individual's space S_i. The group stimulus space may thus be viewed as the space in which the k stimuli can be represented as points in such a way that inter-point distance reflects the 'group' or 'consensus' view of the corresponding inter-stimulus dissimilarity. Given this group stimulus space and the complete set of w_{ij}, it is then possible to generate immediately the stimulus configurations in each subject's private space S_i. This configuration is such that the inter-point distance reflects the corresponding inter-stimulus configuration by multiplying coordinates of all points on the jth axis by $\sqrt{w_{ij}}$. If $w_{ij} = 1$, therefore, there is no change to these coordinates; in this case the jth axis for subject i is exactly the same as the jth axis of the group stimulus space. If $w_{ij} < 1$, then the ith subject contracts the jth axis when perceiving differences between the stimuli, while if $w_{ij} > 1$ then the ith subject expands the jth axis. Another comparison that can be made using the INDSCAL model is thus between the w_{ij} for different subjects. In fact, if we let $\mathbf{w}_i' = (w_{i1}, w_{i2}, \ldots, w_{ip})$, then $\mathbf{w}_i$ can be represented as a point in a p-dimensional space with coordinates given by the complete set of squared weights for the ith subject. This can thus be termed a 'subject space' W; two subjects who assign very similar sets of weights to the p axes will be represented by two points that are close together in this space, while two distant points indicate that the corresponding subjects differ markedly in their assignment of weights.

To illustrate these concepts, consider the hypothetical two-dimensional example of Fig. 6.1, reproduced from Wish and Carroll (1971) with permission from Edinburgh University Press. Part (a) shows the group stimulus space in which nine stimuli are represented by points, while part (b) shows the subject space for nine subjects. Subjects 1, 2, and 6 have higher weights on dimension 1 than on dimension 2, their points falling below the 45° line emanating from the origin. Thus their private spaces will stretch dimension 1 relative to dimension 2, as is illustrated for subject 2 in part (c). Subjects 4, 5, and 8, on the other hand, lie above

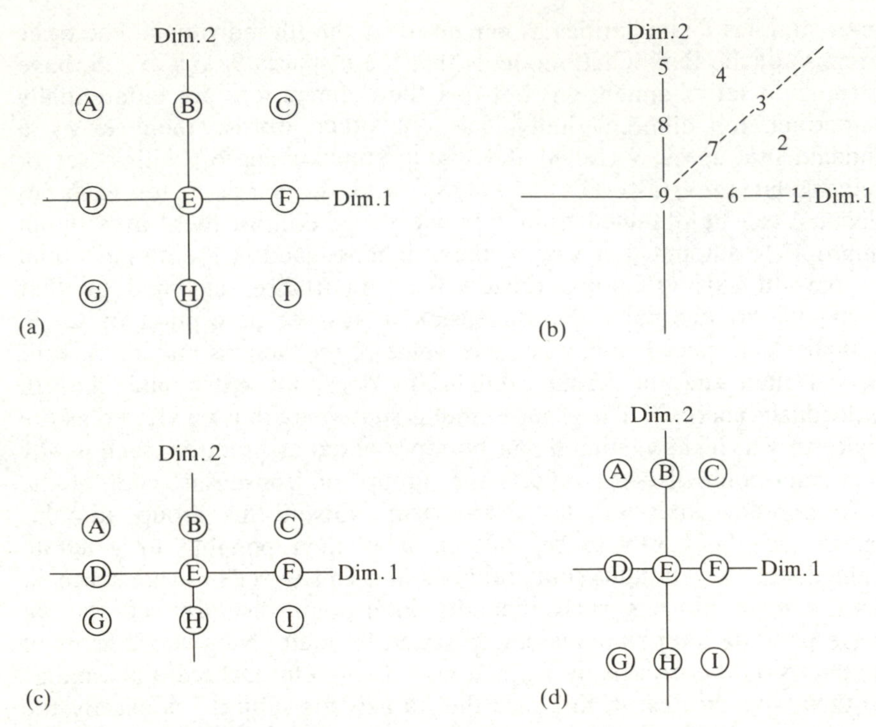

Fig. 6.1 Hypothetical two-dimensional INDSCAL configuration; nine subjects, each comparing nine stimuli. (a) Group stimulus space. (b) Subject space. (c) Private space for subject 2. (d) Private space for subject 4. Reproduced from Wish and Carroll (1971) with permission from Edinburgh University Press

the 45° line so have higher weights on dimension 2 than dimension 1; their private spaces will stretch dimension 2 relative to dimension 1, as illustrated for subject 4 in part (d). Subjects 3 and 7 lie on the 45° line, so have equal weights for dimensions 1 and 2. Their private spaces will thus be similar to the group stimulus space (although of course *both* axes may be equally stretched or shrunk depending on the actual values w_{ij}). Finally, subject 9 has zero weights on both dimensions, so none of this subject's data can be represented in these two dimensions; a higher-dimensional configuration would have to be sought in practice to include this subject satisfactorily. In fact, the further a subject's point is from the origin of the subject space, the better his or her data can be accounted for by the dimensions of the group stimulus space. Thus if Fig. 6.1 represented real data, then from part (b) we could conclude that subjects 1 to 5 provide better representations for the inter-stimulus dissimilarities than do the other subjects.

Having described the underlying concepts of the model, it now remains to consider the method of analysis. We are provided, as input, with n separate dissimilarity matrices between k stimuli. Let δ_{irs} denote the input dissimilarity between the rth and sth stimuli for the ith individual $(r, s = 1, \ldots, k; i = 1, \ldots, n)$. As output, we require the coordinates of the k points in the group stimulus space representing the k stimuli, and the weights w_{ij} for each subject and each dimension. Let us suppose that we require the dimensionality of the group stimulus space to be p, and that x_{rt} is the coordinate of the point representing the rth stimulus on the tth dimension in this space $(r = 1, \ldots, k;\ t = 1, \ldots, p)$. Then the distance d_{rs} between the rth and sth points in the group stimulus space is given by

$$d_{rs}^2 = \sum_{t=1}^{p} (x_{rt} - x_{st})^2 \tag{6.1}$$

and the distance d_{irs} between the rth and sth points in the private space S_i of the ith individual is given by

$$d_{irs}^2 = \sum_{t=1}^{p} w_{it}(x_{rt} - x_{st})^2. \tag{6.2}$$

The objective of an INDSCAL analysis is thus to find the values of x_{rt} and w_{it} $(r = 1, \ldots, k; t = 1, \ldots, p; i = 1, \ldots, n)$ so that d_{irs}^2 matches δ_{irs}^2 as closely as possible for all i, r, and s. Note that in this formulation INDSCAL therefore is a generalization of *metric* scaling.

If we let y_{irt} be the coordinate of the point representing the rth stimulus on the tth dimension of the space S_i, then

$$d_{irs}^2 = \sum_{t=1}^{p} (y_{irt} - y_{ist})^2 \tag{6.3}$$

and a comparison of (6.3) with (6.2) shows that $y_{irt} = x_{rt}\sqrt{w_{it}}$. This reflects the assertion in the introduction to the technique, in which it was stated that coordinates in the private spaces could be obtained from those in the group stimulus space by multiplication by $\sqrt{w_{ij}}$. It also shows that while the square roots of the weights are important in coordinate calculations, the weights themselves are important in *distance* calculations. Hence a plot of the weights, rather than their square roots, is meaningful; this plot yields the subject space W already described.

Standard computer packages such as MDS(X) (see Coxon, 1982) contain routines for performing an INDSCAL analysis, and these routines will output the quantities x_{rt} and w_{it} $(r = 1, \ldots, k;\ t = 1, \ldots, p;$ $i = 1, \ldots, n)$. For completeness, we now give some mathematical and computational details relating to the derivation of these quantities and to the INDSCAL model in general. These details may be omitted by readers primarily interested in the application of the model.

First let us focus on the private space S_i for the ith individual. The input dissimilarities are δ_{irs} $(r, s = 1, \ldots, k)$, and we seek a set of coordinates y_{irt} $(r = 1, \ldots, k;\ t = 1, \ldots, p)$ such that d_{irs}^2 of (6.3) matches δ_{irs}^2 for all r and s. Initially, we can think of this as a single metric scaling problem and use the methods of Section 3.2. Write $\mathbf{Y}_i = (y_{irt})$, the $(k \times p)$ matrix containing the final coordinates of the stimuli. Then if we use eqn (3.12) on the δ_{irs}^2 we obtain

$$q_{irs} = -\tfrac{1}{2}(\delta_{irs}^2 - \delta_{ir.}^2 - \delta_{i.s}^2 + \delta_{i..}^2). \tag{6.4}$$

Write $\mathbf{Q}_i = (q_{irs})$, the $(k \times k)$ symmetric matrix containing q_{irs} as (r, s)th element. Then metric scaling theory shows that the solution $\mathbf{Y}_i$ must satisfy

$$\mathbf{Q}_i = \mathbf{Y}_i\mathbf{Y}_i'. \tag{6.5}$$

This must be true for all $i = 1, \ldots, n$. However, the INDSCAL model imposes the constraint $y_{irt} = x_{rt}\sqrt{w_{it}}$, as derived above. Write $\mathbf{X} = (x_{rt})$, the $(k \times p)$ matrix containing the coordinates of the k stimuli on the p dimensions of the group stimulus space, and $\mathbf{W}_i = \text{diag}(w_{i1}, w_{i2}, \ldots, w_{ip})$, the diagonal matrix containing the ith subject's weights for each axis along the diagonal. Clearly, therefore, $\mathbf{W}_i^{\frac{1}{2}} = \text{diag}(\sqrt{w_{i1}}, \sqrt{w_{i2}}, \ldots, \sqrt{w_{ip}})$ while $y_{irt} = x_{rt}\sqrt{w_{it}}$ for all r, t is equivalent to

$$\mathbf{Y}_i = \mathbf{X}\mathbf{W}_i^{\frac{1}{2}}. \tag{6.6}$$

Substitution of (6.6) into (6.5) thus gives

$$\mathbf{Q}_i = \mathbf{X}\mathbf{W}_i^{\frac{1}{2}}\mathbf{W}_i^{\frac{1}{2}}\mathbf{X}' = \mathbf{X}\mathbf{W}_i\mathbf{X}'. \tag{6.7}$$

The INDSCAL problem is therefore to find, for a given dimensionality p, the group stimulus coordinates $\mathbf{X}$ and individual subject weights $\mathbf{W}_i$ such that $\mathbf{X}\mathbf{W}_i\mathbf{X}$ approximates as closely as possible to $\mathbf{Q}_i$ (derived from the input dissimilarities via (6.4)) for all i $(i = 1, \ldots, n)$. Writing this in element-by-element fashion, we thus require x_{rt} and w_{it} such that q_{irs} is appproximated as closely as possible by $\sum_{t=1}^{p} w_{it}x_{rt}x_{st}$ for all i, r, s $(r, s = 1, \ldots, k;\ i = 1, \ldots, n)$, i.e. such that

$$q_{irs} = \sum_{t=1}^{p} w_{it}x_{rt}x_{st} + \varepsilon_{irs} \tag{6.8}$$

where ε_{irs} is a small 'error'.

A more general version of this problem was treated by Eckart and Young (1936), who sought values a_{it}, b_{rt}, and c_{st} such that a set of known values z_{irs} can be approximated by $\sum_{t=1}^{p} a_{it}b_{rt}c_{st}$, i.e. such that

$$z_{irs} = \sum_{t=1}^{p} a_{it}b_{rt}c_{st} + \varepsilon_{irs}. \tag{6.9}$$

An appropriate set of values a_{it}, b_{rt}, and c_{st} can be found using the principle of least squares, and minimizing $V = \sum_i \sum_r \sum_s \varepsilon_{irs}^2 = \sum_i \sum_r \sum_s \left\{z_{irs} - \sum_{t=1}^{p} a_{it}b_{rt}c_{st}\right\}^2$. This minimization can be achieved by an iterative procedure as follows. Set the b_{rt} and c_{st} to some arbitrary initial values, and find the values of a_{it} minimizing V. Taking these values of the a_{it}, and retaining the initial values of c_{st}, next find the values of b_{rt} minimizing V. Finally, for these computed a_{it} and b_{rt}, find the values of c_{st} minimizing V. This constitutes one cycle of the procedure. This whole cycle is repeated, improving the estimate of one set of parameters using the values of the other sets of parameters obtained in the previous iteration, until a complete cycle is performed with no change in any of the parameters (to within a pre-set level of accuracy). The final set of values constitutes the least-squares solution. The INDSCAL problem as expressed by (6.8) is a special case of this procedure, with $z_{irs} = q_{irs}$, $a_{it} = w_{it}$, and $b_{rt} = c_{rt} = x_{rt}$, but if the general procedure were applied without constraint, two separate and possibly unequal $\mathbf{X}$ matrices would be obtained. In practice, due to the basic symmetry of the problem, these two matrices will normally be very similar to each other. However, a final cycle is commonly added to the general procedure for INDSCAL. The $\mathbf{X}$ matrix computed last, after convergence, is taken to be the correct one and the other one is set equal to it; the $\mathbf{W}$ matrix is then recomputed using this joint value of $\mathbf{X}$.

The solution as outlined above is one which minimizes $V = \sum_i \sum_r \sum_s \varepsilon_{irs}^2$, where ε_{irs} is defined by (6.8). Now if we set $\sum_r \sum_s \varepsilon_{irs}^2 = E_i^2$, the 'residual sum of squares' due to the ith individual, then $V = \sum_i E_i^2$. If the δ_{irs}^2 are of differing magnitudes for each individual i, then the solution will favour those individuals with larger δ_{irs}^2 values in the sense that these individuals will be fitted more accurately. This is because otherwise these individuals will contribute large values of E_i^2 to V. In order to give each individual equal weight in the analysis and equalize the influence of all E_i^2 the matrix $\mathbf{Q}_i$ is customarily normalized at the outset of the analysis by rescaling all q_{irs} so that $\sum_r \sum_s q_{irs}^2 = 1$ for each i. Similarly, once a solution has been obtained, it is open to arbitrary translation and scaling. It is customary to ensure that the origin of the group stimulus space is at the centroid of the configuration, and the sum of squared projections of each point on each dimension is unity. This normalization leads to the property mentioned in the introductory description of the technique, that the square of the distance of an individual from the origin of the subject space reflects the goodness-of-fit of the model for that individual. This is because this squared distance measures $(1 - E_i^2)$ for the ith individual.

An important point worth noting is that, despite INDSCAL being a metric technique, it does not have the metric scaling property that the best r-dimensional representation is given by the first r dimensions of the best $s(>r)$-dimensional representation. Rather, it has the feature observed in non-metric scaling that the whole fitting process has to be repeated whenever the dimensionality of the fitted configuration is changed. This is because the process relies on an iterative computational scheme, rather than on exact algebraic solution.

An excellent illustration of a successful INDSCAL analysis has been

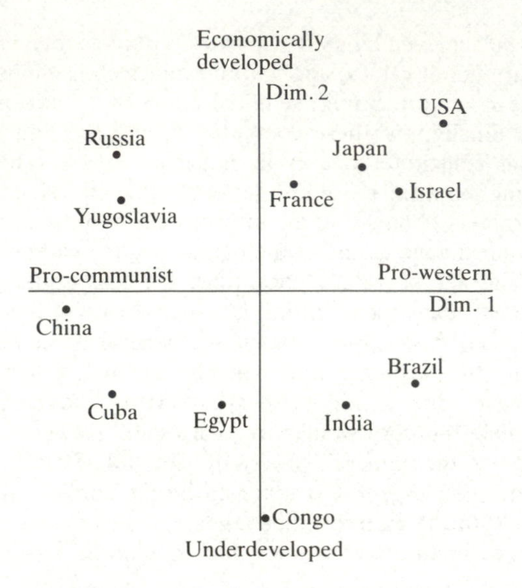

Fig. 6.2 Group stimulus space for similarities between nations. Reproduced from Wish and Carroll (1982) with permission from North Holland Publishing Company

provided by Wish and Carroll (1982). The data relate to students' perceptions of similarity among nations, and are described in Wish *et al.* (1970). The eighteen students involved made pairwise ratings of similarities for a set of twelve nations, and also gave their views about US policy in Vietnam. The set of eighteen 12×12 matrices was then analysed by INDSCAL, and the results are given in Figs 6.2–6.5 (reproduced from Wish and Carroll (1982) with permission from North Holland Publishing Company). The (two-dimensional) group stimulus space is given in Fig. 6.2, the subject space in Fig. 6.3, and two of the private spaces in Figs 6.4 and 6.5.

It has been found empirically that the axes of an unrotated INDSCAL solution can often be interpreted directly, and so it proved in the present example. The major axis of Fig. 6.2 was interpreted as measuring 'political alignment', with pro-western nations featuring at one extreme and pro-communist nations at the other, while 'economic development' constituted the minor axis interpretation, developed countries being marked positive on this axis and underdeveloped countries negative. The subject space, given in Fig. 6.3, revealed a further interesting feature. All those students whose views about US policy in Vietnam labelled them as 'Doves' (D) fell *above* the 45° line in this space, while all those whose views labelled them as 'Hawks' (H) fell *below* the 45° line. The 'Doves'

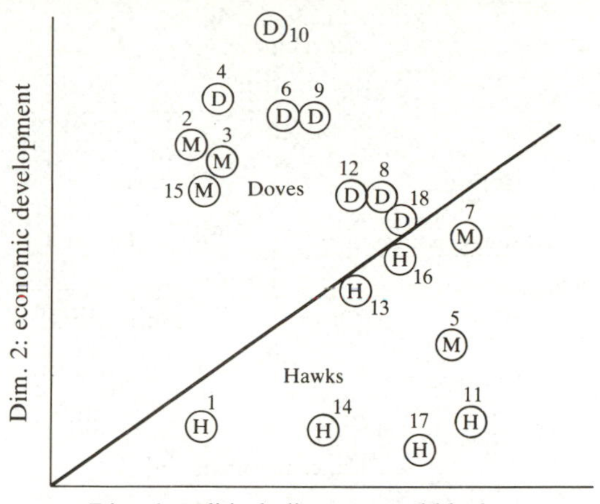

Fig. 6.3 Subject space for similarities between nations. Reproduced from Wish and Carroll (1982) with permission from North Holland Publishing Company

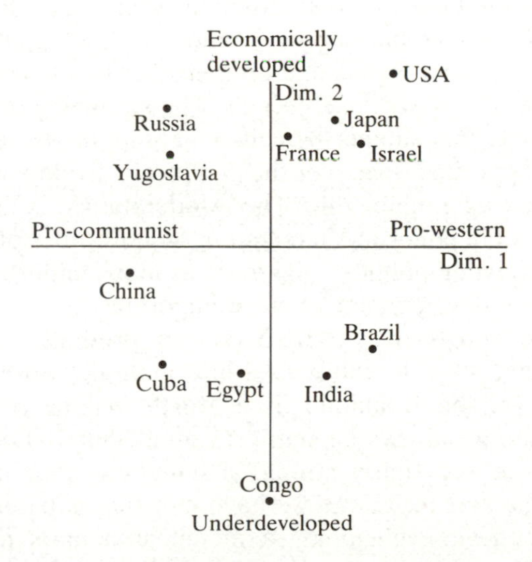

Fig. 6.4 Private space (individual No. 4) for similarities between nations. Reproduced from Wish and Carroll (1982) with permission from North Holland Publishing Company

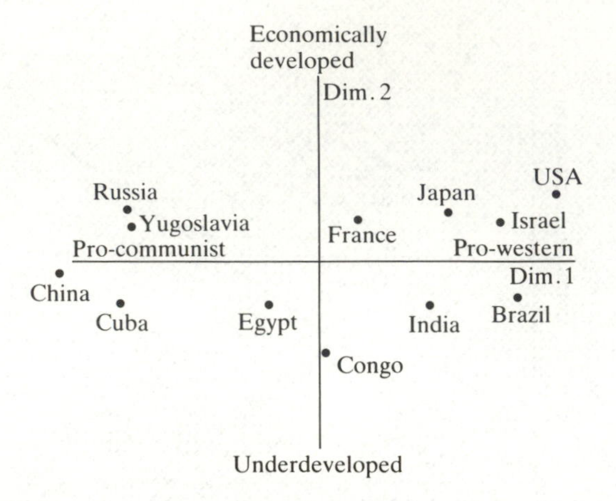

Fig. 6.5 Private space (individual No. 17) for similarities between nations. Reproduced from Wish and Carroll (1982) with permission from North Holland Publishing Company

therefore place greater emphasis on axis 2 (economic development) in their perception of similarities between nations, while the 'Hawks' place greater emphasis on axis 1 (political alignment) in their perception of similarities between nations. Figure 6.4 shows the private space for one of the 'Doves', subject 4. The stretching of the vertical axis shows that this subject views nations as more dissimilar when they differ markedly in terms of their economic development, and that their political alignment has much less influence. Figure 6.5, on the other hand, shows the private space for one of the 'Hawks', subject 17. The stretching of the horizontal axis indicates that this subject perceives nations to be very different if they fall on opposite sides of the political divide, with economic development playing a minor role. The 'Moderate' (M) subjects in regard to their views on US policy in Vietnam fall on both sides of the 45° line in Fig. 6.3, some viewing political alignment as more important, and others viewing economic development as more important.

This example shows that INDSCAL can produce very convincing analyses that relate well to extraneous information. However, in closing this discussion on the technique, it is worth making some cautionary comments. It should always be borne in mind that INDSCAL depends for its success on the (fairly strong) assumptions that it makes being satisfied. It is the first technique we have met that is based on a definite model, and whenever we employ a model as a basis for analysis we should question carefully its applicability to the data in hand. The main assumption in INDSCAL is that the same fundamental dimensions govern all subjects' perceptions of the set of stimuli, with just the weight allocated to each dimension differing between subjects. In the case of the nations example above, this is indeed a plausible assumption; political alignment is clearly a major consideration (particularly among American subjects!), and economic development is also a recognizable factor that would be at the forefront of most people's minds in judging such similarities. Whether the INDSCAL assumption would be as defensible in, say, analysing different people's perception of the similarities between various wines (particularly if the subjects included both connoisseurs and novices), or in analysing the effect of different similarity measures on a set of taxonomic data, is certainly open to question.

Now, while INDSCAL is based on a model, this model could be described as deterministic or descriptive rather than statistical or stochastic, in the sense that no distributional assumptions are made about the 'errors' ε_{irs} of (6.8) or (6.9). Consequently, no objective assessment can be made as to whether the overall model fits a set of data or not; the best one can do is to weigh up the results and make subjective judgements about the success or otherwise of the representations (based on whether or not the interpretation of the configurations is sensible). INDSCAL can therefore be viewed as closer in spirit to the constrained description of other graphical summaries than to a true model-based analysis: it is essentially a descriptive technique, but the model imposes sufficient constraints to enable a concise output to be obtained. The role of the least-squares equations is in fitting the parameters optimally under these constraints (which can be contrasted with the role of least squares in, say, non-metric scaling, where the constraint arose because exact matching of dissimilarities and distances was not possible in the given number of dimensions). We will go on in the next section to a consideration of the imposition of *distributional* assumptions in a model. This will open up a whole new vista of possibilities in both model formulation and model validation. To conclude the present section, however, we may note that there are currently many variants of the INDSCAL model in existence, in both metric and non-metric versions. For example, the ALSCAL program mentioned in Section 3.3 permits individual difference analysis. Space considerations preclude any further discussion of these models, however; the interested reader is referred to Wish and Carroll (1982) for a brief survey and for further references.

6.3 Models for multivariate populations

In the first part of this chapter, it was emphasized that most statistical models consist of some combination of systematic and random components. The systematic part of the model specifies the behaviour that is to be expected 'on average' of the variable or variables under observation, while the random part of the model is concerned with the fluctuation to be expected (about this average) in the individuals on which the variables have been measured. It thus provides a means of describing fairly closely the population from which the sample has been drawn; we now focus on this random component.

Any multivariate data set is obtained as a consequence of measuring the values of p different variables on a set of individuals. Let us denote these variables by $x_1, x_2, \ldots, x_p$. We can put these variables together into a (column) vector, and refer to them collectively as the (random) vector $\boldsymbol{x}$. Thus $\boldsymbol{x}' = (x_1, x_2, \ldots, x_p)$. If we now observe this vector on a set of n individuals, we can denote the values observed on the ith individual by $\boldsymbol{x}_i' = (x_{i1}, x_{i2}, \ldots, x_{ip})$. The observation x_{ij} is thus recognized as the (i, j)th element of the data matrix $\mathbf{X}$; hence this data matrix $\mathbf{X}$ can also be written as $\mathbf{X} = (\boldsymbol{x}_1, \boldsymbol{x}_2, \ldots, \boldsymbol{x}_n)'$ or

$$\mathbf{X} = \begin{pmatrix} \boldsymbol{x}_1' \\ \boldsymbol{x}_2' \\ \vdots \\ \boldsymbol{x}_n' \end{pmatrix}.$$

Note that here we have denoted by lower-case letters both the random vector ($\boldsymbol{x}$) and the values observed for this vector ($\boldsymbol{x}_i$). A common convention is to denote random vectors by upper-case letters and their values by lower-case letters; this convention was in fact used in Chapter 2. However, if we were to adopt this convention more generally then we would find that $\mathbf{X}$ denoted both a random vector and a data matrix, and this would in turn become a source of confusion. Since a statistical sample is often thought of as a collection of independent identically distributed variables, it seems preferable in the remainder of this book to use the lower-case letter as the symbol for the random vector.

Suppose, therefore, that a sample of n individuals has yielded the values $\boldsymbol{x}_1, \boldsymbol{x}_2, \ldots, \boldsymbol{x}_n$ of a random vector $\boldsymbol{x}$, and that these values have been written as the data matrix $\mathbf{X}$. The sample will often have been collected from some *target population* of interest, and it is the features of this population that we wish to describe. Populations can in general be one of two kinds: finite or infinite. An example of the former would be the population comprising all members of the electoral roll of a particular city; while an example of the latter would be the population of all possible blood samples taken from cows during lactation. Valid application of statistical distribution theory requires the samples to have been taken from infinite populations, but in practice the theory is often also applied to finite populations providing that they are sufficiently large.

If each of the variables in $\boldsymbol{x}$ is discrete, then each has only a countable number of possible values. Hence there is only a countable number (say s) of p-tuples $(x_1, x_2, \ldots, x_p)$ that can be formed from the p responses observed on each sample member. The relative frequency f_i/n with which the ith such p-tuple occurs in the sample can thus be evaluated. Now consider making the sample size n larger and larger. In the limit, the sample thus approaches the whole population, and, by the relative frequency interpretation of probability, f_i/n tends to the limiting value p_i. This is the probability that the ith p-tuple will be observed when an individual $\boldsymbol{x}$ is selected at random from the population. A probability model for a discrete random vector $\boldsymbol{x}$ can thus be specified by enumerating the possible p-tuples that $\boldsymbol{x}$ can assume, together with their associated probabilities p_i. This is known as the *probability distribution* of $\boldsymbol{x}$, and is just the familiar definition of the probability distribution of a discrete univariate random variable x but applied to the random vector $\boldsymbol{x}$.

Now suppose that the variables in $\boldsymbol{x}$ are all continuous. We have already seen in Chapter 2 (and subsequently) how a sample of n observations on $\boldsymbol{x}$ can be conceptualized as n points in p-dimensional space. Again imagine increasing the sample size until, in the limit, all members of the population are included. Each member of this sample can now assume one of an infinite set of p-tuples $(x_1, x_2, \ldots, x_p)$, so the population will be represented by a dense swarm of points, infinite in number, in p dimensions.

Each axis of this representation can be divided into intervals, thus partitioning the whole p-dimensional space into a grid or mesh, and the relative frequency of occurrence of sample values in each sub-area of this mesh can be computed. These relative frequencies may be thought of as values in a further dimension, the $(p+1)$th. When the sample size is small the mesh has to be coarse to allow reasonable computation of relative frequencies in each sub-area, and the values in the $(p+1)$th dimension are discontinuous. As the sample size is increased, however, the mesh can be made increasingly finer without decreasing the information content in each sub-area, and the values in the $(p+1)$th dimension become increasingly smoother. In the limit as the sample size becomes infinite, therefore, and the whole population is represented, the relative frequencies generate a surface in the $(p+1)$th dimension. Corresponding to each possible value of the random vector $\boldsymbol{x} = (x_1, x_2, \ldots, x_p)'$ is a point in p dimensions and a value on this surface. Denote by $\mathrm{f}(\boldsymbol{x})$ or $\mathrm{f}(x_1, \ldots, x_p)$ the value on the surface corresponding to the point $\boldsymbol{x}$. If $\boldsymbol{x}$ is surrounded by an infinitesimally small volume $\mathrm{d}V$ in p dimensions, then $\mathrm{f}(\boldsymbol{x})$ represents the relative frequency with which observations falling in this volume occur in the population. The function $\mathrm{f}(\boldsymbol{x})$ or $\mathrm{f}(x_1, \ldots, x_p)$ is thus the p-variate *probability density function* for the population.

Clearly, therefore, $f(\boldsymbol{x}) \geqslant 0$ at all points $\boldsymbol{x}$. Note that the case of discrete $\boldsymbol{x}$ can be subsumed within this general framework, since each observation of $\boldsymbol{x}$ is also representable by a point in p dimensions. However, now only a finite number s of points in this space are allowed as realizations of $\boldsymbol{x}$. Hence the population is represented by s multiply repeated points, and in place of a probability density surface we have simply s 'spikes' of probability $p_1, p_2, \ldots, p_s$.

Since any observation on $\boldsymbol{x}$ constitutes a point in p-dimensional space, it is natural to ask what the probability is that a randomly chosen individual from the population will be represented by a point which falls in a given *region* R of this space. This probability can be denoted by $\text{pr}(\boldsymbol{x} \in R)$, and clearly the required value will be given by the sum of the probabilities of all possible points in R. If $\boldsymbol{x}$ is discrete then simple addition gives $\text{pr}(\boldsymbol{x} \in R) = \sum_{i \in R} p_i$. If $\boldsymbol{x}$ is continuous, however, then we must appeal to the mathematical definition of integration and we obtain $\text{pr}(\boldsymbol{x} \in R) = \int_R f(x_1, x_2, \ldots, x_p)\, dx_1 dx_2 \ldots dx_p$, often written in shorthand as $\text{pr}(\boldsymbol{x} \in R) = \int_R f(\boldsymbol{x})\, d\boldsymbol{x}$. If R is the whole space, therefore, we must have $\int_R f(\boldsymbol{x})\, d\boldsymbol{x} = 1$ or $\sum_{i \in R} p_i = 1$.

Let us now concentrate on the case of continuous $\boldsymbol{x}$, as this will be the main concern in the remainder of this book. In view of the definition and explanation of $f(\boldsymbol{x})$ given above, the 'height' of the surface $f(\boldsymbol{x})$ in the $(p+1)$th dimension is clearly a measure of the 'denseness' of points in the small volume dV around $\boldsymbol{x}$ in the original p dimensions (since the denser the points in a region, the more likely is an observation from that region to be selected at random from the population). If p-dimensional 'slices' are taken through this surface at progressively lower positions along the $(p+1)$th axis, they will map out increasingly broader contours in p dimensions. These contours mark out areas of progressively lower density of points in the population and are termed *contours of equal probability*. What shape might we expect these contours to have? This will depend on the associations or correlations that exist between the variables $x_1, x_2, \ldots, x_p$. Since *all* variables are measured on each sample or population member, it is reasonable to anticipate associations existing among them. Most readers will be familiar with the usual diagrammatic representation of correlation. If two variables have been measured on each of n sample members, and these sample members are plotted as a scatter diagram in two dimensions, then increasing correlation is evidenced by increasing tendency of the points to lie on a straight line. A sample in which there is no discernible correlation between the variables is characterized by a roughly circular scatter diagram in which no straight line fits the points better than any other straight line. A sample in which there is perfect correlation between the variables (whether positive or negative) is characterized by a scatter diagram in which all points lie exactly on a straight line. The nearer the points are to lying on a straight line, the higher is the (absolute value of the) correlation between the variables.

Now let us consider all possible two-dimensional subspaces of the p-dimensional space formed by taking two coordinate axes at a time, and attempt to reconstruct the overall p-dimensional scatter from such fragmentary pictures. A population in which there are no correlations between any of the variables, will appear as a roughly circular blob of points on all possible such two-dimensional subspaces, and hence will have the appearance of a hyperspherical swarm of points in p dimensions. Introducing correlations between pairs of variables has the effect of 'flattening' the circular scatter in the corresponding subspaces, thus making the swarm of points ellipsoidal in shape in these subspaces. Many continuous populations encountered in practice are roughly symmetrical about a central value for each observed variable, and have non-zero correlations between pairs of variables. Such populations can thus be conceptualized by swarms of points which form hyperellipsoids in p dimensions. The 'narrower' these hyperellipsoids are in particular projections on to two coordinate axes, the greater are the correlations between the corresponding pairs of variables. In the limit, when all pairs of variables are perfectly correlated, the population is represented by points which lie on a single line in p dimensions. Increasing correlation between variables is thus characterized by a reduction in size of some of the principal axes of the hyperellipsoid. The degenerate case when all variables are perfectly correlated is clearly a limiting case of a hyperellipsoid in which all principal axes bar one have zero size. These principal axes will be considered further in the next chapter.

In order to specify the random component of a model, it is necessary to specify a model for the fluctuations to be expected in the observations from individual to individual when sampling from a given population. This is often abbreviated to saying that it is necessary to specify a model for the population, and the most obvious way of doing this is to specify a probability density function for the variable(s) under consideration. Various standard mathematical functions exist, any of which can be chosen as a possible probability density function. Either the general characteristics of the sample that has been obtained, or the nature of the variable(s) being measured, can be used to guide the choice of appropriate function. The reader will be aware of the common univariate density functions such as the normal, exponential, gamma, beta, t, χ^2, and F distributions for continuous variables, and the Bernoulli, binomial, Poisson, geometric, and hypergeometric distributions for discrete variables. A summary of the characteristics and chief properties of these distributions can be found in most texts covering first courses in statistics (see, e.g., Mood *et al.* 1974). Multivariate generalizations exist for most of these univariate distributions; a full description may be found in Johnson and Kotz (1969, 1972). However, in practice the use of many of these distributions is severely restricted because of mathematical complexity, and the development of multivariate analysis has been built almost exclusively on the multivariate normal distribution as basis. There are probably two main reasons for this. The first is mathematical simplicity: the multivariate normal distribution has many attractive properties, which make it particularly easy to use in practice. The second is the existence of a multivariate central limit theorem. This states in essence that if a random vector $\boldsymbol{x}$ is composed of the sum of n independent and identically distributed components $\boldsymbol{x}_i$ (i.e. $\boldsymbol{x} = \boldsymbol{x}_1 + \boldsymbol{x}_2 + \ldots + \boldsymbol{x}_n$), then the distribution of $\boldsymbol{x}$ tends to the multivariate normal distribution as n becomes large irrespective of the distribution of the individual $\boldsymbol{x}_i$. Now, many variates that are observed in practice are measurements that can be considered to be made up of such additive components, and hence the multivariate normal distribution provides a natural model for such populations.

We shall be devoting most of our attention to multivariate techniques that use the multivariate normal distribution as the random component of the model. Accordingly, we next give a description of the multivariate normal distribution and distributions following directly from it, and discuss their chief properties. We shall first summarize the basic concepts of multivariate distribution theory, assuming that the reader has complete familiarity with the basic concepts of univariate distribution theory.

7
Distribution theory

7.1 Basic concepts of multivariate distributions

A p-dimensional vector random variable, or *random vector*, $\boldsymbol{x}$ is defined to be the vector $\boldsymbol{x}' = (x_1, x_2, \ldots, x_p)$ whose elements x_i are unidimensional random variables. As was indicated in Section 6.3, the rows of a typical data matrix can be viewed as independent realizations of such a random vector. As the majority of techniques to be discussed are appropriate for continuous numerical data, we shall restrict our attention in this chapter mainly to random vectors composed of continuous elements. By extension, an $(m \times p)$ *random matrix* $\mathbf{Y}$ is defined to be the matrix $\mathbf{Y}$ whose elements y_{ij} are unidimensional random variables.

The *joint distribution function* of $\boldsymbol{x}$ is $F(a_1, \ldots, a_p) = \text{pr}(x_1 \leqslant a_1, \ldots, x_p \leqslant a_p)$. A continuous distribution results from a direct assignment of probability in the space of vector values by means of a *density function* $\text{f}(x_1, \ldots, x_p)$. As was indicated in Section 6.3, such a function is non-negative, its integral over the whole p-dimensional space is unity, and the probability that $\boldsymbol{x}$ falls in a region R is the integral of the density function over that region. It thus follows that

$$F(a_1, \ldots, a_p) = \int_{-\infty}^{a_p} \cdots \int_{-\infty}^{a_1} \text{f}(x_1, \ldots, x_p)\,\text{d}x_1 \ldots \text{d}x_p.$$

Conversely, given a distribution function for $\boldsymbol{x}$ that is a continuous function with a sufficiently smooth pth-order mixed partial derivative with respect to the x_i, then the density function is obtained as

$$\text{f}(x_1, \ldots, x_p) = \frac{\partial^p}{\partial x_1 \ldots \partial x_p} \text{F}(x_1, \ldots, x_p).$$

Marginal distributions are distributions of specified elements of $\boldsymbol{x}$ *ignoring the other elements*. Thus the marginal distribution functions are

$$F_{x_i}(u) = \text{pr}(x_i \leqslant u)$$
$$F_{x_i, x_j}(u, v) = \text{pr}(x_i \leqslant u, x_j \leqslant v) \text{ etc.}$$

To obtain these marginal distribution functions from the joint distribution function, we need only set a_k to ∞ in the latter for all elements x_k not included in the marginal distribution. Thus, for example, $F_{x_i,x_j}(u, v) = F(\infty, \ldots, \infty, u, \infty, \ldots, \infty, v, \infty, \ldots, \infty)$, the u and v appearing in the ith and jth positions respectively in $F(.)$. The labels x_i, x_j, etc. in the marginal distribution function notation are generally omitted if no ambiguity results. Marginal density functions are obtained by integrating the joint density function over all those elements that are not required, between $-\infty$ and $+\infty$. Thus $\mathrm{f}(x_i, x_j) = \int_{-\infty}^{+\infty} \ldots \int_{-\infty}^{+\infty} \mathrm{f}(x_1, \ldots, x_p)$ $\mathrm{d}x_1 \ldots \mathrm{d}x_{i-1}\, \mathrm{d}x_{i+1} \ldots \mathrm{d}x_{j-1}\, \mathrm{d}x_{j+1} \ldots \mathrm{d}x_p$.

Conditional distributions are distributions of specified elements of $\boldsymbol{x}$ *for given values of the other elements*. Elementary probability theory tells us that the conditional probability of event A given that event B has occurred is the ratio of the joint probability of both A and B to the marginal probability of B. The definition of a conditional distribution is obtained if we extend this reasoning to the corresponding density functions. If we write $\mathrm{f}(x_1, \ldots, x_k \mid x_{k+1}, \ldots, x_p)$ to denote the conditional probability density of $x_1, \ldots, x_k$ for given values of $x_{k+1}, \ldots, x_p$, then

$$\mathrm{f}(x_1, \ldots, x_k \mid x_{k+1}, \ldots, x_p) = \frac{\mathrm{f}(x_1, \ldots, x_p)}{\mathrm{f}(x_{k+1}, \ldots, x_p)}$$

and $x_{k+1}, \ldots, x_p$ in the resulting expression are treated as the given values (rather than as random variables).

The *expected value* $\boldsymbol{\mu}$ of the random vector $\boldsymbol{x}$ is defined to be the vector of expected values of elements of $\boldsymbol{x}$. Thus $\mathrm{E}(\boldsymbol{x}') = \boldsymbol{\mu}' = \{\mathrm{E}(x_1), \ldots, \mathrm{E}(x_p)\}$. Similarly, the expected value of the random matrix $\mathbf{Y}$ is the matrix of expected values of elements of $\mathbf{Y}$. Thus if $\mathbf{Y} = (y_{ij})$ then $\mathrm{E}(\mathbf{Y}) = (\mathrm{E}(y_{ij}))$.

The *variance–covariance* matrix of $\boldsymbol{x}$ (often called simply the *covariance* matrix of $\boldsymbol{x}$ or the *dispersion* matrix of $\boldsymbol{x}$) is the matrix $\boldsymbol{\Sigma}$ which has elements $\sigma_{ij} = \begin{cases} \mathrm{cov}(x_i, x_j) & (i \neq j) \\ \mathrm{var}(x_i) & (i = j). \end{cases}$ Since $\mathrm{cov}(x_i, x_j) = \mathrm{E}[\{x_i - E(x_i)\}\{x_j - E(x_j)\}]$ and $\mathrm{var}(x_i) = \mathrm{E}[\{x_i - \mathrm{E}(x_i)\}^2]$ then on using the definitions of vector and matrix expectation given above it readily follows that

$$\boldsymbol{\Sigma} = \mathrm{E}[\{\boldsymbol{x} - \mathrm{E}(\boldsymbol{x})\}\{\boldsymbol{x} - \mathrm{E}(\boldsymbol{x})\}'].$$

Given any function g of p variables, the function $\mathrm{g}(x_1, \ldots, x_p)$ of the p elements of the random vector $\boldsymbol{x}$ defines a new random variable, z say, which has density function $\mathrm{f}(z)$. The expected value of z is clearly $\int_{-\infty}^{+\infty} \ldots \int_{-\infty}^{+\infty} \mathrm{g}(u_1, \ldots, u_p)\mathrm{f}(u_1, \ldots, u_p)\, \mathrm{d}u_1 \ldots \mathrm{d}u_p$. The latter integral can be written $\int_{-\infty}^{+\infty} \ldots \int_{-\infty}^{+\infty} \mathrm{g}(\boldsymbol{u})\, \mathrm{f}(\boldsymbol{u})\, \mathrm{d}\boldsymbol{u}$ in shorthand notation.

We can obviously generate any number of such functions of p

variables. In particular, suppose we generate p new variables $z_1, \ldots, z_p$ by

$$z_1 = g_1(x_1, \ldots, x_p)$$
$$\vdots$$
$$z_p = g_p(x_1, \ldots, x_p).$$

The *Jacobian* of this transformation, J, is the determinant of partial derivatives,

$$J = \begin{vmatrix} \dfrac{\partial z_1}{\partial x_1} \cdots \dfrac{\partial z_p}{\partial x_1} \\ \vdots \\ \dfrac{\partial z_1}{\partial x_p} \cdots \dfrac{\partial z_p}{\partial x_p} \end{vmatrix}.$$

If this Jacobian is non-zero in some region R of the $\boldsymbol{x}$-space, there is an inverse transformation defined on the image of R in the $\boldsymbol{z}$-space such that

$$x_1 = G_1(z_1, \ldots, z_p)$$
$$\vdots$$
$$x_p = G_p(z_1, \ldots, z_p)$$

and the Jacobian of this inverse transformation is $1/J = J^{-1}$. To obtain the probability density function of $\boldsymbol{z}$ from that of $\boldsymbol{x}$, we simply need to replace each x_i in the latter by $G_i(z_1, \ldots, z_p)$ and multiply the resulting expression by $|J|^{-1}$. The range of variation of the z_i must also be specified.

In the majority of problems, however, we shall only be concerned with a special type of transformation, and instead of having to derive a probability density function we often only require the moments of the transformed variables. The special type of transformation is the *linear* transformation, in which

$$z_1 = a_{11}x_1 + \ldots + a_{1p}x_p$$
$$\vdots$$
$$z_m = a_{m1}x_1 + \ldots + a_{mp}x_p$$

for arbitrary (constant) coefficients a_{ij}. We can write this transformation succinctly in matrix and vector notation as $\boldsymbol{z} = \mathbf{A}\boldsymbol{x}$, where $\boldsymbol{x}' =$

$(x_1, \ldots, x_p)$; $\boldsymbol{z}' = (z_1, \ldots, z_m)$ and

$$\mathbf{A} = (a_{ij}) = \begin{pmatrix} a_{11} \ldots a_{1p} \\ \vdots \\ a_{m1} \ldots a_{mp} \end{pmatrix}.$$

For such a transformation it is evident that $\dfrac{\partial z_i}{\partial x_j} = a_{ij}$.

If $m = p$ and $\mathbf{A}$ is non-singular, then $\boldsymbol{x} = \mathbf{A}^{-1}\boldsymbol{z}$, $J = |\mathbf{A}|$, and we can obtain the probability density function of $\boldsymbol{z}$. Otherwise, however, the Jacobian vanishes, no inverse transformation exists, and the probability density of $\boldsymbol{z}$ can not be found directly. The *moments* of $\boldsymbol{z}$, however, can always be found directly, even if the distribution of $\boldsymbol{z}$ is difficult to obtain.

Let the mean vectors of $\boldsymbol{x}$ and $\boldsymbol{z}$ be $\boldsymbol{\mu}' = (\mu_1, \ldots, \mu_p)$ and $\boldsymbol{\nu}' = (\nu_1, \ldots, \nu_m)$ respectively, and their variance–covariance matrices be $\boldsymbol{\Sigma} = (\sigma_{ij})$ and $\boldsymbol{\Omega} = (\omega_{ij})$ respectively.

Now

$$\begin{aligned} \mathrm{E}\{z_i\} &= \mathrm{E}\{a_{i1}x_1 + \ldots + a_{ip}x_p\} \\ &= a_{i1}\mathrm{E}(x_1) + \ldots + a_{ip}\mathrm{E}(x_p) \\ &= \sum_{j=1}^{p} a_{ij}\mu_j \\ &= i\text{th element of } \mathbf{A}\boldsymbol{\mu}. \end{aligned}$$

Thus

$$\boldsymbol{\nu} = \mathbf{A}\boldsymbol{\mu}. \tag{7.1}$$

$$\begin{aligned} \mathrm{var}(z_i) &= \mathrm{var}\{a_{i1}x_1 + \ldots + a_{ip}x_p\} \\ &= a_{i1}^2\,\mathrm{var}(x_1) + \ldots + a_{ip}^2\,\mathrm{var}(x_p) + 2a_{i1}a_{i2}\,\mathrm{cov}(x_1, x_2) + \ldots \\ &\quad + 2a_{i,p-1}a_{ip}\,\mathrm{cov}(x_{p-1}, x_p) \\ &= \sum_{j=1}^{p} a_{ij}^2\sigma_{jj} + \sum_{\substack{j=1 \\ j \neq k}}^{p} \sum_{k=1}^{p} a_{ij}a_{ik}\sigma_{jk} \\ &= (i, i)\text{th element of } \mathbf{A\Sigma A}'. \end{aligned} \tag{7.2}$$

$$\begin{aligned} \mathrm{cov}(z_i, z_j) &= \mathrm{cov}\left\{\sum_{k=1}^{p} a_{ik}x_k, \sum_{l=1}^{p} a_{jl}x_l\right\} \\ &= \sum_{k=1}^{p} a_{ik}a_{jk}\sigma_{kk} + \sum_{\substack{k=1 \\ k \neq l}}^{p} \sum_{l=1}^{p} (a_{ik}a_{jl} + a_{il}a_{jk})\sigma_{kl} \\ &= (i, j)\text{th element of } \mathbf{A\Sigma A}'. \end{aligned} \tag{7.3}$$

On combining (7.2) and (7.3), therefore, we obtain

$$\boldsymbol{\Omega} = \mathbf{A}\boldsymbol{\Sigma}\mathbf{A}'. \tag{7.4}$$

Finally, since addition of a constant to a random variable does not affect variances and covariances, we deduce that for the most general linear transformation $\boldsymbol{z} = \mathbf{A}\boldsymbol{x} + \boldsymbol{c}$, where $\boldsymbol{c}$ is a constant vector, we have

$$\boldsymbol{v} = \mathbf{A}\boldsymbol{\mu} + \boldsymbol{c} \tag{7.5}$$

and

$$\boldsymbol{\Omega} = \mathbf{A}\boldsymbol{\Sigma}\mathbf{A}'. \tag{7.6}$$

A common operation on random variables is the process of *standardization,* i.e. reducing them to random variables which all have zero mean and unit variance. If the z_i are standardized versions of the x_i, then $z_i = \dfrac{x_i - \mu_i}{\sqrt{\sigma_{ii}}}$ $(i = 1, \ldots, p)$. Standardization can thus be viewed as a linear transformation, since these p transformations can be written as

$$\boldsymbol{z} = \mathbf{A}\boldsymbol{x} + \boldsymbol{c}$$

where

$$\mathbf{A} = \begin{pmatrix} \frac{1}{\sqrt{\sigma_{11}}} & 0 & \cdots & 0 \\ 0 & \frac{1}{\sqrt{\sigma_{22}}} & & \cdot \\ 0 & \cdot & & \cdot \\ 0 & \cdot & & \cdot \\ \vdots & \cdot & & 0 \\ 0 & \cdots & 0 & \frac{1}{\sqrt{\sigma_{pp}}} \end{pmatrix}$$

and

$$\boldsymbol{c} = \begin{pmatrix} -\mu_1/\sqrt{\sigma_{11}} \\ -\mu_2/\sqrt{\sigma_{22}} \\ \vdots \\ -\mu_p/\sqrt{\sigma_{pp}} \end{pmatrix}.$$

Using (7.5) shows that, as expected, $\boldsymbol{v} = \boldsymbol{0}$, while using (7.6) gives

$$\boldsymbol{\Omega} = \mathbf{A}\boldsymbol{\Sigma}\mathbf{A}' = \begin{pmatrix} 1 & \rho_{12} & \cdots & \rho_{1p} \\ \rho_{21} & 1 & \cdots & \rho_{2p} \\ \vdots & & & \\ \rho_{p1} & \rho_{p2} & \cdots & 1 \end{pmatrix}$$

where $\rho_{ij} = \dfrac{\sigma_{ij}}{\sqrt{(\sigma_{ii}\sigma_{jj})}}$. Thus the covariances of pairs of z_i are simply the correlations of corresponding pairs of x_i. The covariance matrix of $\boldsymbol{z}$ is thus the *correlation* matrix of $\boldsymbol{x}$.

We now turn from these general background concepts to consider some specific distributions that are used frequently in multivariate analysis.

7.2 The multivariate normal distribution

There are various ways of introducing the multivariate normal distribution, some mathematically more rigorous than others. For a fully rigorous definition of the distribution, starting from first principles, the reader must refer to a text on mathematical statistics. We shall content ourselves with the simplest possible introduction that will be sufficient to establish the most important properties for subsequent use.

If $\boldsymbol{u} = (u_1, \ldots, u_m)'$ is distributed such that the u_i are all independent standard normal univariate random variables, then $\boldsymbol{u}$ is said to have a *standardized multivariate normal distribution.* By the definitions of Section 7.1 above, the expected value of $\boldsymbol{u}$ is the zero vector $\boldsymbol{0}$ and the covariance matrix of $\boldsymbol{u}$ is the $(m \times m)$ identity matrix $\mathbf{I}_m$. (Independent u_i implies that covariances between pairs of u_i are zero, while standardization implies that variances of the u_i are all 1.) The distribution of $\boldsymbol{u}$ is denoted by $\mathrm{N}_m(\boldsymbol{0}, \mathbf{I}_m)$, indicating an m-component multivariate normal distribution with mean vector $\boldsymbol{0}$ and covariance matrix $\mathbf{I}_m$. If $\mathbf{A}$ is any $(p \times m)$ matrix of rank $m \leqslant p$, and $\boldsymbol{\mu}$ is any p-vector, then the random vector $\boldsymbol{x} = \boldsymbol{\mu} + \mathbf{A}\boldsymbol{u}$ is said to have a general multivariate normal distribution. Use of (7.5) and (7.6) shows that the expected value of $\boldsymbol{x}$ is $\boldsymbol{\mu} + \mathbf{A}\boldsymbol{0} = \boldsymbol{\mu}$, and the covariance matrix of $\boldsymbol{x}$ is $\mathbf{A}\mathbf{I}_m\mathbf{A}' = \mathbf{A}\mathbf{A}'$. Thus if we write $\boldsymbol{\Sigma} = \mathbf{A}\mathbf{A}'$ then $\boldsymbol{x}$ has a p-component multivariate normal distribution with mean $\boldsymbol{\mu}$ and covariance matrix $\boldsymbol{\Sigma}$, denoted by $\mathrm{N}_p(\boldsymbol{\mu}, \boldsymbol{\Sigma})$. We shall write $\boldsymbol{x} \sim \mathrm{N}_p(\boldsymbol{\mu}, \boldsymbol{\Sigma})$ to indicate that '$\boldsymbol{x}$ has the $\mathrm{N}_p(\boldsymbol{\mu}, \boldsymbol{\Sigma})$ distribution'.

If $m < p$, then $\boldsymbol{\Sigma}$ is not of full rank, and the distribution is degenerate in that all the probability lies in a subspace of dimension m. This situation *can* be handled, but once again requires a journey into the higher realms of mathematics. We shall assume for our development that $m = p$, and that the distribution is well behaved (as it will be in most cases of practical interest). We first derive the probability density function for $\boldsymbol{x}$.

Since $u_1, \ldots, u_p$ are independent standard normal variates, their joint probability density function is

$$\mathrm{f}(\boldsymbol{u}) = (2\pi)^{-p/2} \exp\left(-\tfrac{1}{2}\sum_{i=1}^{p} u_i^2\right) = (2\pi)^{-p/2} \exp(-\tfrac{1}{2}\boldsymbol{u}'\boldsymbol{u}). \qquad (7.7)$$

Now we transform from $\boldsymbol{u}$ to $\boldsymbol{x} = \boldsymbol{\mu} + \mathbf{A}\boldsymbol{u}$. From the general theory of linear transformations given above, the Jacobian J is $|\mathbf{A}|$. Hence $1/J = 1/|\mathbf{A}| = |\mathbf{A}^{-1}|$ by the general theory of determinants. Also, since $\mathbf{A}$ is a square matrix of full rank, $\boldsymbol{u} = \mathbf{A}^{-1}(\boldsymbol{x} - \boldsymbol{\mu})$. Substituting this in (7.7) and multiplying by $1/J$ yields the joint probability density function of $\boldsymbol{x}$ as

$$\mathrm{f}(\boldsymbol{x}) = (2\pi)^{-p/2}\,|\mathbf{A}^{-1}|\exp\{-\tfrac{1}{2}(\boldsymbol{x} - \boldsymbol{\mu})'(\mathbf{A}^{-1})'(\mathbf{A}^{-1})(\boldsymbol{x} - \boldsymbol{\mu})\}. \quad (7.8)$$

Now $(\mathbf{A}^{-1})'(\mathbf{A}^{-1}) = (\mathbf{A}')^{-1}(\mathbf{A}^{-1}) = (\mathbf{AA}')^{-1}$ from general matrix theory (Appendix A, Section A3). Since $\boldsymbol{\Sigma} = \mathbf{AA}'$ then $(\mathbf{A}^{-1})'(\mathbf{A}^{-1}) = \boldsymbol{\Sigma}^{-1}$. Also, $|\boldsymbol{\Sigma}| = |\mathbf{AA}'| = |\mathbf{A}|\,|\mathbf{A}'| = |\mathbf{A}|^2$ from general determinant theory (Appendix Section A3), so $|\mathbf{A}| = |\boldsymbol{\Sigma}|^{\frac{1}{2}}$. Hence $|\mathbf{A}^{-1}| = |\mathbf{A}|^{-1} = |\boldsymbol{\Sigma}|^{-\frac{1}{2}}$. Substituting into (7.8) thus yields, finally,

$$\mathrm{f}(\boldsymbol{x}) = (2\pi)^{-p/2}\,|\boldsymbol{\Sigma}|^{-\frac{1}{2}}\exp\{-\tfrac{1}{2}(\boldsymbol{x} - \boldsymbol{\mu})'\boldsymbol{\Sigma}^{-1}(\boldsymbol{x} - \boldsymbol{\mu})\} \quad (7.9)$$

as the probability density function for a $\mathrm{N}(\boldsymbol{\mu}, \boldsymbol{\Sigma})$ distribution. Note therefore that knowledge of the means and variances of each variate, and all covariances between pairs of variates, specifies a multivariate normal distribution completely.

Contours of equi-probability of this distribution in the p-dimensional space of which $\boldsymbol{x}$ is a point are given by $\mathrm{f}(\boldsymbol{x}) = C$, for some constant C. Now since $\boldsymbol{\Sigma}$ is specified and constant for a given distribution, these contours must satisfy $(\boldsymbol{x} - \boldsymbol{\mu})'\boldsymbol{\Sigma}^{-1}(\boldsymbol{x} - \boldsymbol{\mu}) = C'$, where C' is also a constant (equal to $-2\log_e\{(2\pi)^{p/2}\,|\boldsymbol{\Sigma}|^{\frac{1}{2}}C\}$). This equation defines a hyper-ellipsoid in this p-dimensional space. The probability density function (7.9) is thus a suitable model for populations of the type discussed in Chapter 6.3.

The multivariate normal distribution has a number of simple properties that make its use as a model for observed data most attractive, and these properties are now summarized. In all the following it is assumed that $\boldsymbol{x}$ has a p-component multivariate normal distribution with a mean of $\boldsymbol{\mu}$ and covariance matrix $\boldsymbol{\Sigma}$.

Property 1

If $\boldsymbol{y} = \mathbf{B}\boldsymbol{x}$ is any linear transformation of $\boldsymbol{x}$, where $\mathbf{B}$ is an $(m \times p)$ matrix of real numbers with $m \leqslant p$ and rank m, then $\boldsymbol{y}$ also has a multivariate normal distribution, with m components, a mean of $\mathbf{B}\boldsymbol{\mu}$, and covariance matrix $\mathbf{B}\boldsymbol{\Sigma}\mathbf{B}'$.

This property follows directly from our definition of the general multivariate normal distribution, because $\boldsymbol{y} = \mathbf{B}\boldsymbol{x}$ and $\boldsymbol{x} = \boldsymbol{\mu} + \mathbf{A}\boldsymbol{u}$ implies that $\boldsymbol{y} = \mathbf{B}\boldsymbol{\mu} + \mathbf{BA}\boldsymbol{u}$. Hence $\mathbf{y} \sim \mathrm{N}(\mathbf{B}\boldsymbol{\mu}, \mathbf{BA}(\mathbf{BA})')$ and $(\mathbf{BA})(\mathbf{BA})' = \mathbf{BA}'\mathbf{B}' = \mathbf{B}\boldsymbol{\Sigma}\mathbf{B}'$ as required.

Property 2

The marginal distribution of *any* subset of the x_i has a multivariate normal distribution, whose mean vector and covariance matrix are obtained as the subvector of $\boldsymbol{\mu}$ and submatrix of $\boldsymbol{\Sigma}$ appropriate to the chosen subset of x_i.

This property can be obtained as a direct consequence of Property 1, by expressing the subset vector of x_i in the form $\mathbf{B}\boldsymbol{x}$ with $\mathbf{B}$ a matrix having all elements zero except for one unity in each row picking out each element of the subset. For example, if the required subset is $x_1, \ldots, x_q$, then we choose the $q \times p$ matrix

$$\mathbf{B} = \begin{pmatrix} 1 & 0 & 0 & \ldots. & 0 & 0 & \ldots & 0 \\ 0 & 1 & 0 & \ldots. & 0 & 0 & \ldots & 0 \\ 0 & 0 & 1 & \ldots. & 0 & 0 & \ldots & 0 \\ \vdots & & & & & & & \\ 0 & 0 & 0 & \ldots. & 1 & 0 & \ldots & 0 \\ 0 & 0 & 0 & \ldots. & 0 & 1 & \ldots & 0 \end{pmatrix}$$

$\mathbf{B}\boldsymbol{x}$ then picks out the first q elements of $\boldsymbol{x}$, $\mathbf{B}\boldsymbol{\mu}$ correspondingly selects the first q elements of $\boldsymbol{\mu}$ and $\mathbf{B\Sigma B}'$ picks out the principal $(q \times q)$ submatrix of $\boldsymbol{\Sigma}$ (corresponding to variances of and covariances between the first q of the x_i).

Property 3

Suppose that $\boldsymbol{x}$ is partitioned into two portions $\boldsymbol{x}_1$ and $\boldsymbol{x}_2$, i.e. $\boldsymbol{x} = \begin{pmatrix} \boldsymbol{x}_1 \\ \boldsymbol{x}_2 \end{pmatrix}$ where $\boldsymbol{x}_1$ has q elements and $\boldsymbol{x}_2$ has $(p-q)$ elements. Partition $\boldsymbol{\mu}$ and $\boldsymbol{\Sigma}$ in the obvious conformable way

$$\boldsymbol{\mu} = \begin{pmatrix} \boldsymbol{\mu}_1 \\ \boldsymbol{\mu}_2 \end{pmatrix} \quad \text{and} \quad \boldsymbol{\Sigma} = \begin{pmatrix} \boldsymbol{\Sigma}_{11} & \boldsymbol{\Sigma}_{12} \\ \boldsymbol{\Sigma}_{21} & \boldsymbol{\Sigma}_{22} \end{pmatrix}.$$

Thus, for example, $\boldsymbol{\mu}_1$ is the mean of the portion $\boldsymbol{x}_1$ of $\boldsymbol{x}$, $\boldsymbol{\Sigma}_{11}$ is the covariance matrix of elements of $\boldsymbol{x}_1$, $\boldsymbol{\Sigma}_{21} = \boldsymbol{\Sigma}'_{12}$ gives the covariances between each element of $\boldsymbol{x}_1$ and each element of $\boldsymbol{x}_2$, and so on. Then the conditional distribution of $\boldsymbol{x}_1$, given that $\boldsymbol{x}_2$ has a known value $\boldsymbol{a}$, is multivariate normal with a mean of $\{\boldsymbol{\mu}_1 + \boldsymbol{\Sigma}_{12}\boldsymbol{\Sigma}_{22}^{-1}(\boldsymbol{a} - \boldsymbol{\mu}_2)\}$ and a covariance matrix $(\boldsymbol{\Sigma}_{11} - \boldsymbol{\Sigma}_{12}\boldsymbol{\Sigma}_{22}^{-1}\boldsymbol{\Sigma}_{21})$. Note that if nothing were known about $\boldsymbol{x}_2$, then Property 2 tells us that $\boldsymbol{x}_1$ has the marginal distribution $\mathrm{N}_q(\boldsymbol{\mu}_1, \boldsymbol{\Sigma}_{11})$. Thus knowing that $\boldsymbol{x}_2 = \boldsymbol{a}$ changes the location of this multivariate normal distribution by addition of $\boldsymbol{\Sigma}_{12}\boldsymbol{\Sigma}_{22}^{-1}(\boldsymbol{a} - \boldsymbol{\mu}_2)$, and changes the dispersion by subtraction of $\boldsymbol{\Sigma}_{12}\boldsymbol{\Sigma}_{22}^{-1}\boldsymbol{\Sigma}_{21}$. The matrix $\boldsymbol{\Sigma}_{12}\boldsymbol{\Sigma}_{22}^{-1}$ is known as the matrix of (population) *regression coefficients.*

Proof of this property is considerably more intricate than proofs of the other two properties, and it is included here primarily for the benefit of more mathematical readers. Fundamentally, however, it requires only the basic ideas of transformations summarized in the previous section together with some algebraic manipulation.

If $\boldsymbol{\Sigma}^{-1}$ is the inverse of $\boldsymbol{\Sigma}$, and $\boldsymbol{\Sigma} = \begin{pmatrix} \boldsymbol{\Sigma}_{11} & \boldsymbol{\Sigma}_{12} \\ \boldsymbol{\Sigma}_{21} & \boldsymbol{\Sigma}_{21} \end{pmatrix}$, then let the corresponding partition of $\boldsymbol{\Sigma}^{-1}$ be $\begin{pmatrix} \boldsymbol{\Sigma}^{11} & \boldsymbol{\Sigma}^{12} \\ \boldsymbol{\Sigma}^{21} & \boldsymbol{\Sigma}^{22} \end{pmatrix}$. Now $\boldsymbol{\Sigma}^{-1}\boldsymbol{\Sigma} = \mathbf{I}_p$, so that $\begin{pmatrix} \boldsymbol{\Sigma}^{11} & \boldsymbol{\Sigma}^{12} \\ \boldsymbol{\Sigma}^{21} & \boldsymbol{\Sigma}^{22} \end{pmatrix}\begin{pmatrix} \boldsymbol{\Sigma}_{11} & \boldsymbol{\Sigma}_{12} \\ \boldsymbol{\Sigma}_{21} & \boldsymbol{\Sigma}_{22} \end{pmatrix} = \begin{pmatrix} \mathbf{I}_q & \boldsymbol{0} \\ \boldsymbol{0} & \mathbf{I}_{p-q} \end{pmatrix}$. Multiplying out these matrices yields

$$\boldsymbol{\Sigma}^{11}\boldsymbol{\Sigma}_{11} + \boldsymbol{\Sigma}^{12}\boldsymbol{\Sigma}_{21} = \mathbf{I}_q \tag{7.10}$$

$$\boldsymbol{\Sigma}^{21}\boldsymbol{\Sigma}_{11} + \boldsymbol{\Sigma}^{22}\boldsymbol{\Sigma}_{21} = \boldsymbol{0} \tag{7.11}$$

$$\boldsymbol{\Sigma}^{11}\boldsymbol{\Sigma}_{12} + \boldsymbol{\Sigma}^{12}\boldsymbol{\Sigma}_{22} = \boldsymbol{0} \tag{7.12}$$

$$\boldsymbol{\Sigma}^{21}\boldsymbol{\Sigma}_{12} + \boldsymbol{\Sigma}^{22}\boldsymbol{\Sigma}_{22} = \mathbf{I}_{p-q}. \tag{7.13}$$

Multiplying (7.12) on the left by $(\boldsymbol{\Sigma}^{11})^{-1}$ and on the right by $\boldsymbol{\Sigma}_{22}^{-1}$ yields

$$(\boldsymbol{\Sigma}^{11})^{-1}\boldsymbol{\Sigma}^{12} = -\boldsymbol{\Sigma}_{12}\boldsymbol{\Sigma}_{22}^{-1}. \tag{7.14}$$

Multiplying (7.10) on the left by $(\boldsymbol{\Sigma}^{11})^{-1}$ yields

$$\boldsymbol{\Sigma}_{11} + (\boldsymbol{\Sigma}^{11})^{-1}\boldsymbol{\Sigma}^{12}\boldsymbol{\Sigma}_{21} = (\boldsymbol{\Sigma}^{11})^{-1},$$

and substituting for $(\boldsymbol{\Sigma}^{11})^{-1}\boldsymbol{\Sigma}^{12}$ from (7.14) gives

$$(\boldsymbol{\Sigma}^{11})^{-1} = \boldsymbol{\Sigma}_{11} - \boldsymbol{\Sigma}_{12}\boldsymbol{\Sigma}_{22}^{-1}\boldsymbol{\Sigma}_{21}. \tag{7.15}$$

From (7.9), the joint probability density function of $\boldsymbol{x}$ is

$$\mathrm{f}(\boldsymbol{x}) = (2\pi)^{-p/2}|\boldsymbol{\Sigma}|^{-\frac{1}{2}}\exp\{-\tfrac{1}{2}(\boldsymbol{x}-\boldsymbol{\mu})'\boldsymbol{\Sigma}^{-1}(\boldsymbol{x}-\boldsymbol{\mu})\}.$$

Writing $\boldsymbol{x}$, $\boldsymbol{\mu}$, and $\boldsymbol{\Sigma}^{-1}$ inside the exponential in their partitioned form and expanding gives

$$\begin{aligned}\mathrm{f}(\boldsymbol{x}) = (2\pi)^{-p/2}\,|\boldsymbol{\Sigma}|^{-\frac{1}{2}}\exp[-\tfrac{1}{2}\{&(\boldsymbol{x}_1-\boldsymbol{\mu}_1)'\boldsymbol{\Sigma}^{11}(\boldsymbol{x}_1-\boldsymbol{\mu}_1) \\ &+ 2(\mathbf{x}_1-\boldsymbol{\mu}_1)'\boldsymbol{\Sigma}^{12}(\mathbf{x}_2-\boldsymbol{\mu}_2) + (\mathbf{x}_2-\boldsymbol{\mu}_2)'\boldsymbol{\Sigma}^{22}(\mathbf{x}_2-\boldsymbol{\mu}_2)\}].\end{aligned}$$

The conditional distribution of $\boldsymbol{x}_1$ given the value of $\boldsymbol{x}_2$ is thus obtained by dividing this density by the marginal density of $\boldsymbol{x}_2$, and treating $\boldsymbol{x}_2$ as constant in the resulting expression (Section 7.1). The only portion of the resultant that is not constant is the portion involving $\boldsymbol{x}_1$. Hence we can easily deduce that

$$\mathrm{f}(\boldsymbol{x}_1 \mid \boldsymbol{x}_2) \propto \exp[-\tfrac{1}{2}\{(\boldsymbol{x}_1-\boldsymbol{\mu}_1)'\boldsymbol{\Sigma}^{11}(\boldsymbol{x}_1-\boldsymbol{\mu}_1) + 2(\boldsymbol{x}_1-\boldsymbol{\mu}_1)'\boldsymbol{\Sigma}^{12}(\boldsymbol{x}_2-\boldsymbol{\mu}_2)\}],$$

the constant of proportionality being obtained from the condition that $\int f(\boldsymbol{x}_1 \mid \boldsymbol{x}_2)\,d\boldsymbol{x}_1 = 1$.

Now,

$$
\begin{aligned}
&(\boldsymbol{x}_1-\boldsymbol{\mu}_1)'\boldsymbol{\Sigma}^{11}(\boldsymbol{x}_1-\boldsymbol{\mu}_1)+2(\boldsymbol{x}_1-\boldsymbol{\mu}_1)'\boldsymbol{\Sigma}^{12}(\boldsymbol{x}_2-\boldsymbol{\mu}_2)\\
&\quad=(\boldsymbol{x}_1-\boldsymbol{\mu}_1)'\boldsymbol{\Sigma}^{11}(\boldsymbol{x}_1-\boldsymbol{\mu}_1)+(\boldsymbol{x}_1-\boldsymbol{\mu}_1)'\boldsymbol{\Sigma}^{12}(\boldsymbol{x}_2-\boldsymbol{\mu}_2)+(\boldsymbol{x}_2-\boldsymbol{\mu}_2)'\boldsymbol{\Sigma}^{21}(\boldsymbol{x}_1-\boldsymbol{\mu}_1)\\
&\quad=\{\boldsymbol{x}_1-\boldsymbol{\mu}_1+(\boldsymbol{\Sigma}^{11})^{-1}\boldsymbol{\Sigma}^{12}(\boldsymbol{x}_2-\boldsymbol{\mu}_2)\}'\boldsymbol{\Sigma}^{11}\{\boldsymbol{x}_1-\boldsymbol{\mu}_1+(\boldsymbol{\Sigma}^{11})^{-1}\boldsymbol{\Sigma}^{12}(\boldsymbol{x}_2-\boldsymbol{\mu}_2)\}\\
&\qquad-(\boldsymbol{x}_2-\boldsymbol{\mu}_2)'\boldsymbol{\Sigma}^{21}(\boldsymbol{\Sigma}^{11})^{-1}\boldsymbol{\Sigma}^{12}(\boldsymbol{x}_2-\boldsymbol{\mu}_2).
\end{aligned}
$$

But since the last term involves only constants (as far as $f(\boldsymbol{x}_1 \mid \boldsymbol{x}_2)$ is concerned), it follows that

$$
\begin{aligned}
f(\boldsymbol{x}_1 \mid \boldsymbol{x}_2) \propto \exp[-\tfrac{1}{2}\{\boldsymbol{x}_1-\boldsymbol{\mu}_1+(\boldsymbol{\Sigma}^{11})^{-1}\boldsymbol{\Sigma}^{12}(\boldsymbol{x}_2-\boldsymbol{\mu}_2)\}'\boldsymbol{\Sigma}^{11}\\
\times\{\boldsymbol{x}_1-\boldsymbol{\mu}_1+(\boldsymbol{\Sigma}^{11})^{-1}\boldsymbol{\Sigma}^{12}(\boldsymbol{x}_2-\boldsymbol{\mu}_2)\}]
\end{aligned}
$$

which is the density function of a multivariate normal distribution that has mean $\boldsymbol{\mu}_1-(\boldsymbol{\Sigma}^{11})^{-1}\boldsymbol{\Sigma}^{12}(\boldsymbol{x}_2-\boldsymbol{\mu}_2)=\boldsymbol{\mu}_1+\boldsymbol{\Sigma}_{12}\boldsymbol{\Sigma}_{22}^{-1}(\boldsymbol{x}_2-\boldsymbol{\mu}_2)$ from (7.14), and dispersion matrix $(\boldsymbol{\Sigma}^{11})^{-1}=\boldsymbol{\Sigma}_{11}-\boldsymbol{\Sigma}_{12}\boldsymbol{\Sigma}_{22}^{-1}\boldsymbol{\Sigma}_{21}$ from (7.15). Setting $\boldsymbol{x}_2=\boldsymbol{a}$ thus establishes the stated property.

These three properties show that if a random vector $\boldsymbol{x}$ has a multivariate normal distribution, then the joint distribution of any set of elements of $\boldsymbol{x}$ also has a multivariate normal distribution (whether or not we know the values of the remaining elements), and the joint distribution of any set of linear transformations of elements of $\boldsymbol{x}$ also has the multivariate normal form. Furthermore, the mean vectors and dispersion matrices of these distributions are readily obtained from those of $\boldsymbol{x}$, so the distributions are completely specified. These properties thus make the multivariate normal distribution extremely attractive as a model for many populations encountered in practice. Furthermore, insight into the geometrical nature of the multivariate normal distribution is also provided readily. Taking up the general discussion of a probability density function from Section 6.3, the multivariate normal density of (7.9) is a surface in $(p+1)$ dimensions. The coordinate axes for p of these dimensions are provided by elements of $\boldsymbol{x}$, while the function $f(\boldsymbol{x})$ is defined along the $(p+1)$th dimension. Slicing this dimension into a p-dimensional subspace at any value of $f(\boldsymbol{x})$ produces the hyperellipsoidal equiprobability contour arising from equation (7.9). Property 1 tells us that if the surface $f(\boldsymbol{x})$ is compressed into any subspace of the p dimensions defined by $\boldsymbol{x}$ (and re-scaled to integrate to unity), it will just be another multivariate normal surface. This follows because any linear transformation of $\boldsymbol{x}$ defines such a subspace. In particular, subsets of coordinate axes can be used to define the subspace: this is the upshot of Property 2. At the extreme of this is the one-dimensional subspace provided by a single coordinate axis. If the surface $f(\boldsymbol{x})$ is compressed on to each of the coordinate axes and re-scaled, therefore, it will have the form of a univariate normal density. Finally, fixing the values of some elements of $\boldsymbol{x}$ amounts to 'slicing' through the $(p+1)$ dimensions at the coordinate values determined by these elements. Property 3 thus says that if such a slice is chosen (and $f(\boldsymbol{x})$ is re-scaled to integrate to unity) then the shape of the surface is still the multivariate normal shape.

Many multivariate techniques involve the calculation and manipulation of sums of squares of, and sums of cross-products between, the constituent variates. If multivariate normality is assumed for these constituent variates, then the joint distribution of the calculated sums of squares and products has a special distribution, which is now briefly described.

7.3 The Wishart distribution

In univariate theory, if $x_1, x_2, \ldots, x_k$ are a sequence of independent identically distributed (i.i.d.) random variables, each having distribution $N(0, \sigma^2)$, then $\frac{1}{\sigma^2}\Sigma x_i^2$ has a chi-squared distribution on k degrees of freedom denoted χ_k^2. This distribution plays an important role in statistical inference based on normal assumptions, since it forms the basis of many hypothesis tests and confidence interval calculations. The multivariate analogue occurs when $\boldsymbol{x}_1, \ldots, \boldsymbol{x}_k$ form a sequence of independent p-variable random vectors each with distribution $N(\boldsymbol{0}, \boldsymbol{\Sigma})$ and the matrix $\mathbf{C}$ is defined by $\mathbf{C} = \sum_{i=1}^{k} \boldsymbol{x}_i\boldsymbol{x}_i'$. Thus the (i, i)th element of $\mathbf{C}$ is the sum of squares of the ith elements of the vectors $\boldsymbol{x}_1, \ldots, \boldsymbol{x}_k$, while the (i, j)th element of $\mathbf{C}$ $(i \neq j)$ is the sum of products of ith and jth elements of these vectors. The joint distribution of all elements of $\mathbf{C}$ (or the distribution of $\mathbf{C}$ for short) is said to be a *Wishart distribution* based on p variables, with k degrees of freedom and parameter $\boldsymbol{\Sigma}$, and is denoted $W_p(k, \boldsymbol{\Sigma})$.

The Wishart distribution can thus be thought of as the multivariate analogue of the χ^2 distribution, and indeed many of its properties are either bound up with the χ^2 distribution or mirror those of the χ^2 distribution. Its probability density function is a complicated mathematical expression and of little practical interest in its own right, so it will not be stated here. Of more concern are its properties, and some of the main ones are now listed.

1. If $p = 1$, then $C = \sum_{i=1}^{k} x_i^2$ where the x_i are i.i.d. $N(0, \sigma^2)$. Hence the $W_1(k, \sigma^2)$ distribution is just σ^2 times the χ_k^2 distribution, so that the χ^2 distribution is a special case of the Wishart distribution.

2. If $\mathbf{C}_1$ and $\mathbf{C}_2$ are independent, with $\mathbf{C}_1 \sim \mathrm{W}_p(k_1, \boldsymbol{\Sigma})$ and $\mathbf{C}_2 \sim \mathrm{W}_p(k_2, \boldsymbol{\Sigma})$, then $\mathbf{C}_1 + \mathbf{C}_2 \sim \mathrm{W}_p(k_1 + k_2, \boldsymbol{\Sigma})$. This property is a simple consequence of the definition of the Wishart distribution, because $\mathbf{C}_1$ can be written as $\sum_{i=1}^{k_1} \boldsymbol{x}_i\boldsymbol{x}_i'$ and $\mathbf{C}_2$ can be written $\sum_{i=1}^{k_2} \boldsymbol{y}_i\boldsymbol{y}_i'$, where $\boldsymbol{x}_1, \ldots, \boldsymbol{x}_{k_1}, \boldsymbol{y}_1, \ldots, \boldsymbol{y}_{k_2}$ are all independent $\mathrm{N}_p(\boldsymbol{0}, \boldsymbol{\Sigma})$ random vectors. Hence $\mathbf{C}_1 + \mathbf{C}_2 = \sum_{i=1}^{k_1} \boldsymbol{x}_i\boldsymbol{x}_i' + \sum_{i=1}^{k_2} \boldsymbol{y}_i\boldsymbol{y}_i'$, so that the distribution of $\mathbf{C}_1 + \mathbf{C}_2$ is $\mathrm{W}_p(k_1 + k_2, \boldsymbol{\Sigma})$ by definition. Now standard univariate theory tells us that if $x \sim \chi^2_{k_1}$ independently of $y \sim \chi^2_{k_2}$, then $x + y \sim \chi^2_{k_1+k_2}$. Thus the Wishart distribution shares the χ^2 additive property.
3. If $\mathbf{B}$ is any $(q \times p)$ matrix of constants and $\mathbf{C} \sim \mathrm{W}_p(k, \boldsymbol{\Sigma})$, then $\mathbf{BCB}' \sim \mathrm{W}_p(k, \mathbf{B\Sigma B}')$. To show this, first note from property 1 of the multivariate normal that if $\boldsymbol{x}_i \sim \mathrm{N}(\boldsymbol{0}, \boldsymbol{\Sigma})$ then $\boldsymbol{u}_i = \mathbf{B}\boldsymbol{x}_i \sim \mathrm{N}(\boldsymbol{0}, \mathbf{B\Sigma B}')$. Now $\mathbf{BCB}' = \mathbf{B}\left(\sum_{i=1}^{k} \boldsymbol{x}_i\boldsymbol{x}_i'\right)\mathbf{B}' = \sum_{i=1}^{k} (\mathbf{B}\boldsymbol{x}_i)(\boldsymbol{x}_i'\mathbf{B}') = \sum_{i=1}^{k} (\mathbf{B}\boldsymbol{x}_i)(\mathbf{B}\boldsymbol{x}_i)' = \sum_{i=1}^{k} \boldsymbol{u}_i\boldsymbol{u}_i'$ which has the required distribution by definition.

   Now set $q = 1$, $\mathbf{B} = \boldsymbol{b}$, a p-vector of constants, and use 1 above to obtain:
4. If $\boldsymbol{b}$ is any p-vector of constants and $\mathbf{C} \sim \mathrm{W}_p(k, \boldsymbol{\Sigma})$, then $\boldsymbol{b}'\mathbf{C}\boldsymbol{b} \sim \sigma^2\chi^2_k$ where $\sigma^2 = \boldsymbol{b}'\boldsymbol{\Sigma}\boldsymbol{b}$.

   Finally, two properties concern correlation coefficients and will have some importance in Part V of this book. In these two properties we let c_{ij} be the (i, j)th element of $\mathbf{C}$, σ_{ij} be the (i, j)th element of $\boldsymbol{\Sigma}$, $r_{ij} = c_{ij}/\sqrt{(c_{ii}c_{jj})}$, and we suppose that $\mathbf{C} \sim \mathrm{W}_p(k, \boldsymbol{\Sigma})$.
5. If $\sigma_{ij} = 0$ then $r_{ij}\sqrt{\left\{\frac{(k-1)}{(1-r_{ij}^2)}\right\}}$ has a Student's t distribution on $(k-1)$ degrees of freedom.
6. $\frac{1}{2}\log_e\left(\frac{1+r_{ij}}{1-r_{ij}}\right)$ is asymptotically (i.e. in large samples) normally distributed with mean $\frac{1}{2}\log_e\left(\frac{1+\rho_{ij}}{1-\rho_{ij}}\right)$ and variance $\frac{1}{k-2}$, where $\rho_{ij} = \sigma_{ij}/\sqrt{(\sigma_{ii}\sigma_{jj})}$.

These properties of the Wishart distribution suffice for immediate purposes, and we shall not discuss this distribution any further. For more details and a comprehensive bibliography, the reader is referred to Johnson and Kotz (1972).

7.4 Elliptic distributions

Although most of classical multivariate analysis is built upon the multivariate normal distribution, there has been some recent interest in developing alternative models for multivariate data. One area of applicability of such models is in the study of the *robustness* of multivariate techniques to departures from multivariate normality in the underlying distributions. The difficulties associated with many alternatives are both theoretical and practical. There is, however, a simple class of distributions which contains distributions that have similar features to the multivariate normal but which exhibit either longer or shorter tails than the normal. Such a class forms an ideal basis for robustness studies, and hence has attracted increasing attention. It is known as the class of *elliptic* distributions, and is characterized by the form of the probability density function possessed by its members. A p-vector random variable $\boldsymbol{x}$ is said to have an elliptic distribution with parameters $\boldsymbol{\mu}' = (\mu_1, \ldots, \mu_p)$ and $\boldsymbol{\Psi}$, a $p \times p$ positive definite matrix, if its density has the form $\mathrm{f}(\boldsymbol{x}) = |\boldsymbol{\Psi}|^{-\frac{1}{2}}\mathrm{h}\{(\mathbf{x}-\boldsymbol{\mu})'\boldsymbol{\Psi}^{-1}(\mathbf{x}-\boldsymbol{\mu})\}$ for some function $h(.)$.

It can thus be immediately seen that the multivariate normal distribution belongs to the elliptic class; in addition, this class contains the multivariate *t,* the multivariate Cauchy, the double exponential, and the logistic distributions. The feature of distributions in this class is that they all have elliptical contours of equiprobability (like the multivariate normal). Many of the properties of the multivariate normal distribution extend also to the class of elliptic distributions. Thus, for example, if a distribution belongs to this class then so do all its marginal distributions, and so do all the conditional distributions of some variables given the values of the others (cf Properties 1 and 3 of the multivariate normal). For a more detailed discussion of elliptic distributions see Kelker (1970) or Muirhead (1982). The latter gives details also of various robustness studies and results involving elliptic distributions; additional references in this vein are Devlin *et al.* (1975) and Mitchell and Krzanowski (1985). Broadly speaking, results obtained under assumptions of multivariate normality remain at least qualitatively unaltered if the true distribution is some other member of the elliptic family.

7.5 Tests for multivariate normality

If an assumption of multivariate normality is to be made as a prelude to a statistical analysis—and many of the techniques to be described in the remainder of this book do indeed make such an assumption—then it is often prudent to check prior to embarking on the analysis whether or not the data conform to such an assumption. If the data do not conform to the assumption, then some transformation of the variates may be appropriate before the analysis is undertaken. Alternatively, the assumption may be violated by the presence of a few 'outliers' in the data, in which case these offending units may best be omitted from the data before the analysis.

There are many methods, formal hypothesis testing procedures and informal graphical techniques, which are now available to investigate the assumption that a set of data vectors can be treated as a set of independent observations from a multivariate normal distribution. Excellent descriptions of these methods may be found in Andrews *et al.* (1973) or Gnanadesikan (1977), so detailed discussion here would be superfluous. We shall content ourselves with a brief description of just one easily implementable informal graphical technique.

Suppose that $\boldsymbol{x}$ has a multivariate normal distribution with mean $\boldsymbol{\mu}$ and dispersion matrix $\boldsymbol{\Sigma}$. From the algebra following eqns (7.7) and (7.8) it can be deduced that

$$(\boldsymbol{x}-\boldsymbol{\mu})'\boldsymbol{\Sigma}^{-1}(\boldsymbol{x}-\boldsymbol{\mu}) = \boldsymbol{u}'\boldsymbol{u}$$

where $\boldsymbol{u} \sim \mathrm{N}(\boldsymbol{0}, \mathbf{I}_p)$. But $\boldsymbol{u}'\boldsymbol{u} = \sum_{i=1}^{p} u_i^2$, the sum of squares of p independent N(0, 1) variates. Hence $z = (\boldsymbol{x}-\boldsymbol{\mu})'\boldsymbol{\Sigma}^{-1}(\boldsymbol{x}-\boldsymbol{\mu})$ has a χ^2 distribution with p degrees of freedom. Thus if $\boldsymbol{x}_1, \ldots, \boldsymbol{x}_n$ form a random sample from a $\mathrm{N}(\boldsymbol{\mu}, \boldsymbol{\Sigma})$ population, then $z_i = (\boldsymbol{x}_i-\boldsymbol{\mu})'\boldsymbol{\Sigma}^{-1}(\boldsymbol{x}_i-\boldsymbol{\mu})$ for $i = 1, \ldots, n$ correspondingly form a random sample from a χ_p^2 population. Let $z_{(1)}, z_{(2)}, \ldots, z_{(n)}$ denote the z_i values arranged in increasing order, $z_{(1)}$ denoting the smallest and $z_{(n)}$ denoting the largest, and suppose that $z_{(i)}$ is plotted against the quantile of a χ_p^2 distribution corresponding to a cumulative probability of $(i-\frac{1}{2})/n$ for $i = 1, \ldots, n$. If the z_i are truly independent observations from a χ_p^2 distribution then such a χ^2 probability plot should yield a straight line, as the order statistics $z_{(i)}$ and the corresponding quantiles are proportional. The χ^2 probability plot is thus a simple graphical means of testing the hypothesis that the observations $\boldsymbol{x}_i$ are independent realizations of a $\mathrm{N}(\boldsymbol{\mu}, \boldsymbol{\Sigma})$ variate: if the plot is approximately linear the hypothesis is tenable, but if there are marked deviations from linearity the hypothesis is not tenable.

The bar to direct use of this technique comes from the parameters $\boldsymbol{\mu}$ and $\boldsymbol{\Sigma}$: in practice these will be unknown, and the z_i are thus not directly computable. The simplest way round the problem is to estimate $\boldsymbol{\mu}$ and $\boldsymbol{\Sigma}$ from the sample observations. Anticipating the discussion in the next chapter, the best estimates will be the sample mean vector $\bar{\boldsymbol{x}}$ and the sample covariance matrix $\mathbf{S} = \frac{1}{n-1}\sum_{i=1}^{n}(\boldsymbol{x}_i-\bar{\boldsymbol{x}})(\boldsymbol{x}_i-\bar{\boldsymbol{x}})'$. Replacing $\boldsymbol{\mu}$ and $\boldsymbol{\Sigma}$ by $\bar{\boldsymbol{x}}$ and $\mathbf{S}$ enables the $z_{(i)}$ to be computed and plotted, but now two further complications have been introduced. First, the z_i are no longer independent of each other, because each z_i involves the common sample values $\bar{\boldsymbol{x}}$ and $\mathbf{S}$. Second, since $\bar{\boldsymbol{x}}$ and $\mathbf{S}$ are only *estimates* of $\boldsymbol{\mu}$ and $\boldsymbol{\Sigma}$, the

z_i computed from them no longer have a χ^2 distribution under the hypothesis of normality of the $\boldsymbol{x}_i$. Fortunately, if the sample size is large, then for practical purposes these complications can be ignored. Once n is greater than about $10p$, the difference between using the true distribution and the χ^2 approximation seems to be negligible, and for samples of such size the departure from independence of the z_i is also insignificant. Hence a χ^2 probability plot will yield a good graphical method of testing multivariate normality. For small samples, however, some modifications need to be made. The main distributional result on which these modifications depend will be derived in the next chapter, and will simply be quoted here. If $\bar{\boldsymbol{x}}$ and $\mathbf{S}$ are respectively the mean vector and covariance matrix of a sample of size n from a $\mathrm{N}_p(\boldsymbol{\mu}, \boldsymbol{\Sigma})$ population, and if $\boldsymbol{x}$ is a further (independent) observation from this population, then $(\boldsymbol{x}-\bar{\boldsymbol{x}})'\mathbf{S}^{-1}(\boldsymbol{x}-\bar{\boldsymbol{x}})$ is distributed as $\dfrac{p(n^2-1)}{n(n-p)}$ times an F variate with p and $(n-p)$ degrees of freedom. Hence the quantiles of this latter distribution must be used in place of the quantiles of the χ_p^2 distribution when constructing the probability plot. Additionally, independence of each $\boldsymbol{x}_i$ from $\bar{\boldsymbol{x}}$ and $\mathbf{S}$ must be ensured. To do this, *jack-knifed* means and covariance matrices can be used instead of the single mean vector $\bar{\boldsymbol{x}}$ and covariance matrix $\mathbf{S}$. The jack-knifed mean $\bar{\boldsymbol{x}}_{(i)}$ and covariance matrix $\mathbf{S}_{(i)}$ for use with the observation $\boldsymbol{x}_i$ are simply the mean vector and covariance matrix respectively of the $(n-1)$ sample members excluding $\boldsymbol{x}_i$. The appropriately adjusted z_i for use in the probability plot are thus $z_i' = (\boldsymbol{x}_i - \bar{\boldsymbol{x}}_{(i)})'\mathbf{S}_{(i)}^{-1}(\boldsymbol{x}_i - \bar{\boldsymbol{x}}_{(i)})$. Excessive computation is avoided by using a well-known matrix identity (Bartlett 1951b), from which it can be shown that

$$\mathbf{S}_{(i)}^{-1} = a\mathbf{S}^{-1} + \frac{ab\mathbf{S}^{-1}(\boldsymbol{x}_i-\bar{\boldsymbol{x}})(\boldsymbol{x}_i-\bar{\boldsymbol{x}})'\mathbf{S}^{-1}}{\{1-b(\boldsymbol{x}_i-\bar{\boldsymbol{x}})'\mathbf{S}^{-1}(\boldsymbol{x}_i-\bar{\boldsymbol{x}})\}}$$

where $a=(n-2)/(n-1)$ and $b=n/(n-1)^2$. Also $\bar{\boldsymbol{x}}_{(i)} = \dfrac{1}{n-1}(n\bar{\boldsymbol{x}} - \boldsymbol{x}_i)$ so that each z_i' can be obtained using just $\bar{\boldsymbol{x}}$, $\mathbf{S}^{-1}$, and $\boldsymbol{x}_i$.

As an illustration of this technique, Fig. 7.1 shows the probability plot obtained from the eight test scores observed for the students at one of the time points in the British Council investigation described in Section 5.3. Although there is some suggestion of curvature, for practical purposes the plot appears to be roughly linear with no obviously discrepant observations so that normality can be assumed for any formal analysis of the data.

It is perhaps worth noting that quantiles of the χ^2-distribution are much easier to obtain than quantiles of the F-distribution, since reliance on

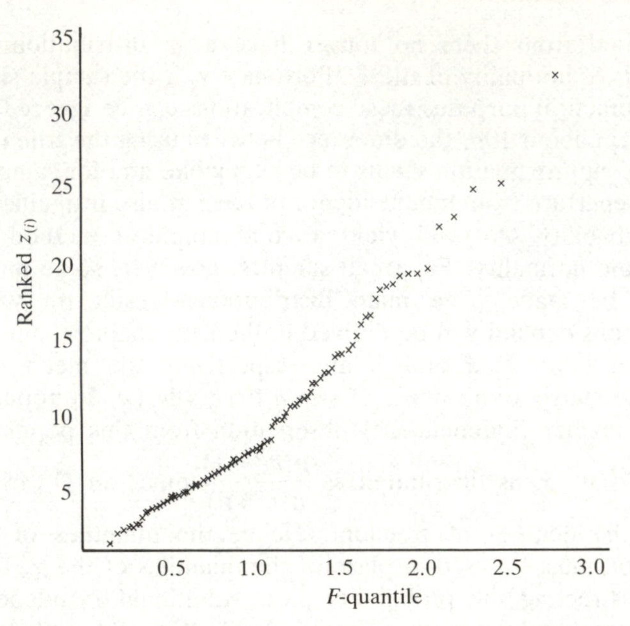

Fig. 7.1 Probability plot to test for multivariate normality

fewer degrees of freedom means that more detailed tabulations of the χ^2-distribution are possible in standard books of tables. Hence the χ^2 probability plots are operationally the easiest; they should be perfectly adequate for all but the smallest sample sizes in practice.

As a final point, it has been established empirically that many of the classical multivariate techniques are relatively unaffected by non-normality, providing the observations come from some other centrally symmetric distribution. Hence, even if the hypothesis of normality is rejected, no special corrective action need be taken as long as the observations are reasonably symmetric. Gnanadesikan (1977) also describes various simple graphical ways of investigating whether or not a data set exhibits symmetry. For this purpose, it is sufficient to consider each variate separately in a multivariate data set, and a very simple method is as follows. Arrange the values of each variate in increasing order. Thus, for the variate x, the ordered values can be denoted $x_{(1)} \leqslant x_{(2)} \leqslant \ldots \leqslant x_{(n-1)} \leqslant x_{(n)}$. Plot the points $\{x_{(1)}, x_{(n)}\}$, $\{x_{(2)}, x_{(n-1)}\}$, $\{x_{(3)}, x_{(n-2)}\}$, . . . , the horizontal axis representing the 'lower half', and the vertical axis the 'upper half' of the data. If the plot is reasonably linear with slope approximately -1, then the observations can be taken as coming from a symmetric distribution. In this case, the centre of symmetry can be deduced as half the value of the intercept of the plot.

However, if the plot shows systematic deviations from linearity then non-symmetry of data is indicated. An upward bow to the plot indicates a longer upper tail, a downward bow a longer lower tail, while individual peculiarities in the plot are a sign of individually aberrant observations.

7.6 Transformations

If preliminary inspection of the sample, or some formal test of hypothesis, suggests that an assumption of either multivariate normality, or at least symmetry of the data, is not appropriate, then direct use of any of the classical multivariate techniques may provoke misleading inferences. Most of these techniques are based on such summary matrices as sample covariance or correlation matrices, and the justification for using such summaries is considerably weakened in the presence of asymmetric distributions. In some cases, modifications exist to make a given technique more robust to departures from normality. We shall see in Chapter 8 how to obtain robust estimates of covariance or correlation. A more general approach, however, is to seek some suitable transformation of the data so that the transformed observations are more nearly normal than the original ones. Even if such a transformation does not exactly accomplish normality, it will often go a long way towards symmetrizing the data. The desired technique can then be applied to the transformed observations with confidence. Results of the application of the technique can be ultimately back-transformed (using the inverse of the original transformation) so that they can be expressed in comparable terms to those of the original observations.

A suitable transformation may be suggested by theoretical or a priori considerations, and many readers will be familiar with such transformations of this type as the angular transformation for binomial data (or proportions), the square root transformation for Poisson data (or counts), the logistic transformation of binary data, and the z-transformation for correlation coefficients. All these transformations can be formally derived with the objective of stabilizing variance, or inducing additivity or normality, in the context of hypothesis testing or analysis of variance. Their justification can be found in most general statistical texts, and we shall not be concerned with them here. Of more concern are transformations that have to be derived from the data, when no a priori considerations exist. Such transformations are often required in multivariate problems, but we shall introduce them by considering first the univariate case.

If a variate x is found to be positively skewed, then often either the transformation $y = \log x$ or the transformation $y = x^{1/n}$ will produce a symmetrical variate y. Both of these transformations have the effect of reducing the tail of the distribution, as the larger the value of x the greater is the reduction applied to it. Box and Cox (1964) developed this idea into a formal procedure for the data-based estimation of the power transformation of a variate which best enhances some specific property of interest. Possible properties might be normality, additivity, or homogeneity of variance; we shall concentrate here on normality, as this is the only property required for use in the classical multivariate context. The family of transformations considered by Box and Cox was

$$y = \begin{cases} (x^\lambda - 1)/\lambda & \lambda \neq 0, \\ \log_e x & \lambda = 0. \end{cases} \tag{7.16}$$

The parameter λ is to be estimated from the data. Note that, formally, $\lim_{\lambda \to 0} \frac{(x^\lambda - 1)}{\lambda} = \log_e x$, and this explains the different form of y at $\lambda = 0$. Box and Cox suggested the use of maximum likelihood for estimation of λ; the essential ideas of the method are as follows.

Given the sample values $x_1, x_2, \ldots, x_n$ *and the value of* λ, we can calculate the transformed values $y_1^{(\lambda)}, y_2^{(\lambda)}, \ldots, y_n^{(\lambda)}$ from (7.16). The superscript λ is appended to each of these values to denote that they depend on the value of λ: a different λ produces a different set of transformed values. Since we aim to transform to normality, we assume that the $y_i^{(\lambda)}$ form a random sample from a normal distribution having some mean μ_λ and some variance σ_λ^2. The likelihood of this sample is

$$L(y_1^{(\lambda)}, \ldots, y_n^{(\lambda)}; \mu_\lambda; \sigma_\lambda^2) = (2\pi\sigma_\lambda^2)^{-n/2} \exp\left\{ -\frac{1}{2\sigma_\lambda^2} \sum_{i=1}^{n} (y_i^{(\lambda)} - \mu_\lambda)^2 \right\} \tag{7.17}$$

and the maximum likelihood estimates of μ_λ and σ_λ^2, from standard theory, are the sample mean $\bar{y}_\lambda$ and the sample variance s_λ^2 respectively. Replacing the parameters in (7.17) by these estimates thus yields the maximized likelihood for given λ as

$$\hat{L}_\lambda = (2\pi s_\lambda^2)^{-n/2} \exp(-n/2). \tag{7.18}$$

The maximum likelihood estimate of λ will therefore be the value that yields the largest possible $\hat{L}_\lambda$. Finding this value is a task clearly suited to the computer, as (7.18) can be used to define the objective function in a function-maximization routine. Alternatively, a range of values of λ can be selected, $\hat{L}_\lambda$ calculated at each and plotted against λ. If these points are connected by a smooth curve, the maximum likelihood estimate of λ will be the value λ at which this curve is maximum.

Turning now to the multivariate case, it seems sensible to allow different parameters λ_i to govern the transformation of each variate in the system. The simplest approach is to apply the Box and Cox procedure to each variate separately, thereby achieving approximate *marginal* normality of the transformed data. However, while the marginal distributions of a multivariate normal distribution are always normal (Property 2, Section 7.2), it is not always the case that the joint distribution of a collection of normal variates is a multivariate normal distribution. For this reason, the Box and Cox method was extended to the multivariate case by Andrews *et al.* (1971), and this method may be used to produce *joint* normality of the transformed data. The computations are fairly involved, so will not be described here; a detailed discussion of the bivariate case may be found in Gnanadesikan (1977). Given the fact that symmetrizing the data is often sufficient for multivariate purposes, however, it seems as if the simple univariate procedure described above may safely be applied to each variate separately in most practical situations. Note, incidentally, that this procedure may easily be converted into a formal *test* of normality of the data, because if the data are already normal then no transformation is needed. Thus, for normal data, $\lambda = 1$ for each variate under consideration, and hence a formal test of normality is equivalent to testing the null hypothesis $\lambda_1 = \lambda_2 = \ldots = \lambda_p = 1$ against the alternative that at least one λ_i is not unity. If each variate is being considered separately, then a likelihood ratio test can be constructed very simply. Let $\hat{L}$ be the maximized likelihood at the selected value of λ for this variate (i.e. the maximum of all $\hat{L}_\lambda$ from (7.18)), and $\hat{L}'$ the maximized likelihood when λ is set to unity (i.e. the value of (7.18) when s_λ^2 is the sample variance of the *untransformed* data). Then the likelihood ratio test statistic is given by $\psi = \hat{L}'/\hat{L}$, and if the null hypothesis (that $\lambda = 1$) is true, $-2\log_e \psi$ asymptotically has a χ^2 distribution on one degree of freedom. Hence if $-2\log_e \psi$ exceeds the upper $100(1-\alpha)$ per cent point of the χ_1^2 distribution, the null hypothesis of normality of the given variate is rejected. This produces a test of marginal normality of the data, suitable for large samples. To obtain a test of joint normality, the extension due to Andrews *et al.* (1971) would have to be used in constructing the likelihood ratio test.

To illustrate the above ideas, we consider some data obtained in a study to develop screening tests for the disease cystic fibrosis (R. N. Curnow, private communication). Test results were obtained on forty-six people known to have the disease, and on 129 controls, known not to have the disease. Four tests were carried out on each subject: OS (salivary chloride), EIL (salivary sodium), OH (sweat chloride, sweat induced by heating), and OPC (sweat chloride, sweat induced by injection). For purposes of analysis, it was desirable that these four variates be distributed normally within each of the two groups of subjects. Consider the OPC data first. Normal probability plots of this

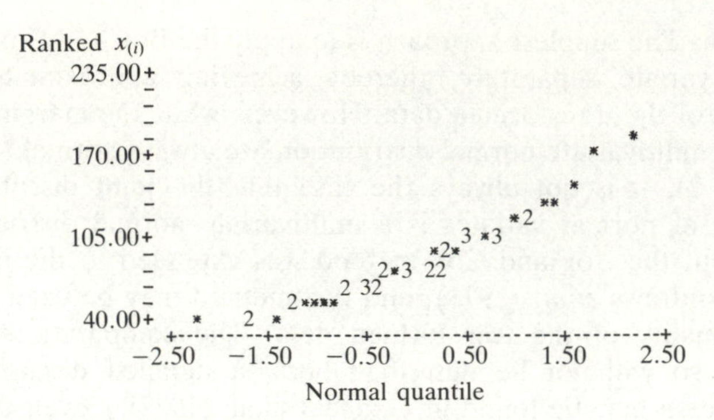

Fig. 7.2 Normal probability plot of raw OPC values for the cystic fibrosis group

variate within the two groups suggested non-normality: Fig. 7.2 shows the plot for the cystic fibrosis group, and there seems to be a suggestion of curvature at each end of the plot. Using the method of Box and Cox, the estimated value of λ for a data-based transformation to normality was $\hat{\lambda} = 0.25$; Fig. 7.3 shows the normal probability plot of the transformed variate for the cystic fibrosis group, and the previous suggestion of curvature seems to have been removed. Applying the same technique to each of the other three variates in turn yielded the following estimated values of λ: OS, $\hat{\lambda} = -0.20$; OH, $\hat{\lambda} = 0.25$; EIL, $\hat{\lambda} = -0.40$. Figures 7.4, 7.5, and 7.6 show the normal probability plots of each of these three variates, after transformation, for the cystic fibrosis group. All three plots appear to be suitably linear (to within computer line-plot accuracy).

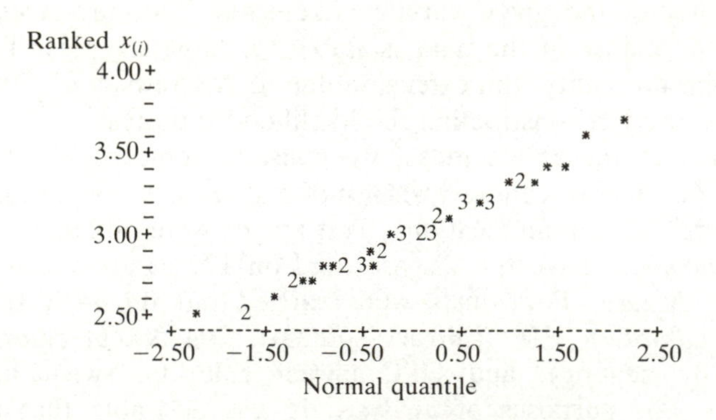

Fig. 7.3 Normal probability plot of transformed OPC values for the cystic fibrosis group

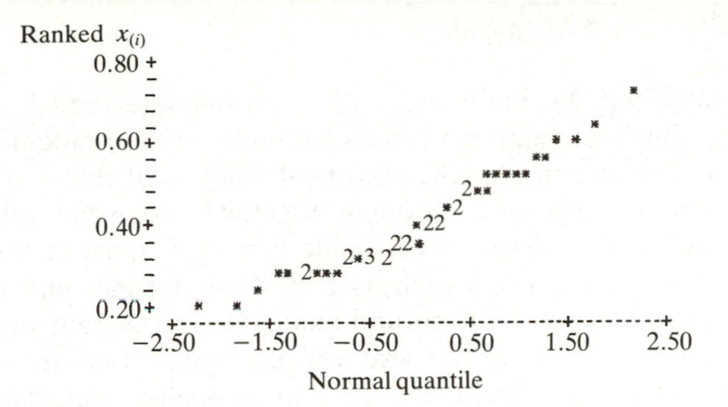

Fig. 7.4 Normal probability plot of transformed EIL values for the cystic fibrosis group

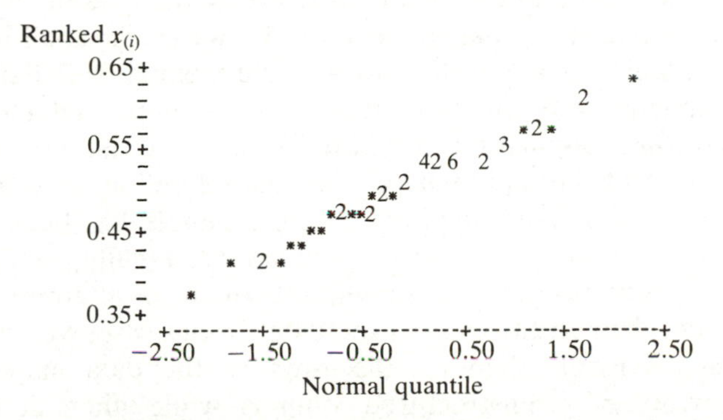

Fig. 7.5 Normal probability plot of transformed OS values for the cystic fibrosis group

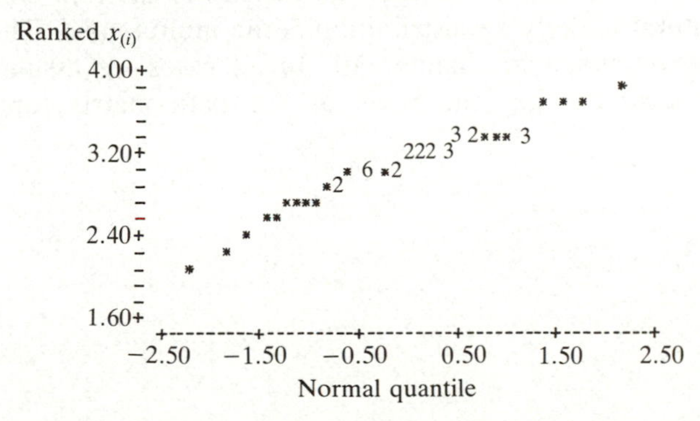

Fig. 7.6 Normal probability plot of transformed OH values for the cystic fibrosis group

7.7 Multivariate analysis

Having established the main ideas of a multivariate model, and summarized the chief distributional results for multivariate random variables, we are now able to consider the statistical analysis of multivariate data. Various different aspects are brought together under the general umbrella of 'statistical analysis'. As is to be expected, most of the familiar univariate inferential aspects such as hypothesis testing, and point and interval estimation, have their natural analogues in the multivariate case. In addition, however, there are also various aspects that are specific to multivariate situations. Reduction of dimensionality, determination of the most effective variables for a given problem, and explanation of observed associations between variables by a suitable set of unobserved (or 'latent') variables are just some of these latter aspects. In keeping with a problem-oriented approach to the subject, we will consider all these aspects according to either the structure of the data or the focus of attention in a particular problem. In Part III, we consider all inferential techniques suitable for application to a single unstructured data matrix, which corresponds to the simplest situation of a sample obtained from a single multivariate population. In Part IV we consider problems that require the rows of the data matrix to be considered as partitioned into groups, which is the case whenever a single sample has been obtained from each of a number of different populations. Finally, in Part V we consider all the questions that arise when the focus of attention is on the variables rather than on the units in the data. In this case, we concentrate on the columns rather than on the rows of the data matrix. Some situations correspond to unstructured columns, while others demand that the columns be considered as partitioned into groups. In the majority of cases we shall assume multivariate normality when setting up models, but some attention will also be devoted to analysis of discrete data in which case the natural underlying distribution is the multinomial. This distribution will be discussed in Chapter 10. In all cases we assume that the individuals sampled, i.e. the rows of the data matrix, are mutually uncorrelated.

Part III: Analysing ungrouped data

8 Estimation and hypothesis testing

8.1 Basic concepts and distributional results

We suppose the data for analysis to be in the form of a single unstructured $(n \times p)$ data matrix $\mathbf{X}$. There are thus n observations of a p-vector $\boldsymbol{x}$, and the observed values $\boldsymbol{x}_1', \boldsymbol{x}_2', \ldots, \boldsymbol{x}_n'$ constitute the rows of $\mathbf{X}$. To effect any form of statistical analysis we need to postulate a model for the behaviour of $\boldsymbol{x}$. Since there is no a priori information (in the form of structure imposed on $\mathbf{X}$), we can at most assume that $\boldsymbol{x}_1, \boldsymbol{x}_2, \ldots, \boldsymbol{x}_n$ are uncorrelated (or independent) observations from a single population. Details of this population can be specified by assuming some distributional form for the random vector $\boldsymbol{x}$. We shall suppose initially that the observations are all *quantitative,* in which case the most practically reasonable assumption to make is that $\boldsymbol{x}$ follows a multivariate normal distribution with some mean $\boldsymbol{\mu}$ and some dispersion matrix $\boldsymbol{\Sigma}$. Situations in which some or all of the observations are qualitative will be considered in Chapter 10.

Given this model, many questions of practical interest may be answered either by *estimating* the unknown parameters $\boldsymbol{\mu}$ and $\boldsymbol{\Sigma}$, or by testing some *hypothesis* about them. To fix ideas, consider a small example. Rao (1948) gave a 28×4 data matrix showing the weights of bark deposits of twenty-eight trees in the four directions, North, East, South, and West. If these trees are taken to be a random sample from the population of all such trees, it would be of interest to determine the vector of mean bark deposits for this population. Estimation of the population dispersion matrix will then give some insight into the possible variation in bark deposits from tree to tree. Finally, the investigator may have some theory about the bark deposits that he or she would like to test: one possible hypothesis is that the mean bark deposit for the population is the same in all four directions. This process of establishing population features from sample data is known as *statistical inference.* The inferential questions cited above for bark deposits are simply extensions to the multivariate case of corresponding univariate problems, and many readers will be familiar already with the approaches taken in the univariate case to deal with such questions. We shall therefore merely summarize briefly the general philosophy behind these approaches, and leave the reader to fill in any necessary details from a standard text on elementary statistical inference.

Since the sample data form the only available information on which to base inferences, any parameter estimate or hypothesis test must be formed from, or based on, some function of sample values. Such a function, which yields a numerical value when sample values are substituted into it, is termed a *statistic.* If this statistic is used to generate an estimate of a parameter it is usually referred to as an *estimator,* while if it is used to test a hypothesis it is called a *test statistic.* Irrespective of the use to which it is to be put, any statistic will show fluctuation in its values from sample to sample. For example, the mean weight of bark deposits in the North direction is a statistic for the data alluded to earlier. The value of this statistic for the twenty-eight trees discussed by Rao was 50.536 centigrams. A further twenty-eight trees might yield a mean of 52.460 centigrams, while a third set of twenty-eight trees could yield a mean of, say, 46.875 centigrams. In general, a *distribution* of values will be generated for any statistic by taking (random) samples of the given size n from the parent population; this distribution is termed the *sampling distribution* of that statistic. In order to assess the worth of a statistic as an estimator of a population parameter, we must judge the behaviour of the statistic over all possible samples that might be drawn from the target population. This behaviour is described by the sampling distribution of the statistic. An estimator of some parameter is usually adjudged to be a good one if its sampling distribution is centred at that parameter's true value, and has as small a spread as possible. Such an estimator is an *unbiased* estimator having *minimum variance.* Its intuitive appeal is that any single value obtained in a practical application should thus be 'reasonably close' to the true value of the parameter. Note that the spread of the sampling distribution gives an indication of the likely fluctuation in the estimator's values from sample to sample, but does not tell us how close the estimate is to the true value in any *particular* single realization. Therefore, instead of obtaining a single ('point') estimate of the unknown parameter with no knowledge of how close it is to the true value, we might prefer to find a set of values ('interval') within which the true parameter is almost certain to lie. The sampling distribution again plays a central role, in setting the end points of this interval. Finally, when testing a hypothesis about the unknown parameter, we wish to determine the likelihood of occurrence of the observed value of the test statistic assuming the hypothesis to be true. This question can also be answered by reference to the test statistic's sampling distribution under this hypothesis.

The above general reasoning is employed in all inferential problems, whether they be univariate or multivariate. The only difference between the two cases is that in the former, all statistics and their sampling distributions are univariate, while in the latter the statistics used are often (but not always) multivariate, so their sampling distributions will also generally be multivariate. However, many of these multivariate statistics and sampling distributions are natural analogues of the corresponding univariate quantities, so a firm knowledge of univariate techniques will ensure a good appreciation of the multivariate counterparts. We therefore summarize the chief sampling distributions that play a major part in inferential statistics based on the normal distribution.

Let us start with the univariate case. Suppose that $x_1, \ldots, x_n$ constitute a random sample of size n from a normal population with mean μ and variance σ^2. The two statistics that summarize all the available sample information about the unknown parameters (i.e. are *sufficient* for μ and σ^2) are the sample mean $\bar{x} = \frac{1}{n} \sum_{i=1}^{n} x_i$ and the sample variance $s^2 = \frac{1}{n-1} \sum_{i=1}^{n} (x_i - \bar{x})^2$. Because of this sufficiency, all optimal procedures concerning μ and σ^2 are based on these two statistics, which are therefore the two most common statistics used in univariate inference. Both of these statistics possess sampling distributions that describe the likely fluctuations in their values from sample to sample, and it is shown in most elementary texts on statistical inference that:

(1) the sampling distribution of $\bar{x}$ is the normal distribution with mean μ and variance $\frac{1}{n}\sigma^2$,
(2) the sampling distribution of $(n-1)s^2$ is σ^2 times a chi-squared distribution on $(n-1)$ degrees of freedom, and
(3) $\bar{x}$ and s^2 are statistically independent in their behaviour.

The central limit theorem furthermore establishes that, if n is large, the sampling distribution of $\bar{x}$ is *approximately* normal $\left(\text{with mean } \mu \text{ and variance } \frac{1}{n}\sigma^2\right)$, *whatever the shape of the parent population of the* x_i. Thus result (1) is applied almost invariably when n is sufficiently large (say $n > 25$). No similar approximation exists for result (2), however, and result (3) is certainly not true for most non-normal distributions.

Now consider the multivariate analogue of the above situation. Here $\boldsymbol{x}_1, \ldots, \boldsymbol{x}_n$ constitute a random sample of p-vectors from a multivariate normal population with mean vector $\boldsymbol{\mu}$ and dispersion matrix $\boldsymbol{\Sigma}$. It can be shown readily that the two statistics containing all the available sample information about the unknown parameters, i.e. that are sufficient for $\boldsymbol{\mu}$ and $\boldsymbol{\Sigma}$, are the sample mean vector $\bar{\boldsymbol{x}} = \frac{1}{n}\sum_{i=1}^{n} \boldsymbol{x}_i$, and the sample covariance matrix $\mathbf{S} = \frac{1}{n-1}\sum_{i=1}^{n} (\boldsymbol{x}_i - \bar{\boldsymbol{x}})(\boldsymbol{x}_i - \bar{\boldsymbol{x}})'$. These two statistics have already been used extensively in the descriptive techniques of Part I. It is worth recollecting that the sample mean vector is just the vector of sample means of each of the variates, while the (j, k)th element of $\mathbf{S}$ contains the sample covariance $\frac{1}{n-1}\sum_{i=1}^{n} (x_{ij} - \bar{x}_j)(x_{ik} - \bar{x}_k)$ between the jth and kth variates (if $j \neq k$) or the sample variance $\frac{1}{n-1}\sum_{i=1}^{n} (x_{ik} - \bar{x}_k)^2$ of the kth variate (if $j = k$). It is also convenient to use the symbol $\mathbf{C}$ to denote the sample sum of squares and products (SSP) matrix, so that $\mathbf{C} = (n-1)\mathbf{S}$. Now, we saw in Chapter 7 that the multivariate normal and Wishart distributions are the multivariate analogues of the univariate normal and chi-squared distributions respectively. Hence it will come as no surprise to find that:

(1) the sampling distribution of $\bar{\boldsymbol{x}}$ is the multivariate normal distribution with mean vector $\boldsymbol{\mu}$ and dispersion matrix $\frac{1}{n}\boldsymbol{\Sigma}$,
(2) the sampling distribution of $\mathbf{C} = (n-1)\mathbf{S}$ is the Wishart distribution with $(n-1)$ degrees of freedom and parameter $\boldsymbol{\Sigma}$, and
(3) $\bar{\boldsymbol{x}}$ and $\mathbf{C}$ are statistically independent in their behaviour.

Furthermore, the multivariate central limit theorem states that if n is large, the sampling distribution of $\bar{\boldsymbol{x}}$ is *approximately* multivariate normal $\left(\text{with mean vector } \boldsymbol{\mu} \text{ and dispersion matrix } \frac{1}{n}\boldsymbol{\Sigma}\right)$ *whatever the shape of the parent population of the* $\boldsymbol{x}_i$. Formal proof of these results is not attempted here, as it is in the province of mathematical statistics. However, the practical implications are that simple extensions of the familiar univariate procedures should also yield good procedures in the multivariate case.

One further multivariate analogue of a univariate distributional result will prove to be of crucial importance in the inferential problems to be tackled subsequently, and can be stated here. Let us first recollect the important univariate result. If the random variable x has a normal distribution with zero mean and variance σ^2, independently of the random variable v which has σ^2 times a chi-squared distribution on k degrees of freedom, then $x \div \sqrt{(v/k)}$ follows a Student's t distribution on k degrees of freedom. Thus $x\sqrt{k} \div \sqrt{v}$ follows a t_k distribution, which is a definition of this distribution. But a standard univariate result tells us that if the random variable z has a t_k distribution then z^2 has an $F_{1,k}$

distribution. Hence kx^2/v has an $F_{1,k}$ distribution. Rewriting this relationship, it follows that $xv^{-1}x$ is distributed as $\frac{1}{k}F_{1,k}$. The expression has been rewritten into this form to enable a direct connection to be made with the corresponding multivariate result. This result states that if $\boldsymbol{x}$ has a multivariate normal distribution with mean vector zero and dispersion matrix $\boldsymbol{\Sigma}$, independently of the matrix $\mathbf{C}$ which has a Wishart distribution with k degrees of freedom and parameter $\boldsymbol{\Sigma}$, then $\boldsymbol{x}'\mathbf{C}^{-1}\boldsymbol{x}$ is distributed as $\frac{p}{k-p+1}$ times an F variable on p and $k-p+1$ degrees of freedom, i.e. $\left(\frac{k-p+1}{p}\right)\boldsymbol{x}'\mathbf{C}^{-1}\boldsymbol{x}$ has an $F_{p,k-p+1}$ distribution.

This last result can be used to justify the modified probability plot of Section 7.5 (dealing with tests of normality). There we used the sample mean $\bar{\boldsymbol{x}}$, the sample covariance matrix $\mathbf{S}$, and a further independent observation $\boldsymbol{x}$, where the parent population is assumed to be $\mathrm{N}(\boldsymbol{\mu}, \boldsymbol{\Sigma})$. Since $\boldsymbol{x}$ and $\bar{\boldsymbol{x}}$ are independent, and the distribution of $\bar{\boldsymbol{x}}$ is $\mathrm{N}\left(\boldsymbol{\mu}, \frac{1}{n}\boldsymbol{\Sigma}\right)$ from above, then $(\boldsymbol{x}-\bar{\boldsymbol{x}})$ has a multivariate normal distribution with mean vector $\boldsymbol{0}$ and dispersion matrix $\left(1+\frac{1}{n}\right)\boldsymbol{\Sigma} = \left(\frac{n+1}{n}\right)\boldsymbol{\Sigma}$. Thus $\left\{\sqrt{\left(\frac{n}{n+1}\right)}\right\}\{\boldsymbol{x}-\bar{\boldsymbol{x}}\}$ is distributed as $N(\boldsymbol{0}, \boldsymbol{\Sigma})$. Also, $\mathbf{S}^{-1} = (n-1)\mathbf{C}^{-1}$ and $\mathbf{C}$ has a $\mathrm{W}(n-1, \boldsymbol{\Sigma})$ distribution. Hence, using the above result with $(n-1)$ in place of k, and $\left\{\sqrt{\left(\frac{n}{n+1}\right)}\right\}\{\boldsymbol{x}-\bar{\boldsymbol{x}}\}$ in place of $\boldsymbol{x}$, it follows that $\left(\frac{n-p}{p}\right)\left(\frac{n}{n+1}\right)(\boldsymbol{x}-\bar{\boldsymbol{x}})'\mathbf{C}^{-1}(\boldsymbol{x}-\bar{\boldsymbol{x}})$ has an $F_{p,n-p}$ distribution. Thus $\left(\frac{n-p}{p}\right)\left(\frac{n}{n+1}\right)\left(\frac{1}{n-1}\right)(\boldsymbol{x}-\bar{\boldsymbol{x}})'\mathbf{S}^{-1}(\boldsymbol{x}-\bar{\boldsymbol{x}})$, i.e. $\frac{n(n-p)}{p(n^2-1)}(\boldsymbol{x}-\bar{\boldsymbol{x}})'\mathbf{S}^{-1}(\boldsymbol{x}-\bar{\boldsymbol{x}})$, has an $F_{p,n-p}$ distribution. Hence $(\boldsymbol{x}-\bar{\boldsymbol{x}})'\mathbf{S}^{-1}(\boldsymbol{x}-\bar{\boldsymbol{x}})$ has $\frac{p(n^2-1)}{n(n-p)}$ times an $F_{p,n-p}$ distribution, as asserted in Section 7.5.

8.2 Estimation of $\boldsymbol{\mu}$ and $\boldsymbol{\Sigma}$

Suppose that $\boldsymbol{x}_1, \boldsymbol{x}_2, \ldots, \boldsymbol{x}_n$ are n independent observations taken from a p-variate population which has mean $\boldsymbol{\mu}$ and dispersion matrix $\boldsymbol{\Sigma} = (\sigma_{ij})$. Then the sample mean vector $\bar{\boldsymbol{x}}$ and the sample covariance matrix $\mathbf{S} = \frac{1}{n-1}\mathbf{C} = \frac{1}{n-1}\sum_{i=1}^{n}(\boldsymbol{x}_i-\bar{\boldsymbol{x}})(\boldsymbol{x}_i-\bar{\boldsymbol{x}})'$ are the 'obvious' estimates of $\boldsymbol{\mu}$ and $\boldsymbol{\Sigma}$. If we further assume some specific distributional form for the

population, we can then apply a general principle such as maximum likelihood to obtain estimates. In particular, under the assumption of normality, maximum likelihood estimates of $\boldsymbol{\mu}$ and $\boldsymbol{\Sigma}$ are $\hat{\boldsymbol{\mu}} = \bar{\mathbf{x}}$ and $\hat{\boldsymbol{\Sigma}} = \frac{1}{n}\mathbf{C} = \frac{n-1}{n}\mathbf{S}$. (For a relatively simple mathematical derivation, see Rayner (1985a), together with the amendments in Carlstein *et al.* (1985) and Rayner (1985b).) Given the assumption of normality, it is relatively straightforward to find the expectation of each element of $\bar{\mathbf{x}}$ and $\mathbf{S}$ under repeated sampling, and to show that $\bar{\mathbf{x}}$ and $\mathbf{S}$ are *unbiased* estimates of $\boldsymbol{\mu}$ and $\boldsymbol{\Sigma}$. The maximum likelihood estimate $\hat{\boldsymbol{\Sigma}}$ is therefore biased, although its bias of $-\frac{1}{n}\boldsymbol{\Sigma}$ is unlikely to be of any practical significance in samples of even moderate size. However, the most commonly used estimates in practice are the unbiased ones, $\bar{\mathbf{x}}$ and $\mathbf{S}$.

Now $\bar{\mathbf{x}}$ is a *point estimate of* $\boldsymbol{\mu}$. Our conceptual image of the population from which the sample is taken is of a dense swarm of points in p-dimensional space centred on the point $\boldsymbol{\mu}$. If the assumption of normality is appropriate, this swarm forms a hyperellipsoid whose density is greatest around the point $\boldsymbol{\mu}$, decreasing progressively as the edges of the hyperellipsoid are approached. $\bar{\mathbf{x}}$ is a point in this space, representing a 'guess' at the location of $\boldsymbol{\mu}$. Because of the vagaries of sampling, we can be almost certain that this guess will be incorrect, so a more appealing procedure is to identify a *region* in the space within which we are reasonably sure that the point $\boldsymbol{\mu}$ is located. Assuming multivariate normality for the observations, we can obtain such a region as follows. Recollect from the previous section that $(\bar{\mathbf{x}} - \boldsymbol{\mu})$ is distributed as $N\left(\mathbf{0}, \frac{1}{n}\boldsymbol{\Sigma}\right)$, independently of $\mathbf{C} = (n-1)\mathbf{S}$ which has a $W(n-1, \boldsymbol{\Sigma})$ distribution. Thus, using the general result quoted in the same section, $n\left(\frac{n-p}{p}\right)(\bar{\mathbf{x}} - \boldsymbol{\mu})'\mathbf{C}^{-1}(\bar{\mathbf{x}} - \boldsymbol{\mu})$ has an $F_{p,n-p}$ distribution. Now $\mathbf{C} = (n-1)\mathbf{S}$, so that $\mathbf{C}^{-1} = \frac{1}{n-1}\mathbf{S}^{-1}$.

Hence $\left\{\frac{n(n-p)}{p(n-1)}\right\}(\bar{\mathbf{x}} - \boldsymbol{\mu})'\mathbf{S}^{-1}(\bar{\mathbf{x}} - \boldsymbol{\mu})$ has an $F_{p,n-p}$ distribution. Let $F^{\alpha}_{p,n-p}$ be the $100(1-\alpha)$ per cent point of the $F_{p,n-p}$ distribution, i.e. the value exceeded by 100α per cent of the $F_{p,n-p}$ distribution. (This value is the one usually found from statistical tables when interpreting, e.g. analysis of variance.) Then

$$\Pr\left[\left\{\frac{n(n-p)}{p(n-1)}\right\}(\bar{\mathbf{x}} - \boldsymbol{\mu})'\mathbf{S}^{-1}(\bar{\mathbf{x}} - \boldsymbol{\mu}) < F^{\alpha}_{p,n-p}\right] = 1 - \alpha$$

i.e.

$$\Pr\left\{(\bar{\boldsymbol{x}}-\boldsymbol{\mu})'\mathbf{S}^{-1}(\bar{\boldsymbol{x}}-\boldsymbol{\mu}) < \frac{p(n-1)}{n(n-p)} F^{\alpha}_{p,n-p}\right\} = 1-\alpha. \tag{8.1}$$

But the expression $(\bar{\boldsymbol{x}}-\boldsymbol{\mu})'\mathbf{S}^{-1}(\bar{\boldsymbol{x}}-\boldsymbol{\mu}) = c$, when considered as a function of $\boldsymbol{\mu}$, defines the boundary of a hyperellipsoid, centred at the point $\bar{\boldsymbol{x}}$, in the p-dimensional space of possible values of $\boldsymbol{\mu}$. Thus $(\bar{\boldsymbol{x}}-\boldsymbol{\mu})'\mathbf{S}^{-1}(\bar{\boldsymbol{x}}-\boldsymbol{\mu}) < \frac{p(n-1)}{n(n-p)} F^{\alpha}_{p,n-p}$ can be intuitively interpreted as the interior of that hyperellipsoid in the population space which is centred at the point $\bar{\boldsymbol{x}}$, and which has a probability $(1-\alpha)$ of containing the population mean $\boldsymbol{\mu}$.

In precise terms, if we were to collect many samples of size n from the same population, obtaining the region $(\bar{\boldsymbol{x}}-\boldsymbol{\mu})'\mathbf{S}^{-1}(\bar{\boldsymbol{x}}-\boldsymbol{\mu}) < \frac{p(n-1)}{n(n-p)} F^{\alpha}_{p,n-p}$ for each sample, then $100(1-\alpha)$ per cent of all these regions would enclose the true population mean $\boldsymbol{\mu}$. Such a region is termed the *100(1 − α) per cent confidence region* for $\boldsymbol{\mu}$. Clearly, the smaller we make α the larger will be the *confidence coefficient* $(1-\alpha)$, and the larger will be the confidence region (since the larger is the region, the more confident we are of capturing the value $\boldsymbol{\mu}$ within it).

There are many ways of calculating confidence regions in any practical situation. If there are several different prescriptions for obtaining a $100(1-\alpha)$ per cent confidence region for a given parameter then, in general, the various regions will enclose different volumes of the underlying space. The 'best' region will be the one enclosing the smallest such volume, because this region will give us the most precise estimate of the location of $\boldsymbol{\mu}$. The region described above for $\boldsymbol{\mu}$ is the best of various regions that could be derived for the mean of a single population. Note that it is based only on the sufficient statistics $\bar{\boldsymbol{x}}$ and $\mathbf{S}$, but requires the assumption of multivariate normality in its derivation.

Corresponding confidence region theory does not exist for the dispersion matrix $\boldsymbol{\Sigma}$ because $\mathbf{S}$, being a $(p \times p)$ matrix of numbers, is no longer a point estimate in the sense of being representable as a point in the underlying p-dimensional population space. It could be represented by a point in some other space, however, and one such possibility is the space of all $(p \times p)$ symmetric positive semi-definite matrices. The constraint of symmetry means that the dimensionality of this space is $\frac{1}{2}p(p+1)$ rather than p^2, and the coordinates of the point representing a particular matrix in this space are given by the entries in the lower triangle of the matrix (including the diagonal). The constraint of positive semi-definiteness further restricts the matrix to lie within only a portion of this $\frac{1}{2}p(p+1)$-space, namely a convex cone. Both $\mathbf{S}$ and $\boldsymbol{\Sigma}$ are represented by points in this cone, and hence **S** could be viewed in this sense as a point estimate of $\boldsymbol{\Sigma}$. In principle, therefore, we could attempt to find a confidence region within this space for estimation of $\boldsymbol{\Sigma}$. In practice, however, the mathematical theory is so complicated, and the resulting interpretation of the region so far removed from reality, as to render the exercise of academic interest only. Some results associated with this space of variance matrices are given by James (1973).

The above theory concerns the standard approach to the one-sample estimation problem, when applied to well-behaved data. In practice, however, various problems may occur that mean the data no longer fall into the standard framework assumed earlier. The most obvious such problem is when some of the values in the data matrix are missing for some reason. Another common problem arises when some of the values in the data matrix are extreme or unusual in some way. For useful practical applications, we therefore need some modifications to the existing theory that will enable both missing values and extreme values (or 'outliers') to be accommodated without unduly influencing the estimates that are derived.

The missing-value problem has already been discussed in other contexts, and the general approach to it in the present case is the same as before. Assuming multivariate normality, the iterative scheme described by Beale and Little (1975) converges to give the maximum likelihood estimates $\hat{\boldsymbol{\mu}}$ and $\hat{\boldsymbol{\Sigma}}$. It is tempting to adopt the simple process of estimating each element of $\boldsymbol{\mu}$ and $\boldsymbol{\Sigma}$ separately, in the standard univariate way from all available single variate or paired variate observations, and then putting the separate estimates together to form $\tilde{\boldsymbol{\mu}}$ and $\tilde{\boldsymbol{\Sigma}}$. However, $\tilde{\boldsymbol{\Sigma}}$ formed in this way need not be positive semi-definite, and this will cause problems if an inverse is subsequently required. Devlin *et al.* (1981) suggest 'shrinking' the elements to achieve positive semi-definiteness. Occasionally the missing values have a special form. One such form that occurs from time to time in practice has been termed the *monotone sample*. This is where all the individuals observed on the ith element of $\boldsymbol{x}$ are also observed on the $(i-1)$th, but only a subset of those observed on the $(i-1)$th are observed on the ith $(i = 1, \ldots, p)$. Thus if there are n_i observations made on the ith element of $\boldsymbol{x}$ $(i = 1, \ldots, p)$, then $n_1 \geqslant n_2 \geqslant n_3 \geqslant \ldots \geqslant n_p$. Afifi and Elashoff (1966) give explicit expressions for the maximum likelihood estimators of $\boldsymbol{\mu}$ and $\boldsymbol{\Sigma}$ in this case.

A more substantial discussion can be allotted to the problem of accommodating untypical observations in the estimation process, generally referred to as *robust estimation*. For the uniresponse situation, problems and methods of robust estimation have received considerable attention. This univariate work has been concerned largely with the problem of obtaining and studying either estimates or test statistics that are insensitive to outliers, and the distributions generally considered as alternatives to the normal have almost always been symmetrical with heavier tails than the normal. Extension to the multivariate situation is still relatively undeveloped, the main problem being that the variety, both in kind and effect, of outliers in multiresponse data can be very large. Observations that are grossly atypical in a univariate context are generally readily detectable simply by inspection. For multivariate data, however, observations are often found to be atypical only when the value for each variable is considered in relation to all the other variables. Even more problematic is the interrelated nature of outlier effects on the location, scale, and orientation of the data. It is often not possible to isolate the effect of outliers on one of these aspects without also considering the other aspects, and adjusting an estimate of location, for example, to correct for the effect of an outlier will inevitably require some adjustment to estimates of dispersion and orientation. Thus an iterative process is strongly suggested. A comprehensive general treatment may be found in Huber (1977), while specifically multivariate techniques are discussed by Campbell (1980b) and Devlin *et al.* (1981). We shall content ourselves with a brief description of one general approach, using the implementation proposed by Campbell.

The idea here is extremely simple. Instead of using the standard estimates of location and dispersion, we use weighted estimates in which outliers or untypical observations are downweighted relative to the majority of the sample members. Such estimators are often called M-estimators, and the basic equations used to define them are

$$\bar{\boldsymbol{x}}_{\mathrm{M}} = \sum_{s=1}^{n} w_s \boldsymbol{x}_s \Big/ \sum_{s=1}^{n} w_s \tag{8.2}$$

$$\mathbf{S}_{\mathrm{M}} = \sum_{s=1}^{n} w_s^2 (\boldsymbol{x}_s - \bar{\boldsymbol{x}}_{\mathrm{M}})(\boldsymbol{x}_s - \bar{\boldsymbol{x}}_{\mathrm{M}})' \Big/ \left(\sum_{s=1}^{n} w_s^2 - 1 \right) \tag{8.3}$$

where w_s is the weight attached to the sth sample member. If we set $w_s = 1$ for all s then all sample members receive the same weight, and eqns (8.2) and (8.3) define the usual estimators of $\boldsymbol{\mu}$ and $\boldsymbol{\Sigma}$. If we set $w_s = 1$ for 'typical' sample members and $w_s = 0$ for 'atypical' members, then we ignore the latter entirely and form estimates of $\boldsymbol{\mu}$ and $\boldsymbol{\Sigma}$ in the usual way from the former members alone. The 'best' approach is generally considered to lie between these two extremes, with each w_s being assigned a number in the range $(0, 1)$. The problem is that the weights can rarely be determined at the outset, but must be assigned on the basis of the corresponding observation's 'atypicality' relative to $\boldsymbol{\mu}$ and $\boldsymbol{\Sigma}$. This dependence makes an iterative approach inevitable. At the first stage there is no prior information about possible outliers, so all w_s are assigned the value 1, and $\boldsymbol{\mu}$, $\boldsymbol{\Sigma}$ are estimated by $\bar{\boldsymbol{x}}$, $\mathbf{S}$. Comparison of each $\boldsymbol{x}_s$ with $\bar{\boldsymbol{x}}$ will give some idea of its 'atypicality' and will hence suggest suitable values of w_s $(s = 1, \ldots, n)$, thus enabling (8.2) and (8.3) to be employed, and $\bar{\boldsymbol{x}}_{\mathrm{M}}$, $\mathbf{S}_{\mathrm{M}}$ to be calculated. Since $\bar{\boldsymbol{x}}_{\mathrm{M}}$ and $\mathbf{S}_{\mathrm{M}}$ are almost certain to differ from $\bar{\boldsymbol{x}}$ and $\mathbf{S}$, however, the 'atypicality' of each sample member will need to be re-assessed and the weights w_s correspondingly recomputed. Thus (8.2) and (8.3) can be re-employed and new values of $\bar{\boldsymbol{x}}_{\mathrm{M}}$ and $\mathbf{S}_{\mathrm{M}}$ obtained. The whole process can be repeated as many times as is necessary for stability in weights and estimates to be achieved; in practice, more than a few iterations are rarely needed before successive estimates agree to within any predetermined level of accuracy.

Before the above scheme can be implemented, however, some method is needed of measuring each sample member's 'atypicality' relative to the others in the sample. Intuitively, it seems sensible to consider those sample members 'close to' the sample mean as being 'typical' of the sample, and those 'far away' from the sample mean as being 'atypical', but how are we to measure these distances? We have already had recourse to a geometrical representation of the sample as n points in p-dimensional space, where each variate measured on the sample members is identified with an orthogonal axis of this space. The sample mean is a point in this space, so that the Euclidean distance $\sqrt{\{(\boldsymbol{x}_s - \bar{\boldsymbol{x}})'(\boldsymbol{x}_s - \bar{\boldsymbol{x}})\}}$ between the points representing the sth sample member $\boldsymbol{x}_s$ and the sample mean $\bar{\boldsymbol{x}}$ suggests itself as a measure of 'atypicality' of the sth sample member. However, there is one major objection to this measure. The Euclidean distance attaches equal weight to all the axes of the representation, but in the presence of differential variances among variates, and of correlations between variates, this may not be a desirable feature. If, for example, variate x_1 has much larger variance than variate x_2 (but x_1 and x_2 are uncorrelated), a scatter plot of a sample of (x_1, x_2) pairs may look like Fig. 8.1. (It is assumed in this diagram that both variates are mean-centered.) An observation can be much further from the sample mean in the x_1 direction than in the x_2 direction without being necessarily more 'atypical'. Putting this another way, the point P looks 'atypical' in the x_2 direction, although its distance from the origin in this direction is less than the distance from the origin in the x_1 direction of many more 'typical' observations. If we now envisage two further variates x_3 and x_4 which have similar variances to x_1 and x_2 respectively but are highly correlated, then a large difference in x_3 values between two sample members will be associated with a large difference in x_4 values. Thus, in a sense, the Euclidean distance between the corresponding points in the x_3x_4 plane overemphasizes the difference between these sample members. More interest would attach to a large

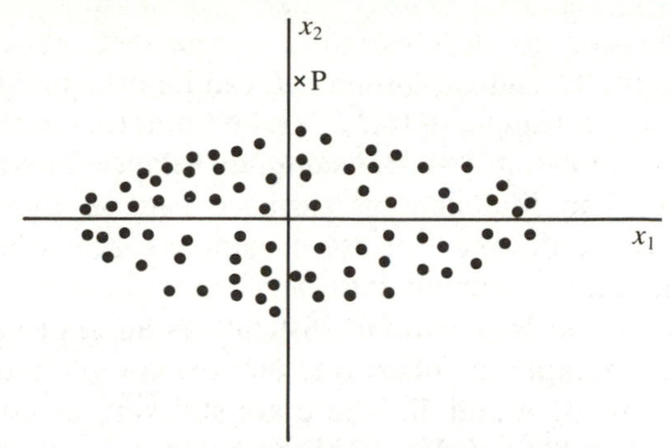

Fig. 8.1 Scatter plot for bivariate sample in which $\text{Var}(x_1) > \text{Var}(x_2)$, and x_1, x_2 are uncorrelated

Euclidean distance if it were in the x_1x_2 plane, where the variates are uncorrelated and where there is no spurious contribution to the difference between the observations.

Thus, for Euclidean distance to be an appropriate measure, the variates used in its computation should be uncorrelated and of equal variance. Given observed variates $\boldsymbol{x}$, let us therefore consider a linear transformation to new variates $\boldsymbol{y} = \mathbf{C}\boldsymbol{x}$ in order to satisfy these requirements. Using the results of Section 7.1, it follows that if the sample dispersion matrix of $\boldsymbol{x}$ is $\mathbf{S}$, then the sample dispersion matrix of $\boldsymbol{y}$ is $\mathbf{CSC}'$. Since we want the variates in $\boldsymbol{y}$ to be uncorrelated and to have equal variance, we thus need to choose $\mathbf{C}$ to satisfy $\mathbf{CSC}' = k\mathbf{I}$ for some k. It is simplest to standardize the new variates, in which case $k = 1$. If we restrict attention to non-singular transformations from p variates x_i to p variates y_i, it follows readily that $\mathbf{C}$ satisfies

$$\mathbf{S} = (\mathbf{C}'\mathbf{C})^{-1}$$

or

$$\mathbf{S}^{-1} = \mathbf{C}'\mathbf{C}. \tag{8.4}$$

Now an observed sample member $\boldsymbol{x}_s$ will have values $\boldsymbol{y}_s = \mathbf{C}\boldsymbol{x}_s$ of the new variables, and the sample mean $\bar{\boldsymbol{x}}$ will be transformed to $\bar{\boldsymbol{y}} = \mathbf{C}\bar{\boldsymbol{x}}$. Thus the 'atypicality' d_s of the sth sample member can be measured by the Euclidean distance between $\boldsymbol{y}_s$ and $\bar{\boldsymbol{y}}$, i.e.

$$\begin{aligned} d_s^2 &= (\boldsymbol{y}_s - \bar{\boldsymbol{y}})'(\boldsymbol{y}_s - \bar{\boldsymbol{y}}) \\ &= (\mathbf{C}\boldsymbol{x}_s - \mathbf{C}\bar{\boldsymbol{x}})'(\mathbf{C}\boldsymbol{x}_s - \mathbf{C}\bar{\boldsymbol{x}}) \\ &= \{\mathbf{C}(\boldsymbol{x}_s - \bar{\boldsymbol{x}})\}'\{\mathbf{C}(\boldsymbol{x}_s - \bar{\boldsymbol{x}})\} \\ &= (\boldsymbol{x}_s - \bar{\boldsymbol{x}})'\mathbf{C}'\mathbf{C}(\boldsymbol{x}_s - \bar{\boldsymbol{x}}) \\ &= (\boldsymbol{x}_s - \bar{\boldsymbol{x}})'\mathbf{S}^{-1}(\boldsymbol{x}_s - \bar{\boldsymbol{x}}) \end{aligned} \tag{8.5}$$

on substituting from (8.4).

The transformation to new variables $\boldsymbol{y}$ described above is a purely notional one, as the Euclidean distance d_s can be calculated directly from the values $\boldsymbol{x}_s$, $\bar{\boldsymbol{x}}$ and $\mathbf{S}$ using eqn (8.5). Viewed in terms of the $\boldsymbol{x}$ variables, this distance d_s is known as the *Mahalanobis* distance between $\boldsymbol{x}_s$ and the sample mean $\bar{\boldsymbol{x}}$. The Mahalanobis distance can be thought of as an appropriate statistical distance for use in sample spaces where there exist differential variances and correlations between variates.

Having specified the Mahalanobis distance as an appropriate measure of atypicality of a sample member, we can now complete the description of the M-estimators of $\boldsymbol{\mu}$ and $\boldsymbol{\Sigma}$. There are still various possible ways of linking d_s of (8.5) to the weights w_s of (8.2) and (8.3); Campbell (1980b) suggests setting

$$w_s = \omega(d_s)/d_s$$

where

$$\omega(d) = \begin{cases} d & \text{if } d \leq d_0 \\ d_0 \exp\{-\frac{1}{2}(d - d_0)^2/\beta\} & \text{if } d > d_0 \end{cases}$$

and

$$d_s = \surd\{(\boldsymbol{x}_s - \bar{\boldsymbol{x}}_{\mathrm{M}})'\mathbf{S}_{\mathrm{M}}^{-1}(\boldsymbol{x}_s - \bar{\boldsymbol{x}}_{\mathrm{M}})\}.$$

This choice ensures that full weight $w_s = 1$ is assigned to those observations within a Mahalanobis distance d_0 of the sample mean, and progressively decreasing weight is assigned to those observations further than this distance away. Campbell reparameterizes d_0 as $(\surd p + \alpha/\surd 2)$. The user thus needs to select values for the two parameters α and β. If the sample is genuinely drawn from a multivariate normal distribution with mean $\boldsymbol{\mu}$ and dispersion matrix $\boldsymbol{\Sigma}$, d^2 approximately follows a χ_p^2 distribution. Fisher's square-root approximation gives $d \sim \mathrm{N}(\surd p, 1/\surd 2)$. A natural choice for α is thus a suitable percentage point of the standard normal distribution, while β controls the rate of decrease of the weight function to zero. Campbell suggests from practical experience that one of the following three selections should prove acceptable in most applications.

(1) $\alpha = \infty$; β immaterial;
(2) $\alpha = 2$; $\beta = \infty$;
(3) $\alpha = 2$; $\beta = 1.25$.

Choice of (1) results in $w_s = 1$ for all s; choice of (2) results in $w_s = 1$ for $d_s < \surd p + \surd 2$ or $w_s = (\surd p + \surd 2)/d_s$ for $d_s > \surd p + \surd 2$, which is thus a step function with two possible values; while choice of (3) gives the full descending set of weights over the whole sample.

Campbell (1980b) describes the robust estimation of location and dispersion for several real data sets, and discusses the weights derived in the analyses in relation to the atypicality of the corresponding observations. He also reports on some Monte Carlo simulation experiments designed to check that the robust estimates obtained by his suggested procedure differ little from the usual estimates when applied to uncontaminated data. He concludes that the robust estimates of variance will generally be within 2–3 per cent of the usual estimates for well-behaved data, and that large sample agreement of the expected value of d^2 under robust and usual estimation is of the same order of magnitude.

8.3 Testing hypotheses about $\boldsymbol{\mu}$ and $\boldsymbol{\Sigma}$

Most of the hypotheses commonly encountered in the univariate case can also be generalized readily to the multivariate case. Thus, given a sample of size n from a p-variate population which has mean $\boldsymbol{\mu}$ and dispersion matrix $\boldsymbol{\Sigma}$, we would typically be interested in testing one of the hypotheses: that $\boldsymbol{\mu}$ had some specified value $\boldsymbol{\mu}_0$, that $\boldsymbol{\Sigma}$ had some specified value $\boldsymbol{\Sigma}_0$, or that $\boldsymbol{\mu}$ and $\boldsymbol{\Sigma}$ jointly had some specified values $\boldsymbol{\mu}_0$ and $\boldsymbol{\Sigma}_0$. Additionally, when testing the individual hypotheses about $\boldsymbol{\mu}$ or $\boldsymbol{\Sigma}$, we might or might not wish to assume that the true value of the other parameter was known. It is assumed that the reader is familiar with the standard theory underlying the general concept of hypothesis testing, and will therefore appreciate that for any tests to be developed it is necessary to assume a definite probability distribution for the random variable defining the population. The results of the present section are all therefore built on the assumption of multivariate normality for the random vector under consideration, for reasons already discussed in Chapter 7. It will also be assumed that the reader is familiar with the details of all the univariate tests of hypothesis that correspond to the situations outlined above.

Now, given that any multivariate observation can be viewed simply as a collection of univariate observations, it might be thought that a multivariate hypothesis could be tested most easily by means of a series of univariate tests. For example, given a sample from a $N(\boldsymbol{\mu}, \boldsymbol{\Sigma})$ population, then to test the hypothesis $\boldsymbol{\mu} = \boldsymbol{\mu}_0$ (with $\boldsymbol{\Sigma}$ unknown) we need only test the p separate hypotheses $\mu_i = \mu_{0i}$, where $\boldsymbol{\mu} = (\mu_1, \mu_2, \ldots, \mu_p)$ and $\boldsymbol{\mu}_0 = (\mu_{01}, \mu_{02}, \ldots, \mu_{0p})$. Since all marginal distributions of a multivariate normal distribution are themselves normal (by Property 2 of Section 7.2), then we could view the multivariate sample as a set of univariate samples, the ith of which was from a $N(\mu_i, \sigma_i^2)$ population, and σ_i^2 was unknown. Standard theory would thus indicate a one-sample t-test for each of these univariate hypotheses.

Such an approach to testing the multivariate hypothesis would be

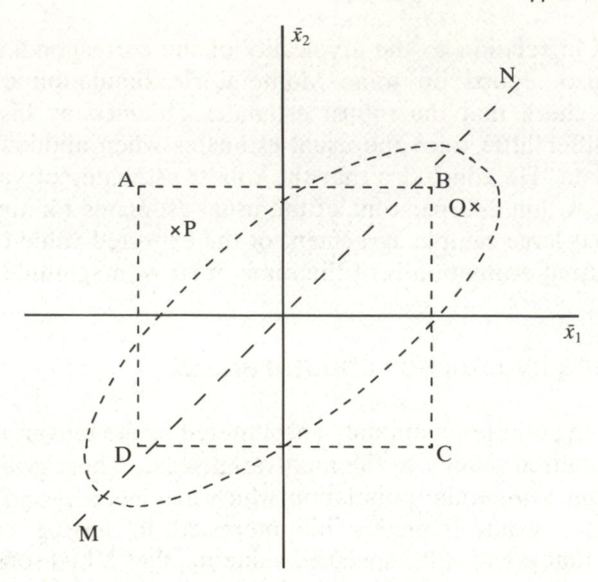

Fig. 8.2 Plot of sample means for a bivariate sample, with regions within which the hypothesis $\boldsymbol{\mu} = \boldsymbol{0}$ would not be rejected when using (1) a multivariate criterion and (2) two univariate criteria

unsatisfactory, however, for two main reasons. Most crucially, the series of univariate tests ignores any dependencies between the variables that will exist in the multivariate sample, and hence may lead to the wrong inferences. An illustration of one such possibility is given in Fig. 8.2. Suppose that we have a bivariate sample of size n, the two variables being denoted by x_1 and x_2, and we wish to test the null hypothesis that $\boldsymbol{\mu} = \boldsymbol{0}$ against the alternative that $\boldsymbol{\mu} \neq \boldsymbol{0}$. Considering first x_1, we therefore require a test of $\mu_1 = 0$ against $\mu_1 \neq 0$, and in the absence of knowledge about the variance we would use the standard t-test. Denoting the sample mean and variance of x_1 by $\bar{x}_1$ and s_1^2, the appropriate test statistic is $t = (\bar{x}_1 - 0) \div s_1/\sqrt{n}$. The hypothesis $\mu_1 = 0$ would be rejected at the α per cent level of significance if $t < -t_{\frac{1}{2}\alpha}$ or $t > t_{\frac{1}{2}\alpha}$, i.e. if $\bar{x}_1$ fell outside the interval $(-s_1 t_{\frac{1}{2}\alpha}/\sqrt{n}, +s_1 t_{\frac{1}{2}\alpha}/\sqrt{n})$ where $t_{\frac{1}{2}\alpha}$ is the $100(1 - \frac{1}{2}\alpha)$ per cent point of the t-distribution on $(n - 1)$ degrees of freedom. Thus the hypothesis would *not* be rejected if $\bar{x}_1$ fell *within* this interval. Similarly, the hypothesis $\mu_2 = 0$ for the variable x_2 would not be rejected if the mean $\bar{x}_2$ of the x_2-observations lay within the interval $(-s_2 t_{\frac{1}{2}\alpha}/\sqrt{n}, +s_2 t_{\frac{1}{2}\alpha}/\sqrt{n})$ where s_2^2 is the variance of the x_2-observations and $t_{\frac{1}{2}\alpha}$, n are as before. The multivariate hypothesis $\boldsymbol{\mu} = \boldsymbol{0}$ would not therefore be rejected if both of these conditions were satisfied. If we were to plot the point $(\bar{x}_1, \bar{x}_2)$ against rectangular axes, the area within which the point could lie and the hypothesis $\boldsymbol{\mu} = \boldsymbol{0}$ would *not* be rejected is given by the rectangle ABCD of Fig. 8.2, where AB and DC are of length $2s_1t_{\frac{1}{2}\alpha}/\sqrt{n}$ while AD and BC are of length $2s_2t_{\frac{1}{2}\alpha}/\sqrt{n}$.

Thus a sample that gave the means $(\bar{x}_1, \bar{x}_2)$ represented by point P would lead to acceptance of the hypothesis $\boldsymbol{\mu} = \boldsymbol{0}$. Suppose, however, that the variables x_1 and x_2 are moderately highly correlated. Then all points (x_1, x_2) and hence also $(\bar{x}_1, \bar{x}_2)$ should lie reasonably close to the straight line MN through the origin, marked on Fig. 8.2. Hence samples consistent with the hypothesis $\boldsymbol{\mu} = \boldsymbol{0}$ should be represented by points $(\bar{x}_1, \bar{x}_2)$ which lie within a region encompassing the line MN. When we take account of the nature of variation of bivariate normal samples that include correlation, this region can be shown to be an ellipse such as the one marked in Fig. 8.2. The point P is *not* consistent with this region, and in fact the hypothesis $\boldsymbol{\mu} = \boldsymbol{0}$ should be rejected for this sample (P being more consistent with an ellipse constructed about a line parallel to MN but above MN on the diagram). Thus the inference drawn from the two separate univariate tests conflicts with the one that would be drawn from a single multivariate test, and is the wrong inference. A sample giving the $(\bar{x}_1, \bar{x}_2)$ values represented by the point Q would give the other type of mistake, where the two univariate tests lead to rejection of the null hypothesis but the correct multivariate inference is that the hypothesis should not be rejected.

The above simple illustration shows that major errors of inference can result from ignoring the dependencies between the variables and using separate univariate test procedures. The second drawback with the latter approach is a consequence of the philosophy of hypothesis testing. When we say that a test is to be conducted at the α per cent level of significance, what we are implicitly accepting is that the null hypothesis will be true in α per cent of all those occasions on which we reject it. Thus we are admitting that we will make a proportion $\alpha/100$ of such 'errors of the first kind'. The more variables in our sample, therefore, the more such 'false significances' are likely to arise when adopting a sequence of univariate tests, and hence the more possibility there is that a confusing pattern of inferences will emerge when we try to summarize our conclusions.

For these reasons, therefore, it seems clear that if we want to make overall inferences about a multivariate system, we need a *multivariate* approach to hypothesis testing. Obviously, if particular interest attaches to the behaviour of some individual variables then separate tests on these particular variables are perfectly in order. However, for general conclusions about the whole *system,* a multivariate procedure has to be used. Sometimes it may be possible to deduce a reasonable multivariate test statistic by extrapolating from the corresponding univariate problem, but such extrapolation generally requires exceptional insight and appreciation of multivariate methods. More commonly, given a particular problem and associated hypothesis test, we need to follow through some general procedure that will provide us with a suitable test statistic from which the hypothesis can be tested. There are two such general procedures in common use, the *likelihood ratio* procedure and the *union–intersection* procedure. We shall briefly describe each of these, and then derive the test statistic for one of the most common hypotheses about the mean of a multivariate population using both approaches. Finally we shall give the test statistics for a range of multivariate hypotheses, and briefly comment on the differences between the two general approaches used in deriving these statistics.

The likelihood ratio procedure is a completely general method of obtaining test statistics in any situation (whether multivariate or not) for which it is possible to write down the likelihood of the observations. It is used in any situation in which one wishes to test a null hypothesis H_0 against a completely general alternative H_1 (in the sense that the union of H_0 and H_1 comprises the complete parameter space), and is particularly appropriate where there are nuisance parameters in the problem (these being unknown parameters not specified in either H_0 or H_1). If Ω denotes the complete parameter space and ω denotes the parameter space restricted by the null hypothesis H_0, then the likelihood ratio test criterion for the null hypothesis H_0 against the alternative hypothesis (not H_0) is

$$\lambda = \frac{\sup_{\omega} L}{\sup_{\Omega} L} \tag{8.6}$$

where L denotes the likelihood of the sample and $\sup_{S} L$ denotes the maximum of this likelihood over the parameter set S.

To illustrate this procedure, consider a simple univariate example. Let $x_1, x_2, \ldots, x_n$ denote a random sample of size n from a $\mathrm{N}(\mu, \sigma^2)$ distribution, and suppose that we wish to test the null hypothesis $H_0: \mu = 0$ against the general alternative $H_1: \mu \neq 0$. Since nothing has been said about the parameter σ^2, it is a nuisance parameter as defined above. Clearly, the whole parameter space Ω is the set of all possible (μ, σ^2) pairs; μ can take any real value, either positive, negative, or zero, while σ^2 can take any positive value. Thus in mathematical set notation, we can write

$$\Omega = \{(\mu, \sigma^2) \mid -\infty < \mu < \infty; 0 < \sigma^2 < \infty\}.$$

Under the null hypothesis H_0, μ is forced to be zero, but σ^2 is unrestricted and can take any positive value. Thus,

$$\omega = \{(0, \sigma^2) \mid 0 < \sigma^2 < \infty\}.$$

Clearly, since H_1 excludes $\mu = 0$ but otherwise allows all possible values of μ and σ^2, the set ω_1 of parameter values under H_1 is such that $\omega \cup \omega_1 = \Omega$. Now the likelihood of the sample is

$$L(\boldsymbol{x}; \mu, \sigma^2) = \{\sigma\sqrt{(2\pi)}\}^{-n} \exp\left\{-\frac{1}{2\sigma^2}\sum_{i=1}^{n}(x_i - \mu)^2\right\} \tag{8.7}$$

while under H_0 μ is forced to be zero, and the likelihood is modified to

$$L(\boldsymbol{x}; \sigma^2) = \{\sigma\sqrt{(2\pi)}\}^{-n} \exp\left\{-\frac{1}{2\sigma^2}\sum_{i=1}^{n} x_i^2\right\}. \tag{8.8}$$

Thus $\sup_{\Omega} L$ is found by maximizing (8.7) over μ and σ^2. The maximum is achieved at the maximum likelihood estimates $\hat{\mu} = \bar{x}$ and $\hat{\sigma}^2 = \frac{1}{n}\sum_{i=1}^{n}(x_i - \bar{x})^2$, these values being found by differentiating the logarithm of (8.7) successively with respect to μ and σ^2, setting the resulting expressions to zero and solving the consequent pair of simultaneous equations. Substituting $\hat{\mu}$ and $\hat{\sigma}$ for μ and σ in (8.7) then yields $\sup_{\Omega} L = \{\hat{\sigma}\sqrt{(2\pi)}\}^{-n} \exp\left(-\frac{n}{2}\right)$. Next, $\sup_{\omega} L$ is found by maximizing (8.8) over σ^2. Another application of straightforward calculus yields the value $\tilde{\sigma}^2$ of σ^2 at the maximum as $\tilde{\sigma}^2 = \frac{1}{n}\sum_{i=1}^{n} x_i^2$, and substituting this value for σ^2 in (8.8) yields $\sup_{\omega} L = \{\tilde{\sigma}\sqrt{(2\pi)}\}^{-n} \exp\left(-\frac{n}{2}\right)$. Hence, finally, the likelihood ratio test criterion is given by

$$\lambda = \frac{\sup_{\omega} L}{\sup_{\Omega} L} = \left(\frac{\tilde{\sigma}^2}{\hat{\sigma}^2}\right)^{-n/2}.$$

(Further algebraic manipulation can be used to show that this ratio is a simple function of the usual t-statistic, but such algebra is not of particular interest here.)

Having obtained the test statistic by the use of (8.6), we then require a critical value for the test of the hypothesis. The logic behind the criterion λ is that, if H_0 is true, then the maximum of the likelihood over Ω should occur at a value of the parameter set consistent with ω. Thus $\sup_{\omega} L$ and

$\sup_{\Omega} L$ should be equal (or at least close to each other in value). On the other hand, if H_0 is not true, then the maximum of the likelihood will occur at a value of Ω that is not in ω. Thus $\sup_{\omega} L$ will be smaller than $\sup_{\Omega} L$, often considerably so. Hence if λ is close to unity then acceptance of H_0 is indicated, while if λ is close to zero then rejection of H_0 is indicated. In general we thus seek a value λ_0 such that H_0 is rejected if $\lambda \leq \lambda_0$. The appropriate value of λ_0 will depend on the required size α of the test, since we must have $\text{pr}\{\lambda \leq \lambda_0\} = \alpha$ whenever H_0 is true. In order to find λ_0, therefore, we must derive the sampling distribution of λ under H_0, but this is often a very difficult task. To simplify matters an asymptotic general result may be used, which says that for large samples the distribution of $-2\log_e \lambda$ is χ_r^2 if H_0 is true, where the number of degrees of freedom r equals the number of independent constraints used in defining H_0. This general result may be used even for n as small as 25 with acceptable accuracy, and will enable the critical value λ_0 to be derived easily from tables of the χ^2 distribution. Thus in the illustrative example above, the null hypothesis $H_0: \mu = 0$ involves one constraint so that $-2\log_e \lambda = n(\log_e \tilde{\sigma}^2 - \log_e \hat{\sigma}^2)$ has a χ_1^2 distribution if H_0 is true.

By contrast with the likelihood ratio procedure, the union–intersection procedure was specifically designed with multivariate problems in mind. The idea here is to turn the multivariate problem into a univariate one by forming a linear combination of the original variables $x_1, \ldots, x_p$. Thus for arbitrary coefficients $a_1, a_2, \ldots, a_p$ we consider the linear combination $y = a_1x_1 + \ldots + a_px_p = \boldsymbol{a}'\boldsymbol{x}$. For given values $a_1, \ldots, a_p$, this process converts the multivariate hypothesis H_0 into a univariate hypothesis $H_0(\boldsymbol{a})$ for which standard univariate theory will provide a test statistic $T(\boldsymbol{a})$ and an associated rejection region $R(\boldsymbol{a})$. By varying the coefficients we are able to generate all possible univariate hypotheses, test statistics and rejection regions. Now H_0 will be true if and only if all the component $H_0(\boldsymbol{a})$ are true so that H_0 should be rejected if *any* of the $H_0(\boldsymbol{a})$ are rejected. Hence $H_0 = \bigcap_{\boldsymbol{a}} H_0(\boldsymbol{a})$, the *intersection* of all component hypotheses, and the multivariate rejection region $R = \bigcup_{\boldsymbol{a}} R(\boldsymbol{a})$, the *union* of the univariate rejection regions. This is the genesis of the name of the technique; its use will be made clear in the derivation below.

We now state appropriate test statistics and their distributions for a range of common hypotheses. To illustrate the use of the likelihood ratio and union–intersection procedures, we will derive the first test statistic using each of these procedures; the other test statistics can all then be derived in analogous manner. For further details, the interested reader is referred to T. W. Anderson (1984) who deals with the likelihood ratio method, Morrison (1976) who deals with the union–intersection method, or Mardia *et al.* (1979) who consider both methods. Less mathematically orientated readers may of course omit all the details of these derivations without detriment.

1. To test the null hypothesis $H_0: \boldsymbol{\mu} = \boldsymbol{\mu}_0$, a specified constant, against the alternative $H_1: \boldsymbol{\mu} \neq \boldsymbol{\mu}_0$ without making any assumptions about $\boldsymbol{\Sigma}$, both the likelihood ratio and the union–intersection procedures lead to the test statistic known as *Hotelling's* T^2,

$$T^2 = n(\bar{\boldsymbol{x}} - \boldsymbol{\mu}_0)'\mathbf{S}^{-1}(\bar{\boldsymbol{x}} - \boldsymbol{\mu}_0) \tag{8.9}$$

where $\bar{\boldsymbol{x}}$ is the sample mean, $\mathbf{S}$ is the unbiased sample covariance matrix $\dfrac{1}{n-1}\sum_{i=1}^{n} (\boldsymbol{x}_i - \bar{\boldsymbol{x}})(\boldsymbol{x}_i - \bar{\boldsymbol{x}})'$ and n is the sample size.

If H_0 is true then $(n-p)T^2/p(n-1)$ has an F distribution on p and $(n-p)$ degrees of freedom, while if H_0 is not true then the distribution is a non-central F. Hence choosing the $100(1-\alpha)$ per cent point F_α of the $F_{p,n-p}$ distribution, and rejecting H_0 whenever $T^2 > p(n-1)F_\alpha/(n-p)$, yields an appropriate test of size α.

We now derive this result, using the likelihood ratio and union–intersection principles in turn.

First consider the likelihood ratio principle. Given a random sample $\boldsymbol{x}_1, \boldsymbol{x}_2, \ldots, \boldsymbol{x}_n$ from $\mathrm{N}(\boldsymbol{\mu}, \boldsymbol{\Sigma})$, the likelihood is obtained by multiplying together the probability density (7.9) for each observation.

Thus

$$L(\boldsymbol{x}_1, \ldots, \boldsymbol{x}_n; \boldsymbol{\mu}, \boldsymbol{\Sigma}) = \{(2\pi)^{p/2}\,|\boldsymbol{\Sigma}|^{\frac{1}{2}}\}^{-n} \exp\left\{-\tfrac{1}{2}\sum_{r=1}^{n} (\boldsymbol{x}_r - \boldsymbol{\mu})'\boldsymbol{\Sigma}^{-1}(\boldsymbol{x}_r - \boldsymbol{\mu})\right\}.$$

The values of $\boldsymbol{\mu}$ and $\boldsymbol{\Sigma}$ which maximize L are the same as the ones which maximize $\log_e L$; maximizing the latter is usually simpler mathematically so we consider

$$\log_e L = -\tfrac{1}{2}\sum_{r=1}^{n} (\boldsymbol{x}_r - \boldsymbol{\mu})'\boldsymbol{\Sigma}^{-1}(\boldsymbol{x}_r - \boldsymbol{\mu}) - \frac{n}{2}\log_e\{(2\pi)^p\,|\boldsymbol{\Sigma}|\}. \tag{8.10}$$

This expression is maximized at the maximum likelihood estimates of $\boldsymbol{\mu}$, $\boldsymbol{\Sigma}$ (by definition), and from Section 8.2 we know these estimates to be $\hat{\boldsymbol{\mu}} = \bar{\boldsymbol{x}}$ and $\hat{\boldsymbol{\Sigma}} = \dfrac{1}{n}\sum_{r=1}^{n} (\boldsymbol{x}_r - \bar{\boldsymbol{x}})(\boldsymbol{x}_r - \bar{\boldsymbol{x}})' = \dfrac{1}{n}\mathbf{C}$. Substituting these values into (8.10) yields

$$\sup_{\Omega}(\log_e L) = -\tfrac{1}{2}\sum_{r=1}^{n} (\boldsymbol{x}_r - \hat{\boldsymbol{\mu}})'\hat{\boldsymbol{\Sigma}}^{-1}(\boldsymbol{x}_r - \hat{\boldsymbol{\mu}}) - \frac{n}{2}\log_e\{(2\pi)^p\,|\hat{\boldsymbol{\Sigma}}|\}.$$

But

$$\begin{aligned}\sum_{r=1}^{n}(\boldsymbol{x}_r-\hat{\boldsymbol{\mu}})'\hat{\boldsymbol{\Sigma}}^{-1}(\boldsymbol{x}_r-\hat{\boldsymbol{\mu}}) &= \text{trace}\left\{\sum_{r=1}^{n}(\boldsymbol{x}_r-\hat{\boldsymbol{\mu}})'\hat{\boldsymbol{\Sigma}}^{-1}(\boldsymbol{x}_r-\hat{\boldsymbol{\mu}})\right\}\\ &= \sum_{r=1}^{n}\text{trace}\{(\boldsymbol{x}_r-\hat{\boldsymbol{\mu}})'\hat{\boldsymbol{\Sigma}}^{-1}(\boldsymbol{x}_r-\hat{\boldsymbol{\mu}})\}\\ &= \sum_{r=1}^{n}\text{trace}\{\hat{\boldsymbol{\Sigma}}^{-1}(\boldsymbol{x}_r-\hat{\boldsymbol{\mu}})(\boldsymbol{x}_r-\hat{\boldsymbol{\mu}})'\}\\ &= \text{trace}\left[\hat{\boldsymbol{\Sigma}}^{-1}\left\{\sum_{r=1}^{n}(\boldsymbol{x}_r-\hat{\boldsymbol{\mu}})(\boldsymbol{x}_r-\hat{\boldsymbol{\mu}})'\right\}\right]\\ &= \text{trace}\{\hat{\boldsymbol{\Sigma}}^{-1}(n\hat{\boldsymbol{\Sigma}})\}\\ &= n\,\text{trace}(\hat{\boldsymbol{\Sigma}}^{-1}\hat{\boldsymbol{\Sigma}})\\ &= n\,\text{trace}(\mathbf{I})\\ &= np.\end{aligned}$$

The above steps follow through on noting that $(\boldsymbol{x}_r-\hat{\boldsymbol{\mu}})'\hat{\boldsymbol{\Sigma}}^{-1}(\boldsymbol{x}_r-\hat{\boldsymbol{\mu}})$ can be viewed as a (1×1) matrix and hence equals its trace, that $\hat{\boldsymbol{\Sigma}}^{-1}\hat{\boldsymbol{\Sigma}}$ yields the $(p\times p)$ identity matrix, and by using standard properties of the trace of a matrix (see Appendix, Section A2).

Thus

$$\sup_{\Omega} L = \exp(-\tfrac{1}{2}np)\Big/\left\{(2\pi)^{np/2}\left|\frac{1}{n}\mathbf{C}\right|^{n/2}\right\}. \tag{8.11}$$

Now, under H_0, $\boldsymbol{\mu}$ is forced to equal $\boldsymbol{\mu}_0$ and the log-likelihood becomes

$$\log_e L = -\tfrac{1}{2}\sum_{r=1}^{n}(\boldsymbol{x}_r-\boldsymbol{\mu}_0)'\boldsymbol{\Sigma}^{-1}(\boldsymbol{x}_r-\boldsymbol{\mu}_0)-\frac{n}{2}\log_e\{(2\pi)^p\,|\boldsymbol{\Sigma}|\}. \tag{8.12}$$

Maximizing with respect to the remaining parameter $\boldsymbol{\Sigma}$ (by using standard calculus) yields the restricted maximum likelihood estimate

$$\begin{aligned}\tilde{\boldsymbol{\Sigma}} &= \frac{1}{n}\sum_{r=1}^{n}(\boldsymbol{x}_r-\boldsymbol{\mu}_0)(\boldsymbol{x}_r-\boldsymbol{\mu}_0)'\\ &= \frac{1}{n}\sum_{r=1}^{n}(\boldsymbol{x}_r-\bar{\boldsymbol{x}}+\bar{\boldsymbol{x}}-\boldsymbol{\mu}_0)(\boldsymbol{x}_r-\bar{\boldsymbol{x}}+\bar{\boldsymbol{x}}-\boldsymbol{\mu}_0)'\\ &= \frac{1}{n}\{\mathbf{C}+n(\bar{\boldsymbol{x}}-\boldsymbol{\mu}_0)(\bar{\boldsymbol{x}}-\boldsymbol{\mu}_0)'\}\quad\text{on expanding the above line.}\end{aligned}$$

Using a similar set of trace operations to those above we find, on substituting $\tilde{\boldsymbol{\Sigma}}$ for $\boldsymbol{\Sigma}$ in (8.12),

$$\sup_{\omega} L = \mathrm{e}^{-\frac{1}{2}np}/\{(2\pi)^{np/2}|\tilde{\boldsymbol{\Sigma}}|^{\frac{1}{2}n}\}. \tag{8.13}$$

The likelihood ratio test statistic is thus given by the ratio of (8.13) to (8.11), i.e.

$$\lambda = \frac{\left|\frac{1}{n}\mathbf{C}\right|^{n/2}}{\left|\frac{1}{n}[\mathbf{C} + n(\bar{\boldsymbol{x}} - \boldsymbol{\mu}_0)(\bar{\boldsymbol{x}} - \boldsymbol{\mu}_0)']\right|^{n/2}}.$$

It is generally sufficient to find the simplest possible monotonic function of λ to serve as the test statistic. We can cancel the factor $\frac{1}{n}$ in the numerator and the denominator, and take the $(2/n)$th power of the expression. Hence the test criterion λ satisfies

$$\lambda^{2/n} = \frac{|\mathbf{C}|}{|\mathbf{C} + \boldsymbol{b}\boldsymbol{b}'|} \tag{8.14}$$

where $\boldsymbol{b} = \sqrt{n}(\bar{\boldsymbol{x}} - \boldsymbol{\mu}_0)$.

Now if we consider the determinant of the partitioned $(p+1) \times (p+1)$ matrix $\begin{pmatrix} 1 & -\boldsymbol{b}' \\ \boldsymbol{b} & \mathbf{C} \end{pmatrix}$, standard theory shows it to be equal to $|\mathbf{C} + \boldsymbol{b}\boldsymbol{b}'|$ on expanding about 1 and equal to $|\mathbf{C}| \cdot |1 + \boldsymbol{b}'\mathbf{C}^{-1}\boldsymbol{b}|$ on expanding about $\mathbf{C}$ (see, e.g. Graybill (1969) Chapter 8). Also, since $\boldsymbol{b}'\mathbf{C}^{-1}\boldsymbol{b}$ is a scalar, $|1 + \boldsymbol{b}'\mathbf{C}^{-1}\boldsymbol{b}| = 1 + \boldsymbol{b}'\mathbf{C}^{-1}\boldsymbol{b}$. Thus $|\mathbf{C} + \boldsymbol{b}\boldsymbol{b}'| = |\mathbf{C}|(1 + \boldsymbol{b}'\mathbf{C}^{-1}\boldsymbol{b})$, and on substituting into (8.14) we find

$$\begin{aligned} \lambda^{2/n} &= (1 + \boldsymbol{b}'\mathbf{C}^{-1}\boldsymbol{b})^{-1} \\ &= \{1 + n(\bar{\boldsymbol{x}} - \boldsymbol{\mu}_0)'\mathbf{C}^{-1}(\bar{\boldsymbol{x}} - \boldsymbol{\mu}_0)\}^{-1} \quad \text{from (8.14)} \\ &= \left\{1 + \frac{n}{n-1}(\bar{\boldsymbol{x}} - \boldsymbol{\mu}_0)'\mathbf{S}^{-1}(\boldsymbol{x} - \boldsymbol{\mu}_0)\right\}^{-1} \quad \text{since } \mathbf{S} = \frac{1}{n-1}\mathbf{C}. \end{aligned}$$

Hence the likelihood ratio test statistic is a monotonic function of $T^2 = n(\bar{\boldsymbol{x}} - \boldsymbol{\mu}_0)'\mathbf{S}^{-1}(\bar{\boldsymbol{x}} - \boldsymbol{\mu}_0)$.

Next, consider the union–intersection approach. If $\boldsymbol{x} \sim \mathrm{N}(\boldsymbol{\mu}, \boldsymbol{\Sigma})$, then Property 1 of Section 7.2 shows that $y = \boldsymbol{a}'\boldsymbol{x} \sim \mathrm{N}(\boldsymbol{a}'\boldsymbol{\mu}, \boldsymbol{a}'\boldsymbol{\Sigma}\boldsymbol{a})$. Thus if we reduce the multivariate problem to a univariate one by forming an arbitrary linear combination $y = \boldsymbol{a}'\boldsymbol{x} = a_1x_1 + \ldots + a_px_p$ of the original variables, the multivariate null hypothesis $H_0\colon \boldsymbol{\mu} = \boldsymbol{\mu}_0$ induces the univariate null hypothesis $H_0(\boldsymbol{a})\colon \boldsymbol{a}'\boldsymbol{\mu} = \boldsymbol{a}'\boldsymbol{\mu}_0$. For specified $\boldsymbol{a}$, this hypothesis is tested against the general alternative $H_1(\boldsymbol{a})\colon \boldsymbol{a}'\boldsymbol{\mu} \neq \boldsymbol{a}'\boldsymbol{\mu}_0$ by the standard one-sample t-test. The sample mean of the derived variable y is $\boldsymbol{a}'\bar{\boldsymbol{x}}$, and since $\mathbf{S}$ is an unbiased estimate of $\boldsymbol{\Sigma}$, $\boldsymbol{a}'\mathbf{S}\boldsymbol{a}$ provides an unbiased estimate of the variance $\boldsymbol{a}'\boldsymbol{\Sigma}\boldsymbol{a}$ of y. Hence the acceptance region for $H_0(\boldsymbol{a})$ is the set of values of $\boldsymbol{x}$ such that $t^2(\boldsymbol{a}) \leq t^2_{\frac{1}{2}\alpha, n-1}$ where $t(\boldsymbol{a})$ is the t statistic given by

$$t(\boldsymbol{a}) = \frac{\boldsymbol{a}'(\bar{\boldsymbol{x}} - \boldsymbol{\mu}_0)\sqrt{n}}{\sqrt{(\boldsymbol{a}'\mathbf{S}\boldsymbol{a})}},$$

and $t_{\frac{1}{2}\alpha, n-1}$ is the $100(1 - \frac{1}{2}\alpha)$ per cent point of the t-distribution on $(n-1)$ degrees of freedom. The union–intersection principle thus tells us to choose as the acceptance region for the multivariate hypothesis H_0 the set of $\boldsymbol{x}$ for which *all* univariate hypotheses would be accepted on varying $\boldsymbol{a}$. Thus the acceptance region should be the set of $\boldsymbol{x}$ for which *all* $t^2(\boldsymbol{a}) \leq t^2_{\frac{1}{2}\alpha,n-1}$ when $\boldsymbol{a}$ is varied.

Now if all possible $t^2(\mathbf{a})$ obtained by varying $\boldsymbol{a}$ are to be less than some fixed constant, then clearly the largest such $t^2(\boldsymbol{a})$ must be less than this constant. Hence a suitable test criterion for the multivariate hypothesis will be provided by $\max_{\boldsymbol{a}} t^2(\boldsymbol{a})$, the maximum value of $t^2(\boldsymbol{a})$ obtained by varying $\boldsymbol{a}$. However, the form of $t(\boldsymbol{a})$ given above shows that it is invariant to multiplication of $\boldsymbol{a}$ by a scalar constant k, i.e. $t(\boldsymbol{a}) = t(k\boldsymbol{a})$ for any $k \neq 0$. Hence for unique maximization we must impose some constraint on the elements of $\boldsymbol{a}$. This constraint may be chosen arbitrarily; the most convenient one is $\boldsymbol{a}'\mathbf{S}\boldsymbol{a} = 1$, as this makes the denominator of $t(\boldsymbol{a})$ unity. Maximization of $t^2(\boldsymbol{a})$ subject to this constraint is achieved by the use of the Lagrange undetermined multiplier λ, and requires maximization of

$$V = \{\boldsymbol{a}'(\bar{\boldsymbol{x}} - \boldsymbol{\mu}_0)\sqrt{n}\}^2 - \lambda(\boldsymbol{a}'\mathbf{S}\boldsymbol{a} - 1).$$

But since $\boldsymbol{a}'(\bar{\boldsymbol{x}} - \boldsymbol{\mu}_0)$ is a scalar and equal to its transpose $(\bar{\boldsymbol{x}} - \boldsymbol{\mu}_0)'\boldsymbol{a}$, we can write

$$V = n\boldsymbol{a}'(\bar{\boldsymbol{x}} - \boldsymbol{\mu}_0)(\bar{\boldsymbol{x}} - \boldsymbol{\mu}_0)'\boldsymbol{a} - \lambda(\boldsymbol{a}'\mathbf{S}\boldsymbol{a} - 1).$$

Thus $\dfrac{\partial V}{\partial \boldsymbol{a}} = 2n(\bar{\boldsymbol{x}} - \boldsymbol{\mu}_0)(\bar{\boldsymbol{x}} - \boldsymbol{\mu}_0)'\boldsymbol{a} - 2\lambda\mathbf{S}\boldsymbol{a}$, so that $\dfrac{\partial V}{\partial \boldsymbol{a}} = \boldsymbol{0}$ yields $\{n(\bar{\boldsymbol{x}} - \boldsymbol{\mu}_0)(\bar{\boldsymbol{x}} - \boldsymbol{\mu}_0)' - \lambda\mathbf{S}\}\hat{\boldsymbol{a}} = \boldsymbol{0}$, where $\hat{\boldsymbol{a}}$ is the value maximizing V. Premultiplication by $\hat{\boldsymbol{a}}'$ shows that

$$\lambda = \frac{n\hat{\boldsymbol{a}}'(\bar{\boldsymbol{x}} - \boldsymbol{\mu}_0)(\bar{\boldsymbol{x}} - \boldsymbol{\mu}_0)'\hat{\boldsymbol{a}}}{\hat{\boldsymbol{a}}'\mathbf{S}\hat{\boldsymbol{a}}} = \max_{\boldsymbol{a}} \{t^2(\boldsymbol{a})\}.$$

Thus the union–intersection test statistic is λ.

Now, $\{n\mathbf{S}^{-1}(\bar{\boldsymbol{x}} - \boldsymbol{\mu}_0)(\bar{\boldsymbol{x}} - \boldsymbol{\mu}_0)' - \lambda\mathbf{I}\}\hat{\boldsymbol{a}} = \boldsymbol{0}$, so that $\hat{\boldsymbol{a}}$ is an eigenvector of $n\mathbf{S}^{-1}(\bar{\boldsymbol{x}} - \boldsymbol{\mu}_0)(\bar{\boldsymbol{x}} - \boldsymbol{\mu}_0)'$ corresponding to the eigenvalue λ. But the rank of $(\bar{\boldsymbol{x}} - \boldsymbol{\mu}_0)(\bar{\boldsymbol{x}} - \boldsymbol{\mu}_0)'$ is 1, which is thus the rank of $n\mathbf{S}^{-1}(\bar{\boldsymbol{x}} - \boldsymbol{\mu}_0)(\bar{\boldsymbol{x}} - \boldsymbol{\mu}_0)'$. Hence this matrix has just one non-zero eigenvalue, which must equal λ. Since the trace of a matrix is equal to the sum of its eigenvalues it thus follows that

$$\begin{aligned}
\lambda &= \text{trace}\{n\mathbf{S}^{-1}(\bar{\boldsymbol{x}} - \boldsymbol{\mu}_0)(\bar{\boldsymbol{x}} - \boldsymbol{\mu}_0)'\}, \\
&= \text{trace}\{n(\bar{\boldsymbol{x}} - \boldsymbol{\mu}_0)'\mathbf{S}^{-1}(\bar{\boldsymbol{x}} - \boldsymbol{\mu}_0)\} \quad \text{since trace}(\mathbf{AB}) = \text{trace}(\mathbf{BA}), \\
&= n(\bar{\boldsymbol{x}} - \boldsymbol{\mu}_0)'\mathbf{S}^{-1}(\bar{\boldsymbol{x}} - \boldsymbol{\mu}_0) \quad \text{since this is scalar}, \\
&= T^2.
\end{aligned}$$

Thus the union–intersection principle also yields T^2 as the test statistic for H_0.

Finally, we require the sampling distribution of T^2, and this follows readily from two results already discussed for the multivariate normal distribution. If H_0 is true, then on using the sampling results of Section 8.1 it follows that $\sqrt{n}(\bar{\boldsymbol{x}} - \boldsymbol{\mu}_0)$ has a $\text{N}(\boldsymbol{0}, \boldsymbol{\Sigma})$ distribution independently of $\mathbf{C}$, which has a $\text{W}_p(n-1, \boldsymbol{\Sigma})$ distribution. Hence, using the final result in Section 8.1, $n\left(\dfrac{n-p}{p}\right)(\bar{\boldsymbol{x}} - \boldsymbol{\mu}_0)'\mathbf{C}^{-1}(\bar{\boldsymbol{x}} - \boldsymbol{\mu}_0)$ has an $F_{p,n-p}$ distribution. Setting $(n-1)\mathbf{S} = \mathbf{C}$ it follows that

$$\left\{\frac{n-p}{p(n-1)}\right\}n(\bar{\boldsymbol{x}}-\boldsymbol{\mu}_0)'\mathbf{S}^{-1}(\bar{\boldsymbol{x}}-\boldsymbol{\mu}_0)\sim F_{p,n-p}$$

i.e. $(n-p)T^2/p(n-1)\sim F_{p,n-p}$ as required.

2. To test the null hypothesis $H_0:\boldsymbol{\mu}=\boldsymbol{\mu}_0$ against the alternative $H_1:\boldsymbol{\mu}\neq\boldsymbol{\mu}_0$ *when* $\boldsymbol{\Sigma}$ *is known,* both the likelihood ratio approach and the union–intersection approach lead to the test statistic

$$Z^2=n(\bar{\boldsymbol{x}}-\boldsymbol{\mu}_0)'\boldsymbol{\Sigma}^{-1}(\bar{\boldsymbol{x}}-\boldsymbol{\mu}_0) \tag{8.15}$$

where $\bar{\boldsymbol{x}}$ is, as usual, the sample mean vector. If H_0 is true then Z^2 has a χ_p^2 distribution, while if H_0 is not true the distribution is non-central χ_p^2. Thus, for a size α test, we reject H_0 if $Z^2>\chi^2_{\alpha,p}$, the $100(1-\alpha)$ per cent point of the χ_p^2 distribution.

3. To test the null hypothesis $H_0:\boldsymbol{\Sigma}=\boldsymbol{\Sigma}_0$ (a specified matrix) against the alternative $H_1:\boldsymbol{\Sigma}\neq\boldsymbol{\Sigma}_0$ without making any assumptions about $\boldsymbol{\mu}$, the likelihood ratio and union–intersection approaches lead to different test statistics. However, both test statistics are functions of the matrix $\mathbf{U}=\boldsymbol{\Sigma}_0^{-1}\mathbf{S}$ (where $\mathbf{S}$ as usual is the unbiased estimate of $\boldsymbol{\Sigma}$) or, more particularly, of the eigenvalues $l_1\geqslant l_2\geqslant\ldots\geqslant l_p$ of $\mathbf{U}$. The likelihood ratio criterion λ satisfies

$$-2\log_e\lambda=n\text{ trace }\mathbf{U}-n\log_e|\mathbf{U}|-np. \tag{8.16}$$

Denote the arithmetic mean of the l_i by a and the geometric mean by g, so that $a=(l_1+l_2+\ldots+l_p)/p$ and $g=(l_1l_2\ldots l_p)^{1/p}$. Then it follows that trace $\mathbf{U}=pa$ and $|\mathbf{U}|=g^p$ so that

$$-2\log_e\lambda=np(a-\log_e g-1). \tag{8.17}$$

The drawback with this statistic is that its null distribution is very complicated, and most results about it in the literature are not easy to use. Recourse must therefore be made to the asymptotic general result cited earlier. Since the specification of $\boldsymbol{\Sigma}_0$ requires $\frac{1}{2}p(p+1)$ separate independent elements to be specified, it follows that the statistic $-2\log_e\lambda$ of (8.16) or (8.17) has a χ^2 distribution on $m=\frac{1}{2}p(p+1)$ degrees of freedom if H_0 is true, and for a size α test we therefore reject H_0 if $-2\log_e\lambda$ is greater than the $100(1-\alpha)$ per cent point of the χ_m^2 distribution.

Whereas the likelihood ratio test is a function of *all* the eigenvalues of $\mathbf{U}$, the union–intersection test depends only on the largest (l_1) and the smallest (l_p) of these eigenvalues. The critical region of the test is such that we reject H_0 if *either* $l_1>c_1$ *or* $l_p<c_2$ where the constants c_1 and c_2 are chosen to make the size of the test α. However, the joint distribution of

(l_1, l_p) is quite complicated, and the critical values required for this test have not yet been tabulated. It is therefore simpler to use the likelihood ratio test in this situation.

4. To test the hypothesis $H_0: \boldsymbol{\Sigma} = k\boldsymbol{\Sigma}_0$, where $\boldsymbol{\Sigma}_0$ is a specified matrix but k is unknown, against the alternative $H_1: \boldsymbol{\Sigma} \neq k\boldsymbol{\Sigma}_0$ without making any assumptions about $\boldsymbol{\mu}$, only the likelihood ratio approach yields a tractable procedure. The likelihood ratio criterion λ satisfies

$$-2 \log_e \lambda = np \log_e(a/g) \tag{8.18}$$

where a and g again denote the arithmetic and geometric means respectively of the eigenvalues of the matrix $\mathbf{U} = \boldsymbol{\Sigma}_0^{-1}\mathbf{S}$. Asymptotically, $-2 \log_e \lambda$ follows a χ^2 distribution on $m = \frac{1}{2}(p-1)(p+2)$ degrees of freedom if H_0 is true, so that for a size α test we reject H_0 if $np \log_e(a/g)$ exceeds the $100(1-\alpha)$ per cent point of the χ^2_m distribution.

A common use that is made of this test is in the so-called *test of sphericity*, $H_0: \boldsymbol{\Sigma} = k\mathbf{I}$, which tests whether all p variates in the sample can be treated as independent with equal variance. In this case $\boldsymbol{\Sigma}_0 = \mathbf{I}$, so that $\mathbf{U} = \mathbf{S}$, and a and g denote the arithmetic and geometric means respectively of the eigenvalues of $\mathbf{S}$.

5. Finally, we may be interested in testing for independence of only *some* of the variates. Suppose that the p variates $\boldsymbol{x}$ have been partitioned into two groups $\boldsymbol{x}_1$ and $\boldsymbol{x}_2$ containing p_1 and p_2 elements respectively, and we wish to test whether the variates in $\boldsymbol{x}_1$ are independent of those in $\boldsymbol{x}_2$. The partition of $\boldsymbol{x}$ induces corresponding partitions of $\boldsymbol{\mu}$ into $\boldsymbol{\mu}_1$ and $\boldsymbol{\mu}_2$, and of $\boldsymbol{\Sigma}$ into $\begin{pmatrix} \boldsymbol{\Sigma}_{11} & \boldsymbol{\Sigma}_{12} \\ \boldsymbol{\Sigma}_{21} & \boldsymbol{\Sigma}_{22} \end{pmatrix}$ where (as in Property 3 of Section 7.2) $\boldsymbol{\Sigma}_{11}$, $\boldsymbol{\Sigma}_{22}$ denote the covariance matrices *within* $\boldsymbol{x}_1$, $\boldsymbol{x}_2$ respectively, and $\boldsymbol{\Sigma}_{12}$ denotes the covariance matrix *between* $\boldsymbol{x}_1$ and $\boldsymbol{x}_2$. The sample statistics $\bar{\boldsymbol{x}}$ and $\mathbf{S}$ can be partitioned similarly. Our hypothesis of interest thus becomes $H_0: \boldsymbol{\Sigma}_{12} = \boldsymbol{0}$ (against the general alternative $H_1: \boldsymbol{\Sigma}_{12} \neq \boldsymbol{0}$) and no assumptions are made about $\boldsymbol{\mu}$ or the other elements of $\boldsymbol{\Sigma}$. If we now denote by $l_1 \geqslant l_2 \geqslant \ldots \geqslant l_k$ the non-zero eigenvalues of $\mathbf{I} - \mathbf{S}_{22}^{-1}\mathbf{S}_{21}\mathbf{S}_{11}^{-1}\mathbf{S}_{12}$, where $\mathbf{S}_{ij}$ is the sample equivalent of $\boldsymbol{\Sigma}_{ij}$ and $k = \min(p_1, p_2)$, then the likelihood ratio criterion λ satisfies

$$-2 \log_e \lambda = -n \sum_{i=1}^{k} \log_e(1 - l_i) \tag{8.19}$$

and if H_0 is true then $-2 \log_e \lambda$ is asymptotically distributed as a $\chi^2_{p_1 p_2}$ variate. The union–intersection approach, on the other hand, leads to rejection of H_0 if the largest eigenvalue l_1 exceeds a constant c_1 chosen to make the size of the test α. Again, however, we need to appeal to some more complicated distributional results to establish the value of c_1, so the likelihood ratio approach will be simpler in practice.

The above cases 1 to 5 cover most of the common situations in practice in which a test of hypothesis about $\boldsymbol{\mu}$ or $\boldsymbol{\Sigma}$ is required when sampling from a single normal population. Note that various other, seemingly unrelated, hypotheses can be reduced to the form required by one of the cases 1–5 if a suitable transformation of variates followed by Properties 1 or 3 of Section 7.2 are employed. For example, consider one of the theories about bark deposits used to introduce the hypothesis testing at the start of this chapter. We have a sample of twenty-eight trees for each of which the weights of bark deposits in the North, East, South, and West directions are recorded. Denote these weights by x_N, x_E, x_S, and x_W; our sample thus consists of twenty-eight observations on the random vector $\boldsymbol{x}' = (x_N, x_E, x_S, x_W)$. We assume this vector to have a normal distribution with mean vector $\boldsymbol{\mu}' = (\mu_N, \mu_E, \mu_S, \mu_W)$ and (4×4) dispersion matrix $\boldsymbol{\Sigma}$. We wish to test the hypothesis that the mean bark deposit is the same in all directions. Thus our null hypothesis will be $H_0: \mu_N = \mu_E = \mu_S = \mu_W = \mu$, say, where μ is unknown, and the alternative will be the general H_1: at least one μ_i differs from the rest. This is a null hypothesis giving relationships among the μ_i, and it apparently has no connection with any of the cases 1–5 above. However, consider rewriting H_0 as follows: $\mu_N - \mu_E = 0$; $\mu_N - \mu_S = 0$; $\mu_N - \mu_W = 0$. If we set $\boldsymbol{v}' = (\mu_N - \mu_E, \mu_N - \mu_S, \mu_N - \mu_W)$, then we have $H_0: \boldsymbol{v} = \boldsymbol{0}$ versus $H_1: \boldsymbol{v} \neq \boldsymbol{0}$, which is of an appropriate form for case 1. Now the function of elements of $\boldsymbol{x}$ corresponding to this form of H_0 is $\boldsymbol{y}' = (x_N - x_E, x_N - x_S, x_N - x_W)$, and if we write $\boldsymbol{y}' = (y_1, y_2, y_3)$ we see that $\boldsymbol{y} = \mathbf{B}\boldsymbol{x}$ where

$$\mathbf{B} = \begin{pmatrix} 1 & -1 & 0 & 0 \\ 1 & 0 & -1 & 0 \\ 1 & 0 & 0 & -1 \end{pmatrix}.$$

By Property 1 of Section 7.2, if $\boldsymbol{x} \sim \mathrm{N}_4(\boldsymbol{\mu}, \boldsymbol{\Sigma})$ then $\boldsymbol{y} \sim \mathrm{N}_3(\mathbf{B}\boldsymbol{\mu}, \mathbf{B}\boldsymbol{\Sigma}\mathbf{B}')$. But $\mathbf{B}\boldsymbol{\mu} = (\mu_N - \mu_E, \mu_N - \mu_S, \mu_N - \mu_W)' = \boldsymbol{v}$, so $\boldsymbol{y} \sim \mathrm{N}_3(\boldsymbol{v}, \boldsymbol{\Omega})$ where $\boldsymbol{\Omega} = \mathbf{B}\boldsymbol{\Sigma}\mathbf{B}'$. If the sample mean and covariance matrix of the $\boldsymbol{x}$ observations are $\bar{\boldsymbol{x}}$, $\mathbf{S}$ respectively, then the sample mean and covariance matrix of the corresponding $\boldsymbol{y}$ observations will be $\mathbf{B}\bar{\boldsymbol{x}}$ and $\mathbf{BSB}'$. Hence we can test H_0 by direct use of case 1, giving the test statistic

$$\begin{aligned} T^2 &= n(\mathbf{B}\bar{\boldsymbol{x}} - \boldsymbol{0})'(\mathbf{BSB})^{-1}(\mathbf{B}\bar{\boldsymbol{x}} - \boldsymbol{0}) \\ &= 28\bar{\boldsymbol{x}}'\mathbf{B}'(\mathbf{BSB}')^{-1}\mathbf{B}\bar{\boldsymbol{x}}. \end{aligned}$$

Note that by effecting the transformation from $\boldsymbol{x}$ to $\boldsymbol{y}$ we have reduced the number of variables from 4 to 3, so when finding the sampling distribution of T^2 we must use $n = 28$, $p = 3$ in the general formula given in 1. Hence we reject H_0 if $T^2 > 3.24F_\alpha$, where F_α is the $100(1-\alpha)$ percent point of the $F_{3,25}$ distribution.

To illustrate the use of these test statistics, let us consider the following simple set of data. The blackbody CIE chromaticity specification for a colour temperature of 4000 K is $x = 0.3804$ and $y = 0.3768$. In ten colour matching trials, one subject had mean chromaticity values $\bar{x} = 0.3745$ and $\bar{y} = 0.3719$, and sample covariance matrix $\mathbf{S} = 10^{-5}\begin{pmatrix} 1.843 & 1.799 \\ 1.799 & 1.836 \end{pmatrix}$.

Example 1

Can the observations be treated as coming from a bivariate normal population with mean given by the 4000 K standards?

Solution

The situation here corresponds exactly with case 1, setting $\bar{\mathbf{x}} = (0.3745, 0.3719)'$; $\boldsymbol{\mu}_0 = (0.3804, 0.3768)$, $n = 10$, and $\mathbf{S}$ as given above.

Thus $(\bar{\mathbf{x}} - \boldsymbol{\mu}_0)' = (-0.0059, \ -0.0049)$ and $\mathbf{S}^{-1} = (10^5 \div 0.14735) \times \begin{pmatrix} 1.836 & -1.799 \\ -1.799 & 1.843 \end{pmatrix}$—since the inverse of the (2×2) matrix $\mathbf{B} = \begin{pmatrix} a & b \\ c & d \end{pmatrix}$ is $(\det \mathbf{B})^{-1}\begin{pmatrix} d & -b \\ -c & a \end{pmatrix}$.

Hence $T^2 = 10(\bar{\mathbf{x}} - \boldsymbol{\mu}_0)'\mathbf{S}^{-1}(\bar{\mathbf{x}} - \boldsymbol{\mu}_0) = 28.1214$. Now, $n = 10$, $p = 2$, so $(n-p)T^2/p(n-1) = 12.5$, which has an $F_{2.8}$ distribution if H_0 is true.

From tables, the 95 per cent point of this distribution is 4.46, the 99 per cent point is 8.65, and the 99.9 per cent point is 18.5.

Hence the observations differ significantly from the 4000 K standards at the 1 per cent level.

Example 2

Can the subject's x and y observations be treated as independent and having equal variance?

Solution

We require the test of sphericity of case 4, for which we require the eigenvalues of $\mathbf{S}$. Now the eigenvalues l_i of $\mathbf{S}$ satisfy $\det(\mathbf{S} - l\mathbf{I}) = 0$, which in the present case means solving $\begin{vmatrix} 1.843 - l & 1.799 \\ 1.799 & 1.836 - l \end{vmatrix} = 0$ (the constant

10^{-5} cancelling throughout), i.e. $(1.843 - l)(1.836 - l) - 1.799^2 = 0$, and $l^2 - 3.679l + 0.14735 = 0$. Hence the eigenvalues are $l_1 = 3.6385$ and $l_2 = 0.0405$. These have arithmetic mean $a = \frac{1}{2}(l_1 + l_2) = 1.8395$, and geometric mean $g = \sqrt{(l_1 l_2)} = 0.3839$. Also $n = 10$, $p = 2$, so the test statistic is

$$-2 \log_e \lambda = np \log_e(a/g) = 10 \times 2 \times \log_e(1.8395/0.3839) = 31.34.$$

Strictly speaking, the sample size of 10 is too small for the asymptotic result to be valid when seeking the critical value for the test, and the exact distribution should be used. However, in this case it is fairly clear that even if constancy of variance is not unreasonable, independence of variates will clearly not be appropriate, and the test statistic should be 'obviously' significant even for such a small sample. The asymptotic distribution of $-2 \log_e \lambda$ is χ^2 on $\frac{1}{2}(p-1)(p+2) = 2$ degrees of freedom if H_0 is true, and this distribution has 95 per cent and 99 per cent points of 5.99 and 9.21. The observed value of 31.34 thus gives overwhelming evidence against the hypothesis of sphericity.

8.4 Some comments on multivariate hypothesis tests

The likelihood ratio and the union–intersection principles have been described in the previous section, and have been used to derive test statistics in a variety of situations. For most hypothesis tests the two principles lead to different test statistics, the only cases in which the two coincide being the T^2 and Z^2 statistics given in 1 and 2. In fact, the only other situations in which the two principles agree are the two-sample analogues of 1 and 2, to be described in Section 12.1, and in all other cases different statistics compete for each possible hypothesis test. While the likelihood-ratio and union–intersection principles are the most popular methods of deriving test statistics, there are other general principles (e.g. invariance) as well as *ad-hoc* approaches that could equally be used; they all tend to lead to different test statistics. In fact, for testing hypotheses about the multivariate general linear model (see Chapter 13), as many as half a dozen test statistics are popularly available in any given situation, with the possibility of conflicting conclusions being reached depending on which statistic is chosen. How has this situation arisen, and how does one choose the most appropriate test statistic in a given situation?

The root of the problem is that a *multivariate* hypothesis is to be tested and, while the null hypothesis H_0 is usually specified very tightly and precisely, departures from H_0 can occur in many ways (or dimensions) and yet still be consistent with the general alternative hypothesis H_1.

Different test statistics—or indeed different general *methods* of constructing test statistics—are sensitive to different departures from H_0. To illustrate this point, it is convenient to look ahead slightly to topics that will be covered in Part IV, and consider testing the equality of means in a set of k populations. To be specific, suppose that we draw a sample from each of a set of populations, the ith sample being of size n_i and coming from the population $N(\boldsymbol{\mu}_i, \boldsymbol{\Sigma})$ $(i = 1, \ldots, k)$. These samples can be visualized as k swarms of points in our usual p-dimensional space, and we can represent the k population means $\boldsymbol{\mu}_1, \boldsymbol{\mu}_2, \ldots, \boldsymbol{\mu}_k$ as k further points in this space. We now wish to test the null hypothesis $H_0: \boldsymbol{\mu}_1 = \boldsymbol{\mu}_2 = \ldots = \boldsymbol{\mu}_k$ against the general alternative that the means are not all equal. The null hypothesis specifies that the k points representing the population means are coincident, which is a very precise specification (the only uncertainty being the exact location of this point). The alternative hypothesis, however, can be satisfied by very many different configurations. For example, we might have $(k-1)$ coincident points and one point very far from this cluster. Or we might have no coincident points, but each point differs equally from all the others on just two of its coordinates. In the first case there is a well-defined 'major direction' of difference in the means (i.e. the line joining the cluster of means to the one different mean), but the simplex bounded by the k points has zero volume as it is degenerate and lies in just one dimension. In the second case there is no 'major direction' of difference in the means if the distance between each pair of means is the same, but the simplex bounded by the k points might have a large volume. Thus a test statistic that aims to detect a departure from H_0 along the 'major direction' of difference between sample means would correctly detect a difference in the first case but possibly not in the second; while a test statistic based on some measure of volume enclosed by the sample means would be more likely to detect a difference with the second configuration than with the first.

An intuitive description of all the common test statistics associated with hypotheses in the general linear model has been given in such geometrical terms by Kenward (1979). The likelihood ratio criterion provides statistics that tend to detect 'overall' departures from H_0, and hence tend to be functions of determinants (or products of all eigenvalues) of certain matrices, while the union–intersection principle provides statistics that tend to detect 'maximum' departures from H_0, and hence tend to be functions of the extreme eigenvalues of these matrices. All test statistics in cases 1 to 5 above conform to this general pattern. Furthermore, when testing a hypothesis about a single mean or a difference of two means, departures from H_0 can be at most one-dimensional (along the line joining the points representing the true and postulated values). Hence 'maximum' and 'overall' differences must be the same, which accounts for the coincidence of likelihood ratio and union–intersection test statistics in these cases.

Whenever there is more than one test statistic available for a multivariate hypothesis test, therefore, some thought should be given to the types of departures from H_0 it is desired to detect, and the choice of test statistic should be made accordingly. In some cases, of course, mathematical difficulty or intractability may limit the initial choice of statistic (as in cases 3 and 4 above). However, it should always be borne in mind that conflicting results may arise if more than one statistic is chosen, and misleading results may be obtained if an inappropriate statistic is chosen. These considerations reinforce the author's view that hypothesis testing, in the multivariate even more than in the univariate case, should be used in an 'informal inference' sense (Chatfield, 1985). By this is meant that the significance levels computed in any practical applications should be used as a guide to action rather than in a rigid decision-making framework. Thus a significance level somewhere in the range 1–10 per cent should be indicative of there being something of interest in the data, which should then be explored in much greater detail via the descriptive methods of Part I and those to be discussed in Parts IV and V. Simply computing the value of a test statistic, declaring the result to be significant if its associated null probability is less than 0.05, and making all future action depend on whether or not the result is significant is poor statistical practice.

9
Reduction of dimensionality: inferential aspects of descriptive methods

We now return to some of the techniques already described in Part I, and consider ways in which their purely descriptive use can be underpinned by statistical theory. Broadly speaking, the majority of these techniques comprise two main operations: finding a suitable small-dimensional space in which the sample can be embedded, and then inspecting the representation of sample members in this space for any evidence of pattern. Both of these operations carry associated statistical questions, which were hardly even alluded to earlier. In the first place, what is the 'correct' number of dimensions in which the sample is located? Given an r-dimensional representation of a p-variate sample, is this only an approximation, or have we identified the 'true' dimensionality? Secondly, having obtained a small-dimensional representation, what effect will sampling variation have on the positions of the various entities in this space? For example, how far apart must be two points (or vectors, or curves) before we can safely assume that the entities they represent are genuinely different? In order to answer any of these questions we cannot rely simply on our geometrical model, but must also assume a statistical model for our initial multivariate sample. Hence consideration of these matters belongs properly to this part of the book. However, at this point we will break slightly with our problem-oriented philosophy. Since we have already devoted much space to the introduction and description of the various techniques earlier, it seems more natural to switch (in this chapter only) to a technique-oriented discussion, in which all aspects associated with statistical inference are summarized for each technique in turn.

While much research effort has been expended on such inferential aspects, some of the results either require advanced mathematical knowledge, or are of limited practical utility. We therefore concentrate only on the more applicable results. In general, questions about the sampling distribution of entities in the representation will require the assumption of (multivariate) normality of sample individuals, with no special restrictions on the mean vector and covariance matrix of this distribution. Questions about the 'true' dimensionality of the sample, on the other hand, will additionally require assumptions about the covariance structure of the sample members. These assumptions will be amplified further in the various techniques discussed below.

9.1 Andrews curves

Here the sample individuals are *always* represented by curves in two dimensions, so there is no question of finding a 'true' dimensionality. On the other hand, since we are dealing with a sample, we may wish to test hypotheses about, or construct confidence intervals for, the curve representing the mean $\boldsymbol{\mu}$ of the population from which the sample was taken. The curve representing the sample member $\boldsymbol{x}' = (x_1, x_2, \ldots, x_p)$ is given by

$$\mathrm{f}_{\boldsymbol{x}}(t) = x_1/\sqrt{2} + x_2 \sin t + x_3 \cos t + x_4 \sin 2t + x_5 \cos 2t + \ldots$$

so the curve representing the population mean is

$$\mathrm{f}_{\boldsymbol{\mu}}(t) = \mu_1/\sqrt{2} + \mu_2 \sin t + \mu_3 \cos t + \mu_4 \sin 2t + \mu_5 \cos 2t + \ldots$$

where μ_i is the population mean for x_i $(i = 1, \ldots, p)$.

We have seen in Section 2.1.3 that if the variates measured in the sample are uncorrelated and have equal variance σ^2, the variance σ_{f}^2 of the function value at t is independent of t if p is odd, and only depends minimally on t if p is even. This variance is $\frac{1}{2}p\sigma^2$ in the former case and lies between $\frac{1}{2}\sigma^2(p-1)$ and $\frac{1}{2}\sigma^2(p+1)$ in the latter, so may be approximated by $\sigma_{\mathrm{f}}^2 = \frac{1}{2}p\sigma^2$ across the whole graph. This approximation enables the following tests to be constructed.

(i) Tests of significance at particular values of t

Sometimes it is possible to identify a priori for an Andrews plot certain values of t at which it is of interest to test the hypothesis that the expectation of $\mathrm{f}_{\boldsymbol{x}}(t)$ is $\mathrm{f}_{\boldsymbol{\mu}_0}(t)$, where $\boldsymbol{\mu}_0$ is a hypothesized known constant. To test this hypothesis, it is necessary to evaluate the significance level of

$$z = \{\mathrm{f}_{\boldsymbol{x}}(t) - \mathrm{f}_{\boldsymbol{\mu}_0}(t)\}/\sigma_{\mathrm{f}}.$$

If the variates $x_1, \ldots, x_p$ are assumed to be independent normal variates then z has a standard normal distribution under the hypothesis $\boldsymbol{\mu} = \boldsymbol{\mu}_0$. This distribution may therefore be used if σ^2 is known, to assess the hypothesis $\boldsymbol{\mu} = \boldsymbol{\mu}_0$, or to construct a $100(1-\alpha)$ per cent confidence interval for $\boldsymbol{\mu}$ as the set of values not rejected by this test at a level α.

Estimation of σ^2 by a pooled estimate $\hat{\sigma}^2$ (across the p variates), and use of $\hat{\sigma}^2$ in σ_f, modifies the standard normal null distribution to the t distribution in the usual way.

(ii) Overall tests

Under the same assumptions of independence and equality of variance of the variates $x_1, \ldots, x_p$, Andrews (1972) established that, if σ^2 is known, with probability $(1-\alpha)$ the function $f_{\boldsymbol{x}}(t)$ lies in the band with fixed width about $f_{\boldsymbol{\mu}}(t)$ given by

$$|f_{\boldsymbol{x}}(t) - f_{\boldsymbol{\mu}}(t)|^2 \leq \tfrac{1}{2}\sigma^2(p+1)\chi^2_{p,\alpha}$$

where $\chi^2_{p,\alpha}$ denotes the $100(1-\alpha)$ per cent point of the χ^2 distribution on p degrees of freedom. If $\boldsymbol{\mu}$ is known, outliers will fall outside this band and will be easily detectable. A band of the same width centred at $f_{\boldsymbol{x}}(t)$ is a $100(1-\alpha)$ per cent confidence region for $f_{\boldsymbol{\mu}}(t)$, so that if $f_{\boldsymbol{\mu}_0}(t)$ falls outside this band there is evidence against the hypothesis $\boldsymbol{\mu} = \boldsymbol{\mu}_0$.

The simple forms of the above test statistics depend heavily on the assumption of equal variance and independence of the variates $x_1, \ldots, x_p$, as under these assumptions the variance of $f_{\boldsymbol{x}}(t)$ will be approximately constant irrespective of the value of t. These are, in general, very unrealistic assumptions. Goodchild and Vijayan (1974) pointed out that if the sample covariance matrix of the x_i is $\mathbf{W}$, then the variance of $f_{\boldsymbol{x}}(t)$ is $\boldsymbol{u}'\mathbf{W}\boldsymbol{u}$ where $\boldsymbol{u}' = (1/\sqrt{2}, \sin t, \cos t, \sin 2t, \ldots)$. Thus, in general, the variability will *not* be constant over the range of the graph. Tests of significance may still be made at specified values of t, however, by modifying Andrews' test statistic to $\{f_{\boldsymbol{x}}(t) - f_{\boldsymbol{\mu}}(t)\}/\sqrt{(\boldsymbol{u}'\mathbf{W}\boldsymbol{u})}$ and referring the calculated value to the t distribution that has degrees of freedom equal to those of $\mathbf{W}$. The overall tests are no longer appropriate under this more general situation.

9.2 Principal components

We saw in Section 2.2 that if $\mathbf{X}$ is an $(n \times p)$ data matrix from which the sample variance–covariance matrix $\mathbf{S}$ is computed, then the measured variables $x_1, x_2, \ldots, x_p$ can be transformed into principal components $y_1, y_2, \ldots, y_p$ by means of the linear transformations $y_i = a_{i1}x_1 + a_{i2}x_2 + \ldots + a_{ip}x_p$ $(i = 1, \ldots, p)$, where $\boldsymbol{a}_i' = (a_{i1}, \ldots, a_{ip})$ is the eigenvector corresponding to the ith largest eigenvalue l_i of $\mathbf{S}$. Now suppose that we consider the rows of the data matrix to be independent realizations of a random vector $\boldsymbol{x}' = (x_1, \ldots, x_p)$ which has a multivariate normal distribution with mean vector $\boldsymbol{\mu}$ and dispersion matrix $\boldsymbol{\Sigma}$. By analogy with the sample case, we can define the *population principal components* to be the new variables $z_1, z_2, \ldots, z_p$ given by $z_i = \alpha_{i1}x_1 + \alpha_{i2}x_2 + \ldots + \alpha_{ip}x_p$ $(i = 1, \ldots, p)$, where $\boldsymbol{\alpha}_i' = (\alpha_{i1}, \ldots, \alpha_{ip})$ is the eigenvector corresponding to the ith largest eigenvalue λ_i of $\boldsymbol{\Sigma}$. We might then ask: what can be inferred about the z_i, λ_i, and $\boldsymbol{\alpha}_i$ from knowledge of the y_i, l_i, and $\boldsymbol{a}_i$? Mathematically, this is a very complicated area, and many of the results that have been derived are beyond the scope of the present book. However, it is possible at least to give some idea of the extent of sampling variability that is to be expected in large samples.

First, let us establish the difference between sample and population principal components in geometric terms. Each individual for which values have been observed on the p variables $x_1, x_2, \ldots, x_p$ is represented by a point in p-dimensional space in the usual fashion. The population is thus represented (cf. Chapter 6) by a dense swarm of points in this space, the swarm being ellipsoidal in shape (on account of the multivariate normality) and centred at the population mean $\boldsymbol{\mu}$. The eigenvectors $\boldsymbol{\alpha}_i$ define the principal axes of this ellipsoid, and the eigenvalues λ_i measure the lengths of these principal axes (since they are measures of the variance of the corresponding population principal components). Now if a sample of size n is taken from the population, then n points are selected from this swarm. These n points are treated as a second swarm of points; the eigenvectors $\boldsymbol{a}_i$ define the principal axes of the smallest ellipsoid that encloses these n points, and the eigenvalues l_i measure the lengths of these principal axes. If the sample is large and representative of the whole population, then we would expect each l_i, $\boldsymbol{a}_i$ pair to be 'close' to its counterpart λ_i, $\boldsymbol{\alpha}_i$. However, it is evident that, because of vagaries of sampling, there could be considerable discrepancy between these quantities, particularly if the sample is small. For example, in Fig. 9.1, the empty circles denote a 'population' of points in two dimensions while the solid circles denotes a sample selected from this population. The population is bounded by the dotted ellipsoid, the sample by the solid ellipsoid. The dotted lines represent the population vectors $\boldsymbol{\alpha}_i$, while the solid lines the sample vectors $\boldsymbol{a}_i$. The difference in ellipsoids, and hence in $(l_i, \boldsymbol{a}_i)$ and $(\lambda_i, \boldsymbol{\alpha}_i)$ pairs is evident.

Different samples from this population will give rise to different $(l_i, \boldsymbol{a}_i)$ pairs $(i = 1, \ldots, p)$, and hence these quantities will possess sampling distributions. While the small sample theory is complex, T. W. Anderson (1963) has established tractable asymptotic results that are useful for large samples. The main results are as follows:

1. The sample eigenvalues l_i are asymptotically normal and independent, with means λ_i and variances $2\lambda_i^2/(n-1)$ for $i = 1, \ldots, p$.

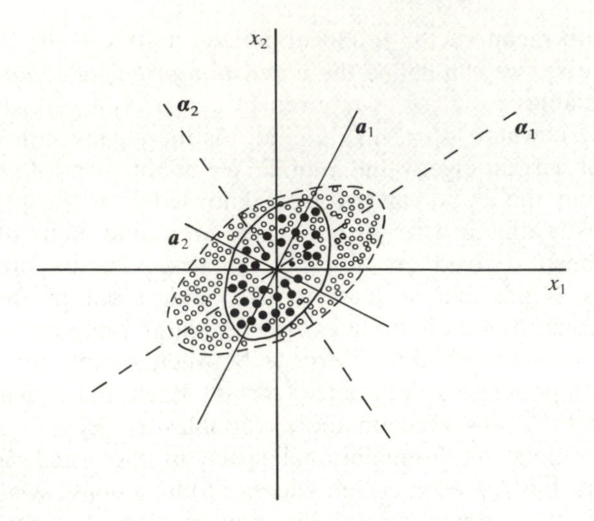

Fig. 9.1 Hypothetical illustration of the difference between population and sample principal components

2. The sample eigenvector $\boldsymbol{a}_i$ is asymptotically normal with mean $\boldsymbol{\alpha}_i$ and covariance matrix $\dfrac{\lambda_i}{(n-1)} \sum_{j \neq i} \dfrac{\lambda_j}{(\lambda_j - \lambda_i)^2} \boldsymbol{\alpha}_j \boldsymbol{\alpha}_j'$ for $i = 1, \ldots, p$.
3. The covariance between a_{is} and a_{jt} is $-\lambda_i \lambda_j \alpha_{it} \alpha_{js} / \{(n-1)(\lambda_i - \lambda_j)^2\}$ for $i \neq j$.
4. The sample eigenvalues are asymptotically independent of the sample eigenvectors.

These results can be used in straightforward fashion to construct asymptotic confidence intervals for the λ_i, or to conduct hypothesis tests about the $\boldsymbol{\alpha}_i$ that will have the desired size if the sample is large. For example, l_1 is asymptotically $\mathrm{N}\{\lambda_1, 2\lambda_1^2/(n-1)\}$ using result 1. Replacing λ_1 by its estimate l_1 in the variance, $(n-1)^{\frac{1}{2}}(l_1 - \lambda_1)/l_1\sqrt{2}$ is asymptotically approximately standard normal. Hence a 95 per cent confidence interval for the largest population eigenvalue λ_1 is given approximately by

$$l_1 - (1.96 l_1)\sqrt{\left(\frac{2}{n-1}\right)} \leq \lambda_1 \leq l_1 + (1.96 l_1)\sqrt{\left(\frac{2}{n-1}\right)}$$

i.e.

$$l_1\left\{1 - \frac{2.77}{\sqrt{(n-1)}}\right\} \leq \lambda_1 \leq l_1\left\{1 + \frac{2.77}{\sqrt{(n-1)}}\right\}.$$

For a sample size of 101, therefore, this interval would be $(0.723 l_1, 1.277 l_1)$.

One question of great practical interest concerns the minimum number of components that are required to represent the sample accurately. Unfortunately it is very difficult even to formulate an adequate hypothesis for this situation, let alone test for it. Suppose that we wish to decide whether a k-dimensional representation of the data is adequate. Such a representation will be identified by the projection of the sample on to the hyperplane defined by the first k principal components, so we are looking for evidence that the remaining $(p-k)$ components can be ignored. A first attempt at formulating the appropriate hypothesis would be that the $(p-k)$ smallest population eigenvalues $\lambda_{k+1}, \lambda_{k+2}, \ldots, \lambda_p$ are all zero. However, if this were the case, then the population would lie in k dimensions instead of p, and hence so would any sample drawn from it. Consequently there would also be $(p-k)$ zero sample eigenvalues. Thus if the sample eigenvalues are not zero, then clearly the population values can not be zero either. We must therefore seek a different way of formulating the hypothesis.

One possibility is to formalize the most popular criterion for choosing k in the descriptive use of principal components. We saw in Section 2.2.5 that k is often chosen as the minimum value for which a given proportion of variance is accounted for by the first k components. Typically, we might choose k as the smallest value such that the first k components account for, say, 80 per cent of the sample variance. To formalize this criterion we need to test the hypothesis that the first k *population* components account for a specified proportion π of the total variance, i.e. we test $H:(\lambda_1+\lambda_2+\ldots+\lambda_k)/(\lambda_1+\lambda_2+\ldots+\lambda_p)=\pi$. Now if we write $\hat{\pi}=(l_1+l_2+\ldots+l_k)/(l_1+l_2+\ldots+l_p)$, the proportion of variance accounted for by the first k sample components, then it can be shown (Mardia *et al.* 1979) that $\hat{\pi}$ is asymptotically normal with mean π and with variance $2\{\text{trace}(\boldsymbol{\Sigma}^2)\}\{\pi^2-2\psi\pi+\psi^2\}/\{(n-1)(\text{trace}\,\boldsymbol{\Sigma})^2\}$ where $\psi=(\lambda_1^2+\lambda_2^2+\ldots+\lambda_k^2)/(\lambda_1^2+\lambda_2^2+\ldots+\lambda_p^2)$. By replacing $\boldsymbol{\Sigma}$ by $\mathbf{S}$ and λ_i by l_i in this variance, an approximate test of the stated hypothesis can be derived. However, even if this approximation is good, there is an unsatisfactory element of arbitrariness in the choice of value for π which makes this test less than useful in practice.

Possibly the most appropriate way of deciding whether the data can be represented adequately in k dimensions is to test that the $(p-k)$ smallest population eigenvalues are equal. Thus the formal hypothesis is

$$H:\lambda_{k+1}=\lambda_{k+2}=\ldots=\lambda_p.$$

If H is true, then there exists no one preferred direction in the subspace spanned by the last $(p-k)$ eigenvectors, so there is no reason to choose any one eigenvector in preference to any of the others. Thus we should either reduce dimensionality to k dimensions, or not reduce dimensionality at all. The likelihood ratio test statistic for H can be found in the same way as was employed in the previous chapter for tests of hypothesis about $\boldsymbol{\Sigma}$. Minus twice the logarithm of the likelihood ratio provides the statistic $W = n(p-k)\log_e(a/g)$, where a is the arithmetic mean of the $(p-k)$ smallest sample eigenvalues (i.e. $a = (l_{k+1} + l_{k+2} + \ldots + l_p)/(p-k)$) and g is the geometric mean of these eigenvalues (i.e. $g = (l_{k+1} \times l_{k+2} \times \ldots \times l_p)^{1/(p-k)}$). Under H, $W = n(p-k)\log_e(a/g)$ has approximately the chi-squared distribution on $(p-k+2)(p-k-1)/2$ degrees of freedom, which enables a test of significance to be made. A slightly better chi-squared approximation is obtained if n is replaced by $n - (2p+11)/6$ in W; this modification gives Bartlett's test of isotropy, which is often printed in the output of standard computer package implementations of principal component analysis. Typical procedure is to test sequentially $k=0$, $k=1$, $k=2$, etc., until H is accepted. This procedure generally leads to selection of values of k much larger than those chosen on the basis of the heuristic methods discussed in Section 2.2.5; Krzanowski (1983a) provides some Monte Carlo simulations to illustrate this feature.

Sampling fluctuations will also play an important role in the comparison of two sets of components via the techniques of Section 5.3, particularly if the comparison is to be viewed inferentially rather than descriptively. Suppose that two independent samples are drawn from the same population, which has population principal components $\boldsymbol{\alpha}_i$ with variances λ_i $(i = 1, \ldots, p)$. Suppose also that the two samples yield principal component solutions l_i, $\boldsymbol{a}_i$ and r_i, $\boldsymbol{b}_i$ respectively $(i = 1, \ldots, p)$. Using the subspace comparison technique of Section 5.3 to compare the first k vectors $\boldsymbol{a}_i$ with the first k vectors $\boldsymbol{b}_i$, how large might the critical angles in this comparison be? In general, we need to know the sampling distributions of these critical angles under both null (i.e. the two samples are from the same population) and alternative (i.e. the two samples are from different populations) hypotheses before they can be used inferentially as test statistics. No analytical results are yet available about these distributions. In fact, the only published results are the limited Monte Carlo simulations reported by Krzanowski (1982a).

We can thus see that inferential aspects of principal component analysis present a rather confusing picture. Small-sample theory is either very complicated or unavailable, and although large-sample theory can lead to relatively simple results it is not clear how large the sample must be before these results are valid. Also, there is ambiguity about the appropriate hypothesis to test. Consequently, it is this writer's view that the most satisfactory uses of principal component analysis are the descriptive ones.

9.3 Multidimensional scaling

Scaling methods are all designed to produce a configuration of points from a matrix of dissimilarities in such a way that the distance between any two points of the configuration approximates as closely as possible the dissimilarity between the entities represented by these points; the different types of scaling described in Chapter 3 can be distinguished by the way in which they measure the goodness-of-fit between distances and dissimilarities.

Given a dissimilarity matrix, every scaling method will provide a configuration as a solution; descriptive use of scaling implies that the input dissimilarities are taken at face value and no allowance is made for any errors that might have arisen in their construction. In the early stages of development, the main thrust of research in multidimensional scaling was in devising and comparing different criteria to be satisfied by the fitted configurations, and in improving the efficiency of the consequent computational algorithms. More recently, however, consideration has been given to the effect of sampling variability in the observed dissimilarities on the recovered configurations and on the magnitude of the goodness-of-fit criteria. Attempts have also been made to derive statistically efficient procedures for estimating a 'true' configuration in the presence of noise-contaminated dissimilarities. We shall therefore outline briefly the work done in each of these three general areas.

Sibson (1978, 1979) adopted a numerical analytic rather than statistical standpoint, and provided a very interesting investigation of the extent to which the accuracy of operation of classical (metric) scaling can be put on to a quantitative footing. To this end, he considered small perturbations of order ε to the elements of the input dissimilarity matrix, and he evaluated the consequent perturbations to the eigenvalues and eigenvectors of the fundamental metric scaling matrix $\mathbf{F}$ (Section 3.2, p. 107). These latter perturbations determine the displacements in the points of the fitted configuration caused by the changes to the input dissimilarities. Comparing the fitted configuration after perturbing dissimilarities with the original fitted configuration by means of procrustes rotation (Section 5.1) yields a single summary measure of the extent to which the perturbation in dissimilarities changes the fitted configuration. Sibson provided expressions for this measure under various specific perturbations to the dissimilarities, and established approximate sampling distributions assuming these perturbations to be from a standard normal distribution. The expressions he derived are simple enough to be of practical utility. He also concluded from this work that metric scaling is robust against errors that leave observed dissimilarities still approximately linearly related to distance.

Turning next to non-metric multidimensional scaling, there have been very many Monte Carlo studies of the performance of Kruskal's (1964a,b) algorithm and measure of goodness-of-fit STRESS. However, many arguments have raged in the Psychometry literature about the appropriateness of design of some of these studies, and applicability of results from them. Consequently it seems inappropriate to devote too much space to a discussion here, beyond stating the main objectives and general structure of the experiments, together with some pertinent references from which the reader may gather much more detail. There have been two main objectives in these studies: establishing sampling distributions for the quantity STRESS (to provide guidance for the user about satisfactory levels of STRESS to achieve in practical applications), and deciding whether random or systematic starting configurations yield the best results (to provide guidance for the user about the best way of operating the computational algorithm). To conduct a Monte Carlo study of either question it is necessary to provide replicate sets of input dissimilarities; to make sure the results of the study are of practical use, the structure of any given input practical set should be roughly identifiable with that of one of these replicate sets. There have been two main designs employed in such studies. In the first design, the input dissimilarities are generated directly by means of a random selection from a uniform distribution, which corresponds to a hypothesis of 'randomness' in which each set of ranked dissimilarities is equally likely. In the second design, the input dissimilarities are obtained by calculation of inter-point distances in a specified configuration, made up of a 'true' configuration on top of which random error is superimposed. Interest then focuses on the effect of variation in factors such as degree of error, number of stimuli, type of metric, and dimensionality. Critics of the former design point to its lack of realism in practice, while critics of the latter emphasize that a 'true' configuration will never be known in practice, and that any practical application is bound to deviate in some unknown way from the idealized Monte Carlo counterpart. To obtain some idea of the available results, together with reasonable bibliographies of previous work, the reader is referred to Arabie (1978) and Spence and Young (1978) on the random versus rational starting configuration questions, and to Levine (1978) on the distribution of STRESS. It is probably fair to say that neither question has yet been fully resolved, and scope exists for further work, although it is certainly clear that Kruskal's original recommendations were over-simplistic.

The third area of statistical interest in multidimensional scaling concerns the modelling of the input dissimilarities as random variables, and is best exemplified by the series of papers by Ramsay culminating in Ramsay (1982). Here it is assumed that the dissimilarity D_{ijr} between objects or stimuli i and j as perceived by subject r, and the inter-point distance d^*_{ijr} between points representing stimuli i and j in the 'true' configuration representing the perception of subject r, are linked by assuming that the D_{ijr} are random variables with distributions including d^*_{ijr} as parameters. From empirical considerations Ramsay suggests that, either in raw form or after suitable transformations, the most appropriate models are either:

(1) the log-normal distribution, $\log_e D_{ijr} \sim N(\log_e d^*_{ijr}, \sigma^2_{ijr})$, or
(2) the normal distribution, $D_{ijr} \sim N\{d^*_{ijr}, (\sigma_{ijr} d^*_{ijr})^{2\alpha}\}$ for some α, or
(3) the inverse normal distribution, with probability density

$$g(D_{ijr}) = (2\pi\sigma^2_{ijr} D^3_{ijr})^{-\frac{1}{2}} \exp\{-(1 - D_{ijr} d^*_{ijr})^2/(2\sigma^2_{ijr} D_{ijr})\}.$$

Further empirical considerations suggest a multiplicative variance components structure imposed on σ^2_{ijr}. Having formulated these models, Ramsay is then able to invoke the maximum likelihood principle in estimating the various parameters (i.e. transformation parameters if necessary, configuration parameters, variance component parameters), as well as establishing methods of interval estimation and hypothesis testing about these parameters. Tests about the appropriate dimensionality of the configuration come within this framework. These latter aspects go beyond the scope of previous results in multidimensional scaling theory. Computational aspects also need to be considered, of course, and Ramsay describes the structure of a computer program to implement these various procedures.

There seems to be much current interest in such model-building approaches to multidimensional scaling, particularly among psychologists. The discussion following Ramsay's paper makes interesting reading, with cases both for and against inferential usage of multidimensional scaling being equally strongly put. Again, it seems that much more work needs to be done in this area before definitive conclusions can be reached, and at this stage the present writer once more comes down in favour of the purely descriptive use of the techniques.

9.4 Cluster analysis

When a clustering algorithm is applied to a set of data, a classification of objects is obtained whether or not the data exhibit a true or 'natural' grouping structure. In Section 3.1.1 we discussed the use of a dendrogram obtainable from a hierarchical cluster analysis as a method of description of a multivariate sample, showing which sample members grouped into relatively homogeneous classes. Used purely descriptively, it does not really matter whether or not the stratification of the sample corresponds to a division into 'real' groups; it is simply a convenient summary of the data. However, in the wider context of cluster analysis the problem is an important one. If genuine groups *do* exist then the botanist or zoologist, say, will require the chosen numerical method to identify them; conversely if no such groups exist then he or she will not want to be misled by arbitrary grouping of the sample. Much effort has therefore also been expended by researchers in this area. However, we have only briefly touched on cluster analysis, and have not discussed this vast subject in any depth. Some more comments are made in Section 11.3, but for the present we only indicate the main statistical lines of research that have been pursued, and leave the reader to follow up any topics of interest in the cited references.

Of most direct relevance to the uses of the dendrogram as envisaged in Section 3.1.1 is the question of how accurately the dendrogram represents the input dissimilarities. To attempt an answer to this question, we first need some way of quantifying the information in the dendrogram. A suitable approach is to consider the clustering criterion value for the lowest node of the dendrogram that links each possible pair of objects in the sample. If the value of this node for objects i and j is denoted by d_{ij}^+, then the set of values d_{ij}^+ $(i=2,\ldots,n;\ j=1,\ldots,i-1)$ forms a set of *ultrametric distances* (i.e. they have the property that $d_{uv}^+ \leqslant \max(d_{uw}^+, d_{vw}^+)$ for all u, v, w), and there is a (1–1) correspondence between the ultrametric distances and the dendrogram itself (see Gower and Banfield 1975, and references therein). The goodness-of-fit of a dendrogram to a set of input dissimilarities can then be assessed by some function $\mathrm{f}(d_{ij}^+, \delta_{ij})$ of all ultrametric distances and corresponding dissimilarities δ_{ij}. Many such goodness-of-fit criteria have been proposed in the literature. For example, three such criteria can be formed using functions already considered in Chapter 3:

(1) $\mathrm{f}(d_{ij}^+, \delta_{ij}) = \sum_{i<j}\sum (\delta_{ij}^2 - d_{ij}^{+2})$ (Principal components),

(2) $\mathrm{f}(d_{ij}^+, \delta_{ij}) = \sum_{i<j}\sum (\delta_{ij} - d_{ij}^+)^2$ (Least squares),

(3) $\mathrm{f}(d_{ij}^+, \delta_{ij}) = \sum_{i<j}\sum (\delta_{ij} - d_{ij}^+)^2/\delta_{ij}^2$ (Non-linear mapping).

Other possibilities include

(4) $\mathrm{f}(d_{ij}^+, \delta_{ij})$ = product-moment correlation between d_{ij}^+ and δ_{ij} (Cophenetic correlation).

(5) As (4), but calculated on ranked d_{ij}^+ and δ_{ij} (Kendall's tau), and many more.

One possible study that can be made of such a goodness-of-fit criterion is into its sampling distribution for data drawn from a population of given structure. By comparing the sampling distribution for data drawn from a homogeneous population with the sampling distribution for data drawn from a population containing clusters, it may be possible to assess whether the criterion value obtained for a real set of data is more likely to have come from the former or latter type of population. Unfortunately, analytical results seem to be difficult to obtain, and recourse must generally be made to Monte Carlo methods. There are also very many different criteria and clustering methods that can be studied. Last, but not least, is the problem of modelling an appropriate 'clustered' population for one of the hypotheses. As with the difficulties discussed in the previous section, the 'true' structure of the data is never known in a real application for comparison with the idealized distribution. Despite all these obstacles, various results do exist in the literature. The reader is referred to Gower and Banfield (1975) for a detailed study of the distribution of seven criteria applied to single-linkage dendrograms obtained from data generated under the null hypothesis (i.e. a homogeneous multivariate normal population). Further studies and more extensive references may be found in Milligan (1981a, b).

Passing to more general aspects, various authors have suggested models for clustered data as a basis for statistical inference in cluster analysis. Suppose that P is the population distribution, $\boldsymbol{x}$ is a typical point in p-dimensional Euclidean space and f is the density of P. The following models form the basis of most of the reported statistical investigations:

1. High-density clusters: the population clusters are the maximal connected subsets of the high-density region $\{\boldsymbol{x} \mid \mathrm{f}(\boldsymbol{x}) \geqslant c\}$ for each c. The family of population clusters forms a tree, in that if two clusters overlap then one includes the other, so this model is suitable for examining hierarchic techniques.
2. Multimodality: the common density $\mathrm{f}(\boldsymbol{x})$ is multimodal, i.e. there exists a finite number $k \geqslant 2$ of distinct points $\boldsymbol{\mu}_1, \ldots, \boldsymbol{\mu}_k$ of the p-dimensional space where f(.) attains a strict relative maximum.
3. Mixture distribution: the common density $\mathrm{f}(\boldsymbol{x})$ is a translation mixture of the form $\mathrm{f}(\boldsymbol{x}) = \sum_{i=1}^{k} p_i \mathrm{h}(\boldsymbol{x} - \boldsymbol{\mu}_i)$ with $k \geqslant 2$ different subpopulation centres $\boldsymbol{\mu}_1, \ldots, \boldsymbol{\mu}_k$, a density $\mathrm{h}(\boldsymbol{x})$ describing the shape of the clusters, and k class proportions $p_1, \ldots, p_k$ satisfying $\sum_{i=1}^{k} p_i = 1$.
(Note that in both 2 and 3, all $\boldsymbol{\mu}_i$, p_i, and k are unknown.)

Good recent summaries of the results obtained from these models, as well as of problems outstanding, are given by Hartigan (1985) and Bock (1985), which include comprehensive lists of references.

9.5 Comment

The development of the descriptive methods of multivariate data display discussed in Part I of this book has been accelerated in recent years by the dramatic increase in computer power that has been achieved in this period, and has also been fuelled by the enormous increase in size of data sets made possible by automatic data recording devices. Research into the possible applications of inferential techniques based upon statistical models, as discussed in the present chapter, has trailed some way behind these developments. However, there is a fundamental conflict between the objectives of the two applications (description and inference), which has caused some statisticians to voice their unease (see the discussion following Ramsay, 1982). Used in their initial descriptive manner, the techniques are all geared towards *generating* hypotheses, by exposing inherent pattern in a set of data and suggesting interesting aspects to be followed up in subsequent experiments or investigations. The inferential applications, on the other hand, are geared towards *testing* hypotheses on the basis of models of data. It is therefore worth keeping in mind, when contemplating the use of one of these techniques, that its descriptive use carries no penalty but an inferential extension may be founded on an inappropriate model and may therefore lead to incorrect conclusions.

10
Discrete data

Most of the problems that have been discussed thus far have assumed the data presented for analysis are quantitative. In many multivariate data sets, however, some or all of the measured variables may be qualitative. For such variables, an assumption of multivariate normality is manifestly inappropriate, and methods of analysis that are radically different from those considered earlier must therefore be sought. Discrete quantitative variables having only a few possible values also come under the same umbrella, and currently popular usage is to term the collection of all such variables as *discrete multivariate data*. We first consider the case where all variables in a data set are discrete, and later turn to the treatment of a mixture of discrete and continuous data.

10.1 Entirely discrete data: summary and model

Whether the data are discrete or continuous, the starting point of any multivariate investigation is the $(n \times p)$ data matrix $\mathbf{X}$. However, when the p variables are discrete, such data presentation is usually wasteful of space. With a restricted range of values for each variable, there will generally be subsets of individuals in the sample having the same pattern of responses over all p variables. Consider for example the data in Table 1.2. If we focus on variables x_3 and x_4 only, we have a bivariate sample of twelve individuals, with only the three values 0, 1, or 2 possible for each variable. We see that the six individuals 1, 2, 4, 5, 6, and 8 all have value 0 for x_3 ('temper') together with value 2 for x_4 ('feelings'). There are additional replicate observations for the variable values $(x_3 = 1, x_4 = 1)$ and $(x_3 = 2, x_4 = 0)$. A much more concise presentation of the data is therefore as the frequency distribution shown at the top of p. 266.

Even more informative will be the arrangement of the data in a two-way array; each row is indexed by the values of one variable (say x_3), each column is indexed by the values of the other variable (x_4), and the entry

Value of x_3	x_4	Sample frequency
0	1	1
0	2	6
1	0	1
1	1	2
2	0	2
	Total	12

in the ith row and jth column gives the number of sample members with values $x_3 = i$ and $x_4 = j$:

x_3 \ x_4	0	1	2	.
0	0	1	6	7
1	1	2	0	3
2	2	0	0	2
.	3	3	6	12

Not only is this a concise arrangement of all possible combinations of (x_3, x_4) values (thus allowing those combinations not present in the sample to be readily highlighted), but by summing the entries in the rows and columns as shown we can also immediately obtain the *marginal* frequency distributions of the separate variables. In this chapter we shall denote summation over an index set by a dot, so these marginal frequencies are indexed by a dot in the appropriate row or column. (Note that the dot notation here differs slightly from its use in Chapter 3, where it denoted averaging rather than summing over a variable.)

Such a table is, of course, the familiar *contingency table* encountered in many simple statistical problems. The arrangement can be extended readily to higher-way tables which display the joint frequencies of 3, 4, 5, or more variables. As an example, consider the three binary variables x_1, x_{11}, and x_{12} of Table 1.2. To avoid confusion with suffixes, denote these three variables for the present by x_1, x_2 and x_3. Then the three-way table formed from them is given at the top of p. 267.

With three variables, there will be three single-variable marginal frequency distributions and three two-variable marginal distributions. Marginal frequency distributions are often simply referred to as margins. The most concise way of incorporating these margins into the table is to add

x_1	x_2	x_3 0	1
0	0	1	0
	1	2	5
1	0	0	1
	1	1	2

one category, indexed by a dot, to the existing categories for each variable and enter frequencies into this category by appropriate summations. Thus the above table will be expanded into the following table on addition of marginal totals:

x_1	x_2	x_3 0	1	·
	0	1	0	1
0	1	2	5	7
	·	3	5	8
	0	0	1	1
1	1	1	2	3
	·	1	3	4
	0	1	1	2
·	1	3	7	10
	·	4	8	12

The x_1, x_2 margin is obtained by summing over values of x_3, i.e. by summing adjacent entries in the two columns headed $x_3 = 0$ and $x_3 = 1$; the x_1, x_3 margin is obtained by summing corresponding entries of each $x_2 = 0$, $x_2 = 1$ row pair; and the x_2, x_3 margin is obtained by summing the corresponding entries of the two 2×2 tables defined by $x_1 = 0$ and $x_1 = 1$. Each marginal distribution can be extracted from the above table and written as a separate table; such a process is often referred to as *collapsing* a table over one or more variables. For example, collapsing the above table over x_1 yields the (x_2, x_3) marginal table:

x_2	x_3 0	1	·
0	1	1	2
1	3	7	10
·	4	8	12

while collapsing it over both x_1 and x_2 yields the x_3 table:

x_3		
0	1	•
4	8	12

It is evident that the inclusion of marginal totals rapidly expands the size of a table, and for this reason margins are generally omitted in tables of order greater than two.

We now give four examples of discrete multivariate data sets, which will be used to motivate and illustrate the methods that follow.

Example 10.1

The following data refer to an experimental investigation into the possible carcinogenic effect of a certain fungicide. Sixteen male mice were fed with the fungicide for the period of the experiment, whilst a control group of seventy-nine male mice were kept without treatment under similar conditions. At the end of the experiment, the following two attributes were recorded for each mouse: x_1, presence or absence of pulmonary tumours; x_2, whether a treated or a control mouse. The data can be summarized in the two-way table:

	x_2		
x_1	Treated	Control	Total
Tumours	4	5	9
No tumours	12	74	86
Total	16	79	95

Example 10.2

In an experiment to investigate the relationship between the nasal carrier rate for *streptococcus pyogenes* and the size of tonsils among school-children, 1398 children were examined, and the following two attributes were recorded on each child: x_1, size of tonsil (normal, large, or very large); x_2, whether or not the child is a carrier. The data can again be summarized in a two-way table:

x_1	x_2 Carrier	Non-carrier	Total
Normal	19	497	516
Large	29	560	589
Very large	24	269	293
Total	72	1326	1398

Example 10.3

The following data were obtained from an experiment designed to investigate the propagation of plum root stocks from root cuttings (Bartlett, 1935). Each cutting could be either long or short (x_3) and was either planted at once or not until spring (x_2). A total of 240 cuttings were taken for each of the four possible combinations of these two conditions, and at the end of the experiment it was recorded whether the plant was alive or dead (x_1). The data can thus be summarized in the following three-way table:

		x_3	
x_1	x_2	Long	Short
Alive	At once	156	107
	In spring	84	31
Dead	At once	84	133
	In spring	156	209

Example 10.4

Kihleberg *et al.* (1964) report the following data, obtained in a study of the relationship between car size and accident injuries. The 'individuals' in the study were accidents to solitary drivers, and the attributes recorded for each individual were: x_1, whether the car was small or large (assessed by weight); x_2, whether the driver was ejected from the car or not; x_3, whether the injuries were severe or not severe; x_4, whether the accident was a collision or a rollover. The data are summarized in the following four-way table.

x_1	x_2	x_3	x_4: Collision	Rollover
Small car	Ejected	Severe	23	80
		Not severe	26	19
	Not ejected	Severe	150	112
		Not severe	350	60
Large car	Ejected	Severe	161	265
		Not severe	111	22
	Not ejected	Severe	1022	404
		Not severe	1878	148

It is evident from these examples that the qualitative variables can be of two types. Some are such that their values for a particular experimental unit are determined *in advance* by the investigator. For example, the experimenter chooses the mice that will be treated and those that will be controls in the set-up described in Example 10.1. Similarly, the cuttings to be planted now and those to be planted in spring, or those to be long and those to be short, are also determined by the investigator. Moreover, the total number of such cuttings, or of the various categories of mice, are decided before the experiment is conducted. Consequently, the appropriate margins of the contingency table are *fixed* at these pre-selected values. Remaining variables, on the other hand, are genuine responses in the sense that their outcomes cannot be predicted in advance. Thus we do not know (before the experiment) how many mice will develop tumours, how many plants will die, how many drivers will be ejected in crashes, and so on. The margins for such variables thus depend on the outcome of the experiment or investigation. Although the *type* of variable is generally not distinguished in the analysis as outlined below, it is important to incorporate into this analysis any experimental constraints which lead to fixed margins.

Having considered some possible methods for displaying discrete multivariate data in Section 4.5, we now turn our attention to analysis of such data. First, a suitable model must be postulated. Now the data for any discrete multivariate problem can be expressed in the form of a contingency table containing a number of mutually exclusive categories or 'cells'. In Example 10.1 above, two binary variables yielded four such cells; in Example 10.2, one binary and one three-state variable yielded six cells; in Example 10.3, three binary variables yielded eight cells; while in Example 10.4 four binary variables yielded sixteen cells. In general, if there are q discrete variables and the ith one has s_i states ($i = 1, \ldots, q$), then it is easy to see that the contingency table formed from all these variables will have $s = s_1 \times s_2 \times \ldots \times s_q = \prod_{i=1}^{q} s_i$ cells. These cells can be labelled $z_1, z_2, \ldots, z_s$ in some specified order. Every sample individual must be placed into one, and only one, such cell. Thus we may denote a typical sample member by z, and the simplest model that can be formulated for the data is to postulate that $\text{Prob}(z = z_i) = p_i$ for some set of probabilities p_i satisfying $\sum_{i=1}^{s} p_i = 1$. Suppose that we have a sample of n individuals, n_i of which are located in cell z_i of the table ($i = 1, \ldots, s$). The above model then implies that the probability of the distribution $(n_1, n_2, \ldots, n_s)$ among cells $(z_1, z_2, \ldots, z_s)$ is given by the *multinomial distribution* as

$$P = \frac{n!}{n_1!\, n_2! \ldots n_s!} p_1^{n_1} p_2^{n_2} \ldots p_s^{n_s} = n! \prod_{i=1}^{s} \frac{1}{n_i!} (p_i)^{n_i}$$

where $\sum_{i=1}^{s} p_i = 1$ and $\sum_{i=1}^{s} n_i = n$. The parameters of the model are thus the cell probabilities p_i, and inference about these probabilities constitutes the analysis of discrete multivariate data (in analogous fashion to the inference about $\boldsymbol{\mu}$ and $\boldsymbol{\Sigma}$ that constituted analysis of continuous multivariate data).

A comprehensive discussion of discrete multivariate data analysis would require a separate volume. Indeed, excellent accounts may already be found in the books by Everitt (1977) and Fienberg (1981), while a comprehensive text book is the one by Bishop *et al.* (1975). Our purpose here is merely to introduce the basic ideas underlying such analysis, and to provide the reader with enough material to enable simple analyses to be done of straightforward data sets. A secondary purpose is to highlight the difference in approach between the analysis of discrete data and the analysis of continuous data, pointing up the essential features of the former.

10.2 Entirely discrete data: analysis

Many of the most important features of discrete data analysis are brought out in the simplest situation, that of bivariate data summarized in a two-way table. We will therefore devote most of our attention to this case. Extension of the results to the multivariate case and higher-order tables will follow readily, without the need for such detail.

10.2.1 Two-dimensional tables

Here we have two discrete variables x_1 and x_2 measured on each of n individuals. Let us suppose that x_1 contains a states and x_2 contains b states, and that n_{ij} denotes the number of sample members that have state i of x_1 and state j of x_2. The full summary of the sample is therefore in the two-way table

	x_2					
x_1	1	2	3	...	b	Totals
1	n_{11}	n_{12}	n_{13}	...	n_{1b}	$n_{1.}$
2	n_{21}	n_{22}	n_{23}	...	n_{2b}	$n_{2.}$
3	n_{31}	n_{32}	n_{33}	...	n_{3b}	$n_{3.}$
$\vdots$	$\vdots$	$\vdots$	$\vdots$		$\vdots$	$\vdots$
a	n_{a1}	n_{a2}	n_{a3}	...	n_{ab}	$n_{a.}$
Totals	$n_{.1}$	$n_{.2}$	$n_{.3}$	...	$n_{.b}$	n

Summation over all states of a variable is indicated by a dot in the appropriate subscript position, so that $n_{i.} = \sum_{j=1}^{b} n_{ij}$ and $n_{.j} = \sum_{i=1}^{a} n_{ij}$. The dots are suppressed here on summing over both variables, as $n = \sum\sum n_{ij}$.

To conform with this arrangement of the data, write p_{ij} as the probability of obtaining a sample member which has state i of x_1 and state j of x_2. Inferences from the data will generally involve some hypothesis about the p_{ij}. For any hypothesis which specifies $p_{ij} = p_{ij}^{(0)} (i = 1, \ldots, a; j = 1, \ldots, b)$, we may calculate the *expected* number $\mu_{ij} = np_{ij}^{(0)}$ of individuals exhibiting state i of x_1 and state j of x_2. Only exceptionally will the hypothesis assign actual numerical values to all p_{ij}, enabling the μ_{ij} to be calculated directly. Most commonly the hypothesis imposes some simple structure on the p_{ij}, in which case the $p_{ij}^{(0)}$ are still unknown (but are usually functions of a number of parameters). In such cases we must *estimate* the expected numbers μ_{ij}. Denote by $\hat{\mu}_{ij}$ the estimate of μ_{ij} under the given hypothesis $(i = 1, \ldots, a; j = 1, \ldots, b)$; such estimates are generally called the *fitted values*. A test of the hypothesis can then be made by assessing whether or not these fitted values are in reasonable agreement with the observed cell frequencies n_{ij}.

Of most general practical interest is the null hypothesis of *no association* between x_1 and x_2. In Example 10.1, we would be interested in testing whether the treated mice are more disposed to developing tumours than those of the control group; if this is not the case, then tumour incidence is independent of group membership, and the two variables thus exhibit no association. In Example 10.2, we might wish to investigate whether the carrier rate is independent of tonsil size, which is again a statement of the 'no association' hypothesis. No association between x_1 and x_2 implies that occurrence of a unit with category i of x_1 is not influenced by its corresponding category j of x_2. Under this hypothesis, the probability of a unit falling in category i of x_1 and category j of x_2 is the product of the marginal probabilities of obtaining category i of x_1 and category j of x_2 (by the laws of simple probability). Thus if we denote by H_0 the null hypothesis of no association, then under H_0:

$$p_{ij} = p_{i.}p_{.j} \tag{10.1}$$

so that

$$\mu_{ij} = \frac{\mu_{i.}\mu_{.j}}{n} \tag{10.2}$$

where $p_{i.} = \sum_j p_{ij}$, $p_{.j} = \sum_i p_{ij}$ are the *marginal probabilities,* $\mu_{ij} = np_{ij}$, and $\mu_{i.} = np_{i.}$, $\mu_{.j} = np_{.j}$ are the *expected marginal totals*. Taking logarithms in eqn (10.2) yields

$$\log_e(\mu_{ij}) = \log_e(\mu_{i.}) + \log_e(\mu_{.j}) - \log_e(n). \tag{10.3}$$

Now let θ denote the average of all the logarithms of the expected cell frequencies, i.e. $\theta = \frac{1}{ab}\sum_{i=1}^{a}\sum_{j=1}^{b}\log_e(\mu_{ij})$. Then summing (10.3) over both i and j and dividing by ab, we see that

$$\theta = \frac{1}{a}\sum_{i=1}^{a}\log_e(\mu_{i.}) + \frac{1}{b}\sum_{j=1}^{b}\log_e(\mu_{.j}) - \log_e(n). \tag{10.4}$$

Further, let α_i denote the difference between the average of the logarithms of the expected frequencies in the b cells for category i of x_1 and the average of all the logarithms of expected cell frequencies. Then

$$\alpha_i = \frac{1}{b}\sum_{j=1}^{b}\log_e(\mu_{ij}) - \frac{1}{ab}\sum_{i=1}^{a}\sum_{j=1}^{b}\log_e(\mu_{ij}) = \frac{1}{b}\sum_{j=1}^{b}\log_e(\mu_{ij}) - \theta,$$

so that $\alpha_i + \theta = \frac{1}{b}\sum_{j=1}^{b}\log_e(\mu_{ij})$. Hence, summing (10.3) over j and dividing by b, we see that

$$\alpha_i + \theta = \log_e(\mu_{i.}) + \frac{1}{b}\sum_{j=1}^{b}\log_e(\mu_{.j}) - \log_e(n),$$

and subtracting (10.4) from this expression yields

$$\alpha_i = \log_e(\mu_{i.}) - \frac{1}{a}\sum_{i=1}^{a} \log_e(\mu_{i.}). \tag{10.5}$$

Next, let β_j denote the difference between the average of the logarithms of the expected cell frequencies in the a cells for category j of x_2 and the average of all the logarithms of expected cell frequencies. Then

$$\beta_j = \frac{1}{a}\sum_{i=1}^{a} \log_e(\mu_{ij}) - \frac{1}{ab}\sum_{i=1}^{a}\sum_{j=1}^{b} \log_e(\mu_{ij}) = \frac{1}{a}\sum_{i=1}^{a} \log_e(\mu_{ij}) - \theta,$$

so that $\beta_j + \theta = \frac{1}{a}\sum_{i=1}^{a} \log_e(\mu_{ij})$. Hence summing (10.3) over i and dividing by a, we see that

$$\beta_j + \theta = \log_e(\mu_{.j}) + \frac{1}{a}\sum_{i=1}^{a} \log_e(\mu_{i.}) - \log_e(n),$$

and subtracting (10.4) from this expression yields

$$\beta_j = \log_e(\mu_{.j}) - \frac{1}{b}\sum_{j=1}^{b} \log_e(\mu_{.j}). \tag{10.6}$$

Hence we can substitute for $\log_e(\mu_{i.})$ and $\log_e(\mu_{.j})$ from (10.5) and (10.6) respectively into (10.3) to yield

$$\log_e(\mu_{ij}) = \alpha_i + \frac{1}{a}\sum_{i=1}^{a} \log_e(\mu_{i.}) + \beta_j + \frac{1}{b}\sum_{j=1}^{b} \log_e(\mu_{.j}) - \log_e(n)$$

which, from (10.4) finally gives

$$\log_e(\mu_{ij}) = \theta + \alpha_i + \beta_j. \tag{10.7}$$

This expression is known as a *log-linear model* for the expected frequencies μ_{ij}, *on the hypothesis that* x_1 *and* x_2 *are independent.*

Consideration of this model shows it to be very similar to simple additive models encountered in the analysis of variance. Furthermore, eqns (10.5) and (10.6) show that $\sum_{i=1}^{a} \alpha_i = 0$ and $\sum_{j=1}^{b} \beta_j = 0$. By analogy with analysis of variance we may therefore term the α_i and the β_j the *main effects* of the categories in the two-way table. The set of main effects α_i $(i = 1, \ldots, a)$ reflect the differences in the marginal totals of the rows of the table, while the set β_j $(j = 1, \ldots, b)$ reflect the differences in the marginal totals of the columns of the table.

Model (10.7) was derived assuming that x_1 and x_2 are independent, and so variation in the α_i and β_j reflects the only possible variation in the table when this constraint is imposed. If x_1 and x_2 are *not* independent then there will be some association between these variables, which implies that patterns of frequencies in the cells of the tables will be affected by the values of the (i, j) combinations of x_1, x_2 categories. Reverting to analysis of variance terminology, this means that there will be an *interaction* between categories of x_1 and categories of x_2. If (10.7) represents a constrained model for the data in the absence of interaction, it seems reasonable to postulate that the full model will be of the form

$$\log_e(\mu_{ij}) = \theta + \alpha_i + \beta_j + \gamma_{ij} \tag{10.8}$$

where γ_{ij} represents the interaction term for categories i, j of x_1, x_2. Further analogy with analysis of variance suggests that these interaction terms will satisfy the constraints $\sum_{i=1}^{a} \gamma_{ij} = \sum_{j=1}^{b} \gamma_{ij} = 0$. Because of these constraints, and the earlier ones for α_i and β_j, model (10.8) contains $1 + (a-1) + (b-1) + (a-1)(b-1) = ab$ unknown parameters. If this model is fitted to an $a \times b$ array of observed frequencies, these ab unknown parameters must be estimated from ab items of information. Consequently a perfect fit is to be anticipated, and (10.8) is termed a *saturated* or *full* model.

To test the null hypothesis of no association between x_1 and x_2, it therefore suffices to test whether $\gamma_{ij} = 0$ for all i and j in (10.8). If γ_{ij} *is* zero for all i, j then (10.8) becomes (10.7). Thus standard statistical practice can be followed by successively fitting (10.8) and (10.7) to the data. If the fit of the latter to the data is significantly worse than the fit of the former then the null hypothesis is rejected, otherwise the null hypothesis is accepted. To make progress, we thus need to be able to estimate the μ_{ij} for any specified model. The theory of maximum likelihood estimation for log-linear models was considered by Birch (1963), and the following two fundamental results provide the cornerstone as regards practical applications.

Result 1: the maximum likelihood estimators of the μ_{ij} are uniquely determined by equating appropriate observed marginal totals to the maximum likelihood estimators of their expectations.

Result 2: if a term is included in a log-linear model, then its presence ensures that the estimated expected marginal totals are equal to the observed marginal totals corresponding to that term.

To clarify these two results, consider estimation of the μ_{ij} using model (10.7), i.e. assuming that there is no association between x_1 and x_2. This model contains terms θ, α_i, and β_j, and the expected marginal totals corresponding to these terms are $\mu_{..}$, $\mu_{i.}$ and $\mu_{.j}$ respectively. Corresponding observed marginal totals are n, $n_{i.}$, and $n_{.j}$, so from Result 2 we have the following maximum likelihood equations;

$$\hat{\mu}_{..} = n \tag{10.9}$$

$$\hat{\mu}_{i.} = n_{i.} \tag{10.10}$$

$$\hat{\mu}_{.j} = n_{.j}. \tag{10.11}$$

Result 1 tells us these equations uniquely determine the $\hat{\mu}_{ij}$. These equations can in fact be solved for the $\hat{\mu}_{ij}$ without first having to obtain the estimators for θ, α_i, and β_j. From (10.7) we have $\log_e(\mu_{ij}) = \theta + \alpha_i + \beta_j$, so that

$$\mu_{ij} = \exp(\theta + \alpha_i + \beta_j) = e^{\theta}e^{\alpha_i}e^{\beta_j}.$$

Thus

$$\mu_{i.} = e^{\theta+\alpha_i}\left(\sum_{j=1}^{b} e^{\beta_j}\right)$$

$$\mu_{.j} = e^{\theta+\beta_j}\left(\sum_{i=1}^{a} e^{\alpha_i}\right)$$

and

$$\mu_{..} = e^{\theta}\left(\sum_{i=1}^{a} e^{\alpha_i}\right)\left(\sum_{j=1}^{b} e^{\beta_j}\right)$$

$$= \frac{\mu_{i.}\mu_{.j}}{\mu_{ij}}$$

on substituting for the two summations. Rearrangement of terms gives $\mu_{ij} = \dfrac{\mu_{i.}\mu_{.j}}{\mu_{..}}$, and the invariance property of maximum likelihood estimators thus yields $\hat{\mu}_{ij} = \dfrac{\hat{\mu}_{i.}\hat{\mu}_{.j}}{\hat{\mu}_{..}}$. By using (10.9), (10.10), and (10.11), it follows that

$$\hat{\mu}_{ij} = \frac{n_{i.}n_{.j}}{n}. \tag{10.12}$$

Application of Results 1 and 2 enables the estimates $\hat{\mu}_{ij}$ to be obtained for *any* log-linear model. Having fitted a model, the next step is to assess how well that model fits the data, i.e. how closely the $\hat{\mu}_{ij}$ match the n_{ij}. There are two widely used summary measures for assessing the overall goodness of fit of a log-linear model. There is the classical Pearson statistic

$$X^2 = \sum_i \sum_j (n_{ij} - \hat{\mu}_{ij})^2/\hat{\mu}_{ij} \tag{10.13}$$

and the log-likelihood ratio statistic

$$D = 2\sum_i \sum_j n_{ij} \log_e(n_{ij}/\hat{\mu}_{ij}). \tag{10.14}$$

Both statistics are asymptotically distributed as χ^2 if the model is 'correct', but the D statistic is to be preferred since it is additive under partitioning while X^2 is not. The statistic D was termed the *deviance* of the model by Nelder and Wedderburn (1972), who established the mathematical theory upon which the results summarized here are based. In fact, if the model is correct then D is asymptotically distributed as χ^2 on v degrees of freedom, where v is the number of cells in the table minus the number of independent parameters fitted by the model. This result enables a test of goodness-of-fit of the model to be made. More informally, since $E(\chi^2_v) = v$, a well-fitting model will have a deviance roughly equal to its number of degrees of freedom. Furthermore, if M_1 and M_2 denote two nested models (i.e. the parameters of M_1 are all included in M_2) with v_1 and v_2 degrees of freedom respectively, then if all extra parameters of M_2 are zero the difference in deviances of the two models has a χ^2 distribution on $v_1 - v_2$ degrees of freedom. We can therefore assess the significance of particular terms by finding the reduction in deviance incurred by their inclusion in a model, and comparing this reduction with the critical values of a $\chi^2_{v_1-v_2}$ distribution. Note that, in particular, the deviance for the full model of equation (10.8) must be zero, since the model will be a perfect fit to the data and $\hat{\mu}_{ij} = n_{ij}$. Thus the deviance for the model corresponding to eqn (10.7) measures the importance of the interaction terms γ_{ij}. This is a test of the hypothesis of no interaction based on $(a-1)(b-1)$ degrees of freedom.

Finally, some comments need to be made about general model-fitting strategies and the imposition of external constraints on a model. From the above discussion it is evident that the way in which particular terms should be tested is by fitting a model with and without the relevant terms, and then assessing the resultant change in deviance. Thus we are led naturally by this process to a nested sequence of hierarchic models. However, it is important to remember that, in fitting such sequences of models, higher-order terms may only be included if the related lower-order terms are also included. Thus γ_{ij} should not be included in a model unless the terms α_i and β_j are already present. This is because if interaction terms are non-negligible then it is of little interest that the main effects may be zero. Such considerations become increasingly important as the number of discrete variables, and hence the complexity of models, increases. The second comment relates to constraints imposed by the design of the investigation, as discussed after Example 10.4 above. Result 2 above says that, if a term is included in a model, then the fitted marginal totals are equal to the observed marginal totals. The corollary to this is that, if a design forces certain marginal totals to be fixed, then *the corresponding terms must be present* in any log-linear model fitted to the data. To illustrate this feature, consider again Example 10.1. Here the marginal totals for the categories of x_2 (i.e. treated/control) are fixed by the design of the experiment. Thus the column totals are fixed, and consequently any model fitted to the data *must* contain β_j. In particular, the *minimal* model that can be fitted to these data will be

$$\log_e \mu_{ij} = \theta + \beta_j.$$

On the other hand, the marginal totals for categories of x_1 are obviously not fixed by the design, so inclusion or otherwise of α_i is optional. In Example 10.2, the only total to be fixed by the design is the total number of children, 1398, used in the study and so here only the θ term is constrained to be included in models fitted to these data.

Analysis of any two-way table constructed from bivariate discrete data can thus be effected by obtaining parameter estimates and deviances as illustrated above. The statistical computing package GLIM (Baker and Nelder 1978) is designed to fit any model of the generalized linear class discussed by Nelder and Wedderburn (1972), which includes the log-linear model, and hence is tailor-made for discrete multivariate data analysis. To illustrate such an analysis, we could construct the following *analysis of deviance* table for the data of Example 10.1. In this example, both x_1 and x_2 have two categories, so that $a = b = 2$ in previous notation. We have already seen that the minimal model must contain $\theta + \beta_j$, to ensure that the marginal totals for x_2 are the same for both the observed and the estimated frequencies. There are four items of data, and two independent parameters (since $\beta_1 + \beta_2 = 0$). Hence the minimal model has two degrees of freedom. Having fitted this model it would then make sense to add the α_i terms (one extra parameter since $\alpha_1 + \alpha_2 = 0$), and finally the γ_{ij} terms (again one extra parameter because of the constraints $\gamma_{11} + \gamma_{12} = \gamma_{21} + \gamma_{22} = 0$). The analysis-of-deviance table is thus:

Model terms fitted	Deviance	Degrees of freedom
$\theta + \beta_j$	76.43	2
$\theta + \alpha_i + \beta_j$	4.27	1
$\theta + \alpha_i + \beta_j + \gamma_{ij}$	0	0

These deviance values would normally be found by putting the observed frequencies plus specified model into a general-purpose package such as GLIM. To illustrate the relevant computations fully, however, we will now derive these deviances. First we consider the minimal model:

$$\log_e(\mu_{ij}) = \theta + \beta_j.$$

Thus $\qquad \mu_{ij} = \exp(\theta + \beta_j)$

so $\qquad \mu_{.j} = 2\exp(\theta + \beta_j) = 2\mu_{ij}.$

$\therefore \qquad \mu_{ij} = \frac{1}{2}\mu_{.j}$

and $\qquad \hat{\mu}_{ij} = \frac{1}{2}\hat{\mu}_{.j} = \frac{1}{2}n_{.j}$ for each i.

$n_{.1} = 16$ and $n_{.2} = 79$, so expected values under this model are thus given in the following table:

x_1	x_2 Treated	Control	Totals
Tumours	8	39.5	47.5
No tumours	8	39.5	47.5
Totals	16	79	95

The corresponding observed values are as given earlier (in the table for Example 10.1) and we note that, as required, the expected marginal totals for x_2 coincide with their observed values. The deviance for this model is thus $D = 2\{4\log_e(4/8) + 5\log_e(5/39.5) + 12\log_e(12/8) + 74\log_e(74/39.5)\} = 76.426$. Next we add the α_i terms and thus have the model $\log_e(\mu_{ij}) = \theta + \alpha_i + \beta_j$. This is in fact model (10.7), and expected values under this model are given by eqn (10.12). For the present set of data these expected values are

x_1	x_2 Treated	Control	Totals
Tumours	1.516	7.484	9.000
No tumours	14.484	71.516	86.000
Totals	16.000	79.000	95.000

Hence the deviance is $D = 2\{4\log_e(4/1.516) + 5\log_e(5/7.484) + 12\log_e(12/14.484) + 74\log_e(74/71.516)\} = 4.266$. Finally we add the terms γ_{ij} to obtain the saturated model $\log_e(\mu_{ij}) = \theta + \alpha_i + \beta_j + \gamma_{ij}$, for which $\hat{\mu}_{ij} = n_{ij}$ and $D = 2\{4\log_e(1) + 5\log_e(1) + 12\log_e(1) + 74\log_e(1)\} = 0$.

The reduction in deviance on including the γ_{ij} terms is 4.27 on 1 degree of freedom, which is significant at the 5 per cent level. We thus conclude that there *is* association between x_1 and x_2, which means the probability of an individual developing a tumour in the treated group is different from that in the control group. Comparing the fitted cell frequencies under the hypothesis of independence with the observed cell frequencies clearly shows that the observed proportion of tumours in the treated group is greater than the expected proportion under the hypothesis of there being no treatment effect.

10.2.2 Three- and higher-dimensional tables

All the ideas and methodology of the two-dimensional analysis can be extended readily to higher dimensional tables. We illustrate this extension with the three-variable case. Suppose that observed cell frequencies n_{ijk} are classified according to the three variables x_1, x_2, and x_3 where i denotes the category of x_1 $(i = 1, \ldots, a)$; j, the category of x_2 $(j = 1, \ldots, b)$; and k, the category of x_3 $(k = 1, \ldots, c)$. An instance of such a table is provided in Example 10.3 above, where $a = b = c = 2$.

An appropriate saturated log-linear model for the data is $\log_e(\mu_{ijk}) = \theta + \alpha_i + \beta_j + \gamma_k + (\alpha\beta)_{ij} + (\beta\gamma)_{jk} + (\alpha\gamma)_{ik} + (\alpha\beta\gamma)_{ijk}$ where, to avoid proliferation of Greek letters, we denote by α_i, β_j, γ_k the *main effects* of x_1, x_2, x_3; by $(\alpha\beta)_{ij}$, $(\beta\gamma)_{jk}$, $(\alpha\gamma)_{ik}$ the *two-factor interactions* between x_1 and x_2, x_2 and x_3, x_1 and x_3 respectively; and by $(\alpha\beta\gamma)_{ijk}$ the *three-factor interaction* between x_1, x_2, and x_3. The model carries the constraints:

$$\sum_i \alpha_i = \sum_j \beta_j = \sum_k \gamma_k = 0,$$

$$\sum_i (\alpha\beta)_{ij} = \sum_j (\alpha\beta)_{ij} = \sum_j (\beta\gamma)_{jk} = \ldots = \sum_k (\alpha\gamma)_{ik} = 0,$$

$$\sum_i (\alpha\beta\gamma)_{ijk} = \sum_j (\alpha\beta\gamma)_{ijk} = \sum_k (\alpha\beta\gamma)_{ijk} = 0.$$

Particular hierarchic models encountered in a log-linear approach to the analysis of three-way tables have the following interpretations.

1. *Three-factor interaction $(\alpha\beta\gamma)_{ijk}$ absent*
   Under this model, each two-factor interaction is unaffected by the category of the third variable. There is therefore *partial association* between each pair of variables.
2. *Three-factor and one two-factor interaction absent*
   There are three versions of this model. If, for example, $(\alpha\beta)_{ij}$ is the absent two-factor interaction, then under the model x_1 and x_2 are independent for every category of x_3, but each is associated with x_3. In

other words, x_1 and x_2 are *conditionally independent,* given the category of x_3.

3. *Three-factor and two two-factor interactions absent*
There are three versions of this model. If, for example, $(\alpha\beta\gamma)_{ijk} = (\alpha\beta)_{ij} = (\alpha\gamma)_{ik} = 0$ for all i, j, k, then x_1 is *completely independent* of both x_2 and x_3, although x_2 and x_3 are associated.
4. *Three-factor and all two-factor interactions absent*
This is the model for *complete independence* of x_1, x_2, x_3.

Isolating an appropriate model from among these eight possibilities can be accomplished by an analysis of deviance. Expected values can be estimated using the same methods and results as given for the two-dimensional case in Section 10.2.1. For example, in the model of complete independence where

$$\log_e(\mu_{ijk}) = \theta + \alpha_i + \beta_j + \gamma_k$$

the $\hat{\mu}_{ijk}$ satisfy

$$\hat{\mu}_{...} = n; \quad \hat{\mu}_{i..} = n_{i..}; \quad \hat{\mu}_{.j.} = n_{.j.}; \quad \hat{\mu}_{..k} = n_{..k}$$

and we find that

$$\hat{\mu}_{ijk} = (\hat{\mu}_{i..}\hat{\mu}_{.j.}\hat{\mu}_{..k})/n^2.$$

For some models, it is not possible to obtain $\hat{\mu}_{ijk}$ in an explicit form. For example, it is not possible to derive the $\hat{\mu}_{ijk}$ explicitly for the partial association model (1). An iterative technique such as Newton–Raphson or iterative proportional scaling is required to fit such models. Details of such schemes are included in Fienberg (1981). GLIM uses Newton–Raphson; an algorithm for implementing iterative proportional scaling is given by Haberman (1972).

Four- and higher-dimensional tables can be analysed similarly, by seeking appropriate log-linear models via an analysis of deviance. No new principles are involved, although the number of possible models rapidly increases with the dimension of the table. Some sort of model selection process will then be useful.

We conclude this section by considering analysis of the data in Example 10.3. In this experiment, the marginal totals corresponding to the four combinations of categories of x_3 (length of cutting) and x_2 (time of planting) were fixed at 240. Consequently, the minimal model must contain the terms $(\beta\gamma)_{jk}$ and hence, by the hierarchic principle, β_j, γ_k, and θ as well. The remaining variable x_1 denotes the condition of the plant (alive/dead). Using GLIM to calculate the expected frequencies and deviances for each model, the following analysis-of-deviance table was constructed.

Terms fitted	Deviance	d.f.
$\theta+\beta_j+\gamma_k+(\beta\gamma)_{jk}$	194.70	4
$+\alpha_i$	151.02	3
$+(\alpha\gamma)_{ik}$	105.18	2
$+(\alpha\beta)_{ij}$	2.29	1
$+(\alpha\beta\gamma)_{ijk}$	0	0

Each row of this table gives the deviance and degrees of freedom for the model containing all the parameters specified up to and including that row. From the results of this analysis, we see that the reduction in deviance (2.29) on introducing the three-factor interaction is not significant (χ^2 on 1 d.f.). Thus the two-factor interaction between x_3 and x_2 does not affect the ultimate condition of the plant. The $(\alpha\gamma)_{ik}$ and $(\alpha\beta)_{ij}$ terms are both highly significant, confirming that both time of planting and length of cutting affect propagation. If we examine the estimated expected frequencies under the model

$$\log_e(\mu_{ijk}) = \theta + \alpha_i + \beta_j + \gamma_k + (\beta\gamma)_{jk}$$

we can identify the direction of association between x_1, x_2, and x_3. We find that to maximize the survival rate, long cuttings should be taken and planted at once.

10.3 Mixed discrete and continuous data

Frequently in practice it happens that only some of the variables measured on each sample member are discrete, the remainder being continuous. For example, in an attitude survey, each individual may receive a score on a set of attitude scales (yielding p continuous values) and at the same time may be classified according to religious and political affiliations, region of domicile, etc, which are all categorical variables. Alternatively, patients attending hospital for treatment may have a set of continuous laboratory measurements made on them, as well as having various signs or symptoms scored as being either present or absent. In order to conduct any analysis of such mixed variable types, we must first formulate a plausible model as a basis for the analysis. The multivariate normal distribution has been used already as a basis for continuous data analysis, while the multinomial distribution was shown earlier in this chapter to be a natural basis for discrete data analysis. Some combination of these distributions should therefore be appropriate as a model for mixed variables. A satisfactory model will be one specifying the joint distribution of all variables, so that the association between discrete and continuous variables can be modelled as readily as the marginal features of each variable type.

Since the joint distribution of a set of variates can be expressed as the product of the marginal distribution of some of the variates, and the conditional distribution of the remainder given the values of these selected ones, the first step towards finding a suitable model is to look for a possible division of the variates into two such sets. The most natural division is into discrete and continuous components, which therefore gives rise to two possible types of model according to the choice of component for the conditioning variables. Olkin and Tate (1961) proposed a model in which the marginal distribution of the discrete variables was multiplied by the conditional distribution of the continuous variables given the values of the discrete ones, while Cox (1970) suggested a model in which the discrete and continuous variables interchanged roles. Since the form of the conditional distribution of the discrete variables given the continuous ones in this latter model is logistic, however, estimation of parameters inevitably incurs iterative computational schemes. The model proposed by Olkin and Tate, on the other hand, can be handled in computationally simpler fashion and has thus been more widely applied in mixed variable problems; it is generally known in the literature as the 'location model', and will be the main focus of our attention.

In the location model it is assumed that the discrete variables are arranged in contingency table form, as in the analyses discussed earlier in this chapter, and thus the pattern of discrete variable values for any sample member determines uniquely a cell (or 'location') in this table. The continuous variables are assumed to follow a multivariate normal distribution, but the parameters of the distribution depend on the contingency table location as defined by the associated discrete variables. The contingency table locations are then identified with the categories of a multinomial distribution, and this completes the specification of the model. In algebraic form, the model is thus built up as follows. Let each sample member have q discrete and p continuous variables observed on it. Denote the discrete variables by $x_1, x_2, \ldots, x_q$ and the continuous variables by $y_1, y_2, \ldots, y_p$. If the ith discrete variable has s_i categories ($i = 1, \ldots, q$) then the contingency table formed from the x_i has $s = s_1 \times s_2 \times \ldots \times s_q$ locations; denote these locations by $z_1, z_2, \ldots, z_s$. Writing $\boldsymbol{x}' = (x_1, \ldots, x_q)$, $\boldsymbol{y}' = (y_1, \ldots, y_p)$, and z for the variable denoting cell membership in the contingency table, the model has the two components:

(1) the conditional distribution of $\boldsymbol{y}$ given that $\boldsymbol{x}$ falls in location z_m is

$$N(\boldsymbol{\mu}^{(m)}, \boldsymbol{\Sigma}) \tag{10.15}$$

and

(2) the marginal distribution of the locations is given by

$$\mathrm{pr}(z = z_{\mathrm{m}}) = p_m, \quad \text{with} \quad \sum_{m=1}^{s} p_m = 1. \tag{10.16}$$

The continuous variables $\boldsymbol{y}$ thus follow a multivariate normal distribution, but the mean of the distribution depends on the associated discrete variables $\boldsymbol{x}$, and this dependency induces correlations between components of $\boldsymbol{x}$ and components of $\boldsymbol{y}$. A more general model could be defined by letting the dispersion matrix of the normal distribution also vary with contingency table location, and Krzanowski (1983b) has included various possibilities of this type. However, such generalizations lead to a very large increase in model parameters. Moreover, patterns of covariances in practice suggest that a common dispersion matrix between locations is not an unduly restrictive assumption (and is compatible with the types of assumptions that will be encountered in Part IV). Finally, satisfactory practical results seem to be obtained in general with the simpler model, and hence (10.15) and (10.16) are usually taken as the defining equations for the location model. Parameters of this model are thus the conditional continuous variable means $\boldsymbol{\mu}^{(m)}$ $(m = 1, \ldots, s)$, the conditional continuous variable dispersion matrix $\boldsymbol{\Sigma}$, and the marginal multinomial probabilities p_m $(m = 1, \ldots, s)$.

Olkin and Tate (1961) considered various implications of this model. In particular, they established relationships linking the conditional means $\boldsymbol{\mu}^{(m)}$ of the continuous variables and the correlations between the discrete and continuous variates. The fundamental quantities in these relationships are the *canonical correlations* between $\boldsymbol{y}$ and $\boldsymbol{x}$. We shall discuss canonical correlations in Section 14.5, so will not enter into technicalities at this stage beyond saying that Olkin and Tate prove that these canonical correlations will be zero if, and only if, all $\boldsymbol{\mu}^{(m)}$ are equal. Furthermore, Olkin and Tate show that the likelihood for a sample of individuals following the location model factorizes into two parts: one part contains the multinomial likelihood with parameters $p_1, p_2, \ldots, p_s$, while the other consists of the product of s multivariate normal likelihoods, the ith of which has parameters $\boldsymbol{\mu}^{(i)}$ and $\boldsymbol{\Sigma}$. The consequence is that the maximum likelihood estimates of $\boldsymbol{\mu}^{(m)}$ and $\boldsymbol{\Sigma}$ are the corresponding sample counterparts (i.e. within-location mean vectors and pooled within-location covariance matrix), and the maximum likelihood estimates of p_m are the corresponding location relative frequencies. Invariance of maximum likelihood estimation then enables the maximum likelihood estimates of any derived quantities such as canonical correlations to be obtained by substituting these sample estimates into the appropriate population formulae.

Using these fundamental results, Olkin and Tate derive various procedures for testing hypotheses about the parameters of the location model. The reader is referred to their paper for technical details. Ways of handling missing values in location model problems using the E–M algorithm (Dempster *et al.* 1977) are described by Little and Schluchter (1985), who also provide some practical examples.

Part IV: Analysing grouped data

11
Incorporating group structure: descriptive methods

A common feature of all the situations considered thus far is an absence of any external information about the sample members prior to analysis of the data at hand. In other words, there is no external structure that must be imposed on the data matrix, so this matrix can be treated at face value for the purposes of analysis. Thus the descriptive methods of Part I presuppose an essentially *homogeneous* sample of individuals, and the same situation is more formally assumed for the inferential procedures of Part III. The possibility of heterogeneity is raised by those techniques in Chapters 2, 3, and 4 that enable a search to be made for any evidence of grouping among the sample members, but once again the starting point for these techniques is equivalent to a null hypothesis of 'no structure'. Indeed, it was emphasized in Chapter 3 that a cluster analysis is *not* appropriate when we *do* have some information about structure of the units. Methods of incorporating such information into the analysis will occupy Part IV of this book.

By far the most common external information about sample members in practice is the knowledge that they have come from a number of different populations. Thus the *rows* of the data matrix can be considered to be divided into a number of *groups,* with the rows in a particular group representing the sample from the corresponding population. It is clearly just as important to take account of this information in any analysis of the data as it is, say, to take proper account of blocking devices in the univariate analysis of a designed experiment. The present chapter will focus on various descriptive techniques that make use of such grouping information, and inferential extensions will be considered in Chapters 12 and 13. Much more complex external information arises when some further structure is imposed on the groups of units. Such further structure is common in designed experiments where, for example, the groups may constitute all possible combinations of a set of treatments in a factorial experiment. If a multivariate response is observed on each unit in such an experiment, then we will require a multivariate generalization of the univariate analysis of a factorial design. Complex group structures can arise in many different areas of application, and techniques that enable such information to be incorporated into a multivariate analysis will also be discussed in Chapter 13.

For the present, let us return to the simple external constraint in which the sample members are assumed to have come from g separate populations, and accordingly the n rows of the data matrix are divided into g separate groups, n_i rows comprising the ith group. We shall concentrate throughout most of the present chapter on the case of *quantitative* data. Hence if p variables have been measured on each sample member, the sample can be modelled by n points in p-dimensional space in the fashion already familiar from Part I. Now, however, we can divide these n points into g distinct subsets, while interest in the statistical analysis of the data will generally focus on some form of comparison between the groups of individuals. We would therefore like to have suitable methods of displaying the data, in order to investigate possible relationships between the corresponding subsets of points. As before, the dimensionality of the data will generally be too large to enable direct data inspection to be carried out, and some low-dimensional approximations to the data must first be sought. The best type of approximation will be dictated by the objective of the analysis, and we can in general distinguish two contrasting objectives in the presence of group structure; highlighting the *differences* or highlighting the *similarities* between the groups. We will consider each of these objectives in turn, and possible descriptive techniques for accomplishing them will be discussed in Sections 11.1 and 11.2 respectively.

It may sometimes happen that we collect a set of data which we strongly believe to have come from g separate groups, but we do not know which individuals belong to each group. This may be thought of as a case of *partial* structure, and the methods of hierarchical cluster analysis described in Chapter 3 can obviously be used in an attempt to identify the group membership of the sample individuals. Such methods are constrained by the hierarchical requirement, however, and are rarely optimal for the stated objective. More appropriate methods will utilize in some way the results developed in Sections 11.1 and 11.2, and such extensions are considered in Section 11.3.

Finally, in Section 11.4, we consider modifications necessary to the analysis when the data deviate from the ideal conditions assumed earlier. In particular, we focus on qualitative data, missing values, and the stability of the techniques in the presence of either high correlation or outliers.

11.1 Highlighting differences between groups: canonical variates

Suppose that we have available an $(n \times p)$ data matrix, in which the n individuals are divided into g different groups with n_i individuals in the ith group $\left(\text{so that } n = \sum_{i=1}^{g} n_i\right)$, and we wish to investigate possible differences between the groups. Since we are assuming the p measured variates to be quantitative, the data matrix can be modelled by n points in p-dimensional space. Furthermore, since the n individuals come from g different groups, it is reasonable to suppose that the n points are composed of g different sets (with n_i points in the ith set). The more separated are these sets in space, the more readily distinguishable are the groups of individuals, and hence the more distinct are the populations from which the individuals were sampled. However, in general p will be larger than 3, and hence we have no direct way of observing the differences between the subsets of points. We must first seek an approximation in several dimensions to the true configuration before we can look at the data. Various ways of constructing approximate configurations were discussed in Chapters 2 to 4, but in none of these approaches was a group structure of units allowed for. It was stressed in Chapter 2 that many different low-dimensional approximations can be obtained to a high-dimensional configuration, depending on the objective of the data inspection. Our major objective in the present instance is to provide a low-dimensional representation of the data that highlights as accurately as possible the true differences existing between the g subsets of points in the full configuration.

Since none of the earlier techniques allowed for a grouping of the units, it follows that none of these techniques will necessarily meet the stated objective. To illustrate how wide of the mark any of these earlier techniques might be, consider the simple two-dimensional example illustrated in Fig. 11.1. Here we have two variables x_1 and x_2 measured on all individuals of two groups, the individuals in one group being marked by crosses and those of the other group by circles. Suppose we would like to produce a one-dimensional approximation to this configuration. One possible way of doing so would be via principal component analysis (Section 2.2). This technique finds the new direction in space (i.e. the linear combination $y = a_1x_1 + a_2x_2$) such that when all the individuals are projected on to this direction they exhibit the maximum spread. This principal component direction y is shown in Fig. 11.1. However, it is evident that, although the overall spread of points

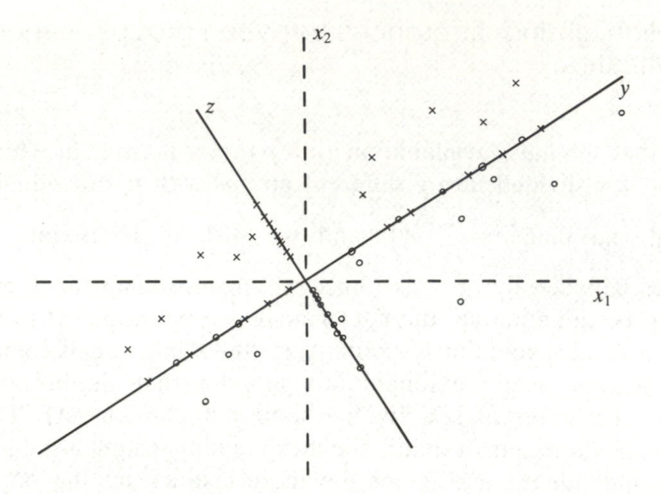

Fig. 11.1 Illustrative two-dimensional example of bivariate data in two groups

along this direction may be maximum, there is no indication of any difference between the two groups along this direction. The projection of all points on to y is shown in Fig. 11.1, and points from both groups are completely intermingled. What is in fact a total separation of the two sets of points in the true configuration is represented as a single homogeneous group of points in the one-dimensional principal component approximation. Principal component analysis would therefore be completely unsuitable as a dimension-reducing technique for highlighting the main features that interest us in this set of data. In fact, the best one-dimensional projection from our point of view would be the direction z illustrated in Fig. 11.1: projection of all points on to z now shows a clear separation between the two sets.

To provide a motivation for, and illustration of, the following theory, let us return to the Fisher iris data already introduced in Table 2.3. Here there are 150 iris plants, divided into three equal groups of fifty plants each of *Iris setosa, Iris versicolor,* and *Iris virginica.* Each plant has four quantitative measurements made on it: sepal length, sepal width, petal length, and petal width. We now wish to investigate possible differences between the three groups. With four-dimensional data, initial approximation is necessary, and Fig. 11.2 shows the projection of the 150 points on to the space of the first two principal components of the data. Some separation of the groups is evident, with the *setosa* group 1 distinct from the other two groups (among which there is a certain amount of overlap). Can we find a better representation for highlighting the differences between the groups?

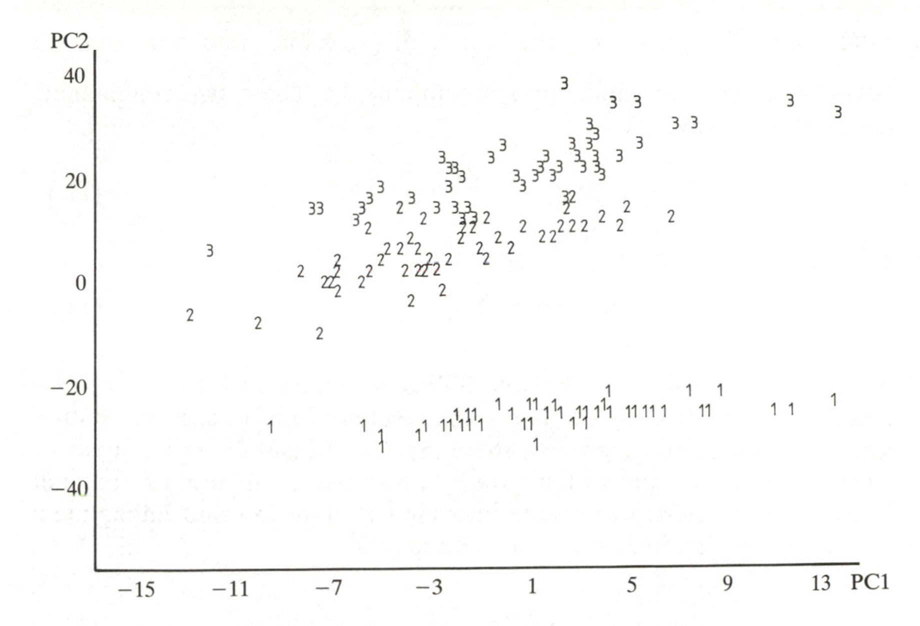

Fig. 11.2 *Iris* data plotted against first two principal components. 1 = *setosa*; 2 = *versicolor*; 3 = *virginica*

Let us tackle this question step by step, and first enquire what would be the best *single* direction (i.e. one-dimensional representation) in which to view the differences between the groups. We consider the general case in which p variables $x_1, x_2, \ldots, x_p$ are measured on each individual, and any direction in the p-dimensional sample space is specified by the p-tuple $(a_1, a_2, \ldots, a_p)$. Focus attention on the linear combination $y = a_1x_1 + \ldots + a_px_p$. *If we specify all the* a_i, we convert each multivariate observation $\boldsymbol{x}_i' = (x_{i1}, x_{i2}, \ldots, x_{ip})$ into a univariate observation $y_i = \boldsymbol{a}'\boldsymbol{x}_i$ where $\boldsymbol{a}' = (a_1, \ldots, a_p)$. The n sample members are thus reduced to n univariate observations $y_1, \ldots, y_n$, which in turn are represented by n points along the direction y in space. However, since the n observations are divided into g groups, it will be convenient to relabel these y values. Let y_{ij} denote the y value for the jth observation in the ith group $(i = 1, \ldots, g; j = 1, \ldots, n_i)$, and denote by $\bar{y}_i$ the mean of the y_{ij} in the ith group, and by $\bar{y}$ the mean of the y_{ij} in the whole sample. Thus $\bar{y}_i = \frac{1}{n_i}\sum_{j=1}^{n_i} y_{ij}$ and $\bar{y} = \frac{1}{n}\sum_{i=1}^{g}\sum_{j=1}^{n_i} y_{ij} = \frac{1}{n}\sum_{i=1}^{g} n_i\bar{y}_i$. If we now wish to investigate whether the y_{ij} show evidence of systematic (mean) differences between the g groups, one-way analysis of variance is an appropriate technique to invoke. In such an analysis, we partition the total sum of squares of the y_{ij}, $\sum_{i=1}^{g}\sum_{j=1}^{n_i}(y_{ij}-\bar{y})^2$, into the sum of between-groups and within-groups components. These two components are given respectively by

$$\text{SSB}(\boldsymbol{a}) = \sum_{i=1}^{g} n_i(\bar{y}_i - \bar{y})^2 \tag{11.1}$$

and

$$\text{SSW}(\boldsymbol{a}) = \sum_{i=1}^{g}\sum_{j=1}^{n_i}(y_{ij} - \bar{y}_i)^2 \tag{11.2}$$

where by writing $(\boldsymbol{a})$ after SSB and SSW it is emphasized that choice of $\boldsymbol{a}$ determines the values y_{ij} and hence the resultant sums of squares. With n sample members and g groups, there are $(g-1)$ and $(n-g)$ degrees of freedom between and within groups respectively. A test of the null hypothesis that there are no differences in (true) mean value among the g groups is obtained from the mean square ratio

$$F = \left\{\frac{1}{(g-1)}\text{SSB}(\boldsymbol{a})\right\} \Big/ \left\{\frac{1}{(n-g)}\text{SSW}(\boldsymbol{a})\right\}. \tag{11.3}$$

The larger the value of this ratio, the more variability is there between groups than within groups and the greater is the evidence against the null hypothesis. This hypothesis is formally rejected at the 100α per cent significance level if F exceeds the $100(1-\alpha)$ percent point of the F distribution on $(g-1)$ and $(n-g)$ degrees of freedom.

Returning now to the original multivariate problem, it is evident that every different choice of coefficients $\boldsymbol{a}' = (a_1, \ldots, a_p)$ will yield a different set of values y_{ij} and hence a different value of F from (11.3). The *best* choice of $\boldsymbol{a}$ will clearly be the one that yields the *largest* F value, as for this choice of $\boldsymbol{a}$ we have the greatest evidence of a difference between the g groups, and hence the resulting values y_{ij} will yield the one-dimensional projection of sample points that shows up differences among groups as much as is possible. The only question that remains to be settled is how to find the optimal coefficients $\boldsymbol{a}$. To do this, we first need to define multivariate analogues of the between-groups and within-groups sums of squares used in the univariate analysis of variance. These are, respectively, the between-groups sum-of-squares and -products matrix $\mathbf{B}_0$ and the within-groups sum-of-squares and -products matrix $\mathbf{W}_0$ defined by

$$\mathbf{B}_0 = \sum_{i=1}^{g} n_i(\bar{\boldsymbol{x}}_i - \bar{\boldsymbol{x}})(\bar{\boldsymbol{x}}_i - \bar{\boldsymbol{x}})',$$

and

$$\mathbf{W}_0 = \sum_{i=1}^{g} \sum_{j=1}^{n_i} (\boldsymbol{x}_{ij} - \bar{\boldsymbol{x}}_i)(\boldsymbol{x}_{ij} - \bar{\boldsymbol{x}}_i)'$$

where the labelling $\boldsymbol{x}_{ij}$ is analogous to that of the y_{ij} earlier, $\bar{\boldsymbol{x}}_i = \frac{1}{n_i} \sum_{j=1}^{n_i} \boldsymbol{x}_{ij}$ is the sample mean vector in the ith group and $\bar{\boldsymbol{x}} = \frac{1}{n} \sum_{i=1}^{g} \sum_{j=1}^{n_i} \boldsymbol{x}_{ij} = \frac{1}{n} \sum_{i=1}^{g} n_i \bar{\boldsymbol{x}}_i$ is the overall sample mean vector. Since $y_{ij} = \boldsymbol{a}'\boldsymbol{x}_{ij}$, it can readily be verified (in similar fashion to the derivation of the sum of squares for principal component analysis, Section 2.2.4) that

$$\text{SSB}(\boldsymbol{a}) = \boldsymbol{a}'\mathbf{B}_0\boldsymbol{a} \quad \text{and } \text{SSW}(\boldsymbol{a}) = \boldsymbol{a}'\mathbf{W}_0\boldsymbol{a}. \tag{11.4}$$

Hence, from (11.3),

$$\begin{aligned} F &= \left\{ \frac{1}{(g-1)} \boldsymbol{a}'\mathbf{B}_0\boldsymbol{a} \right\} \Big/ \left\{ \frac{1}{(n-g)} \boldsymbol{a}'\mathbf{W}_0\boldsymbol{a} \right\} \\ &= \frac{\boldsymbol{a}'\mathbf{B}\boldsymbol{a}}{\boldsymbol{a}'\mathbf{W}\boldsymbol{a}} \end{aligned} \tag{11.5}$$

where $\mathbf{B} = \frac{1}{(g-1)} \mathbf{B}_0$ is the *between-groups covariance matrix* and $\mathbf{W} = \frac{1}{(n-g)} \mathbf{W}_0$ is the *within-groups covariance matrix*.

Thus the best linear combination $y = \boldsymbol{a}'\boldsymbol{x}$ of the original variables on which to highlight differences between the g groups has coefficients $\boldsymbol{a}$ which maximize (11.5). Maximizing $F = \boldsymbol{a}'\mathbf{B}\boldsymbol{a}/\boldsymbol{a}'\mathbf{W}\boldsymbol{a}$ with respect to $\boldsymbol{a}$ is a straightforward exercise in calculus. Differentiating F with respect to $\boldsymbol{a}$ and setting the result to zero yields $\mathbf{B}\boldsymbol{a} - \left(\frac{\boldsymbol{a}'\mathbf{B}\boldsymbol{a}}{\boldsymbol{a}'\mathbf{W}\boldsymbol{a}}\right)\mathbf{W}\boldsymbol{a} = \boldsymbol{0}$. But at the maximum of F, $\frac{\boldsymbol{a}'\mathbf{B}\boldsymbol{a}}{\boldsymbol{a}'\mathbf{W}\boldsymbol{a}}$ must be a constant l (equal to max(F)), so the required value of $\boldsymbol{a}$ must satisfy

$$(\mathbf{B} - l\mathbf{W})\boldsymbol{a} = \boldsymbol{0}. \tag{11.6}$$

This equation can be written $(\mathbf{W}^{-1}\mathbf{B} - l\mathbf{I})\boldsymbol{a} = \boldsymbol{0}$, so l must be an eigenvalue, and $\boldsymbol{a}$ must be an eigenvector, of $\mathbf{W}^{-1}\mathbf{B}$. Furthermore, since l is the maximum of F, $\boldsymbol{a}$ must be the eigenvector corresponding to the *largest* eigenvalue of $\mathbf{W}^{-1}\mathbf{B}$. This eigenvector determines the required linear combination $y = \boldsymbol{a}'\boldsymbol{x}$. Note that l measures the ratio (between group variance/within group variance) for this combination, so if there is clear separation of groups then l will be substantially greater than unity.

So far, we have concentrated on finding the *single* direction in the multivariate space in which to examine differences between the g groups. However, if g is large, or if the original dimensionality of the space is large, a single direction will provide a gross over-simplification of the true multivariate configuration, and between-group differences may still be obscured. In that case, we really need to find a suitable two-, three-, or even higher-dimensional space for an adequate representation. Fortunately, the solution to the one-dimensional approach given above extends directly to the higher-dimensional representation, as eqn (11.5) will generally possess more than one eigenvalue/eigenvector pair. Suppose in fact that $l_1 > l_2 > \ldots l_s > 0$ are the eigenvalues of $\mathbf{W}^{-1}\mathbf{B}$ with associated eigenvectors $\boldsymbol{a}_1, \boldsymbol{a}_2, \ldots, \boldsymbol{a}_s$. It has already been shown that $\boldsymbol{a}_1$ gives the direction in the p-dimensional data space along which the between-group variability is greatest relative to the within-group variability, and an individual $\boldsymbol{x}_{ij}$ has score $\boldsymbol{a}_1'\boldsymbol{x}_{ij}$ in this direction. Extending the argument (in analogous fashion to that for principal component analysis in Section 2.2.4) it follows that $\boldsymbol{a}_2, \boldsymbol{a}_3, \boldsymbol{a}_4, \ldots, \boldsymbol{a}_s$ give the directions along which between-group variability relative to within-group is second, third, fourth, ..., sth greatest. If we define new variates $y_1, y_2, \ldots$ by $y_i = \boldsymbol{a}_i'\boldsymbol{x}$, then the y_i are termed *canonical variates*; the best r-dimensional representation of the differences between the groups is obtained by plotting the sample individuals against the first r canonical variates as axes.

Figure 11.3 shows the *Iris* individuals plotted against the first two canonical variates as axes, and this representation may be contrasted with the principal component representation of Fig. 11.2. Apart from the trivial difference of a rotation through 90°, the division between the *I. versicolor* and *I. virginica* individuals seems to be more clear-cut, and overall the groups have a more compact representation than before.

The derivation of canonical variates outlined above might lead one to suppose that there is a close connection between these variates and principal components. In fact, although some of their uses and interpretations follow the lines outlined earlier for principal components, there are also some profound differences between the two techniques. Some consideration of the properties of canonical variates is necessary in order to establish these similarities and differences. These properties will necessarily involve some matrix manipulation, which will therefore be kept within the main text as it is felt to be central to the technique rather than merely a mathematical aside.

First, eqn (11.6) has $s = \min(p, g-1)$ non-zero eigenvalues in general; if the number of groups is less than or equal to the number of original variates the matrix $\mathbf{B}$ is not of full rank, and there will be $(p - g + 1)$ zero eigenvalues. The maximum dimensionality for a canonical variate representation is thus s, and the relevant canonical variates are $y_i = \boldsymbol{a}_i'\boldsymbol{x}$

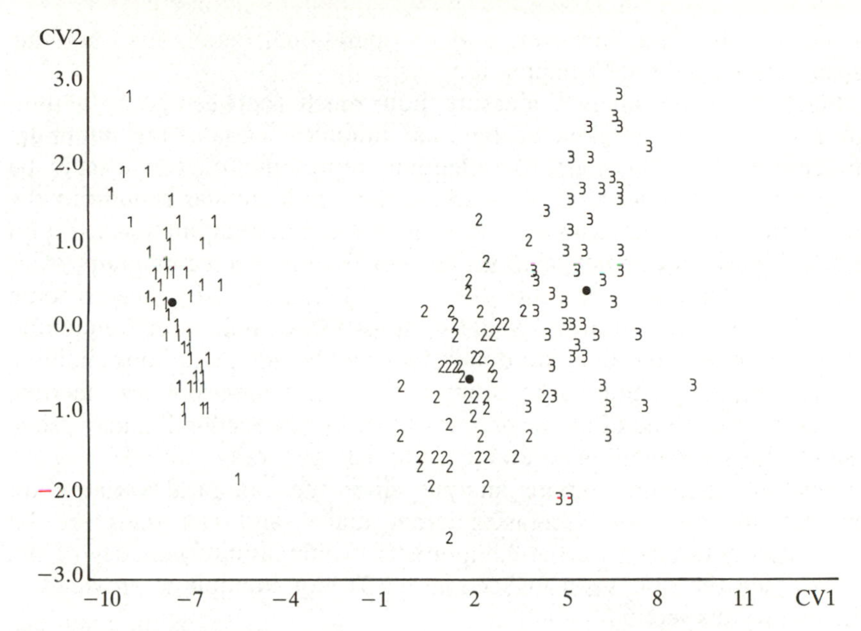

Fig. 11.3 *Iris* data plotted against first two canonical variates. 1 = *setosa*; 2 = *versicolor*; 3 = *virginica*. Solid circles denote group means

$(i = 1, \ldots, s)$. Write $\boldsymbol{y}' = (y_1, y_2, \ldots, y_s)$, and gather all eigenvalues l_i and eigenvectors $\boldsymbol{a}_i$ together so that $\boldsymbol{a}_i$ is the ith column of a $(p \times s)$ matrix $\mathbf{A}$, while l_i is the ith diagonal element of the $(s \times s)$ diagonal matrix $\mathbf{L}$. Then in matrix terms, eqn (11.6) may be written $\mathbf{BA} = \mathbf{WAL}$, and the collection of canonical variates is given by $\boldsymbol{y} = \mathbf{A}'\boldsymbol{x}$. The space of all vectors $\boldsymbol{y}$ is termed the *canonical variate space*. In this space, the mean of the ith group of individuals is $\bar{\boldsymbol{y}}_i = \mathbf{A}'\bar{\boldsymbol{x}}_i$. Plotting the group means on the first few canonical variates as axes will thus exhibit as clearly as possible the group separations in the chosen dimensionality: the means of the three iris groups are shown as solid circles in Fig. 11.3. Note that for the iris data there are $s = \min(4, 2) = 2$ canonical variates, so Fig. 11.3 depicts the whole space. In this example, it is possible to plot *all* iris individuals on the diagram without undue confusion. Where the total number of sample observations is very large, or there are many groups, or there is considerable overlap between the individuals of different groups, plotting all individuals may not give a very clear picture. In such circumstances, a plot of just the group means will convey the main features that are sought, and this is the form of representation most often used in practice. Indication of sampling variability may be additionally incorporated by surrounding each group mean point on the diagram by a confidence or tolerance region; such regions are based on a statistical model for the data, however, and so operational details for obtaining them are delayed until Chapter 13.

Since the eigenvalues l_i measure how much (between-group/within-group) variability is taken up by each canonical variate, the minimum dimensionality r necessary for adequate representation can usually be judged from consideration of these eigenvalues in similar fashion to the determination of dimensionality in principal component analysis (Section 2.2.5). The most popular methods are either from a scree diagram, or as the value r for which the ratio $(l_1 + \ldots + l_r)/(l_1 + \ldots + l_s)$ exceeds some suitably large (but arbitrary) value such as 0.75 or 0.8. Most frequently, of course, one hopes that two dimensions will be adequate, thus enabling a direct representation to be drawn. If higher dimensions are needed, however, then some of the techniques described in Section 2.1 may prove useful. In particular, Andrews curves will generally provide a good adjunct to canonical variate analysis since the canonical variates are ordered in terms of decreasing 'importance' and can therefore be associated naturally with the components of increasing frequency of the curves (see the discussion in Section 2.1.3). An example is provided at the end of this section.

Interpretation of a particular canonical variate y_i may be attempted in similar vein to interpretation of a principal component, by considering the coefficients $\boldsymbol{a}_i$ and identifying those variates with large coefficients as 'important' in distinguishing between the groups. However, the matter is now complicated by the fact that between-group variability is being assessed relative to within-group variability, and large coefficients may be a reflection *either* of large between-group variability *or* of small within-group variability in the corresponding variate. For interpretation, therefore, it is better to consider the modified coefficients $\boldsymbol{a}_i^* = (a_{i1}^*, \ldots, a_{ip}^*)$ which are appropriate when the original variates are each standardized to have unit within-group variance. If w_{ii} is the ith diagonal element of $\mathbf{W}$, the modified coefficients a_{ij}^* are given by $a_{ij}^* = a_{ij}\sqrt{w_{jj}}$. This standardization puts all variates on a comparable footing as regards within-group variability, and allows the modified canonical variates to be interpreted in the manner suggested above.

Having outlined the various similarities between canonical variate and principal component analysis, we now turn to the fundamental differences between the two techniques and consider the resulting implications. From (11.6), any two particular eigenvalue/eigenvector pairs $(l_i, \boldsymbol{a}_i)$ and $(l_j, \boldsymbol{a}_j)$ satisfy

$$\mathbf{B}\boldsymbol{a}_i = l_i \mathbf{W} \boldsymbol{a}_i \tag{11.7}$$

and

$$\mathbf{B}\boldsymbol{a}_j = l_j \mathbf{W} \boldsymbol{a}_j. \tag{11.8}$$

Pre-multiplying (11.7) by $\boldsymbol{a}_j'$, and (11.8) by $\boldsymbol{a}_i'$ yields

$$\boldsymbol{a}_j'\mathbf{B}\boldsymbol{a}_i = l_i\boldsymbol{a}_j'\mathbf{W}\boldsymbol{a}_i$$

and

$$\boldsymbol{a}_i'\mathbf{B}\boldsymbol{a}_j = l_j\boldsymbol{a}_i'\mathbf{W}\boldsymbol{a}_j.$$

But $\boldsymbol{a}_j'\mathbf{B}\boldsymbol{a}_i = \boldsymbol{a}_i'\mathbf{B}\boldsymbol{a}_j$, since both expressions are scalars and $\mathbf{B}$ is symmetric, so it follows that

$$l_i\boldsymbol{a}_j'\mathbf{W}\boldsymbol{a}_i = l_j\boldsymbol{a}_i'\mathbf{W}\boldsymbol{a}_j. \tag{11.9}$$

However, $\boldsymbol{a}_j'\mathbf{W}\boldsymbol{a}_i = \boldsymbol{a}_i'\mathbf{W}\boldsymbol{a}_j$ for the same reasons, so that if $l_i \neq l_j$, the only way in which (11.9) can be satisfied is when $\boldsymbol{a}_i'\mathbf{W}\boldsymbol{a}_j = \boldsymbol{a}_j'\mathbf{W}\boldsymbol{a}_i = 0$. Clearly these arguments hold for all $i \neq j$, which establishes the fundamental property of the coefficients $\boldsymbol{a}_i$, that $\boldsymbol{a}_i'\mathbf{W}\boldsymbol{a}_j = 0$ for all $i \neq j$. Considering all the vectors $\boldsymbol{a}_i$ together as columns of the matrix $\mathbf{A}$, this property is equivalent to $\mathbf{A}'\mathbf{W}\mathbf{A}$ being a diagonal matrix. To overcome arbitrary scaling of $\boldsymbol{a}_i$, it is usual to adopt the normalization

$$\mathbf{A}'\mathbf{W}\mathbf{A} = \mathbf{I}. \tag{11.10}$$

With this normalization, the canonical variates are arranged to be uncorrelated and of equal variance *within groups*. Since $\mathbf{BA} = \mathbf{WAL}$, then $\mathbf{A}'\mathbf{BA} = \mathbf{A}'\mathbf{WAL} = \mathbf{L}$. Thus the canonical variates are also uncorrelated *between groups,* but now they occur in decreasing order of between-group variance. Finally, since the total sum-of-squares and products matrix is the sum of within-group and between-group components, the canonical variates will indeed be uncorrelated over the whole sample.

We may now contrast this with the corresponding result for principal components. If $\boldsymbol{a}_i$ contains the coefficients of the ith principal component, and if $\mathbf{A} = (\boldsymbol{a}_1, \boldsymbol{a}_2, \ldots, \boldsymbol{a}_p)$, then the conditions $\boldsymbol{a}_i'\boldsymbol{a}_i = 1$ and $\boldsymbol{a}_i'\boldsymbol{a}_j = 0$ $(i \neq j = 1, \ldots, p)$ established in Section 2.2.4 show that the principal components are also uncorrelated over the whole sample, but that the appropriate normalization is $\mathbf{A}'\mathbf{A} = \mathbf{I}$. Thus the principal component transformation from the original variates $\boldsymbol{x}$ to the new variates $\boldsymbol{y}$ is *orthogonal,* but reference to (11.10) shows that the canonical variate transformation from $\boldsymbol{x}$ to $\boldsymbol{y}$ is *not* orthogonal. In geometric terms, this means that the principal component axes are all at right angles to each other, and the frame of reference $Oy_1, Oy_2, \ldots, Oy_p$ for the principal component space is obtained by a *rigid rotation* of the original frame of reference $Ox_1, Ox_2, \ldots, Ox_p$. On the other hand, the canonical variate axes are *not* at right angles to each other (in general), and the frame of reference $Oy_1, Oy_2, \ldots, Oy_s$ for the canonical variate space involves some deformation of the original frame of reference, with some axes being pressed closer to each other and others pulled further apart. When producing a canonical variate diagram, however, we construct orthogonal axes from the canonical variates. A justification for such a procedure stems from the objective of representing *distances* between groups as accurately as possible in the canonical variate space.

Consider two groups that differ in their means, but which have comparable covariance structures. If all variates have equal variance and are mutually uncorrelated within groups, then the Euclidean distance between the group means will be an accurate representation of the dissimilarity of (or equivalently, distance between) the two groups. If some variates have much larger within-group variance than the rest, however, then the influence of these variates should be lessened when computing the distance (since even individuals from the same group are likely to differ appreciably on these variates). Similarly, if two variates are highly correlated within groups, then including both equally in the computation will distort the calculated distance between the two groups, as the values of the group mean will tend to be either small or large together for the two variates. (Corresponding arguments have already been propounded in Section 8.2 regarding distance between individual points in data space.) Thus, rather than using the Euclidean distance $\{(\bar{\boldsymbol{x}}_i - \bar{\boldsymbol{x}}_j)'(\bar{\boldsymbol{x}}_i - \bar{\boldsymbol{x}}_j)\}^{\frac{1}{2}}$ between the group means, a more appropriate measure would be $\{(\bar{\boldsymbol{x}}_i - \bar{\boldsymbol{x}}_j)'\mathbf{M}(\bar{\boldsymbol{x}}_i - \bar{\boldsymbol{x}}_j)\}^{\frac{1}{2}}$ for some suitable matrix $\mathbf{M}$ that modifies the influence of each variate in the computation.

An early such modification was proposed by Pearson (1926) with the coefficient of racial likeness, which took account of differential variances within groups but ignored covariances between variates. Pearson's coefficient used a diagonal matrix $\mathbf{M}$, in which the ith diagonal entry was the inverse of the within-group variance for the ith variate. Subsequently, a more thorough analysis was undertaken by the Indian statistician P. C. Mahalanobis, who proposed as the matrix $\mathbf{M}$ the inverse of the pooled within-groups dispersion matrix (Mahalanobis 1936). This choice ensures that full account is taken of covariances between variables as well as of differential variances, and it is now the accepted measure of distance between two (quantitative) multivariate populations. In the context of the present chapter, the Mahalanobis squared distance between the ith and jth groups is given by

$$D^2 = (\bar{\boldsymbol{x}}_i - \bar{\boldsymbol{x}}_j)'\mathbf{W}^{-1}(\bar{\boldsymbol{x}}_i - \bar{\boldsymbol{x}}_j). \tag{11.11}$$

Recollect that we have already met the Mahalanobis distance in Section 8.2, in the form $d^2 = (\boldsymbol{x}_i - \bar{\boldsymbol{x}})'\mathbf{S}^{-1}(\boldsymbol{x}_i - \bar{\boldsymbol{x}})$ for the distance between an individual point $\boldsymbol{x}_i$ and the mean $\bar{\boldsymbol{x}}$ of a single sample. In Section 8.2 the sample was a homogeneous one with no grouping, and hence the overall variance/covariance matrix **S** played the role of the pooled within-groups matrix **W** of the present chapter.

Suppose that $s = p$, i.e. that the dimensionalities of the canonical variate space and of the original space are the same. If we construct orthogonal axes from the canonical variates to form the canonical variate space, and we use the resulting space to represent differences between the groups, then we are tacitly accepting Euclidean distance as the appropriate measure in this space (since this is the measure with which we are familiar in everyday life). The group means in this space have been earlier defined as $\bar{\mathbf{y}}_i = \mathbf{A}'\bar{\mathbf{x}}_i$ $(i = 1, \ldots, g)$. Hence, the squared Euclidean distance between the ith and jth groups in the canonical variate space is given by

$$\begin{aligned} d^2 &= (\bar{\mathbf{y}}_i - \bar{\mathbf{y}}_j)'(\bar{\mathbf{y}}_i - \bar{\mathbf{y}}_j) \\ &= (\bar{\mathbf{x}}_i - \bar{\mathbf{x}}_j)'\mathbf{A}\mathbf{A}'(\bar{\mathbf{x}}_i - \bar{\mathbf{x}}_j), \end{aligned}$$

on substituting for the $\bar{\mathbf{y}}_i$ and $\bar{\mathbf{y}}_j$.

But from (11.10),

$$\mathbf{I} = \mathbf{A}'\mathbf{W}\mathbf{A}.$$

Pre-multiplying this equation by **A** and postmultiplying it by **A**′ thus yields

$$\mathbf{A}\mathbf{A}' = (\mathbf{A}\mathbf{A}')\mathbf{W}(\mathbf{A}\mathbf{A}').$$

which is an equation satisfied by $\mathbf{A}\mathbf{A}' = \mathbf{W}^{-1}$.

Thus substituting for $\mathbf{A}\mathbf{A}'$ above yields $d^2 = (\bar{\mathbf{x}}_i - \bar{\mathbf{x}}_j)'\mathbf{W}^{-1}(\bar{\mathbf{x}}_i - \bar{\mathbf{x}}_j)$. Hence by constructing the canonical variate diagram in the way described earlier, we produce a diagram in which the ordinary Euclidean distance between points representing group means is equal to the statistical (i.e. Mahalanobis) distance between the corresponding groups. Of course this is only exactly true if the diagram has s dimensions; if we reduce dimensionality to r then we are, as usual, providing an approximation to the true configuration. Also, the above justification is no longer valid when $s < p$, as the matrix $\mathbf{A}\mathbf{A}'$ is then of less than full rank and non-invertible. In this case, recourse must be made to a justification based on generalized inverses. See Gower (1966a) for the use of generalized inverses in this context.

One question may be raised as to the exact nature of the approximation that is obtained by taking only the first r of the s canonical variates. The above development has shown that this will be the best approximation in the sense of maximizing between-group relative to within-group variability. When canonical variates are being used primarily for descriptive purposes, it may be more appropriate to seek a low-dimensional representation in which the total squared distance between every pair of sample (or group) means is maximized. It can be shown (Ashton *et al.* 1957), that in this case we need to replace the matrix **B** in (11.6) by the *unweighted* between-groups covariance matrix $\mathbf{B}_u = \frac{1}{g-1} \sum_{i=1}^{g} (\bar{\boldsymbol{x}}_i - \bar{\boldsymbol{x}})(\bar{\boldsymbol{x}}_i - \bar{\boldsymbol{x}})'$. Furthermore, Gower (1966b) showed that this latter solution can be obtained equivalently by computing the matrix of Mahalanobis D^2 values between every pair of groups (using eqn 11.11), and then subjecting this matrix to a metric scaling (Section 3.2). Note also that if all group sizes n_i are equal, this latter solution will be the same as the one obtained from the weighted between-groups matrix **B**.

To illustrate these ideas, consider one of the analyses conducted by Ashton *et al.* (1957), who compared measurements on the teeth of fossils with those of various different races of men and species of apes. Ashton *et al.* analysed several teeth, but Andrews (1972) re-examined the data for just one of the teeth (the lower first premolar) so we will restrict attention to this subset of the data. Eight measurements (1, trans.-bas. lab.; 2, trans.-mid. lab.; 3, trans.-max. lab.; 4, ht.-lab.; 5, ht.-prox. seg.; 6, thick. bas.; 7, thick.-mid. lab.; 8, thick.-max.) were available on each individual. Samples were taken from nine populations:

(1) West African (WA)
(2) British (B)
(3) Australian aboriginal (AA)
(4) Gorilla, male (GM)
(5) Gorilla, female (GF)
(6) Orang-utan, male (OM)
(7) Orang-utan, female (OF)
(8) Chimpanzee, male (CM)
(9) Chimpanzee, female (CF)

and additionally measurements were obtained for six individual fossils:

(1) *Pithecanthropus pekinensis* (PPA)
(2) *Pithecanthropus pekinensis* (PPB)
(3) *Paranthropus robustus* (PR)
(4) *Paranthropus crassidens* (PC)
(5) *Meganthropus palaeojavanicus* (MP)
(6) *Proconsul africanus* (PA)

First, Ashton *et al.* computed canonical variates from the nine groups of individuals, and the canonical variate coefficients are shown in Table 11.1. The constant terms ensure that each canonical variate is centred at the origin. Next, group means for each of the nine groups, and individual fossil values for each of the six fossils, were obtained on each of the eight canonical variates (note that $s = \min(p, g-1) = \min(8, 8) = 8$); these

Table 11.1 Coefficients of canonical variates for eight-variable data on nine human and ape populations (from Ashton *et al.* 1957)

Variable (x)	CV1	CV2	CV3	CV4	CV5	CV6	CV7	CV8
1	+10.2	+3.7	−6.6	−18.1	−6.7	+19.3	−6.4	+9.9
2	−1.4	+0.9	+0.1	+14.7	−16.4	−0.8	+13.8	+19.7
3	+24.6	−3.8	+2.7	+24.2	+18.0	−6.0	−0.7	−30.5
4	+0.8	+0.9	−8.1	−1.8	+5.0	−9.6	−5.9	+7.5
5	−5.0	+2.4	+2.8	+4.6	+2.1	+8.9	−2.8	0.0
6	−2.0	+14.0	−5.3	−8.4	+17.7	+3.3	+25.6	−1.3
7	−13.1	+11.5	−19.9	+2.7	−10.3	−1.2	−7.6	−13.5
8	−1.2	+0.6	+42.0	−10.7	−15.5	−15.5	−19.4	+7.7
Constant	−31.78	−54.53	−16.55	−14.23	+10.98	+7.97	+7.76	+3.16

values are shown in Table 11.2. The group means were then plotted against the first two canonical variates as axes, and the fossil values were added to the diagram and assessed relative to the group means. The resulting picture is shown in Fig. 11.4.

From this plot it is evident that the human populations are well separated from the apes along the first canonical variate axis, and that the

Table 11.2 Group mean and fossil values on canonical variates (from Ashton *et al.* 1957)

	CV1	CV2	CV3	CV4	CV5	CV6	CV7	CV8
(a) Means of human and ape groups								
WA	−8.09	+0.49	+0.18	+0.75	−0.06	−0.04	+0.04	+0.03
B	−9.37	−0.68	−0.44	−0.37	+0.37	+0.02	−0.01	+0.05
AA	−8.87	+1.44	+0.36	−0.34	−0.29	−0.02	−0.01	−0.05
GM	+6.28	+2.89	+0.43	−0.03	+0.10	−0.14	+0.07	+0.08
GF	+4.82	+1.52	+0.71	−0.06	+0.25	+0.15	−0.07	−0.10
OM	+5.11	+1.61	−0.72	+0.04	−0.17	+0.13	+0.03	+0.05
OF	+3.60	+0.28	−1.05	+0.01	−0.03	−0.11	−0.11	−0.08
CM	+3.46	−3.37	+0.33	−0.32	−0.19	−0.04	+0.09	+0.09
CF	+3.05	−4.21	+0.17	+0.28	+0.04	+0.02	−0.06	−0.06
(b) Fossils								
PPA	−6.73	+3.63	+1.14	+2.11	−1.90	+0.24	+1.23	−0.55
PPB	−5.90	+3.95	+0.89	+1.58	−1.56	+1.10	+1.53	+0.58
PR	−7.56	+6.34	+1.66	+0.10	−2.23	−1.01	+0.68	−0.23
PC	−7.79	+4.33	+1.42	+0.01	−1.80	−0.25	+0.04	−0.87
MP	−8.23	+5.03	+1.13	−0.02	−1.41	−0.13	−0.28	−0.13
PA	+1.86	−4.28	−2.14	−1.73	+2.06	+1.80	+2.61	+2.48

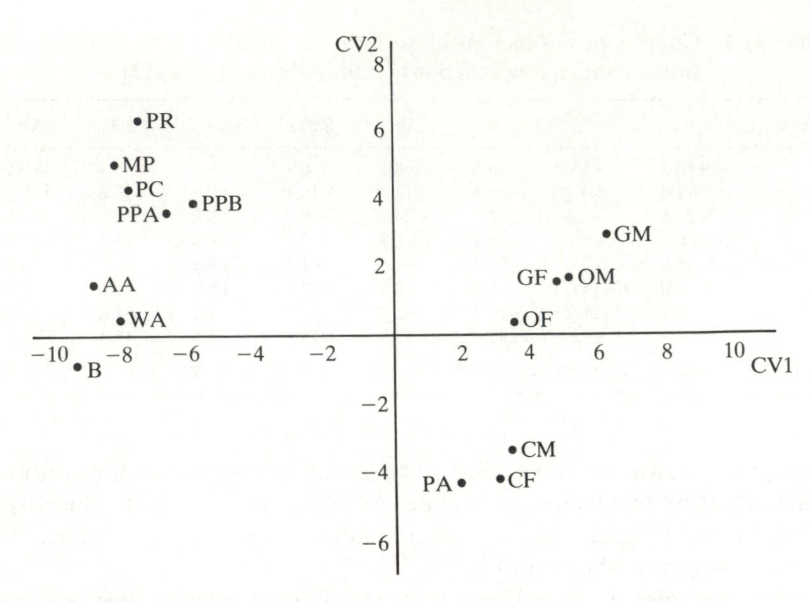

Fig. 11.4 Human and ape group means and fossil values plotted against first two canonical variates as axes, for data from Ashton *et al.* (1957)

second canonical variate axis discriminates among the latter groups. Considering next the fossils, the authors concluded that *Proconsul africanus* is very like a chimpanzee, while the other fossils are more like humans. However, no indication was given by Ashton *et al.* as to the size of the eight eigenvalues, and hence it is not clear how adequate the two-dimensional approximation is to the full eight-dimensional space. Andrews (1972) pointed out that some group means have some moderately large values of the third and fourth canonical variates, while large values occur for all eight variates for the fossils. Plotting *all* the variates would thus permit the examination of the effect of these large values, and this can be achieved by the technique of Andrews curves. Use of the ith canonical variate value for the x_i in $f_{\mathbf{x}}(t)$ of Section 2.1.3 yields the appropriate set of functions. Figure 11.5 shows the group means of all the canonical variates, and demonstrates the clear separation between the human and the ape populations, with chimpanzees standing out from the gorillas and orang-utans among the latter. The human populations have a very precise function value at t_2 and t_4, very different from the corresponding ape values; the point t_1 similarly distinguishes chimpanzees from gorillas and orang-utans, while t_3 is the point at which all nine groups are maximally separated. When the fossils are drawn to the same scale, Fig. 11.6 is obtained. *Proconsul africanus* seems to be very

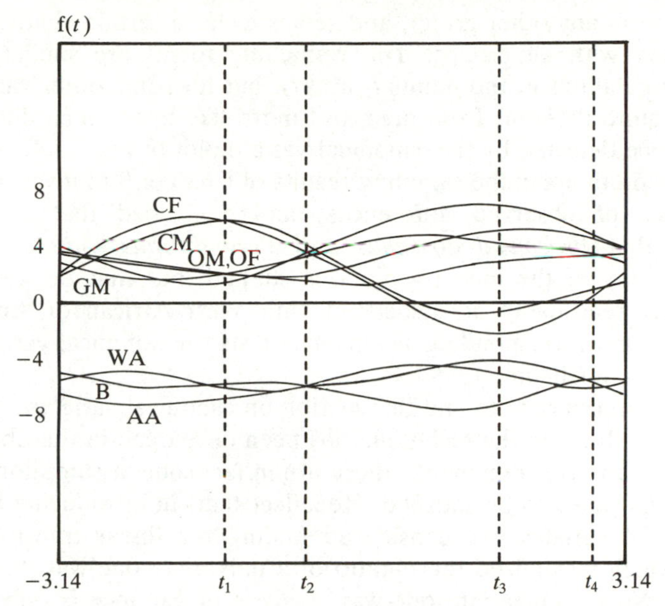

Fig. 11.5 Andrews curves for human/ape group means on canonical variates, as given in Table 11.2(a). Reproduced from Andrews (1972) with permission from the Biometric Society

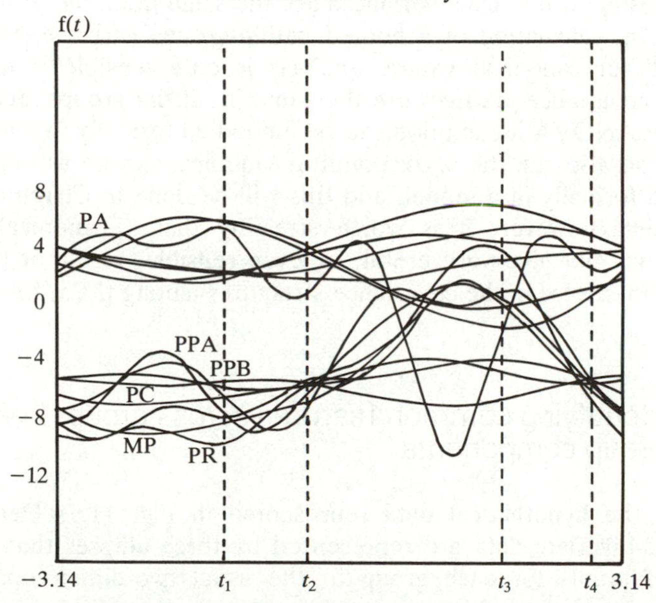

Fig. 11.6 Andrews curves for fossil values on canonical variates (Table 11.2(b)) superimposed on Fig. 11.5. Reproduced from Andrews (1972) with permission from the Biometric Society

different from any other group, and seems to have certain characteristics in common with all groups. The remaining fossils are similar to the human populations at the points t_2 and t_4, but for some other values of t they are quite different from man and more like apes. Such differences could not be detected by the canonical variate plot of Fig. 11.4. Andrews then went on to apply the sampling results of Chapter 9 to investigate the significance of observed differences, and concluded that there was evidence that *Proconsul africanus* and *Paranthropus robustus* do not belong to any of the nine populations studied, but that the remaining fossils may reasonably be associated with West African or Australian humans. Some corresponding significance tests for canonical variates will be discussed in Chapter 13.

It seems appropriate to end this section on canonical variates on a note of caution. Although the technique has been developed in this chapter in a purely descriptive framework, there are in fact some assumptions about the data that need to be satisfied. Recollect that, in introducing the idea of canonical variates, we considered finding the linear transformation $y = \boldsymbol{a}'\boldsymbol{x}$ which maximized the F-ratio of a univariate one-way analysis of variance. Now a univariate one-way analysis of variance is only strictly applicable if there is variance homogeneity among all the groups being compared. In other words, calculation of a pooled within-groups variance is only sensible if the 'true' variances are the same in all the groups. In a similar vein, calculation of a pooled within-groups variance–covariance matrix $\mathbf{W}$ for canonical variate analysis is only sensible if the 'true' variance–covariance matrices are the same in all the groups. Just as the univariate ANOVA assumption can be embodied formally in a model for the data, so also can the corresponding canonical variate assumption be embodied formally in a model, and this will be done in Chapter 13. For the present, however, it is worth stressing that a canonical variate analysis may not give very useful, or even sensible, results if there are obvious differences in the covariance structures among the x_i for different groups.

11.2 Identifying common features across groups: within-group components

Consider the hypothetical data represented in Fig. 11.7. Here, three groups of bivariate data are represented by three ellipses that give the scatter of points for each group in the usual two-dimensional representation. In attempting to summarize these data unidimensionally, we are now in a position to produce several different indices. The first canonical variate, as considered in the previous section, will be the best

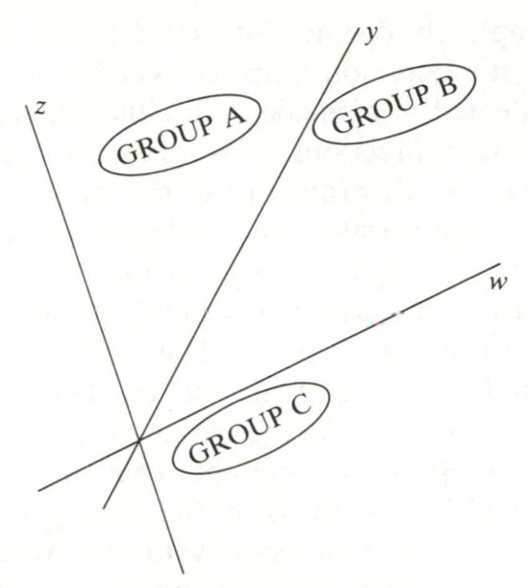

Fig. 11.7 Hypothetical two-dimensional representation illustrating different components of interest: between-group component indicated by z; overall component indicated by y; within-group component indicated by w

single index for showing up the group differences, and will be in the approximate direction z shown in the diagram. Alternatively, the first principal component of the whole data set will be the best single index for representing the scatter of all points (ignoring groups): this will be approximately in the direction y shown. There is a particular feature of this set of data, and of many sets encountered in practice, however, that will be obscured by either of these indices. It frequently happens that, while different groups exhibit different *average* values of the measured variates, the variances and covariances between the variates are roughly similar in each group. This similarity will be exhibited, as in Fig. 11.7, by sets of points that form ellipsoids of roughly comparable sizes and orientations. Even if variances differ between groups but all correlations are roughly comparable between groups, the orientations of the ellipsoids will all be approximately the same. In such circumstances the major sources of variation between individuals will be similar from group to group, and determining these sources is of interest in any description of the data. For example, we may be interested in determining the common sources of variation among students' examination marks across colleges, among people's attitudes to work across different socio-economic groupings, among subjects' responses to psychological stimuli across different treatment groups, and so on.

The main such source of variation that is common to all groups is shown as the direction w in Fig. 11.7, this being the common direction of the major axis of each ellipse. The three directions y, z, and w may be expected to be completely distinct for any data set. More generally, of course, *all* axes of the common ellipsoid will be of interest, and these axes will not be expected to coincide with either the canonical variate or the principal component directions. It is relatively straightforward to see how these axes may be determined. If the only differences between groups are in the average variate values, while the scatter of points about these average values is comparable from group to group, then the best representation of this scatter will be obtained by superimposing all group centroids before plotting the points. This operation removes all extraneous effects, and leaves us with the residual variation. The best estimate of the latter is clearly **W**, the within-groups variance–covariance matrix defined in the previous section. Thus the major axes of the common ellipsoids will be identified by the principal components of **W**. These components may be contrasted with the overall principal components, which are the principal components of the *total* variance–covariance matrix **T**. To illustrate the large differences that can occur, the two sets of components are presented in Table 11.3 for the Fisher iris data, and it can be seen that only the fourth components in each set are at all comparable. In fact, using the subspace comparison technique of Section 5.3, the critical angles between the planes defined by the first two components of each type are 34.2° and 47.3°. Thus the two main sources of overall variability are rather different from the two main sources of within-group variability.

We shall return to this topic in Chapter 13, where we shall consider some more sophisticated models for identifying common sources of variability between groups. In passing, we may also note that the canonical variate analysis of the previous section may be obtained by a

Table 11.3 Comparison of within-group principal components and overall principal components for the Fisher iris data of Table 2.3

Component		Variate			
		Sepal length	Sepal width	Petal length	Petal width
Within-group	1	0.74	0.32	0.57	0.16
	2	0.63	−0.18	−0.58	−0.47
	3	0.06	−0.87	0.46	−0.15
	4	0.23	−0.32	−0.35	0.85
Overall	1	0.36	−0.08	0.86	0.36
	2	0.66	0.73	−0.17	−0.07
	3	−0.58	0.60	0.08	0.55
	4	0.32	−0.32	−0.48	0.75

two-stage principal component analysis as follows. First, a principal component analysis is conducted on the within-groups variance–covariance matrix **W**, and the original individual data values are converted into scores on each of these principal components. The scores on each principal component are standardized to unit variance, and a second principal component analysis is conducted on the group means computed from these standardized scores. The canonical variate coefficients for the original data are obtained by a simple transformation of these latter principal components. Full details, and a geometrical demonstration of equivalence, are provided by Campbell and Atchley (1981).

11.3 Partial information: identifying group membership

It sometimes happens that the investigator has collected an ostensibly unstructured sample, but in fact suspects strongly that the n sample members have come from g different populations. For example, a psychologist may believe that the n subjects in his experiment are of g different psychological types; a botanist may believe that the n plants have g different origins; a zoologist may believe that the n marine organisms are of g different species, and so on. It is then of considerable interest to find the partition of the sample that will produce the g most distinct groups (as judged by the data alone). Since the data are initially considered to be unstructured, a common practice is to plot the data against the first few principal components as axes and to look for the most obvious division of the n points into g groups from this plot. Having made the division, canonical variate analysis is often invoked to confirm the chosen grouping. However, there are various dangers in such a procedure. Earlier discussion in this chapter should have made evident that the first few principal components need not necessarily lie in the directions of natural group differences, and in fact such differences may often be hidden in the resultant plot. Chang (1983) has provided a quantitative analysis highlighting this problem and warning against the use of principal components in such circumstances. The opposite problem can also occur, in that 'natural' groupings can seem to appear even in completely random data. Subsequent confirmatory canonical variate analysis will almost inevitably strengthen opinions about the grouping, and thus lead to false conclusions. It is evident, therefore, that the ideas behind canonical variate analysis must play a much more central role in any satisfactory resolution of the problem.

Suppose that the n sample members have been divided arbitrarily into g groups. Canonical variate analysis requires computation of the between-group and within-group covariance matrices **B** and **W**, followed by solution of eqn (11.6). If the groups are distinct and compact then group differences should show up well on the canonical variates. We have seen in Section 11.1 that the eigenvalues of $\mathbf{W}^{-1}\mathbf{B}$ are interpreted as measures of between-group relative to within-group variability. Consequently, among a set of possible partitions of the sample, the most distinct and compact groups will be indicated by the partition with largest such eigenvalues. Combining all the eigenvalues into a single expression in some way will thus furnish a criterion which gives a measure of separation and compactness of the g groups in any partition, and hence a measure of 'goodness' of that partition. The best partition will be the one optimizing the criterion. Two possible ways of combining all the eigenvalues of a matrix are through the determinant of the matrix (product of the eigenvalues) or the trace of the matrix (sum of the eigenvalues). Various different criteria have been suggested since the pioneering work of Friedman and Rubin (1967), who looked for criteria based on the eigenvalues of $\mathbf{W}^{-1}\mathbf{B}$ that are invariant under non-singular linear transformations of the data. Probably the simplest and most common of these criteria that are in common use are:

$$V_1 = \text{trace}(\mathbf{W}), \tag{11.12}$$

$$V_2 = \det(\mathbf{W}), \tag{11.13}$$

$$V_3 = \sum_i \sum_j (\bar{\mathbf{x}}_i - \bar{\mathbf{x}}_j)'\mathbf{W}^{-1}(\bar{\mathbf{x}}_i - \bar{\mathbf{x}}_j). \tag{11.14}$$

Since compactness of groups implies 'small' **W**, the best partition of the sample is the one *minimizing* (11.12) or (11.13). V_3 is the sum of Mahalanobis D^2 values between all pairs of groups, so the best partition is the one *maximizing* (11.14). For a list, and brief description, of a range of other possible criteria, see Milligan and Cooper (1985).

Having selected one of these criteria, in principle all that is necessary is to compute its value for every possible partition of the n objects into g groups and to select the partition that has the optimum value. In practice, of course, the number of possible partitions may be computationally prohibitive for even moderate combinations of n and g, and we must seek some alternative strategy. A reasonable approximation to the global optimum may often be found by the following *iterative relocation scheme*. First, choose some initial partition of the data. This can be done randomly or in some systematic fashion. An alternative is to do a preliminary hierarchic cluster analysis and take the g group division as the initial partition. Sample members are then systematically moved between groups in order to try and improve the chosen criterion, and the optimal partition of the sample is taken to be the one for which no further relocations are possible. A suitable strategy for doing the relocations, and a FORTRAN listing of the complete algorithm, is given by Banfield and Bassill (1977).

The procedure is commonly extended to include also the choice of the best value of g, by repeating the whole iterative relocation process for a range of values of g, and plotting the optimal value of the criterion against g. As in a scree diagram for principal component analysis, the 'best' value of g is usually taken to be the value at which an 'elbow' occurs in the plot.

Various drawbacks may be evident from the above description. There is no guarantee that the iterative relocation algorithm will yield the partition corresponding to the optimal criterion value, and indeed different starting configurations may result in different final partitions for the same set of data. It is generally recommended to try several different starting configurations and to choose the best of the resulting partitions. However, the problems here are similar to the ones encountered in iterative computer methods for obtaining multidimensional scaling solutions; see the discussion in Section 9.3 and references cited therein. Secondly, relatively few objective guidelines exist in the literature about methods of choosing the best value of g. Marriott (1971) has studied the $\det(\mathbf{W})$ criterion (11.13), and suggests that optimal choice of g is the one minimizing $g^2 \det(\mathbf{W})$. Some recent research on the $\text{trace}(\mathbf{W})$ criterion (11.12) is reported by Krzanowski and Lai (1988). The other criteria have not proved amenable to such specific study, and for them the scree diagram appears to be the only available general method but success or otherwise with this method is unpredictable. See also Milligan and Cooper (1985) for a study of the performances of a range of criteria on simulated data. Perhaps most fundamentally, all criteria such as (11.12)–(11.14) assume a common covariance structure within groups, and hence may lead to inappropriate choices of groups if this assumption is unwarranted. For some insight into the characteristics of partitions produced by different criteria, see Marriott (1982). Finally, even though the iterative relocation method is meant to be a feasible approximation to the 'all partitions' solution, it can still be expensive in terms of computer time, and large data sets can prove troublesome to partition. In this respect, the trace criterion (11.12) is computationally fastest (but may not be the most appropriate from other considerations).

It should thus be clear that there are many questions still unanswered in this particular area, and many applications of the methodology are at best exploratory. One such example is provided by Grime *et al.* (1987). Other *ad hoc* modifications of the methods, or individually tailored suggestions, have been made to meet particular requirements. For example, external constraints acting on the data may prevent a free application of the data partition. The most obvious instance of such constraints occurs when the data have a spatial structure, and only contiguous points are allowed to be grouped. For example, a soil scientist who studies the properties of soil profiles sampled along a line transect may wish to partition the whole area into sections of land within which the variation in soil properties is small. Each section should thus comprise a set of *contiguous* sites with similar properties. Webster and McBratney (1981) suggest setting up a moving window along the transect, calculating the Mahalanobis D^2 value between points in either half of the window, and plotting D^2 against window mid-point position. Natural soil boundaries will correspond to maxima of this D^2 plot. Much work still remains to be done on evaluation of methods such as these, and on the provision of objective criteria for their interpretation.

11.4 Miscellaneous topics

11.4.1 Categorical data

The computation and interpretation of canonical variates as described in Section 11.1 has tacitly assumed that the data for analysis are continuous, complete, and exhibit no obviously discrepant values. We now consider what can be done if one or more of these requirements is not satisfied.

Of course, the mathematical derivation of canonical variates can be carried through formally on *any* type of numerical data. Thus discrete and even binary data, where the two possible states are given an arbitrary scoring such as 0 or 1, can always be subjected to canonical variate analysis. Moreover, any categorical variable can be re-expressed as a set of binary variables; if the variable has c categories then $(c-1)$ binary variables will do the job. For example, the three-state variable 'colour' with states red, white, and pink can be replaced by the two binary variables x_1 and x_2 which take values $x_1 = 0$, $x_2 = 0$ if colour is red; $x_1 = 1$, $x_2 = 0$ if colour is white; and $x_1 = 0$, $x_2 = 1$ if colour is pink. Thus it follows that canonical variates can be formally obtained from any type of data whatsoever. The only question on which the user needs to satisfy himself/herself is whether the results will be meaningful and useful.

Claringbold (1958) has set out the standard theory, but in the context of multivariate quantal analysis. Here it is assumed that g groups of individuals are observed, and q quantal (i.e. binary) variates are recorded on each individual. The vector of variates $\boldsymbol{x}' = (x_1, x_2, \ldots, x_q)$ is thus specified by $x_i = 1$ or 0, according to whether or not the ith response is observed on a given individual. Canonical variates can then be computed in the usual fashion and tested for significance (see Chapter 13), to find the number of scores necessary to distinguish between the vectors of responses in the g groups. Such data frequently occur in bioassay and in toxicological investigations. Here the g groups either consist of g different applied doses of chemical, or of $g_1 = g/c$ different doses applied under each of c external conditions. Of prime interest in such work is the detection of a single score which maximizes the dependence of response on dose, and Claringbold shows how the usual canonical variate equation can be modified to meet this objective. He presents a numerical example in which two canonical variates are needed to distinguish between eighty-one groups of mice arranged in a 3^4 factorial pattern. Each mouse had three vaginal smears taken at fixed times after the administration of oestrogens, and the cytological picture of each smear was scored quantally to yield a trivariate quantal response. The two significant canonical variates had coefficient vectors (1.66, 1.07, −1.15) and (−0.19, 1.35, 1.99). One of the four three-level factors was log(dose), integrally spaced on the $\log_2$ scale, and the linear combination that maximized dependence between response and log(dose) had coefficient vector $(0.45, 0.69, 0.56)$. This vector has correlation 0.99 with the vector $(1, 1, 1)$, and this latter vector together with $(-1, 0, 1)$ span virtually the same space as the two canonical vectors given earlier. It was therefore concluded that the two independent aspects of the response that accounted fully for the differences between the eighty-one groups were $y_1 = x_1 + x_2 + x_3$ and $y_2 = x_3 - x_1$. The first derived variate is the total number of positive responses, which is related to duration of response, while the second derived variate is related to the time of onset of the response. Additionally, y_1 is the derived variate most closely associated with the applied dose. Thus in this case canonical variate analysis applied to 0/1 data yielded useful, and physiologically sensible, results.

A second example of the application of canonical variate analysis to binary variables is that given by Corbet *et al.* (1970), in the study of British water voles, genus *Arvicola,* already discussed in Section 3.2. Information on the presence or absence of thirteen characteristics was obtained for each of about 300 voles subdivided into fourteen groups, and a summary of the resulting values has been given in Table 3.2. The dissimilarity matrix between every pair of *Arvicola* populations, using a dissimilarity coefficient based on these summary values, was also given in Table 3.3 and a metric scaling of this dissimilarity matrix in Figure 3.7. Now since the raw data consisted of absence/presence (i.e. 0/1) values for *each* vole on each of the thirteen characteristics, it was also possible to compute the Mahalanobis D^2 between every pair of populations using eqn (11.11) on these 0/1 values. The resulting matrix of D^2 values is given in Table 11.4. Metric scaling of these D^2 values is equivalent to

Table 11.4 Mahalanobis D^2 matrix for *Arvicola* populations

	Surrey	Shrops	Yorks	Perths	Aberd.	Eilean Gamhna	Alps	Yugo.	Germ.	Norway	Pyrenees I	II	North Spain
Shropshire	2.309												
Yorkshire	0.615	2.036											
Perthshire	5.066	4.162	4.230										
Aberdeen	2.924	4.163	2.254	3.012									
Eilean Gamhna	3.368	1.884	2.376	4.137	1.985								
Alps	8.358	6.244	7.519	9.936	6.273	3.683							
Yugoslavia	11.212	9.402	9.976	9.405	6.685	6.056	2.187						
Germany	2.250	2.639	2.111	3.665	2.103	2.225	4.498	5.025					
Norway	6.085	3.767	5.591	8.619	5.404	4.641	5.510	9.517	5.696				
Pyrenees I	6.424	7.561	5.792	10.606	5.495	4.579	4.568	4.773	3.871	8.606			
Pyrenees II	14.031	13.847	15.112	14.279	13.930	13.165	8.634	10.836	11.004	13.401	11.819		
North Spain	10.864	11.664	11.448	14.191	12.804	12.774	9.416	12.530	8.292	11.875	13.104	4.344	
South Spain	12.516	13.525	13.687	14.326	12.566	13.737	10.002	12.558	9.313	11.503	14.257	3.844	1.212

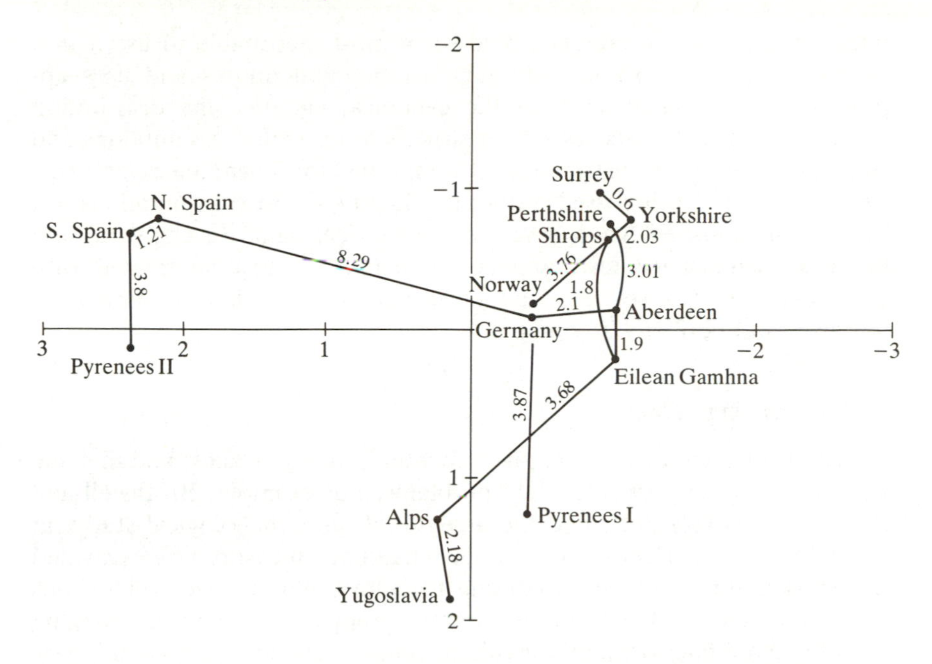

Fig. 11.8 *Arvicola* population means plotted on the first two canonical coordinate axes, with superimposed minimum spanning tree. Reproduced from Corbet *et al.* (1970) with permission from the Zoological Society of London

canonical variate analysis of the original 0/1 data, as has been explained in Section 11.1; a plot of the *Arvicola* populations against the first two canonical variates as axes is given in Fig. 11.8 (reproduced from Corbet *et al.* (1970) with permission from the Zoological Society of London). The minimum spanning tree is superimposed on this diagram, in the same way as for Fig. 3.7.

Comparison of Fig. 11.8 with Fig. 3.7 shows a high level of similarity; only the points representing the Pyrenees I population seem to be in appreciably different positions in the two diagrams (although the two minimum spanning trees have rather different shapes). Since the dissimilarity measure used in the computation of Table 3.3 makes no allowance for correlations between variables while Mahalanobis D^2 does, one inference is that correlations between characters were small in this data set. However, this might be a consequence of the binary nature of the data, since correlations are generally underemphasized in such data. It can certainly be concluded that application of canonical variate analysis to this set of data also gave both meaningful and sensible results.

As a final comment on the application of canonical variate analysis to binary or categorical data, it would seem most reasonable to use it in a purely descriptive fashion. Although Claringbold quotes and uses approximate significance tests for the canonical variates, the distribution theory behind such tests rests on multivariate normality assumptions and consequent Wishart distribution assumptions for covariance matrices. This theory will be discussed further in Chapter 13. Such assumptions will be far from satisfied when binary or categorical data are analysed, and hence all significance tests applied to such data must be treated with great caution. Use of the canonical variates for data summary is, however, still perfectly acceptable.

11.4.2 Missing values

Missing values are a common phenomenon in many studies, and in some situations they can provide acute problems. For example, Brothwell and Krzanowski (1974) described the analysis of an archaeological study, in which forty-eight craniometric variables had been measured on excavated British skeletons classified according to culture phase. Over 2000 skulls were examined, and sixty-three different groups were identified ranging from the Neolithic to Medieval culture phases. The aim of the study was to explore any group differences that might exist on the basis of the series of measurements taken; so, canonical variate analysis was appropriate. In measuring excavated skeletal material (particularly the skull), however, it is often found that post-mortem damage and erosion prevents the recording of more than a small number of the definable craniometric measurements. In the present study, many of the forty-eight variables were completely riddled with missing values across the sample, and use of these variables in analysis was seriously questionable. After careful preliminary inspection, eleven of the forty-eight variables were selected as worthy of inclusion in an analysis on account of their lowest incidence of missing values. It turned out that these eleven were all skull vault measurements, presumably because the bones there are stronger than those of the face and are therefore more likely to withstand post-mortem decay. Nevertheless, there was still an appreciable number of missing values, even in this carefully selected subset of the variables.

Various strategies are possible for handling missing values in a canonical variate analysis. One possibility is to ignore completely those individuals that have *any* missing values, and to base the analysis solely on those individuals that possess a complete set of measurements. In the study under discussion, inspection of the data showed that, while there was an appreciable number of missing values, it frequently happened that only one or two measurements were missing on a particular skull. Restricting analysis to those individuals that had no missing values would therefore reduce sample sizes dramatically, and would throw away an intolerably large amount of valuable information, so this option was rejected. A second general possibility is to impute values where measurements are missing. The simplest imputation is the insertion of variate group means for missing values; in the present instance this was felt to be too crude. In particular, where there are missing values, such imputation will lead to serious under-estimation of variances and corresponding over-estimation of covariances. At the other extreme, the imputation scheme of Beale and Little (1975), already discussed in other contexts, would have been computationally prohibitive with so many missing values. In addition, it was felt that this scheme would be over-sophisticated, in that completing the data is only essential if we require detailed analytical results for each skull in the sample. In the present case, interest centred mainly on relationships between *groups,* so a computationally simpler solution was sought.

Adopting Gower's (1966b) approach to canonical variate analysis, we require the matrix of Mahalanobis D^2 values as given by eqn (11.11) between every pair of groups, which in turn requires the mean values of each measurement in every group and the inverse of the within-groups dispersion matrix. We could utilize all available measurements to best advantage in calculating these quantities, and also avoid the use of 'fictitious' values, as follows. Each variate group mean could be computed separately from all available measurements, and each element of the within-groups dispersion could be found separately from the maximum number of data values possible. Thus the variance of each measurement could be calculated from all its available values, and the covariance between two measurements could be calculated from all those individuals for which both these measurements are available. However, it is easy to see that in general with this procedure each mean, variance, and covariance will be based on a different number of individuals when there are many missing values in the data. While properties of the mean vectors are unaffected adversely by this aspect, the within-groups dispersion matrix need not necessarily be positive definite as a consequence. If the matrix is not positive definite, its inverse may not exist (see Appendix A, Section A4). Computation of the matrix of Mahalanobis D^2 values might then run into difficulties. The only way to guarantee positive definiteness is to compute the within-groups dispersion matrix from complete cases. Hence a final compromise possibility is to compute means separately from all available data, but to select only those individuals which have no missing values for calculation of the within-groups dispersion matrix.

Brothwell and Krzanowski (1974) encountered the problem of non-positive definiteness, so they settled for the last-mentioned compromise

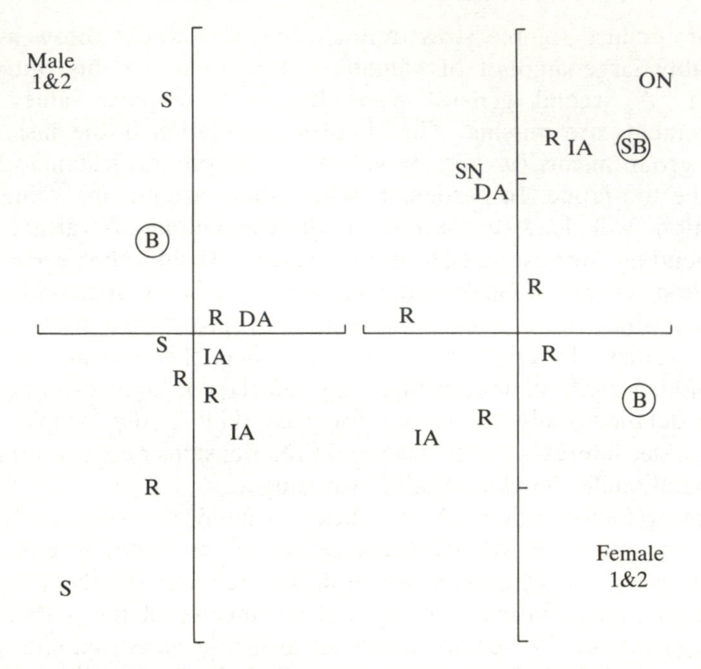

Fig. 11.9 The positioning of the Roman and Iron Age series for axes 1 and 2. A few Saxon samples (S) are indicated, as well as the Brandon series (B), a Shetland (SN) and Orkney (ON) Norse series, and a southern English Dark Age (DA) sample. Reproduced from Brothwell and Krzanowski (1974) with permission from Academic Press

solution. To avoid diagrams with a confusingly large number of points, they plotted group means against canonical variates as axes separately for the earlier prehistoric groups, then for the Iron Age and Roman series, and finally for Saxon, Medieval, and contemporary groups. A representation of some of the samples on the first four canonical variate axes has already been shown in Fig. 2.4. As a further example, Fig. 11.9 (reproduced with permission from Academic Press) presents the Roman and Iron Age series, plotted separately for males and females, with some other individual group means superimposed. (For a full discussion of the analyses, see Brothwell and Krzanowski (1974).) Further investigation of the estimation of variances and covariances separately from all available information has been made in the context of regression analysis by Haitovsky (1968).

11.4.3 Outliers and robustness

The final problem area concerns the stability of canonical variate coefficients in the presence of non-ideal data. Since the relative magnitudes of the canonical variate coefficients for the variables standardized within groups are often used to identify important variables and contrasts between variables, accurate and stable determination of these coefficients is important. Two distinct situations can be isolated in which there is potential instability of coefficients. One is when a number of untypical or outlying observations are contained in the data set, and these observations exert undue influence on the canonical variate coefficients. The second is when the within-groups covariance matrix **W** has a small eigenvalue, and the between-groups sum-of-squares in the direction of the corresponding eigenvector is also small. It is well known that, in regression analysis, the presence of high correlation between a pair of regressor variables leads to instability in the corresponding regression coefficients, reflected in large standard errors. Small eigenvalues of **W** correspond to high within-group correlations among the variables in a canonical variate analysis, and the strong formal connections between this technique and regression analysis (see Chapter 15, in particular, Section 15.3) therefore suggest potential instability of coefficients and hence possibly misleading conclusions. In both situations, modified estimates of the canonical variate coefficients are needed, to guard against these deficiencies. *Robust* estimates are designed to downweight the influence of outliers, while *shrunken* estimates will produce greater stability in the face of high correlations. Campbell (1980a and 1982) has proposed possible schemes for deriving each type of estimate, and these schemes are now described briefly.

First, consider the robust procedure. Campbell (1982) extended his robust covariance matrix M-estimation scheme, as described in Section 8.2, to encompass canonical variates as follows. Each observation again has a weight associated with it; since we are now denoting by $\boldsymbol{x}_{ij}$ the jth observation in the ith group ($i = 1, \ldots, g; j = 1, \ldots, n_i$), we can write the associated weights as v_{ij}. For a *given* set of weights, the effective number of observations is $m_i = \sum_{j=1}^{n_i} v_{ij}$ in the ith group, and $m = \sum_{i=1}^{g} m_i$ altogether, while the effective number of degrees of freedom is

$$f = \sum_{i=1}^{g} \left(\sum_{j=1}^{n_i} v_{ij}^2 - 1 \right).$$

We then define the group means by

$$\bar{\boldsymbol{x}}_i = m_i^{-1} \sum_{j=1}^{n_i} v_{ij} \boldsymbol{x}_{ij}, \tag{11.15}$$

the overall mean by

$$\bar{\boldsymbol{x}} = m^{-1} \sum_{i=1}^{g} m_i \bar{\boldsymbol{x}}_i, \tag{11.16}$$

the within-groups sum-of-squares and products matrix by

$$\mathbf{W}_0 = \sum_{i=1}^{g}\sum_{j=1}^{n_i} v_{ij}^2(\boldsymbol{x}_{ij} - \bar{\boldsymbol{x}}_i)(\boldsymbol{x}_{ij} - \bar{\boldsymbol{x}}_i)', \tag{11.17}$$

and the between-groups sum-of-squares and products matrix by

$$\mathbf{B}_0 = \sum_{i=1}^{g} m_i(\bar{\boldsymbol{x}}_i - \bar{\boldsymbol{x}})(\bar{\boldsymbol{x}}_i - \bar{\boldsymbol{x}})'. \tag{11.18}$$

Note that these are the weighted equivalents of the corresponding vectors and matrices of Section 11.1. Also, eqn (11.15) is the ith group version of eqn (8.2), while eqn (11.17) is the pooled within-groups extension of eqn (8.3).

Now suppose we require the first c robust canonical variates. Their coefficients, as might be expected, are given by the eigenvectors corresponding to the c largest eigenvalues of $\mathbf{W}_0^{-1}\mathbf{B}_0$. Let $\mathbf{D}$ be the diagonal $(c \times c)$ matrix containing these eigenvalues, and $\mathbf{L}$ the $(p \times c)$ matrix containing the corresponding eigenvectors in its columns. It only now remains for the weights v_{ij} to be specified, and the complete solution is then defined. Campbell (1982) suggests that the weight v_{ij} be determined from the sample Mahalanobis distance between $\boldsymbol{x}_{ij}$ and its estimated (robust) group mean.

Setting

$$d_{ij}^2 = (\boldsymbol{x}_{ij} - \boldsymbol{a}_i)'\mathbf{C}^{-1}(\boldsymbol{x}_{ij} - \boldsymbol{a}_i),$$

where

$$\boldsymbol{a}_i = \bar{\boldsymbol{x}} + f^{-1}\mathbf{W}_0\mathbf{L}\mathbf{L}'(\bar{\boldsymbol{x}}_i - \bar{\boldsymbol{x}})$$

and

$$m\mathbf{C} = \mathbf{W}_0 + \mathbf{B}_0 - \mathbf{B}_0\mathbf{L}\mathbf{D}^{-1}\mathbf{L}'\mathbf{B}_0 m^{-1},$$

the weight v_{ij} is given by

$$v_{ij} = w(d_{ij})/d_{ij}$$

with the influence function defined, as in Section 8.2, by

$$w(d_{ij}) = \begin{cases} d_{ij} & \text{if } d_{ij} \leqslant d_0 \\ d_0 \exp\{-\frac{1}{2}(d_{ij} - d_0)^2/b_2^2\} & \text{if } d_{ij} > d_0 \end{cases}$$

for $d_0 = \{\surd(2p-1) + b_1\}/\surd 2$.

Choice of the parameters b_1, b_2 in this influence function is discussed by Campbell; (the discussion in Section 8.2 is also relevant in this context). Choice of the pair $b_1 = 2.25$, $b_2 = 1.25$ seems to give ease of computation and good properties, and from some simulation results Campbell concludes that with this choice a weight of less than 0.3 will be associated with an atypical observation. The whole process is clearly iterative, as the weights depend on the robust estimates, and vice versa. For a detailed practical example and further discussion, see Campbell (1982).

Passing on to problems of instability caused by high within-group correlations, it is first worth noting that, in the corresponding regression situation, generalized ridge estimation procedures will produce more stable estimates. The interested reader is referred to Alldredge and Gilb (1976) for a comprehensive bibliography of this area. Campbell (1980a) adapts the generalized ridge estimation philosophy to canonical variate coefficient estimation. Denoting, as usual, the within-groups covariance matrix by **W**, and the between-groups covariance matrix by **B**, let the eigenvalues and eigenvectors of **W** form the elements of the diagonal matrix **E** and the columns of the matrix **U** respectively. Furthermore, let **K** denote the diagonal matrix of positive constants $k_1, k_2, \ldots, k_p$. Then Campbell's shrunken canonical variates $\mathbf{c}_i$ are given by

$$\mathbf{c}_i = \mathbf{U}(\mathbf{E}+\mathbf{K})^{-\frac{1}{2}}\mathbf{a}_i,$$

where $\mathbf{a}_i$ is the eigenvector corresponding to the ith largest eigenvalue of $(\mathbf{E}+\mathbf{K})^{-\frac{1}{2}}\mathbf{U}'\mathbf{B}\mathbf{U}(\mathbf{E}+\mathbf{K})^{-\frac{1}{2}}$. The optimum values of the constants k_i are estimated by conducting the analysis for a range of values of k_i, plotting the canonical variate coefficients against each k_i, and choosing the values of k_i at which the canonical variate coefficients stabilize, in the same way that the constants in a generalized ridge regression are estimated from the so-called 'ridge trace'. Indeed, it can be seen that the above shrunken estimates bear exactly analogous relationships to the ordinary coefficients as the generalized ridge estimates do to ordinary least-squares regression coefficients. Campbell (1980a) considers mean square error aspects of the estimates, and provides some detailed practical examples.

11.5 The need for inferential methods

The techniques discussed in this chapter have all been concerned with the description of a multivariate data set in which there exists an a-priori grouping of the units. The techniques are therefore appropriate if the data set itself is the sole focus of interest, and there is no desire to extrapolate the results beyond this data set. Very often in practice, however, the data set consists of random samples from a number of different populations, and the investigator wishes to explore the relationships between (or test hypotheses about) these populations. Mere description of the samples is not enough: inferences about the populations are required. This mirrors the objectives of Chapters 8 and 10, where inferences about a single population were sought using a single (homogeneous) sample from this population. Appropriate results were obtained by first formulating a probability model for the population, and then expressing the desired inferences in terms of the parameters of this model. Thus parameters could be estimated by maximizing the likelihood of the sample assuming this model, and tests of hypotheses about the parameters could be derived by application of the likelihood ratio or union–intersection principles.

Similar considerations apply when there is a-priori grouping of units, and each group of units comprises a random sample from a given population. We may wish to describe each of the populations, or perhaps test the hypothesis that the random samples have come from the same populations. To attempt such description or hypothesis testing, we can again formulate appropriate probabilistic models for the populations and then apply maximum likelihood, union–intersection or likelihood ratio methods. Details of all these methods follow in the next two chapters.

As well as these extensions to familiar single-population inferential problems, however, a problem peculiar to multiple-group situations will also be discussed in some detail. This is the problem of allocating an individual of unknown provenance to one of several existing groups or populations. Consider the following medical problem. Patients suffering from jaundice are admitted to a hospital, and appropriate treatment has to be determined. The condition of jaundice can be associated with a number, say g, of different complaints. Some of these complaints can be treated successfully by medication, while others can only be treated by surgery. Is it possible to determine (from p external signs, symptoms, or laboratory measurements) to which category any particular patient belongs? Can the patient then be successfully matched with his/her complaint on the basis of these p measurements alone? Correct diagnosis could possibly be achieved by operating on every patient admitted with jaundice, but this would place an intolerable financial and logistic strain on the hospital as well as subjecting the 'non-surgical' patients to totally unnecessary risks. The intuitive feeling is that, if sufficiently many measurements are taken on each patient, there should be enough information available from the hospital records of previous jaundice patients whose complaint was subsequently correctly identified (either because they were treated successfully or from post-mortem examination) to enable a rule to be devised that will classify future patients to a complaint category with small probability of error.

In view of the wide range of problems that need to be tackled, consideration of the inferential aspects of grouped data is most conveniently dealt with in two stages: Chapter 12 will deal with all problems involving just two groups of individuals, while Chapter 13 will deal with all problems encompassing more than two groups.

12 Inferential aspects: the two-group case

We suppose that the data matrix $\mathbf{X}$ is partitioned into two sections: one section, comprising n_1 rows, constitutes a random sample from a p-variate population π_1, while the second section, comprising n_2 rows, constitutes a random sample from another p-variate population π_2. Most of the methods in this chaper and the next will rest upon the assumption of probabilistic models for π_1 and π_2. Each of these π_i is a single population, and hence the discussions and models of Chapters 8 and 10 are appropriate. In particular, we will adopt the categorization of data types from those chapters, namely that a data matrix can be classified according to whether its variates are all quantitative (continuous), all qualitative (discrete), or a mixture of the two, and we will take as suitable probability models the multivariate normal for the first type, the multinomial for the second type, and the location model for the last type. Again, the various discussions of Part III regarding these models are relevant in the present context. This chapter is divided into five sections. In the first section we consider estimation and hypothesis testing, in the second we derive model-based classification rules and their error rates, in the third we consider other approaches to the derivation of a classification rule, in the fourth we address the problem of data-based assessment of a classification rule, and a practical example is given in the final section. Where appropriate, a section is subdivided according to the type of data under consideration.

12.1 Estimation and hypothesis testing

12.1.1 The multivariate normal model

Suppose that $\boldsymbol{x}_1, \boldsymbol{x}_2, \ldots, \boldsymbol{x}_{n_1}$ are n_1 independent observations taken from a multivariate normal distribution with mean $\boldsymbol{\mu}_1$ and dispersion matrix $\boldsymbol{\Sigma}_1$, while $\boldsymbol{y}_1, \boldsymbol{y}_2, \ldots, \boldsymbol{y}_{n_2}$ are n_2 independent observations taken from a multivariate normal distribution with mean $\boldsymbol{\mu}_2$ and dispersion matrix $\boldsymbol{\Sigma}_2$. Associating the $\boldsymbol{x}_i$ with those rows of the data matrix that belong to one group and the $\boldsymbol{y}_i$ with those rows that belong to the other group leads to a suitable model for the grouped data matrix if all observations are continuous and quantitative, although perhaps some preliminary transformation of the variables to normality may first be necessary (see Section 7.6). Assume also that the two samples are independent of each other. Results of Section 8.2 show that unbiased estimates of the population parameters $\boldsymbol{\mu}_1$, $\boldsymbol{\mu}_2$, $\boldsymbol{\Sigma}_1$, and $\boldsymbol{\Sigma}_2$ are $\bar{\boldsymbol{x}}$, $\bar{\boldsymbol{y}}$, $\frac{1}{(n_1-1)}\sum_{i=1}^{n_1}(\boldsymbol{x}_i-\bar{\boldsymbol{x}})(\boldsymbol{x}_i-\bar{\boldsymbol{x}})'$ and $\frac{1}{(n_2-1)}\sum_{i=1}^{n_2}(\boldsymbol{y}_i-\bar{\boldsymbol{y}})(\boldsymbol{y}_i-\bar{\boldsymbol{y}})'$ respectively. Write $\mathbf{C}_1=\sum_{i=1}^{n_1}(\boldsymbol{x}_i-\bar{\boldsymbol{x}})(\boldsymbol{x}_i-\bar{\boldsymbol{x}})'$ and $\mathbf{C}_2=\sum_{i=1}^{n_2}(\boldsymbol{y}_i-\bar{\boldsymbol{y}})(\boldsymbol{y}_i-\bar{\boldsymbol{y}})'$ for brevity. Confidence regions for $\boldsymbol{\mu}_1$ and $\boldsymbol{\mu}_2$ may also be found, by using the results of Section 8.2.

Of more interest, however, is estimation of the *difference* between the means $\boldsymbol{\mu}_1$ and $\boldsymbol{\mu}_2$, because this difference gives a measure of divergence between the populations. In this case we require an estimate of $\boldsymbol{v}=\boldsymbol{\mu}_1-\boldsymbol{\mu}_2$. A point estimate of $\boldsymbol{v}$ is clearly given by $\bar{\boldsymbol{x}}-\bar{\boldsymbol{y}}$, but by itself this point estimate is of little use as we have no way of assessing the importance of calculated values. Without some indication of sampling variability, we have no way of deciding whether a calculated mean difference is 'small' or 'large'. A confidence region for $\boldsymbol{v}$ is a much more satisfactory way of estimating the difference, but in trying to derive such a region we encounter a severe theoretical difficulty. Since $\bar{\boldsymbol{x}}-\bar{\boldsymbol{y}}$ is the best point estimate of $\boldsymbol{v}$, it should form the basis for calculation of the confidence region. The sampling distribution of $\bar{\boldsymbol{x}}-\bar{\boldsymbol{y}}$ is $\mathrm{N}\left(\boldsymbol{\mu}_1-\boldsymbol{\mu}_2, \frac{1}{n_1}\boldsymbol{\Sigma}_1+\frac{1}{n_2}\boldsymbol{\Sigma}_2\right)$. The dispersion matrix of this distribution is estimated by $\frac{1}{n_1(n_1-1)}\mathbf{C}_1+\frac{1}{n_2(n_2-1)}\mathbf{C}_2$, but there is no simple function of this estimate that will combine with $\bar{\boldsymbol{x}}-\bar{\boldsymbol{y}}$ to yield a statistic having a known tabulated distribution from which confidence limits can be obtained. The root cause of the problem is the assumption of different dispersion matrices $\boldsymbol{\Sigma}_1$ and $\boldsymbol{\Sigma}_2$ in the two populations. This is the multivariate analogue of a problem long familiar in univariate analysis, the *Behrens–Fisher* problem. Approximate confidence regions, and mathematically more complicated ways of overcoming the problem, do exist; the reader is referred for details to T. W. Anderson (1984) or Muirhead (1982). In practice the simplest way round the problem is to assume a common dispersion matrix in the two populations, thereby setting $\boldsymbol{\Sigma}_1=\boldsymbol{\Sigma}_2=\boldsymbol{\Sigma}$. Such an assumption leads to a more restrictive model but is often perfectly acceptable, and is the natural extension of the assumption of common within-population variance that is usually made in univariate problems. It can be justified on the grounds that differences between populations will generally be manifested in differences of location, while the nature of the variables and measuring techniques will ensure roughly comparable variances and covariances from population to population. However, if any doubts exist about such justification, the reasonableness or otherwise of the assumption can be tested formally in any situation by means of the test given later in this section.

Assuming therefore a common dispersion matrix in the two populations, $\boldsymbol{x}_1, \boldsymbol{x}_2, \ldots, \boldsymbol{x}_{n_1}$ are n_1 independent observations from a $\mathrm{N}(\boldsymbol{\mu}_1, \boldsymbol{\Sigma})$ distribution, while $\boldsymbol{y}_1, \boldsymbol{y}_2, \ldots, \boldsymbol{y}_{n_2}$ are n_2 independent observations from a $\mathrm{N}(\boldsymbol{\mu}_2, \boldsymbol{\Sigma})$ distribution. Maximum likelihood estimates of $\boldsymbol{\mu}_1$ and $\boldsymbol{\mu}_2$ are still $\bar{\boldsymbol{x}}$ and $\bar{\boldsymbol{y}}$ respectively, while the maximum likelihood estimate of $\boldsymbol{\Sigma}$ now becomes

$$\hat{\boldsymbol{\Sigma}} = \frac{1}{n_1 + n_2}(\mathbf{C}_1 + \mathbf{C}_2) \tag{12.1}$$

where $\mathbf{C}_1$ and $\mathbf{C}_2$ are as defined earlier. Formal details of the derivation of $\hat{\boldsymbol{\Sigma}}$ may be found in Mardia *et al.* (1979) for example, and will not be repeated here. The sampling distribution of $\hat{\boldsymbol{\Sigma}}$ may be deduced readily, however, by employing the additive property of the Wishart distribution (property (2), Section 7.3). Since the $\boldsymbol{x}_i$ are independent observations from a $\mathrm{N}(\boldsymbol{\mu}_1, \boldsymbol{\Sigma})$ distribution while the $\boldsymbol{y}_i$ are independent observations from a $\mathrm{N}(\boldsymbol{\mu}_2, \boldsymbol{\Sigma})$ distribution then $\mathbf{C}_1$ has a $\mathrm{W}(n_1 - 1, \boldsymbol{\Sigma})$ distribution while $\mathbf{C}_2$ has a $\mathrm{W}(n_2 - 1, \boldsymbol{\Sigma})$ distribution (Section 8.1). Furthermore, the $\boldsymbol{x}_i$ are independent of the $\boldsymbol{y}_i$, so that $\mathbf{C}_1$ is independent of $\mathbf{C}_2$. Thus $\mathbf{C}_1 + \mathbf{C}_2$ has a $\mathrm{W}(n_1 + n_2 - 2, \boldsymbol{\Sigma})$ distribution, using result (2) of Section 7.3. Hence $(n_1 + n_2)\hat{\boldsymbol{\Sigma}}$ has a $\mathrm{W}(n_1 + n_2 - 2, \boldsymbol{\Sigma})$ distribution. Using this result it can be established that $\hat{\boldsymbol{\Sigma}}$ is a biased estimator, as was the maximum likelihood estimator of $\boldsymbol{\Sigma}$ in the single-population case; the corresponding unbiased estimator of $\boldsymbol{\Sigma}$ is

$$\begin{aligned}\mathbf{W} &= \frac{1}{n_1 + n_2 - 2}(\mathbf{C}_1 + \mathbf{C}_2) \\ &= \frac{1}{n_1 + n_2 - 2}\left\{\sum_{i=1}^{n_1}(\boldsymbol{x}_i - \bar{\boldsymbol{x}})(\boldsymbol{x}_i - \bar{\boldsymbol{x}})' + \sum_{i=1}^{n_2}(\boldsymbol{y}_i - \bar{\boldsymbol{y}})(\boldsymbol{y}_i - \bar{\boldsymbol{y}})'\right\}. \end{aligned} \tag{12.2}$$

This matrix $\mathbf{W}$ is the within-groups covariance matrix already encountered in Chapter 11, but restricted here to the special case of two groups. It is often referred to as the *pooled* covariance matrix, and readers will of course recognize it as the multivariate analogue of the pooled estimate of variance in two-sample univariate problems (in which population variances are assumed equal). Using this matrix in conjunction with the distributional result of Section 8.1 enables a confidence region for $\boldsymbol{\mu}_1 - \boldsymbol{\mu}_2$ to be derived as follows. Since the two population dispersion matrices are equal, we see from above that $\boldsymbol{z} = (\bar{\boldsymbol{x}} - \bar{\boldsymbol{y}}) - (\boldsymbol{\mu}_1 - \boldsymbol{\mu}_2)$ has a $\mathrm{N}\left\{\boldsymbol{0}, \left(\frac{1}{n_1} + \frac{1}{n_2}\right)\boldsymbol{\Sigma}\right\}$ distribution. Also from above, $(n_1 + n_2 - 2)\mathbf{W}$ has a $\mathrm{W}(n_1 + n_2 - 2, \boldsymbol{\Sigma})$ distribution. Finally, result (3) in Section 8.1 shows that $\bar{\boldsymbol{x}}$ is independent of $\mathbf{C}_1$, and $\bar{\boldsymbol{y}}$ is independent of $\mathbf{C}_2$. Since the $\boldsymbol{x}_i$ and $\mathbf{y}_i$ are mutually independent it is also clear that $\bar{\boldsymbol{x}}$ is independent of $\mathbf{C}_2$, and $\bar{\boldsymbol{y}}$ is independent of $\mathbf{C}_1$. Thus $\boldsymbol{z}$ is independent of $\mathbf{W}$. Hence we can apply the distributional result of Section 8.1 and deduce that $\left(\frac{n_1 n_2}{n_1 + n_2}\right)\boldsymbol{z}'\{(n_1 + n_2 - 2)\mathbf{W}\}^{-1}\boldsymbol{z}$ is distributed as $\frac{p}{n_1 + n_2 - p - 1}$ times an F variable on p and $(n_1 + n_2 - p - 1)$ degrees of freedom. Let $F^{\alpha}_{p,n_1+n_2-p-1}$ be the $100(1-\alpha)$ per cent point of the F_{p,n_1+n_2-p-1} distribution, i.e. the value exceeded by 100α per cent of this distribution. Then

$$\mathrm{pr}\left[\{(\bar{\boldsymbol{x}} - \bar{\boldsymbol{y}}) - (\boldsymbol{\mu}_1 - \boldsymbol{\mu}_2)\}'\mathbf{W}^{-1}\{(\bar{\boldsymbol{x}} - \bar{\boldsymbol{y}}) - (\boldsymbol{\mu}_1 - \boldsymbol{\mu}_2)\} < \frac{p(n_1 + n_2)(n_1 + n_2 - 2)}{n_1 n_2 (n_1 + n_2 - p - 1)} F^{\alpha}_{p,n_1+n_2-p-1}\right] = 1 - \alpha. \quad (12.3)$$

Thus

$$\{(\bar{\boldsymbol{x}} - \bar{\boldsymbol{y}}) - \boldsymbol{\nu}\}'\mathbf{W}^{-1}\{(\bar{\boldsymbol{x}} - \bar{\boldsymbol{y}}) - \boldsymbol{\nu}\} < \frac{p(n_1 + n_2)(n_1 + n_2 - 2)}{n_1 n_2 (n_1 + n_2 - p - 1)} F^{\alpha}_{p,n_1+n_2-p-1}$$

defines the interior of a hyperellipsoid in the population space, centred at $(\bar{\boldsymbol{x}} - \bar{\boldsymbol{y}})$, which has probability $1 - \alpha$ of containing the difference in population means $\boldsymbol{\nu} = \boldsymbol{\mu}_1 - \boldsymbol{\mu}_2$. This hyperellipsoid defines the $100(1-\alpha)$ per cent confidence region for $\boldsymbol{\mu}_1 - \boldsymbol{\mu}_2$.

While it is the author's view that estimation, as exemplified by the above results, should provide the main approach in multivariate problems, situations nevertheless arise where a formal test of some hypothesis is demanded. We therefore conclude this section by providing suitable tests of the two most common hypotheses in two-group multivariate situations, and illustrate all the ideas by means of an example. The first hypothesis is one of equality of dispersion matrices in two multivariate normal populations, irrespective of the values of their means. This hypothesis may need to be tested prior to the common dispersion assumption made in the estimation models above. The second hypothesis is then of equality of means in two multivariate normal populations that have equal (but unspecified) dispersion matrices.

1. Suppose that $\boldsymbol{x}_1, \boldsymbol{x}_2, \ldots, \boldsymbol{x}_{n_1}$ are independent observations from a

$N(\boldsymbol{\mu}_1, \boldsymbol{\Sigma}_1)$ distribution, while $\boldsymbol{y}_1, \boldsymbol{y}_2, \ldots, \boldsymbol{y}_{n_2}$ are independent observations from a $N(\boldsymbol{\mu}_2, \boldsymbol{\Sigma}_2)$ distribution, and it is desired to test the null hypothesis $H_0: \boldsymbol{\Sigma}_1 = \boldsymbol{\Sigma}_2 = \boldsymbol{\Sigma}$ against the alternative $H_1: \boldsymbol{\Sigma}_1 \neq \boldsymbol{\Sigma}_2$. No simple union–intersection test seems to be available, but the likelihood-ratio test does have a simple form. Write $\hat{\boldsymbol{\Sigma}}_1$ and $\hat{\boldsymbol{\Sigma}}_2$ for the maximum likelihood estimates of the separate dispersion matrices, and $\hat{\boldsymbol{\Sigma}}$ for the maximum likelihood estimate of the common dispersion matrix under H_0. From above, employing standard terminology for the sum-of-squares and -products matrices, we have $\hat{\boldsymbol{\Sigma}}_1 = \frac{1}{n_1}\mathbf{C}_1$; $\hat{\boldsymbol{\Sigma}}_2 = \frac{1}{n_2}\mathbf{C}_2$; $\hat{\boldsymbol{\Sigma}} = \frac{1}{n_1+n_2}(\mathbf{C}_1 + \mathbf{C}_2)$. Then the likelihood ratio test statistic λ is given by

$$-2\log_e \lambda = (n_1 + n_2)\log_e |\hat{\boldsymbol{\Sigma}}| - n_1 \log_e |\hat{\boldsymbol{\Sigma}}_1| - n_2 \log_e |\hat{\boldsymbol{\Sigma}}_2|. \quad (12.4)$$

Using the general asymptotic result given in Section 8.3, this quantity has an asymptotic chi-squared distribution with $\frac{1}{2}p(p+1)$ degrees of freedom if H_0 is true. When n_1 and n_2 are small, a modification due to Box (1949) will ensure a good approximation to this chi-squared distribution. If $\mathbf{S}_1 = \frac{1}{(n_1 - 1)}\mathbf{C}_1$ and $\mathbf{S}_2 = \frac{1}{(n_2-1)}\mathbf{C}_2$ are the unbiased estimates of $\boldsymbol{\Sigma}_1$ and $\boldsymbol{\Sigma}_2$ respectively, Box's criterion is given by

$$M = k\{(n_1 + n_2 - 2)\log_e |\mathbf{W}| - (n_1 - 1)\log_e |\mathbf{S}_1| - (n_2 - 1)\log_e |\mathbf{S}_2|\}$$

where

$$k = 1 - \frac{2p^2 + 3p - 1}{6(p+1)}\left\{\frac{1}{(n_1 - 1)} + \frac{1}{(n_2 - 1)} - \frac{1}{(n_1 + n_2 - 2)}\right\}.$$

2. Suppose that $\boldsymbol{x}_1, \boldsymbol{x}_2, \ldots, \boldsymbol{x}_{n_1}$ are independent observations from a $N(\boldsymbol{\mu}_1, \boldsymbol{\Sigma})$ distribution while $\boldsymbol{y}_1, \boldsymbol{y}_2, \ldots, \boldsymbol{y}_{n_2}$ are independent observations from a $N(\boldsymbol{\mu}_2, \boldsymbol{\Sigma})$ distribution, and it is desired to test the null hypothesis $H_0: \boldsymbol{\mu}_1 = \boldsymbol{\mu}_2$ against the alternative $H_1: \boldsymbol{\mu}_1 \neq \boldsymbol{\mu}_2$. Then both the union–intersection and the likelihood ratio principles lead to the test statistic

$$T^2 = \left(\frac{n_1 n_2}{n_1 + n_2}\right)(\bar{\boldsymbol{x}} - \bar{\boldsymbol{y}})'\mathbf{W}^{-1}(\bar{\boldsymbol{x}} - \bar{\boldsymbol{y}}) \quad (12.5)$$

where $\mathbf{W} = \frac{1}{n_1 + n_2 - 2}(\mathbf{C}_1 + \mathbf{C}_2)$ as before. The quantity T^2 of (12.5) is known as Hotelling's two-sample T^2 statistic, and under the null hypothesis $\frac{(n_1 + n_2 - p - 1)}{(n_1 + n_2 - 2)p}T^2$ follows an F distribution on p and $n_1 + n_2 - p - 1$ degrees of freedom. Thus we can obtain the critical value for an exact size α test from tables of the F distribution.

We now illustrate these various procedures by means of a simple example. Measurements of cranial length and cranial breadth gave the following summary statistics (Seal 1964, p. 106) for samples of male and female frogs:

$$\text{Male sample: } n_1 = 14; \quad \bar{\mathbf{x}} = \begin{pmatrix} 21.821 \\ 22.843 \end{pmatrix}; \quad \mathbf{C}_1 = \begin{pmatrix} 240.226 & 248.234 \\ 248.234 & 269.822 \end{pmatrix}$$

$$\text{Female sample: } n_2 = 35; \quad \bar{\mathbf{y}} = \begin{pmatrix} 22.860 \\ 24.397 \end{pmatrix}; \quad \mathbf{C}_2 = \begin{pmatrix} 601.230 & 689.850 \\ 689.850 & 829.850 \end{pmatrix}.$$

To test the hypothesis of equality of dispersion matrices in the male and female populations, we require:

$$\hat{\mathbf{\Sigma}}_1 = \frac{1}{14}\mathbf{C}_1 \quad = \begin{pmatrix} 17.159 & 17.731 \\ 17.731 & 19.273 \end{pmatrix}, \quad \text{so } |\hat{\mathbf{\Sigma}}_1| = 16.3170;$$

$$\hat{\mathbf{\Sigma}}_2 = \frac{1}{35}\mathbf{C}_2 \quad = \begin{pmatrix} 17.178 & 19.710 \\ 19.710 & 23.710 \end{pmatrix}, \quad \text{so } |\hat{\mathbf{\Sigma}}_2| = 18.8063;$$

$$\hat{\mathbf{\Sigma}} = \frac{1}{49}(\mathbf{C}_1 + \mathbf{C}_2) = \begin{pmatrix} 17.173 & 19.145 \\ 19.145 & 22.442 \end{pmatrix}, \quad \text{so } |\hat{\mathbf{\Sigma}}| = 18.8654.$$

From (12.4) we thus find that

$$-2 \log_e \lambda = 49 \log_e(18.8654) - 14 \log_e(16.317) - 35 \log_e(18.8063).$$

Hence $-2 \log_e \lambda = 2.142$. The asymptotic null distribution of $-2 \log_e \lambda$ is chi-squared on $\frac{1}{2} \times 2 \times 3 = 3$ degrees of freedom, and the observed value is less than the expectation of this distribution. Thus, even though sample sizes are relatively small, we may safely assume a common dispersion matrix in the two frog populations. (The reader may like to verify that the value of Box's criterion for these data is 1.889.)

Next we might wish to test equality of mean vectors in the two populations. For this test we require:

$$(\bar{\mathbf{x}} - \bar{\mathbf{y}}) = \begin{pmatrix} -1.039 \\ -1.554 \end{pmatrix}$$

and

$$\mathbf{W} = \frac{1}{47}(\mathbf{C}_1 + \mathbf{C}_2) = \begin{pmatrix} 17.903 & 19.959 \\ 19.959 & 23.397 \end{pmatrix}.$$

Hence

$$\mathbf{W}^{-1} = \begin{pmatrix} 1.1405 & -0.9729 \\ -0.9729 & 0.8727 \end{pmatrix}.$$

Thus

$$T^2 = \left(\frac{35 \times 14}{49}\right)(-1.039, -1.554)\begin{pmatrix} 1.1405 & -0.9729 \\ -0.9729 & 0.8727 \end{pmatrix}\begin{pmatrix} -1.039 \\ -1.554 \end{pmatrix}$$

$$= 10(0.3269, -0.3453)\begin{pmatrix} -1.039 \\ -1.554 \end{pmatrix} = 10 \times 0.1969 = 1.969.$$

Hence

$$\frac{(n_1 + n_2 - p - 1)}{(n_1 + n_2 - 2)p} T^2 = \frac{46}{2 \times 47} \times 1.969 = 0.964,$$

which is clearly not significant when referred to the F distribution on 2 and 46 degrees of freedom. Thus the mean vectors of male and female frog populations are not significantly different. Hence, the two hypothesis tests suggest that male and female frog populations can be treated as a single population, as regards measurements of cranial length and breadth. If the bivariate confidence region given by (12.3) were calculated, the point (0, 0) would lie well within its boundary.

12.1.2 The multinomial model

The multinomial model is an appropriate model for the analysis of discrete multivariate data, and such analysis has already been discussed in Chapter 10. Although this chapter was located in the section of the book dealing with analysis of data from a single population, the methods described there are perfectly general and can be applied directly to the two-population (or even multi-population) case. In fact, some of the examples shown in Section 10.1 were already of this type. For instance, one of the variables in Example 10.1 denoted 'group membership' of the mice: whether a mouse belonged to the treated or to the control group. Testing for presence of interaction between x_1 and x_2 is thus equivalent to testing for a difference between the treated and control 'populations' of mice in their tumour incidence. Similarly, interaction between x_1 and x_2 in Example 10.2 can be viewed as evidence of difference between 'carriers' and 'non-carriers' in size of tonsils.

Thus if p discrete variables $x_1, x_2, \ldots, x_p$ are observed on each individual, and individuals are sampled from two populations, then a test of difference between populations can be conducted by indicating an individual's 'population membership' as an extra binary variable x_{p+1}, and by employing the analysis of Chapter 10 on the contingency table formed from all $p + 1$ variables. In this case the x_{p+1} margin is fixed, so any log-linear model fitted to the data must include the 'population' term (α_{p+1}, say). Clearly, the interaction terms of most interest will be those between x_{p+1} and each of the other variables, as presence of such interactions implies differences between the populations. Apart from these considerations, however, analysis is exactly as described in Chapter 10.

12.1.3 The location model

The location model was introduced in Section 10.3 as a suitable model for use in one-population situations when some of the variables measured on each sample member are discrete and the remainder are continuous. This model was extended to two-population situations by Afifi and Elashoff (1969) for the case of mixed dichotomous and continuous variables, and by Krzanowski (1980) for more general discrete/continuous mixtures. Since the latter extension was with particular reference to discrimination and classification, however, it will be deferred until later in the chapter, and we content ourselves here with a brief summary of the former extension alone.

Afifi and Elashoff (1969) adopted the location model as specified in eqns (10.15) and (10.16), allowing different values for the continuous variable location means $\boldsymbol{\mu}_i^{(m)}$ and multinomial probabilities p_{im} ($m = 1, \ldots, s$) in the two populations ($i = 1, 2$) but constraining the conditional continuous variable dispersion matrix $\boldsymbol{\Sigma}$ to be constant over all locations and over both populations. This latter constraint is no more restrictive than the common dispersion matrix assumption of Section 12.1.1, or the corresponding dispersion assumption made in the more complex linear model structures to be discussed in Chapter 13, and the model appears to be widely applicable in practice. Afifi and Elashoff cited three medical studies in which the model provided acceptable fit to the data. They then considered testing the null hypothesis of equality of all corresponding parameters between the two populations against the general alternative that at least one parameter differed between populations, i.e. $H_0 : \boldsymbol{\mu}_1^{(m)} = \boldsymbol{\mu}_2^{(m)}$ and $p_{1m} = p_{2m}$ for all $m = 1, \ldots, s$ versus H_1: not H_0. They first showed that if the dichotomous variables are scored 0 or 1, and their discrete nature is ignored, then the two-sample Hotelling T^2 test of (12.5) is not a consistent test. They then went on to derive an information-theoretic and also a likelihood-ratio test for the given hypothesis, but neither test has a simple sampling distribution. The interested reader is referred to their paper for full details, and also for references to applications of their test procedures.

12.1.4 Discrimination and classification

In many of the situations considered above, the null hypothesis is a statement about equality of two populations, or about some aspect of them. Generally speaking this is a rather uninteresting hypothesis, because the mere presence of two populations suggests that there must be *some* difference between them (especially if many variables have been measured). A more fruitful endeavour than the testing for equality of two populations might therefore be a search for the best way in which the populations can be distinguished. In statistics this type of analysis is known as *discriminant analysis,* and a *discriminant function* is some combination of the observed variables that achieves this objective of population separation. The discriminant function may thus be used to describe and interpret the differences between the populations, which is a much more positive approach than a test of a null hypothesis.

Sometimes there is a much more specific, predictive, reason for wishing to describe the difference between two populations. Suppose that an individual $\boldsymbol{x}$ is observed, and we wish to decide from which of two populations $\boldsymbol{x}$ has come. Knowledge of the differences between the two populations will then be necessary for the construction of a suitable *classification* (or *allocation*) *rule* for $\boldsymbol{x}$. The 'best' such rule will be the one that leads to the smallest probability of misclassifying $\boldsymbol{x}$, and in statistical terms this is equivalent to the rule that leads to the smallest *error rate* of all future allocations of individuals from these populations.

That there is a very close connection between discrimination and classification should already be evident from the above brief introduction. A discriminant function for a pair of populations can obviously double as a classification rule for allocating an individual to one of the two populations, and the better the discrimination between the populations the smaller will be the error rate given by the corresponding classification. However, it is important to recognize the difference in objectives: discriminant functions aim to maximize separation between groups of (available) individuals, while classification rules aim to minimize the misclassification rate over all possible (future) allocations, and two different functions may be necessary to achieve these objectives (even when other conditions do not alter).

What is clear, however, is that construction of such functions is a necessary activity in many branches of science; even when such construction is not explicitly demanded, it will generally provide a much more positive analysis than will mere hypothesis testing. We therefore examine the topic in greater detail in the following sections. There has been much vigorous research in this area since the pioneering work of Fisher (1936), and it is not possible to cover this work exhaustively here. Our objective is to provide an overview of the main ideas, and to supply sufficient working details to enable the user to put the ideas into practice. For more comprehensive treatment, the reader is referred to specialist texts (e.g. Lachenbruch (1975), Goldstein and Dillon (1978), and Hand (1981)).

12.2 Classification rules based on probability models

12.2.1 Fundamental principles

Let us suppose it is required to allocate an individual to one of two populations, on the basis of a set of p measurements that have been made on it. For example, we might wish to deduce from which of two Bronze Age populations an excavated skeleton came, on the basis of p skull measurements; or to which of two vole populations an unidentified individual belongs, on the basis of p presence/absence characteristics; or from which of two literary schools a newly discovered manuscript might have come, on the basis of p stylistic variables. Probabilistic classification rules are founded on the premise that a large number of individuals will need to be classified in the future, and hence the classification rule should be chosen in such a way as to minimize the expected consequences of mistakes made in this series of allocations. Mistakes will arise because virtually any of the possible sets of p values that constitute the p-dimensional sample space R could plausibly be observations from *either* population (although, in general, each possible observation vector is more likely to come from one of the two populations than from the other). Thus probability models are central not only to a description of the populations but also to an assessment of the performance of the classification rule.

To fix ideas, we suppose that $\boldsymbol{v}$ denotes a p-component random vector of observations made on any individual, $\boldsymbol{v}_0$ denotes a particular observed value of $\boldsymbol{v}$, and π_1, π_2 denote the two populations involved in the problem. The basic assumption is that $\boldsymbol{v}$ has different probability distributions in π_1 and π_2 (as otherwise the two populations cannot be distinguished). Let the probability density of $\boldsymbol{v}$ be $\mathrm{f}_1(\boldsymbol{v})$ in π_1, and $\mathrm{f}_2(\boldsymbol{v})$ in π_2. Since we envisage a possibly infinite sequence of individuals that will require classification, we must be prepared for *any* point in the sample space R to occur as the value of an individual. A classification rule can therefore be defined by a partition of R into two exhaustive and mutually exclusive regions R_1 and R_2, together with the decision rule that assigns to π_1 individuals falling in R_1, and to π_2 individuals falling in R_2.

(Note that the above procedure corresponds to *forced* classification, whereby we require a definite decision about population membership to be made for each individual that is considered. It may be felt advantageous to divide R instead into three regions R_1, R_2, and R_3, such that individuals are allocated to π_1 if they fall in R_1, to π_2 if they fall in R_2, and are left unclassified if they fall in R_3. R_3 is then a region of 'doubt', and further information is sought on individuals that fall in it before a final allocation is made. However, if this further information is in the form of q extra variables, then we can view both stages together as forced classification on $(p+q)$ variables. Hence we will restrict attention throughout to forced classification alone.)

How should the two regions R_1 and R_2 be chosen? The simplest intuitive argument suggests that $\boldsymbol{v}_0$ should be allocated to π_1 whenever it has greater probability of coming from π_1 than from π_2, to π_2 whenever these probabilities are reversed, and arbitrarily to π_1 or π_2 whenever these probabilities are equal. Since the decision is arbitrary in the last situation, we can choose one of the two populations and always allocate to this population whenever the probabilities are equal. Adopting this argument yields R_1 as the set of points $\boldsymbol{v}$ for which $\mathrm{f}_1(\boldsymbol{v}) > \mathrm{f}_2(\boldsymbol{v})$, and R_2 as the set of points for which $\mathrm{f}_1(\boldsymbol{v}) \leqslant \mathrm{f}_2(\boldsymbol{v})$. It is easy to see that R_1 and R_2 chosen in this way are mutually exclusive and hence form a disjoint partition of the sample space R. In set theory terms, these conditions are given by $R_1 \cap R_2 = \phi$ and $R = R_1 \cup R_2$. Rewriting the above slightly, we thus arrive at the classification rule:

$$\text{Allocate } \boldsymbol{v} \text{ to } \pi_1 \text{ if } \mathrm{f}_1(\boldsymbol{v})/\mathrm{f}_2(\boldsymbol{v}) > 1,$$
$$\text{and to } \pi_2 \text{ if } \mathrm{f}_1(\boldsymbol{v})/\mathrm{f}_2(\boldsymbol{v}) \leqslant 1. \qquad (12.6)$$

The above allocation rule is often termed the *likelihood ratio* rule. While it is intuitively reasonable, however, it nevertheless fails to take account of several factors that may be important in practice. These factors are: differential prior probabilities of observing individuals from the two populations, and differential costs incurred by misclassification. The influence of these factors can best be illustrated by considering a value $\boldsymbol{v}_0$ at which $\mathrm{f}_2(\boldsymbol{v}_0)$ is less than, but very nearly equal to, $\mathrm{f}_1(\boldsymbol{v}_0)$. Using (12.6), therefore, $\boldsymbol{v}_0$ would be allocated to π_1. Suppose it is known, however, that individuals from π_1 are observed very rarely in practice while individuals from π_2 are observed quite frequently. For example, π_1 might denote the population of individuals suffering from active tuberculosis, while π_2 might denote the population of individuals suffering from bronchitis. Despite the fact that $\boldsymbol{v}_0$ is more likely to occur in population π_1 than in π_2, our prior knowledge of the incidences of π_1 and π_2 would persuade us to ignore (12.6) and allocate $\boldsymbol{v}_0$ to π_2. This is because of the relative closeness of $\mathrm{f}_1(\boldsymbol{v}_0)$ and $\mathrm{f}_2(\boldsymbol{v}_0)$. The probability density $\mathrm{f}_1(\boldsymbol{v}_0)$ would need to be considerably in excess of $\mathrm{f}_2(\boldsymbol{v}_0)$ before the evidence became sufficiently persuasive for us to disregard the prior information and allocate $\boldsymbol{v}_0$ to π_1. Similar considerations would apply if there were grossly disparate costs incurred in making a mistake in classification. For example, suppose π_1 is the population of individuals that require medical treatment for jaundice, while π_2 is the population of individuals that require surgical treatment. Viewed from the patient's standpoint, being allocated to π_1 when he or she should be allocated to π_2 is a very serious mistake (as this incorrect classification could easily result in death). The reverse misclassification runs some risk caused by unnecessary surgery, but has far greater chance of eventual correct diagnosis, treatment, and recovery. Thus the 'cost' of misallocating to π_1 is far higher than the 'cost' of misclassifying to π_2. With $f_1(\boldsymbol{v}_0)$ only slightly greater than $f_2(\boldsymbol{v}_0)$, therefore, it would make sense to 'play safe' by ignoring (12.6) and classifying to π_2. In this case, $f_1(\boldsymbol{v}_0)$ would again need to be considerably larger than $f_2(\boldsymbol{v}_0)$ before allocation to π_1 was seriously considered.

In both of these cases, the allocation rule being suggested is of the form:

$$\text{Allocate } \boldsymbol{v} \text{ to } \pi_1 \text{ if } f_1(\boldsymbol{v})/f_2(\boldsymbol{v}) > k$$
$$\text{and to } \pi_2 \text{ if } f_1(\boldsymbol{v})/f_2(\boldsymbol{v}) \leq k \qquad (12.7)$$

where k is greater than 1. Reversing the populations of the two examples would lead to the same form of rule but now with k less than 1. Can such a rule be justified theoretically, and if it can, then how should k be chosen? This is within the province of statistical *decision theory*; both of these questions can be answered by electing to choose the allocation rule that will minimize the expected cost incurred by future misallocations.

In order to derive the rule formally, we must quantify the various factors involved. First, we will assume that q_1 is the prior probability that an observed value of $\boldsymbol{v}$ is from π_1, and that q_2 is the corresponding prior probability from π_2 (with $q_1 + q_2 = 1$). Secondly, we will assume that $c(1 \mid 2)$ is the cost incurred whenever an individual from π_2 is incorrectly allocated to π_1, and that $c(2 \mid 1)$ is the cost incurred whenever an individual from π_1 is incorrectly allocated to π_2. The prior probabilities may be known from extensive past history or ancillary knowledge of the populations considered. For example, if a screening test is being devised for detecting the condition cystic fibrosis in children, the two populations to be distinguished are the population of children suffering from cystic fibrosis and the population of 'normal' children not suffering from the condition. It is known that the incidence rate of the disease in Great Britain is approximately 1 in every 2500 births, so appropriate values of the prior probabilities would be $q_1 = 1/2500$ and $q_2 = 1 - q_1$. On the other hand, the costs $c(1 \mid 2)$ and $c(2 \mid 1)$ present more of a problem. In some situations the costs can be evaluated simply in monetary terms, but often other considerations must be brought into play. For example, how does one evaluate the cost of wrong medical treatment of a patient, with all the associated dangers to, and hardships suffered by, the individual? One possibility here is to repeat the statistical analysis for a *range* of possible cost values, from the relatively small to the severe, and to investigate the stability of the derived solution. An example of such a procedure is presented in Section 12.5. For the present, however, we assume that the q_i and the $c(i \mid j)$ are fixed and known.

Recollect from Section 7.1 that if $\boldsymbol{v}$ has probability density function $f(\boldsymbol{v})$, then the probability that an observed value falls in a region ω of the sample space is $\int_\omega f(\boldsymbol{v})\,d\boldsymbol{v}$. Now if $\boldsymbol{v}$ comes from population π_1, then it has probability density function $f_1(\boldsymbol{v})$. It will be misclassified if it is allocated to π_2, and this will happen if it falls in region R_2 of the sample space. Given that $\boldsymbol{v}$ comes from π_1, therefore, the probability that it is misallocated is $\int_{R_2} f_1(\boldsymbol{v})\,d\boldsymbol{v}$. But $\boldsymbol{v}$ will come from π_1 with probability q_1, so the probability that $\boldsymbol{v}$ comes from π_1 *and* is misallocated is $p(2 \mid 1) = q_1 \int_{R_2} f_1(\boldsymbol{v})\,d\boldsymbol{v}$. By similar argument, the probability that $\boldsymbol{v}$ comes from π_2 and is misallocated is $p(1 \mid 2) = q_2 \int_{R_1} f_2(\boldsymbol{v})\,d\boldsymbol{v}$. The costs associated with each of these mistakes are $c(2 \mid 1)$ and $c(1 \mid 2)$ respectively, so the expected cost due to misallocation is given by

$$\begin{aligned} C &= c(2 \mid 1)p(2 \mid 1) + c(1 \mid 2)p(1 \mid 2) \\ &= c(2 \mid 1)q_1 \int_{R_2} f_1(\boldsymbol{v})\,d\boldsymbol{v} + c(1 \mid 2)q_2 \int_{R_1} f_2(\boldsymbol{v})\,d\boldsymbol{v}. \end{aligned} \tag{12.8}$$

The best allocation rule is the one that yields minimum expected cost due to misallocation, and this rule will be obtained by finding the regions R_1 and R_2 minimizing C in (12.8).

Since R is the complete sample space, $\int_R f_1(\boldsymbol{v})\,d\boldsymbol{v} = \int_R f_2(\boldsymbol{v})\,d\boldsymbol{v} = 1$ (see Section 7.1). But $R = R_1 \cup R_2$ and $R_1 \cap R_2 = \phi$ from above, so that $\int_R f_1(\boldsymbol{v})\,d\boldsymbol{v} = \int_{R_1} f_1(\boldsymbol{v})\,d\boldsymbol{v} + \int_{R_2} f_1(\boldsymbol{v})\,d\boldsymbol{v}$ (since the integral of a function over a region must equal the sum of the integrals of the function over any disjoint subsets into which the region can be partitioned). Hence $\int_{R_1} f_1(\boldsymbol{v})\,d\boldsymbol{v} + \int_{R_2} f_1(\boldsymbol{v})\,d\boldsymbol{v} = 1$, or $\int_{R_2} f_1(\boldsymbol{v})\,d\boldsymbol{v} = 1 - \int_{R_1} f_1(\boldsymbol{v})\,d\boldsymbol{v}$. Substituting for $\int_{R_2} f_1(\boldsymbol{v})\,d\boldsymbol{v}$ in (12.8) yields

$$\begin{aligned} C &= c(2 \mid 1)q_1 - c(2 \mid 1)q_1 \int_{R_1} f_1(\boldsymbol{v})\,d\boldsymbol{v} + c(1 \mid 2)q_2 \int_{R_1} f_2(\boldsymbol{v})\,d\boldsymbol{v} \\ &= c(2 \mid 1)q_1 + \int_{R_1} \{c(1 \mid 2)q_2 f_2(\boldsymbol{v}) - c(2 \mid 1)q_1 f_1(\boldsymbol{v})\}\,d\boldsymbol{v}. \end{aligned}$$

It is required to minimize C. From the definition of integration, the integral above can be (loosely) viewed as the sum of contributions to the expression in curly brackets from all possible points $\boldsymbol{v}$ in the region R_1. Since $c(2 \mid 1)q_1$ is a constant, therefore, C will be minimized by choosing the region R_1 to be the set of all those points, and only those points, that give a negative contribution to the expression $c(1 \mid 2)q_2 f_2(\boldsymbol{v}) - c(2 \mid 1)q_1 f_1(\boldsymbol{v})$. This is because, with this choice, the largest possible amount will be subtracted from $c(2 \mid 1)q_1$ to yield C. Hence the optimal decision-theoretic rule is associated with the region R_1 composed of all those points $\boldsymbol{v}$ for which $c(1 \mid 2)q_2 f_2(\boldsymbol{v}) - c(2 \mid 1)q_1 f_1(\boldsymbol{v}) < 0$, i.e. for which $f_1(\boldsymbol{v})/f_2(\boldsymbol{v}) > c(1 \mid 2)q_2/c(2 \mid 1)q_1$, while the remaining points in the sample space constitute the region R_2. The optimal rule is thus:

$$\text{Allocate } \boldsymbol{v} \text{ to } \pi_1 \text{ if } f_1(\boldsymbol{v})/f_2(\boldsymbol{v}) > c(1 \mid 2)q_2/c(2 \mid 1)q_1,$$
$$\text{and to } \pi_2 \text{ if } f_1(\boldsymbol{v})/f_2(\boldsymbol{v}) \leq c(1 \mid 2)q_2/c(2 \mid 1)q_1. \qquad (12.9)$$

This argument justifies theoretically a rule of the form (12.7), and establishes the appropriate value of k therein. Within the scope of general decision theory this rule is in fact a *Bayes procedure,* and is sometimes referred to by this name.

It has been remarked above that assigning values to $c(1 \mid 2)$ and $c(2 \mid 1)$ is sometimes very difficult. If this is the case in practice, and there is no reason to penalize one type of mistake more than the other, then it is convenient to set $c(1 \mid 2) = c(2 \mid 1)$. There is then no need to set an actual value for this cost, because reference to (12.9) shows that the rule now becomes:

$$\text{Allocate } \boldsymbol{v} \text{ to } \pi_1 \text{ if } f_1(\boldsymbol{v})/f_2(\boldsymbol{v}) > q_2/q_1,$$
$$\text{and to } \pi_2 \text{ if } f_1(\boldsymbol{v})/f_2(\boldsymbol{v}) \leq q_2/q_1. \qquad (12.10)$$

This rule could have been derived equivalently by minimizing P, where P is given by C of (12.8) but with $c(2 \mid 1)$ and $c(1 \mid 2)$ removed. Thus P is the *total probability of misallocation* of an individual, given by

$$P = q_1 \int_{R_2} f_1(\boldsymbol{v})\, d\boldsymbol{v} + q_2 \int_{R_1} f_2(\boldsymbol{v})\, d\boldsymbol{v} \qquad (12.11)$$

and the rule (12.10) is the one which minimizes this total probability of misallocation. Note that if q_1 and q_2 are also set equal, then (12.10) reduces to the likelihood ratio rule (12.6). Thus (12.9), (12.10), and (12.6) form a nested set of allocation rules, and are all of the common form (12.7). The most useful is arguably (12.10), as it does not require any difficult questions regarding costs due to misallocation to be resolved yet it permits differential prior probabilities of population membership.

Rule (12.10) has a further attractive property, in that it is equivalent to the allocation rule derived by maximizing the posterior probability of population membership. Suppose, as before, that $f_1(\boldsymbol{v})$, $f_2(\boldsymbol{v})$ are the probability densities of $\boldsymbol{v}$ in π_1, π_2 respectively, and that q_1, q_2 are the prior probabilities of an individual coming from π_1, π_2 respectively. Then the posterior probability that an individual with observed vector $\boldsymbol{v}_0$ comes from population π_i is obtained from Bayes' theorem as

$$q(\pi_i \mid \boldsymbol{v}_0) = q_i f_i(\boldsymbol{v}_0)/(q_1 f_1(\boldsymbol{v}_0) + q_2 f_2(\boldsymbol{v}_0)). \qquad (12.12)$$

(See Seber (1984) for justification.) A reasonable classification criterion would be to assign $\boldsymbol{v}$ to the population with the larger such posterior probability, that is:

$$\text{Allocate } \boldsymbol{v} \text{ to } \pi_1 \text{ if } q(\pi_1 \mid \boldsymbol{v}) > q(\pi_2 \mid \boldsymbol{v}),$$
$$\text{and to } \pi_2 \text{ if } q(\pi_1 \mid \boldsymbol{v}) \leq q(\pi_2 \mid \boldsymbol{v}). \qquad (12.13)$$

Substituting from (12.12) into (12.13) and tidying up the algebra yields (12.10).

Thus, most reasonable criteria for obtaining an allocation rule lead to one of the form (12.7), and the form (12.10) is the rule of this type that has the widest appeal. To operate any of these rules, it is necessary to find the likelihood ratio $f_1(\boldsymbol{v})/f_2(\boldsymbol{v})$, but the $f_i(\boldsymbol{v})$ are of course probability *models* for the populations, and they will not be known precisely except under the most unusual circumstances. In practice, the only information about the populations π_1 and π_2 which is usually available comes from two initial samples: $\boldsymbol{v}_1^{(1)}, \boldsymbol{v}_2^{(1)}, \ldots, \boldsymbol{v}_{n_1}^{(1)}$ known to come from π_1, and $\boldsymbol{v}_1^{(2)}, \boldsymbol{v}_2^{(2)}, \ldots, \boldsymbol{v}_{n_2}^{(2)}$ known to come from π_2. Such samples are often referred to as *training sets* in the literature. In the archaeological example at the beginning of this section the training sets would be samples of skull measurements made on previously identified members of the two Bronze Age populations; in the zoological example they would be samples of presence/absence characteristics on known species of voles; and in the literature example they would be values of the stylistic variables made on manuscripts of known provenance. In such circumstances, estimation is necessary in order to make any progress. Various different approaches have been proposed, each approach involving a different level of assumptions.

If no assumptions at all are made about the form of the densities $f_i(\boldsymbol{v})$, then these densities must be estimated from the training sets by some non-parametric method. A comprehensive bibliography of such methods is provided by Wertz and Schneider (1979), but detailed consideration is beyond the scope of the present volume. A recently popularized non-parametric method of density estimation is the kernel method. This method can be used to estimate $f_1(\boldsymbol{v})$ and $f_2(\boldsymbol{v})$ from the training sets, and the estimates can then be substituted into (12.6), (12.9), or (12.10) to obtain a usable allocation rule. Full details of such an approach are given in the book by Hand (1982).

Despite the recent interest in these non-parametric methods, however, the overwhelming number of applications of discrimination and classification still rely on various parametric assumptions. This is due to the relatively heavy demands made on the computer by the non-parametric methods, and to the lack of widely available software for doing the computations. Parametric methods, on the other hand, require much less computation, and are represented in all major statistical software packages. They are also the most powerful and efficient methods if the probability model is a suitable one for the data. We shall therefore concentrate almost exclusively on such methods in this chapter.

The standard parametric approach is to assume that the probability density functions $f_1(\boldsymbol{v})$ and $f_2(\boldsymbol{v})$ are given by familiar expressions which involve unknown parameters, $\boldsymbol{\theta}$, say. We shall consider the multivariate normal and multinomial forms in detail below, but in principle any parametric density functions could be used. Given parametric forms for $f_1(\boldsymbol{v})$ and $f_2(\boldsymbol{v})$, the ratio $f_1(\boldsymbol{v})/f_2(\boldsymbol{v})$ is also a function of the unknown parameters $\boldsymbol{\theta}$, and hence can be written $G(\boldsymbol{v}, \boldsymbol{\theta})$. The simplest way of proceeding is to estimate $\boldsymbol{\theta}$ from the training sets, and to replace $\boldsymbol{\theta}$ by this estimate $\hat{\boldsymbol{\theta}}$ wherever necessary. This procedure yields the *estimative* (or *plug-in*) allocation rule, using $G(\boldsymbol{v}, \hat{\boldsymbol{\theta}})$ in the left-hand side of (12.6), (12.9), or (12.10). Thus the estimative allocation rule is of the form:

$$\text{Allocate } \boldsymbol{v} \text{ to } \pi_1 \text{ if } G(\boldsymbol{v}, \hat{\boldsymbol{\theta}}) > k, \quad \text{and to } \pi_2 \text{ if } G(\boldsymbol{v}, \hat{\boldsymbol{\theta}}) \leq k, \tag{12.14}$$

where k is chosen suitably. This estimative rule is very easy to implement computationally in a large variety of circumstances, and hence is by far the most popular rule in practical applications. However, it is open to criticism in that $\hat{\boldsymbol{\theta}}$ is used *as if it were the true value* $\boldsymbol{\theta}$, whereas in many cases $\hat{\boldsymbol{\theta}}$ may be far from the true value $\boldsymbol{\theta}$ (particularly if the training sets are small). Consequently, the allocation rule that is used may be nowhere near the optimum rule.

An alternative approach to estimation of $G(\boldsymbol{v}, \boldsymbol{\theta})$ is provided by Bayesian methods. This approach requires a prior distribution $p(\boldsymbol{\theta})$ to be specified for the unknown parameters. The training sets are then used to derive the posterior distribution $p(\boldsymbol{\theta} \mid \boldsymbol{v}_i^{(1)}, \boldsymbol{v}_i^{(2)})$ of the unknown parameters, and the *predictive* allocation rule is formed by integrating $G(\boldsymbol{v}, \boldsymbol{\theta})$ over this distribution and substituting the result in the left-hand side of (12.6), (12.9), or (12.10). Thus the predictive rule is of the form:

$$\text{Allocate } \boldsymbol{v} \text{ to } \pi_1 \text{ if } \hat{G}(\boldsymbol{v}) > k, \quad \text{and to } \pi_2 \text{ if } \hat{G}(\boldsymbol{v}) \leq k \tag{12.15}$$

where $\hat{G}(\boldsymbol{v}) = \int G(\boldsymbol{v}, \boldsymbol{\theta}) p(\boldsymbol{\theta} \mid \boldsymbol{v}_i^{(1)}, \boldsymbol{v}_i^{(2)})\, d\boldsymbol{\theta}$, and k is chosen suitably. This predictive rule therefore averages $G(\boldsymbol{v}, \boldsymbol{\theta})$ over the *whole range* of possible values of $\boldsymbol{\theta}$, weighting each value $G(\boldsymbol{v}, \boldsymbol{\theta})$ by the posterior probability of the corresponding $\boldsymbol{\theta}$ value. It therefore avoids the previous criticisms of the estimative rule. Many statisticians, however, feel unable to make the required assumptions about the prior distribution $p(\boldsymbol{\theta})$, and the predictive allocation rule is not perhaps used as widely as it should be.

A third parametric approach sidesteps both of these problems, by ignoring the decision-theoretic considerations and adopting instead a hypothesis-testing standpoint. The allocation rule is given simply by the generalized likelihood ratio test statistic for the null hypothesis $H_0: \boldsymbol{v}, \boldsymbol{v}_1^{(1)}, \boldsymbol{v}_2^{(1)}, \ldots, \boldsymbol{v}_{n_1}^{(1)} \in \pi_1; \boldsymbol{v}_1^{(2)}, \boldsymbol{v}_2^{(2)}, \ldots, \boldsymbol{v}_{n_2}^{(2)} \in \pi_2$ versus the alternative $H_1: \boldsymbol{v}_1^{(1)}, \boldsymbol{v}_2^{(1)}, \ldots, \boldsymbol{v}_{n_1}^{(1)} \in \pi_1; \boldsymbol{v}, \boldsymbol{v}_1^{(2)}, \boldsymbol{v}_2^{(2)}, \ldots, \boldsymbol{v}_{n_2}^{(2)} \in \pi_2$. The resulting allocation rule has been termed the *hypothesis-testing* (or *testimative*) rule. It is a fully parametric rule, because parametric density functions must be specified to enable the likelihood to be formed. Also, in the more common situations in practice, its form has close affinity with that of the predictive rule. However, both hypothesis-testing and predictive rules run into a number of analytical and computational problems in all but the most common practical applications, and these problems have also contributed to the relative lack of usage of the rules. We shall therefore concentrate mainly on the estimative rules in our detailed consideration below of allocation rules for different data types, simply quoting relevant results from the other approaches and leaving more mathematical readers to fill in details from cited references. As before, we consider the multivariate normal model to be appropriate for quantitative data, the multinomial model for qualitative data, and the location model for mixed data.

12.2.2 The multivariate normal model

The most general form of the model is to assume that π_i is a multivariate normal population with mean $\boldsymbol{\mu}_i$ and dispersion matrix $\boldsymbol{\Sigma}_i$ for $i = 1, 2$. Thus $f_i(\boldsymbol{v}) = (2\pi)^{-p/2} |\boldsymbol{\Sigma}_i|^{-\frac{1}{2}} \exp\{-\frac{1}{2}(\boldsymbol{v} - \boldsymbol{\mu}_i)' \boldsymbol{\Sigma}_i^{-1} (\boldsymbol{v} - \boldsymbol{\mu}_i)\}$, so that (after a little bit of algebra) we obtain

$$f_1(\boldsymbol{v})/f_2(\boldsymbol{v}) = |\boldsymbol{\Sigma}_2|^{\frac{1}{2}} |\boldsymbol{\Sigma}_1|^{-\frac{1}{2}} \exp[-\tfrac{1}{2}\{\boldsymbol{v}'(\boldsymbol{\Sigma}_1^{-1} - \boldsymbol{\Sigma}_2^{-1})\boldsymbol{v} - 2\boldsymbol{v}'(\boldsymbol{\Sigma}_1^{-1}\boldsymbol{\mu}_1 - \boldsymbol{\Sigma}_2^{-1}\boldsymbol{\mu}_2) + \boldsymbol{\mu}_1'\boldsymbol{\Sigma}_1^{-1}\boldsymbol{\mu}_1 - \boldsymbol{\mu}_2'\boldsymbol{\Sigma}_2^{-1}\boldsymbol{\mu}_2\}].$$

Now an allocation rule of the form (12.7) is exactly equivalent to the rule:

$$\text{Allocate } \boldsymbol{v} \text{ to } \pi_1 \text{ if } \log_e\{f_1(\boldsymbol{v})/f_2(\boldsymbol{v})\} > \log_e k, \text{ and to } \pi_2 \text{ if } \log_e\{f_1(\boldsymbol{v})/f_2(\boldsymbol{v})\} \leq \log_e k. \quad (12.16)$$

Hence, on taking logarithms above, we find that the optimal allocation rule for this model is:

$$\text{Allocate } \boldsymbol{v} \text{ to } \pi_1 \text{ if } \tfrac{1}{2}\log_e\{|\boldsymbol{\Sigma}_2| \div |\boldsymbol{\Sigma}_1|\} - \tfrac{1}{2}\{\boldsymbol{v}'(\boldsymbol{\Sigma}_1^{-1} - \boldsymbol{\Sigma}_2^{-1})\boldsymbol{v} - 2\boldsymbol{v}'(\boldsymbol{\Sigma}_1^{-1}\boldsymbol{\mu}_1 - \boldsymbol{\Sigma}_2^{-1}\boldsymbol{\mu}_2) + \boldsymbol{\mu}_1'\boldsymbol{\Sigma}_1^{-1}\boldsymbol{\mu}_1 - \boldsymbol{\mu}_2'\boldsymbol{\Sigma}_2^{-1}\boldsymbol{\mu}_2\} > \log_e k, \text{ and otherwise to } \pi_2$$

or,

$$\text{Allocate } \boldsymbol{v} \text{ to } \pi_1 \text{ if } \mathrm{Q}(\boldsymbol{v}) > \log_e k, \text{ and otherwise to } \pi_2 \quad (12.17)$$

where $\mathrm{Q}(\boldsymbol{v})$ is the discriminant function $\frac{1}{2}\log_e\{|\boldsymbol{\Sigma}_2| \div |\boldsymbol{\Sigma}_1|\} - \frac{1}{2}(\boldsymbol{v}'(\boldsymbol{\Sigma}_1^{-1} - \boldsymbol{\Sigma}_2^{-1})\boldsymbol{v} - 2\boldsymbol{v}'(\boldsymbol{\Sigma}_1^{-1}\boldsymbol{\mu}_1 - \boldsymbol{\Sigma}_2^{-1}\boldsymbol{\mu}_2) + \boldsymbol{\mu}_1'\boldsymbol{\Sigma}_1^{-1}\boldsymbol{\mu}_1 - \boldsymbol{\mu}_2'\boldsymbol{\Sigma}_2^{-1}\boldsymbol{\mu}_2\}$. Since the terms in $\mathrm{Q}(\boldsymbol{v})$ include the quadratic form $\boldsymbol{v}'(\boldsymbol{\Sigma}_1^{-1} - \boldsymbol{\Sigma}_2^{-1})\boldsymbol{v}$, which will be a function of the squares of elements of $\boldsymbol{v}$ and cross-products between pairs of them, this discriminant function is known as the *quadratic discriminant function*. Given training sets from each of π_1 and π_2, the parameters $\boldsymbol{\mu}_1$, $\boldsymbol{\mu}_2$, $\boldsymbol{\Sigma}_1$, and $\boldsymbol{\Sigma}_2$ can be estimated by the means and covariance matrices of these training sets: $\bar{\boldsymbol{v}}_1$, $\bar{\boldsymbol{v}}_2$, $\mathbf{S}_1$, and $\mathbf{S}_2$. Replacing the unknown parameters in (12.17) by these estimates then yields an allocation rule based on the function $\hat{\mathrm{Q}}(\boldsymbol{v})$ which can be fully evaluated for any observation $\boldsymbol{v}$.

Although this rule is based on the most general multivariate normal model, it is not one that tends to be used very often in practical applications. The reason for this is that considerable simplification can be achieved by making one more assumption. We have already seen in this chapter that the presence of two different population dispersion matrices renders difficult the testing of hypotheses about the population mean vectors, and it has been argued that the assumption $\boldsymbol{\Sigma}_1 = \boldsymbol{\Sigma}_2 = \boldsymbol{\Sigma}$ is a reasonable one in many practical situations. Similar arguments can be brought to bear in the present case also (and the hypothesis of equality $\boldsymbol{\Sigma}_1$ and $\boldsymbol{\Sigma}_2$ can be tested by using (12.4) on the training sets, if doubt exists about the a priori justification of the assumption). The practical benefits of making this assumption are that the discriminant function and allocation rule become very simple indeed. For if $\boldsymbol{\Sigma}_1 = \boldsymbol{\Sigma}_2 = \boldsymbol{\Sigma}$, then

$$\mathrm{f}_i(\boldsymbol{v}) = (2\pi)^{-p/2}|\boldsymbol{\Sigma}|^{-\frac{1}{2}}\exp\{-\tfrac{1}{2}(\boldsymbol{v} - \boldsymbol{\mu}_i)'\boldsymbol{\Sigma}^{-1}(\boldsymbol{v} - \boldsymbol{\mu}_i)\} \qquad (i = 1, 2)$$

and

$$\mathrm{f}_1(\boldsymbol{v})/\mathrm{f}_2(\boldsymbol{v}) = \exp\{(\boldsymbol{\mu}_1 - \boldsymbol{\mu}_2)'\boldsymbol{\Sigma}^{-1}\boldsymbol{v} - \tfrac{1}{2}(\boldsymbol{\mu}_1 - \boldsymbol{\mu}_2)'\boldsymbol{\Sigma}^{-1}(\boldsymbol{\mu}_1 + \boldsymbol{\mu}_2)\}$$

so that the allocation rule (12.17) reduces to the rule:

$$\text{Allocate } \boldsymbol{v} \text{ to } \pi_1 \text{ if } \mathrm{L}(\boldsymbol{v}) > \log_e k, \text{ and otherwise to } \pi_2 \quad (12.18)$$

where $\mathrm{L}(\boldsymbol{v}) = (\boldsymbol{\mu}_1 - \boldsymbol{\mu}_2)'\boldsymbol{\Sigma}^{-1}\{\boldsymbol{v} - \frac{1}{2}(\boldsymbol{\mu}_1 + \boldsymbol{\mu}_2)\}$.

No quadratic terms now exist in the discriminant function $\mathrm{L}(\boldsymbol{v})$, which is therefore called the *linear discriminant function*. In fact, $(\boldsymbol{\mu}_1 - \boldsymbol{\mu}_2)'\boldsymbol{\Sigma}^{-1}$ is a row vector, $\boldsymbol{\alpha} = (\alpha_1, \alpha_2, \ldots, \alpha_p)$ say. Thus if $v_1, v_2, \ldots, v_p$ denote the elements of $\boldsymbol{v}$, $\mathrm{L}(\boldsymbol{v}) = \alpha_0 + \alpha_1 v_1 + \alpha_2 v_2 + \ldots + \alpha_p v_p$ where $\alpha_0 = -\frac{1}{2}(\boldsymbol{\mu}_1 - \boldsymbol{\mu}_2)'\boldsymbol{\Sigma}^{-1}(\boldsymbol{\mu}_1 + \boldsymbol{\mu}_2)$.

Before looking at estimation of this linear discriminant function, let us consider its performance in a little more detail. In particular, let us determine the frequency of mistakes that the rule (12.18) will give rise to.

The first kind of error arises whenever we classify a π_2 individual to π_1, and we denote this error by $p(1 \mid 2)$. Now if $\boldsymbol{v}$ is a π_2 individual, then $\boldsymbol{v} \sim N(\boldsymbol{\mu}_2, \boldsymbol{\Sigma})$. But since $L(\boldsymbol{v})$ is just a linear transformation of $\boldsymbol{v}$, we can use Property 1 of Section 7.2 to establish that $L(\boldsymbol{v})$ has a univariate normal distribution with mean $-\frac{1}{2}\Delta^2$ and variance Δ^2, where $\Delta^2 = (\boldsymbol{\mu}_1 - \boldsymbol{\mu}_2)'\boldsymbol{\Sigma}^{-1}(\boldsymbol{\mu}_1 - \boldsymbol{\mu}_2)$. Since the individual $\boldsymbol{v}$ is classified to π_1 whenever $L(\boldsymbol{v}) > \log_e k$, it follows that

$$\begin{aligned} p(1 \mid 2) &= \text{pr}\{L(\boldsymbol{v}) > \log_e k \mid \boldsymbol{v} \in \pi_2\} \\ &= \text{pr}\{L(\boldsymbol{v}) > \log_e k \mid L(\boldsymbol{v}) \sim N(-\tfrac{1}{2}\Delta^2, \Delta^2)\} \\ &= \text{pr}\left[\frac{1}{\Delta}\{L(\boldsymbol{v}) + \tfrac{1}{2}\Delta^2\} > \frac{1}{\Delta}(\log_e k + \tfrac{1}{2}\Delta^2) \mid L(\boldsymbol{v}) \sim N(-\tfrac{1}{2}\Delta^2, \Delta^2)\right]. \end{aligned}$$

But if $L(\boldsymbol{v}) \sim N(-\frac{1}{2}\Delta^2, \Delta^2)$ then $z = \dfrac{1}{\Delta}\{L(\boldsymbol{v}) + \frac{1}{2}\Delta^2\}$ has a $N(0, 1)$ distribution. If $\Phi(\cdot)$ denotes the cumulative distribution function of a $N(0, 1)$ distribution, then $\text{pr}\{z > a\} = \text{pr}\{z \leq -a\} = \Phi(-a)$. Thus

$$p(1 \mid 2) = \Phi\left\{-\frac{1}{\Delta}(\log_e k + \tfrac{1}{2}\Delta^2)\right\}. \tag{12.19}$$

The second kind of error arises whenever we classify a π_1 individual to π_2, and we denote this error by $p(2 \mid 1)$. If $\boldsymbol{v}$ is a π_1 individual then $\boldsymbol{v} \sim N(\boldsymbol{\mu}_1, \boldsymbol{\Sigma})$, and similar argument to that used above shows that $L(\boldsymbol{v})$ then has a univariate normal distribution with mean $\frac{1}{2}\Delta^2$ and variance Δ^2. Thus

$$\begin{aligned} p(2 \mid 1) &= \text{pr}\{L(\boldsymbol{v}) \leq \log_e k \mid \boldsymbol{v} \in \pi_1\} \\ &= \text{pr}\left[\frac{1}{\Delta}\{L(\boldsymbol{v}) - \tfrac{1}{2}\Delta^2\} \leq \frac{1}{\Delta}[\log_e k - \tfrac{1}{2}\Delta^2] \mid L(\boldsymbol{v}) \sim N(\tfrac{1}{2}\Delta^2, \Delta^2)\right] \\ &= \Phi\left\{\frac{1}{\Delta}(\log_e k - \tfrac{1}{2}\Delta^2)\right\}. \end{aligned} \tag{12.20}$$

Choice of k is governed by whether we wish to use (12.6), (12.9), or (12.10) as the allocation rule. If we use the likelihood ratio rule (12.6) then $k = 1$, so that $\log_e k = 0$. In this case, therefore, the two kinds of error have the same probability: $p(1 \mid 2) = p(2 \mid 1) = \Phi(-\frac{1}{2}\Delta)$. Introducing differential prior probabilities and/or costs due to misclassification creates an imbalance in the two error probabilities.

The two probabilities (12.19) and (12.20) may be termed the *optimal error rates* for discriminating between two multivariate normal populations with equal dispersion matrices. Even if we know all the characteristics of the two populations, and choose the best possible allocation rule in the given circumstances, we will still misallocate future individuals from each population, at a rate given by (12.19) and (12.20). This is because the two populations overlap, and however great the probability is that a particular value $\boldsymbol{v}_0$ has come from one population, there will still be *some* probability that it came from the other population. Probabilities of misallocation will only be eliminated if every possible value $\boldsymbol{v}_0$ of the sample space has zero (or negligible) probability of having come from one of the populations. This situation corresponds to a total separation of π_1 and π_2. In practice this is virtually impossible to achieve, but obviously the greater the separation the smaller will be the error rates. Now we see from (12.19) and (12.20) that both error rates are monotonically decreasing functions of Δ, i.e. the greater the value of Δ the smaller are the error rates. Thus Δ provides a possible measure of separation of π_1 and π_2. It is, in fact, the *Mahalanobis distance* between π_1 and π_2, being defined by $\Delta^2 = (\boldsymbol{\mu}_1 - \boldsymbol{\mu}_2)'\boldsymbol{\Sigma}^{-1}(\boldsymbol{\mu}_1 - \boldsymbol{\mu}_2)$, and is the population counterpart of the distance between individuals and samples already introduced in Chapters 8 and 11 respectively.

Figure 12.1 illustrates the various facets of the linear discriminant function $L(\boldsymbol{v})$ discussed above. Plotted are the probability densities of $L(\boldsymbol{v})$ in each of the two populations. These densities are centred at $-\frac{1}{2}\Delta^2$ and $\frac{1}{2}\Delta^2$ for π_2, π_1 respectively, and have the same shape (normal) and size (since the variance is the same in the two populations). An individual is allocated according to (12.18), with k chosen by the user. If the likelihood ratio rule (12.6) is employed, then $k = 1$, and the cut-off point ($\log_e k$) is zero. It can be seen from Fig. 12.1 that this occurs at the intersection of the two probability density curves (because of the equal size and shape of the curves and their equidistant displacement from zero). Misclassification probabilities are given by the areas in the tails of the curves cut off at zero: $p(1 \mid 2)$ is the area under the π_2 curve to the right of zero, and $p(2 \mid 1)$ is the area under the π_1 curve to the left of zero. By symmetry, these areas are clearly equal, agreeing with the

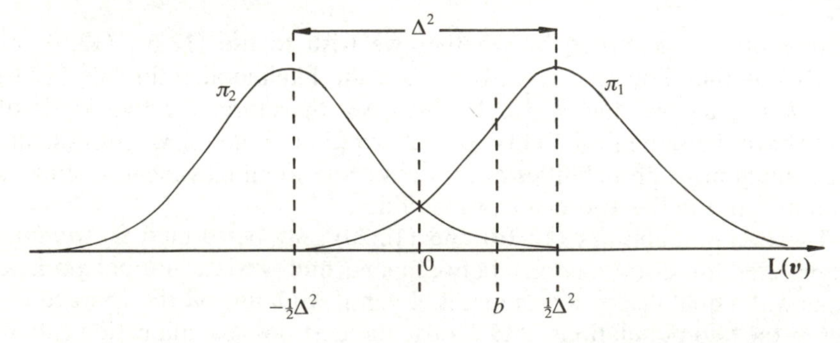

Fig. 12.1 Linear discriminant analysis between populations.

result given above. Now, if the prior probability of π_2 were much greater than that for π_1, and allocation rule (12.10) were employed, $k = q_2/q_1$ would be much greater than 1, and $\log_e k$ would be much greater than zero. The cut-off point would therefore be shifted appreciably to the right, say to the position b in the figure. The effect of this is to make the area under the π_2 curve to the right of the cut-off point much smaller than before, while area under the π_1 curve to the left of the cut-off point becomes much larger than before. Our prior knowledge about the populations has made us much less willing to allocate an individual to π_1. This is evidenced by the reduced region in which allocation is made to π_1, and the consequence is that π_2 individuals have greatly reduced (conditional) probability of misclassification while π_1 individuals have greatly increased (conditional) probability of misclassification. Of course, if the incidence of π_1 individuals is indeed low, then the *overall* (i.e. unconditional) error rate will be maintained at about its former level because the larger conditional error rate is appropriate much more rarely than the smaller one. Similar considerations apply with differential costs due to misallocation, and k chosen from allocation rule (12.9) will induce some other shift of the cut-off point. Finally, we note that the two population means are a distance Δ^2 apart. However, each population has variance Δ^2, i.e. standard deviation Δ, so the separation of the two means is Δ when measured in units of one standard deviation. Thus the distance between the populations can be taken to be the Mahalanobis distance Δ. Increasing this distance will reduce the overlap between the two density curves, and hence will reduce the areas in the tails of the curves to the right or left of any cut-off point. Hence probabilities of misclassification will be reduced whichever allocation rule (i.e. value of k) is chosen.

Of course, to use the allocation rule (12.18) we must assume that π_1 and π_2 are $N(\boldsymbol{\mu}_1, \boldsymbol{\Sigma})$ and $N(\boldsymbol{\mu}_2, \boldsymbol{\Sigma})$ populations respectively, and that *all the population parameters are known.* In any practical application, the most that can be said is that $N(\boldsymbol{\mu}_1, \boldsymbol{\Sigma})$ and $N(\boldsymbol{\mu}_2, \boldsymbol{\Sigma})$ are probability *models* for π_1 and π_2, and the parameters will not be known in these distributions. Given two training sets, $\boldsymbol{v}_1^{(1)}, \ldots, \boldsymbol{v}_{n_1}^{(1)}$ from π_1, and $\boldsymbol{v}_1^{(2)}, \ldots, \boldsymbol{v}_{n_2}^{(2)}$ from π_2, however, we can estimate $\boldsymbol{\mu}_1$, $\boldsymbol{\mu}_2$, and $\boldsymbol{\Sigma}$ in the usual way by $\bar{\boldsymbol{v}}^{(1)} = \frac{1}{n_1}\sum_{i=1}^{n_1} \boldsymbol{v}_i^{(1)}$, $\bar{\boldsymbol{v}}^{(2)} = \frac{1}{n_2}\sum_{i=1}^{n_2} \boldsymbol{v}_i^{(2)}$ and the pooled covariance matrix

$$\mathbf{S} = \frac{1}{n_1 + n_2 - 2}\left\{\sum_{i=1}^{n_1} (\boldsymbol{v}_i^{(1)} - \bar{\boldsymbol{v}}^{(1)})(\boldsymbol{v}_i^{(1)} - \bar{\boldsymbol{v}}^{(1)})' + \sum_{i=1}^{n_2} (\boldsymbol{v}_i^{(2)} - \bar{\boldsymbol{v}}^{(2)})(\boldsymbol{v}_i^{(2)} - \bar{\boldsymbol{v}}^{(2)})'\right\}.$$

(We use the symbol **S** in place of the earlier **W** for the pooled covariance matrix in the remainder of this chapter, as **S** is the usual symbol in the discrimination and classification literature.) The estimative approach is then to substitute these estimates for the unknown parameters in allocation rule (12.18), thereby obtaining the rule:

$$\text{Allocate } \boldsymbol{v} \text{ to } \pi_1 \text{ if } \hat{\mathrm{L}}(\boldsymbol{v}) > \log_e k, \text{ and otherwise to } \pi_2 \qquad (12.21)$$

where $\hat{\mathrm{L}}(\boldsymbol{v}) = (\bar{\boldsymbol{v}}^{(1)} - \bar{\boldsymbol{v}}^{(2)})'\mathbf{S}^{-1}\{\boldsymbol{v} - \frac{1}{2}(\bar{\boldsymbol{v}}^{(1)} + \bar{\boldsymbol{v}}^{(2)})\}$. The function $\hat{\mathrm{L}}$ is known as the *sample linear discriminant function,* or sometimes as Anderson's classification statistic after T. W. Anderson (1951). Furthermore, if the prior probabilities q_1, q_2 of populations π_1 and π_2 that are necessary for calculation of k are not known, but the training sets are obtained from a *mixture* of π_1 and π_2 (i.e. each population was not sampled separately), then we can additionally estimate q_1 and q_2 by $n_1/(n_1 + n_2)$ and $n_2/(n_1 + n_2)$ for use in the right-hand side of (12.21). This is because the rates of incidence of each population in the mixture sample are determined by q_1 and q_2. If training sets of sizes n_1 and n_2 are obtained from each population separately, however, then other information must be used to estimate q_1 and q_2.

Assessment of the performance of this rule can be tackled on several different levels. At the simplest level, we may assume that a rule has been formulated once-for-all from a fixed pair of training sets, and this rule will then be used on all individuals presented for classification in the future. For example, past hospital records are used to construct a linear discriminant function between 'medical' and 'surgical' jaundice patients, and all future patients suffering from jaundice are allocated to one of these groups by using this discriminant function. In this case, the values $\bar{\boldsymbol{v}}^{(1)}$, $\bar{\boldsymbol{v}}^{(2)}$, and **S** are *fixed* (because the same values are used in all future classifications) and hence can be treated as *known constants.* Thus the only random variable in $\hat{\mathrm{L}}(\boldsymbol{v})$ is again of the linear form $a_0 + a_1v_1 + a_2v_2 + \ldots + a_pv_p$ where the a_i are all constants. Hence Property 1 of Section 7.2 can be invoked again to show that $\hat{\mathrm{L}}(\boldsymbol{v})$ has a normal distribution in both π_1 and π_2, with mean $(\bar{\boldsymbol{v}}^{(1)} - \bar{\boldsymbol{v}}^{(2)})'\mathbf{S}^{-1}\{\boldsymbol{\mu}_i - \frac{1}{2}(\bar{\boldsymbol{v}}^{(1)} + \bar{\boldsymbol{v}}^{(2)})\}$ in π_i $(i = 1, 2)$ and variance $(\bar{\boldsymbol{v}}^{(1)} - \bar{\boldsymbol{v}}^{(2)})'\mathbf{S}^{-1}\boldsymbol{\Sigma}\mathbf{S}^{-1}(\bar{\boldsymbol{v}}^{(1)} - \bar{\boldsymbol{v}}^{(2)})$ in both π_1 and π_2. Now $\mathrm{p}(1 \mid 2) = \mathrm{pr}\{\hat{\mathrm{L}}(\boldsymbol{v}) > \log_e k \mid \boldsymbol{v} \in \pi_2\}$ and $\mathrm{p}(2 \mid 1) = \mathrm{pr}\{\hat{\mathrm{L}}(\boldsymbol{v}) \leqslant \log_e k \mid \boldsymbol{v} \in \pi_1\}$ so by using the above distributions we can show that

$$\mathrm{p}(1 \mid 2) = \Phi([(\bar{\boldsymbol{v}}^{(1)} - \bar{\boldsymbol{v}}^{(2)})'\mathbf{S}^{-1}\{\boldsymbol{\mu}_2 - \tfrac{1}{2}(\bar{\boldsymbol{v}}^{(1)} + \bar{\boldsymbol{v}}^{(2)})\} - \log_e k]/\{(\bar{\boldsymbol{v}}^{(1)} - \bar{\boldsymbol{v}}^{(2)})'\mathbf{S}^{-1}\boldsymbol{\Sigma}\mathbf{S}^{-1}(\bar{\boldsymbol{v}}^{(1)} - \bar{\boldsymbol{v}}^{(2)})\}^{\frac{1}{2}}),$$

$$\mathrm{p}(2 \mid 1) = \Phi([\log_e k - (\bar{\boldsymbol{v}}^{(1)} - \bar{\boldsymbol{v}}^{(2)})'\mathbf{S}^{-1}\{\boldsymbol{\mu}_1 - \tfrac{1}{2}(\bar{\boldsymbol{v}}^{(1)} + \bar{\boldsymbol{v}}^{(2)})\}]/\{(\bar{\boldsymbol{v}}^{(1)} - \bar{\boldsymbol{v}}^{(2)})'\mathbf{S}^{-1}\boldsymbol{\Sigma}\mathbf{S}^{-1}(\bar{\boldsymbol{v}}^{(1)} - \bar{\boldsymbol{v}}^{(2)})\}^{\frac{1}{2}}). \qquad (12.22)$$

The probabilities in (12.22) are known as the *actual error rates,* because they give the true probabilities of misclassifying $\boldsymbol{v}$ with the given $\hat{\mathrm{L}}(\boldsymbol{v})$. However, these error rates are of little practical use, as they involve the unknown parameters $\boldsymbol{\mu}_1$, $\boldsymbol{\mu}_2$, and $\boldsymbol{\Sigma}$, and hence will be unknown themselves. Invoking the estimative principle once again, we can replace these parameters by $\bar{\boldsymbol{v}}^{(1)}$, $\bar{\boldsymbol{v}}^{(2)}$, and $\mathbf{S}$ in (12.22). When this done, considerable simplification occurs with the expressions inside the function Φ and we obtain

$$\mathrm{p}(1 \mid 2) = \Phi\{-(\log_e k + \tfrac{1}{2}D^2)/D\}$$

and (12.23)

$$\mathrm{p}(2 \mid 1) = \Phi\{(\log_e k - \tfrac{1}{2}D^2)/D\}$$

where $D^2 = (\bar{\boldsymbol{v}}^{(1)} - \bar{\boldsymbol{v}}^{(2)})'\mathbf{S}^{-1}(\bar{\boldsymbol{v}}^{(1)} - \bar{\boldsymbol{v}}^{(2)})$ is the squared Mahalanobis distance between the training samples. If, moreover, we have used the likelihood ratio rule (12.6), so that $\log_e k = 0$, then $\mathrm{p}(1 \mid 2) = \mathrm{p}(2 \mid 1) = \Phi(-\tfrac{1}{2}D)$. These probabilities are called *estimated error rates:* note that they can be obtained by substituting estimates for all unknown parameters *either* in the actual error rates (12.22) *or* in the optimum error rates (12.19) and (12.20).

At a more complicated level, we can treat (12.21) as a prescription for future action. In other words, every time we have a classification problem involving the two populations π_1 and π_2 we are to find two training sets; obtain $\bar{\boldsymbol{v}}^{(1)}$, $\bar{\boldsymbol{v}}^{(2)}$, and $\mathbf{S}$; compute $\hat{\mathrm{L}}(\bar{\boldsymbol{v}})$; and finally classify the individual. If we adopt this viewpoint, then $\bar{\boldsymbol{v}}^{(1)}$, $\bar{\boldsymbol{v}}^{(2)}$, and $\mathbf{S}$ are now no longer fixed but are themselves random variables (whose possible values are the set of all possible means and covariance matrices computed from samples of size n_1 and n_2 taken from π_1 and π_2). Hence if we want to assess the performance of (12.21) as a *prescription* for discriminating between π_1 and π_2, $\hat{\mathrm{L}}(\boldsymbol{v})$ has a much more complicated distribution than before. In particular, the actual error rates (12.22) have probability distributions induced by the sampling distributions of $\bar{\boldsymbol{v}}^{(1)}$, $\bar{\boldsymbol{v}}^{(2)}$, and $\mathbf{S}$. The means of these probability distributions are known as the *expected error rates,* and can be found by integrating the actual error rates over the joint probability density of $\bar{\boldsymbol{v}}^{(1)}$, $\bar{\boldsymbol{v}}^{(2)}$, and $\mathbf{S}$. This is a highly mathematical area of research, and one in which there are still many unanswered questions, so it will not be considered any further here. Some relevant references are John (1961), Okamoto (1963), and Moran (1975).

The relationships between the optimal, actual, estimated, and expected error rates have been discussed by Hills (1966). Taking a fairly pragmatic viewpoint, the optimal error rates are the error rates associated with the best possible allocation rule that could be used (if all assumptions made are appropriate), and hence are an assessment of the *best* that the user could hope to do; the actual rates are an assessment of how well the user will *actually* do with a given classification rule; the estimated rates are an assessment of how well the user *thinks* he is doing with his classification rule; while the expected error rates are an assessment of how well the *prescription* will work *on average*. Intuitively, one would expect the optimum error rates to be the smallest, with the others all larger either because estimation is involved or because averaging over a distribution is involved. The potential user may be surprised to find, therefore, that the estimated rates can turn out to be smaller. This is because $\hat{L}(\boldsymbol{v})$ is formed in such a way as to maximize the separation between the training sets, and then *the same* parameter estimates are substituted into either the optimal or the actual error rates to obtain the estimated rates. Such estimation presents an over-optimistic view, and future allocations are rarely made with the accuracy that is implied by these estimated rates. In technical terms, the sample Mahalanobis distance is a biased estimate of the population distance, consistently overestimating this distance. Consequently, the error rates will be underestimated. Although a bias correction can be applied, it can sometimes overcompensate and produce a negative squared distance. This line will not be pursued any further here, because the reader has probably already formed a further criticism of the estimated error rate as an assessment of the allocation rule, namely its heavy dependence on normality of the observations. Of course, normality has been assumed in deriving the allocation rule. However, whether or not normality is justified, classification based on the linear function $\hat{L}(\boldsymbol{v})$ may still be perfectly reasonable. On the other hand, if the data are not normal then the estimated error rates (12.23) may be wildly inaccurate. Hence, some data-based estimates of error rates would be more sensible. As this is a topic of general application, however, it is deferred to Section 12.4 to allow for various other classification rules to be discussed first.

Another matter that deserves some comment is the choice between a linear and a quadratic discriminant function. The quadratic function $Q(\boldsymbol{v})$ allows for different dispersions in the two populations. Viewing the populations as swarms of points in multidimensional space, the curved function $Q(\boldsymbol{v})$ can provide an effective boundary between two swarms that have different shapes and orientations. For example, in the simple illustration of Fig. 12.2, the swarm of crosses is separated totally from the swarm of circles by the solid curved function. The linear function $L(\boldsymbol{v})$, on the other hand, provides only a straight-line boundary in two dimensions (or a hyperplane in p dimensions), so that the best dotted linear function in Fig. 12.2 still misclassifies some of the circles.

It may thus be supposed that the more general quadratic function $Q(\boldsymbol{v})$ should always provide a preferable allocation rule, as it should be able to

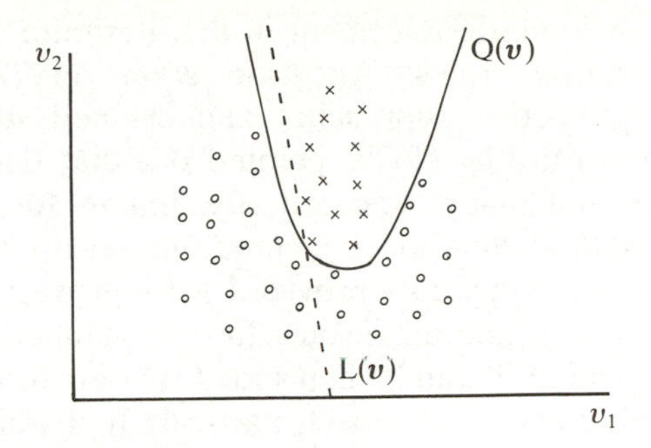

Fig. 12.2 Two-variable example in which a quadratic function (solid line) separates the two groups (crosses and circles) completely, while the best linear function (dotted line) misclassifies an appreciable number of individuals of one group.

cope better with *any* situation one is likely to encounter. This would indeed be the case if we dealt with populations rather than samples. In practice we always have to estimate our allocation rules from training sets, and work with $\hat{Q}(\boldsymbol{v})$ and $\hat{L}(\boldsymbol{v})$ instead of $Q(\boldsymbol{v})$ and $L(\boldsymbol{v})$. Estimation affects the two functions very differently. $\hat{L}(\boldsymbol{v})$ is fairly stable and not subject to wild fluctuation even in fairly small samples. $\hat{Q}(\boldsymbol{v})$, on the other hand, requires estimation of many more parameters, is notoriously unstable, and often gives very poor classification performance in small samples. Indeed, it sometimes happens that the linear function gives better performance even when the population dispersion matrices cannot be assumed equal. If in doubt, both functions can be tried and the methods of Section 12.4 used to determine which of the two is the better.

Having concentrated throughout on the estimative approach in the derivation of the sample allocation rules, we conclude this section with a brief review of the alternative approaches. The equal dispersion assumption $\boldsymbol{\Sigma}_1 = \boldsymbol{\Sigma}_2 = \boldsymbol{\Sigma}$ is necessary in all of these. Using $\bar{\boldsymbol{v}}^{(1)}$, $\bar{\boldsymbol{v}}^{(2)}$, $\mathbf{S}$ as defined earlier, and writing $\mathbf{C} = (n_1 + n_2 - 2)\mathbf{S}$, it can be shown that the predictive, hypothesis testing, and a best invariant estimate due to Murray (1977) all lead to the same form of allocation rule; this is to:

$$\text{Allocate } \boldsymbol{v} \text{ to } \pi_1 \text{ if } \frac{1 + \{n_2/(n_2+1)\}(\boldsymbol{v} - \bar{\boldsymbol{v}}^{(2)})'\mathbf{C}^{-1}(\boldsymbol{v} - \bar{\boldsymbol{v}}^{(2)})}{1 + \{n_1/(n_1+1)\}(\boldsymbol{v} - \bar{\boldsymbol{v}}^{(1)})'\mathbf{C}^{-1}(\boldsymbol{v} - \bar{\boldsymbol{v}}^{(1)})} \geqslant d, \quad \text{and otherwise to } \pi_2, \tag{12.24}$$

where d is a suitably chosen constant. Details of the derivations of the rules and values of the constant d are provided by Han (1979), who also discusses the interpretation of the rules in both Bayesian and frequentist contexts.

There has been some disagreement in the literature about the relative merits of the various rules. Aitchison *et al.* (1977) compared the estimative and predictive approaches and claimed superiority for the latter. Moran and Murphy (1979) pointed out that the comparison had been misleading, and that if standard adjustments for bias are made in the estimative method then more comparable results are obtained with the two approaches. They also provided a frequentist interpretation of the predictive method, and drew some further parallels between the two approaches. In general, it can be concluded that an allocation rule of the form (12.24) does have some appeal, particularly if training sample sizes n_1 and n_2 are unequal (John 1963), and could be tried routinely alongside (12.18) whenever classification is required. Once again, the methods of Section 12.4 can be used to decide between the rules.

12.2.3 The multinomial model

Suppose now that the elements $v_1, v_2, \ldots, v_p$ of the random vector $\boldsymbol{v}$ are all discrete or qualitative random variables, and that the ith element v_i can have a value in one of s_i possible categories or states $(i = 1, \ldots, p)$. Unless each s_i is very large, the multivariate normal distribution will not be suitable for such a random vector. Instead, the multinomial distribution will be the most appropriate model that can be used. The sample space consists of $s = \prod_{j=1}^{p} s_j$ possible values or 'cells', these being all the possible combinations of discrete variable categories that can be exhibited by the individual v_i. The most general form of model that can be postulated is to assume simply that the probability of obtaining an observation in cell i is p_{1i} in π_1 and p_{2i} in π_2 $(i = 1, \ldots, s)$. If z is a random variable denoting the cell membership of an individual with values $v_1, v_2, \ldots, v_p$ of the original p qualitative variables, then z has a multinomial distribution with parameters p_{1i} in π_1 and p_{2i} in π_2. Given that z falls in cell r, classification rule (12.7) becomes:

$$\text{Allocate } \boldsymbol{v} \text{ to } \pi_1 \text{ if } p_{1r} > kp_{2r}, \text{ and otherwise to } \pi_2, \tag{12.25}$$

where k is chosen to be one of the values in (12.6), (12.9), or (12.10) as appropriate. Thus the regions R_1 and R_2 of the sample space (in which classification to π_1 and π_2 respectively is indicated) are given by:

$$R_1 = \{r \mid p_{1r} > kp_{2r}\}$$

and

$$R_2 = \{r \mid p_{1r} \leqslant kp_{2r}\}.$$

It follows immediately that the conditional probabilities of misclassification are

$$p(1 \mid 2) = \sum_{r \in R_1} p_{2r} \qquad \text{and } p(2 \mid 1) = \sum_{r \in R_2} p_{1r}, \tag{12.26}$$

and the total (unconditional) error rate is $q_1 \sum_{r \in R_2} p_{1r} + q_2 \sum_{r \in R_1} p_{2r}$.

In order to be able to use allocation rule (12.25), we must first estimate the probabilities p_{ij} ($i = 1, 2; j = 1, \ldots, s$). On the face of it, this would seem to be an easy matter. The training sets, of size n_1 from π_1 and n_2 from π_2, yield the observed incidences n_{1j} ($j = 1, \ldots, s$) and n_{2j} ($j = 1, \ldots, s$) in each of the cells. Maximum likelihood estimates of the probabilities p_{ij} are thus given by $\hat{p}_{1j} = n_{1j}/n_1$ and $\hat{p}_{2j} = n_{2j}/n_2$, and these estimates can be substituted for the true values in (12.25). If the training sets were obtained by sampling a mixture of π_1 and π_2 then prior probability estimates are $\hat{q}_1 = n_1/(n_1 + n_2)$ and $\hat{q}_2 = n_2/(n_1 + n_2)$; otherwise ancillary information must be invoked to specify these prior probabilities. Thus in principle a sample version of (12.25) can be obtained easily.

Such an approach to the estimation of the probabilities p_{ij} is not very satisfactory in the majority of practical applications, however, because it is not clear what should be done when one or more of the multinomial cells are empty in the training sets. Such empty cells will occur quite frequently, particularly if the training sets are not large. For example, five qualitative variables $v_1, v_2, \ldots, v_5$ each of which has three possible states, generates a sample space of 243 cells. There are bound to be empty cells, therefore, if the size of the training set is less than 243, and empty cells are still quite likely to occur even with a sample size as large as 1000. What should therefore be done about classifying an individual that falls in a cell that is empty in both of the training sets? It is clearly unsatisfactory to allocate the individual on the basis of some random outcome such as the spin of a coin. Even more problematic is the situation in which a cell is empty in one training set (say the one from π_1) but not empty in the other training set. Maximum likelihood estimates would thus be $\hat{p}_{1r} = 0$ and $\hat{p}_{2r} = b > 0$ for this cell, and any future individual falling in this cell would be forcibly allocated to π_2. However, the empty cell in the training set from π_1 might be entirely a chance event and the forced allocation would therefore be misleading. In other cases, empty cells in one population might be more informative and might mean something entirely different from empty cells in the other population. Maximum likelihood estimates from a full multinomial model are subject to sampling variation, and lead to problems such as those mentioned above. When the data exhibit signs of sparseness, it is therefore preferable to look for *smoothed* estimates of the probabilities in (12.25).

A very simple *ad hoc* smoothing procedure was suggested by Hills (1967) for the special case of dichotomous variables. If each v_i can take just two values, then every cell of the multinomial z can be represented by a string of 0s and 1s. Suppose, for example, that $p = 5$. Fix ideas by considering the allocation of an individual that falls into the particular cell defined by the string 11010. Assume that this is cell number 1. Then the full multinomial estimate of the probability in this cell for population π_1, say, is the number of individuals n_{11} that exhibit the values $v_1 = 1$, $v_2 = 1$, $v_3 = 0$, $v_4 = 1$, $v_5 = 0$ in the training set from π_1, divided by the total number n_1 of individuals in this training set. Hills defined the *nearest-neighbour procedure of order r* by estimating this probability as $\left(\sum_{i \in D} n_{1i}\right) \Big/ n_1$, where D is the set of cells whose binary string differs from 11010 in no more than r symbols. Thus the nearest neighbour estimate of order 0 is just n_{11}/n_1, as the cell defined by the string 11010 is the only cell in D. For the nearest-neighbour estimate of order 1, the cells in D are the ones defined by the strings 01010, 10010, 11110, 11000, 11011 (these being the ones differing from 11010 in just one symbol) plus the original cell defined by 11010. If the numbers of individuals in the cells 01010, 10010, 11110, 11000, and 11011 are n_{12}, n_{13}, n_{14}, n_{15}, and n_{16} respectively, then the order-1 nearest-neighbour estimate of the probability is $\left(\sum_{i=1}^{6} n_{1i}\right) \Big/ n_1$. The nearest-neighbour estimate of order 2 would be the sum of the above six incidences plus the ones in all cells differing from 11010 in just two symbols, divided by n_1. Similarly for nearest neighbour estimates of higher orders. The idea here is that cells defined by binary strings which have many symbols in common should be relatively 'similar', so amalgamating their incidences should provide a larger data set for estimating the probabilities of obtaining a given cell in the two populations. In this way the zero cell incidence problem is obviated, and the sampling error of the full multinomial estimate is simultaneously reduced. It should be unneccessary to require a nearest-neighbour estimate of order greater than 2 in most practical circumstances.

The nearest-neighbour estimates, however, are *ad hoc* ones for which there is no theoretical or model-based justification. Indeed, there may often be some sound reason why the cell defined by the string 01010, say, will *not* be 'similar' to the one defined by 11010. A more satisfactory approach in this case is to define a parametric model for the probabilities, to estimate the parameters in the model by fitting it in turn to the two training sets, and then to obtain estimates of the probabilities as expected values under the model. Since the training sets define contingency tables in the presence of qualitative variables, we have already considered a class of suitable models in the form of the log-linear models of Chapter 10. Furthermore, these models are appropriate for *any* number of states of each qualitative variable v_i and are not merely restricted to dichotomous variables as is the case with the nearest-neighbour approach. Now we saw in Chapter 10 that the logarithms of expected cell values in a contingency table constructed from p variables can be fully modelled by a linear function of the main effects of each variable plus all possible interactions up to order p between the variables. This full or 'saturated' model is exactly equivalent to the full multinomial model, and hence fitting it to each of the two training sets will yield expected cell incidences that match observed cell incidences exactly. Taking the ratio of expected cell incidence and training sample size thus recovers the maximum likelihood estimates $\hat{p}_{1i} = n_{1i}/n_1$ and $\hat{p}_{2i} = n_{2i}/n_2$ obtainable from the multinomial model. To obtain smoothed estimates, we must ignore some of the terms in the saturated model, and fit a reduced model to each training set in the manner explained in Chapter 10. A first-order model is one in which all terms except the main effects are ignored; a second-order model is one in which only these main effects plus all interactions between *pairs* of variables are included; a third-order model is one in which only main effects plus all interactions between pairs and triples of variables are included; and so on. Fitting the model to a training set is done in the manner outlined in Chapter 10, by applying the iterative scaling procedure after fixing suitable margins of the contingency table: for a first-order model, the one-way margins of variable state totals are fixed; for a second-order model, all one-way and two-way margins are fixed; for a third-order model, all one-, two-, and three-way margins are fixed; and so on. Having fitted the chosen model and obtained the expected cell frequencies for each training set, the probabilities for use in (12.25) are estimated simply by dividing the relevant expected cell frequencies by the corresponding training sample size. Allocation rule (12.25) is then used directly. The 'best' model to use can be determined by successively testing the various terms for significance as in Chapter 10, and choosing the model that has fewest parameters yet fits the training sets adequately. Alternatively, an a priori 'blanket' choice can be made by selecting one of the first-, second-, third-, or higher-order models; in practice a second-order model is often found to be adequate.

Various other parametric models have been suggested for multinomial data, but will not be discussed here, as the above models provide sufficient armoury for all practical applications. (The reader who wants a more extensive discussion of discrete variable discrimination and allocation is referred to Goldstein and Dillon (1978).) We conclude this section by giving a simple illustration, to highlight the differences between all the

Table 12.1 Multinomial classification using different models on fictitious data. (The higher probability is underlined for each cell and each method)

Multinomial cells			Data		Multinomial probabilities		First-order log-linear probabilities		Second-order log-linear probabilities		Nearest-neighbour probabilities	
v_1	v_2	v_3	π_1	π_2	π_1	π_2	π_1	π_2	π_1	π_2	π_1	π_2
0	0	0	12	0	0.24	0.00	0.12	0.14	0.24	0.05	0.48	0.64
1	0	0	8	12	0.16	0.24	0.11	0.15	0.16	0.19	0.48	0.44
0	1	0	4	12	0.08	0.24	0.13	0.09	0.08	0.19	0.52	0.32
0	0	1	0	8	0.00	0.16	0.13	0.15	0.3×10^{-4}	0.11	0.52	0.44
1	1	0	0	0	0.00	0.00	0.12	0.10	0.3×10^{-4}	0.05	0.40	0.52
1	0	1	4	10	0.08	0.20	0.12	0.16	0.08	0.25	0.32	0.44
0	1	1	10	4	0.20	0.08	0.14	0.10	0.20	0.13	0.52	0.56
1	1	1	12	4	0.24	0.08	0.13	0.11	0.24	0.03	0.52	0.36

above approaches. In the first column of Table 12.1 are the binary strings defining the eight multinomial cells obtained from three dichotomous variables v_1, v_2, and v_3 (each variable taking possible values 0 or 1), and in the next two columns are the numbers of observations falling in each of these cells when samples of size 50 are drawn from each of the two populations π_1 and π_2. (These data are fictitious, and are merely intended to provide an example of the various methods discussed earlier.) In the following columns of Table 12.1 are the cell probabilities for each population, estimated from:

(1) the full multinomial model;
(2) the log-linear model including main effects only;
(3) the log-linear model including main effects and first-order interactions;
(4) the nearest-neighbour order-1 method.

For each estimation scheme, the higher of the two probabilities is underlined in each cell. A future individual falling in a given cell would be allocated to the population with the underlined probability in that cell (if prior probabilities and costs due to misallocation were to be ignored).

The multinomial probabilities are obtained by dividing cell incidence by sample size. Thus $12/50 = 0.24$, $8/50 = 0.16$, and so on. The log-linear probabilities are obtained by the iterative scheme outlined in Chapter 10. However, the log-linear model that includes main effects only is equivalent to an assumption of independence among the v_i, and hence the probabilities under this model can also be obtained by multiplication of the marginal probabilities of the v_i. For example, the marginal probability that $v_1 = 0$ in π_1 is estimated by summing the incidences in all cells that have $v_1 = 0$ in π_1 and dividing by 50. This estimate is $(12 + 4 + 0 + 10)/50 = 0.52$. Similarly, the marginal probabilities that $v_2 = 0$ and $v_3 = 0$ in π_1 are estimated by $(12 + 8 + 0 + 4)/50 = 0.48$ and $(0 + 12 + 12 + 0)/50 = 0.48$ respectively. Multiplying these three probabilities together yields $0.52 \times 0.48 \times 0.48 = 0.12$. This is the probability of the cell $(0, 0, 0)$ in π_1 assuming independence of the v_i, i.e. the estimated probability in the first cell of π_1 under the first-order log-linear model. The other probabilities in this column of the table can be obtained in similar fashion. The first-order nearest-neighbour probabilities are obtained by summing the incidence in all cells whose binary strings do not differ by more than one symbol from that of the cell under consideration. For example, if we wish to estimate the probabilities in cell $(0, 0, 0)$ we must sum the incidences in cells $(0, 0, 0)$, $(1, 0, 0)$, $(0, 1, 0)$, and $(0, 0, 1)$. Thus in π_1 we obtain $(12 + 8 + 4 + 0)/50 = 0.48$, and in π_2 we obtain $(0 + 12 + 12 + 8)/50 = 0.64$. Similarly for all other cells.

There are a number of features of interest in Table 12.1. First, the $(1, 1, 0)$ cell has no observations falling in it from either training set. The multinomial estimated probabilities are both zero, therefore, and no allocation decision can be made for future observations that might fall in this cell. All the other estimation methods yield non-zero probabilities, however, and *do* lead to an allocation decision. The second-order log-linear estimated probabilities lead to the same allocation rule as is given by the full multinomial estimates for all the other cells, but the two remaining methods give a considerable difference in their allocation rules. The first-order log-linear model (i.e. independent v_i) yields the same allocation rule for five of the remaining seven cells, but the first-order nearest-neighbour procedure yields the same allocation rule as the multinomial for only two of the remaining seven cells. A choice between the various methods would require an estimate of the error rate associated with each. Since the various probabilities shown in Table 12.1 are computed under different assumptions, they cannot be used directly in assessing the error rate for each method. In particular, the 'probabilities' in the nearest-neighbour column do not form a probability distribution over cells, and are therefore not directly usable in any error rate calculation. A comparison of the methods can only be effected by a data-based estimation of error rates, and a suitable method of obtaining such an estimate will be given in Section 12.4. On intuitive grounds, however, high-order log-linear models seem to be the most preferable, because they provide enough smoothing to remove estimated zero probabilities while not over-smoothing and thereby disturbing the optimal multinomial decision rule too much.

12.2.4 The location model

We now turn to mixtures of discrete and continuous variables, and first state more formally the two-group location model that was described in Section 12.1.3. Suppose that c continuous variables $\boldsymbol{y}$ and d discrete variables $\boldsymbol{x}$ are measured on each individual, and the d discrete variables define a multinomial table containing s cells. The location model then assumes that:

(1) the probability of obtaining an observation in the jth cell of the multinomial table is p_{ij} in population π_i $(i = 1, 2; j = 1, 2, \ldots, s)$;
(2) if the discrete variables have located an individual in cell j, the continuous variables $\boldsymbol{y}$ have a multivariate normal distribution with mean $\boldsymbol{\mu}_i^{(j)}$ and dispersion matrix $\boldsymbol{\Sigma}$ in population π_i $(i = 1, 2; j = 1, 2, \ldots, s)$.

From (2), the conditional probability density of $\boldsymbol{y}$, given that the discrete variables locate the individual in cell j, is

$$\frac{1}{(2\pi)^{c/2}\,|\boldsymbol{\Sigma}|^{\frac{1}{2}}}\exp\{-\tfrac{1}{2}(\boldsymbol{y}-\boldsymbol{\mu}_i^{(j)})'\boldsymbol{\Sigma}^{-1}(\boldsymbol{y}-\boldsymbol{\mu}_i^{(j)})\}$$

in π_i $(i = 1, 2)$. Thus the joint probability density of obtaining the individual in cell j *and* observing the continuous variable values $\boldsymbol{y}$ is

$$\frac{p_{ij}}{(2\pi)^{c/2}\,|\boldsymbol{\Sigma}|^{\frac{1}{2}}}\exp\{-\tfrac{1}{2}(\boldsymbol{y}-\boldsymbol{\mu}_i^{(j)})'\boldsymbol{\Sigma}^{-1}(\boldsymbol{y}-\boldsymbol{\mu}_i^{(j)})\}$$

in π_i $(i = 1, 2)$. Inserting these two joint probability densities into (12.6), and tidying up the expression by algebraic manipulation yields the allocation rule:

> Allocate the individual $\boldsymbol{v}' = (\boldsymbol{y}', \boldsymbol{x}')$ to π_1 if the discrete variables $\boldsymbol{x}$ correspond to the jth multinomial cell and
>
> $(\boldsymbol{\mu}_1^{(j)} - \boldsymbol{\mu}_2^{(j)})'\boldsymbol{\Sigma}^{-1}\{\boldsymbol{y} - \tfrac{1}{2}(\boldsymbol{\mu}_1^{(j)} + \boldsymbol{\mu}_2^{(j)})\} > \log_e(p_{2j}/p_{1j})$;
>
> otherwise allocate $\boldsymbol{v}$ to π_2. (12.27)

The rule obtained from either (12.9) or (12.10) would be identical, except that the right-hand side of the inequality in (12.27) would contain $\log_e\{c(1 \mid 2)q_2 p_{2j}/c(2 \mid 1)q_1 p_{1j}\}$ in the former case, and $\log_e(q_2 p_{2j}/q_1 p_{1j})$ in the latter. For simplicity we will retain the form in (12.27) throughout. Comparison of (12.27) with (12.18), therefore, shows that the allocation rule requires a choice to be made of one out of s linear discriminant functions. The choice is governed by the multinomial cell occupied by the individual to be allocated and is determined by the individual's discrete variable values, while the linear discriminant function is applied to the individual's continuous variable values. Given that the location model is appropriate, probabilities of misallocation from populations π_1 and π_2 are easily shown to be

$$p(2 \mid 1) = \sum_{j=1}^{s} p_{1j}\Phi[\{\log_e(p_{2j}/p_{1j}) - \tfrac{1}{2}\Delta_j^2\}/\Delta_j]$$

and

$$p(1 \mid 2) = \sum_{j=1}^{s} p_{2j}\Phi[\{-\log_e(p_{2j}/p_{1j}) - \tfrac{1}{2}\Delta_j^2\}/\Delta_j] \qquad (12.28)$$

where $\Delta_j^2 = (\boldsymbol{\mu}_1^{(j)} - \boldsymbol{\mu}_2^{(j)})'\boldsymbol{\Sigma}^{-1}(\boldsymbol{\mu}_1^{(j)} - \boldsymbol{\mu}_2^{(j)})$ is the Mahalanobis squared distance between π_1 and π_2 *in cell j of the multinomial table,* and $\Phi(\cdot)$ is the cumulative normal distribution function.

Suppose now that there are two training sets, $\boldsymbol{v}_1^{(1)}, \ldots, \boldsymbol{v}_{n_1}^{(1)}$ from π_1, and $\boldsymbol{v}_1^{(2)}, \ldots, \boldsymbol{v}_{n_2}^{(2)}$ from π_2, and that population parameters in (12.27) and (12.28) are unknown. The discrete variable components of these training sets define two contingency tables, giving incidences $n_{11}, n_{12}, \ldots, n_{1s}$ in the multinomial cells of π_1, and incidences $n_{21}, n_{22}, \ldots, n_{2s}$ in those of π_2 (with $n_{11} + \ldots + n_{1s} = n_1$, and $n_{21} + \ldots + n_{2s} = n_2$). Maximum likelihood estimates of all the population parameters are thus:

$\hat{p}_{ij} = n_{ij}/n_i$ $(i = 1, 2; j = 1, \ldots, s)$,
$\hat{\boldsymbol{\mu}}_i^{(j)} = \bar{\boldsymbol{y}}_i^{(j)}$, the mean vector of the continuous variable values for the individuals in the jth contingency table cell of π_i, while $\hat{\boldsymbol{\Sigma}}$ is the covariance matrix pooled within cells and training sets.

However, these maximum likelihood estimates suffer from the same drawbacks as the full multinomial probability estimates of Section 12.2.3. Indeed, the problem is even more acute in the present instance. Suppose, for example, that a particular contingency table cell has no entry in one or other (or both) of the training sets. Then not only is there no estimate of p_{ij} for this cell, but furthermore there are no individuals from which $\boldsymbol{\mu}_i^{(j)}$ can be estimated. Thus, if a future individual occurs in this cell, then no estimate is available of the linear discriminant function (12.27) that is to be used for classifying this individual. The problem can be circumvented once more, as in 12.2.3, by using smoothed estimates of the parameters. The parameters p_{ij} can be estimated by fitting a log-linear model of suitable order to the contingency tables formed from the discrete variables, in exactly the same manner as for the multinomial case discussed in 12.2.3. A similar linear expression can then be employed to model $\boldsymbol{\mu}_i^{(j)}$ in terms of main effects and low-order interactions, and the parameters in this expression can be estimated by means of multivariate regression (Section 15.3). Given these estimated parameters, $\boldsymbol{\mu}_i^{(j)}$ can be predicted from the linear expression even for those cells that have no observations in the training sets. Hence an allocation rule can be implemented for use in practice, and the error rates (12.28) can also be estimated. Full details of the estimation procedures may be found in Krzanowski (1975, 1980), together with examples of their use. Methods of handling missing values for such data have been given by Little and Schluchter (1985). Furthermore, an allocation rule using the hypothesis testing principle has also been derived from the location model by Krzanowski (1982b), but this rule will not be discussed here.

12.3 Other classification rules

The classification rules of the previous section were all founded on the basic equation (12.7), assuming different parametric forms of probability density function. The theoretical justification that led up to this equation was first set out by Welch (1939), following the work done in the 1930s by Neyman and Pearson on testing statistical hypotheses. Over the years, however, there have been many alternative principles put forward for the calculation of allocation rules, and some of these principles and rules have now become firmly established methods. For an overview of these various approaches, and a review of the methodology up to 1972, the reader is referred to Das Gupta (1972). Here we will outline just three possible alternative approaches and the rules that result from them, starting with the oldest and most popular of these rules.

12.3.1 Fisher's linear discriminant function

In contrast to the probabilistic approach of Section 12.2, Fisher (1936) tackled discrimination from a purely data-based standpoint. He supposed that one was presented with independent random samples, of sizes n_1 and n_2 respectively, from each of two p-variate populations, and that a method of best distinguishing between these samples was required. The only assumption that he made was that the dispersion matrices in these two populations were equal; otherwise, the populations were completely unspecified. With this assumption, the data can be summarized by computing the sample mean vectors $\bar{\boldsymbol{v}}^{(1)}$ and $\bar{\boldsymbol{v}}^{(2)}$, and the pooled within-sample covariance matrix $\mathbf{S}$. Fisher then looked for the linear combination $w = \boldsymbol{a}'\boldsymbol{v}$ of responses that gave maximum separation of the group means, when measured relative to the within-group variance of the data. This linear combination he found, by maximizing $\{\boldsymbol{a}'(\bar{\boldsymbol{v}}^{(1)} - \bar{\boldsymbol{v}}^{(2)})\}^2/\boldsymbol{a}'\mathbf{S}\boldsymbol{a}$, to be $w = (\bar{\boldsymbol{v}}^{(1)} - \bar{\boldsymbol{v}}^{(2)})'\mathbf{S}^{-1}\boldsymbol{v}$. Given the group separation is maximized by this function, then a sensible allocation rule can be constructed by allocating $\boldsymbol{v}$ to π_1 if $(\bar{\boldsymbol{v}}^{(1)} - \bar{\boldsymbol{v}}^{(2)})'\mathbf{S}^{-1}\boldsymbol{v}$ is greater than some constant k, and otherwise to π_2. Note that this function is of exactly the same form as (12.21), since the portion $\frac{1}{2}(\bar{\boldsymbol{v}}^{(1)} - \bar{\boldsymbol{v}}^{(2)})'\mathbf{S}^{-1}(\bar{\boldsymbol{v}}^{(1)} + \bar{\boldsymbol{v}}^{(2)})$ of the latter is merely a sample-based constant and can be absorbed into the right-hand side of the inequality. The function $(\bar{\boldsymbol{v}}^{(1)} - \bar{\boldsymbol{v}}^{(2)})'\mathbf{S}^{-1}\boldsymbol{v}$ is generally known as *Fisher's linear discriminant function* (LDF). Because of its non-probabilistic derivation, Fisher's LDF should provide a useful tool for discrimination under wide distributional conditions (but may nevertheless be quite unsuitable for allocating a particular observation to one of two populations that are not multivariate normal). Furthermore, Fisher (1938) went on to show that if a dummy dependent variable w is defined to have one value for each individual in the sample from π_1 and another (different) value for each individual in the sample from π_2, then regressing w on the variables $\boldsymbol{v}$ using both samples yields $w = a + (\bar{\boldsymbol{v}}^{(1)} - \bar{\boldsymbol{v}}^{(2)})'\mathbf{S}^{-1}\boldsymbol{v}$ as the regression equation (for some constant a). This can be interpreted as saying that Fisher's LDF is the *linear* function that best predicts group membership of a sample individual (although, for non-normal distributions, the best function for predicting group membership according to Welch's (1939) approach will not be linear).

In view of these two non-distributional justifications, Fisher's LDF has been applied to a wide variety of discrimination problems and has not been restricted only to those situations in which normality can be assumed. Frequently, it has produced very good discrimination performance. Because of its ease of computation, moreover, there is a temptation to apply it blindly in all circumstances when discrimination between two groups is required. A number of studies have shown, however, that its performance may be less than satisfactory in certain circumstances. Lachenbruch *et al.* (1973) considered continuous but non-normal data. They showed that Fisher's LDF can be greatly affected by non-normality: error rates can be imbalanced, with those for one population being greater than the optimum values, while those for the other population are less than the optimum values, and the sum of the two error rates increases for some distributions. They recommend transforming to approximate normality before using the LDF. Moore (1973) studied the use of the LDF on binary data, where the values are coded 0 or 1, and found certain situations in which it performed badly. In particular, if the true log likelihood ratio for the two populations is plotted against the number of variables having the value 1, then in some populations this log likelihood ratio will not increase monotonically and is said to undergo a 'reversal'. Fisher's LDF, on the other hand, may increase monotonically for these populations; if it does increase monotonically then it will not be able to follow such 'reversals'. In such circumstances its performance will be much worse than that of the multinomial allocation rule. Krzanowski (1977) considered the application of Fisher's LDF to mixtures of binary and continuous data, and also found circumstances in which it performed poorly. Broadly speaking, positively correlated binary variables or sign reversals in the binary/continuous correlations from one population to the other are warning signs that the LDF may not do well, and that the location model would be preferable. Further evidence was provided by Knoke (1982), while Vlachonikolis and Marriott (1982) suggested modifications that could be made to the LDF to improve its performance in this situation. The message from all these studies is clear: do not trust to the sole use of the LDF if the data show gross departures from multivariate normality. The safest procedure is to try several allocation rules, and compare their performance using the methods to be described in Section 12.4.

12.3.2 Logistic discrimination

The probabilistic methods of Section 12.2 require parametric specification of each density function $f_i(\boldsymbol{v})$, followed by estimation of $f_1(\boldsymbol{v})/f_2(\boldsymbol{v})$ from the training samples and evaluation of this ratio for the individual $\boldsymbol{v}$ to be allocated. An alternative basis for discrimination and classification is a parametric specification of the posterior probabilities $q(\pi_1 \mid \boldsymbol{v})$ and $q(\pi_2 \mid \boldsymbol{v})$ of (12.12). Parameters in these expressions can again be estimated from the training sets, and then allocation rule (12.13) can be used directly. A possible model for the posterior probabilities is the logistic model:

$$q(\pi_1 \mid \boldsymbol{v}) = \frac{\exp\{\alpha_0 + \boldsymbol{\alpha}'\boldsymbol{v}\}}{1 + \exp\{\alpha_0 + \boldsymbol{\alpha}'\boldsymbol{v}\}}; \qquad q(\pi_2 \mid \boldsymbol{v}) = \frac{1}{1 + \exp\{\alpha_0 + \boldsymbol{\alpha}'\boldsymbol{v}\}}. \tag{12.29}$$

With this model, the posterior log odds-ratio is a linear function of the observed variables, i.e.

$$\log_e\{q(\pi_1 \mid \boldsymbol{v})/q(\pi_2 \mid \boldsymbol{v})\} = \alpha_0 + \boldsymbol{\alpha}'\boldsymbol{v}$$

and hence application of the allocation rule based on this model has close affinity with Fisher's LDF used via (12.10). However, the great advantage of the logistic approach is that only the $(p+1)$ parameters α_0 and $\boldsymbol{\alpha}$ need to be estimated from the sample data, whereas the other approaches require specification of $f_i(\boldsymbol{v})$ and estimation of many more parameters (e.g. means, variances, and covariances over all p variables). Moreover, J. A. Anderson (1972) has pointed out that the log odds-ratio is linear in $\boldsymbol{v}$ for a range of different assumptions about the $f_i(\boldsymbol{v})$ (e.g. for multivariate normal densities with common dispersion parameters; for independent binary variables; for multivariate discrete distributions following the log-linear model with the same interactions in each population), so that the logistic model will be optimal under a wide range of data types. Hence a direct assumption of such a logistic form for the posterior probabilities should yield a reasonable allocation rule in general; this approach is termed *logistic discrimination.*

This approach was suggested by Cox (1966) and Day and Kerridge (1967). It was developed in a series of papers by J. A. Anderson, commencing with Anderson (1972). A general review of all this work is provided by Anderson (1982), and the reader is referred to these papers and to Seber (1984, Chapter 6) for details of parameter estimation via iterative Newton–Raphson procedures. In view of its close affinity with the LDF, the performance of logistic discrimination has been studied in relation to that of linear discriminant analysis by a number of authors. The general consensus is that logistic discrimination is to be preferred when the distributions are clearly non-normal or the dispersion matrices are clearly unequal. Otherwise, the results of the two methods are likely to be very similar. Interested readers are referred for details to the papers by Halperin *et al.* (1971), Press and Wilson (1978), Crawley (1979), and Byth and McLachlan (1980).

12.3.3 Distance-based discrimination

The two training sets in any discrimination problem may be thought of as two swarms of points in p-dimensional space; the greater is the difference between the two populations π_1 and π_2, the greater will be the separation between the two swarms. An individual $\boldsymbol{v}$ to be allocated to one of π_1, π_2 may then be thought of as a single point in this space, and an intuitively attractive procedure would be to allocate $\boldsymbol{v}$ to the population to whose training set it is 'nearer'. This approach requires a definition of distance between the single observation $\boldsymbol{v}$ and each training sample. One possibility, already encountered in Chapter 8, is to define the squared distance by the Mahalanobis quantity $D_i^2 = (\boldsymbol{v} - \bar{\boldsymbol{v}}^{(i)})'\mathbf{S}^{-1}(\boldsymbol{v} - \bar{\boldsymbol{v}}^{(i)})$, where $\bar{\boldsymbol{v}}^{(i)}$ is the mean of the ith training set ($i = 1, 2$), and $\mathbf{S}$ is the covariance matrix pooled within the two training sets. Allocation of $\boldsymbol{v}$ would then to be π_1 if $D_1^2 < D_2^2$, and to π_2 if $D_1^2 \geq D_2^2$. Some simple algebraic manipulation establishes that this rule is exactly the same as (12.21) with k set to zero, i.e. Fisher's LDF assuming equal prior probabilities and equal costs.

On a more formal level, a number of authors have proposed definitions of distance between two populations characterized by density functions $f_1(\boldsymbol{v})$ and $f_2(\boldsymbol{v})$. (For a list of definitions and references, see Krzanowski (1983b).) A single point $\boldsymbol{v}$ can be represented by a degenerate population whose mass $f(\boldsymbol{v})$ is all concentrated at that point. Adopting any of these definitions, therefore, enables the distance between $\boldsymbol{v}$ and $f_i(\boldsymbol{v})$ to be evaluated in terms of the parameters of $f_i(\boldsymbol{v})$. Estimating these parameters from the training sets and substituting all unknown parameters in the distance by their estimates will then yield a distance between the individual and each training set.

The problem with this approach is to decide on the distance function to be used. Ali and Silvey (1966) showed that all the proposed distance functions are monotonic functions of the Mahalanobis distance when $f_1(\boldsymbol{v})$ and $f_2(\boldsymbol{v})$ are multivariate normal with common dispersion matrix. Hence the distance measure D_i^2 given above will be suitable for normal data, and the distance principle will yield an allocation rule equivalent to the maximum likelihood rule (12.6). For other distributions $f_i(\boldsymbol{v})$, different distance measures will yield different rules. However, if the distance function of Matusita (1964) is used, it can be shown that the distance-based allocation rule based on the above principle is equivalent to the maximum likelihood rule (12.6) for the multinomial model (Dillon and Goldstein, 1978) and for the location model (Krzanowski, 1986) as well. Since the distance-based rule is equivalent to the maximum likelihood rule for the multinomial model, however, it will exhibit that model's unsatisfactory features when there are empty cells in the contingency tables formed from the training sets. In an attempt to obviate these drawbacks, Dillon and Goldstein proposed an alternative distance-based allocation principle. They suggested calculating the distance between the two training sets twice: first after including the object $\boldsymbol{v}$ to be allocated with the training set from π_1 (yielding distance $D^{(1)}$), and then after including it with the training set from π_2 (yielding distance $D^{(2)}$). The individual $\boldsymbol{v}$ is then allocated to π_1 if $D^{(1)} > D^{(2)}$, to π_2 if $D^{(1)} < D^{(2)}$, and randomly if $D^{(1)} = D^{(2)}$. Dillon and Goldstein showed that this procedure overcame the empty-cell problem in certain circumstances, and hence was useful with discrete data. It could be used, in principle, with any other type of data and model, but once again the problem is to find first a suitable distance function. As yet, this aspect does not appear to have been studied.

12.4 Evaluating the performance of an allocation rule

It should be evident from the previous two sections and from the vast literature on discriminant analysis that very many allocation rules have been defined, and the user is at liberty to select any of these rules for a particular application. How can the user decide whether a chosen rule is a good one? This situation is somewhat analogous to that of point estimation: very many point-estimators exist for any given distributional parameter, and to decide between them the user must have access to some measure of their sampling behaviour (such as bias and/or standard error). In the case of an allocation rule, the most obvious measure of performance is an estimate of the proportion of individuals that will be misallocated by the rule. In other words, we require to find the *estimated error rate* of any allocation rule that is used in practice, to give an idea of its precision and to enable it to be compared with other allocation rules. We have derived various error rates already in Section 12.2, but these are all error rates computed from theoretical models for the data and they assume the optimal allocation rule for the given model is being employed. There are two major objections or drawbacks to the use of such theoretically based error rates in practice. First, the distributional model may be inappropriate for the data at hand. For example, allocation rule (12.21) and its associated estimated error rates (12.23) are derived by assuming the data come from multivariate normal populations having equal dispersion matrices. Even if these assumptions are invalid in a given case, there is nothing to prevent the user from employing the rule (12.21). However, if the data are grossly non-normal then (12.23) will provide very poor estimates of the performance of this rule on such data. The second major drawback is that sometimes a rule may be too complicated for the error rates to be found theoretically, even under the most straightforward probabilistic assumptions about the data. Thus if we wish to compare two or more allocation rules (which may be of very different appearances) with respect to their expected performance on a particular set of data, we require a *data-based* method of estimating their error rates.

The simplest data-based method is to apply the given allocation rule to the two training sets, and to estimate the error rate for each population by the proportion of individuals that are misclassified in the training set from that population. These estimates are often called the *apparent error rates,* and this method of estimation is generally referred to as the *resubstitution* method (because the individuals of the training sets are used to find the allocation rule and are then resubstituted into it to estimate its performance). Although a very simple procedure, the resubstitution method is a biased method of estimating error rates. The reason for this is that the *same* individuals are used to determine the allocation rule as well as to evaluate its performance. Now, since, by definition, any estimated allocation rule will be based on a function that maximally separates the two training sets, then the individuals of these training sets will be precisely those individuals (of all the ones in populations π_1 and π_2) that have least chance of being misallocated by the given rule. Hence the resubstitution method will provide an over-optimistic assessment of the success rate of the allocation rule on future individuals from π_1 and π_2: the smaller are the sample sizes of the training sets, the more extreme is this over-optimism likely to be. The resubstitution method may therefore give very misleading results (unless sample sizes are indeed large). Some authors accept that the resulting estimated error rates are unreliable in absolute terms, but maintain that they are useful in relative terms (for comparing the worth of several different allocation rules on a given set of data). The problem here is that different rules may be subject to different amounts of resubstitution bias because of their form, or because of the number of their parameters, or perhaps because of the configuration of the training data, and hence the results may still be grossly misleading. Consequently, the resubstitution method should be used with great caution.

It is evident, therefore, that a reliable estimate of the error rate incurred by an allocation rule will only be obtained if *different* data are used in the assessment of the rule from the data that are used in the formulation of the rule. This is essentially the principle of *cross-validation*. The simplest implementation of this principle is to split each training set randomly into two portions, and then to use one portion of each training set for estimation of the allocation rule itself, and the other portion to assess its performance by finding the proportion of individuals misallocated by the rule. This approach is also known as *sample-splitting*. It will provide unbiased estimates of the error rates, but also has a number of practical drawbacks. The chief of these is that, unless initial sample sizes are very large, either the estimation of the allocation rule or the assessment of its performance (or both) will be based on small samples and hence will be subject to large sampling fluctuations. A second drawback is that any future allocations will be made according to a rule based on the *whole* of the training sets, not just on a random portion of them. Hence the rule whose performance is being assessed by sample-splitting is not the rule that will be used in the future (unless sample sizes are very large indeed). The only situation in which the sample-splitting method can be used with complete confidence is when two additional samples from π_1 and π_2 become available *after* the allocation rule has been formulated.

To overcome most of the problems inherent in these two previous methods, Lachenbruch and Mickey (1968) proposed the *leave-one-out* method, now popularly referred to as the *cross-validation* method. This technique consists of determining the allocation rule using the sample data minus one observation, and then using the consequent rule to classify the omitted observation. Repeating this procedure by omitting each of the individuals in the two training sets in turn yields, as estimates of the error rates, the proportions of misclassified individuals in the two training sets. This approach thus circumvents the drawbacks of the resubstitution method in that the individual being classified each time has not been used in the formulation of the rule. It also circumvents the drawbacks of the sample-splitting method, in that the rule being assessed is based on all individuals bar one in the two training sets, and hence it will differ only minimally from the rule to be used on future individuals. Its chief practical drawback is the amount of computation necessary, as a new allocation rule must be calculated each time a new individual is omitted from the training sets. However, various algebraic identities exist which enable some of the standard allocation rules to be updated very easily for each unit omission, given the allocation rule based on the complete data. Lachenbruch and Mickey (1968) give simple formulae for updating Fisher's LDF on omitting an individual from the sample, Krzanowski (1975) considers similarly the location model rule for mixed binary and continuous data, while Campbell (1985) provides analogous formulae for updating the Mahalanobis distance. Where such formulae do not exist, moreover, suitable arrangement of the computations can minimize the computer effort. For example, the starting values for the iterative numerical estimation of logistic discrimination parameters can be set to the values of these parameters obtained for the full data set, and computation should be minimized. The leave-one-out method is thus a viable method for all allocation rules. It should be said, however, that whereas the resubstitution method can have large bias, the cross-validation method can have large variance and thus can still produce poor estimates in individual situations. Nevertheless is does seem to be the preferred method in much current work.

One final method that may be mentioned is also heavily computer-orientated, but in a slightly different manner from the Lachenbruch and Mickey proposal. This is Efron's (1979, 1981) *bootstrap* correction to the resubstitution method. In this technique (see also Section 17.2) a new sample is taken from each of π_1 and π_2 by sampling each of the original training sets *with replacement*. These two new samples are used to derive a new allocation rule, and the proportion of misallocations in each of π_1 and π_2 is noted when this new rule is applied first to the original training sets and then to the new training sets. Denote the differences between these two proportions by d_1 in π_1, and by d_2 in π_2. The resampling of the training sets is repeated a large number of times, and the averages $\bar{d}_1$ and $\bar{d}_2$ over these realizations are estimates of the bias in the resubstitution error rates. The estimated error rate in π_i is then the ordinary resubstitution error rate plus $\bar{d}_i$ $(i = 1, 2)$.

For further details of all the above techniques, and other data-based methods of error rate estimation, see the review by McLachlan (1986).

12.5 A practical example

Fibrocystic disease of the pancreas, commonly known as cystic fibrosis (CF), is a serious disease. It is an hereditary disease transmitted by a recessive gene that leads to a biochemical defect, namely a mucous abnormality of the exocrine function. This causes progressive lung damage and impairment of the pancreatic function. Some cases are not recognized until late childhood or adult life, while others are severely affected from birth. The pancreas malfunction is present in about 85 per cent of cases and can be effectively treated; most of the deaths caused by CF result from the lung damage. Early diagnosis is therefore very important, but exact diagnostic tests are very expensive. Various simple screening tests have been proposed, however, to enable large numbers of people to be tested. The idea is that anyone with an 'abnormal' value on these tests should be given one of the exact tests, while 'normality' on the simple tests should be taken to mean that the subject is free of the disease. An investigation into four such screening tests was undertaken by biochemists at the Royal Berkshire Hospital in Reading, to determine their usefulness in discriminating between CF-afflicted and normal subjects.

Test results were obtained on forty-six people known to have CF, and on 129 controls, known not to have CF. (These results have already been considered briefly in Section 7.6, to illustrate the Box and Cox transformation to normality.) Four tests were carried out on each subject: Orion Salivary (OS), E.I.L Sodium (EIL), Orion Heat (OH), and Orion Pilocarpine (OPC). Means and standard deviations of each test in each training set are given in Table 12.2.

It was thought that the test results might depend on age of subject, so simple linear regressions of results on age were first done, within each of the groups, for all tests. In the fibrocystic group, only EIL appeared to depend on age ($t = 2.27$), while in the control group OH and OPC gave significant regressions ($t = 3.05$ and 2.38 respectively). However, none of the tests exhibited a similar relationship with age in both groups. Since the age distribution of people was similar in the two groups, it was decided to ignore age effects. (If the effect of age on one variable is the same in both groups then ignoring the age effect will affect both misclassification rates equally. If the age effect differs between groups then to allow for it would require knowing into which group the person fell before correcting for age.) Although the age effects could be ignored, the data exhibited evidence of non-normality and heteroscedasticity. The method of Box and Cox (1964), as described in Section 7.6, was applied to each variable separately in an attempt to transform to approximate normality and to achieve equal variances for the two groups. The

Table 12.2 Means and standard deviations of raw data for CF and control groups in cystic fibrosis study

	CF group		Control group	
Test	Mean	s.d.	Mean	s.d.
OS	37.2	22.3	23.3	12.4
OH	97.1	42.0	30.9	21.3
OPC	85.2	34.0	18.0	13.0
EIL	17.0	15.8	7.2	5.7

maximum likelihood estimates of the Box–Cox parameter λ are given in Section 7.6 for each of the tests. The data were transformed using these estimates, and a summary of the transformed data is given in Table 12.3.

It is evident from Figs 7.3–7.6 of Chapter 7 that the transformed data adequately display the features of normality, while a comparison of the standard deviations between the CF and control groups of Table 12.3 also demonstrates that the heteroscedasticity has been removed satisfactorily. A linear discriminant function should therefore be suitable for use on the transformed data. This function can be calculated in various ways, and here the discriminant function was first tackled by the multiple regression method described in Section 12.3.1. To use this approach, a dummy dependent variable y must first be defined. In general, if n_1 and n_2 are the sample sizes in the two training sets respectively, then the two dummy variable values $n_2/(n_1 + n_2)$ for individuals in the first set, and $-n_1/(n_1 + n_2)$ for individuals in the second set, are the most suitable ones to choose, as then the average of the dummy variable over the whole data set is zero; a future individual is then allocated to π_1 or π_2 according to

Table 12.3 Means and standard deviations of transformed data for CF and control groups in cystic fibrosis study

	CF group		Control group	
Test	Mean	s.d.	Mean	s.d.
OS	0.50	0.05	0.55	0.05
OH	3.08	0.32	2.27	0.34
OPC	3.00	0.28	1.99	0.30
EIL	0.39	0.12	0.51	0.13

whether its predicted y value is positive or negative. In the present case these two values were $129/175 = 0.737$ and $-46/175 = -0.263$ respectively for the CF and control groups, and the fitted regression equation was

$$y = 0.562(\text{OPC}) + 0.127(\text{OH}) - 0.245(\text{EIL}) - 0.316(\text{OS}) - 1.296.$$

This regression had a multiple correlation coefficient of 0.852, indicating good ability to predict group membership. One major benefit of using the multiple regression approach is that well-established techniques associated with multiple regression can be used in any way that might be helpful. One such technique is variable subset selection, which can be employed to see if comparable discrimination can be achieved with fewer tests than all four. Regressing y on each single test, and then on each pair of tests, showed that the best single regressor variable was OPC, while the best two variables for inclusion together were OPC and OH. These gave multiple correlation coefficients of 0.840 and 0.846 respectively. Thus inclusion of all four tests gives only small improvement over the use of just OPC and OH. This latter pair was also the one chosen as the best pair by a standard stepwise variable selection program.

The above regression approach takes no account of either prior probabilities or differing costs due to misallocation; it simply produces the best linear functions from a straightforward discrimination point of view. In order to subject the four tests to a much more rigorous examination, therefore, it was decided to compute the LDF of (12.21) with k chosen as in (12.9), for each of the four tests individually, for all possible pairs of tests, and for all four tests together. To assess each of these functions the error rates would be estimated by the leave-one-out method. One problem had to be overcome, however, before this objective could be achieved. While the incidence of CF in the population is believed to be about 1 in 2500 births, so that 1/2500 can be taken as the prior probability q_1 of membership of the CF group, and 2499/2500 as the prior probability q_2 of membership of the control group, what should be the values $c(1 \mid 2)$ and $c(2 \mid 1)$ for the costs due to misallocating an individual? Clearly $c(2 \mid 1)$, the cost of misclassifying a fibrocystic patient as normal, will be greater than $c(1 \mid 2)$, the cost of misclassifying a normal person as fibrocystic, but by how much? The best solution, it was felt, would be to compute the error rates of the LDF for various values of r, the ratio $c(2 \mid 1)/c(1 \mid 2)$ of these two costs. The results are given in Table 12.4, where $P(C \mid N)$ denotes the probability of declaring a normal person to be fibrocystic, while $P(N \mid C)$ denotes the probability of declaring a fibrocystic patient normal.

It can be seen that when the two costs $c(1 \mid 2)$ and $c(2 \mid 1)$ are considered equal, the two error rates are very imbalanced for all test

Table 12.4 Leave-one-out error rates for various test combinations and various ratios of the two costs due to misallocation

Test combination	Ratio $r = c(2 \mid 1)/c(1 \mid 2)$ 1:1 P(C | N)	P(N | C)	10:1 P(C | N)	P(N | C)	100:1 P(C | N)	P(N | C)	1000:1 P(C | N)	P(N | C)	2500:1 P(C | N)	P(N | C)
OS	0.0	1.0	0.0	1.0	0.0	1.0	0.031	0.761	0.380	0.326
OH	0.0	1.0	0.0	0.913	0.008	0.609	0.085	0.239	0.163	0.130
OPC	0.008	0.674	0.008	0.413	0.008	0.174	0.023	0.087	0.031	0.0
EIL	0.0	1.0	0.0	1.0	0.0	1.0	0.078	0.717	0.325	0.261
OS + OH	0.0	1.0	0.0	0.935	0.008	0.565	0.085	0.217	0.124	0.152
OS + OPC	0.008	0.674	0.008	0.413	0.008	0.174	0.016	0.0	0.023	0.0
OS + EIL	0.0	1.0	0.0	1.0	0.0	1.0	0.085	0.630	0.279	0.261
OH + OPC	0.008	0.674	0.008	0.348	0.008	0.174	0.015	0.065	0.039	0.021
OH + EIL	0.0	1.0	0.0	0.891	0.0	0.5	0.109	0.217	0.140	0.152
OPC + EIL	0.008	0.674	0.008	0.457	0.008	0.196	0.023	0.065	0.031	0.022
All four	0.008	0.674	0.008	0.326	0.008	0.152	0.023	0.065	0.023	0.022

combinations: the chance of a false positive result (i.e. P(C | N)) is negligible, while the chance of a false negative result is very high. This is because the prior probability q_1 of obtaining a fibrocystic patient is so small that the cut-off value k of the LDF is exceptionally high. Thus nearly all the fibrocystic training set members are misclassified, while at most one of the control group individuals is misclassified (and in six of the test combinations none is misclassified). The imbalance is gradually evened out as r increases, and by the time r reaches the value 2500, most of the probabilities of a false negative are lower than their corresponding probabilities of a false positive. This is the state cf affairs we prefer, as we do not mind a false positive (since the individual is then subjected to a more rigorous test) but wish to avoid a false negative (since the fibrocystic patient does not then have a further chance to be detected). Clearly, we could force the opposite imbalance to the one in the first column, by making r extremely large. However, the main features and trends are evident from the table as it stands. The best single test for all values of r is OPC, which agrees with the previous regression result. However, while OPC + OH is the (equally) best two-test combination at lower values of r, OPC + OS is clearly best for the higher values of r. The great advantage here over the regression approach is the assessment of worth of each combination through the leave-one-out error rates.

13
Inferential aspects: more than two groups

We now generalize the situation postulated at the start of Chapter 12, and suppose that the data matrix $\mathbf{X}$ is partitioned into g (>2) sections, the ith of which comprises n_i rows and constitutes a random sample from a p-variate population π_i $(i = 1, \ldots, g)$. Inferences are sought about these populations, using the sample data in $\mathbf{X}$ as basis. Most of the techniques that are currently available require the assumption of multivariate normality, and are therefore only strictly applicable to continuous (or at least quantitative) data. In view of this restriction, we shall concentrate throughout the chapter on the multivariate normal model; a summary of available techniques for non-normal data will be given at the end of the chapter.

13.1 The multivariate normal model

Denote by $\boldsymbol{x}_{ij}$ the vector of p values observed on the jth individual of the ith group $(i = 1, \ldots, g; j = 1, \ldots, n_i)$. Then $\boldsymbol{x}_{i1}, \boldsymbol{x}_{i2}, \ldots, \boldsymbol{x}_{in_i}$ is assumed to be a random sample from a multivariate normal population whose mean vector is $\boldsymbol{\mu}_i$ and whose dispersion matrix is $\boldsymbol{\Sigma}_i$ $(i = 1, \ldots, g)$. This is the most general form of the model, allowing different means and dispersion matrices in each of the populations. Writing $\bar{\boldsymbol{x}}_i = \frac{1}{n_i}\sum_{j=1}^{n_i} \boldsymbol{x}_{ij}$ for the sample mean, and $\mathbf{C}_i = \sum_{j=1}^{n_i} (\boldsymbol{x}_{ij} - \bar{\boldsymbol{x}}_i)(\boldsymbol{x}_{ij} - \bar{\boldsymbol{x}}_i)'$ for the corrected sum of squares and products matrix in the ith sample, results of Section 8.2 show that maximum likelihood estimators of $\boldsymbol{\mu}_i$ and $\boldsymbol{\Sigma}_i$ are $\hat{\boldsymbol{\mu}}_i = \bar{\boldsymbol{x}}_i$ and $\hat{\boldsymbol{\Sigma}}_i = \frac{1}{n_i}\mathbf{C}_i$, while an unbiased estimator of $\boldsymbol{\Sigma}_i$ is given by $\mathbf{S}_i = \frac{1}{(n_i - 1)}\mathbf{C}_i$ $(i = 1, \ldots, g)$.

However, for many purposes, this model is *too* general. It contains very many dispersion parameters, and consequently very large data matrices are required for stable parameter estimates to be achieved. In practice, each group will usually be relatively small, and hence estimates of the individual $\mathbf{\Sigma}_i$ will be poor. Moreover, assumption of different $\mathbf{\Sigma}_i$ in each population leads to generalizations of the Behrens–Fisher problem already discussed in Chapter 12. A far more stable and satisfactory model is obtained by setting all the $\mathbf{\Sigma}_i$ equal to a common value $\mathbf{\Sigma}$, and hence assuming that $\boldsymbol{x}_{i1}, \boldsymbol{x}_{i2}, \ldots, \boldsymbol{x}_{in_i}$ is a random sample from a $\mathrm{N}(\boldsymbol{\mu}_i, \mathbf{\Sigma})$ population ($i = 1, \ldots, g$). This model can be justified by an extension of the argument used in Section 12.1.1 for the two-group case. However, if there is any doubt about the validity of this justification in a particular case, we can formally test the null hypothesis $H_0: \mathbf{\Sigma}_1 = \mathbf{\Sigma}_2 = \ldots = \mathbf{\Sigma}_g = \mathbf{\Sigma}$ against the alternative H_1: at least one $\mathbf{\Sigma}_i$ differs from the rest. The likelihood ratio principle yields the test statistic

$$-2 \log_e \lambda = n \log_e |\hat{\mathbf{\Sigma}}| - \sum_{i=1}^{g} n_i \log_e |\hat{\mathbf{\Sigma}}_i| \tag{13.1}$$

where $n = \sum_{i=1}^{g} n_i$ and $\hat{\mathbf{\Sigma}} = \frac{1}{n} \sum_{i=1}^{g} \mathbf{C}_i$ is the maximum likelihood estimate of the common dispersion matrix $\mathbf{\Sigma}$ under H_0. For large n, $-2 \log_e \lambda$ has a chi-squared distribution on $\frac{1}{2}p(p+1)(g-1)$ degrees of freedom if H_0 is true, and this distribution can be used to set critical values for a size α test. Note that (12.4) is just a special case of the above test when $g = 2$; the correction due to Box (1949) can be also used in the present circumstances.

13.2 Differences between the groups: canonical variates

We now adopt the more parsimonious model of the previous section, and assume that $\boldsymbol{x}_{i1}, \boldsymbol{x}_{i2}, \ldots, \boldsymbol{x}_{in_i}$ is a random sample from a $\mathrm{N}(\boldsymbol{\mu}_i, \mathbf{\Sigma})$ population ($i = 1, \ldots, g$). Thus we have g separate random samples, the ith of which has size n_i. Maximum likelihood estimators of $\boldsymbol{\mu}_i$ and $\mathbf{\Sigma}$ are $\hat{\boldsymbol{\mu}}_i = \bar{\boldsymbol{x}}_i$ and $\hat{\mathbf{\Sigma}} = \frac{1}{n} \sum_{i=1}^{g} \mathbf{C}_i = \frac{1}{n} \sum_{i=1}^{g} \sum_{j=1}^{n_i} (\boldsymbol{x}_{ij} - \bar{\boldsymbol{x}}_i)(\boldsymbol{x}_{ij} - \bar{\boldsymbol{x}}_i)'$ respectively. The $\hat{\boldsymbol{\mu}}_i$ are unbiased, but $\hat{\mathbf{\Sigma}}$ is biased. A suitable unbiased estimator is $\mathbf{W} = \frac{1}{n-g} \sum_{i=1}^{g} \mathbf{C}_i = \frac{1}{n-g} \sum_{i=1}^{g} \sum_{j=1}^{n_i} (\boldsymbol{x}_{ij} - \bar{\boldsymbol{x}}_i)(\boldsymbol{x}_{ij} - \bar{\boldsymbol{x}}_i)'$, the pooled within-group covariance matrix. Following the notation of Section 11.1, write $\mathbf{W}_0 = (n-g)\mathbf{W} = \sum_{i=1}^{g} \sum_{j=1}^{n_i} (\boldsymbol{x}_{ij} - \bar{\boldsymbol{x}}_i)(\boldsymbol{x}_{ij} - \bar{\boldsymbol{x}}_i)'$ for the pooled within-group sum-of-squares and products (SSP) matrix, and $\mathbf{B}_0 = \sum_{i=1}^{g} n_i(\bar{\boldsymbol{x}}_i - \bar{\boldsymbol{x}})(\bar{\boldsymbol{x}}_i - \bar{\boldsymbol{x}})'$ for the between-group SSP matrix (where $\bar{\boldsymbol{x}}$ is the mean of all n observations).

The first thing we might wish to ascertain is whether or not these g

random samples can be treated as a single sample of size $n = \sum n_i$ from one homogeneous population. This can be decided by testing the null hypothesis $H_0: \boldsymbol{\mu}_1 = \boldsymbol{\mu}_2 = \ldots = \boldsymbol{\mu}_g (= \boldsymbol{\mu})$ against the alternative H_1: at least one $\boldsymbol{\mu}_i$ differs from the rest. A suitable test statistic can be derived either by using one of the general principles of test construction described in Section 8.3, or by formulating a statistic that has clearly distinguishable behaviour under the two hypotheses H_0 and H_1.

Consider first the union–intersection test. Here we reduce the p-variate observation vector $\boldsymbol{x}$ to a univariate one by taking a linear combination $y = \boldsymbol{a}'\boldsymbol{x}$. For a specified value of the vector $\boldsymbol{a}$, this process yields g univariate samples with the jth observation in the ith sample denoted by $y_{ij} = \boldsymbol{a}'\boldsymbol{x}_{ij}$ $(i = 1, \ldots, g;\ j = 1, \ldots, n_i)$. Since the $\boldsymbol{x}_{ij}$ are independent observations from a $\mathrm{N}(\boldsymbol{\mu}_i, \boldsymbol{\Sigma})$ population, it follows from Property 1 of the multivariate normal distribution (Section 7.2) that the y_{ij} are independent observations from a univariate normal distribution with mean $\boldsymbol{a}'\boldsymbol{\mu}_i$ and variance $\boldsymbol{a}'\boldsymbol{\Sigma}\boldsymbol{a}$. Unbiased estimators of these univariate parameters are given by $\boldsymbol{a}'\bar{\boldsymbol{x}}_i$ and $\boldsymbol{a}'\mathbf{W}\boldsymbol{a}$ respectively ($i = 1, \ldots, g$). Also, the multivariate hypothesis H_0 is converted into the univariate hypothesis $H_0(\boldsymbol{a}): \boldsymbol{a}'\boldsymbol{\mu}_1 = \boldsymbol{a}'\boldsymbol{\mu}_2 = \ldots = \boldsymbol{a}'\boldsymbol{\mu}_g\ (= \boldsymbol{a}'\boldsymbol{\mu})$. Given the above assumptions, if $\boldsymbol{a}$ is known then $H_0(\boldsymbol{a})$ can be tested by means of a (univariate) one-way analysis of variance in which the ratio of between-group to within-group mean squares is compared with critical values given in F-tables. In terms of the vector $\boldsymbol{a}$, the test statistic is thus a simple multiple of $F = \boldsymbol{a}'\mathbf{B}_0\boldsymbol{a}/\boldsymbol{a}'\mathbf{W}_0\boldsymbol{a}$. By the union–intersection principle, therefore, the test statistic for the multivariate H_0 is obtained as the maximum of F over all possible vectors $\boldsymbol{a}$, and H_0 is rejected for large values of this maximum. Using exactly the same argument as in Section 11.1 (preceding eqn (11.6)), the maximum value of F is equal to the maximum eigenvalue of $\mathbf{W}_0^{-1}\mathbf{B}_0$. Thus the union–intersection test statistic for the hypothesis H_0 is θ_1, where $\theta_1 \geqslant \theta_2 \geqslant \ldots \geqslant \theta_p$ are the eigenvalues of $\mathbf{W}_0^{-1}\mathbf{B}_0$. For further details of this test statistic, along with references and tables of critical values, see Morrison (1976).

Consider next the generalized likelihood ratio test of H_0. For this test, we must maximize the joint likelihood of the g samples $\boldsymbol{x}_{ij}$ $(i = 1, \ldots, g;\ j = 1, \ldots, n_i)$ twice. The first maximization is done assuming that H_0 is true, and leads to the estimates

$$\bar{\boldsymbol{x}} = \frac{1}{n}\sum_{i=1}^{g}\sum_{j=1}^{n_i}\boldsymbol{x}_{ij} \quad \text{and} \quad \mathbf{T} = \frac{1}{n}\sum_{i=1}^{g}\sum_{j=1}^{n_i}(\boldsymbol{x}_{ij} - \bar{\boldsymbol{x}})(\boldsymbol{x}_{ij} - \bar{\boldsymbol{x}})'$$

for the common mean vector $\boldsymbol{\mu}$ and dispersion matrix $\boldsymbol{\Sigma}$. The second maximization is done without imposing any constraints on the parameters, and leads to the estimates $\bar{\boldsymbol{x}}_i$ for each $\boldsymbol{\mu}_i$ and $\hat{\boldsymbol{\Sigma}} = \frac{1}{n}\mathbf{W}_0$ for the common dispersion matrix $\mathbf{\Sigma}$. The generalized likelihood ratio test statistic for H_0 is the ratio of the maximized likelihoods under each of these two schemes. After some algebraic manipulation, it can be shown that the likelihood ratio test statistic is of the form

$$\Lambda = \frac{|\mathbf{W}_0|}{|\mathbf{B}_0 + \mathbf{W}_0|}. \tag{13.2}$$

This ratio of determinants is known as *Wilks' lambda*. Using various properties of determinants, it can further be shown that $\Lambda = \prod_{i=1}^{p} (1 + \theta_i)^{-1}$ where the θ_i are the eigenvalues of $\mathbf{W}_0^{-1}\mathbf{B}_0$ as defined above. The exact distribution of Λ under H_0 is discussed by Rao (1973), and for more general multivariate situations by T. W. Anderson (1984). For certain special cases it reduces to an F distribution, but in general its use is quite complicated as it is a function of three parameters (number p of variables, number b of degrees of freedom for $\mathbf{B}_0$, and number w of degrees of freedom for $\mathbf{W}_0$). However, various approximations to its distribution have been obtained and are useful in practice. Bartlett (1947) showed that $-\left(w - \frac{p - b + 1}{2}\right) \log_e \Lambda$ has approximately a χ^2 distribution on pb degrees of freedom if H_0 is true. This is an approximation that can be used in a variety of circumstances; in the present case $b = g - 1$ and $w = n - g$. A more detailed asymptotic expression is given by T. W. Anderson (1984), but Bartlett's approximation should be adequate in practice. An alternative approximation to the null distribution of Λ is the F approximation given by Rao (1973).

It can be seen from above that the union–intersection and likelihood-ratio principles lead to different test statistics, but both statistics are functions of the eigenvalues θ_i of $\mathbf{W}_0^{-1}\mathbf{B}_0$. In fact, T. W. Anderson (1984) shows that if a test statistic for $H_0: \boldsymbol{\mu}_1 = \ldots = \boldsymbol{\mu}_g$ is to be invariant under choice of origin and scale of the data, it *must* be a function of these eigenvalues. Thus whenever we seek a test statistic on heuristic grounds for H_0 versus H_1, we must restrict ourselves to functions of the θ_i if the statistic is to satisfy these sensible requirements. A number of such test statistics have been suggested for testing H_0 versus H_1, (see, e.g. Kenward (1979)). Two that merit some mention are the Lawley–Hotelling trace, $T_{(1)} = \text{trace}(\mathbf{W}_0^{-1}\mathbf{B}_0) = \sum_{i=1}^{p} \theta_i$, and the Pillai trace, $T_{(2)} = \text{trace}\{(\mathbf{W}_0 + \mathbf{B}_0)^{-1}\mathbf{B}_0\} = \sum_{i=1}^{p} \theta_i(1 + \theta_i)^{-1}$. These two criteria are probably the most-used alternatives to either Wilks' Λ or the largest-root criterion. Critical values for both may be found in appendix D of Seber (1984). $T_{(1)}$ is often called Hotelling's generalized T^2.

We saw in Chapters 8 and 12 that the two general principles of test construction lead to the *same* test statistic (Hotelling's T^2) when only one or two groups are present in the data, and indeed Hotelling's T^2 is also the end product of most heuristic approaches in these cases. On the other hand, each approach in the g-group case leads to a *different* test statistic. General reasons for this behaviour have already been outlined in Section 8.4. More precisely, when $g = 2$ the rank of $\mathbf{B}_0$ is 1, so there is only one non-zero eigenvalue θ_1 of $\mathbf{W}_0^{-1}\mathbf{B}_0$ and $\theta_2 = \theta_3 = \ldots = \theta_p = 0$. Hence in this case all the different test statistics given above reduce to θ_1 (or to a simple function of θ_1), which can be shown to equal T^2. The problem of which statistic to choose in the g-group case can sometimes be resolved by considering the structure of the alternative hypothesis; see Section 8.4 in particular, and also Kenward (1979) for further discussion. Unfortunately, studies of power and robustness have not given any clearcut guidelines; see Chatfield and Collins (1980) for additional references. In the end a decision between the statistics may have to be made according to ease of computation and access to suitable tables of critical values. In this respect Wilks' Λ is probably the most popular, if only because of the asymptotic chi-squared approximation to its null distribution given above. This approximation makes the statistic very easy to use, and provides a suitable guideline to the appropriate inference even when the sample size is quite small.

If the null hypothesis $H_0: \boldsymbol{\mu}_1 = \boldsymbol{\mu}_2 = \ldots = \boldsymbol{\mu}_g$ is rejected, the next step is clearly an investigation of the data to pinpoint the differences that exist among the groups. This investigation can be effected by means of canonical variate analysis, as described in Section 11.1. In addition to the eigenvalues of $\mathbf{W}_0^{-1}\mathbf{B}_0$, which are used in testing H_0, we now also need the eigenvectors (the canonical variates) ranked in decreasing order of the corresponding eigenvalues. (Note that $\mathbf{W}^{-1}\mathbf{B} = \{(n-g)/(g-1)\}\mathbf{W}_0^{-1}\mathbf{B}_0$. Hence eigenvectors of $\mathbf{W}_0^{-1}\mathbf{B}_0$ are the same as those of $\mathbf{W}^{-1}\mathbf{B}$, but any eigenvalue of $\mathbf{W}_0^{-1}\mathbf{B}_0$ is $(g-1)/(n-g)$ times the corresponding eigenvalue of $\mathbf{W}^{-1}\mathbf{B}$.) The best r-dimensional representation of the differences between the groups is obtained by plotting the individuals and the group means against the first r canonical variates as axes. This representation was derived in Section 11.1 from a purely descriptive standpoint, but we are now able to invoke the multivariate normal model to add inferential power to the technique. Return to the geometrical data representation as n points in p dimensions. The points in this space corresponding to the sample means are the best estimates of the positions of the unknown $\boldsymbol{\mu}_i$. When a canonical variate analysis is done, and points are plotted against the first few canonical variates as axes, the resulting diagram is simply a projection of the original p-dimensional space into a chosen subspace. Moreover, the points representing the sample means in this subspace are still the best estimates of the positions of the population means in the same subspace. However, since the $\bar{\boldsymbol{x}}_i$ are subject to sampling variability, it is important to be able to show on the canonical variate diagram the extent of the uncertainty in the positions of the $\boldsymbol{\mu}_i$ caused by sampling fluctuation. This will enable a judgement to be made as to whether an observed difference between two canonical variate means is indicative of a real difference between the corresponding population means or not.

Since the canonical variates in any given application are fixed linear functions of the original variates, and since the latter follow multivariate normal distributions, then the canonical variates will also be jointly normal. Furthermore, eqn (11.10) shows that *within each group,* the canonical variates are uncorrelated and have unit variance. Coupled with normality, this implies within-group independence and unit variance for the canonical variates. Thus if $\bar{\boldsymbol{y}}_i$ is the mean of the ith sample on a canonical variate diagram of dimension r, and n is large enough for $\mathbf{W}$ to be sufficiently close to $\boldsymbol{\Sigma}$, then $\bar{\boldsymbol{y}}_i \sim \mathrm{N}_r\left(\boldsymbol{v}_i, \frac{1}{n_i}\mathbf{I}\right)$, where $\boldsymbol{v}_i$ is the ith population mean on this diagram. Hence $(\bar{\boldsymbol{y}}_i - \boldsymbol{v}_i)\surd(n_i) \sim \mathrm{N}(\boldsymbol{0}, \mathbf{I})$, so that $n_i \sum_{j=1}^{r} (\bar{y}_{ij} - v_{ij})^2$ has a χ_r^2 distribution (where $\bar{y}_{ij}$, v_{ij} denote the jth elements of $\bar{\boldsymbol{y}}_i$ and $\boldsymbol{v}_i$ respectively). Thus if $\chi^2_{\alpha,r}$ is the value exceeded by 100α per cent of a χ^2 distribution on r degrees of freedom (obtainable from χ^2 tables), then

$$\mathrm{pr}\left\{\sum_{j=1}^{r} (\bar{y}_{ij} - v_{ij})^2 < \chi^2_{\alpha,r} \div n_i\right\} = 1 - \alpha.$$

But $\sum_{j=1}^{r} (\bar{y}_{ij} - z_j)^2 < c^2$ is the interior of an r-dimensional hypersphere centred at $\bar{\boldsymbol{y}}_i$ and having radius c, referred to axes $z_1, z_2, \ldots, z_r$. Hence a $100(1-\alpha)$ per cent confidence region for the true mean $\boldsymbol{v}_i$ (relative to the fixed canonical variate axes) is given by the interior of the hypersphere of radius $(\chi^2_{\alpha,r}/n_i)^{\frac{1}{2}}$ centred at $\bar{\boldsymbol{y}}_i$. In particular, if the canonical variate diagram is in two dimensions, then the $100(1-\alpha)$ per cent confidence region is a circle centred at $\bar{\boldsymbol{y}}_i$ and having radius $(\chi^2_{\alpha,2}/n_i)^{\frac{1}{2}}$. By similar reasoning, the $100(1-\alpha)$ per cent tolerance region for π_i (i.e. the region within which $100(1-\alpha)$ per cent of the *whole population* π_i is expected to lie) is given by the hypersphere centred at $\bar{\boldsymbol{y}}_i$ and having radius $(\chi^2_{\alpha,r})^{\frac{1}{2}}$. In two dimensions, this region is again a circle, and its radius is $(\chi^2_{\alpha,2})^{\frac{1}{2}}$.

For an illustration, let us return to the data of Ashton *et al.* (1957) illustrated in the two-dimensional canonical diagram of Fig. 11.4. The size of the eigenvalues of $\mathbf{W}^{-1}\mathbf{B}$ leads to rejection of the null hypothesis of equality of population means. The canonical variate means of Fig. 11.4 indicate that differences do exist between the groups, but which apparent differences are real and which ones are simply within the limits of sampling variability? Also, which of the fossils can be associated with which of the populations? From tables of the χ^2 distribution we see that the upper 90 per cent point of the χ^2 distribution on 2 degrees of freedom is 4.61. Thus a 90 per cent tolerance region for a population will be a circle with radius $\sqrt{4.61} = 2.15$, while a 90 per cent confidence region for a population mean will be a circle with radius $2.15 \div \sqrt{n}$, where n is the size of the sample taken from that population. Figure 13.1 shows the canonical variate diagram of Fig. 11.4, with 90 per cent tolerance regions constructed round each sample mean.

It is thus evident that the human populations are well separated from the ape-like populations (but not from each other), while the chimpanzee populations seem separated from the gorilla and orang-utan populations. H_0 is thus rejected in favour of a grouping of the population means. Furthermore, the point representing *Proconsul africanus* (PA) falls within both chimpanzee tolerance regions, thus reinforcing the conclusion of Ashton *et al.* that PA was like a chimpanzee. The remaining fossils were all thought to be more like man than ape, since they all fell on the

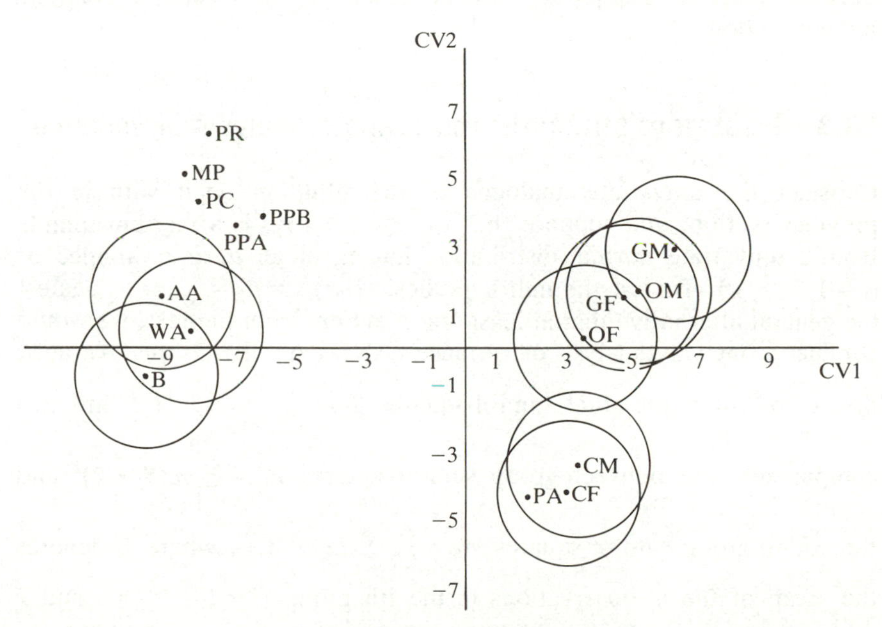

Fig. 13.1 Group means and fossil values plotted against first two canonical variates as axes, as in Fig. 11.4 but with 90 per cent tolerance regions added for each population

left-hand side of the canonical variate diagram. However, all the fossil points fall *outside* the tolerance regions for the human populations, and hence the fossils cannot be considered as part of the human population with any degree of certainty.

Finally, it is worth noting that if the difference between two specified groups is of a priori interest, then we can always conduct a Hotelling's T^2 test on their means (as described in Section 12.1.1). This procedure is equivalent to the testing of specified comparisons or contrasts among the means in a univariate analysis of variance, and similar strictures apply to the multivariate case as to the univariate one. In particular, testing differences between groups which appear to be different from a canonical variate diagram will lead to a significance rate much higher than the nominal level. This is because the Euclidean distance between the group means on the canonical variate diagram is equal to the Mahalanobis distance between the groups in the full data space (Section 11.1), and the Mahalanobis distance D is related to the two-sample Hotelling's T^2 statistic by the simple equation $(n_1+n_2)T^2 = n_1 n_2 D^2$. Hence a large distance between two points on the canonical variate diagram is guaranteed to produce a large T^2 value. For a genuine size α test, therefore, we should restrict ourselves to testing those differences between pairs of groups that are proposed *before* a canonical variate analysis is done.

13.3 Treatment structure: multivariate analysis of variance

Consider the univariate analogue of the situation dealt with in the previous section, and suppose that $x_{i1}, x_{i2}, \ldots, x_{in_i}$ is a random sample from a univariate normal distribution having mean μ_i and variance σ^2 $(i = 1, \ldots, g)$. To test the null hypothesis $H_0: \mu_1 = \mu_2 = \ldots = \mu_g$ against the general alternative that at least one μ_i differs from the rest, we would conduct a one-way analysis of variance (ANOVA). To do this, we need first to partition the total sum-of-squares $T_0 = \sum_{i=1}^{g} \sum_{j=1}^{n_i} (x_{ij} - \bar{x})^2$ into two components: the between-group sum-of-squares $B_0 = \sum_{i=1}^{g} n_i(\bar{x}_i - \bar{x})^2$ and the within-group sum-of-squares $W_0 = \sum_{i=1}^{g} \sum_{j=1}^{n_i} (x_{ij} - \bar{x}_i)^2$, where $\bar{x}_i$ denotes the mean of the n_i observations in the ith group $(i = 1, \ldots, g)$, and $\bar{x}$ denotes the mean of all $n = \sum_{i=1}^{g} n_i$ observations. The test is completed by setting out the following ANOVA table:

Source	Sum of squares	Degrees of freedom	Mean square	Ratio
Between groups	B_0	$g-1$	$B = B_0/(g-1)$	$F = B/W$
Within groups	W_0	$n-g$	$W = W_0/(n-g)$	
Total	T_0	$n-1$		

If H_0 is true, the ratio B/W is an observation from an F distribution on $(g-1)$ and $(n-g)$ degrees of freedom, while if H_0 is not true it is an observation from a non-central F distribution on the same degrees of freedom. Observations from the latter distribution are in general larger than ones from the former. Hence if a calculated value of the ratio in a given application exceeds the $100(1-\alpha)$ per cent point of the $F_{g-1,n-g}$ distribution, we reject H_0 in favour of the general alternative at the 100α per cent significance level. This is because the probability of obtaining such an extreme value under H_0 is less than α, while such large values are more common if H_0 is not true.

Now let us turn to the multivariate case, in which a p-variate vector $\boldsymbol{x}_{ij}$ replaces each univariate observation x_{ij}. The matrices $\mathbf{B}_0$ and $\mathbf{W}_0$ of Sections 13.2 and 11.1 are the natural analogues of the sums-of-squares B_0 and W_0 defined above. In fact, the elements of these two matrices can be obtained by successive application of the B_0 and W_0 formulae to individual variables or pairs of variables. Thus the p diagonal elements of $\mathbf{B}_0$ and $\mathbf{W}_0$ are simply the p values of B_0 and W_0 evaluated for each variable in the multivariate vector. To obtain the elements in the off-diagonal positions of $\mathbf{B}_0$ and $\mathbf{W}_0$, we must convert the sums-of-squares B_0 and W_0 into sums of products. For example, let $x_{ij}^{(k)}$ denote the kth variate value for the jth observation in the ith group $(i = 1, \ldots, g;$ $j = 1, \ldots, n_i$; $k = 1, \ldots, p)$, and identify the kth elements of the mean vectors $\bar{\boldsymbol{x}}_i$ and $\bar{\boldsymbol{x}}$ by $\bar{x}_i^{(k)}$ and $\bar{x}^{(k)}$ respectively. Then the element in the (k, l)th position $(k \neq l)$ of $\mathbf{B}_0$ is given by the formula for B_0 but with every square $(\bar{x}_i - \bar{x})^2$ replaced by the product $(\bar{x}_i^{(k)} - \bar{x}^{(k)})(\bar{x}_i^{(l)} - \bar{x}^{(l)})$. Similarly, the element in the (k, l)th position $(k \neq l)$ of $\mathbf{W}_0$ is given by the formula for W_0 but with every square $(x_{ij} - \bar{x}_i)^2$ replaced by the product $(x_{ij}^{(k)} - \bar{x}_i^{(k)})(x_{ij}^{(l)} - \bar{x}_i^{(l)})$. Thus, denoting these elements by $(\mathbf{B}_0)_{kl}$ and $(\mathbf{W}_0)_{kl}$ respectively, we have

$$(\mathbf{B}_0)_{kl} = \sum_{i=1}^{g} n_i(\bar{x}_i^{(k)} - \bar{x}^{(k)})(\bar{x}_i^{(l)} - \bar{x}^{(l)})$$

and

$$(\mathbf{W}_0)_{kl} = \sum_{i=1}^{g} \sum_{j=1}^{n_i} (x_{ij}^{(k)} - \bar{x}_i^{(k)})(x_{ij}^{(l)} - \bar{x}_i^{(l)}).$$

(Readers thoroughly familiar with univariate analysis of variance will recognize these modifications as precisely those introduced in an analysis of covariance.)

Furthermore, the ratio B_0/W_0 of the univariate analysis can be viewed as the product $\mathbf{W}_0^{-1}\mathbf{B}_0$ in the multivariate case. Direct analogy with the univariate ANOVA table thus provides us with the following multivariate analysis of variance (MANOVA) table:

Source	Sum-of-squares and products	Degrees of freedom	Mean squares and products
Between groups	$\mathbf{B}_0$	$g-1$	$\mathbf{B}=\mathbf{B}_0/(g-1)$
Within groups	$\mathbf{W}_0$	$n-g$	$\mathbf{W}=\mathbf{W}_0/(n-g)$
Total	$\mathbf{T}_0$	$n-1$	

The null hypothesis of equality of population means can be tested by using any of the statistics described in the previous section. These statistics are all functions of the eigenvalues θ_i of $\mathbf{W}_0^{-1}\mathbf{B}_0$. If the null hypothesis is rejected, then the reasons for its rejection can be investigated by plotting group means against the first few canonical variates, i.e. eigenvectors of $\mathbf{W}_0^{-1}\mathbf{B}_0$.

Frequently in practice there is further structure in the groups, beyond the simple fact that samples have been taken from g different populations. For example, if we consider the six species of apes on which the teeth measurements in Ashton *et al.* (1957) were recorded, we find that the six groups break down into a (3×2) arrangement (males and females for each of chimpanzee, gorilla, and orang-utan). This is a simple *factorial* arrangement, and interest in the corresponding univariate case would focus on determining whether there was evidence of any factorial *main effects* or *interactions*. A main effect for species would mean that chimpanzees, gorillas, and orang-utans exhibited differences in their responses when averaged over sex. A main effect for sex would mean that males and females exhibited differences in their responses when averaged over species. An interaction between sex and species would mean that the difference between male and female responses varied across the species. Thus if there is evidence of difference among the six groups, a search for the reason for this difference can be systematized by considering the two main effects and the interaction. If the factorial arrangement is *balanced,* in the sense that equal numbers of observations have been made in each of the six groups, then the univariate sum-of-squares between groups can be directly partitioned into a sum-of-squares for species, a sum-of-squares for sex, and a sum-of-squares for the species-by-sex interaction. The sum-of-squares within groups is then generally termed the *residual* sum-of-squares. Associated with each sum-of-squares is a value for degrees of freedom, and division by this value converts a sum-of-squares to a mean square. The significance or otherwise of each factorial effect is then tested by taking the ratio of the mean square for that effect to the residual mean square, and comparing the resulting value with critical values of the F distribution on effect and residual degrees of freedom respectively. If the computed value exceeds the critical value then the given effect is deemed to be significant at the chosen level of significance.

There are many possible *designs* in which structure is imposed on groups of units. It is not the purpose of this book to review these designs or the analysis of data arising within them, as the subject has been covered exhaustively in the univariate case by a number of authors who give precise details of each design and step-by-step methods of calculating the sums-of-squares for the corresponding ANOVA table: a classic reference is Cochran and Cox (1957), while a more recent one is Box *et al.* (1978). The sole aim here is to point out that any design or arrangement for which a univariate ANOVA exists (complete with formulae for sums-of-squares) can be generalized readily to a multivariate MANOVA in which sums-of-squares become sum-of-squares and -products matrices but degrees of freedom remain unaltered. The general principle described above is adhered to: diagonal elements of each sum-of-squares and -products matrix are obtained by applying the relevant univariate sum-of-squares formula to each variate in turn, while off-diagonal elements are obtained by converting the sum-of-squares formula into a sum-of-products, and applying it to the relevant pair of variates. Of course, in practice the computations would be done on a computer using a specially written statistical software package such as BMDP or SAS. However, the reader who wishes to follow some step-by-step calculations of this type is referred to the examples in Chatfield and Collins (1980, Chapters 8 and 9).

A typical MANOVA table will thus take on the following form:

Source	Sum of squares and products	Degrees of freedom	Mean squares and products
Factor A	$\mathbf{S}_A$	$a-1$	$\mathbf{S}_A \div (a-1)$
Factor B	$\mathbf{S}_B$	$b-1$	$\mathbf{S}_B \div (b-1)$
$\vdots$	$\vdots$	$\vdots$	
Factor Z	$\mathbf{S}_Z$	$z-1$	$\mathbf{S}_Z \div (z-1)$
Residual	$\mathbf{R}$	$r-1$	$\mathbf{R} \div (r-1)$
Total	$\mathbf{T}$	$n-1$	

Significance of a given factor, say Factor I, is tested by computing any of the test statistics described in Section 13.2 using the eigenvalues of $\mathbf{R}^{-1}\mathbf{S}_I$. If the factor is deemed to be significant, then reasons for significance can be investigated by plotting group means against canonical variates (eigenvectors of $\mathbf{R}^{-1}\mathbf{S}_I$). If Bartlett's asymptotic approximation to the distribution of Wilks' Λ is used, the appropriate degrees of freedom are $b = i - 1$ and $w = r - 1$.

To illustrate the general procedure, consider a two-factor analysis given in Johnson and Wichern (1982, pp. 274–7). (The following details are reprinted by permission of Prentice–Hall, Inc., Englewood Cliffs, New Jersey.) The data come from an experiment investigating the optimum conditions for extruding plastic film. The factors were *rate of extrusion* and *amount of an additive;* each factor had two levels, a high and a low level, and five replicate observations of plastic film were taken at each combination of the factor levels. Three responses were measured on each individual film: x_1(tear resistance), x_2(gloss), and x_3(opacity). The data are given in Table 13.1.

This is a (2×2) factorial arrangement with five observations in each of the four factor combinations, so we have sufficient data to extract information on the main effects of each factor and also the interaction between the two factors. Let us first consider three separate univariate analyses on each of x_1, x_2, and x_3 in turn. Formulae for calculation of factorial effects and sums of squares are given in Chapter 5 of Cochran and Cox (1957); they yield the ANOVA tables given in Table 13.2.

Critical values of the F distribution on 1 and 16 degrees of freedom are 4.49 for 5 per cent significance and 8.53 for 1 per cent significance. Thus we see that the main effect of rate is significant at the 1 per cent level for x_1 and at the 5 per cent level for x_2, while the main effect of additive is significant at the 5 per cent level for x_1. No other effects are significant. In the present case, the conclusions overall seem fairly clear: no evidence of interaction (hence the effects of the factors combine additively) and both main effects show significance to some extent. In many multivariate problems, however, conducting separate univariate analyses may lead to confusing or contradictory results, with conflicting patterns of significances among the different variates. A single overall MANOVA will take account of correlations between the variates and will give a much clearer view. For the plastic film data, the MANOVA table is given in Table 13.3. Note that the sums-of-squares from Table 13.2 now form the diagonal elements of the respective SSP matrices, while the off-diagonal elements are obtained by converting the sums-of-squares formulae to sums-of-products. (Only the lower half of each symmetric matrix is given, and we denote the residual SSP matrix by $\mathbf{S}_E$ rather than $\mathbf{R}$.)

Table 13.1 Data from an experiment on plastic film, from Johnson and Wichern, *Applied Multivariate Statistical Analysis*, p. 275. © 1982, Prentice-Hall Inc., reprinted by permission of Prentice-Hall, Inc., Englewood Cliffs, New Jersey

Rate of extrusion	Amount of additive	Replicate	x_1	x_2	x_3
Low	Low	1	6.5	9.5	4.4
		2	6.2	9.9	6.4
		3	5.8	9.6	3.0
		4	6.5	9.6	4.1
		5	6.5	9.2	0.8
	High	1	6.9	9.1	5.7
		2	7.2	10.0	2.0
		3	6.9	9.9	3.9
		4	6.1	9.5	1.9
		5	6.3	9.4	5.7
High	Low	1	6.7	9.1	2.8
		2	6.6	9.3	4.1
		3	7.2	8.3	3.8
		4	7.1	8.4	1.6
		5	6.8	8.5	3.4
	High	1	7.1	9.2	8.4
		2	7.0	8.8	5.2
		3	7.2	9.7	6.9
		4	7.5	10.1	2.7
		5	7.6	9.2	1.9

The logical way of proceeding with any analysis of variance (whether univariate or multivariate) is by testing for the presence of interactions first. If interactions are absent (i.e. not significant) then it is valid to average over levels of any particular factor for summary purposes, and hence tests of main effects are of interest. If interactions are present (i.e. significant), then averaging over levels of any factor included in a significant interaction is misleading and should not be done. In that case, tests of main effects are irrelevant, and data summary should be by way of higher-order tables of mean.

For the plastic film data, therefore, we start by testing for the presence of a rate–additive interaction. To do this we can calculate one of the

Table 13.2 Univariate analyses of variance for the plastic film data of Table 13.1

Source	Sum of squares	D.F.	Mean square	Ratio
ANOVA for x_1				
Main effect of rate	1.7405	1	1.7405	15.78
Main effect of additive	0.7605	1	0.7605	6.89
Interaction	0.0005	1	0.0005	$\ll 1$
Residual	1.7640	16	0.1103	
Total	4.2655			
ANOVA for x_2				
Main effect of rate	1.3005	1	1.3005	7.92
Main effect of additive	0.6125	1	0.6125	3.73
Interaction	0.5445	1	0.5445	3.31
Residual	2.6280	16	0.1643	
Total	5.0855			
ANOVA for x_3				
Main effect of rate	0.4205	1	0.4205	0.10
Main effect of additive	4.9005	1	4.9005	1.21
Interaction	3.9605	1	3.9605	0.98
Residual	64.9240	16	4.0578	
Total	74.2055			

following test statistics:

(1) Wilks' lambda, $\Lambda = \dfrac{|\mathbf{S}_E|}{|\mathbf{S}_I + \mathbf{S}_E|}$ = determinant of $(\mathbf{S}_E + \mathbf{S}_I)^{-1}\mathbf{S}_E$;
(2) maximum eigenvalue of $\mathbf{S}_E^{-1}\mathbf{S}_I$;
(3) trace of $\mathbf{S}_E^{-1}\mathbf{S}_I$;
(4) trace of $(\mathbf{S}_E + \mathbf{S}_I)^{-1}\mathbf{S}_I$;

and refer the calculated value to an appropriate chart or table. Johnson and Wichern (1982) give step-by-step details of the calculation of Wilks' lambda, and find that $\Lambda = 0.7771$. This can be converted to a value $F = 1.34$, which comes from an F distribution on 3 and 14 degrees if the null hypothesis of no interaction is true. The 95 per cent point of this F distribution is 3.34, and the calculated value is well below this critical value. Hence there is no significant evidence of interaction between the

Table 13.3 Multivariate analysis of plastic film data of Table 13.1

MANOVA

Source	SSP	D.F.
Main effect of rate	$\mathbf{S}_R = \begin{pmatrix} 1.7405 & & \\ -1.5045 & 1.3005 & \\ 0.8555 & -0.7395 & 0.4205 \end{pmatrix}$	1
Main effect of additive	$\mathbf{S}_A = \begin{pmatrix} 0.7605 & & \\ 0.6825 & 0.6125 & \\ 1.9305 & 1.7325 & 4.9005 \end{pmatrix}$	1
Interaction	$\mathbf{S}_I = \begin{pmatrix} 0.0005 & & \\ 0.0165 & 0.5445 & \\ 0.0445 & 1.4685 & 3.9605 \end{pmatrix}$	1
Residual	$\mathbf{S}_E = \begin{pmatrix} 1.7640 & & \\ 0.0200 & 2.6280 & \\ -3.0700 & -0.5520 & 64.9240 \end{pmatrix}$	16
Total	$\begin{pmatrix} 4.2655 & & \\ -0.7855 & 5.0855 & \\ -0.2395 & 1.9095 & 74.2055 \end{pmatrix}$	

Mean responses

		x_1	x_2	x_3
Factor 1 (rate)	High level	7.08	9.06	4.08
	Low level	6.49	9.57	3.79
Factor 2 (additive)	High level	6.98	9.49	4.43
	Low level	6.59	9.14	3.44

two factors. If we do not have access to tables of Wilks' Λ, we could use Bartlett's χ^2 approximation. Even though the sample size is not large, the result using this approximation should provide useful inferential guidelines. From Section 13.2, $W = -\left\{w - \frac{(p-b+1)}{2}\right\} \log_e \Lambda$ approximately has a χ^2 distribution on pb degrees of freedom if H_0 (no interaction) is true, where p is the number of variables (3), w is the number of residual degrees of freedom (16), and b is the number of hypothesis, i.e. interaction, degrees of freedom (1). With $\Lambda = 0.7771$ we thus find

$W = 3.66$; if H_0 is true this value should be commensurate with observations from a χ^2 distribution on 3 degrees of freedom. Since the observed value is only slightly greater than the expected value of this distribution the result is not significant, and we therefore reach the same conclusion as with the exact test. This means that the effect of the first factor—rate—is the same at each of the two levels of the second factor—additive—and vice-versa. Consequently, the effects of either factor can be averaged over the levels of the other factor for summary purposes.

To find out which of the factors is important, we next examine the two main effects. To test for the main effect of the first factor we simply replace $\mathbf{S}_I$ by $\mathbf{S}_R$ in our chosen test statistic, and to test for the main effect of the second factor we likewise replace $\mathbf{S}_I$ by $\mathbf{S}_A$. Johnson and Wichern again give full details for Wilks' lambda, from which statistic it can be concluded that both factors have significant main effects at the 5 per cent level. This means that the response at the high level of each factor is significantly different from the response at the low level, when averaged over the levels of the other factor. In general, we would examine the reasons for significance by plotting the means for each level of a factor against the canonical variates for that factor. Thus we would plot the means for each level of the factor 'rate' against the first few eigenvectors of $\mathbf{S}_E^{-1}\mathbf{S}_R$ as axes, and the means for each level of the factor 'additive' against the first few eigenvectors of $\mathbf{S}_E^{-1}\mathbf{S}_A$ as axes. However, since in this case there are only two levels for each factor, these plots would only contain two points each, and would not be any more informative than simply quoting the relevant mean vectors. These vectors are given below the MANOVA table in Table 13.3.

13.4 Univariate repeated measurements

A common practice in some areas of biological, agricultural, or psychological research is to set up experiments in which the same variable is measured on each experimental unit at a number of different occasions, the experimental design remaining unchanged between measurements. Such experiments are often said to be of 'repeated measurement design', and are particularly important when it is required to investigate the behaviour of some physiological or psychological process over a period of time. For example, blood haemoglobin might be measured at weekly intervals for a period of, say, twelve weeks after calving for a number of cows. The cows might be in several treatment groups according to diet, and it is of interest to investigate differences between the treatment groups over the whole period of the experiment. In an experiment on the efficacy of a vaccine for foot-and-mouth disease, a herd of cattle might be subdivided into males and females comprising calves, yearlings, and adults, vaccinated at the start of the experiment, and then antibody measurements are made on each animal every thirty days for a period of one year. It is then of interest to investigate differences between sexes and age groups in respect of the response over time to the vaccine. In a psychological experiment, a certain stimulus might be applied to groups of children of different ages and different home backgrounds, and the reaction times to the stimulus noted. If the same stimulus is applied to the same children on a number of separate occasions, the differences in 'learning effect' between ages and backgrounds can be studied.

A naive approach to the analysis of all these experiments might be to consider the time points at which measurements were taken as levels of a treatment factor to rank alongside the other treatment factors such as age, sex, diet, etc., and then to conduct a univariate analysis of variance of the responses with respect to all these factors. However, such an analysis assumes independent and equally variable responses that are normally distributed. While the normality and equal variance may be quite reasonable assumptions, independence most certainly is not. Observations that are made at different times on the same experimental unit (i.e. same cow, same child, etc.) will always show some correlation. This correlation will not generally be predictable (although it is reasonable to suppose that observations made close together in time will be more highly correlated than ones taken far apart in time). Analyses developed for split-plot experiments (Cochran and Cox 1957, Chapter 7) might be thought to be appropriate for repeated measurements, as such analyses allow for correlation between sub-plots (i.e. time points) within the whole plots (i.e. experimental units). However, such analyses assume equal correlation among all pairs of sub-plots, and random allocation of factor levels among sub-plots. In the repeated measure context we have already mentioned the intuitive impression of decreasing correlation with time, and hence violation of the equal correlation assumption, while randomization of factor levels is impossible if those levels are time points. The split-plot type of analysis is perhaps just about tenable if only a few, well-spaced, time points are used in the experiment (and see also Huyn and Feldt (1970) for a relaxation of the equal correlation assumption); in other circumstances a more general analysis is needed which allows for arbitrary correlation structure.

The way forward in general is to treat the data as if they were multivariate, in spite of the univariate nature of the response. Suppose that we have a simple one-way arrangement with g groups of experimental units (i.e. cows, children, etc.) and observations are taken at p time points. Let there be n_i experimental units in the ith group and let $x_{ij}^{(k)}$

denote the response to the variable of interest observed for the jth experimental unit in the ith group at the kth time point ($i=1, \ldots, g$; $j=1, \ldots, n_i$, $k=1, \ldots, p$). Then denote by $\boldsymbol{x}_{ij}$ the vector of observations made on the jth individual of the ith group, the elements of this vector being that individual's response at the p time-points, i.e. $\boldsymbol{x}'_{ij} = (x_{ij}^{(1)}, \ldots, x_{ij}^{(p)})$. Since individuals are independent, but responses at different time points for the same individual are correlated, we can assume that $\boldsymbol{x}_{i1}, \boldsymbol{x}_{i2}, \ldots, \boldsymbol{x}_{in_i}$ forms a random sample from a $\mathrm{N}(\boldsymbol{\mu}_i, \boldsymbol{\Sigma})$ distribution where $\boldsymbol{\Sigma}$ is an arbitrary (unknown) covariance matrix ($i=1, \ldots, g$). The data and assumptions are now precisely the same as those for the one-way arrangement in Section 13.3 above, and we can therefore use multivariate analysis of variance to test for differences among the group means $\boldsymbol{\mu}_i$. Such an analysis thus tests for differences in the responses between the groups, simultaneously for the p time points. Moreover, the logic remains the same irrespective of the experimental design employed, providing this design remains constant over the time points. For example, the g groups might be the 2 'sexes' $\times$ 3 'ages' arrangement of the foot-and-mouth vaccine experiment above, in which case we can use the general scheme of Section 13.3 to obtain main effects of both sex and age as well as sex–age interaction, simultaneously for all p time points.

Basing the analysis of repeated measurement data on multivariate analysis of variance in this way has the advantage of not requiring any assumption or information about the structure of covariances or correlations among the repeated measures. However, there are some disadvantages that must also be borne in mind. The chief difficulty arises if the number of repeated measurements p exceeds the number of experimental units. More precisely, the test statistics for multivariate analysis of variance described in Section 13.3 (such as Wilks' lambda, maximum eigenvalue, and trace) require that the number of degrees of freedom for residual be greater than p. If this is not the case then the residual matrix $\mathbf{R}$ is singular, and the analysis cannot be done. The only solution to this problem is to analyse a subset of the repeated measurements. A secondary drawback is that the canonical variates produced for a detailed examination of the components of the null hypothesis are often meaningless in terms of the aims of the experiment. Nevertheless, in such areas as psychology where there is no reason to expect a systematic relationship among means at each time point, the multivariate analysis of variance approach can prove to be most useful. (For some further technical details see, e.g. Bock (1975).)

In some areas such as biology, agriculture, or medicine, however, a further complication is introduced in that the primary interest is not in looking at between-group differences for isolated means at different time points, but in these between-group differences for the general response over time. Typically, responses are measures of size of some organism or unit (e.g. weights of cows, heights of plants, dry weights of roots). Since such responses tend to increase with time, repeated-measures data in this context are often referred to as *growth-curve* data, and their analysis as growth-curve analysis. What is usually necessary with such data is to fit a model that adequately represents the response over time, and then test for differences in the parameters of this model across the different treatment groups. The chosen model should be some convenient mathematical function that summarizes the essential aspects of the response using a relatively small number of parameters; an obvious pragmatic choice (to obtain a smooth function over time) is a low-order polynomial. In essence, therefore, the p unrelated group means of the repeated-measures analysis are replaced by p or fewer more meaningful measures representing the relationships among means over time. This process is equivalent to applying a set of constraints to the means across time. The parameters of the derived measures may then be tested for equality across treatment groups, or some parameters may be tested to see if they can be omitted and the model simplified, and so on.

The complicating feature in the analysis is the fact that the repeated measures are correlated but the underlying covariance structure is unknown. The general model formulation for the problem, and a suggested method of analysis, was first proposed by Potthof and Roy (1964). This method includes an arbitrary feature in the selection of a covariance matrix. An alternative approach using analysis of covariance was suggested independently by Rao (1966, 1967) and Khatri (1966). The Potthof–Roy or Rao–Khatri methods form the basis of most growth-curve analyses. A description of either would require some detailed and heavy algebraic manipulation, and so is not presented here. The interested reader is referred either to the sources cited above, or to Seber (1984, Section 9.7).

13.5 Similarities between the groups: common principal components

Sections 13.2 to 13.4 have been concerned with the testing of hypotheses about means of different populations, and all test statistics have been derived under the assumption of a common dispersion matrix for observations in each population. Returning to the models and notation of Section 13.1, the observations $\boldsymbol{x}_{i1}, \boldsymbol{x}_{i2}, \ldots, \boldsymbol{x}_{in_i}$ form a random sample from a $\mathrm{N}(\boldsymbol{\mu}_i, \boldsymbol{\Sigma}_i)$ population ($i = 1, \ldots, g$), and for the techniques of

Sections 13.2 to 13.4 to be applicable, we must assume

$$H_a : \boldsymbol{\Sigma}_1 = \boldsymbol{\Sigma}_2 = \ldots = \boldsymbol{\Sigma}_g \quad (= \boldsymbol{\Sigma}, \text{ say}).$$

Hypothesis H_a can be tested formally using (13.1), and sometimes this test will lead to rejection of H_a. It is then concluded that differences exist between the groups (since their dispersion matrices differ), and, hence, further tests on means are unnecessary. Often, however, it may be suspected that there is *some* similarity between the $\boldsymbol{\Sigma}_i$, but stopping short of actual equality among them. Formalizing and testing this similarity would be useful in giving a broad description of the similarities between the groups, in improving the precision of other parameter estimates, and in use for subsequent analysis.

One characterization of similarity among the sample covariance matrices $\mathbf{S}_i$ which has great potential for practical application is that the $\boldsymbol{\Sigma}_i$ share common principal axes although the size of each axis, and hence its relative importance, may differ from population to population. Following the discussions in Chapter 9, this characterization effectively states that the populations share a common set of principal components, but the variance of each component (and hence its ranking in the set) may differ from population to population. We are of course restricted to the sample data $\boldsymbol{x}_{ij}$, from which we can only compute sample principal components (Section 2.2). We therefore require to deduce from these sample components whether or not the populations share the same set of 'true' components. We have already considered a descriptive technique in Chapter 5 for comparing sets of principal components between groups, by looking at the critical angles between the subspaces they span in the common data space. The difference in the present instance is that we are considering the g complete sets of principal components, and asking whether they could have come from g populations having common population components.

Before considering the test of this hypothesis, we need to discuss its practical implications. One of the main uses of principal component analysis is in identifying the major sources of variation between individuals in a sample and thereby effecting data reduction by isolating minor (discardable) sources of variation. Interpretation of the components in terms meaningful to the experimenter additionally enables labels to be attached to the sources of variation, and it may give further insight into the situation under study. If all $\boldsymbol{\Sigma}_i$ are assumed to be different, then we are assuming that sources of variation are different from population to population, and no further structure is imposed. If all $\boldsymbol{\Sigma}_i$ are assumed equal to $\boldsymbol{\Sigma}$, then we are assuming that sources of variation and their order of importance are the same from population to population. If the $\boldsymbol{\Sigma}_i$ are assumed to share common principal axes but the size of each axis may vary from population to population, then we are assuming that *sources of variation* are the same from population to population, but that their *order of importance* may differ between populations. This last assumption may be a perfectly reasonable one in many practical situations. For example, if the same set of examinations is taken by each student in a number of different schools or colleges (as in the data set of Section 5.3), it is not unreasonable to assume that the basic sources of variation between students are similar in each college, but that differences in teaching practice between colleges will place different emphases on these sources. Thus two typical sources of variation between students might be arithmetic ability and language skills. A college which specializes in language teaching would tend to reduce the variation between students in language skills, perhaps at the expense of variation in arithmetic ability, while a college which specializes in teaching numerical skills might show the reverse features. Thus the same two sources of variation are present in both colleges, but in the reverse order of importance. Similarly, one might postulate the same sources of variation in people's attitude to work but differing in importance between the various socio-economic groups; or between subjects' responses to psychological stimuli, with differing levels of importance between adults and children; and so on.

The hypothesis that the $\boldsymbol{\Sigma}_i$ share common principal axes, but possibly of different sizes, is equivalent to the requirement that the $\boldsymbol{\Sigma}_i$ are all diagonalizable by the *same* orthogonal matrix, i.e.

$$H_{\mathrm{b}}: \mathbf{L}'\boldsymbol{\Sigma}_i\mathbf{L} = \boldsymbol{\Lambda}_i \quad (i = 1, \ldots, g)$$

where $\mathbf{L}$ is a $(p \times p)$ orthogonal matrix, and the $\boldsymbol{\Lambda}_i$ are all diagonal matrices. (But note that since the principal axes may be of different sizes in the different populations, the elements of $\boldsymbol{\Lambda}_i$ need not have the same rank order for each $i = 1, \ldots, g$.) Hypothesis H_{b} has been studied comprehensively by Flury (1984), who gives a maximum likelihood scheme for estimating $\mathbf{L}$ and $\boldsymbol{\Lambda}_i$ $(i = 1, \ldots, g)$ and develops the likelihood ratio test of H_{b}. He also discusses various properties of the common principal components, and analyses a number of data sets in some detail. The interested reader is referred to this paper for full details. Here we merely give some very simple intuitive estimators of $\mathbf{L}$ and $\boldsymbol{\Lambda}_i$, and provide an informal way of assessing the reasonableness of H_{b} which does not require assumptions of normality. These simple methods are given in Krzanowski (1984b).

If H_{b} is true, then $\mathbf{L}'\boldsymbol{\Sigma}_i\mathbf{L} = \boldsymbol{\Lambda}_i$ for $i = 1, \ldots, g$ so that $\mathbf{L}'(\boldsymbol{\Sigma}_1 + \ldots + \boldsymbol{\Sigma}_g)\mathbf{L} = \boldsymbol{\Lambda}_1 + \boldsymbol{\Lambda}_2 + \ldots + \boldsymbol{\Lambda}_g$. Thus writing $\boldsymbol{\Omega} = \boldsymbol{\Sigma}_1 + \ldots + \boldsymbol{\Sigma}_g$ and $\boldsymbol{\Lambda} = \boldsymbol{\Lambda}_1 + \ldots + \boldsymbol{\Lambda}_g$, then $\boldsymbol{\Lambda}$ is a diagonal matrix and $\mathbf{L}'\boldsymbol{\Omega}\mathbf{L} = \boldsymbol{\Lambda}$. This is just the spectral decomposition of $\boldsymbol{\Omega}$ (Appendix A, Section A5), and therefore shows that the columns of $\mathbf{L}$ are the eigenvectors (i.e. principal component coefficients) of $\boldsymbol{\Omega}$ corresponding to the eigenvalues given by the diagonal elements of $\boldsymbol{\Lambda}$. Given the usual (unbiased) estimator $\mathbf{S}_i$ of $\boldsymbol{\Sigma}_i$, where $(n_i - 1)\mathbf{S}_i = \sum_{j=1}^{n_i} (\boldsymbol{x}_{ij} - \bar{\boldsymbol{x}}_i)(\boldsymbol{x}_{ij} - \bar{\boldsymbol{x}}_i)'$, then an (unbiased) estimator of $\boldsymbol{\Omega}$ is $\mathbf{V} = \mathbf{S}_1 + \ldots + \mathbf{S}_g$. Hence simple estimators $\hat{\mathbf{L}}$, $\hat{\boldsymbol{\Lambda}}$ of $\mathbf{L}$, $\boldsymbol{\Lambda}$ are given by $\hat{\mathbf{L}}'\mathbf{V}\hat{\mathbf{L}} = \hat{\boldsymbol{\Lambda}}$. Thus the jth column of $\hat{\mathbf{L}}$ is the jth set of principal component coefficients in a principal component analysis of $\mathbf{V}$, and the principal component variances are the diagonal elements of $\hat{\boldsymbol{\Lambda}}$. Furthermore, having obtained $\hat{\mathbf{L}}$, the diagonal matrices $\boldsymbol{\Lambda}_i$ are estimated simply. Let the jth column of $\hat{\mathbf{L}}$ be written $\hat{\mathbf{l}}_j$. Then, if we set $\hat{\lambda}_{ij} = \hat{\mathbf{l}}_j'\mathbf{S}_i\hat{\mathbf{l}}_j$, $\boldsymbol{\Lambda}_i$ is estimated by $\hat{\boldsymbol{\Lambda}}_i = \text{diag}(\hat{\lambda}_{i1}, \hat{\lambda}_{i2}, \ldots, \hat{\lambda}_{ip})$.

If the sample sizes n_i are not all equal, then an informal method of assessing the reasonableness of H_b can be obtained by a similar argument. If H_b is true, then $\mathbf{L}'\boldsymbol{\Sigma}_i\mathbf{L} = \boldsymbol{\Lambda}_i$ for $i = 1, \ldots, g$ so that $\mathbf{L}'(c_i\boldsymbol{\Sigma}_i)\mathbf{L} = c_i\boldsymbol{\Lambda}_i$ for $i = 1, \ldots, g$, and any constants c_i. Thus $\mathbf{L}'[c_1\boldsymbol{\Sigma}_1 + \ldots + c_g\boldsymbol{\Sigma}_g]\mathbf{L} = c_1\boldsymbol{\Lambda}_1 + \ldots + c_g\boldsymbol{\Lambda}_g$. Let us now choose $c_i = (n_i - 1)/(n - g)$ where $n = \sum_{i=1}^{g} n_i$, and write $\boldsymbol{\Psi} = \{(n_1 - 1)\boldsymbol{\Sigma}_1 + \ldots + (n_g - 1)\boldsymbol{\Sigma}_g\}/(n - g)$, $\boldsymbol{\Lambda}_0 = \{(n_1 - 1)\boldsymbol{\Lambda}_1 + \ldots + (n_g - 1)\boldsymbol{\Lambda}_g\}/(n - g)$. Then $\mathbf{L}'\boldsymbol{\Psi}\mathbf{L} = \boldsymbol{\Lambda}_0(\text{diag})$. Hence the columns of $\mathbf{L}$ are given by the eigenvectors of $\boldsymbol{\Psi}$ corresponding to the eigenvalues in the diagonal positions of $\boldsymbol{\Lambda}_0$. Now an unbiased estimator of $\boldsymbol{\Psi}$ is given by $\{(n_1 - 1)\mathbf{S}_1 + \ldots + (n_g - 1)\mathbf{S}_g\}/(n - g)$, which is just the pooled within-group matrix $\mathbf{W}$ defined at the beginning of the chapter. Thus an estimate $\tilde{\mathbf{L}}$ of $\mathbf{L}$ is given by $\tilde{\mathbf{L}}'\mathbf{W}\tilde{\mathbf{L}} = \tilde{\boldsymbol{\Lambda}}_0$, i.e. by a principal component analysis of $\mathbf{W}$. Hence, if H_b is true, principal component analyses of $\mathbf{V}$ and $\mathbf{W}$ should both yield (as principal component coefficients) estimates of the same quantities $\mathbf{L}$. A comparison of the two sets of coefficients (either visually or by using the critical angles of Section 5.3) should therefore show up the reasonableness or otherwise of H_b. Similar sets of coefficients indicate that H_b is tenable, whereas very different sets indicate that it is not.

Note that if the n_i are all equal to m, then the above method is not applicable as $\mathbf{W} = (m - 1)(\mathbf{S}_1 + \ldots + \mathbf{S}_g)/(mg - g) = (1/g)\mathbf{V}$ and the two sets of coefficients are bound to agree. In this case, we can approximate the likelihood ratio test statistic for H_b (Flury, 1984) by $-2\log_e \lambda_b = \sum_{i=1}^{g} n_i \log_e |\mathbf{S}_i^{-1}\hat{\boldsymbol{\Sigma}}_i|$, where $\hat{\boldsymbol{\Sigma}}_i = \hat{\mathbf{L}}'\hat{\boldsymbol{\Lambda}}_i\hat{\mathbf{L}}$. If H_b is true, $-2\log_e \lambda_b$ should be approximately a χ^2 variable on $(g - 1)p(p - 1)/2$ degrees of freedom, so that critical values for testing H_b can be obtained from χ^2 tables.

As an example of common principal components, Flury (1984) re-examined the data on bone dimensions of the North American marten discussed by Jolicoeur (1963). The data comprised the four variables: x_1, log(length of humerus); x_2, log(width of humerus); x_3, log(length of femur); x_4, log(width of femur)—all measured on each of $n_1 = 92$(male), and $n_2 = 47$(female) individuals of the species *Martes americana*. The hypothesis H_a of equal dispersion matrices in the two populations was rejected, but principal component analyses performed separately in the two groups showed a similar pattern of the two orthogonal matrices. Flury therefore fitted common principal components to the two groups by maximum likelihood, and the likelihood ratio statistic for testing H_b yielded a value of 8.34. If H_b is true, this value should be an observation from a χ^2 distribution on 6 degrees of freedom. Since 8.34 is consistent with observations from such a distribution, the hypothesis of common principal axes in the two populations seems quite plausible. From Flury's analysis, the ordering of the eigenvalues in $\hat{\mathbf{\Lambda}}_1$ and $\hat{\mathbf{\Lambda}}_2$ was the same, but there was difference in the magnitude of the respective eigenvalues. This latter was therefore the reason for the rejection of hypothesis H_a. The maximum likelihood estimates of the common principal component coefficients are given in Table 13.4 in the lines labelled M. For direct comparison, the estimated coefficients are given from principal component analyses of matrices **V** and **W** respectively, in the lines labelled V and W.

Table 13.4 Common principal components of American marten data. Principal component coefficients estimated by maximum likelihood (M), principal component analysis of **V**, and principal component analysis of **W** respectively

		Coefficients			
Component	Estimation method	x_1	x_2	x_3	x_4
1	M	0.73	−0.14	−0.66	0.09
	V	0.73	−0.11	−0.67	0.08
	W	0.72	−0.26	−0.62	0.16
2	M	0.39	0.57	0.39	0.61
	V	0.38	0.57	0.40	0.61
	W	0.41	0.58	0.39	0.59
3	M	0.49	−0.58	0.63	−0.19
	V	0.51	−0.56	0.62	−0.20
	W	0.47	−0.48	0.68	−0.30
4	M	−0.28	−0.57	−0.08	0.77
	V	−0.24	−0.60	−0.08	0.77
	W	−0.30	−0.61	0.10	0.74

Comparison of the M and V lines shows that the simple method based on principal component analysis of **V** yields virtually identical coefficients to those from the full (iterative) maximum likelihood scheme. Furthermore, since $n_1 \neq n_2$ here, we can use the simple comparison between coefficients calculated from **V** and those calculated from **W** to assess informally the reasonableness of hypothesis H_b. Comparison of the V and W lines shows the two sets of coefficients to be acceptably close, and hence the hypothesis H_b to be reasonable. This is of course confirmed by the likelihood ratio test conducted by Flury. If the approximation to the likelihood ratio statistic suggested above is used, we obtain a value of 9.39 instead of 8.34; this is still consistent with an observation from a χ_6^2 distribution. Flury additionally compares the common principal components with Jolicoeur's original components given for each sex separately. He shows that the original interpretation of the components applies equally well to the common components, but that it is simplified in that only one set of transformations is needed for both groups. He further argues that the common principal component model has biological meaning and validity for a range of related biometrical problems.

For further studies relating principal component coefficients across groups, see Flury (1983, 1985, 1987).

13.6 Discrimination and classification

We now suppose that it is required to allocate an individual to one of g populations, on the basis of a set of p measurements that have been made on it. Thus we might wish to know to which of g Bronze Age populations an excavated skull belonged on the basis of p skull measurements, from which of g literary schools a newly discovered manuscript might have come on the basis of p stylistic variables, and so on. Most of the methods of Sections 12.2 and 12.3 can be generalized readily from the two-group to the g-group case to deal with the problem. We shall focus on just three of the methods that can be applied to continuous data via the multivariate normal model, starting with the generalization of the fundamental principles of Section 12.2.

13.6.1 Sample space partition

We suppose that $\boldsymbol{v}$ denotes a p-component random vector of observations made on any individual, $\boldsymbol{v}_0$ denotes a particular observed value of $\boldsymbol{v}$, and $\pi_1, \pi_2, \ldots, \pi_g$ denote the g populations involved in the problem. The basic assumption is that $\boldsymbol{v}$ has a different probability distribution in each of $\pi_1, \pi_2, \ldots, \pi_g$ (as otherwise some or all of the populations are indistinguishable); let the probability density of $\boldsymbol{v}$ be $f_i(\boldsymbol{v})$ in π_i $(i = 1, \ldots, g)$. As in Section 12.2.1, we envisage a possibly infinite sequence of future individuals that will require classification so that we must be prepared for any point in the sample space R to occur as the value of an individual. A classification rule can therefore be defined by a partition of the sample space into g exhaustive and mutually exclusive regions $R_1, R_2, \ldots, R_g$, together with the decision rule that assigns to π_i those individuals that fall in R_i $(i = 1, \ldots, g)$. Our objective is to determine the optimum partition of R.

As before, let q_i denote the prior probability of observing an individual from π_i $(i = 1, \ldots, g)$, and suppose that the R_i have been fixed in the above decision rule. Given that an individual $\boldsymbol{v}$ has come from population π_i, therefore, the probability that it will be allocated to population π_j is $\int_{R_j} f_i(\boldsymbol{v})\, d\boldsymbol{v}$ for all i and j. Thus the probability of observing $\boldsymbol{v}$ from π_i *and* allocating it to π_j is $p(j \mid i) = q_i \int_{R_j} f_i(\boldsymbol{v})\, d\boldsymbol{v}$. When $i = j$ this is the probability of correct classification, but when $i \neq j$ it is the probability of one of the $(g - 1)$ possible misclassifications for individuals from π_i. Clearly, therefore, the total probability of observing and misclassifying an individual from π_i is $p(\bar{i} \mid i) = \sum_{j \neq i} p(j \mid i)$, while the probability of observing and correctly classifying an individual from π_i is $p(i \mid i) = 1 - p(\bar{i} \mid i) = 1 - \sum_{j \neq i} p(j \mid i)$. As before, denote by $c(j \mid i)$ the cost associated with misallocating an individual from π_i into π_j. If we conventionally assume that $c(i \mid i) = 0$ for all i, then the total expected cost due to misallocation is given by

$$C = \sum_{i=1}^{g} \sum_{j=1}^{g} c(j \mid i) p(j \mid i)$$

$$= \sum_{i=1}^{g} \sum_{j=1}^{g} c(j \mid i) q_i \int_{R_j} f_i(\boldsymbol{v})\, d\boldsymbol{v} \qquad (13.3)$$

and the optimum partition of the sample space is that choice of $R_1, \ldots, R_g$ which minimizes (13.3).

Although a partition minimizing (13.3) can be obtained without too much difficulty (see, e.g. Seber (1984), exercise (6.13)), differential costs due to misallocation prove to be very troublesome to deal with in practice. Consequently it is generally assumed that all costs are equal, and the partition minimizing (13.3) is then equivalent to the partition which minimizes the total probability of misclassification,

$$P = 1 - \sum_{i=1}^{g} p(i \mid i) = 1 - \sum_{i=1}^{g} q_i \int_{R_i} f_i(\boldsymbol{v})\, d\boldsymbol{v}. \qquad (13.4)$$

This partition is given (Seber, 1984) by

$$R_i = \{\boldsymbol{v} \mid \mathrm{q}(\pi_i \mid \boldsymbol{v}) \geq \mathrm{q}(\pi_j \mid \boldsymbol{v}), \quad j = 1, \ldots, g\} \tag{13.5}$$

where $\mathrm{q}(\pi_i \mid \boldsymbol{v})$ is the posterior probability of membership of population π_i given the observation $\boldsymbol{v}$. Generalizing (12.12) to g populations, this posterior probability is given by

$$\mathrm{q}(\pi_i \mid \boldsymbol{v}) = q_i \mathrm{f}_i(\boldsymbol{v}) \Big/ \sum_{i=1}^{g} q_i \mathrm{f}_i(\boldsymbol{v}). \tag{13.6}$$

The classification rule associated with partition (13.5) thus allocates an individual $\boldsymbol{v}$ to the population π_i for which it has highest posterior probability, which is a direct generalization of allocation rule (12.13) to the g-group case. Use of (13.6) in (13.5) followed by algebraic manipulation yields the optimal allocation rule:

$$\text{Allocate } \boldsymbol{v} \text{ to } \pi_i \text{ if } \mathrm{f}_i(\boldsymbol{v})/\mathrm{f}_j(\boldsymbol{v}) \geq q_j/q_i \text{ for all } j = 1, \ldots, g. \tag{13.7}$$

Hence the g-group allocation can be effected by means of pairwise comparisons of all possible two-group allocation rules (12.10). Furthermore, if information about the prior probabilities q_i is unavailable then they may all be set equal to $1/g$ and (13.7) becomes:

$$\text{Allocate } \boldsymbol{v} \text{ to } \pi_i \text{ if } \mathrm{f}_i(\boldsymbol{v}) \geq \mathrm{f}_j(\boldsymbol{v}) \text{ for all } j = 1, \ldots, g. \tag{13.8}$$

In this case $\boldsymbol{v}$ is allocated to the population for which it has the highest likelihood, which is a direct generalization of the likelihood ratio rule (12.6) to the g-group case. In all the above optimal rules, unknown population parameters can be replaced by their estimates from g training sets to provide the corresponding estimative rules.

Now suppose that the π_i are multivariate normal populations, with different mean vectors $\boldsymbol{\mu}_i$ ($i = 1, \ldots, g$) but the same dispersion matrix $\boldsymbol{\Sigma}$ in each. Then

$$\mathrm{f}_i(\boldsymbol{v}) = \{(2\pi)^p \, |\boldsymbol{\Sigma}|\}^{\frac{1}{2}} \exp\{-\tfrac{1}{2}(\boldsymbol{v} - \boldsymbol{\mu}_i)'\boldsymbol{\Sigma}^{-1}(\boldsymbol{v} - \boldsymbol{\mu}_i)\}$$

so that

$$\begin{aligned} \log_e\{q_i \mathrm{f}_i(\boldsymbol{v})\} &= \log_e q_i + \tfrac{1}{2}\log_e\{(2\pi)^p \, |\boldsymbol{\Sigma}|\} - \tfrac{1}{2}(\boldsymbol{v} - \boldsymbol{\mu}_i)'\boldsymbol{\Sigma}^{-1}(\boldsymbol{v} - \boldsymbol{\mu}_i) \\ &= k + \log_e q_i + \boldsymbol{\mu}_i'\boldsymbol{\Sigma}^{-1}(\boldsymbol{v} - \tfrac{1}{2}\boldsymbol{\mu}_i) \end{aligned} \tag{13.9}$$

where $k = \frac{1}{2}\log_e((2\pi)^p \, |\boldsymbol{\Sigma}|) - \frac{1}{2}\boldsymbol{v}'\boldsymbol{\Sigma}^{-1}\boldsymbol{v}$. Allocating $\boldsymbol{v}$ to the population for which it has highest posterior probability is equivalent to allocating it to the population for which $\log_e\{q_i \mathrm{f}_i(\boldsymbol{v})\}$ is highest. Since k has the same value for all populations π_i, the use of (13.9) leads to the optimal rule:

Allocate $\boldsymbol{v}$ to the population π_i for which

$$L_i(\boldsymbol{v}) = \log_e q_i + \boldsymbol{\mu}_i'\boldsymbol{\Sigma}^{-1}(\boldsymbol{v} - \tfrac{1}{2}\boldsymbol{\mu}_i) \text{ is greatest.} \tag{13.10}$$

If population parameters are unknown but training sets are available from each population, then we may estimate the population means $\boldsymbol{\mu}_i$ by the training set means $\bar{\boldsymbol{v}}_i$, and the common dispersion matrix $\boldsymbol{\Sigma}$ by the usual pooled within-groups covariance matrix $\mathbf{W}$. We then obtain the estimative rule:

Allocate $\boldsymbol{v}$ to the population π_i for which

$$\hat{L}_i(\boldsymbol{v}) = \log_e q_i + \bar{\boldsymbol{v}}_i'\mathbf{W}^{-1}\{\boldsymbol{v} - \tfrac{1}{2}\bar{\boldsymbol{v}}_i\} \text{ is greatest.} \quad (13.11)$$

If prior probabilities are assumed to be equal, then the term $\log_e q_i$ is removed from both (13.10) and (13.11).

The differences $D_{ij}(\boldsymbol{v}) = L_i(\boldsymbol{v}) - L_j(\boldsymbol{v})$ (or $\hat{D}_{ij}(\boldsymbol{v}) = \hat{L}_i(\boldsymbol{v}) - \hat{L}_j(\boldsymbol{v})$) can be used to set up the optimal partition of the sample space, since $D_{ij}(\boldsymbol{v}) = 0$ (or $\hat{D}_{ij}(\boldsymbol{v}) = 0$) defines the hyperplane in the sample space that separates group i from group j. Probabilities of misallocation can again be estimated by cross-validation, as in the two-group case. If the dispersion matrices in the g populations are allowed to be different then the linear functions $L_i(\boldsymbol{v})$ and $\hat{L}_i(\boldsymbol{v})$ become quadratic functions, whose form is easily derivable from the corresponding two-group case, and the separating hyperplanes become curved hypersurfaces.

To illustrate the above ideas, consider the bivariate observations given by Lubischew (1962) that are shown in Table 13.5. These observations concern the variables: v_1, the maximal width of the aedeagus in the forepart (in microns), and v_2, the front angle of the aedeagus (in units of 7.5°) which were measured on specimens of three species (*Chaetocnema concinna, Chaetocnema heikertingeri,* and *Chaetocnema heptapotamica*) of male flea beatles. Seber (1984) gives the following sample mean vectors and pooled covariance matrix, rounded to two decimal places:

$$\bar{\boldsymbol{v}}_1 = \begin{pmatrix} 146.19 \\ 14.10 \end{pmatrix}, \quad \bar{\boldsymbol{v}}_2 = \begin{pmatrix} 124.65 \\ 14.29 \end{pmatrix}, \quad \bar{\boldsymbol{v}}_3 = \begin{pmatrix} 138.27 \\ 10.09 \end{pmatrix},$$

and

$$\mathbf{W} = \begin{pmatrix} 23.02 & -0.56 \\ -0.56 & 1.01 \end{pmatrix}.$$

In the absence of knowledge about the prior probabilities, we set $q_1 = q_2 = q_3 = \frac{1}{3}$, and hence can ignore the terms $\log_e q_i$ in the $\hat{L}_i(\boldsymbol{v})$. Use of these estimates in (13.11) gives the three functions

$$\hat{L}_1(\boldsymbol{v}) = 6.78v_1 + 17.64v_2 - 619.75$$
$$\hat{L}_2(\boldsymbol{v}) = 5.83v_1 + 17.31v_2 - 487.28$$
$$\hat{L}_3(\boldsymbol{v}) = 6.33v_1 + 13.44v_2 - 505.62.$$

Table 13.5 Bivariate observations on three species of male flea beatles (*Chaetocnema* spp.), reproduced from Lubischew (1962) with permission from The Biometric Society

C. concinna		*C. heikertingeri*		*C. heptapotamica*	
v_1	v_2	v_1	v_2	v_1	v_2
150	15	120	14	145	8
147	13	123	16	140	11
144	14	130	14	140	11
144	16	131	16	131	10
153	13	116	16	139	11
140	15	122	15	139	10
151	14	127	15	136	12
143	14	132	16	129	11
144	14	125	14	140	10
142	15	119	13	137	9
141	13	122	13	141	11
150	15	120	15	138	9
148	13	119	14	143	9
154	15	123	15	142	11
147	14	125	15	144	10
137	14	125	14	138	10
134	15	129	14	140	10
157	14	130	13	130	9
149	13	129	13	137	11
147	13	122	12	137	10
148	14	129	15	136	9
		124	15	140	10
		120	13		
		119	16		
		119	14		
		133	13		
		121	15		
		128	14		
		129	14		
		124	13		
		129	14		

A new observation $\boldsymbol{v} = \begin{pmatrix} v_1 \\ v_2 \end{pmatrix}$ is allocated to the species with the largest $\hat{L}_i(\boldsymbol{v})$. Alternatively, we can calculate the functions $\hat{D}_{ij}(\boldsymbol{v}) = \hat{L}_i(\boldsymbol{v}) - \hat{L}_j(\boldsymbol{v})$ giving

$$\begin{aligned} \hat{D}_{12}(\boldsymbol{v}) &= 0.95v_1 + 0.33v_2 - 132.47 \\ \hat{D}_{23}(\boldsymbol{v}) &= -0.50v_1 + 3.87v_2 + 18.34 \\ \hat{D}_{13}(\boldsymbol{v}) &= 0.45v_1 + 4.20v_2 - 114.13. \end{aligned}$$

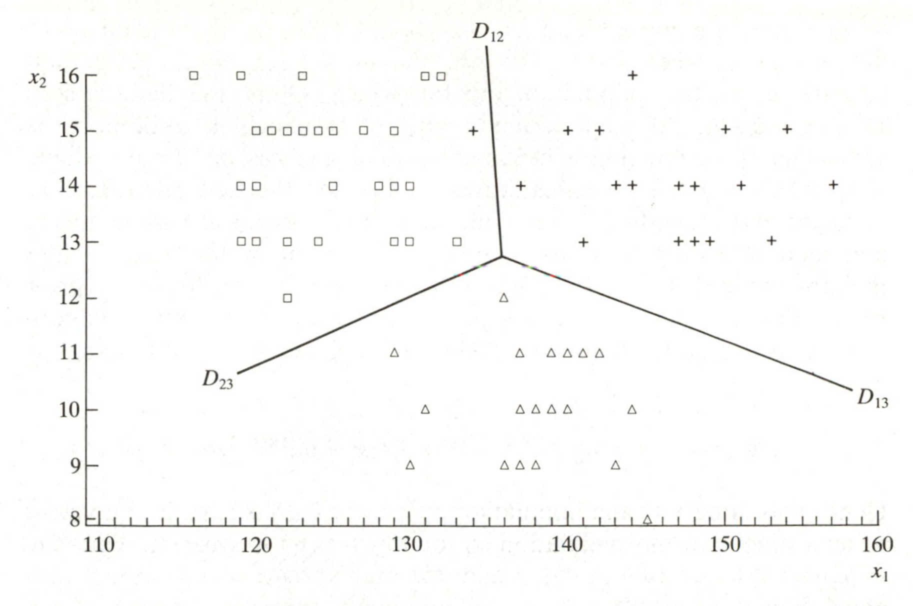

Fig. 13.2 Sample space for data of Table 13.5, with superimposed boundaries of classification regions. From *Multivariate Observations* by G. A. F. Seber, copyright © 1984 John Wiley & Sons Inc., reprinted by permission of John Wiley & Sons Inc

The allocation rule can be obtained by deducing the implications for the $\hat{D}_{ij}(\boldsymbol{v})$ of a particular $\hat{L}_i(\boldsymbol{v})$ being maximum. Thus, for example, $\hat{L}_1(\boldsymbol{v})$ is the greatest of the three $\hat{L}_i(\boldsymbol{v})$ if, and only if, $\hat{L}_1(\boldsymbol{v}) > \hat{L}_2(\boldsymbol{v})$ and $\hat{L}_1(\boldsymbol{v}) > L_3(\boldsymbol{v})$, which is the case if, and only if, $D_{12}(\boldsymbol{v}) > 0$ and $D_{13}(\boldsymbol{v}) > 0$. Also, note that $\hat{D}_{ij}(\boldsymbol{v}) = -\hat{D}_{ji}(\boldsymbol{v})$. Hence we deduce the rule:

$$\begin{aligned} \text{Allocate } \boldsymbol{v} \text{ to } \pi_1 \quad &\text{if} \quad \hat{D}_{12}(\boldsymbol{v}) > 0 \text{ and } \hat{D}_{13}(\boldsymbol{v}) > 0, \\ \text{to } \pi_2 \quad &\text{if} \quad \hat{D}_{12}(\boldsymbol{v}) < 0 \text{ and } \hat{D}_{23}(\boldsymbol{v}) > 0, \\ \text{and to } \pi_3 \quad &\text{otherwise.} \end{aligned}$$

Figure 13.2 shows the partition of the bivariate sample space obtained by drawing in the boundaries $\hat{D}_{ij}(\boldsymbol{v}) = 0$. It is evident that these three training sets show good separation, and in fact only one of the data points falls on the wrong side of a dividing boundary. (Note, however, that a more realistic assessment of future error rates by cross-validation would yield a greater number of misclassifications among the training samples.)

13.6.2 Distance-based allocation rule

Direct generalization of the idea in Section 12.3.3 suggests that if an individual $\boldsymbol{v}$ is to be allocated to one of g populations from each of which training sets are available, then we should compute the Mahalanobis distance D_i between $\boldsymbol{v}$ and the ith training set for $i=1,\ldots,g$, and allocate $\boldsymbol{v}$ to the population π_j for which D_j is the least. Here $D_i^2=(\boldsymbol{v}-\bar{\boldsymbol{v}}_i)'\mathbf{W}^{-1}(\boldsymbol{v}-\bar{\boldsymbol{v}}_i)$ where $i=1,\ldots,g$. Such a procedure is equivalent to conducting a canonical variate analysis on the g training sets, calculating the canonical variate score for the new individual $\boldsymbol{v}$, including this individual in the (full dimension) canonical variate space, and then allocating it to its nearest group mean in this space. Note therefore that this approach makes no provision for inclusion of prior probabilities q_i. Note also that if population parameters in the common-dispersion multivariate normal model are replaced by their training-set estimates, then

$$\log_e\{\hat{\mathrm{f}}_i(\boldsymbol{v})\}=\tfrac{1}{2}\log_e\{(2\pi)^p\,|\mathbf{W}|\}-\tfrac{1}{2}(\boldsymbol{v}-\bar{\boldsymbol{v}}_i)'\mathbf{W}^{-1}(\boldsymbol{v}-\bar{\boldsymbol{v}}_i).$$

Hence allocating $\boldsymbol{v}$ to the population π_j for which D_j is least is equivalent to allocating it to the population π_j for which $\hat{\mathrm{f}}_j(\boldsymbol{v})$ is greatest. Thus the minimum-distance rule in the case of the multivariate normal model with equal dispersion matrices is equivalent to the estimative version of the maximum likelihood rule (13.8), i.e. to rule (13.11) with all prior probabilities set equal.

13.6.3 Logistic discrimination

The ideas of Section 12.3.2 admit direct generalization to the g-group case. Equation (12.29), which provided a model for the posterior probabilities in the two-group case, can now be extended to:

$$\mathrm{q}(\pi_i\,|\,\boldsymbol{v})=\exp(\alpha_{0i}+\boldsymbol{\alpha}_i'\boldsymbol{v})\mathrm{q}(\pi_g\,|\,\boldsymbol{v})\quad\text{where } i=1,\ldots,g-1$$
$$\mathrm{q}(\pi_g\,|\,\boldsymbol{v})=1\Big/\left\{1+\sum_{i=1}^{g-1}\exp(\alpha_{0i}+\boldsymbol{\alpha}_i'\boldsymbol{v})\right\}.\tag{13.12}$$

Once again, parameters in this model can be estimated by iterative Newton–Raphson procedures, and allocation of individual $\boldsymbol{v}$ is made to the population for which it has highest estimated posterior probability. For details see the references cited in Section 12.3.2.

Note that with this model, the posterior log odds-ratio between any two of the g populations is again a linear function, namely

$$\begin{aligned}\log_e\{\mathrm{q}(\pi_i\,|\,\boldsymbol{v})/\mathrm{q}(\pi_j\,|\,\boldsymbol{v})\}&=(\alpha_{0i}-\alpha_{0j})+(\boldsymbol{\alpha}_i-\boldsymbol{\alpha}_j)'\boldsymbol{v}\quad(i\neq j\neq g)\\&=\beta_{0ij}+\boldsymbol{\beta}_{ij}'\boldsymbol{v},\quad\text{say,}\end{aligned}$$

and

$$\log_e\{q(\pi_i \mid \boldsymbol{v})/q(\pi_g \mid \boldsymbol{v})\} = \alpha_{0i} + \boldsymbol{\alpha}_i'\boldsymbol{v}.$$

Thus, application of the allocation rule based on this model has close affinity with the rules (13.10) and (13.11) obtained from the sample space partition principle as applied to the common-dispersion multivariate normal model. The argument in favour of logistic discrimination, as before, is that the model is appropriate under a much wider set of distributions than just the multivariate normal. Moreover, even when common-dispersion multivariate normality is an appropriate assumption, logistic discrimination may have the edge over the other methods because the parameters of the linear functions are estimated directly. Set against this, however, is the fact that iterative numerical procedures rather than exact algebraic expressions have to be employed, so that, for example, estimation of error rates by cross-validation can become prohibitively expensive in terms of computer time. Again, reference may be made to the work cited in Section 12.3.2 for comparative assessments.

13.7 Non-normal data

All the techniques discussed in this chapter have assumed that the multivariate normal model is appropriate, and moreover the assumption of equal dispersion matrices in the g populations has also been necessary in the majority of cases. Relatively little has either been done, or needs to be said, about other types of data.

For discrete data, obvious extensions of the methods given in Chapters 10 and 12 will yield the desired results. This is because division into g populations can be introduced by means of an extra g-state categorical variable denoting 'population membership', and all the ideas of Chapter 12 can be applied with this g-state variable replacing the 2-state one of that chapter. The fundamental log-linear models, with associated fitting procedures, are exactly as described in Chapter 10.

For mixed discrete/continuous data, the situation at present is less clear. The location model itself is readily generalizable to g groups, but no hypothesis-testing procedures to parallel those of the multivariate analysis of variance (described in Sections 13.2 and 13.3) are as yet available. The only g-group procedure at present available is the minimum-distance classification procedure discussed by Krzanowski (1986), using the distance measure of Krzanowski (1983b). If p_{im} is the probability of obtaining the mth pattern of discrete variable values in π_i, and the conditional distribution of the continuous variables is $N(\boldsymbol{\mu}_i^{(m)}, \boldsymbol{\Sigma})$ for this pattern of discrete variable values, then an individual whose discrete variables exhibit this pattern, and whose continuous variable vector is $\boldsymbol{y}$, is allocated to the population π_j for which

$$p_j = \{(2\pi)^p\,|\boldsymbol{\Sigma}|\}^{-\frac{1}{4}}(p_{jm})^{\frac{1}{2}}\exp\{-\tfrac{1}{4}(\boldsymbol{y}-\boldsymbol{\mu}_j^{(m)})'\boldsymbol{\Sigma}^{-1}(\boldsymbol{y}-\boldsymbol{\mu}_j^{(m)})\}$$

is maximum. Parameter estimation using training sets is a direct generalization of the 2-group procedure of 12.2.4, and the estimative rule follows directly. For details and examples, see Krzanowski (1986).

Part V: Analysing association among variables

Attention so far has been concentrated almost exclusively on the individuals in a multivariate sample, as represented by the rows of the data matrix. Techniques have therefore been concerned either with the highlighting of patterns among these individuals (or groups of individuals), or with the testing of hypotheses about them. The objective in each case has generally been achieved by finding some appropriate (linear) combination of variate values for each sample individual. Since each variate is either measured or observed on each individual in the sample, associations exist among the variates. These associations are exploited when deriving those linear combinations of the variates that optimize some criterion. Thus associations among the variates have played a fundamental role in most of the techniques discussed in earlier chapters, but the variates themselves have rarely been referred to explicitly (at least not until interpretation of results is required).

We now turn the spotlight on the variates, and for the last part of the book consider questions that may be asked about the columns of the data matrix. At the most basic level, we wish to have some mechanism for calculating the association between two variables or between two sets of variables. Having calculated an association, we might then wish to decide whether the calculated value was indicative of association between those variables in the whole population, or whether it would have arisen by chance in a random sample between two variables that are independent in the whole population. If there are many variables in a sample then there will be very many pairwise associations. Hence graphical techniques may convey the overall patterns among the variables much more readily than will a large matrix of numbers. Also, high association between two given variables may only exist because of their joint association with a third variable. Hence methods of correcting an observed association for the influence of other variables are of considerable practical use. All of these aspects are covered in Chapter 14.

Having measured and perhaps interpreted a set of associations, the next stage in any deeper analysis of a particular problem is usually some form of model-building. Models are generally constructed whenever we want to gain some insight into the mechanism of a system, or whenever we want to predict future behaviour in the most efficient manner possible. We may therefore want either to *use* the observed associations in a model, or to *explain* the observed associations by a fitted model. Multivariate models can involve either *manifest* or *latent* variables. The former type of model links the observed variables to each other, through either a multivariate regression or a multivariate functional relationship; this case is treated in Chapter 15. Latent-variable models, on the other hand, attempt to explain associations between variables by assuming that these variables are (linear) functions of other, unobservable, variables; such models are considered in Chapter 16.

14 Measuring and interpreting association

14.1 Measuring association between two variables

Denote two variables x_1 and x_2. Suppose x_1 and x_2 are both measured on each of n individuals, and the resulting values are (x_{11}, x_{12}), $(x_{21}, x_{22}), \ldots, (x_{n1}, x_{n2})$. These vectors are, of course, just the rows of the $(n \times 2)$ data matrix $\mathbf{X}$. As in previous chapters, we distinguish the three broad characterizations: (1) x_1 and x_2 both quantitative, (2) x_1 and x_2 both qualitative, (3) one of x_1, x_2 quantitative and the other qualitative.

Consider first case (1). 'Association' here means that the values x_{i1} bear some clear relationship to the corresponding values x_{i2} ($i = 1, \ldots, n$). A natural way of investigating whether association exists is therefore to plot a scatter diagram of the sample, i.e. to plot n points in two dimensions in such a way that the ith point has coordinates (x_{i1}, x_{i2}) referred to two orthogonal axes. This produces the familiar geometrical representation of the sample that we have used throughout the book when dealing with quantitative data. In the present case, of course, the data matrix has just two columns so the representation is exactly two-dimensional, and no dimensionality reduction is needed. The *strength* of association between x_1 and x_2 is shown by the closeness with which the points lie to a curve in the diagram, while the *nature* of the association is determined by the mathematical form of this curve. Figure 14.1 shows four possible examples of such a scatter diagram. Diagrams (a) and (b) illustrate *linear* associations, while (c) and (d) illustrate *non-linear* associations (the former being exponential and the latter parabolic). Furthermore, (b) and (c) are *strong* associations, the points lying close to a line and to an exponential curve respectively, while (a) and (d) are *weak* associations, the points being scattered quite widely about a line and a parabola respectively.

Since there are so many possible varieties of non-linear association, the term 'association' is restricted almost exclusively to mean 'linear association'. Hence if we want to measure the association between x_1 and x_2, we have to measure in some way the closeness with which the points on the

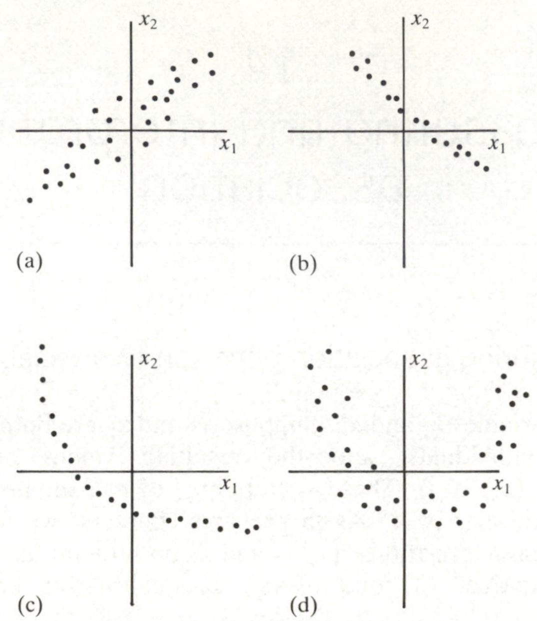

Fig. 14.1 Four examples of bivariate scatter diagrams, illustrating different natures and strengths of association between x_1 and x_2

scatter diagram lie to a straight line. A fundamental measure of linear association between x_1 and x_2 is the (sample) covariance between them:

$$s_{12} = \frac{1}{n-1} \sum_{i=1}^{n} (x_{i1} - \bar{x}_1)(x_{i2} - \bar{x}_2).$$

If x_1 and x_2 tend to increase and decrease together, then x_{i1} will tend to be greater than $\bar{x}_1$ when x_{i2} is greater than $\bar{x}_2$, and lower than $\bar{x}_1$ when x_{i2} is lower than $\bar{x}_2$. Hence $(x_{i1} - \bar{x}_1)$ and $(x_{i2} - \bar{x}_2)$ will tend to be positive or negative together, and their product will generally be positive. Thus a positive covariance means that x_1 and x_2 increase or decrease together (see Fig. 14.1(a)). On the other hand, if x_1 increases when x_2 decreases, and vice-versa, then one of $(x_{i1} - \bar{x}_1)$, $(x_{i2} - \bar{x}_2)$ will tend to be positive when the other is negative. Thus the product will tend to be negative. A negative covariance therefore means that x_1 increases as x_2 decreases, and vice-versa (see Fig. 14.1(b)). If there is no apparent linear trend in the data, the positive and negative terms of s_{12} will tend to cancel, giving a covariance near zero.

The unsatisfactory feature of covariance as a measure of association is that it is scale dependent: it is influenced by the spread of values in x_1 and/or x_2. This is corrected for by standardizing x_1 and x_2, i.e. dividing each set of values by the respective standard deviations. This adjustment converts the covariance between x_1 and x_2 into the correlation between x_1 and x_2:

$$r_{12} = \sum_{i=1}^{n} (x_{i1} - \bar{x}_1)(x_{i2} - \bar{x}_2) \Big/ \sqrt{\left\{ \sum_{i=1}^{n} (x_{i1} - \bar{x}_1)^2 \sum_{i=1}^{n} (x_{i2} - \bar{x}_2)^2 \right\}}$$
$$= s_{12}/(s_1 s_2) \tag{14.1}$$

where $s_i^2 =$ (sample) variance of x_i.

Thus the correlation always lies between +1 and −1: a value of +1 implies a perfect positive linear association between x_1 and x_2, a value of −1 implies a perfect negative linear association between x_1 and x_2, while a value of zero implies no association between x_1 and x_2. It is important to remember, though, that 'no association' means 'no linear association', so that a scatter diagram such as in Fig. 14.1(d) would yield a correlation near zero despite there being some evident non-linear association between x_1 and x_2.

The correlation coefficient of (14.1) is sometimes referred to as the *product-moment* correlation coefficient. If the values x_{i1} are replaced by their rank ordering, and if the values x_{i2} are similarly replaced by their rank ordering, then (14.1) applied to the two sets of ranks results in *Spearman's rank correlation coefficient*. This measure is most useful when the raw data are themselves in the form of ranks (e.g. ranking of n students on each of two examinations). If the x_{i1} and x_{i2} are replaced by the first n integers (i.e. the ranks of the individuals), then it is easy to show that (14.1) reduces to

$$r_{12}^{(s)} = 1 - \left\{ 6 \sum_{i=1}^{n} d_i^2 \right\} \Big/ \{n(n^2 - 1)\} \tag{14.2}$$

where d_i is the difference in ranks between x_1 and x_2 for individual i. However, a correction is necessary if ties occur in one or both of the series of ranks, because then the values are not all the integers between 1 and n. The tied observations are given the mean value of the ranks which they cover, and the correction is to add $(t^3 - t)/12$ to $\sum d_i^2$ *for each set of t tied observations*. (An example of this calculation is given at the end of this section.)

Turn next to case (2), where x_1 and x_2 are both qualitative or categorical. Suppose that x_1 has r categories, and x_2 has c categories. Then the sample space for (x_1, x_2) pairs consists of the rc possible combinations of these categories, and the sample of n individuals can be summarized in an $(r \times c)$ table showing the incidences of each of these combinations. The data can therefore be set out as in Table 14.1 (in the same manner as the tables of Chapter 10). Thus there are n_{ij} individuals

Table 14.1 Contingency table showing incidences of each combination of r categories of x_1 and c categories of x_2 in a sample of n individuals

	x_2				
x_1	1	2	...	c	Total
1	n_{11}	n_{12}	...	n_{1c}	$n_{1.}$
2	n_{21}	n_{22}	...	n_{2c}	$n_{2.}$
⋮	⋮	⋮		⋮	
r	n_{r1}	n_{r2}	...	n_{rc}	$n_{r.}$
Total	$n_{.1}$	$n_{.2}$	...	$n_{.c}$	n

in the sample exhibiting category i of x_1 and category j of x_2, $n_{i.} = \sum_{j=1}^{c} n_{ij}$ individuals that have category i of x_1 irrespective of the category of x_2, and $n_{.j} = \sum_{i=1}^{r} n_{ij}$ individuals that have category j of x_2 irrespective of the category of x_1.

By 'association' in this context is meant the tendency of individuals to exhibit different patterns among the categories of x_2 for each category of x_1. At its extreme this tendency would be manifested in a pattern where individuals fell into only one category of x_2 for each category of x_1, and these x_2 categories differed between the x_1 categories. The converse (i.e. no association) is equivalent to the usual definition of independence between variates: the pattern of incidences across categories of x_2 is the same for each category of x_1. Most readers familiar with elementary statistics will have encountered the χ^2 test of no-association in a contingency table. The test statistic is given by

$$X^2 = \sum_{i=1}^{r} \sum_{j=1}^{c} (n_{ij} - \hat{n}_{ij})^2 / \hat{n}_{ij} \tag{14.3}$$

where $\hat{n}_{ij} = n_{i.} n_{.j} / n$, the estimate of n_{ij} assuming no association between x_1 and x_2. The larger the value of X^2, the greater is the deviation from independence and hence the less likely is the hypothesis of no-association to be true. Consequently it is tempting to treat X^2 itself as a measure of association between x_1 and x_2. However, X^2 suffers from an analogous problem to that of covariance discussed earlier, namely that it is affected by the 'scales' of x_1 and x_2 (i.e. by the values r and c). It is also affected by the sample size n. To be a useful measure of association it must first be 'standardized', and two possible coefficients are:

$$r_{12}^{(t)} = X \div [n\sqrt{\{(r-1)(c-1)\}}]^{\frac{1}{2}} \tag{14.4}$$

or

$$r_{12}^{(c)} = X \div \{n \min(r-1, c-1)\}^{\frac{1}{2}}. \tag{14.5}$$

If x_1 and x_2 are both dichotomous variables then $r = c = 2$, and both (14.4) and (14.5) reduce to $r_{12} = X/\sqrt{n}$, which was used in the dissimilarity measure (xii) between two dichotomous variables as given in Section 1.4.

Two other popular measures of association between dichotomous variables, particularly in epidemiological work, are the *odds ratio* $(n_{11}n_{22})/(n_{12}n_{21})$ and the *relative risk* $(n_{11}n_{2.})/(n_{21}n_{1.})$. However, as usage of these measures is more specialized they will not be considered further here. For up-to-date work and references on them, see Jewell (1986) and Kraemer (1986).

If the categories of x_1 and x_2 are ordered, then they could each be assigned a numerical value, and the contingency table would thus become a bivariate grouped frequency table. Association between x_1 and x_2 can then be measured by the correlation coefficient (14.1), since x_1 and x_2 have both been rendered quantitative. The problem here is in the assignment of the numerical values to the categories, which has the danger of being arbitrary or inappropriate (but see the discussion of correspondence analysis in Section 4.5). A much more formal approach to the problem was taken by psychometric research workers in the 1940s and 1950s, who adopted the view that bivariate ordered categories can be treated as categorized bivariate normal observations. The best measure of association for the categorical data is then given by the best estimate of the correlation coefficient between the underlying bivariate normal observations. If the variates x_1 and x_2 are both dichotomous (so that $r = c = 2$), such an estimate is provided by the *tetrachoric* correlation coefficient. If one of x_1, x_2 is dichotomous but the other has more than two categories, the estimate is the *biserial* coefficient. If both r and c are greater than 2, the best estimate is the *polychoric* correlation coefficient. Calculations for these coefficients are complicated; for the simplest approaches the reader is referred to Hayes (1946), Hamilton (1948), and Lancaster and Hamdan (1964).

Consider finally the case (3) where one of x_1 and x_2 is quantitative while the other is categorical. This is the least satisfactory situation, for which the least guidance is available in the literature. Two possible approaches are to categorize the quantitative variate and treat association under case (2), or to assign numerical values to the categories of the qualitative variate and treat under case (1). Neither approach is entirely satisfactory: in the first approach information is lost in the categorization, while in the second approach the assignment of numerical values is arbitrary and may be inappropriate. If the categorical variable is dichotomous, then the *biserial* coefficient may be obtained directly (see Kendall and Stuart, 1967, Chapter 26). This coefficient assumes that one of the variates in a bivariate normal sample has been dichotomized. Alternatively, the two categories of the dichotomous variate can be assigned the values 0 and 1 respectively, and the correlation coefficient (14.1) can then be computed [illegible]tween the resulting two quantitative variates. Such a correlation [illegible] known as the *point biserial coefficient.*

To illustrate some of the [illegible]onsider the (fictitious) data of Table 14.2 giving results of [illegible]ons obtained by each of twenty students. The first two colu[illegible], give the actual marks obtained on the scale 0–100; the next [illegible]nns, y_1 and y_2, give the ranks of each set of marks (using the mean of the t spanned ranks where there are t tied observations); the following two columns, z_1 and z_2, give two qualitative variables obtained by a division of the mark scale into the four

Table 14.2 Results for twenty students on each of two examinations

	Marks		Ranks		Categorized		Dichotomized	
Student	x_1	x_2	y_1	y_2	z_1	z_2	w_1	w_2
1	64	72	5	4	C	C	Pass	Pass
2	65	63	4	8	C	C	Pass	Pass
3	31	55	17.5	13	F	P	Fail	Pass
4	40	27	14.5	20	P	F	Pass	Fail
5	12	58	20	11	F	P	Fail	Pass
6	77	82	1	1	D	D	Pass	Pass
7	50	50	10	15	P	P	Pass	Pass
8	63	78	6	2	C	D	Pass	Pass
9	44	56	11	12	P	P	Pass	Pass
10	36	59	16	10	F	P	Fail	Pass
11	55	67	7	5	P	C	Pass	Pass
12	31	49	17.5	16.5	F	P	Fail	Pass
13	42	31	12	18	P	F	Pass	Fail
14	54	49	8.5	16.5	P	P	Pass	Pass
15	54	53	8.5	14	P	P	Pass	Pass
16	40	63	14.5	8	P	C	Pass	Pass
17	75	77	2	3	D	D	Pass	Pass
18	41	63	13	8	P	C	Pass	Pass
19	22	28	19	19	F	F	Fail	Fail
20	68	66	3	6	C	C	Pass	Pass

categories 'Fail' (F, 0–39 marks), 'Pass' (P, 40–59 marks), 'Credit' (C, 60–74 marks) and 'Distinction' (D, 75–100 marks); while the final two columns, w_1 and w_2, give the dichotomous variables obtained by combining the P, C, and D categories of z_1 and z_2 into a single 'Pass' category.

To obtain the correlation coefficient (14.1) between x_1 and x_2 we require the following summary statistics:

$$\sum x_{i1}^2 = 52252, \qquad \sum x_{i2}^2 = 70264, \qquad \sum x_{i1}x_{i2} = 58496,$$

$$\bar{x}_1 = 48.2, \qquad \bar{x}_2 = 57.3.$$

Thus covariance between x_1 and x_2 is

$$s_{12} = (58496 - 20 \times 48.2 \times 57.3)/19 = 171.5158;$$

variance of x_1 is

$$s_1^2 = (52252 - 20 \times 48.2 \times 48.2)/19 = 304.5895,$$

so

$$s_1 = 17.4525;$$

variance of x_2 is

$$s_2^2 = (70264 - 20 \times 57.3 \times 57.3)/19 = 242.0105,$$

so

$$s_2 = 15.5567.$$

Hence correlation r_{12} between x_1 and x_2 is $s_{12}/(s_1 s_2) = 0.6317$.

Next we compute Spearman's rank correlation between x_1 and x_2. This could be obtained by calculating the product-moment correlation between y_1 and y_2, but it is more convenient to use (14.2). For this calculation we need the absolute difference in ranks, $|d_i|$, for each student, and these twenty values are as follows: 1, 4, 4.5, 5.5, 9, 0, 5, 4, 1, 6, 2, 1, 6, 8, 5.5, 6.5, 1, 5, 0, 3. Thus $\sum d_i^2 = 439$. There are three ties of two observations each in the marks x_1, and one tie of three observations and one of two observations in the marks x_2. Hence four corrections of $\dfrac{2^3 - 2}{12} = 0.5$ and one of $\dfrac{3^3 - 3}{12} = 2$ must be added to $\sum d_i^2$, and the corrected value is 443.

Thus $r_{12}^{(s)} = 1 - (6 \times 443)/(20 \times 399) = 1 - 0.3331 = 0.6669$.

Next consider the association between the categorized variates z_1 and z_2. We first write these into the form of Table 14.1, and the resulting contingency table is given in Table 14.3. The entries in parentheses in the body of the table are the expected values under the hypothesis of

Table 14.3 Contingency table formed from the categorical variables z_1 and z_2 of Table 14.2; expected values under the hypothesis of independence are given in parentheses in the body of the table

		z_2 F	P	C	D	Total
z_1	F	1 (0.75)	4 (2.0)	0 (1.5)	0 (0.75)	5
	P	2 (1.35)	4 (3.6)	3 (2.7)	0 (1.35)	9
	C	0 (0.6)	0 (1.6)	3 (1.2)	1 (0.6)	4
	D	0 (0.3)	0 (0.8)	0 (0.6)	2 (0.3)	2
Total		3	8	6	3	20

independence, i.e. $\hat{n}_{ij} = n_{i.}n_{.j}/n$. Thus

$$\begin{aligned} X^2 &= (0.25^2/0.75) + (0.65^2/1.35) + (0.6^2/0.6) + (0.3^2/0.3) \\ &\quad + (2.0^2/2.0) + (0.4^2/3.6) + (1.6^2/1.6) + (0.8^2/0.8) \\ &\quad + (1.5^2/1.5) + (0.3^2/2.7) + (1.8^2/1.2) + (0.6^2/0.6) \\ &\quad + (0.75^2/0.75) + (1.35^2/1.35) + (0.4^2/0.6) + (1.7^2/0.3) \\ &= 22.5741. \end{aligned}$$

$r = 4$ and $c = 4$, so that $\sqrt{\{(r-1)(c-1)\}} = \min(r-1, c-1) = 3$. Hence $r_{12}^{(t)} = r_{12}^{(c)} = \sqrt{\{22.5741/(20 \times 3)\}} = 0.6134$.

Turning to the association between the dichotomized variates w_1 and w_2, we first form the analogous table to the one in Table 14.3. This is given in Table 14.4, with the same explanation as before for the terms in parentheses. Thus $X^2 = (0.25^2/0.75) + (0.25^2/2.25) + (0.25^2/4.25) + (0.25^2/12.75) = 0.1307$ so that $r_{12}^{(t)} = r_{12}^{(c)} = \sqrt{\{0.11307/20\}} = 0.0808$.

Finally, suppose that the results of the first examination were available

Table 14.4 Contingency table formed from the dichotomised variables w_1 and w_2 of Table 14.2; expected values under the hypothesis of independence are given in parentheses in the body of the table

		w_2 Fail	Pass	Total
w_1	Fail	1 (0.75)	4 (4.25)	5
	Pass	2 (2.25)	13 (12.75)	15
	Total	3	17	20

only in Pass/Fail form (i.e. as the column w_1), but the full marks of the second examination were available. If we allocate the value 0 for 'Fail' and 1 for 'Pass' in w_1, application of (14.1) to w_1 and x_2 will yield the point biserial coefficient between the two examinations. Calculation of (14.1) requires the summary statistics:

$$\sum w_{i1}^2 = 15, \qquad \sum x_{i2}^2 = 70264, \qquad \sum w_{i1}x_{i2} = 897, \qquad \bar{w}_1 = 0.75,$$
$$\bar{x}_2 = 57.3.$$

Thus covariance between w_1 and x_2 is

$$s_{12} = (897 - 20 \times 0.75 \times 57.3)/19 = 1.9737;$$

variance of w_1 is

$$s_1^2 = (15 - 20 \times 0.75 \times 0.75)/19 = 0.1974,$$

so that

$$s_1 = 0.4443;$$

variance of x_2 is

$$s_2^2 = 242.0105$$

from previous calculation, so that

$$s_2 = 15.5567.$$

Hence the point biserial correlation is $1.9737/(0.4443 \times 15.5567) = 0.2856$.

The values computed above exhibit a pattern which can be accounted for readily. The product-moment correlation coefficient between x_1 and x_2 of 0.6317 is the most accurate measure of association between the two examinations, because it is based on the most detailed data available. When the students are ranked on each examination the overall pattern of the relationship is maintained, but the strength of relationship is slightly *enhanced* as the correlation rises to 0.6669. This is because considerable deviations from linearity can occur in two sets of marks without the rank ordering being disturbed; hence the rank correlation will tend to overestimate the true correlation. When the variables are both categorized into four categories each, the overall pattern of relationship is again maintained, but the strength of relationship is now slightly *reduced* as the measure of association drops to 0.6134. Information has been lost by grouping the values, and this has diluted the association that was present. However, with four categories per variable the spread is wide enough to capture the essential features of the association. On the other hand, dichotomization of either one or both of the variables has proved to be too violent a reduction in information (particularly the Pass/Fail dichotomization, since there were so few failures in the set of students).

Consequently, the two correlation measures involving dichotomized variables are completely different from the other values. Looking at Table 14.4 it is evident that the Pass/Fail results for w_2 are very comparable for each Pass/Fail category of w_1, so that the (2×2) table is consistent with independence between w_1 and w_2. With relatively few individuals in the 'Fail' category, it is also difficult to detect any difference in the x_2 pattern of values for each of these categories of w_1, and hence the point-biserial coefficient is also much closer to zero than to the true correlation value between x_1 and x_2.

14.2 Interpreting association between two variables

Having obtained a value of the association between two variables from a sample of n individuals, how should this value be interpreted? The most common approach, as with other inferential statistical procedures, is to relate it to the value of association in the whole population of individuals from which the n were a sample, either by means of hypothesis tests or by confidence-interval construction. We summarize here the main results needed for such inferences, and again we adopt the categorization of data types used in the previous section.

The main body of theory exists for case (1), where both variables x_1 and x_2 are quantitative. Let us suppose for generality that p variables have been observed on each sample individual, and that the resulting $(n \times p)$ data matrix $\mathbf{X}$ has already been mean-centered. Then the covariance between x_1 and x_2 is given by the element in the first row and second column of $\frac{1}{n-1}\mathbf{X}'\mathbf{X}$, while the variances of x_1 and x_2 are in the first two diagonal positions respectively of this same matrix. We suppose that the correlation between x_1 and x_2 in the whole population is ρ, and this is the quantity about which we wish to draw some inference from the calculated sample correlation coefficient r. In order to make any inference about ρ we must assume a distributional form for the population, and in the case of quantitative variables it is appropriate to assume that the observation vectors have a p-variate normal distribution with mean $\boldsymbol{\mu}$ and dispersion matrix $\boldsymbol{\Sigma}$. If this assumption is made, then the theory of Section 8.1 shows that the corrected sum-of-squares and -products matrix $\mathbf{C} = \mathbf{X}'\mathbf{X}$ has a Wishart distribution with $k = (n-1)$ degrees of freedom, and parameter $\boldsymbol{\Sigma}$. The following two properties of this distribution were listed in Section 7.3, and are given again here using current notation.

Property 5: if $\rho = 0$ then $r\surd\{(k-1)/(1-r^2)\}$ has a t distribution on $(k-1)$ degrees of freedom.

Property 6: as the sample size n increases, the distribution of $\frac{1}{2}\log_e\{(1+r)/(1-r)\}$ approaches the normal distribution with a mean of $\frac{1}{2}\log_e\{(1+\rho)/(1-\rho)\}$ and variance $1/(k-2)$.

These two properties can be used to test hypotheses about, and construct confidence intervals for, ρ. We illustrate them by reference to the examination data of the previous section.

First, we note that the correlation between the two sets of marks x_1 and x_2 was 0.6317 for the twenty students shown in Table 14.2. Is this evidence that there is a positive correlation between all possible marks in these two examinations? To answer this question we must test the null hypothesis H_0: $\rho=0$ against the alternative H_1: $\rho>0$, where ρ is the population correlation between the two examinations. Using Property 5 with $k=19$ and $r=0.6317$,

$$r\sqrt{\{(k-1)/(1-r^2)\}}=0.6317\sqrt{\{18/0.601\}}=3.46.$$

If H_0 is true and $\rho=0$, then this should be an observation from a t distribution on 18 degrees of freedom. But the upper 5 per cent point of this t distribution is 1.734, and since the test is one-tailed we therefore reject H_0 at the 5 per cent significance level. In fact, the upper 1 per cent and upper 0.1 per cent points of the distribution are 2.101 and 2.878, so there is very highly significant evidence in favour of the alternative hypothesis, $\rho>0$. Next, therefore, we would like to find a 95 per cent confidence interval for ρ. To do this we need to use Property 6, but we notice that the sample size of 20 is not very large, so the calculated confidence interval using normality assumptions will be at best only an approximate one.

Now, $\frac{1}{2}\log_e\{(1+r)/(1-r)\}=\frac{1}{2}\log_e\{4.4304\}=0.7442$, and $1/\sqrt{(k-2)}=1/\sqrt{17}=0.2425$. But if $\frac{1}{2}\log_e\{(1+r)/(1-r)\}\sim N[\frac{1}{2}\log_e\{(1+\rho)/(1-\rho)\}, 1/(k-2)]$, then

$$\text{pr}\{-1.96\leqslant\sqrt{(k-2)}[\tfrac{1}{2}\log_e\{(1+r)/(1-r)\} -\tfrac{1}{2}\log_e\{(1+\rho)/(1-\rho)\}]\leqslant 1.96\}=0.95,$$

so that

$$\text{pr}[\tfrac{1}{2}\log_e\{(1+r)/(1-r)\}-1.96/\sqrt{(k-2)}\leqslant\tfrac{1}{2}\log_e\{(1+\rho)/(1-\rho)\} \leqslant\tfrac{1}{2}\log_e\{(1+r)/(1-r)\}+1.96/\sqrt{(k-2)}]=0.95.$$

Using the two values above, therefore,

$$\text{pr}[0.7442-1.96\times0.2425\leqslant\tfrac{1}{2}\log_e\{(1+\rho)/(1-\rho)\} \leqslant 0.7442+1.96\times0.2425]=0.95,$$

i.e. $\text{pr}[0.2688\leqslant\frac{1}{2}\log_e\{(1+\rho)/(1-\rho)\}\leqslant 1.2196]=0.95$,

i.e. $\text{pr}[0.5376\leqslant\log_e\{(1+\rho)/(1-\rho)\}\leqslant 2.4392]=0.95$,

i.e. $\text{pr}[1.7746\leqslant(1+\rho)/(1-\rho)\leqslant 11.4639]=0.95$ on exponentiating throughout.

Now if $(1+\rho)/(1-\rho) \geqslant c$ (and $\rho \neq 1$), then $(1+\rho) \geqslant c(1-\rho)$ so that $\rho \geqslant (c-1)/(c+1)$; similarly $(1+\rho)/(1-\rho) \leqslant c$ implies $\rho \leqslant (c-1)/(c+1)$. Setting c successively equal to 1.7746 and 11.4639, we therefore obtain

$$\mathrm{pr}(0.2792 \leqslant \rho \leqslant 0.8395) = 0.95.$$

Hence a 95 per cent confidence interval for the population correlation coefficient is (0.2792, 0.8395).

Turning to case (2), where both variables x_1 and x_2 are qualitative, we can appeal to simple non-parametric statistical theory which tells us that if x_1 and x_2 are independent (i.e. there is no association between them), then X^2 of eqn (14.3) is an observation from a χ^2 distribution on $(r-1)(c-1)$ degrees of freedom. This enables a χ^2 test of the null hypothesis of no-association to be conducted. However, the condition required for this test to be valid is that the expected value $\hat{n}_{ij}$ in each cell of the contingency table must not be 'small': in practical terms each $\hat{n}_{ij}$ should be greater than about 3. Categories of x_1 or x_2 should be pooled if this condition is not satisfied, and pooling should be continued until it is satisfied. Unfortunately, this condition precludes the use of the χ^2 test on the categorized data of the previous section, because in all tables so formed there is at least one cell with very small expected frequency, and no pooling can be found to remove this feature. (This is because of the very small sample size and the nature of the categories.)

The point-biserial correlation coefficient of case (3) affords ready interpretation in terms of other familiar statistics. Here, one of the two variables is a dichotomy, while the other is quantitative. Independence in such circumstances means that the distribution of the quantitative variable is the same for the two states of the dichotomous variable, while increasing correlation indicates an increasing difference between the two distributions. Thus if normality is assumed for the quantitative variable, then the model for the data consists of two normal populations, one for each state of the dichotomous variable. Independence between the two variables in the population is characterized by two coincident normal distributions, while dependence between the variables is characterized by two different normal distributions. If the two normal distributions are furthermore assumed to have the same variance, then the difference between them resides purely in the difference between their means. Thus testing for a significant point-biserial correlation is exactly equivalent to testing for a difference in these two normal distribution means. In fact, if r_{pb} is the point-biserial correlation coefficient in a sample where x_1 is dichotomous with n_1, n_2 observations in its two categories, x_2 is quantitative, and t is the two-sample t statistic for a difference in means computed from the x_2 values in the two categories of x_1, then it can be shown that

$$r_{\mathrm{pb}}^2/(1-r_{\mathrm{pb}}^2) = t^2/(n_1+n_2-2). \tag{14.6}$$

Thus testing for significance of the point-biserial coefficient can be reduced to a t-test on $(n_1 + n_2 - 2)$ degrees of freedom. Note that as r_{pb}^2 approaches zero then so does t^2 (indicating no difference between the two population means), while as r_{pb}^2 approaches unity, t^2 becomes increasingly large (indicating difference between the two population means).

For the data of Table 14.2, we saw that the point-biserial coefficient between the Pass/Fail categorization of x_1 (i.e w_1) and the x_2 marks was $r_{pb} = 0.2856$. Thus

$$t^2 = (n_1 + n_2 - 2)r_{pb}^2/(1 - r_{pb}^2) = 18 \times 0.2856^2/(1 - 0.2856^2) = 1.5986.$$

Thus $t = 1.264$, which is consistent with an observation from the t-distribution on 18 degrees of freedom, the upper 5 per cent value of this distribution being 1.734. Hence the point-biserial correlation is not significant, which is equivalent to non-significance of difference between the distributions of x_2 under the Pass and Fail categories respectively of x_1.

Sampling distributions of the biserial, tetrachoric, and polychoric coefficients are either very complex or not yet derived, so these coefficients will not be considered further here.

14.3 Graphical investigation of many associations

For the remainder of this chapter, and also for most of the next, we focus on the one form of data most prevalent in practical applications: p *quantitative* variates observed on each of n individuals. For such data, the association between any pair of variates is given by the correlation coefficient (14.1). There are $\frac{1}{2}p(p-1)$ pairs of variates between which association can be calculated, and the resulting values of association will clearly be non-independent. If the overall multivariate system is being studied, then this potentially large number of associations must be jointly appraised. In this section we shall consider graphical methods of displaying many interrelated associations, in such a way as best to highlight their important features; in the following sections we shall consider methods for simplifying the information content in many interrelated associations so that deeper analysis of the whole multivariate system can be achieved. Before proceeding on to such methods of display and analysis, however, we need to establish some notation and basic background.

Since we are dealing with numerical data, the fundamental quantities involved in the calculation of any variabilities or associations are the variances s_{ii} of the p variates, and the covariances s_{ij} between the $\frac{1}{2}p(p-1)$ pairs of the variates $(i \neq j = 1, \ldots, p)$. These quantities are all put together in the $(p \times p)$ symmetric sample variance–covariance matrix

$\mathbf{S}$ whose (i, j)th element is s_{ij}. The sample correlation between the ith and jth variates is given by (14.1) as $r_{ij} = s_{ij}/\sqrt{(s_{ii}s_{jj})}$. By definition, the sample correlation between any variate and itself is unity so that $r_{ii} = 1$ for all i. The sample correlation matrix $\mathbf{R}$ is defined to be the $(p \times p)$ symmetric matrix with r_{ij} as its (i, j)th element. If we let $\mathbf{D}$ denote the $(p \times p)$ matrix with the p sample standard deviations down its diagonal and zeros elsewhere, then $\mathbf{D} = \text{diag}(\sqrt{s_{11}}, \sqrt{s_{22}}, \ldots, \sqrt{s_{pp}})$ and $\mathbf{D}^{-1} = \text{diag}(1/\sqrt{s_{11}}, 1/\sqrt{s_{22}}, \ldots, 1/\sqrt{s_{pp}})$. It is then easy to verify that the sample variance–covariance and correlation matrices are related by the equations:

$$\mathbf{R} = \mathbf{D}^{-1}\mathbf{S}\mathbf{D}^{-1} \tag{14.7}$$

and

$$\mathbf{S} = \mathbf{D}\mathbf{R}\mathbf{D}. \tag{14.8}$$

For inferential purposes, the n individuals are assumed to be a random sample from a population of individuals for each of which there are p observable quantitative variables. Let the population variance of the ith variate be σ_{ii}, and the population covariance between the ith and jth variates be σ_{ij}. The population correlation between the ith and jth variates is defined to be $\rho_{ij} = \sigma_{ij}/\sqrt{(\sigma_{ii}\sigma_{jj})}$, and $\rho_{ii} = 1$ for all i. Thus the population variance–covariance matrix is the $(p \times p)$ matrix $\boldsymbol{\Sigma}$ with (i, j)th element σ_{ij}, and the population correlation matrix is the $(p \times p)$ matrix $\mathbf{P}$ with (i, j)th element ρ_{ij}. Furthermore, if we write $\boldsymbol{\Delta} = \text{diag}(\sqrt{\sigma_{11}}, \sqrt{\sigma_{22}}, \ldots, \sqrt{\sigma_{pp}})$ then it is again easy to verify that $\boldsymbol{\Sigma}$ and $\mathbf{P}$ are related by the equations;

$$\mathbf{P} = \boldsymbol{\Delta}^{-1}\boldsymbol{\Sigma}\boldsymbol{\Delta}^{-1} \tag{14.9}$$

$$\boldsymbol{\Sigma} = \boldsymbol{\Delta}\mathbf{P}\boldsymbol{\Delta}. \tag{14.10}$$

If the divisor n is used in each s_{ij}, then s_{ij} is the maximum likelihood estimator of σ_{ij}. Invoking the invariance property of maximum likelihood estimation, therefore, r_{ij} is the maximum likelihood estimator of ρ_{ij}. Inferential procedures concerning single ρ_{ij} were discussed in the previous section; we now turn to simultaneous comparisons, and discuss some graphical methods described in two papers.

Corsten and Gabriel (1976) adapted the idea of biplots (Section 4.2) to arrive at graphical approximate displays of variance–covariance matrices which they termed *h-plots*. The h-plot takes the geometric representation of the p variates as p vectors in n-dimensional space (the 'variable space' of Section 2.2.1) and projects it on to the plane of least-squares best fit. In the n-dimensional representation the length of each vector is equal to the standard deviation of the corresponding variate, and the cosine of the angle between two vectors is equal to the correlation between the two corresponding variates (Section 2.2.2). In the h-plot, therefore, the vector lengths and angular separations yield the best approximations to these standard deviations and inverse cosines of correlation that can be obtained in a single (two-dimensional) diagram. Inspection of a single h-plot thus allows quick appraisal of the main features of variability and correlation of any one set of variates, while superimposing or comparing two or more h-plots permits the comparison of variabilities and associations that are exhibited in a number of separate samples.

Construction of an h-plot is very simple. Let $\mathbf{S}$ be the $(p \times p)$ variance–covariance matrix to be represented, and let λ_1, λ_2 denote the two largest eigenvalues of $\mathbf{S}$ with corresponding eigenvectors $\boldsymbol{q}_1$, $\boldsymbol{q}_2$. Construct the $(p \times 2)$ matrix $\mathbf{H} = (\boldsymbol{q}_1\sqrt{\lambda_1}, \boldsymbol{q}_2\sqrt{\lambda_2})$, and plot the p rows $\boldsymbol{h}_i'$ $(i = 1, \ldots, p)$ of $\mathbf{H}$ as p points in two dimensions. The ith variate is then represented by an arrow from the origin to the point corresponding to $\boldsymbol{h}_i$. Note from the results of Sections 4.1 and 4.2 that this is just the representation of the ith column effect in a biplot of the original data matrix $\mathbf{X}$. Also, since $\boldsymbol{q}_i\sqrt{\lambda_i}$ is in fact the set of loadings of the p variates on the ith principal component derived from $\mathbf{S}$ (Section 2.2.5), the h-plot is simply the plot of principal component loadings against the first two principal components as axes. In the h-plot, the length of the arrow from the origin to the vertex at the point $\boldsymbol{h}_i$ approximates the standard deviation of the ith variate, and the cosine of the angle between this arrow and the one with its vertex at the point $\boldsymbol{h}_j$, i.e. $\cos(\boldsymbol{h}_i, \boldsymbol{h}_j)$, approximates the correlation r_{ij} between the ith and jth variates. Moreover, if further arrows are constructed by the rules of vector addition and scalar multiplication (Section 1.3), their lengths and angular separations also approximate the standard deviations and inverse cosines of correlation for the corresponding linear combinations of variates. Thus, for example, the length of the arrow which starts at the vertex $\boldsymbol{h}_i$ and ends at the vertex $\boldsymbol{h}_j$ is an approximation to the standard deviation of the difference between the ith and jth variates.

Of course, since the h-plot is a projection of n-dimensional space into two dimensions, all of these distances and angles are approximations, and the h-plot representation will be useful only when these approximations are close enough to the true values. Corsten and Gabriel (1976) define the overall measure of goodness-of-fit of the representation to be $1 - [\mathrm{trace}\{(\mathbf{S} - \mathbf{HH}')'(\mathbf{S} - \mathbf{HH}')\}/\mathrm{trace}(\mathbf{S}'\mathbf{S})]$, which is equal to $(\lambda_1^2 + \lambda_2^2)/\sum_{i=1}^{p} \lambda_i^2$ where $\lambda_1 \geq \lambda_2 \geq \ldots \geq \lambda_p \geq 0$ are all the eigenvalues of $\mathbf{S}$. A high measure of goodness-of-fit confirms that all variances are well approximated by squared lengths $\boldsymbol{h}_i'\boldsymbol{h}_i$, and all covariances are well approximated by inner products $\boldsymbol{h}_i'\boldsymbol{h}_j$. Note, however, that the approximation of r_{ij} by $\cos(\boldsymbol{h}_i, \boldsymbol{h}_j)$ may not necessarily be as good. This is because $\cos(\boldsymbol{h}_i, \boldsymbol{h}_j) = \boldsymbol{h}_i'\boldsymbol{h}_j/\sqrt{\{(\boldsymbol{h}_i'\boldsymbol{h}_i)(\boldsymbol{h}_j'\boldsymbol{h}_j)\}}$ (Section 1.3), and each term in this ratio is approximated in the h-plot. The consequent ratio of approximations may not be close to the true value of r_{ij}.

As an example of the use of h-plots, Corsten and Gabriel (1976) displayed the covariance matrices obtained from data on daily rainfall in eight regions of Israel. The regions (which play the role of variates in this example) are:

Nc North, coastal
Ni North, interior
Ne North, eastern
B Buffer, between North and Centre
Cc Centre, coastal
Ci Centre, interior
Ce Centre, eastern
S Southern.

Each original observation was a rainy day's precipitation averaged over several rain gauges in one region. Observations were taken on 845 rainy days in the period 1949–1960. A rainfall stimulation experiment was then conducted in the period 1961–1967; on each experimental day cloud-seeding was randomly assigned to the North or Centre and carried out upwind of those regions. A further 208 observations were taken on rainy days with seeding in the North, and 183 observations on rainy days with seeding in the Centre. Variance–covariance and correlation matrices were computed separately for the pre-experimental observations ('Natural'), the North-seeded observations, and the Centre-seeded observations. These matrices are shown in Table 14.5, and h-plots of the three variance–covariance matrices are displayed in Fig. 14.2. The goodness-of-fit of each plot was excellent, being 98.79%, 98.53%, and 98.73% respectively. Corsten and Gabriel's interpretation of the plot was as follows.

'Natural' variability in rainfall (Fig. 14.2(a)) is much the same in all regions, except that the North and Buffer have slightly smaller standard deviations. The correlations exhibit a clear geographical pattern: a highly correlated cluster of Northern regions and the Buffer, little correlation of this cluster with the dry regions of the South and Centre-east; while the Central coast and interior are in-between but more highly correlated with the North than with the South.

For the North-seeded days (Fig. 14.2(b)), the Centre-coast and Centre-interior are less highly correlated with the North and Buffer than under 'Natural' conditions. Also, North and Buffer rainfall variabilities appear to have increased under North-seeding, and the seeded areas are more highly correlated with each other. For the Centre-seeded days (Fig. 14.2(c)), there is a reduction in the Centre–North correlations, an

Table 14.5 Variances (along main diagonals), covariances (above main diagonals), and correlations (below main diagonals) of daily rainfall in Israel. (Regions: N, North; B, Buffer; C, Centre; S, South; c, Coastal; i, Interior; e, Eastern.) Reproduced from Corsten and Gabriel (1976) with permission from the Biometric Society

	Nc	Ni	Ne	B	Cc	Ci	Ce	S
(a) *Natural*								
Nc	86.68	80.49	75.50	61.16	80.57	62.93	47.29	39.38
Ni	0.8804	96.43	90.29	61.43	88.35	75.90	63.01	45.64
Ne	0.7966	0.9032	103.63	61.34	93.63	81.24	67.44	47.59
B	0.7802	0.7429	0.7156	70.90	86.35	64.44	41.45	31.40
Cc	0.6591	0.6852	0.7005	0.7811	172.40	126.88	96.76	73.52
Ci	0.6317	0.7224	0.7458	0.7153	0.9031	114.49	101.74	74.38
Ce	0.4229	0.5343	0.5516	0.4099	0.6136	0.7917	144.24	109.64
S	0.3714	0.4080	0.4104	0.3274	0.4916	0.6103	0.8015	129.73
(b) *North-seeded*								
Nc	133.17	145.30	90.70	116.91	70.73	75.97	44.00	26.82
Ni	0.9184	187.96	123.21	142.93	91.51	100.77	63.77	32.27
Ne	0.7955	0.9096	97.61	92.01	58.55	65.32	42.77	21.66
B	0.7805	0.8032	0.7175	168.48	91.62	90.07	43.84	23.25
Cc	0.5980	0.6512	0.5782	0.6886	105.06	113.33	72.43	52.29
Ci	0.5622	0.6277	0.5646	0.5926	0.9442	137.12	87.93	62.26
Ce	0.3786	0.4619	0.4299	0.3354	0.7017	0.7457	101.40	82.16
S	0.2317	0.2347	0.2186	0.1786	0.5086	0.5301	0.8135	100.60
(c) *Centre-seeded*								
Nc	132.48	112.52	81.10	110.77	101.18	105.88	56.06	23.96
Ni	0.7980	150.06	109.72	121.99	114.43	134.08	80.60	34.97
Ne	0.7370	0.9379	91.20	87.58	92.03	107.76	66.80	29.84
B	0.7097	0.7344	0.6763	183.87	134.69	139.22	71.85	36.38
Cc	0.6195	0.6583	0.6791	0.7000	201.36	206.96	115.05	55.96
Ci	0.5826	0.6932	0.7146	0.6502	0.9237	249.32	145.29	71.20
Ce	0.4009	0.5415	0.5757	0.4361	0.6673	0.7573	147.62	90.02
S	0.2234	0.3063	0.3353	0.2879	0.4231	0.4838	0.7950	86.86

increase in Centre variability, and an increase in the correlations between different regions of the Centre. Thus both Figs 14.2(b) and (c) are consistent, with both suggesting that seeding increases variability and correlations among seeded areas, but that it reduces correlations between seeded and unseeded areas. One surprising feature was in Fig. 14.2(c), where there is apparent an increase in variability in the North coastal, North interior, and Buffer regions over the corresponding values under 'Natural' conditions. This observation led to some hypotheses about variability of rainfall being suggested by the h-plots, and Corsten and Gabriel (1976) then went on to develop simultaneous tests on variances using graphical h-plot approximations. The interested reader is referred to their paper for further details.

A different objective was adopted by Hills (1969), who recognized that

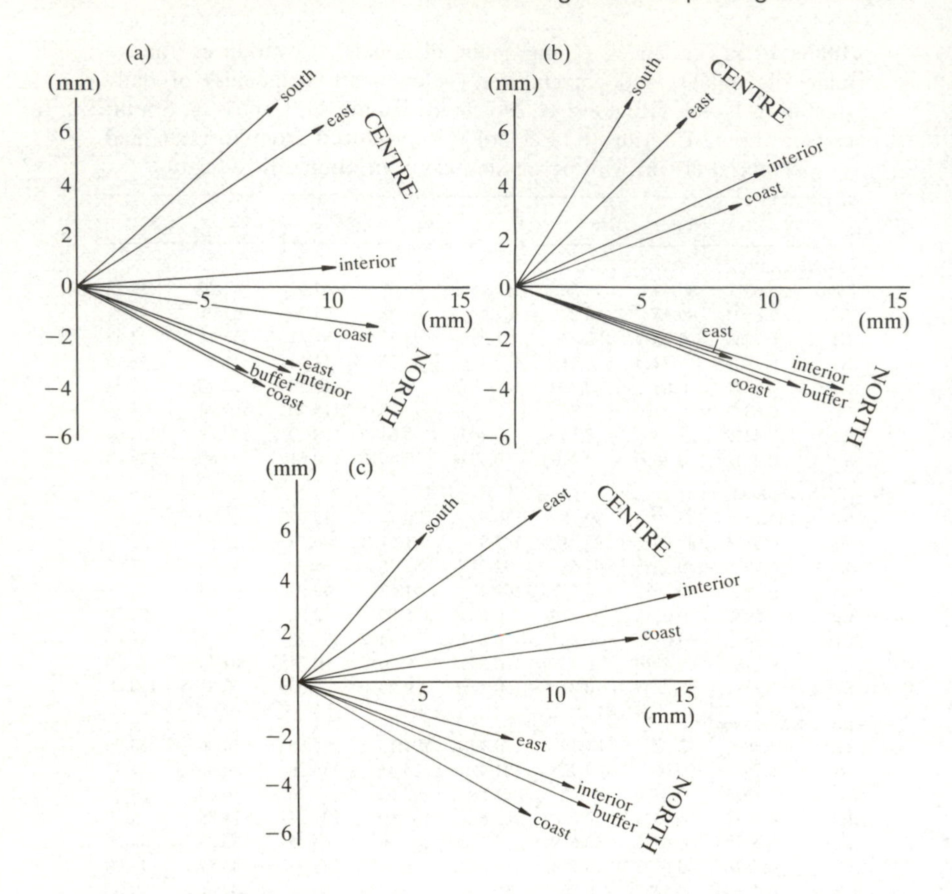

Fig. 14.2 *h*-plots of variance–covariance matrices of Table 14.5. (a) Natural; (b) North-seeded; (c) Centre-seeded. Reproduced from Corsten and Gabriel (1976) with permission from The Biometric Society

in practice correlation matrices are often large and hence difficult to interpret directly. He remarked that the first, and sometimes the only, impression gained from looking at a large correlation matrix is its largeness, and hence was led to consider simple graphical techniques which might help to uncover some structure in the matrix. Specifically, he looked for appropriate graphical techniques to answer the following two questions.

1. Which of the correlation coefficients is large enough (in absolute magnitude) to be considered as arising from a population coefficient different from zero?

2. Are there any groups of variables such that members of the same group are all fairly highly positively correlated with each other and behave similarly in their correlations with other variables?

To answer question 1, Hills suggested first transforming all the correlations r_{ij} $(i<j)$ by means of the transformation $z_{ij}=\frac{1}{2}\log_e\{(1+r_{ij})/(1-r_{ij})\}$. With p variables measured on each of n individuals, there will be $q=\frac{1}{2}p(p-1)$ distinct values z_{ij}, and if $\rho_{ij}=0$, then by Property (6) of Section 14.2 each z_{ij} will be approximately normal with mean zero and variance $\sigma^2=1/(n-3)$. Of course, all the z_{ij} values for $i<j$ are not independent, but if this fact is ignored then the values z_{ij} may be plotted in a half-normal plot to see which are too large numerically to be consistent with a random sample from a $N(0, \sigma^2)$ distribution. The plot is very simple to do. Rank the absolute values of the z_{ij} as $|z_{(1)}|\leqslant|z_{(2)}|\leqslant\ldots\leqslant|z_{(q)}|$, and from appropriate tables of the normal distribution find the q (positive) values z_i' such that $G(z_i')=(i-\frac{1}{2})/q$ for $i=1,\ldots,q$ where $G(c)=\frac{1}{\sqrt{(2\pi)}}\int_{-c}^{+c}e^{-\frac{1}{2}u^2}\,du$. The half-normal plot is then effected by plotting $|z_{(i)}|$ against z_i' for $i=1,\ldots,q$; if the z_{ij} have the distribution $N(0, \sigma^2)$ then the points should lie about a straight line through the origin, with slope σ. This is because the proportion of z_{ij} numerically less than $|z_{(i)}|$ is approximately $(i-\frac{1}{2})/q$ (where the $\frac{1}{2}$ is introduced to avoid the values 0 and 1 at the ends), while $G(z_i')$ gives the proportion of the $N(0, 1)$ distribution numerically less than z_i'. Thus if the z_{ij} come from the distribution $N(0, \sigma^2)$ then $|z_{(i)}|$ should be approximately equal to $\sigma z_i'$.

To illustrate this technique, Hills (1969) considered the correlation matrix given in Table 14.6. This matrix was taken from a study of the physiological effects of examination strain on forty-eight medical students, in which the variables of interest were differences in levels between a normal period of time and the examination period. Differences were obtained for thirteen physiological measurements. Thus $n=48$, $p=13$, and $q=78$. The half-normal plot for these correlations is shown in Fig. 14.3. The points lie very closely about a straight line up to the value $|z|=0.3$, after which they deviate considerably. The slope of this line, however, differs from the theoretical value $\sigma=\sqrt{(1/45)}=0.1491$ because the seventy-eight z_{ij} values are obviously not a random sample from a $N(0, 0.1491^2)$ distribution. Hills re-plotted the sixty-two values of $|z|$ up to the value 0.3 using $(i-\frac{1}{2})/62$ in place of $(i-\frac{1}{2})/78$. This plot is shown in Fig. 14.4, and the previously noted discrepancy has disappeared. Hills concluded from the two figures that correlations with $|z|$ greater than 0.3 are too large to have come from a population with $\rho=0$. He noted also that the z point at which a correlation coefficient would be declared significantly different from zero at the 5 per cent level was

Table 14.6 Matrix of correlations between differences in thirteen physiological measurements. The differences are between normal values for the subject and values at the time of the examination. Reproduced from Hills (1969) with permission from the Biometrika Trustees

1	2	3	4	5	6	7	8	9	10	11	12	13
(SYS)	(DIA)	(P.P)	(PUL)	(CORT)	(U.V.)	(TOT/100)	(ADR/100)	(NOR/100)	(ADR/TOT)	(TOT/HR)	(ADR/HR)	(NOR/HR)
1.00	0.60	0.80	0.20	0.12	−0.20	0.07	0.32	−0.03	−0.03	−0.09	0.14	−0.13
—	1.00	0.00	0.28	0.13	−0.22	0.07	0.19	0.01	−0.05	0.04	0.10	0.01
—	—	1.00	0.04	0.06	−0.09	0.04	0.26	−0.04	0.00	−0.14	0.10	−0.17
—	—	—	1.00	0.15	−0.16	−0.07	0.20	−0.13	−0.12	−0.16	0.08	−0.18
—	—	—	—	1.00	−0.22	−0.01	0.07	−0.03	−0.01	−0.17	−0.03	−0.16
—	—	—	—	—	1.00	−0.13	−0.17	−0.08	0.04	0.24	0.20	0.18
—	—	—	—	—	—	1.00	0.19	0.96	−0.38	0.77	−0.03	0.75
—	—	—	—	—	—	—	1.00	−0.10	0.43	−0.04	0.75	−0.25
—	—	—	—	—	—	—	—	1.00	−0.51	0.79	−0.25	0.84
—	—	—	—	—	—	—	—	—	1.00	−0.49	0.50	−0.62
—	—	—	—	—	—	—	—	—	—	1.00	0.03	0.96
—	—	—	—	—	—	—	—	—	—	—	1.00	−0.26
—	—	—	—	—	—	—	—	—	—	—	—	1.00

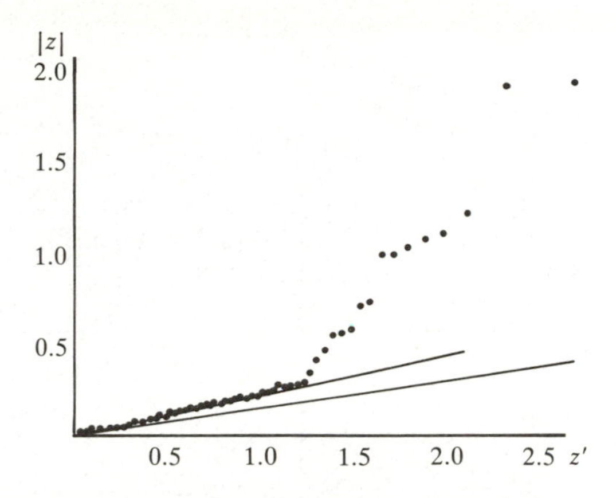

Fig. 14.3 Half-normal plot of the seventy-eight correlations in Table 14.6; line with slope 0.1491 is shown below the fitted line. Reproduced from Hills (1969) with permission from the Biometrika Trustees

$0 + 1.96 \times 0.1491 = 0.29$. Thus, in the present example, similar conclusions would be reached by both significance test and half-normal plot, but this will not always be the case in practice.

To answer question (2) stated above, Hills suggested a simple application of metric scaling (Section 3.2) in which each variable is represented by a point in Euclidean space. Since the objective is to find clusters of variables that are highly *positively* correlated, a suitable measure of distance between the points P_i and P_j representing variables i and j in this diagram is $d_{ij} = 2(1 - r_{ij})$. Thus d_{ij} increases monotonically from 0 when $r_{ij} = 1$ to 4 when $r_{ij} = -1$. Applying this technique to the correlations of Table 14.6 yielded the configuration of Fig. 14.5. The first two principal coordinates are plotted for the thirteen points, and the third coordinate is added in parentheses after each point.

This plot suggests that there exist two clusters containing highly positively intercorrelated variables: variables 7, 9, 11, 13, and variables 8,

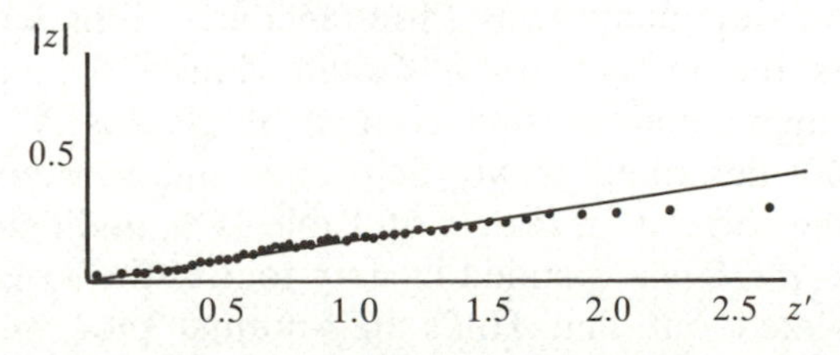

Fig. 14.4 Half-normal plot of the sixty-two correlations whose z values are numerically less than 0.3. Fitted line has slope 0.1491. Reproduced from Hills (1969) with permission from the Biometrika Trustees

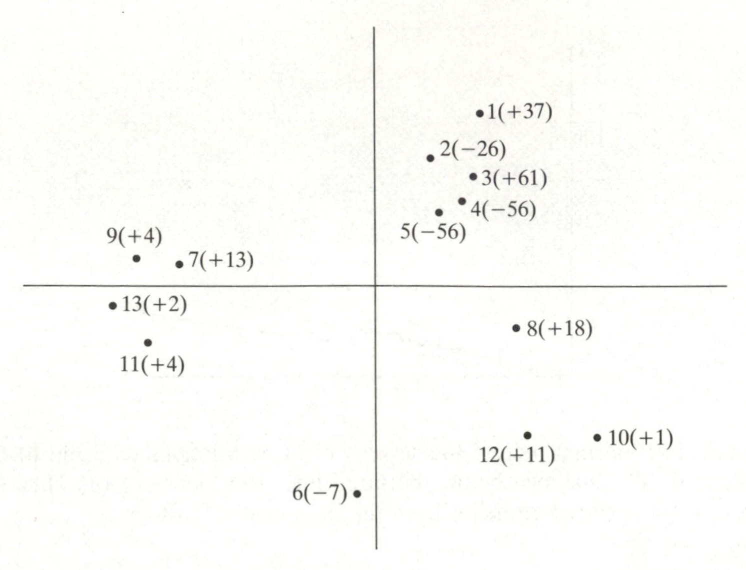

Fig. 14.5 Two-dimensional configuration representing the variables of Table 14.6. Numbers in parentheses are the coordinates of the points in the third dimension. Reproduced from Hills (1969) with permission from the Biometrika Trustees

10, 12. These clusters make quite good physiological sense. Also, if only variable 9 from the first cluster, and variable 8 from the second cluster, are included in the whole correlation matrix, it begins to look more like a diagonal matrix. A half-normal plot verifies that all consequent correlations apart from those of variable 1 with variables 2 and 3 may be taken as coming from populations with $\rho = 0$. Thus a considerable simplification of structure has been achieved. (Table 14.6 and Figs 14.3–14.5 are reproduced from Hills (1969) with permission from the Biometrika Trustees.)

In some related work, Schweder and Spjøtvoll (1982) proposed a plot of p-values to assess many tests simultaneously. This technique can be used to assess the test of $\rho = 0$ simultaneously for each entry in a correlation matrix, and in this context it is closely related to the half-normal plot described above. Schweder and Spjøtvoll applied their technique to the correlation matrix of Table 14.6, and concluded that the null hypothesis $\rho = 0$ was justified in sixty-four of the seventy-eight cases. They also pointed out that Hill's half-normal plot was more widely applicable than simply for correlation matrices (e.g. for plotting comparisons among treatment means in a designed experiment), but that plots of p-values would be more natural in most such contexts.

14.4 Correcting correlations for effects of extraneous variables

Large correlations are often picked out from a correlation matrix as being of special interest, and the variables that exhibit such large correlations become the focus of further attention. Much effort is often expended on explaining, interpreting, and investigating the causes of these large correlations. However, picking out isolated entries from a correlation matrix can be misleading and can sometimes promote incorrect inferences and conclusions, because all the entries in the matrix are interrelated. This non-independence has already been noted in the previous section; it arises because all p variables in the system are measured on the *same* n individuals, and taking the variables in pairwise fashion induces connections between the resulting correlations. For example, if the two pairs of variables (x_1, x_2) and (x_2, x_3) both exhibit high positive association, then the two corresponding scatter plots are very close to two straight lines, each having positive slope. Thus any individual's x_2 value is well predicted from the corresponding x_1 value by $x_2 = a + bx_1$ $(b > 0)$, and the same individual's x_3 value is well predicted from this x_2 value by $x_3 = c + dx_2$ $(d > 0)$. On substituting from the former equation into the latter it follows that the x_3 value will be well predicted from the x_1 value by $x_3 = e + fx_1$, where $e = c + ad$ and $f = bd > 0$. Hence the scatter plot for (x_1, x_3) will be also very close to a straight line with positive slope, and the correlation between x_1 and x_3 will be also large and positive. Thus if x_i and x_j are both highly associated with x_k in the sample, then they will inevitably be associated with each other.

When we try to interpret or explain a high correlation between two variables x_i and x_j, therefore, we must be aware of the possibility that the high correlation has arisen because of the mutual association of x_i and x_j with some other variable(s). For the correlation between x_i and x_j to be of intrinsic interest, therefore, it must remain high when the effects of these extraneous variables have been removed. Now if a variable x_k induces a high correlation between x_i and x_j, it must be because there is a considerable variation in the observed values of x_k, and over this range there are strong linear relationships between x_k and each of x_i, x_j. To examine the 'true' association between x_i and x_j, therefore, we must compute the correlation between x_i and x_j *with the value of* x_k held fixed. That way, x_k has no means of influencing the correlation. Similarly, if a set of variables $x_k, \ldots, x_l$ is thought to be inducing a high correlation between x_i and x_j, then the 'true' association is obtained by computing the correlation between x_i and x_j when the values of *all* variables $x_k, \ldots, x_l$ are held fixed. In order to be able to calculate such a correlation explicitly for sample data, we would need to have sets of replicate observations at each of a number of fixed values of $x_k, \ldots, x_l$. Thus if we had n_1 observations each with $x_k = a_{1k}, \ldots, x_l = a_{1l}; n_2$ observations each with $x_k = a_{2k}, \ldots, x_l = a_{2l}; \ldots; n_s$ observations each with $x_k = a_{sk}, \ldots, x_l = a_{sl}$, where the $(a_{ik}, \ldots, a_{il})$ are fixed constants for $i = 1, \ldots, s$, then the required correlation between x_i and x_j would be the one computed *within* these s sets of observations.

Since such replicate observations are extremely rare in practice (except in very special circumstances), recourse must be made to estimation under a well-defined model. We assume that the rows $\boldsymbol{x}_i'$ of the $(n \times p)$ data matrix $\mathbf{X}$ are independent observations of a p-variate random vector $\boldsymbol{x}$ which has a normal distribution with a mean of $\boldsymbol{\mu}$ and dispersion matrix $\boldsymbol{\Sigma}$. The elements σ_{ij} of $\boldsymbol{\Sigma}$ are of course the population variances and covariances of the variables, estimated by the elements s_{ij} of the sample covariance matrix $\mathbf{S}$. Suppose now that the random vector $\boldsymbol{x}$ is partitioned into two portions, $\boldsymbol{x}' = (\boldsymbol{x}_1', \boldsymbol{x}_2')$ where $\boldsymbol{x}_1$ has q elements, and $\boldsymbol{x}_2$ has the remaining $(p-q)$ elements; let $\boldsymbol{\mu}'$ and $\boldsymbol{\Sigma}$ be partitioned conformally as $(\boldsymbol{\mu}_1', \boldsymbol{\mu}_2')$ and $\begin{pmatrix} \boldsymbol{\Sigma}_{11} & \boldsymbol{\Sigma}_{12} \\ \boldsymbol{\Sigma}_{21} & \boldsymbol{\Sigma}_{22} \end{pmatrix}$. (Thus $\boldsymbol{\Sigma}_{11}$ is the dispersion matrix for $\boldsymbol{x}_1$, $\boldsymbol{\Sigma}_{22}$ the dispersion matrix for $\boldsymbol{x}_2$, and $\boldsymbol{\Sigma}_{12}$ gives the population covariances between elements of $\boldsymbol{x}_1$ and elements of $\boldsymbol{x}_2$.) Then, from Property 3 of Section 7.2, the conditional distribution of $\boldsymbol{x}_1$, given that $\boldsymbol{x}_2$ has the known value $\boldsymbol{a}$, is multivariate normal with a mean of $\boldsymbol{\mu}_1 + \boldsymbol{\Sigma}_{12}\boldsymbol{\Sigma}_{22}^{-1}(\boldsymbol{a} - \boldsymbol{\mu}_2)$, and a dispersion matrix $\boldsymbol{\Sigma}_{1.2} = \boldsymbol{\Sigma}_{11} - \boldsymbol{\Sigma}_{12}\boldsymbol{\Sigma}_{22}^{-1}\boldsymbol{\Sigma}_{21}$. The elements of $\boldsymbol{\Sigma}_{1.2}$ are termed the *partial variances and covariances* of $\boldsymbol{x}_1$ given $\boldsymbol{x}_2 = \boldsymbol{a}$. They indicate the values that the variances and covariances among $\boldsymbol{x}_1$ would take if $\boldsymbol{x}_2$ were constrained to have the value $\boldsymbol{a}$. Since these elements do not depend on $\boldsymbol{a}$, however, they in fact indicate the values that the variances and covariances among $\boldsymbol{x}_1$ would take if $\boldsymbol{x}_2$ were held constant at *any* value. If we set $\boldsymbol{\Delta}_{1.2}$ to be the diagonal matrix containing the square roots of the diagonal elements of $\boldsymbol{\Sigma}_{1.2}$, and standardize the matrix $\boldsymbol{\Sigma}_{1.2}$ in the manner of (14.9) to $\mathbf{P}_{1.2} = \boldsymbol{\Delta}_{1.2}^{-1}\boldsymbol{\Sigma}_{1.2}\boldsymbol{\Delta}_{1.2}^{-1}$, then the elements of $\mathbf{P}_{1.2}$ are the *partial correlations* among the variables in $\boldsymbol{x}_1$ for fixed value of $\boldsymbol{x}_2$.

Given the sample of n observations $\boldsymbol{x}_i'$, we can partition the sample covariance matrix $\mathbf{S}$ into $\begin{pmatrix} \mathbf{S}_{11} & \mathbf{S}_{12} \\ \mathbf{S}_{21} & \mathbf{S}_{22} \end{pmatrix}$, and can estimate $\boldsymbol{\Sigma}_{1.2}$ by $\mathbf{S}_{1.2} = \mathbf{S}_{11} - \mathbf{S}_{12}\mathbf{S}_{22}^{-1}\mathbf{S}_{21}$. Writing $\mathbf{D}_{1.2}$ as the diagonal matrix containing the square roots of the diagonal elements of $\mathbf{S}_{1.2}$, and standardizing $\mathbf{S}_{1.2}$ in the manner of (14.7) to $\mathbf{R}_{1.2} = \mathbf{D}_{1.2}^{-1}\mathbf{S}_{1.2}\mathbf{D}_{1.2}^{-1}$, we obtain the matrix of sample partial correlations. The (i, j)th element of this matrix is an estimate of what the correlation between the ith and jth variates of $\boldsymbol{x}_1$ would be, if only observations with a fixed value of $\boldsymbol{x}_2$ were selected from the population. This therefore removes the effect of the variables in $\boldsymbol{x}_2$ on the relationship between x_i and x_j, and gives an estimate of the intrinsic association between x_i and x_j.

Two further ways of calculating a partial correlation are available. Since correlation is a measure of linear dependence, then removal of the effect of some variables on the correlation between others must in fact be removal of the *linear* effect. Hence the partial correlation between x_i and x_j on fixing the variables in $\boldsymbol{x}_2$ can be obtained equivalently by calculating the correlation between the two sets of residuals from the regressions of x_i and x_j respectively on variables in $\boldsymbol{x}_2$. Alternatively, the partial correlations can be built up using a sequence of recurrence relationships. Denote by $r_{ij.k}$ the (first-order) partial correlation between x_i and x_j on fixing x_k; by $r_{ij.kl}$ the (second-order) partial correlation between x_i and x_j on fixing both x_k and x_l; and so on. Then a sequential application of the equations for $\mathbf{S}_{1.2}$ above yields the relationships:

$$r_{ij.k} = (r_{ij} - r_{ik}r_{jk})/\surd\{(1 - r_{ik}^2)(1 - r_{jk}^2)\} \qquad (14.11)$$

$$r_{ij.kl} = (r_{ij.k} - r_{il.k}r_{jl.k})/\surd\{(1 - r_{il.k}^2)(1 - r_{jl.k}^2)\}. \qquad (14.12)$$

This pattern is continued for higher-order partial correlations, and enables a partial correlation of qth order to be obtained from those of $(q-1)$th order. This method is most useful when low-order partial correlations are required.

To illustrate the calculations, consider the correlation matrix

$$\begin{pmatrix} 1.0000 & 0.6162 & 0.8267 \\ 0.6162 & 1.0000 & 0.7321 \\ 0.8267 & 0.7321 & 1.0000 \end{pmatrix}$$

obtained from observations on the intelligence (x_1), weight (x_2), and age (x_3) of schoolchildren (Mardia *et al.* (1979), p. 170). This matrix suggests that there is a high association (0.6162) between weight and intelligence, but the dependence is likely to be spurious and induced by the age–weight and age–intelligence associations. To investigate this supposition, we calculate the partial correlation between weight and intelligence on fixing age. Partitioning $\boldsymbol{x}$ into the two subvectors (x_1, x_2) and (x_3), we partition the correlation matrix correspondingly into

$$\mathbf{S}_{11} = \begin{pmatrix} 1.0000 & 0.6162 \\ 0.6162 & 1.0000 \end{pmatrix}, \quad \mathbf{S}_{22} = (1.0000),$$

and

$$\mathbf{S}_{12} = \begin{pmatrix} 0.8267 \\ 0.7321 \end{pmatrix}.$$

Hence

$$\mathbf{S}_{1.2} = \begin{pmatrix} 1.0000 & 0.6162 \\ 0.6162 & 1.0000 \end{pmatrix} - \begin{pmatrix} 0.8267 \\ 0.7321 \end{pmatrix} (0.8267 \quad 0.7321)$$

$$= \begin{pmatrix} 1.0000 & 0.6162 \\ 0.6162 & 1.0000 \end{pmatrix} - \begin{pmatrix} 0.6834 & 0.6052 \\ 0.6052 & 0.5360 \end{pmatrix}$$

$$= \begin{pmatrix} 0.3166 & 0.0110 \\ 0.0110 & 0.4640 \end{pmatrix}.$$

$$\mathbf{D}_{1.2} = \begin{pmatrix} \sqrt{0.3166} & 0 \\ 0 & \sqrt{0.4640} \end{pmatrix} = \begin{pmatrix} 0.5627 & 0 \\ 0 & 0.6812 \end{pmatrix}$$

$$\mathbf{D}_{1.2}^{-1} = \begin{pmatrix} 1.7772 & 0 \\ 0 & 1.4681 \end{pmatrix}$$

$$\mathbf{R}_{1.2} = \begin{pmatrix} 1.7772 & 0 \\ 0 & 1.4681 \end{pmatrix} \begin{pmatrix} 0.3166 & 0.0110 \\ 0.0110 & 0.4640 \end{pmatrix} \begin{pmatrix} 1.7772 & 0 \\ 0 & 1.4681 \end{pmatrix}$$

$$= \begin{pmatrix} 1.0000 & 0.0287 \\ 0.0287 & 1.0000 \end{pmatrix}.$$

This shows that, when age is held constant, the correlation between weight and intelligence does drop considerably to 0.0287, confirming that nearly all of the previous association is explained by the age–weight and age–intelligence associations. The partial correlation could have been obtained in this case more directly by applying (14.11) to the entries in the correlation matrix:

$$\begin{aligned} r_{12.3} &= (0.6162 - 0.8267 \times 0.7321)/\sqrt{\{(1-0.8267^2)(1-0.7321^2)\}} \\ &= 0.0110/0.3833 \\ &= 0.0287, \text{ as before.} \end{aligned}$$

In similar fashion, the reader can verify that the high age–intelligence correlation of 0.8267 is only reduced slightly to 0.701 when weight is held constant.

The next question to be raised by these calculations is what significance and/or interpretation can be attached to such partial correlation values? In Section 14.2 we saw that, under assumptions of normality, tests of hypotheses and confidence-interval calculations can be effected for ordinary correlation coefficients. These tests and confidence intervals stem from the fact that the sample sum-of-squares and -products matrix **C** has a Wishart distribution with parameter $\boldsymbol{\Sigma}$ and $(n-1)$ degrees of freedom if the data come from a normal distribution. But Property 3 of Section 7.2 has shown that the conditional distribution of $\boldsymbol{x}_1$ given $\boldsymbol{x}_2 = \boldsymbol{a}$ is then also normal, with dispersion matrix $\boldsymbol{\Sigma}_{1.2}$. Partitioning **C** conformally into $\begin{pmatrix} \mathbf{C}_{11} & \mathbf{C}_{12} \\ \mathbf{C}_{21} & \mathbf{C}_{22} \end{pmatrix}$, it can further be shown that $\mathbf{C}_{1.2} = \mathbf{C}_{11} - \mathbf{C}_{12}\mathbf{C}_{22}^{-1}\mathbf{C}_{21}$ has a Wishart distribution with parameter $\boldsymbol{\Sigma}_{1.2}$ and $\{n-1-(p-q)\}$ degrees of freedom if $\boldsymbol{x}_1$ has q elements and $\boldsymbol{x}_2$ has $(p-q)$ elements. Thus the distribution theory underlying partial correlations is the same as that underlying ordinary correlations, apart from an adjustment to the degrees of freedom. The important consequence is that all the usual hypothesis tests and confidence-interval calculations for an ordinary correlation also apply to a partial correlation, the only modification necessary being a reduction in the number of degrees of freedom of any statistic used. The appropriate modification is accomplished by replacing n in any degrees of freedom for ordinary correlations by $\{n-(p-q)\}$ in the corresponding degrees of freedom for partial correlations (where $p-q$ is the number of variables held fixed). Thus, holding the values constant of some variables can be viewed as reducing the effective sample size; otherwise inferential procedures are the same as for ordinary correlations.

Suppose, for example, the correlation matrix given earlier had been calculated from a sample of $n = 30$ schoolchildren, and we want to test for the presence of correlation between weight and intelligence in the whole population of schoolchildren. From Property 5 of Section 14.2, the appropriate test statistic is $t = r\surd\{(k-1)/(1-r^2)\}$ where k is the number of degrees of freedom of the underlying Wishart distribution. If the population correlation is zero, then this statistic follows a t-distribution on $(k-1)$ degrees of freedom. Taking first the ordinary correlation between weight and intelligence, we have $r = 0.6162$ and $k = n - 1 = 29$. Thus $t = 0.6162\surd\{28/(1-0.6162^2)\} = 4.14$, which is considerably greater than the upper 1 per cent value (2.76) of the t-distribution on 28 degrees of freedom. Hence we conclude that the population correlation between weight and intelligence is different from zero. On eliminating the effect of age, however, we found the partial correlation to be $r = 0.0287$. In this case just one variable has been held constant, so $k = n - 1 - 1 = 28$. Thus $t = 0.0287\surd\{27/(1-0.0287^2)\} = 0.149$, which is close to the mean value (i.e. the 50 per cent point) of the t-distribution on 27 degrees of freedom. Hence we may assume the population correlation between weight and intelligence, after allowing for age, to be zero.

Now consider the partial correlation of 0.701 between age and intelligence, holding weight constant. This value provides significant evidence of a non-zero population partial correlation ρ, so let us find a 99 per cent confidence interval for ρ. For this calculation we need to use Property 6 of Section 14.2 with $k = n - 1 - 1 = 28$ and $r = 0.701$. Hence $z = \frac{1}{2}\log_e\{(1+0.701)/(1-0.701)\} = 0.8693$, and this is an observation from a normal distribution with mean $\frac{1}{2}\log_e\{(1+\rho)/(1-\rho)\}$ and variance $1/(k-2)=1/26$.

Thus

$$\text{pr}[0.8693-(2.576/\sqrt{26})\leq\tfrac{1}{2}\log_e\{(1+\rho)/(1-\rho)\}\leq 0.8693 + (2.576/\sqrt{26})]=0.99,$$

i.e. $\text{pr}[0.3641\leq\frac{1}{2}\log_e\{(1+\rho)/(1-\rho)\}\leq 1.3745]=0.99$,

i.e. $\text{pr}\{2.0713\leq(1+\rho)/(1-\rho)\leq 15.6270\}=0.99$,

i.e. $\text{pr}(0.3489\leq\rho\leq 0.8797)=0.99$ (as in the calculation of Section 14.2).

Thus a 99% confidence interval for the population correlation between age and intelligence, corrected for weight, is (0.349, 0.880).

Although we are now concentrating almost exclusively on quantitative data, it is worth pointing out that allowance for the effect of an extraneous variable on the association between two variables can be made relatively easily when the data are qualitative, because in this case replication *is* generally available. With two qualitative variables x_1 and x_2, we construct an $(r\times c)$ contingency table T, and use one of the measures of association (14.4) or (14.5) that are based on the χ^2 test (14.3) of no-association. If the extraneous variable x_3 whose effect on this association we wish to examine has t categories, then we merely need to construct t contingency tables from the table T (one for each category of x_3) and compute the association between x_1 and x_2 for each of these tables using (14.4) or (14.5). If these values are all substantially in agreement with the value obtained from T, then x_3 has no effect on the association between x_1 and x_2. If they are in substantial agreement with each other but *not* with the one from T, then x_3 does have an effect on the association between x_1 and x_2 in the way indicated. One final possibility is that all the values are diverse, which suggests the presence of an *interaction* between x_1, x_2, and x_3. The data should then be subjected to further scrutiny to determine the nature of this interaction.

14.5 Measuring association between two sets of variables

It often happens that the variables in a multivariate system may be divided a priori into two sets, with each set relating to a particular component of the system and with some idea required of the association between these components. For example, an agronomist may have taken p measurements relating to the yield of plants (e.g. height, dry weight, number of leaves) at each of n sites in a region, and at the same time he or she may have recorded q variables relating to the weather conditions at these sites (e.g. average daily rainfall, humidity, hours of sunshine). The whole system thus consists of n units on each of which $(p+q)$ variables have been measured, and the agronomist is interested in measuring the association between 'yield' and 'weather' as a prelude to modelling the whole process. A medical researcher might take p laboratory measurements (e.g. blood sugar content, chemical constituents of urine) on n patients suffering from some complaint, and at the same time might record some simple measurements that could be made in the doctor's surgery (e.g. temperature, blood pressure, presence or absence of external symptoms) for the same n patients. The researcher would then be interested in measuring any general association that exists between the simple (i.e. doctor's) measurements and the complex (i.e. laboratory) measurements, with the possible aim of establishing a diagnostic system based on simple signs and symptoms. An educational researcher might administer a battery of tests to n schoolchildren, of which p tests are traditional academic ones such as comprehension, grammar, arithmetic, and spelling, while q are geometric/pictorial ones such as picture completion, block diagrams, mazes, and geometrical drawing. The researcher might then be interested in exploring any association that broadly exists between academic achievement and 'pictorial insight', with possible implications for future teaching policy and methods.

In all of these examples, the division of $(p+q)$ variables into two sets of p and q variables respectively is made on external grounds, taking the prior objectives of the study into consideration, and the aim is to measure in some general sense the association between the two *collections* of variables. The overall $(p+q)\times(p+q)$ correlation matrix contains all the information on associations between pairs of variables in the system, but attempting to extract from this matrix some idea of the association between the two sets of variables is a difficult task. This is because the correlations between the two sets may not have a consistent pattern, and these between-set correlations must in any case be adjusted somehow for the within-set correlations. The following simple example will help to fix ideas and highlight some of these points. Table 14.7 presents data from Frets (1921) (analysed also in Rao (1952), Mardia *et al.* (1979), T. W. Anderson (1984), and Seber (1984)) giving head measurements on first and second adult sons in twenty-five families. The variables are:

x_1 head length of first son;

x_2 head breadth of first son;

x_3 head length of second son;

x_4 head breadth of second son.

The sum-of-squares and -products matrix **C**, (unbiased) sample covariance matrix **S**, and correlation matrix **R** for these data are as follows

Table 14.7 Head measurements on the first and second adult sons in twenty-five families, from Frets (1921)

Head length of first son (x_1)	Head breadth of first son (x_2)	Head length of second son (x_3)	Head breadth of second son (x_4)
191	155	179	145
195	149	201	152
181	148	185	149
183	153	188	149
176	144	171	142
208	157	192	152
189	150	190	149
197	159	189	152
188	152	197	159
192	150	187	151
179	158	186	148
183	147	174	147
174	150	185	152
190	159	195	157
188	151	187	158
163	137	161	130
195	155	183	158
186	153	173	148
181	145	182	146
175	140	165	137
192	154	185	152
174	143	178	147
176	139	176	143
197	167	200	158
190	163	187	150

(lower portions only given):

$$\mathbf{C} = \begin{pmatrix} 2287.04 & & & \\ 1268.84 & 1304.64 & & \\ 1671.88 & 1231.48 & 2419.36 & \\ 1106.68 & 841.28 & 1356.96 & 1080.56 \end{pmatrix},$$

$$\mathbf{S} = \begin{pmatrix} 95.293 & & & \\ 52.868 & 54.362 & & \\ 69.662 & 51.312 & 100.807 & \\ 46.112 & 35.053 & 56.540 & 45.023 \end{pmatrix},$$

$$\mathbf{R} = \begin{pmatrix} 1.0000 & & & \\ 0.7346 & 1.0000 & & \\ 0.7108 & 0.6932 & 1.0000 & \\ 0.7040 & 0.7086 & 0.8392 & 1.0000 \end{pmatrix}.$$

Interest focuses here on the association between the head measurements for the first son and those for the second son. Thus the vector of measurements $\boldsymbol{x}' = (x_1, x_2, x_3, x_4)$ can be partitioned into the two sets $\boldsymbol{x}_1' = (x_1, x_2)$ and $\boldsymbol{x}_2' = (x_3, x_4)$, and the matrices $\mathbf{C}$, $\mathbf{S}$, and $\mathbf{R}$ can be partitioned conformally into $\begin{pmatrix} \mathbf{C}_{11} & \mathbf{C}_{12} \\ \mathbf{C}_{21} & \mathbf{C}_{22} \end{pmatrix}$, $\begin{pmatrix} \mathbf{S}_{11} & \mathbf{S}_{12} \\ \mathbf{S}_{21} & \mathbf{S}_{22} \end{pmatrix}$, and $\begin{pmatrix} \mathbf{R}_{11} & \mathbf{R}_{12} \\ \mathbf{R}_{21} & \mathbf{R}_{22} \end{pmatrix}$. The sub-matrix $\mathbf{R}_{21} = \begin{pmatrix} 0.7108 & 0.6932 \\ 0.7040 & 0.7086 \end{pmatrix}$ gives the four correlations between each variable of $\boldsymbol{x}_1$ and each variable of $\boldsymbol{x}_2$. In the present case the correlations *are* consistent, indicating high positive association between every pair of inter-set variables. However, how many 'dimensions' of association are needed to explain these correlations fully; how do the within-set correlations (0.7346 for x_1, x_2, and 0.8392 for x_3, x_4) modify the $\mathbf{R}_{12}$ entries; and how can we quantify the inter-set association? We somehow need to 'clean-up' the rather confusing and interrelated set of entries in $\mathbf{R}$ to extract the main between-set information before we can answer these questions.

In fact, the questions as posed above bring to mind the similar objectives of canonical variate analysis (Section 11.1), although in the latter case the partition of the data was by units rather than by variables. In canonical variate analysis we aim to find the smallest number of dimensions for displaying adequately the between-group differences, while taking into account the within-group differences in the units. The present objectives transfer attention to the variables but retain the elements of dimensionality exploration and of balancing between-group and within-group features. Hence the solution may be expected to show some connection with canonical variate analysis, and this does indeed turn out to be the case. However, we defer establishing this connection until later; to begin with, we approach the problem in the most direct manner.

Suppose quite generally that the $(p+q)$-element observation vector $\boldsymbol{x}$ is partitioned into the two sets $\boldsymbol{x}_1$ and $\boldsymbol{x}_2$; assume without loss of generality that the first p variables $x_1, \ldots, x_p$ comprise $\boldsymbol{x}_1$, and the remaining q variables $x_{p+1}, \ldots, x_{p+q}$ comprise $\boldsymbol{x}_2$. The matrices $\mathbf{C}$, $\mathbf{S}$, and $\mathbf{R}$ can then be partitioned conformally, as in the example above, with $\mathbf{C} = \begin{pmatrix} \mathbf{C}_{11} & \mathbf{C}_{12} \\ \mathbf{C}_{21} & \mathbf{C}_{22} \end{pmatrix}$, $\mathbf{S} = \begin{pmatrix} \mathbf{S}_{11} & \mathbf{S}_{12} \\ \mathbf{S}_{21} & \mathbf{S}_{22} \end{pmatrix}$, and $\mathbf{R} = \begin{pmatrix} \mathbf{R}_{11} & \mathbf{R}_{12} \\ \mathbf{R}_{21} & \mathbf{R}_{22} \end{pmatrix}$. Before we can proceed any further, we must somehow quantify the association between the two sets of variables $\boldsymbol{x}_1$ and $\boldsymbol{x}_2$. The most natural approach, assuming that the variables are either numerical or can have numerical values sensibly assigned to them, is to take as the association between $\boldsymbol{x}_1$ and $\boldsymbol{x}_2$ the largest correlation that can be found between two single variables derived from $\boldsymbol{x}_1$ and $\boldsymbol{x}_2$ respectively. Furthermore, since correlation is a linear concept, we can restrict these derived variables to be linear combinations of $\boldsymbol{x}_1$ elements and $\boldsymbol{x}_2$ elements respectively; this restriction is compatible with those of other dimensionality-reduction techniques such as principal component analysis and canonical variate analysis. We are therefore led to looking for the largest possible correlation between a linear combination $u = \boldsymbol{a}'\boldsymbol{x}_1$ of the first set of variables and a linear combination $v = \boldsymbol{b}'\boldsymbol{x}_2$ of the second set. The necessary mathematics to enable us to do this now follows, but less mathematical readers may omit these details, and pass without detriment straight to the result.

From (14.1),

$$\text{corr}(u, v) = \text{cov}(u, v)/\surd[\{\text{var}(u)\}\{\text{var}(v)\}].$$

Hence the correlation between u and v is maximized by maximizing the covariance between u and v subject to the constraints $\text{var}(u) = \text{var}(v) = 1$. But the covariances between elements of $\boldsymbol{x}_1$ and those of $\boldsymbol{x}_2$ are in $\mathbf{S}_{12}$, while the variances and covariances within elements of $\boldsymbol{x}_1$ and $\boldsymbol{x}_2$ are in $\mathbf{S}_{11}$ and $\mathbf{S}_{22}$ respectively. Thus, since $u = \boldsymbol{a}'\boldsymbol{x}_1$ and $v = \boldsymbol{b}'\boldsymbol{x}_2$ then:

$$\text{cov}(u, v) = \boldsymbol{a}'\mathbf{S}_{12}\boldsymbol{b},$$

$$\text{var}(u) = \boldsymbol{a}'\mathbf{S}_{11}\boldsymbol{a},$$

and

$$\text{var}(v) = \boldsymbol{b}'\mathbf{S}_{22}\boldsymbol{b}.$$

Hence we require the maximum of $\boldsymbol{a}'\mathbf{S}_{12}\boldsymbol{b}$ subject to $\boldsymbol{a}'\mathbf{S}_{11}\boldsymbol{a} = \boldsymbol{b}'\mathbf{S}_{22}\boldsymbol{b} = 1$. Using Lagrange multipliers λ_1 and λ_2, we thus require the maximum of $V = \boldsymbol{a}'\mathbf{S}_{12}\boldsymbol{b} - \lambda_1(\boldsymbol{a}'\mathbf{S}_{11}\boldsymbol{a} - 1) - \lambda_2(\boldsymbol{b}'\mathbf{S}_{22}\boldsymbol{b} - 1)$. Now,

$$\frac{\partial V}{\partial \boldsymbol{a}} = \mathbf{S}_{12}\boldsymbol{b} - 2\lambda_1\mathbf{S}_{11}\boldsymbol{a},$$

and

$$\frac{\partial V}{\partial \boldsymbol{b}} = \mathbf{S}_{21}\boldsymbol{a} - 2\lambda_2\mathbf{S}_{22}\boldsymbol{b} \qquad (\text{since } \mathbf{S}_{21} = \mathbf{S}'_{12}).$$

For maximum V, we set $\frac{\partial V}{\partial \boldsymbol{a}}$ and $\frac{\partial V}{\partial \boldsymbol{b}}$ to zero and solve the resulting equations:

$$\mathbf{S}_{12}\boldsymbol{b} - 2\lambda_1\mathbf{S}_{11}\boldsymbol{a} = \boldsymbol{0} \qquad (14.13)$$

$$\mathbf{S}_{21}\boldsymbol{a} - 2\lambda_2\mathbf{S}_{22}\boldsymbol{b} = \boldsymbol{0}. \qquad (14.14)$$

Pre-multiplying (14.13) by $\boldsymbol{a}'$, and remembering that $\boldsymbol{a}'\mathbf{S}_{11}\boldsymbol{a} = 1$, yields

$$2\lambda_1 = \boldsymbol{a}'\mathbf{S}_{12}\boldsymbol{b};$$

while pre-multiplying (14.14) by $\boldsymbol{b}'$, and similarly remembering that $\boldsymbol{b}'\mathbf{S}_{22}\boldsymbol{b} = 1$, yields

$$2\lambda_2 = \boldsymbol{b}'\mathbf{S}_{21}\boldsymbol{a}.$$

But $\boldsymbol{a}'\mathbf{S}_{12}\boldsymbol{b} = \boldsymbol{b}'\mathbf{S}_{21}\boldsymbol{a}$ since each is a scalar and the transpose of the other. Hence $2\lambda_1 = 2\lambda_2 = \boldsymbol{a}'\mathbf{S}_{12}\boldsymbol{b} = R$ (say), the correlation between u and v. Thus, from (14.13),

$$R\mathbf{S}_{11}\boldsymbol{a} = \mathbf{S}_{12}\boldsymbol{b} \qquad (14.15)$$

while from (14.14)

$$R\mathbf{S}_{22}\boldsymbol{b} = \mathbf{S}_{21}\boldsymbol{a}. \qquad (14.16)$$

Hence from (14.15) $\boldsymbol{a} = \frac{1}{R}\mathbf{S}_{11}^{-1}\mathbf{S}_{12}\boldsymbol{b}$, and when this is substituted in (14.16) we obtain $R\mathbf{S}_{22}\boldsymbol{b} = \frac{1}{R}\mathbf{S}_{21}\mathbf{S}_{11}^{-1}\mathbf{S}_{12}\boldsymbol{b}$, or

$$(\mathbf{S}_{21}\mathbf{S}_{11}^{-1}\mathbf{S}_{12} - R^2\mathbf{S}_{22})\boldsymbol{b} = \boldsymbol{0}. \qquad (14.17)$$

Similarly from (14.16) $\boldsymbol{b} = \frac{1}{R}\mathbf{S}_{22}^{-1}\mathbf{S}_{21}\boldsymbol{a}$, and when this is substituted in (14.15) we obtain $R\mathbf{S}_{11}\boldsymbol{a} = \frac{1}{R}\mathbf{S}_{12}\mathbf{S}_{22}^{-1}\mathbf{S}_{21}\boldsymbol{a}$, or

$$(\mathbf{S}_{12}\mathbf{S}_{22}^{-1}\mathbf{S}_{21} - R^2\mathbf{S}_{11})\boldsymbol{a} = \boldsymbol{0}. \qquad (14.18)$$

From (14.17), therefore, R^2 is an eigenvalue of $\mathbf{S}_{22}^{-1}\mathbf{S}_{21}\mathbf{S}_{11}^{-1}\mathbf{S}_{12}$, and $\boldsymbol{b}$ is its corresponding eigenvector; while from (14.18) R^2 is an eigenvalue of $\mathbf{S}_{11}^{-1}\mathbf{S}_{12}\mathbf{S}_{22}^{-1}\mathbf{S}_{21}$, and $\boldsymbol{a}$ is its corresponding eigenvector. But since $R = \boldsymbol{a}'\mathbf{S}_{12}\boldsymbol{b}$, the quantity that we set out to maximize, then the required eigenvalue of each of these matrices must be the *largest* eigenvalue.

The maximum correlation R_1 between $u = \boldsymbol{a}'\boldsymbol{x}_1$ and $v = \boldsymbol{b}'\boldsymbol{x}_2$ is thus obtained as the square root of the largest eigenvalue of *either* the $(p \times p)$ matrix $\mathbf{E}_1 = \mathbf{S}_{11}^{-1}\mathbf{S}_{12}\mathbf{S}_{22}^{-1}\mathbf{S}_{21}$ *or* the $(q \times q)$ matrix $\mathbf{E}_2 = \mathbf{S}_{22}^{-1}\mathbf{S}_{21}\mathbf{S}_{11}^{-1}\mathbf{S}_{12}$. Furthermore, the vector of coefficients $\boldsymbol{a}$ is the eigenvector of $\mathbf{E}_1$ corresponding to this eigenvalue, while the vector of coefficients $\boldsymbol{b}$ is the eigenvector of $\mathbf{E}_2$ corresponding to this eigenvalue. However, each of these matrices will have more than one eigenvalue/eigenvector pair; in fact, the number of non-zero eigenvalues in each will be $s = \text{rank}(\mathbf{S}_{12}) \leq \min(p, q)$. In practice these s eigenvalues will be distinct, so we can denote them by $R_1^2 > R_2^2 > \ldots > R_s^2$, and we can let $\boldsymbol{a}_i$, $\boldsymbol{b}_i$ denote the eigenvectors of $\mathbf{E}_1$, $\mathbf{E}_2$ respectively corresponding to eigenvalue R_i^2. This shows that a single pair of variables u and v is not sufficient to quantify the association between the two sets $\boldsymbol{x}_1$ and $\boldsymbol{x}_2$, but rather we need s pairs $(u_1, v_1), \ldots, (u_s, v_s)$ to do so. The required pairs are given by

$$u_i = \boldsymbol{a}_i'\boldsymbol{x}_1; \quad v_i = \boldsymbol{b}_i'\boldsymbol{x}_2 \; (i = 1, \ldots, s)$$

and are termed the *canonical variables* (*variates*) of the system, while the values R_i are termed the *canonical correlations* of the system.

We next develop some mathematical properties of these canonical variables, but again, less mathematical readers can omit these details and pass without any detriment straight to the summary and statistical implications.

We first establish that the matrices $\mathbf{E}_1$ and $\mathbf{E}_2$ do indeed have the same non-zero eigenvalues, and we derive a relationship between their eigenvectors. Suppose that a matrix of the form $\mathbf{AB}$ has eigenvalue $\lambda \neq 0$ and corresponding eigenvector $\boldsymbol{x}$. Then $\mathbf{AB}\boldsymbol{x} = \lambda\boldsymbol{x}$. Suppose also that the product $\mathbf{BA}$ is defined.

Then

$$\mathbf{BAB}\boldsymbol{x} = \mathbf{B}(\lambda\boldsymbol{x}) = \lambda\mathbf{B}\boldsymbol{x}$$

i.e.

$$\mathbf{BA}\boldsymbol{y} = \lambda\boldsymbol{y} \quad \text{where} \quad \boldsymbol{y} = \mathbf{B}\boldsymbol{x}.$$

Hence $\lambda \neq 0$ is also an eigenvalue of the matrix $\mathbf{BA}$, with eigenvector $\boldsymbol{y} = \mathbf{B}\boldsymbol{x}$. Now if we set $\mathbf{A} = \mathbf{S}_{11}^{-1}\mathbf{S}_{12}$ and $\mathbf{B} = \mathbf{S}_{22}^{-1}\mathbf{S}_{21}$ then we see that $\mathbf{E}_1 = \mathbf{AB}$ and $\mathbf{E}_2 = \mathbf{BA}$. Thus by the above argument we see that the non-zero eigenvalues of $\mathbf{E}_1$ and $\mathbf{E}_2$ are the same. Furthermore, the eigenvectors $\boldsymbol{a}_i$ of $\mathbf{E}_1$ and $\boldsymbol{b}_i$ of $\mathbf{E}_2$ must be related by $\boldsymbol{b}_i = \mathbf{S}_{22}^{-1}\mathbf{S}_{21}\boldsymbol{a}_i$.

Next, consider the pairs $u_i = \boldsymbol{a}_i'\boldsymbol{x}_1$, $v_i = \boldsymbol{b}_i'\boldsymbol{x}_2$, and $u_j = \boldsymbol{a}_j'\boldsymbol{x}_1$, $v_j = \boldsymbol{b}_j'\boldsymbol{x}_2$ where $i \neq j$. Then

$$\text{cov}(u_i, u_j) = \boldsymbol{a}_i'\mathbf{S}_{11}\boldsymbol{a}_j$$

$$\text{cov}(v_i, v_j) = \boldsymbol{b}_i'\mathbf{S}_{22}\boldsymbol{b}_j$$

and

$$\text{cov}(u_j, v_i) = \boldsymbol{b}_i'\mathbf{S}_{12}\boldsymbol{a}_j.$$

Now, from (14.18),

$$R_j^2\mathbf{S}_{11}\boldsymbol{a}_j = \mathbf{S}_{12}\mathbf{S}_{22}^{-1}\mathbf{S}_{21}\boldsymbol{a}_j$$

so

$$\begin{aligned} R_j^2\boldsymbol{a}_i'\mathbf{S}_{11}\boldsymbol{a}_j &= \boldsymbol{a}_i'\mathbf{S}_{12}\mathbf{S}_{22}^{-1}\mathbf{S}_{21}\boldsymbol{a}_j \\ &= (\boldsymbol{a}_j'\mathbf{S}_{12}\mathbf{S}_{22}^{-1}\mathbf{S}_{21}\boldsymbol{a}_i)' \\ &= (\boldsymbol{a}_j'R_i^2\mathbf{S}_{11}\boldsymbol{a}_i)' && \text{from (14.18)} \\ &= R_i^2\boldsymbol{a}_i'\mathbf{S}_{11}\boldsymbol{a}_j. \end{aligned}$$

Thus, if $i \neq j$ then $R_i \neq R_j$ so we must have $\boldsymbol{a}_i'\mathbf{S}_{11}\boldsymbol{a}_j = 0$ and $\text{cov}(u_i, u_j) = 0$. From (14.17),

$$R_j^2\mathbf{S}_{22}\boldsymbol{b}_j = \mathbf{S}_{21}\mathbf{S}_{11}^{-1}\mathbf{S}_{12}\boldsymbol{b}_j$$

so

$$\begin{aligned} R_j^2\boldsymbol{b}_i'\mathbf{S}_{22}\boldsymbol{b}_j &= \boldsymbol{b}_i'\mathbf{S}_{21}\mathbf{S}_{11}^{-1}\mathbf{S}_{12}\boldsymbol{b}_j \\ &= (\boldsymbol{b}_j'\mathbf{S}_{21}\mathbf{S}_{11}^{-1}\mathbf{S}_{12}\boldsymbol{b}_i)' \\ &= (\boldsymbol{b}_j'R_i^2\mathbf{S}_{22}\boldsymbol{b}_i)' && \text{from (14.17)} \\ &= R_i^2\boldsymbol{b}_i'\mathbf{S}_{22}\boldsymbol{b}_j. \end{aligned}$$

Thus if $i \neq j$ then $R_i \neq R_j$ so we must have $\boldsymbol{b}_i'\mathbf{S}_{22}\boldsymbol{b}_j = 0$ and $\text{cov}(v_i, v_j) = 0$. From (14.18)

$$\mathbf{S}_{11}^{-1}\mathbf{S}_{12}\mathbf{S}_{22}^{-1}\mathbf{S}_{21}\boldsymbol{a}_j = R_j^2\boldsymbol{a}_j$$

so

$$\boldsymbol{b}_i'\mathbf{S}_{21}\mathbf{S}_{11}^{-1}\mathbf{S}_{12}\mathbf{S}_{22}^{-1}\mathbf{S}_{21}\boldsymbol{a}_j = R_j^2\boldsymbol{b}_i'\mathbf{S}_{21}\boldsymbol{a}_j. \tag{14.19}$$

From (14.17)

$$\mathbf{S}_{22}^{-1}\mathbf{S}_{21}\mathbf{S}_{11}^{-1}\mathbf{S}_{12}\boldsymbol{b}_i = R_i^2\boldsymbol{b}_i$$

so

$$\boldsymbol{a}_j'\mathbf{S}_{12}\mathbf{S}_{22}^{-1}\mathbf{S}_{21}\mathbf{S}_{11}^{-1}\mathbf{S}_{12}\boldsymbol{b}_i = R_i^2\boldsymbol{a}_j'\mathbf{S}_{12}\boldsymbol{b}_i. \tag{14.20}$$

But the left-hand sides of (14.19) and (14.20) are equal, because one is the transpose of the other, and they are both scalars.

Hence $R_j^2\boldsymbol{b}_i'\mathbf{S}_{21}\boldsymbol{a}_j = R_i^2\boldsymbol{a}_j'\mathbf{S}_{12}\boldsymbol{b}_i$, and $\boldsymbol{b}_i'\mathbf{S}_{21}\boldsymbol{a}_j = \boldsymbol{a}_j'\mathbf{S}_{12}\boldsymbol{b}_i$.

Thus if $i \neq j$ then $R_i \neq R_j$ so we must have $\boldsymbol{b}_i'\mathbf{S}_{21}\boldsymbol{a}_j = 0$ and $\text{cov}(u_j, v_i) = 0$.

The canonical variates u_i, v_i $(i = 1, \ldots, s)$ are thus such that:

(1) the u_i are mutually uncorrelated ($\text{cov}(u_i, u_j) = 0$ for $i \neq j$);
(2) the v_i are mutually uncorrelated ($\text{cov}(v_i, v_j) = 0$ for $i \neq j$);
(3) the correlation between u_i and v_i is R_i for $i = 1, \ldots, s$;
(4) the u_i are uncorrelated with all v_j except v_i ($\text{cov}(u_i, v_j) = 0$ for $i \neq j$).

The technique of *canonical correlation analysis* thus transforms the p variates in $\boldsymbol{x}_1$, and the q variates in $\boldsymbol{x}_2$ to s pairs of variates $(u_1, v_1), \ldots, (u_s, v_s)$. The $n \times (p + q)$ data matrix is thereby transformed to a new $n \times (2s)$ data matrix. Writing $\boldsymbol{u}' = (u_1, \ldots, u_s)$, $\boldsymbol{v}' = (v_1, \ldots, v_s)$, and $\boldsymbol{y}' = (\boldsymbol{u}', \boldsymbol{v}')$, this transformation reduces the sample covariance matrix $\mathbf{S} = \begin{pmatrix} \mathbf{S}_{11} & \mathbf{S}_{12} \\ \mathbf{S}_{21} & \mathbf{S}_{22} \end{pmatrix}$ of the x_i to one of the form

$$\begin{pmatrix}
1 & 0 & \ldots & 0 & R_1 & 0 & \ldots & 0 \\
0 & 1 & \ldots & 0 & 0 & R_2 & \ldots & 0 \\
\vdots & & & & \vdots & & & \\
0 & 0 & \ldots & 1 & 0 & 0 & \ldots & R_s \\
R_1 & 0 & \ldots & 0 & 1 & 0 & \ldots & 0 \\
0 & R_2 & \ldots & 0 & 0 & 1 & \ldots & 0 \\
\vdots & & & & \vdots & & & \\
0 & 0 & \ldots & R_s & 0 & 0 & \ldots & 1
\end{pmatrix}$$

for the y_i. The s canonical correlations are thus the pure expression of association between the sets $\boldsymbol{x}_1$ and $\boldsymbol{x}_2$ of original variables, and all within-set correlation has been removed in calculating them. The technique can be used, in similar descriptive fashion to other related 'linear transformation' techniques such as principal component analysis or canonical variate analysis, to effect reduction of dimensionality (by choosing only as many of the u_i, v_i pairs as their corresponding canonical correlations seem 'important'), to gain insight into the system (by interpreting the linear combinations that make up the canonical variates), and to inspect the data for any irregularities or outliers (by plotting the data using the u_i, v_i pairs as axes). Interpretation of the linear combinations is aided by rescaling the variates to be expressed relative to their standard deviations, as in canonical variate analysis. At a practical level, it is immaterial whether we base a canonical correlation analysis on the sum-of-squares and -products matrix **C**, the unbiased covariance matrix **S**, or the maximum likelihood covariance matrix $\{(n-1)/n\}\mathbf{S}$, as the constant factors cancel out. Furthermore, the canonical correlations are unaffected even if we work with the correlation matrix **R** (although the canonical variates will be altered in this case).

Let us now turn to the head measurements data of Table 14.7 to illustrate the computations and interpretations. Here $p=q=2$, so that $s=2$ also. The canonical correlations can be found from the eigenvalues of any of the following matrices: $\mathbf{C}_{11}^{-1}\mathbf{C}_{12}\mathbf{C}_{22}^{-1}\mathbf{C}_{21}$, $\mathbf{C}_{22}^{-1}\mathbf{C}_{21}\mathbf{C}_{11}^{-1}\mathbf{C}_{12}$, $\mathbf{S}_{11}^{-1}\mathbf{S}_{12}\mathbf{S}_{22}^{-1}\mathbf{S}_{21}$, $\mathbf{S}_{22}^{-1}\mathbf{S}_{21}\mathbf{S}_{11}^{-1}\mathbf{S}_{12}$, $\mathbf{R}_{11}^{-1}\mathbf{R}_{12}\mathbf{R}_{22}^{-1}\mathbf{R}_{21}$, or $\mathbf{R}_{22}^{-1}\mathbf{R}_{21}\mathbf{R}_{11}^{-1}\mathbf{R}_{12}$. Each of these matrices is (2×2) and hence gives rise to just two eigenvalues, $R_1^2=0.6218$ and $R_2^2=0.0029$. Hence the canonical correlations are $R_1=0.7885$ and $R_2=0.0537$. The canonical variate coefficients $\boldsymbol{a}_i$ for the first adult sons' measurements $\boldsymbol{x}_1'=(x_1, x_2)$ are given by the eigenvectors of $\mathbf{C}_{11}^{-1}\mathbf{C}_{12}\mathbf{C}_{22}^{-1}\mathbf{C}_{21}$ or $\mathbf{S}_{11}^{-1}\mathbf{S}_{12}\mathbf{S}_{22}^{-1}\mathbf{S}_{21}$, and turn out to be $\boldsymbol{a}_1=\begin{pmatrix}0.0577\\0.0722\end{pmatrix}$ and $\boldsymbol{a}_2=\begin{pmatrix}0.1429\\-0.1909\end{pmatrix}$. The canonical variate coefficients $\boldsymbol{b}_i$ for the second adult sons' measurements $\boldsymbol{x}_2'=(x_3, x_4)$ are given by the eigenvectors of $\mathbf{C}_{22}^{-1}\mathbf{C}_{21}\mathbf{C}_{11}^{-1}\mathbf{C}_{12}$ or $\mathbf{S}_{22}^{-1}\mathbf{S}_{21}\mathbf{S}_{11}^{-1}\mathbf{S}_{12}$, and turn out as $\boldsymbol{b}_1=\begin{pmatrix}0.0512\\0.0722\end{pmatrix}$ and $\boldsymbol{b}_2=\begin{pmatrix}0.1796\\-0.2673\end{pmatrix}$. Thus the canonical variates are

$$
\begin{aligned}
u_1 &= 0.0577x_1 + 0.0722x_2\\
u_2 &= 0.1429x_1 - 0.1909x_2\\
v_1 &= 0.0512x_3 + 0.0722x_4\\
v_2 &= 0.1796x_3 - 0.2673x_4.
\end{aligned}
$$

For interpretation, however, it is preferable to express these canonical variates in terms of standardized x_i. The standard deviation s_i of each x_i is given by the square root of the corresponding diagonal entry of **S**. Thus $s_1=\sqrt{95.293}=9.7618$, $s_2=\sqrt{54.362}=7.3731$, $s_3=\sqrt{100.807}=10.0403$,

$s_4 = \sqrt{45.023} = 6.7099$; and the standardized variates are $x_1^* = x_1/s_1$, $x_2^* = x_2/s_2$, $x_3^* = x_3/s_3$, and $x_4^* = x_4/s_4$. Now if $u_i = a_{i1}x_1 + a_{i2}x_2$ then $u_i = (s_1 a_{i1})x_1^* + (s_2 a_{i2})x_2^*$, and similarly for v_i. Thus on multiplying the canonical variate coefficients by the appropriate standard deviations we obtain:

$$u_1 = 0.563x_1^* + 0.532x_2^*$$
$$u_2 = 1.395x_1^* - 1.408x_2^*$$
$$v_1 = 0.514x_3^* + 0.484x_4^*$$
$$v_2 = 1.803x_3^* - 1.794x_4^*.$$

The first canonical variates u_1, v_1 can therefore be interpreted as 'girth' measurements (being approximately proportional to the *sum* of the standardized length and breadth for each brother), while the second canonical variates u_2, v_2 can be interpreted as 'shape' measurements (being approximately proportional to the *difference* of the standardized length and breadth for each brother). From the canonical correlations $R_1 = 0.7885$ and $R_2 = 0.0537$, it is evident that the association between $\boldsymbol{x}_1$ and $\boldsymbol{x}_2$ is entirely channeled through the 'girth' measurements u_1, v_1, and that the two 'shape' measurements u_2, v_2 are virtually uncorrelated. The association between the two sets of variables $\boldsymbol{x}_1$, $\boldsymbol{x}_2$ is thus essentially one-dimensional, and well specified by the single correlation value 0.7885. (Note that this value is larger than any of the entries in $\boldsymbol{R}_{12}$.) It only remains to check that this value represents a genuine association and has not been produced as an artefact of some eccentric grouping or outliers in the data, and this can be done by plotting the twenty-five pairs of (u_1, v_1) values in a scatter plot. The plot is shown in Fig. 14.6, and indicates that the association is indeed genuine and there are no anomalous units present in the sample.

In all the above, canonical correlation analysis has been treated purely from the descriptive point of view, namely for the specification of association between two sets of variables in an observed set of data. Its use can easily be extended to inferential applications by specifying and developing a natural statistical model for the data at hand. If the variables are all continuous (or at least quantitative) then we can assume that the vector $\boldsymbol{x}$ of $(p + q)$ variables measured on each individual (or some appropriate transformation of this vector) comes from a normal distribution with mean vector $\boldsymbol{\mu}$ and dispersion matrix $\boldsymbol{\Sigma}$, and the rows of the data matrix are independent observations from this distribution. Also, since the vector $\boldsymbol{x}$ is partitioned into the two components $\boldsymbol{x}' = (\boldsymbol{x}_1', \boldsymbol{x}_2')$ where $\boldsymbol{x}_1$ has p elements, and $\boldsymbol{x}_2$ has q elements, then the mean vector and dispersion matrix can be partitioned conformally as $\boldsymbol{\mu}' = (\boldsymbol{\mu}_1', \boldsymbol{\mu}_2')$ and

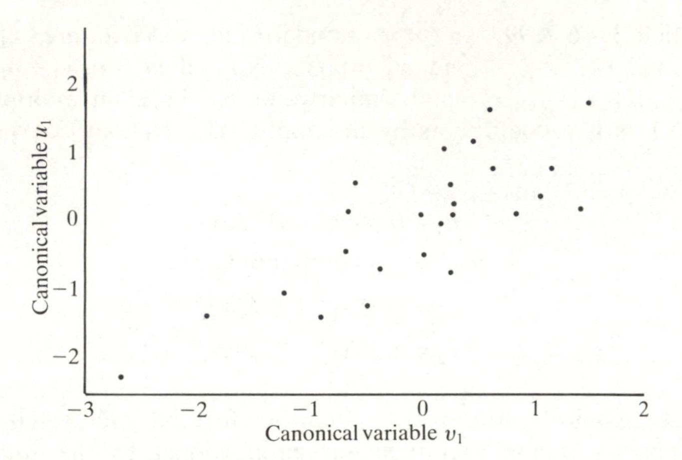

Fig. 14.6 Scatter plot of the first two canonical variates for the data of Table 14.7. Redrawn from *Multivariate Observations* by G. A. F. Seber, copyright © 1984 John Wiley & Sons Inc. Reprinted by permission of John Wiley & Sons Inc

$\boldsymbol{\Sigma} = \begin{pmatrix} \boldsymbol{\Sigma}_{11} & \boldsymbol{\Sigma}_{12} \\ \boldsymbol{\Sigma}_{21} & \boldsymbol{\Sigma}_{22} \end{pmatrix}$. We can therefore look for the transformations $u = \boldsymbol{\alpha}'\boldsymbol{x}_1$ and $v = \boldsymbol{\beta}'\boldsymbol{x}_2$ of the original variables that have maximum *population* correlation. The mathematics in this case is exactly the same as it was before, except that the population dispersion $\boldsymbol{\Sigma}$ replaces the sample dispersion $\mathbf{S}$, and hence we are led to consider the eigenvalues and eigenvectors of the matrices $\boldsymbol{\Gamma}_1 = \boldsymbol{\Sigma}_{11}^{-1}\boldsymbol{\Sigma}_{12}\boldsymbol{\Sigma}_{22}^{-1}\boldsymbol{\Sigma}_{21}$ and $\boldsymbol{\Gamma}_2 = \boldsymbol{\Sigma}_{22}^{-1}\boldsymbol{\Sigma}_{21}\boldsymbol{\Sigma}_{11}^{-1}\boldsymbol{\Sigma}_{12}$. The $s = \min(p, q)$ non-zero eigenvalues of the two matrices are the same, and if they are distinct can be written $\rho_1^2 > \rho_2^2 > \ldots > \rho_s^2$. Denote the eigenvectors corresponding to these eigenvalues by $\boldsymbol{\alpha}_1, \boldsymbol{\alpha}_2, \ldots, \boldsymbol{\alpha}_s$ for $\boldsymbol{\Gamma}_1$, and $\boldsymbol{\beta}_1, \boldsymbol{\beta}_2, \ldots, \boldsymbol{\beta}_s$ for $\boldsymbol{\Gamma}_2$. Then the *population canonical variates* are given by $u_i = \boldsymbol{\alpha}_i'\boldsymbol{x}_1$ and $v_i = \boldsymbol{\beta}_i'\boldsymbol{x}_2$ for $i = 1, \ldots, s$. They are such that, in the population,

(1) the u_i are mutually uncorrelated;
(2) the v_i are mutually uncorrelated;
(3) the correlation between u_i and v_i is ρ_i for $i = 1, \ldots, s$;
(4) the u_i are uncorrelated with all v_j except v_i;

providing that the ρ_i are all distinct. The ρ_i $(i = 1, \ldots, s)$ are the *population canonical correlations.*

In the model thus specified, the canonical correlations ρ_i represent the 'true' associations that exist between the two sets of variables $\boldsymbol{x}_1$ and $\boldsymbol{x}_2$, through the canonical variates $\boldsymbol{\alpha}_i'\boldsymbol{x}_1$ and $\boldsymbol{\beta}_i'\boldsymbol{x}_2$ $(i = 1, \ldots, s)$. Given a random sample from the population, therefore, the sample canonical correlations R_i can be viewed as estimates of ρ_i, and the sample canonical variate coefficients $\boldsymbol{a}_i$, $\boldsymbol{b}_i$ as estimates of $\boldsymbol{\alpha}_i$, $\boldsymbol{\beta}_i$ $(i = 1, \ldots, s)$. Moreover, given invariance of maximum likelihood estimates and the fact that the maximum likelihood estimate of $\boldsymbol{\Sigma}$ can be used in the canonical correlation calculations, it follows that R_i, $\boldsymbol{a}_i$, and $\boldsymbol{b}_i$ are the maximum likelihood estimates of ρ_i, $\boldsymbol{\alpha}_i$, and $\boldsymbol{\beta}_i$ (provided again that the ρ_i are distinct). Statistical inference is then required to answer questions of interest about the ρ_i, $\boldsymbol{\alpha}_i$, and $\boldsymbol{\beta}_i$ from knowledge of the R_i, $\boldsymbol{a}_i$, and $\boldsymbol{b}_i$. Central to such applications are the sampling distributions of the quantities R_i, $\boldsymbol{a}_i$, and $\boldsymbol{b}_i$ $(i = 1, \ldots, s)$. These distributions give information about the behaviour of these quantities under repeated sampling from the same population. Exact sampling distributions are available, but are mathematically complicated and intractable for most practical purposes, so recourse must usually be made to asymptotic theory. A summary of the main results is given by Muirhead (1982, Chapter 11). In practice, the only inferential procedure that is of direct relevance to our descriptive approach is a test of the hypothesis that some or all of the ρ_i are zero. If we can establish that some of the population canonical correlations are zero, then we have achieved dimensionality reduction because we can focus only on those canonical variates whose population correlations are greater than zero. This is a more satisfactory procedure than disregarding in some arbitrary fashion those canonical variates whose sample correlations R_i are 'small'. However, we will delay a discussion of the relevant test statistics until the next section, where they can be given systematically along with other test statistics for correlation matrices.

We now return to the connection with the canonical variate analysis of Section 11.1 that was mentioned at the beginning of the present section. It may seem confusing to have canonical variates in two very different contexts: the grouping of units as in Section 11.1, and the grouping of variables as in the present Section. However, it turns out that the two applications are fundamentally the same. Consider again the situation discussed in Section 11.1 where p variables $\boldsymbol{x}' = (x_1, \ldots, x_p)$ are assumed to be measured on each of n units, and these units have been partitioned into g groups. To display the data in such a way as to emphasize the differences between the group means as much as possible, we need to find the first two canonical variate coefficient vectors $\boldsymbol{a}_1$, $\boldsymbol{a}_2$, and then display the group means using the canonical variates $y_1 = \boldsymbol{a}_1'\boldsymbol{x}$, $y_2 = \boldsymbol{a}_2'\boldsymbol{x}$ as axes. The canonical variate coefficient vectors $\boldsymbol{a}_i$ were obtained by solution of eqn (11.6), i.e. $(\mathbf{B}-l\mathbf{W})\boldsymbol{a}=\mathbf{0}$ where $\mathbf{B}$ and $\mathbf{W}$ are the between-groups and within-groups covariance matrices defined in (11.5). Suppose instead we define g extra dummy 0/1 variables $x_{p+1}, x_{p+2}, \ldots, x_{p+g}$ for each sample individual as follows: if the individual belongs to the ith group then set $x_{p+i}=1$ and $x_{p+j}=0$ $(j \neq i = 1, \ldots, g)$. Thus all dummy variables except one have value zero for each sample individual; the non-zero variable has value 1, and indicates the group membership of that individual. Now set $\boldsymbol{x}' = (x_1, x_2, \ldots, x_p, x_{p+1}, \ldots, x_{p+g})$, $\boldsymbol{x}_1' = (x_1, \ldots, x_p)$ and $\boldsymbol{x}_2' = (x_{p+1}, \ldots, x_{p+g})$. Then the canonical variate coefficients $\boldsymbol{a}_i$ appropriate to the variable set $\boldsymbol{x}_1$, as obtained from a canonical correlation analysis between $\boldsymbol{x}_1$ and $\boldsymbol{x}_2$, are precisely the same as the coefficients obtained from solution of (11.6), while the canonical variates $u_i = \boldsymbol{a}_i'\boldsymbol{x}$ obtained from the canonical correlation analysis are precisely the same as the canonical variates y_i that are used for the plotting of group means. (See Mardia *et al.* (1979), Exercise 11.5.4.) Since the dummy variables in $\boldsymbol{x}_2$ define group membership of the units, it is intuitively reasonable that the canonical variate $u_1 = \boldsymbol{a}_1'\boldsymbol{x}_1$ which has maximum correlation with $v_1 = \boldsymbol{b}_1'\boldsymbol{x}_2$ is the derived variate that best shows up differences between the group means (because all v_i are 'between-group' variates). Readers familiar with univariate design and analysis of experiments will recognize these dummy variables as constituting the *design matrix* of a simple one-way arrangement. Hence the canonical variates for inspecting group differences can be obtained from a canonical correlation analysis between the data and design matrices, and this connection also remains true for more complex designs.

Another connection with analysis of variance (albeit the univariate variety) is provided by the special case $p=1$. Here there is just a single variate x_1 in the first set, and hence a canonical correlation analysis will yield just the single linear combination $\boldsymbol{b}'\boldsymbol{x}_2$ of the second set that has maximum correlation with x_1. Since $p=1$, $\mathbf{S}_{11}$ is just the single value s_1^2 (giving the sample variance of x_1), while $\mathbf{S}_{12}$ is a $(1 \times q)$ vector. Thus $\mathbf{S}_{11}^{-1}\mathbf{S}_{12}\mathbf{S}_{22}^{-1}\mathbf{S}_{21}$ is the scalar $\mathbf{S}_{12}\mathbf{S}_{22}^{-1}\mathbf{S}_{21}/s_1^2$, and therefore must equal its own eigenvalue. Hence the only non-zero canonical correlation is the value $R = \frac{1}{s_1}\sqrt{(\mathbf{S}_{12}\mathbf{S}_{22}^{-1}\mathbf{S}_{21})}$: this is known as the (sample) *multiple correlation* between x_1 and the set $\boldsymbol{x}_2$. It is an estimate of the *population multiple correlation* $\rho = \frac{1}{\sigma_1}\sqrt{(\boldsymbol{\Sigma}_{12}\boldsymbol{\Sigma}_{22}^{-1}\boldsymbol{\Sigma}_{21})}$ for the model in which the observation vector $\boldsymbol{x}' = (x_1, \boldsymbol{x}_2')$ is assumed to have a multivariate normal distribution with mean vector $\boldsymbol{\mu}' = (\mu_1, \boldsymbol{\mu}_2')$ and dispersion matrix $\boldsymbol{\Sigma} = \begin{pmatrix} \sigma_1^2 & \boldsymbol{\Sigma}_{12} \\ \boldsymbol{\Sigma}_{21} & \boldsymbol{\Sigma}_{22} \end{pmatrix}$.

Here σ_1^2 is the (marginal) variance of x_1. But from Property 3 of Section 7.2, the conditional variance of x_1 when the variables in $\boldsymbol{x}_2$ are fixed is given by $\sigma^2 = \sigma_1^2 - \boldsymbol{\Sigma}_{12}\boldsymbol{\Sigma}_{22}^{-1}\boldsymbol{\Sigma}_{21}$. Thus, from above, $\sigma_1^2\rho^2 = \boldsymbol{\Sigma}_{12}\boldsymbol{\Sigma}_{22}^{-1}\boldsymbol{\Sigma}_{21} = \sigma_1^2 - \sigma^2$, so that

$$\rho^2 = (\sigma_1^2 - \sigma^2)/\sigma_1^2. \tag{14.21}$$

Now $\sigma_1^2 - \sigma^2$ is the difference between the marginal variance of x_1 and the conditional variance of x_1 for fixed values of $\boldsymbol{x}_2$, and hence must represent the variance of x_1 caused by its relationship with elements of $\boldsymbol{x}_2$. Furthermore, ρ is the correlation between x_1 and its closest 'predictor' $\boldsymbol{\beta}'\boldsymbol{x}_2$. Hence, in the sample case, $\boldsymbol{\beta}$ can be estimated not only by the sample canonical variate coefficient vector $\boldsymbol{b}$, but also by the set of regression coefficients in a least-squares regression of x_1 on elements of $\boldsymbol{x}_2$. Thus canonical correlation analysis when $p = 1$ is equivalent to multiple regression analysis. Furthermore, R is not only the sample canonical correlation, but also the correlation between the observed values of x_1 and their predictions from the fitted regression. Associated with the regression will be an analysis of variance breaking down the total sum of squares (of x_1) into a regression sum-of-squares and a residual sum-of-squares. Thus using (14.21), a yet further estimate of ρ^2 can be given by $R^2 =$ (regression sum-of-squares)/(total sum-of-squares), and this is the quantity R^2 which is usually provided in any computer package for multiple regression analysis.

Finally, we note that the above discussion of canonical correlation analysis has assumed a division of the observed variates into *two* sets. It may happen that the variates can be divided into more than two sets, and some general between-set measure of association is required. Various possible definitions of such association may now be made, and the ideas of canonical correlation analysis may be generalized in various ways. The mathematics becomes more involved in these circumstances, and it is not clear how practically useful the resulting techniques are likely to be. The interested reader is directed to Gnanadesikan (1977) for a brief overview, and to Kettenring (1971) for full details.

14.6 Testing hypotheses about sets of associations

In Section 14.2 we considered inferential procedures about single correlation coefficients. In the present section we collect together a number of useful test procedures which involve more than a single measure of association. Because most such tests are based on multivariate normality, the tests given here will again be relevant only for quantitative data. It is assumed that the observation vector $\boldsymbol{x}$ comes from a multivariate normal population having mean $\boldsymbol{\mu}$, dispersion matrix $\boldsymbol{\Sigma}$ and correlation matrix $\mathbf{P}$. Rows of the data matrix are independent observations of this vector $\boldsymbol{x}$, and the sample mean, covariance matrix and correlation matrix are given by $\bar{\boldsymbol{x}}$, $\mathbf{S}$ and $\mathbf{R}$ respectively. Where necessary, these vectors and matrices are partitioned in the usual manner.

14.6.1 Test that all population correlations are zero (mutual independence of all variables)

Here we wish to test the null hypothesis H_0: $\mathbf{P} = \mathbf{I}$, against the alternative H_1: $\mathbf{P} \neq \mathbf{I}$. An equivalent statement of these hypotheses is H_0: $\boldsymbol{\Sigma}$ is diagonal, and H_1: $\boldsymbol{\Sigma}$ is not diagonal. The likelihood ratio test statistic λ for this situation may be obtained from a generalization of (8.16) (see (3) of Section 8.3) as $-2 \log_e \lambda = n \operatorname{trace} \mathbf{U} - n \log_e |\mathbf{U}| - np$ where $\mathbf{U} = \hat{\boldsymbol{\Sigma}}^{-1}\mathbf{S}$ and $\hat{\boldsymbol{\Sigma}}$ is the maximum likelihood estimator of $\boldsymbol{\Sigma}$ under H_0. If H_0 is true then the mean and variance of each variable are estimated separately, so that $\hat{\boldsymbol{\mu}} = \bar{\boldsymbol{x}}$, and $\hat{\boldsymbol{\Sigma}} = \operatorname{diag}(s_{11}, s_{22}, \ldots, s_{pp})$. Thus $\hat{\boldsymbol{\Sigma}}^{-1} = \operatorname{diag}(1/s_{11}, 1/s_{22}, \ldots, 1/s_{pp})$, and so the diagonal elements of $\hat{\boldsymbol{\Sigma}}^{-1}\mathbf{S}$ are all unity. Also $|\hat{\boldsymbol{\Sigma}}^{-1}\mathbf{S}| = |\hat{\boldsymbol{\Sigma}}^{-\frac{1}{2}}\hat{\boldsymbol{\Sigma}}^{-\frac{1}{2}}\mathbf{S}| = |\hat{\boldsymbol{\Sigma}}^{-\frac{1}{2}}\mathbf{S}\hat{\boldsymbol{\Sigma}}^{-\frac{1}{2}}|$, since the determinants of $\mathbf{AB}$ and $\mathbf{BA}$ are equal when $\mathbf{A}$ and $\mathbf{B}$ are square matrices. Thus $|\hat{\boldsymbol{\Sigma}}^{-1}\mathbf{S}| = |\mathbf{R}|$. Hence

$$-2 \log_e \lambda = np - n \log_e |\mathbf{R}| - np,$$

i.e.

$$-2 \log_e \lambda = -n \log_e |\mathbf{R}|. \tag{14.22}$$

The exact distribution of λ is difficult to obtain, so recourse must be made to the asymptotic general result of Section 8.3. Since H_0 specifies that all off-diagonal elements of $\mathbf{P}$ are zero, there are $\frac{1}{2}p(p-1)$ constraints in the statement of H_0. Thus if H_0 is true, $-2 \log_e \lambda$ of (14.22) has an asymptotic χ^2 distribution on $\frac{1}{2}p(p-1)$ degrees of freedom. Box (1949) showed that the χ^2 approximation is improved if n is replaced by $n' = n - \frac{1}{6}(2p + 11)$, and hence the best statistic to use for testing H_0 against H_1 is $-n' \log_e |\mathbf{R}|$.

Note that this test statistic provides a formal means of testing the null hypothesis that was investigated graphically by Hills (1969) and by Schweder and Spjøtvoll (1982). These graphical techniques were discussed in Section 14.3 above, so let us return to the correlation matrix of Table 14.6 and subject it to the more formal procedure. Here $n = 48$ and $p = 13$, so that $\frac{1}{2}p(p-1) = 78$ and $n' = 41.83$. We therefore need to calculate the determinant of the correlation matrix to complete the specification of the test statistic. Calculation of a determinant can sometimes be a bit problematic, especially if the matrix is nearly singular. In the present case a number of problems arise: each entry is only given to two decimal places, so accuracy may not be very good; the high correlation of 0.96 between variables 7 and 9 is a warning of near-singularity; and rounding errors in the numerical procedure may exacerbate both of these aspects. The determinant of a matrix equals the product of its eigenvalues, so the latter were calculated for the matrix of Table 14.6. Because of a combination of the factors mentioned above, two of these eigenvalues were small and negative, which makes the quantity in (14.22) undefined. To get round this problem, the determinant was approximated by taking the absolute value of these negative eigenvalues, and then $-n' \log_e|\mathbf{R}| = 988.21$. The 95 per cent point of the χ^2_{78} distribution is 94.4, so the evidence against the null hypothesis of mutual independence is overwhelming. This evidence is of course consistent with the sixteen 'significant' correlations detected by Hills (1969), and the fourteen detected by Schweder and Spjøtvoll (1982).

In the subsequent metric scaling of this matrix, as reported in Section 14.3, Hills suggested removing variables 7, 10, 11, 12, and 13. The remaining eight variables then produce a correlation matrix that is very much more like a diagonal one. Only two 'large' correlations remain, those between variable 1 and each of variables 2 and 3. Applying the likelihood ratio test to this reduced (8×8) matrix, we find $-n' \log_e|\mathbf{R}| = 440.32$ with $\frac{1}{2}p(p-1) = 28$. The observed value of the test statistic is still far in excess of the 95 per cent point of the χ^2_{28} distribution, so we still reject the null hypothesis of mutual independence. However, if variable 1 is additionally removed, then the remaining (7×7) matrix has $-n' \log_e|\mathbf{R}| = 17.65$ with $\frac{1}{2}p(p-1) = 21$. Thus the observed value of the test statistic is less than its expectation of 21 under the null hypothesis, and hence the variable subset 2, 3, 4, 5, 6, 8, and 9 can be taken as having an identity population correlation matrix.

14.6.2 Test that all population canonical correlations are zero

The test that all population canonical correlations are zero is a test of no-association between the two sets of variables $\boldsymbol{x}_1$ and $\boldsymbol{x}_2$, which in turn is equivalent to the test that $\boldsymbol{\Sigma}_{12} = \boldsymbol{0}$ if the sample and population covariance matrices are partitioned in the usual way. An appropriate test has therefore already been discussed (in (5) of Section 8.3); the likelihood ratio test statistic is given in eqn (8.19). However, to make the connection with canonical correlations, this test statistic must be rewritten into the form of Section 14.5. Denoting the canonical correlations by $R_1, \ldots, R_s$ (where $s = \min(p, q)$ for p, q elements in $\boldsymbol{x}_1$, $\boldsymbol{x}_2$ respectively), then the likelihood ratio test statistic λ is given by

$$-2 \log_e \lambda = -n \sum_{i=1}^{s} \log_e(1 - R_i^2). \qquad (14.23)$$

There are pq constraints in the specification of H_0, so if H_0 is true then $-2 \log_e \lambda$ has an asymptotic χ^2 distribution on pq degrees of freedom. Again, the χ^2 approximation may be improved, this time on replacing n by $n' = n - \frac{1}{2}(p + q + 3)$.

As an illustration, consider the head measurements data of Table 14.7, where we are interested in associations between the head measurements of the two brothers. In Section 14.5 we found the two sample canonical correlations to be $R_1^2 = 0.6218$ and $R_2^2 = 0.0029$. Also $p = q = 2$ and $n = 25$. Thus $n' = 21.5$ and $-2 \log_e \lambda = -21.5\{\log_e(0.3782) + \log_e(0.9971)\} = 20.97$. The 95 per cent point of the χ^2 distribution on $pq = 4$ degrees of freedom is 9.49. Hence we clearly reject the null hypothesis that both population canonical correlations are zero, and conclude that some association exists between the two sets of measurements (i.e. the head measurements of one brother are *not* independent of those of the other).

14.6.3 Test that some population canonical correlations are zero

When using canonical correlation analysis for data reduction, it is necessary to be able to determine the smallest number k of 'important' canonical correlations and associated pairs of canonical variables. This reduces the dimensionality of the $\boldsymbol{x}_1 : \boldsymbol{x}_2$ association from s to k. Rather than relying on an arbitrary and heuristic way of choosing k, it would be preferable to base the choice on some inferential considerations. Bartlett (1939) proposed a statistic similar to the one in (14.23), to test the null hypothesis that the first k population canonical correlations are non-zero and the remaining $(s - k)$ are zero. The statistic is $-n' \sum_{i=k+1}^{s} \log_e(1 - R_i^2)$, where $n' = n - \frac{1}{2}(p + q + 3)$ again. If the null hypothesis is true, then this statistic has an asymptotic χ^2 distribution on $(p - k)(q - k)$ degrees of freedom.

For the head measurements data, since the hypothesis in Section 14.6.2 was rejected, it only remains to test whether the first population canonical correlation is non-zero, and the second one is zero. Here $n' = 21.5$ as before, so the test statistic is $-21.5 \log_e(0.9971) = 0.062$. This value is clearly non-significant when tested against the χ^2 distribution on $(2 - 1)(2 - 1) = 1$ degree of freedom. Hence the hypothesis of a single dimension for association between the two sets of head measurements is tenable, which lends inferential justification to the descriptive analysis of the data in Section 14.5.

15
Exploiting observed associations: manifest-variable models

15.1 Motivation

The presence of associations between the variables of a multivariate system implies the existence of *relationships* among the variables, and the logical follow-on from the measurement of the associations is an investigation of these relationships. Often there is an implied *predictive* aim in the investigation, and formulating appropriate and parsimonious relationships among the variables is necessary for prediction. Various examples were cited in Chapter 14 to illustrate the different techniques concerned with measurement of association, and each one could give rise to a possible exercise in prediction. Thus we can readily envisage wanting to predict either the rainfall in the Northern region of Israel from knowledge of the rainfall in the other regions, or a person's response on some physiological variables given his/her responses on other such variables (Section 14.3); to predict a child's intelligence from his/her age and weight (Section 14.4); and to predict the head measurements of the second son in a family given the first son's measurements (Section 14.5).

Such predictive studies are among the most common sources of statistical involvement in scientific work, and examples abound in the literature: prediction of yield variables for crops from weather variables (Agriculture); prediction of physiological or biochemical responses from simple medical signs or symptoms (Medicine); prediction of criminological variables from socio-economic ones (Sociology). Even if prediction is not a primary concern, an appropriate expression of the relationships among the variables will assist in gaining a deeper understanding of, or extra insight into, the mechanism of the system. It is nevertheless convenient to adopt the standpoint of prediction in motivating and developing the theory. Typically, of all variables that have been observed in a system, some will be ones for which a prediction or explanation is sought, while the remainder are to be used in assisting with this prediction or explanation. The former are the *dependent* variables, while the latter are the *explanatory* variables. Let us suppose that there are p dependent and q explanatory variables in a system of $(p+q)$ variables. Establishing the precise relationships between these two sets of manifest variables for an observed data set will lead to a predictive model, and this in turn will be a suitable model from which to examine the mechanism of the system if prediction is not of prime concern.

Before passing to details, we must first decide on the types of systems that will be considered. Since this book is devoted to multivariate analysis, we shall naturally concentrate on multivariate systems. However, there can be some confusion as to what constitutes a multivariate system in the context of predictive models. We shall assume that the dimensionality is governed by the number p of *dependent variables*. This is because, in general, a natural predictor is the mean of some specified distribution but the explanatory variables are considered to be *fixed* when effecting a prediction for the dependent variables. Hence the only distributions that play an important role in the theory are the joint distributions of the dependent variables. These are the distributions that are needed to test hypotheses or construct confidence intervals for predicted values, assess adequacy of fitted models, and so on. Consequently any system in which there is just one dependent variable $(p=1)$ should be treated as a *univariate* system, as the only distribution involved is the distribution of this single dependent variable about a mean value specified by the explanatory variables. Appropriate methods for handling such systems are provided either by simple regression or by multiple regression (for the cases $q=1$ and $q>1$ respectively); we shall assume that the reader is familiar with the basic ideas of these techniques. Suitable textbooks which cover the essential material are those by Draper and Smith (1981), or by Montgomery and Peck (1982). We shall focus exclusively on multivariate systems in which $p>1$.

To emphasize this non-symmetrical role of the two sets of variables, it is convenient to modify our general notation and (for this chapter) to denote the dependent variables by $y_1, \ldots, y_p$, and the explanatory variables by $x_1, \ldots, x_q$. It may be appropriate sometimes to consider the ensemble of each of these sets, in which case we collect them together into the vectors $\boldsymbol{y}' = (y_1, \ldots, y_p)$ and $\boldsymbol{x}' = (x_1, \ldots, x_q)$. If each of these $(p+q)$ variables is measured on each of n sample individuals, then we can denote the vectors of values for the ith individual by $\boldsymbol{y}_i' = (y_{i1}, \ldots, y_{ip})$ and $\boldsymbol{x}_i' = (x_{i1}, \ldots, x_{iq})$. The full data matrix thus has n rows and $(p+q)$ columns and can be written as $\mathbf{Z} = (\mathbf{Y}:\mathbf{X})$, where $\mathbf{Y}$ is the $(n \times p)$ matrix having ith row $\boldsymbol{y}_i'$, while $\mathbf{X}$ is the $(n \times q)$ matrix having ith row $\boldsymbol{x}_i'$ $(i=1, \ldots, n)$. Let $\bar{\boldsymbol{y}}$ and $\bar{\boldsymbol{x}}$ denote the sample mean vectors obtained from $\mathbf{Y}$ and $\mathbf{X}$ respectively. Our objective is to derive models that will enable each of the values $y_{j1}, \ldots, y_{jp}$ to be predicted, given the values $x_{j1}, \ldots, x_{jq}$ of some future individual.

It is first worth noting that some of the techniques described in previous chapters can be employed to achieve related but more limited objectives. First, consider a principal component analysis of the raw data matrix $\mathbf{Z}$. This analysis is effected by computing the eigenvalues and eigenvectors of the covariance matrix $\mathbf{S}$ constructed from $\mathbf{Z}$. Usage of principal component analysis as discussed in Chapter 2 is concerned with data representation and dimensionality reduction, and focuses on those linear combinations of the measured variables whose coefficients are elements of eigenvectors corresponding to the largest k eigenvalues of $\mathbf{S}$. Consider instead the *smallest* eigenvalue l_{p+q} of $\mathbf{S}$ and its corresponding eigenvector $\boldsymbol{a}_{p+q}$. For convenience, drop the subscripts temporarily and denote this eigenvalue/eigenvector pair by l and $\boldsymbol{a} = (a_1, \ldots, a_{p+q})'$. Then, from the definition of principal components, the linear combination

$$w = a_1 y_1 + \ldots + a_p y_p + a_{p+1} x_1 + \ldots + a_{p+q} x_q \tag{15.1}$$

is the one among all linear combinations of $y_1, \ldots, y_p, x_1, \ldots, x_q$ that has *smallest* sample variance, and this sample variance is l. Hence the linear combination w is the one (among all linear combinations) whose n sample values (principal component scores) are *most nearly constant*. Departure from constancy is measured by the variance l of these principal component scores. Thus if l is sufficiently close to zero, we might choose to view the departure from constancy as simply a reflection of errors in the measurements $y_1, \ldots, y_p, x_1, \ldots, x_q$. In this case we can postulate the relationship

$$a_1 y_1 + \ldots + a_p y_p + a_{p+1} x_1 + \ldots + a_{p+q} x_q = \text{constant},$$

or

$$a_1 y_1 + \ldots + a_p y_p = \text{constant} - a_{p+1} x_1 - \ldots - a_{p+q} x_q. \tag{15.2}$$

Equation (15.2) can thus be used as a predictive model for future values of $a_1 y_1 + \ldots + a_p y_p$ given the values $x_1, \ldots, x_q$, the constant in (15.2) being estimated by the average of the w values over the available n individuals. Furthermore, if k eigenvalues $l_{p+q-k+1}, \ldots, l_{p+q}$ are all very close to zero, then k relationships of the form of (15.2) can be deduced from the eigenvectors corresponding to these eigenvalues, and k such predictive models can be constructed. However, all this is of severely limited practical utility, because the linear combination $a_1 y_1 + \ldots + a_p y_p$ of (15.2) is totally data-dependent, and there is no guarantee that the combination obtained in any given circumstance will be of any interest whatsoever for future prediction (or for explanation of the system).

Linear combinations of potentially greater intrinsic interest are those obtained from a canonical correlation analysis of the two sets of variables. In such an analysis (Section 14.5), let $\boldsymbol{a}_i$ and $\boldsymbol{b}_i$ denote the vectors of canonical vector coefficients for the $\boldsymbol{y}$ and $\boldsymbol{x}$ variables, respectively, corresponding to a canonical correlation of R_i ($i = 1, \ldots, s$ where $s = \min(p, q)$). Thus $u_i = \boldsymbol{a}_i'\boldsymbol{y}$, $v_i = \boldsymbol{b}_i'\boldsymbol{x}$ and $\text{corr}(u_i, v_i) = R_i$, where it is assumed as usual that the R_i are ranked $R_1 \geqslant R_2 \geqslant \ldots \geqslant R_s$. Greatest interest attaches to the canonical vectors u_1, v_1 corresponding to R_1, because these are the linear combinations of the $\boldsymbol{y}$ and $\boldsymbol{x}$ variables that are most highly correlated with each other in the sample. Thus among all (u_i, v_i) pairs, prediction of u_1 from v_1 should be best. Moreover, the canonical correlation analysis may have yielded interpretable vectors u_1, v_1 which carry some substantive importance, thereby making prediction of u_1 from v_1 worthwhile. Since the canonical variate coefficients $\boldsymbol{a}_1$ and $\boldsymbol{b}_1$ convert an individual's multivariate observations $\boldsymbol{y}_i$ and $\boldsymbol{x}_i$ to the two univariate quantities $u_{i1} = \boldsymbol{a}_1'\boldsymbol{y}_i$ and $v_{i1} = \boldsymbol{b}_1'\boldsymbol{x}_i$ (that individual's *scores* on the first canonical variates), prediction can be effected by means of simple regression analysis. If we write $\boldsymbol{u}_1 = (u_{11}, u_{21}, \ldots, u_{n1})'$ and $\boldsymbol{v}_1 = (v_{11}, v_{21}, \ldots, v_{n1})'$, then $\boldsymbol{u}_1 = \mathbf{Y}\boldsymbol{a}_1$ and $\boldsymbol{v}_1 = \mathbf{X}\boldsymbol{b}_1$ are the vectors of scores on the first pair of canonical variates for the n sample individuals. The constraints imposed in the canonical correlation analysis (Section 14.5) ensure that the values in each of these two vectors have unit variance. Simple linear regression of the u_{i1} values on the v_{i1} values thus yields the predictive equation

$$\hat{u}_{j1} = R_1(v_{j1} - \boldsymbol{b}_1'\bar{\boldsymbol{x}}) + \boldsymbol{a}_1'\bar{\boldsymbol{y}} \qquad (15.3)$$

for a new individual's u_1 value given his/her value $v_{j1} = \boldsymbol{b}_1'\boldsymbol{x}_j$. Furthermore, R_1^2 represents the proportion of variance of the u_{i1} explained by the regression on the v_{i1}, and hence gives a measure of usefulness of the predictive equation.

An equation of the form (15.3) can be obtained between the canonical variates corresponding to *any* of the canonical correlations R_j ($j = 1, \ldots, s$). However, as the canonical correlations decrease in size so the relationships between the corresponding canonical variates become weaker and the consequent predictions become less accurate. Also, canonical variates corresponding to small canonical correlations are less likely to carry substantive interpretation than those corresponding to large canonical correlations. For these reasons, a predictive equation of the form of (15.3) is only likely to be of practical interest if it links canonical variates that correspond to large R_i.

Although the above two techniques may be of some interest in the study of relationships between the two sets of variables, and may even provide some useful predictive models, it can be seen that their scope is very limited. This is because they only predict *linear combinations* of the y_i, and furthermore the linear combinations that they predict are determined by the data and are not under the control of the investigator.

In practice, we would want to predict either a priori specifiable linear combinations or, even more specifically, the *individual* y_i. Indeed, if we have a prediction of each individual y_i then we can derive a prediction of any linear combination formed from them. Hence the latter objective encompasses the former, as well as all previous models, and this objective is what we will now concentrate on. How do we formulate an appropriate predictive model? One possibility when faced with random variation is to model the distribution of the random variables involved; the best prediction of a future observation is the mean of this distribution. We shall first assume that all variables are quantitative, and that the joint distribution of all y_i and x_i can be modelled adequately by a multivariate normal distribution (perhaps after some preliminary transformation).

Write $\boldsymbol{z}' = (\boldsymbol{y}', \boldsymbol{x}')$, so that $\boldsymbol{z}_i'$ is the ith row of the full data matrix $\mathbf{Z}$, and it represents the complete set of $(p+q)$ variable values for the ith individual ($i = 1, \ldots, n$). Then the $\boldsymbol{z}_i$ can be assumed to be independent observations from a multivariate normal distribution with mean $\boldsymbol{\mu}$ and dispersion matrix $\boldsymbol{\Sigma}$. Since $\boldsymbol{z}$ is partitioned into two portions $\boldsymbol{y}$ and $\boldsymbol{x}$, $\boldsymbol{\mu}$ and $\boldsymbol{\Sigma}$ can be partitioned conformally as $\begin{pmatrix}\boldsymbol{\mu}_y \\ \boldsymbol{\mu}_x\end{pmatrix}$ and $\begin{pmatrix}\boldsymbol{\Sigma}_{yy} & \boldsymbol{\Sigma}_{yx} \\ \boldsymbol{\Sigma}_{xy} & \boldsymbol{\Sigma}_{xx}\end{pmatrix}$ respectively. From Property 2 of Section 7.2, it follows that the marginal distribution of $\boldsymbol{y}$ is multivariate normal with mean vector $\boldsymbol{\mu}_y$ and dispersion matrix $\boldsymbol{\Sigma}_{yy}$. Thus, *in the absence of knowledge about* $\boldsymbol{x}$, we would predict a new individual's $\boldsymbol{y}$ value by its population mean $\boldsymbol{\mu}_y$ if this mean is known, and by the estimate $\bar{\boldsymbol{y}}$ from the sample data otherwise. However, the covariances in $\boldsymbol{\Sigma}_{xy}$ indicate that elements of $\boldsymbol{x}$ are correlated with those of $\boldsymbol{y}$, so if an individual's $\boldsymbol{x}$ value is known to be $\boldsymbol{x}_0$ then we ought to be able to use this information to improve the prediction of its $\boldsymbol{y}$ value. Property 3 of Section 7.2 establishes that the distribution of $\boldsymbol{y}$ for known $\boldsymbol{x} = \boldsymbol{x}_0$ is multivariate normal with mean $\boldsymbol{\mu}_{y\,|\,x} = \boldsymbol{\mu}_y + \boldsymbol{\Sigma}_{yx}\boldsymbol{\Sigma}_{xx}^{-1}(\boldsymbol{x}_0 - \boldsymbol{\mu}_x)$ and dispersion matrix $\boldsymbol{\Sigma}_{y\,|\,x} = \boldsymbol{\Sigma}_{yy} - \boldsymbol{\Sigma}_{yx}\boldsymbol{\Sigma}_{xx}^{-1}\boldsymbol{\Sigma}_{xy}$. Thus the effect of knowing the individual's $\boldsymbol{x}$ value is to modify the $\boldsymbol{y}$ prediction from $\boldsymbol{\mu}_y$ to $\boldsymbol{\mu}_{y\,|\,x}$.

In general, of course, all quantities except for $\boldsymbol{x}_0$ in $\boldsymbol{\mu}_{y\,|\,x}$ will be unknown and will have to be estimated. One possible approach is to replace all unknown quantities by their maximum likelihood estimates assuming joint multivariate normality of $\boldsymbol{y}$ and $\boldsymbol{x}$ (Section 8.2). However, it is often appropriate to assume that the conditional mean of $\boldsymbol{y}$ given $\boldsymbol{x} = \boldsymbol{x}_0$ has a form similar to $\boldsymbol{\mu}_{y\,|\,x}$ above, even when the $\boldsymbol{x}$ values do not themselves have a multivariate normal distribution. This would be the case, for example, if the $\boldsymbol{x}_i$ come from a member of the family of elliptic distributions (Section 7.4). Often, moreover, the elements of $\boldsymbol{x}$ need not have a distribution at all, but are simply predetermined constants. The conditional mean of $\boldsymbol{y}$ given $\boldsymbol{x} = \boldsymbol{x}_0$ can be empirically assumed to have the linear form of $\boldsymbol{\mu}_{y\,|\,x}$, but a quantity such as $\boldsymbol{\Sigma}_{xx}$ is meaningless in this context. Since $\boldsymbol{\Sigma}_{yx}\boldsymbol{\Sigma}_{xx}^{-1}$ is a $(p \times q)$ matrix, then in this case we can simply assume the conditional mean of $\boldsymbol{y}$ given $\boldsymbol{x} = \boldsymbol{x}_0$ to be $\boldsymbol{c} + \mathbf{B}_0\boldsymbol{x}_0$ where the p-component vector $\boldsymbol{c}$ and $(p \times q)$ matrix $\mathbf{B}_0$ both contain unknown constants to be estimated from the sample data. Moreover, by adjoining $\boldsymbol{c}$ to $\mathbf{B}_0$ and creating the $p \times (q+1)$ matrix $\mathbf{B}$, and writing the x_i values in the $(q+1)$ vector $\boldsymbol{x}' = (1, x_1, x_2, \ldots, x_q)$, this conditional mean becomes $\mathbf{B}\boldsymbol{x}_0$. Hence the most general predictive model is obtained by assuming only that the $\boldsymbol{y}$ values have a multivariate normal distribution, whose mean, $\mathbf{B}\boldsymbol{x}$, is determined by the associated $\boldsymbol{x}$ values, and whose dispersion matrix is arbitrary. We will henceforth assume that $\boldsymbol{x}$ has been augmented to a $(q+1)$ component vector by the addition of 1 in the first position, and the data matrix $\mathbf{X}$ has correspondingly been augmented to size $n \times (q+1)$ by the addition of a column of 1s, on the left.

The remainder of this chapter is concerned with a study of this predictive model.

15.2 Multivariate regression

15.2.1 Introduction

The data in Table 15.1 are taken from Chatfield and Collins (1980) and show the four measurements: chest circumference (CC), mid-upper arm circumference (MUAC), height, and age (in months) for a sample of nine young girls. (A small data set is provided deliberately, to enable

Table 15.1 Four measurements taken on each of nine young girls

Individual	Chest circumference (cm) y_1	Mid-upper arm circumference (cm) y_2	Height (cm) x_1	Age (months) x_2
1	58.4	14.0	80	21
2	59.2	15.0	75	27
3	60.3	15.0	78	27
4	57.4	13.0	75	22
5	59.5	14.0	79	26
6	58.1	14.5	78	26
7	58.0	12.5	75	23
8	55.5	11.0	64	22
9	59.2	12.5	80	22

step-by-step calculations to be followed easily.) Summary statistics are:

$$\text{Mean vector} = (58.4 \quad 13.5 \quad 76.0 \quad 24.0).$$

$$\text{Covariance matrix} = \begin{pmatrix} 1.9700 & & & \\ 1.4563 & 1.8125 & & \\ 5.6375 & 4.3125 & 24.5000 & \\ 2.0375 & 2.4375 & 2.5000 & 6.0000 \end{pmatrix}.$$

$$\text{Correlation matrix} = \begin{pmatrix} 1.0000 & & & \\ 0.7707 & 1.0000 & & \\ 0.8115 & 0.6472 & 1.0000 & \\ 0.5926 & 0.7391 & 0.2062 & 1.0000 \end{pmatrix}.$$

One obvious practical objective would be to develop a predictive model for CC and MUAC from knowledge of height and age. The correlation matrix shows this to be a sensible objective, as each of CC and MUAC have suitably large correlations with each of height and age (suggesting that height and age should be reasonably effective explanatory variables), while height has relatively small correlation with age (suggesting that *both* of these variables are worth considering as explanatory variables). Moreover, CC and MUAC are highly correlated with each other, so should be incorporated into a single multivariate predictive model for maximum efficiency (as conducting, e.g. multiple regression analyses for each variate separately will ignore this correlation in any associated hypothesis tests or confidence intervals). Accordingly we designate CC and MUAC by y_1 and y_2 respectively, while height and age are denoted by x_1 and x_2 respectively. We consider prediction of $\boldsymbol{y}' = (y_1, y_2)$ from $\boldsymbol{x}' = (1, x_1, x_2)$, and we write $\boldsymbol{y}_i, \boldsymbol{x}_i$ for the ith individual's values of $\boldsymbol{y}, \boldsymbol{x}$ respectively ($i = 1, \ldots, n$). The data matrix $\mathbf{Y}$ is given by the first two columns of Table 15.1, while the data matrix $\mathbf{X}$ is a column of 1s followed by the last two columns of Table 15.1.

Now return from this example to the general case. Following the argument of the previous section, it is assumed that if $\boldsymbol{x}$ takes value $\boldsymbol{x}_0$ then $\boldsymbol{y}$ has a multivariate normal distribution with mean $\mathbf{B}\boldsymbol{x}_0$ and dispersion matrix $\boldsymbol{\Sigma}$. Thus the rows of $\mathbf{Y}$ are assumed to be independent observations from such normal distributions, and means of these distributions are determined by the corresponding rows of $\mathbf{X}$. Hence the model for the observed data can be written in the form

$$\mathbf{Y} = \mathbf{XB} + \mathbf{E} \tag{15.4}$$

where $\mathbf{Y}$ is the $n \times p$ matrix whose ith row is $\boldsymbol{y}_i'$, $\mathbf{X}$ is the $n \times (q+1)$ matrix whose ith row is $\boldsymbol{x}_i'$, $\mathbf{B}$ is the $(q+1) \times p$ matrix of unknown parameters, and $\mathbf{E}$ is an $n \times p$ matrix of random variables whose rows are independent observations from a multivariate normal distribution with mean zero and dispersion matrix $\boldsymbol{\Sigma}$. Actually, assumption of multivariate normality is only necessary if hypotheses are to be tested or confidence regions are to be constructed (either for the parameters or for the predicted values). For other aspects it is sufficient to assume that rows of $\mathbf{E}$ are mutually uncorrelated and have mean zero and dispersion matrix $\boldsymbol{\Sigma}$. However, we will in general keep the additional normality assumption, as most applications will involve a mixture of point estimation, confidence region construction, and hypothesis testing.

First, let us relate model (15.4) to the familiar multiple regression model for a single dependent variable. Suppose, for example, that only y_1 (i.e. CC) is to be predicted from height and age in the example above. Then the multiple regression model for this would be $y_{i1} = \beta_{01} + \beta_{11}x_{i1} + \beta_{21}x_{i2} + e_{i1}$ $(i = 1, \ldots, n)$. Here, of course, y_{i1} is the ith observation on y_1, x_{i1} is the ith observation on x_1, and so on; e_{i1} is the deviation (departure, or error) of the ith observation y_{i1} from its mean $\beta_{01} + \beta_{11}x_{i1} + \beta_{21}x_{i2}$. In general linear model form this set of n equations can be written compactly as

$$\boldsymbol{y}^{(1)} = \mathbf{X}\boldsymbol{\beta}_1 + \boldsymbol{e}_1$$

where $\boldsymbol{y}^{(1)}$ is the $n \times 1$ vector of values of y_1, $\mathbf{X}$ is the $n \times (q+1)$ matrix of explanatory variable values, $\boldsymbol{\beta}_1$ is the $(q+1) \times 1$ vector of parameter values, and $\boldsymbol{e}_1$ is the $n \times 1$ vector of random variables whose elements are independent observations from a univariate normal distribution with mean zero and variance σ_1^2. (The vector $\boldsymbol{y}^{(1)}$ containing the n values of y_1 should be carefully distinguished from the earlier vector $\boldsymbol{y}_1$ containing the values of $y_1, y_2, \ldots, y_p$ for the first individual.)

For the girls' measurement data, $n = 9$, $\boldsymbol{y}^{(1)}$ is the first column of Table 15.1, $\boldsymbol{\beta}_1 = (\beta_{01}, \beta_{11}, \beta_{21})'$ and $\mathbf{X}$ is as before. In general, if any single variable y_j is to be predicted from all the explanatory variables $x_1, x_2, \ldots, x_q$ in the system, then the multiple regression model can be written

$$\boldsymbol{y}^{(j)} = \mathbf{X}\boldsymbol{\beta}_j + \boldsymbol{e}_j \tag{15.5}$$

where $\boldsymbol{y}^{(j)}$ is the jth column of $\mathbf{Y}$, $\boldsymbol{\beta}_j = (\beta_{0j}, \beta_{1j}, \ldots, \beta_{qj})'$ and $\boldsymbol{e}_j$ has analogous definition to the one above. Note, therefore, that the parameter vectors $\boldsymbol{\beta}_j$ and error vectors $\boldsymbol{e}_j$ are specific to the chosen dependent variable $\boldsymbol{y}^{(j)}$, and hence are different for each j, but *the same* $\mathbf{X}$ appears in all models. This is of course because the *same* explanatory variables are being used in each separate predictive model. Now $\mathbf{Y} = (\boldsymbol{y}^{(1)}, \boldsymbol{y}^{(2)}, \ldots, \boldsymbol{y}^{(p)})$, i.e. $\mathbf{Y}$ is formed by placing the columns $\boldsymbol{y}^{(j)}$ side by side in a matrix. We can similarly put the $\boldsymbol{\beta}_j$ side by side to form $\mathbf{B} = (\boldsymbol{\beta}_1, \boldsymbol{\beta}_2, \ldots, \boldsymbol{\beta}_p)$, and the $\boldsymbol{e}_j$ side by side to form $\mathbf{E} = (\boldsymbol{e}_1, \boldsymbol{e}_2, \ldots, \boldsymbol{e}_p)$. Then $\mathbf{X}\boldsymbol{\beta}_j$ is just the jth column of $\mathbf{XB}$, and hence the multiple regression model (15.5) is obtained by extracting the jth columns of each matrix in the multivariate regression model (15.4).

One final simplification can be made, in analogous fashion to that in multiple regression analysis, and this involves a small modification to the parameters of the model. The multiple regression model (15.5) is the matrix statement of

$$y_{ij} = \beta_{0j} + \beta_{1j}x_{i1} + \beta_{2j}x_{i2} + \ldots + \beta_{qj}x_{iq} + e_{ij} \qquad (i = 1, \ldots, n)$$

for regression of the jth dependent variable on the explanatory variables. It is often more convenient computationally to reparameterize this equation, by measuring each of the explanatory variables about its mean (i.e. by mean-centering all columns of the **X** matrix except the first). Thus the ith individual's value on the kth explanatory variable becomes $x_{ik} - \bar{x}_k$ where $\bar{x}_k$ is the sample mean of this variable. The multiple regression model thus becomes

$$y_{ij} = \beta^*_{0j} + \beta_{1j}(x_{i1} - \bar{x}_1) + \beta_{2j}(x_{i2} - \bar{x}_2) + \ldots + \beta_{qj}(x_{iq} - \bar{x}_q) + e_{ij} \qquad (i = 1, \ldots, n)$$

where the coefficients $\beta_{1j}, \beta_{2j}, \ldots, \beta_{qj}$ are as before but $\beta^*_{0j} = \beta_{0j} + \beta_{1j}\bar{x}_1 + \beta_{2j}\bar{x}_2 + \ldots + \beta_{qj}\bar{x}_q$. Model (15.5) remains unchanged, provided that **X** in it is the *mean-centered* version, and that the first element of $\boldsymbol{\beta}_j$ is understood to be β^*_{0j}. One implication of this re-parameterization is that the first column of **X** is orthogonal to the other columns, which not only has computational benefits but which also ensures that the estimate of β^*_{0j} is uncorrelated with the estimates of the other β_{kj} $(k = 1, \ldots, q)$. Since the same **X** is used in all the separate regressions (15.5) for $j = 1, \ldots, p$, it follows that mean-centering **X** also has corresponding benefits in the multivariate regression model (15.4). Henceforth we will assume that all the explanatory variables have been measured about their means, and all columns except the first of **X** have therefore been mean-centered. Also, the asterisk in β^*_{0j} can be dropped without confusion, and the corresponding parameters will be written simply as β_{0j}.

Mean-centering the explanatory variables of Table 15.1 requires subtraction of 76.0 from each x_1 value, and subtraction of 24.0 from each x_2 value. The resultant mean-centered matrix is given in Table 15.2 (although note of course that the first column of 1s remains unchanged).

15.2.2 Fitting the model

The first stage in a multivariate regression analysis is the fitting of model (15.4) to the observed data. This requires the estimation of the unknown parameters **B** and **Σ**, which can be done by maximum likelihood if normality is assumed for the error matrix **E** or by least squares if no

Table 15.2 Mean-centered matrix **X** obtained from data of Table 15.1

$$\mathbf{X} = \begin{pmatrix} 1 & 4.0 & -3.0 \\ 1 & -1.0 & 3.0 \\ 1 & 2.0 & 3.0 \\ 1 & -1.0 & -2.0 \\ 1 & 3.0 & 2.0 \\ 1 & 2.0 & 2.0 \\ 1 & -1.0 & -1.0 \\ 1 & -12.0 & -2.0 \\ 1 & 4.0 & -2.0 \end{pmatrix}$$

distributional assumptions are made. Fortunately, both approaches lead to the same estimates so the actual assumptions that are made are not critical to the outcome of the analysis. For standard application of the theory, it is necessary for n to be greater than $(p+q)$ and for the matrix $\mathbf{X}$ to be of full rank $(q+1)$ (so that the inverse $(\mathbf{X}'\mathbf{X})^{-1}$ exists). If these conditions are met, then the estimators $\hat{\mathbf{B}}$ and $\hat{\boldsymbol{\Sigma}}$ of $\mathbf{B}$ and $\boldsymbol{\Sigma}$ are given by

$$\hat{\mathbf{B}} = (\mathbf{X}'\mathbf{X})^{-1}\mathbf{X}'\mathbf{Y} \tag{15.6}$$

and

$$\hat{\boldsymbol{\Sigma}} = \frac{1}{n}(\mathbf{Y} - \mathbf{X}\hat{\mathbf{B}})'(\mathbf{Y} - \mathbf{X}\hat{\mathbf{B}}). \tag{15.7}$$

Mathematical derivation of these results is of secondary importance, so will not be presented here. For details of the maximum likelihood derivation the reader is referred to Mardia *et al.* (1979), while the least-squares approach is covered by Rao (1973).

Expressions (15.6) and (15.7) bear a very close resemblance to the corresponding estimates in the multiple regression model (15.5). For this latter model,

$$\hat{\boldsymbol{\beta}}_j = (\mathbf{X}'\mathbf{X})^{-1}\mathbf{X}'\boldsymbol{y}^{(j)} \tag{15.8}$$

and

$$\hat{\sigma}_j^2 = \frac{1}{n}(\boldsymbol{y}^{(j)} - \mathbf{X}\hat{\boldsymbol{\beta}}_j)'(\boldsymbol{y}^{(j)} - \mathbf{X}\hat{\boldsymbol{\beta}}_j). \tag{15.9}$$

Thus the estimates in multivariate regression are obtained from the multiple regression ones simply by replacing the vectors $\boldsymbol{y}^{(j)}$ and $\hat{\boldsymbol{\beta}}_j$ by the matrices $\mathbf{Y}$ and $\hat{\mathbf{B}}$ respectively. But if we place the p column vectors $\hat{\boldsymbol{\beta}}_j$ of (15.8) side by side to form a matrix, we obtain

$$\begin{aligned}(\hat{\boldsymbol{\beta}}_1, \hat{\boldsymbol{\beta}}_2, \ldots, \hat{\boldsymbol{\beta}}_p) &= (\mathbf{X}'\mathbf{X})^{-1}\mathbf{X}'\{\boldsymbol{y}^{(1)}, \boldsymbol{y}^{(2)}, \ldots, \boldsymbol{y}^{(p)}\} \\ &= (\mathbf{X}'\mathbf{X})^{-1}\mathbf{X}'\mathbf{Y},\end{aligned}$$

which is the right-hand side of (15.6). Hence it follows that the multivariate parameter estimate $\hat{\mathbf{B}}$ of (15.6) is simply obtained by amalgamating the p separate vector estimates $\hat{\boldsymbol{\beta}}_j$ of (15.8) into a single matrix. Thus the multivariate estimation is achieved by p independent univariate estimations, so the inter-correlations among the y_i do not influence the estimation of the β_{ij}. On the other hand, the separate multiple regression estimates $\hat{\sigma}_j^2$ of error variance given in (15.9) can be confirmed to be just the diagonal elements of $\hat{\boldsymbol{\Sigma}}$ in (15.7). The off-diagonal elements of $\hat{\boldsymbol{\Sigma}}$ can therefore *only* be obtained by conducting the full multivariate regression analysis, and these off-diagonal elements play an important role in testing hypotheses about elements of $\mathbf{B}$. Hence if correlations among the y_i are ignored, and p separate multiple regressions are done on the columns of $\mathbf{Y}$, then the correct parameter estimates will be obtained but incorrect inferences may be drawn about these parameters.

Inferential aspects are considered below. First we illustrate the above expressions by applying them to the girls' measurement data. Since all explanatory variates are assumed to be mean-centered, we use the $\mathbf{X}$ matrix given in Table 15.2. Thus

$$\mathbf{X}'\mathbf{X} = \begin{pmatrix} 1 & 1 & 1 & 1 & 1 & 1 & 1 & 1 & 1 \\ 4.0 & -1.0 & 2.0 & -1.0 & 3.0 & 2.0 & -1.0 & -12.0 & 4.0 \\ -3.0 & 3.0 & 3.0 & -2.0 & 2.0 & 2.0 & -1.0 & -2.0 & -2.0 \end{pmatrix} \times \begin{pmatrix} 1 & 4.0 & -3.0 \\ 1 & -1.0 & 3.0 \\ 1 & 2.0 & 3.0 \\ 1 & -1.0 & -2.0 \\ 1 & 3.0 & 2.0 \\ 1 & 2.0 & 2.0 \\ 1 & -1.0 & -1.0 \\ 1 & -12.0 & -2.0 \\ 1 & 4.0 & -2.0 \end{pmatrix}$$

$$= \begin{pmatrix} 9 & 0 & 0 \\ 0 & 196 & 20 \\ 0 & 20 & 48 \end{pmatrix}.$$

Mean-centering ensures orthogonality of column 1 with each of columns 2 and 3, as evidenced by the zeros in all positions (except the first) of the first row and column. Also, mean-centering yields the corrected sums of squares and products among the x_i in the remainder of the matrix (hence dividing the lower right 2×2 portion by $(n-1) = 8$ produces the sample variances of x_1, x_2, and the sample covariance between x_1 and x_2).

However, these are subsidiary points of interest; the main interest is in the inverse of $\mathbf{X}'\mathbf{X}$. Orthogonality between the first column and the others means that to invert $\mathbf{X}'\mathbf{X}$ we need merely to invert separately the single element 9 and the 2×2 matrix $\begin{pmatrix} 196 & 20 \\ 20 & 48 \end{pmatrix}$. The determinant of the latter is $196 \times 48 - 20^2 = 9008$, so its inverse is

$$\frac{1}{9008}\begin{pmatrix} 48 & -20 \\ -20 & 196 \end{pmatrix} = \begin{pmatrix} 0.00533 & -0.00222 \\ -0.00222 & 0.02176 \end{pmatrix}.$$

Hence

$$(\mathbf{X}'\mathbf{X})^{-1} = \begin{pmatrix} 0.11111 & 0 & 0 \\ 0 & 0.00533 & -0.00222 \\ 0 & -0.00222 & 0.02176 \end{pmatrix}.$$

Next,

$$\mathbf{X}'\mathbf{Y} = \begin{pmatrix} 1 & 1 & 1 & 1 & 1 & 1 & 1 & 1 & 1 \\ 4.0 & -1.0 & 2.0 & -1.0 & 3.0 & 2.0 & -1.0 & -12.0 & 4.0 \\ -3.0 & 3.0 & 3.0 & -2.0 & 2.0 & 2.0 & -1.0 & -2.0 & -2.0 \end{pmatrix} \times \begin{pmatrix} 58.4 & 14.0 \\ 59.2 & 15.0 \\ 60.3 & 15.0 \\ 57.4 & 13.0 \\ 59.5 & 14.0 \\ 58.1 & 14.5 \\ 58.0 & 12.5 \\ 55.5 & 11.0 \\ 59.2 & 12.5 \end{pmatrix} = \begin{pmatrix} 525.6 & 121.5 \\ 45.1 & 34.5 \\ 16.3 & 19.5 \end{pmatrix}.$$

(Note that the two columns here are the separate values $\mathbf{X}'\boldsymbol{y}^{(1)}$ and $\mathbf{X}'\boldsymbol{y}^{(2)}$.)

Finally, therefore,

$$\begin{aligned} \hat{\mathbf{B}} &= (\mathbf{X}'\mathbf{X})^{-1}(\mathbf{X}'\mathbf{Y}) \\ &= \begin{pmatrix} 0.11111 & 0 & 0 \\ 0 & 0.00533 & -0.00222 \\ 0 & -0.00222 & 0.02176 \end{pmatrix}\begin{pmatrix} 525.6 & 121.5 \\ 45.1 & 34.5 \\ 16.3 & 19.5 \end{pmatrix} \\ &= \begin{pmatrix} 58.4 & 13.5 \\ 0.2042 & 0.1406 \\ 0.2546 & 0.3477 \end{pmatrix}. \end{aligned}$$

(The first element of each column will always be the mean of the corresponding dependent variable because the row of ones in $\mathbf{X}'$ produces the totals of the dependent variable values in the first row of $\mathbf{X}'\mathbf{Y}$, the $\frac{1}{n}$ in the top left-hand corner of $(\mathbf{X}'\mathbf{X})^{-1}$ reduces each total to the corresponding mean, and the zeros in the rest of the first row of $(\mathbf{X}'\mathbf{X})^{-1}$ ensure that there is no other contribution to the first row of $\hat{\mathbf{B}}$.)

Thus the appropriate regression equations are

$$\begin{aligned} y_1 &= 58.4 + 0.2042(x_1 - 76.0) + 0.2546(x_2 - 24.0) + e_1 \\ y_2 &= 13.5 + 0.1406(x_1 - 76.0) + 0.3477(x_2 - 24.0) + e_2 \end{aligned} \qquad (15.10)$$

or

$$\begin{aligned} y_1 &= 36.77 + 0.2042x_1 + 0.2546x_2 + e_1 \\ y_2 &= -5.53 + 0.1406x_1 + 0.3477x_2 + e_2. \end{aligned} \qquad (15.11)$$

The *fitted values* are given by applying these equations to the individual x_i values, using either (15.10) on the mean-centered values of Table 15.2, or (15.11) on the raw values of Table 15.1. If we stay with the mean-centered version throughout, the fitted values can be written as $\hat{\mathbf{Y}} = \mathbf{X}\hat{\mathbf{B}}$ and we therefore obtain

$$\hat{\mathbf{Y}} = \begin{pmatrix} 1 & 4.0 & -3.0 \\ 1 & -1.0 & 3.0 \\ 1 & 2.0 & 3.0 \\ 1 & -1.0 & -2.0 \\ 1 & 3.0 & 2.0 \\ 1 & 2.0 & 2.0 \\ 1 & -1.0 & -1.0 \\ 1 & -12.0 & -2.0 \\ 1 & 4.0 & -2.0 \end{pmatrix} \begin{pmatrix} 58.4 & 13.5 \\ 0.2042 & 0.1406 \\ 0.2546 & 0.3477 \end{pmatrix} = \begin{pmatrix} 58.4530 & 13.0193 \\ 58.9596 & 14.4025 \\ 59.5722 & 14.8243 \\ 57.6866 & 12.6640 \\ 59.5218 & 14.6172 \\ 59.3176 & 14.4766 \\ 57.9412 & 13.0117 \\ 55.4404 & 11.1174 \\ 58.7076 & 13.3670 \end{pmatrix}.$$

The *residuals* are the differences between the observed and fitted values. The matrix of residuals is an estimate of the error matrix $\mathbf{E}$, and can be written $\hat{\mathbf{E}} = \mathbf{Y} - \hat{\mathbf{Y}}(= \mathbf{Y} - \mathbf{X}\hat{\mathbf{B}})$. For the above data we thus have

$$\hat{\mathbf{E}} = (\mathbf{Y} - \hat{\mathbf{Y}}) = \begin{pmatrix} -0.0530 & 0.9807 \\ 0.2404 & 0.5975 \\ 0.7278 & 0.1757 \\ -0.2866 & 0.3360 \\ -0.0218 & -0.6172 \\ -1.2176 & 0.0234 \\ 0.0588 & -0.5117 \\ 0.0596 & -0.1174 \\ 0.4924 & -0.8670 \end{pmatrix}.$$

Each column of the residual matrix should sum to zero (as the residuals denote the deviations from the fitted regressions) and this is indeed the case with the matrix above. From (15.7), therefore, the estimate of the dispersion matrix $\boldsymbol{\Sigma}$ is given by $\hat{\boldsymbol{\Sigma}} = \frac{1}{n}\hat{\mathbf{E}}'\hat{\mathbf{E}}$, i.e. the matrix of raw sums of squares and products of the columns of $\hat{\mathbf{E}}$, divided by n. In the present case we obtain

$$\hat{\boldsymbol{\Sigma}} = \begin{pmatrix} 0.2672 & -0.0395 \\ -0.0395 & 0.3190 \end{pmatrix} \tag{15.12}$$

which shows that the two columns of residuals have a small negative correlation ($r = -0.135$).

Thus we conclude that the best predictive models for y_1 and y_2 given values of x_1 and x_2 are those of (15.10) or, equivalently, (15.11), while (15.12) gives an estimate of the joint dispersion of points about these fitted models. Recollect that at the beginning of Section 15.2.1 the covariance matrix of all four variables was given as

$$\begin{pmatrix} 1.9700 & & & \\ 1.4563 & 1.8125 & & \\ 5.6375 & 4.3125 & 24.5000 & \\ 2.0375 & 2.4375 & 2.5000 & 6.0000 \end{pmatrix}.$$

This was the unbiased estimate with divisor $n - 1 = 8$. Adjusting the values so that they become the maximum likelihood ones with divisor n, and extracting the upper (2×2) submatrix for y_1 and y_2, we obtain $\begin{pmatrix} 1.7511 & 1.2945 \\ 1.2945 & 1.6111 \end{pmatrix}$. This is an estimate of the dispersion matrix for the unconditional (i.e. marginal) distribution of (y_1, y_2), ignoring any relationship that the y_i have with the x_i. On the other hand, (15.12) gives the estimate of the dispersion matrix for the conditional distribution of (y_1, y_2) given the value of (x_1, x_2). That the entries are considerably smaller than in the unconditional distribution testifies to the fact that a large amount of the variation in (y_1, y_2) is attributable to the relationship between (y_1, y_2) and (x_1, x_2), which in turn suggests that the predictive model is justified. To examine this suggestion more formally, we now turn to inferential aspects of the technique.

15.2.3 Inference about model parameters

The parameter estimates $\hat{\mathbf{B}}$ and $\hat{\boldsymbol{\Sigma}}$ of (15.6) and (15.7) are sample statistics and are subject to sampling fluctuation like any other statistic. If we wish to make inferences about the true model parameters $\mathbf{B}$ and $\boldsymbol{\Sigma}$, we must first derive the sampling distributions of $\hat{\mathbf{B}}$ and $\hat{\boldsymbol{\Sigma}}$. To do this, considerable mathematical and algebraic manipulation is necessary. We shall therefore content ourselves with a brief summary of the main results (as they affect practical applications); full mathematical details may be found in more theoretical text-books such as T. W. Anderson (1984) or Mardia *et al.* (1979).

First, if all the assumptions of model (15.4) are obeyed, the following results can be established concerning the sampling distributions of $\hat{\mathbf{B}}$ and $\hat{\boldsymbol{\Sigma}}$:

1. The joint distribution of all elements $\hat{\beta}_{ij}$ of $\hat{\mathbf{B}}$ is multivariate normal; the expected value of $\hat{\mathbf{B}}$ is $\mathbf{B}$, and the covariance between $\hat{\beta}_{ij}$ and $\hat{\beta}_{kl}$ is $\sigma_{ik}u_{jl}$, where $\boldsymbol{\Sigma} = (\sigma_{ij})$, and $(\mathbf{X}'\mathbf{X})^{-1} = \mathbf{U} = (u_{ij})$.
2. $n\hat{\boldsymbol{\Sigma}}$ has a Wishart distribution with $(n - q - 1)$ degrees of freedom and parameter $\boldsymbol{\Sigma}$.
3. $\hat{\mathbf{B}}$ and $\hat{\boldsymbol{\Sigma}}$ are independent.

From (1) we see that $\hat{\mathbf{B}}$ is an unbiased estimator of $\mathbf{B}$. However, from (2) we can deduce that $\hat{\boldsymbol{\Sigma}}$ is a biased estimator of $\boldsymbol{\Sigma}$; the unbiased version is $\left(\frac{n}{n-q-1}\right)\hat{\boldsymbol{\Sigma}}$. (This follows the usual trend with maximum likelihood estimation: unbiased estimators of location parameters but biased estimators of dispersion parameters.) For example, in the data of the previous section, $q + 1 = 3$ so that $n/(n - q - 1) = 9 \div 6$, and an unbiased estimate of $\boldsymbol{\Sigma}$ is $1.5\hat{\boldsymbol{\Sigma}} = \begin{pmatrix} 0.4008 & -0.0593 \\ -0.0593 & 0.4785 \end{pmatrix}$ from (15.12).

A further distributional result that can be derived readily using properties of the multivariate normal distribution is that the joint distribution of elements $\hat{e}_{ij}$ of $\hat{\mathbf{E}}$ is multivariate normal with mean zero. This result is of practical importance when examining the adequacy of the fitted model (see Section 15.2.4 below).

Next, let us consider testing hypotheses about the regression parameters $\mathbf{B}$. Again, when faced with multivariate hypothesis tests, we have available a range of possible approaches and consequent test statistics (as in Chapters 8, 12, and 13). This time we will focus attention on the likelihood ratio approach, and merely indicate other possibilities as appropriate. The likelihood ratio test requires that the likelihood of the sample be maximized twice, once under the null hypothesis and once without restriction. The ratio of these two maxima then constitutes the required test statistic. Since the regression parameters relate to the *mean* of the multivariate normal distribution, the analogous mathematical argument to that employed in Section 8.3 will lead to the maxima

$$\sup_{\Omega} L = (2\pi)^{-\frac{1}{2}np}\, |\hat{\boldsymbol{\Sigma}}|^{-\frac{1}{2}n} \exp(-\tfrac{1}{2}np)$$

and

$$\sup_{\omega} L = (2\pi)^{-\frac{1}{2}np} |\tilde{\boldsymbol{\Sigma}}|^{-\frac{1}{2}n} \exp(-\tfrac{1}{2}np).$$

Here Ω is the complete parameter space for unrestricted maximization, ω is the parameter space constrained to satisfy the null hypothesis, $\hat{\boldsymbol{\Sigma}}$ is the maximum likelihood estimator (m.l.e.) of $\boldsymbol{\Sigma}$ under the full model and given by (15.7) and $\tilde{\boldsymbol{\Sigma}}$ is the m.l.e. of $\boldsymbol{\Sigma}$ when the null hypothesis is imposed. Thus the likelihood ratio test statistic will be the ratio of determinants, i.e. Wilks' Λ, given by $\Lambda = |\hat{\boldsymbol{\Sigma}}|/|\tilde{\boldsymbol{\Sigma}}|$. It is thus only necessary to find $\tilde{\boldsymbol{\Sigma}}$ in any given situation. We will consider several hypothesis tests that occur most commonly in practice, and obtain the appropriate test statistics; the general procedure should then be evident from these special cases.

The most important test in practice is of the significance of regression: do the explanatory variables help at all in predicting the dependent variables, or would simple prediction from the unconditional mean of $\boldsymbol{y}$ be just as accurate? The difference between the conditional and unconditional dispersion matrices of CC and MUAC at the end of Section 15.2.2 suggest that height and age *are* important in prediction, but can we establish this objectively by a hypothesis test? The null hypothesis in this case is that there is no relationship between $\boldsymbol{y}$ and $\boldsymbol{x}$, while the alternative hypothesis is that some relationship does exist. Thus the model for the data under the alternative hypothesis is the full model $\mathbf{Y} = \mathbf{XB} + \mathbf{E}$ and $n\hat{\boldsymbol{\Sigma}} = \hat{\mathbf{E}}'\hat{\mathbf{E}}$, as given in (15.5) and (15.7). Under the null hypothesis, however, all those elements β_{ij} of $\mathbf{B}$ that multiply any of the explanatory variables must be set to zero. Thus the model for the data in this case is $\mathbf{Y} = \mathbf{1}\boldsymbol{\mu}' + \mathbf{E}$, where $\boldsymbol{\mu}$ is the population mean vector of the dependent variable vector and $\mathbf{1}$ is an n-vector of ones. (This model just expresses each row of $\mathbf{Y}$ as a common constant mean $\boldsymbol{\mu}'$ plus a departure $\boldsymbol{e}_i'$.) The maximum likelihood estimator of $\boldsymbol{\mu}$ is $\bar{\boldsymbol{y}}$, the sample mean vector of the dependent variates. Hence the estimate of $\boldsymbol{\Sigma}$ under the null hypothesis is just the ordinary corrected sum-of-squares and -products matrix among the y_i, divided by n. This estimate can be written either as $n\tilde{\boldsymbol{\Sigma}} = \mathbf{Y}'\mathbf{Y} - n\bar{\boldsymbol{y}}\bar{\boldsymbol{y}}'$, or as $n\tilde{\boldsymbol{\Sigma}} = \mathbf{Y}'\mathbf{Y}$ where the matrix $\mathbf{Y}$ has been mean-centered.

Thus Wilks' Λ can be calculated easily. However, to give a pointer to the calculation of the other possible test statistics, as well as to that of the asymptotic approximation to Λ (outlined in Section 13.2), it is convenient to set out a multivariate analysis of variance table in analogous fashion to the analysis of variance table for ordinary multiple regression. The fundamental breakdown that governs this table is

$$(\mathbf{Y}'\mathbf{Y} - n\bar{\boldsymbol{y}}\bar{\boldsymbol{y}}') = (\hat{\mathbf{Y}}'\hat{\mathbf{Y}} - n\bar{\boldsymbol{y}}\bar{\boldsymbol{y}}') + \hat{\mathbf{E}}'\hat{\mathbf{E}}$$

i.e. the total sum-of-squares and -products matrix among the y_i is equal to the 'fitted' sum-of-squares and -products matrix plus the residual sum-of-squares and -products matrix. The proof of this breakdown is relatively simple and is now given. (It is not essential to the development, and can be passed over without any detriment.)

$$\begin{aligned}\hat{\mathbf{E}}'\hat{\mathbf{E}} &= (\mathbf{Y}-\hat{\mathbf{Y}})'(\mathbf{Y}-\hat{\mathbf{Y}}) \\ &= \mathbf{Y}'\mathbf{Y} - \hat{\mathbf{Y}}'\mathbf{Y} - \mathbf{Y}'\hat{\mathbf{Y}} + \hat{\mathbf{Y}}'\hat{\mathbf{Y}}. \qquad (15.13)\end{aligned}$$

But

$$\hat{\mathbf{Y}}'\hat{\mathbf{Y}} = (\mathbf{X}\hat{\mathbf{B}})'(\mathbf{X}\hat{\mathbf{B}}) = \hat{\mathbf{B}}'(\mathbf{X}'\mathbf{X})\hat{\mathbf{B}},$$

and

$$(\mathbf{X}'\mathbf{X})\hat{\mathbf{B}} = \mathbf{X}'\mathbf{Y} \quad \text{from (15.6)},$$

so

$$\hat{\mathbf{Y}}'\hat{\mathbf{Y}} = \hat{\mathbf{B}}'\mathbf{X}'\mathbf{Y} = (\mathbf{X}\hat{\mathbf{B}})'\mathbf{Y} = \hat{\mathbf{Y}}'\mathbf{Y}.$$

Hence

$$\begin{aligned}\hat{\mathbf{E}}'\hat{\mathbf{E}} &= \mathbf{Y}'\mathbf{Y} - \mathbf{Y}'\hat{\mathbf{Y}} \text{ on substituting into (15.13)} \\ &= \mathbf{Y}'\mathbf{Y} - \mathbf{Y}'\mathbf{X}\hat{\mathbf{B}} \\ &= \mathbf{Y}'\mathbf{Y} - \hat{\mathbf{B}}'\mathbf{X}'\mathbf{X}\hat{\mathbf{B}} \text{ using } \mathbf{Y}'\mathbf{X} = \hat{\mathbf{B}}'(\mathbf{X}'\mathbf{X}) \\ &= \mathbf{Y}'\mathbf{Y} - \hat{\mathbf{Y}}'\hat{\mathbf{Y}}.\end{aligned}$$

Thus

$$\mathbf{Y}'\mathbf{Y} = \hat{\mathbf{Y}}'\hat{\mathbf{Y}} + \hat{\mathbf{E}}'\hat{\mathbf{E}},$$

i.e.

$$(\mathbf{Y}'\mathbf{Y} - n\bar{\boldsymbol{y}}\bar{\boldsymbol{y}}') = (\hat{\mathbf{Y}}'\hat{\mathbf{Y}} - n\bar{\boldsymbol{y}}\bar{\boldsymbol{y}}') + \hat{\mathbf{E}}'\hat{\mathbf{E}}.$$

We can set out these matrices in a multivariate analysis of variance table. Writing $\mathbf{H} = (\hat{\mathbf{Y}}'\hat{\mathbf{Y}} - n\bar{\boldsymbol{y}}\bar{\boldsymbol{y}}')$ as the 'fitted' matrix, we have

MANOVA

Source	SSP matrix	Degrees of freedom
Multivariate regression	$\mathbf{H}$	q
Residual	$\hat{\mathbf{E}}'\hat{\mathbf{E}}$	$n-q-1$
Total (corrected)	$\mathbf{Y}'\mathbf{Y} - n\bar{\boldsymbol{y}}\bar{\boldsymbol{y}}'$	$n-1$

Comparison of this table with that of Section 13.3 shows that significance of the multivariate regression can be tested by computing any of the test statistics described there. These statistics all depend on the eigenvalues $\theta_1, \theta_2, \ldots, \theta_p$ of $(\hat{\mathbf{E}}'\hat{\mathbf{E}})^{-1}\mathbf{H}$; in particular, Wilks' Λ is given by $\Lambda = \prod_{i=1}^{p} (1+\theta_i)^{-1}$. Bartlett's statistic based on Λ is $-\{n-q-1-\frac{1}{2}(p-q+1)\}\log_e \Lambda$, and asymptotically (i.e. for large n) this statistic has approximately a χ^2 distribution on pq degrees of freedom if H_0 is true (i.e. if the explanatory variables $\boldsymbol{x}$ do not affect prediction of $\boldsymbol{y}$). Thus an exact test can be conducted by referring any of the calculated test statistics to appropriate tables or charts (Mardia *et al.* (1979), Morrison (1976), T. W. Anderson (1984), Seber (1984)), while a quick but approximate test can be done by calculating Bartlett's statistic and using chi-squared tables.

We illustrate the calculations on the girls' measurement data. Note that in this case $n=9$, which is far from asymptotic! However, Bartlett's approximation should nevertheless yield useful indication of the extent of the influence of the explanatory variables, and can be used in an informal inference sense to guide logical model-building. For this set of data, $\mathbf{Y'Y} - n\bar{\mathbf{y}}\bar{\mathbf{y}}'$ is 9 times the unconditional dispersion estimate, while $\hat{\mathbf{E}}'\hat{\mathbf{E}}$ is $9\hat{\boldsymbol{\Sigma}}$ where $\hat{\boldsymbol{\Sigma}}$ is given in eqn (15.12).

Thus

$$\hat{\mathbf{E}}'\hat{\mathbf{E}} = \begin{pmatrix} 2.4048 & -0.3555 \\ -0.3555 & 2.8710 \end{pmatrix},$$

so

$$(\hat{\mathbf{E}}'\hat{\mathbf{E}})^{-1} = 0.14754\begin{pmatrix} 2.8710 & 0.3555 \\ 0.3555 & 2.4048 \end{pmatrix} = \begin{pmatrix} 0.4236 & 0.0525 \\ 0.0525 & 0.3548 \end{pmatrix}.$$

Also

$$\mathbf{Y'Y} - n\bar{\mathbf{y}}\bar{\mathbf{y}}' = \begin{pmatrix} 15.7600 & 11.6504 \\ 11.6504 & 14.5000 \end{pmatrix},$$

so

$$\mathbf{H} = (\mathbf{Y'Y} - n\bar{\mathbf{y}}\bar{\mathbf{y}}') - \hat{\mathbf{E}}'\hat{\mathbf{E}} = \begin{pmatrix} 13.3552 & 12.0059 \\ 12.0059 & 11.6290 \end{pmatrix}.$$

Thus

$$(\hat{\mathbf{E}}'\hat{\mathbf{E}})^{-1}\mathbf{H} = \begin{pmatrix} 6.2876 & 4.9608 \\ 5.6962 & 4.7563 \end{pmatrix}.$$

Since the eigenvalues of a matrix $\mathbf{A} = (a_{ij})$ are the roots of $|\mathbf{A} - \lambda\mathbf{I}| = 0$, the eigenvalues of the (2×2) matrix $\mathbf{A}$ can be found easily by solving the quadratic equation $(a_{11} - \lambda)(a_{22} - \lambda) - a_{12}a_{21} = 0$. Hence we obtain the eigenvalues 10.8926 and 0.1513 for $(\hat{\mathbf{E}}'\hat{\mathbf{E}})^{-1}\mathbf{H}$, and from above it follows that Wilks' Λ is given by $\Lambda = (11.8926 \times 1.1513)^{-1} = 0.073$. Bartlett's statistic is thus $-\{6 - \frac{1}{2}(2 - 2 + 1)\}\log_e(0.073) = 14.39$. If all regression parameters are zero, this should be commensurate with values from the chi-squared distribution on $2 \times 2 = 4$ degrees of freedom. The 95 per cent point of this distribution is 9.49, well exceeded by 14.39. We therefore conclude that the regression is significant, and that use of the variables height and age will improve the prediction of CC and MUAC.

Having established that a relationship *does* exist between $\boldsymbol{y}$ and $\boldsymbol{x}$ for predictive purposes, the next objective often in model-building is to derive the most parsimonious relationship, i.e. the one that has good predictive accuracy but that involves the fewest number of explanatory variables. In this respect it is of interest to test the null hypothesis that a specified subset of explanatory variables provides no predictive information if the remaining explanatory variables are included in the relationship. This hypothesis is equivalent to the hypothesis that all regression parameters β_{ij} corresponding to this subset of variables are zero. Suppose (without loss of generality, as variables can always be permuted) that we wish to test the null hypothesis that the last q_2 explanatory variables are not important for prediction if the first q_1 are present in the fitted model ($q_1 + q_2 = q$). Then the models for the data under null and alternative hypothesis are both given by (15.4), where $\mathbf{X}$ is the full $n \times (q+1)$ matrix when no constraint is imposed but under the null hypothesis it is the $n \times (q_1 + 1)$ matrix made up of the first $(q_1 + 1)$ columns only. Hence $\hat{\mathbf{\Sigma}}$ is obtained exactly as before, i.e. from (15.7) using (15.6) and the full $\mathbf{X}$ matrix, while $\tilde{\mathbf{\Sigma}}$ is given by (15.7) using (15.6) and the reduced $\mathbf{X}$ matrix. If we write $\tilde{\mathbf{E}}$ as the matrix of residuals obtained from the constrained (reduced) model, then $n\tilde{\mathbf{\Sigma}} = \tilde{\mathbf{E}}'\tilde{\mathbf{E}}$, and the SSP matrix 'due to fitting just the first q_1 explanatory variables' is given by $\mathbf{H}_{(q_1)} = (\mathbf{Y}'\mathbf{Y} - n\bar{\mathbf{y}}\bar{\mathbf{y}}') - \tilde{\mathbf{E}}'\tilde{\mathbf{E}}$. Hence the SSP matrix 'due to fitting the last q_2 explanatory variables, after making allowance for the first q_1 explanatory variables' is $\mathbf{H}_{(q_2 \mid q_1)} = \mathbf{H} - \mathbf{H}_{(q_1)}$. Thus $\mathbf{H}_{(q_2 \mid q_1)}$ is a measure of the extra predictive information contained in the last q_2 explanatory variables if they are added to a model already containing the first q_1 explanatory variables. This matrix is assessed relative to the overall error matrix $\hat{\mathbf{E}}'\hat{\mathbf{E}}$, to determine the extra significance of these q_2 last variables. This procedure is thus the matrix extension of the 'extra-sum-of-squares principle' used in multiple regression analysis; a multivariate analysis of variance table setting out these matrices is given as follows.

MANOVA

Source	SSP matrix	Degrees of freedom
Multivariate regression (all q variables)	$\mathbf{H}$	q
First q_1 variables ignoring last q_2	$\mathbf{H}_{(q_1)} = \mathbf{T} - \tilde{\mathbf{E}}'\tilde{\mathbf{E}}$	q_1
Last q_2 variables eliminating first q_1	$\mathbf{H}_{(q_2 \mid q_1)} = \mathbf{H} - \mathbf{H}_{(q_1)}$	q_2
Residual	$\hat{\mathbf{E}}'\hat{\mathbf{E}}$	$n - q - 1$
Total	$\mathbf{T} = \mathbf{Y}'\mathbf{Y} - n\bar{\mathbf{y}}\bar{\mathbf{y}}'$	$n - 1$

The significance of the last q_2 explanatory variables over and above the first q_1 is thus tested by any of the test statistics based on eigenvalues $\theta_1, \theta_2, \ldots, \theta_p$ of $(\hat{\mathbf{E}}'\hat{\mathbf{E}})^{-1}\mathbf{H}_{(q_2 \mid q_1)}$. In particular, Wilks' Λ is again given by $\Lambda = \prod_{i=1}^{p} (1 + \theta_i)^{-1}$. Bartlett's statistic is $-\{n - q - 1 - \frac{1}{2}(p - q_2 +$

1)} $\log_e \Lambda$; if n is large, and if the last q_2 variables do not contribute any further predictive information, then this statistic has a χ^2 distribution on pq_2 degrees of freedom.

To illustrate the procedure, let us return to the girls' measurements and test whether age is an important predictive variable if height is already being used in the prediction. We already have the matrices $\mathbf{H}$, $(\hat{\mathbf{E}}'\hat{\mathbf{E}})$, and $(\hat{\mathbf{E}}'\hat{\mathbf{E}})^{-1}$ from the previous example, so we only need to find $\mathbf{H}_{(q_1)}$ and hence $\mathbf{H}_{(q_2|q_1)}$. For $\mathbf{H}_{(q_1)}$ we need to find $\tilde{\mathbf{E}}'\tilde{\mathbf{E}}$, where $\tilde{\mathbf{E}}$ is the matrix of residuals obtained from the model which includes only height (x_1). Denote this model by $\mathbf{Y} = \mathbf{X}_{(1)}\mathbf{B}_{(1)} + \mathbf{E}_{(1)}$. Then $\mathbf{X}_{(1)}$ is the matrix composed of the first two columns of Table 15.2. Hence $\mathbf{X}'_{(1)}\mathbf{X}_{(1)}$ is simply the top (2×2) portion of the previous $\mathbf{X}'\mathbf{X}$ matrix, while $\mathbf{X}'_{(1)}\mathbf{Y}$ is the matrix composed of the first two rows of $\mathbf{X}'\mathbf{Y}$, i.e.

$$\mathbf{X}'_{(1)}\mathbf{X}_{(1)} = \begin{pmatrix} 9 & 0 \\ 0 & 196 \end{pmatrix}, \text{ and } \mathbf{X}'_{(1)}\mathbf{Y} = \begin{pmatrix} 525.6 & 121.5 \\ 45.1 & 34.5 \end{pmatrix}.$$

Thus, the parameter estimates are

$$\hat{\mathbf{B}}_{(1)} = \begin{pmatrix} 1/9 & 0 \\ 0 & 1/196 \end{pmatrix}\begin{pmatrix} 525.6 & 121.5 \\ 45.1 & 34.5 \end{pmatrix} = \begin{pmatrix} 58.4 & 13.5 \\ 0.2301 & 0.1760 \end{pmatrix}.$$

(Note that the estimates of the regression coefficients for x_1 have changed on removal of x_2 from the model.)

Next, the fitted values are

$$\tilde{\mathbf{Y}} = \mathbf{X}_{(1)}\hat{\mathbf{B}}_{(1)} = \begin{pmatrix} 1 & 4.0 \\ 1 & -1.0 \\ 1 & 2.0 \\ 1 & -1.0 \\ 1 & 3.0 \\ 1 & 2.0 \\ 1 & -1.0 \\ 1 & -12.0 \\ 1 & 4.0 \end{pmatrix}\begin{pmatrix} 58.4 & 13.5 \\ 0.2301 & 0.1760 \end{pmatrix} = \begin{pmatrix} 59.3204 & 14.204 \\ 58.1699 & 13.324 \\ 58.8602 & 13.852 \\ 58.1699 & 13.324 \\ 59.0903 & 14.028 \\ 58.8602 & 13.852 \\ 58.1699 & 13.324 \\ 55.6388 & 11.388 \\ 59.3204 & 14.204 \end{pmatrix}$$

so that the residual matrix $\tilde{\mathbf{E}}$ is

$$\mathbf{Y} - \tilde{\mathbf{Y}} = \begin{pmatrix} -0.9204 & -0.204 \\ 1.0301 & 1.676 \\ 1.4398 & 1.148 \\ -0.7699 & -0.324 \\ 0.4097 & -0.028 \\ -0.7602 & 0.648 \\ -0.1699 & -0.824 \\ -0.1388 & -0.388 \\ -0.1204 & -1.705 \end{pmatrix}$$

and

$$\tilde{\mathbf{E}}'\tilde{\mathbf{E}} = \begin{pmatrix} 5.3824 & 3.7115 \\ 3.7115 & 8.4273 \end{pmatrix}.$$

From before

$$\mathbf{T} = \mathbf{Y}'\mathbf{Y} - n\bar{\mathbf{y}}\bar{\mathbf{y}}' = \begin{pmatrix} 15.7600 & 11.6504 \\ 11.6504 & 14.5000 \end{pmatrix}.$$

Thus

$$\mathbf{H}_{(q_1)} = \mathbf{T} - \tilde{\mathbf{E}}'\tilde{\mathbf{E}} = \begin{pmatrix} 10.3776 & 7.9389 \\ 7.9389 & 6.0727 \end{pmatrix}.$$

From before,

$$\mathbf{H} = \begin{pmatrix} 13.3552 & 12.0059 \\ 12.0059 & 11.6290 \end{pmatrix},$$

so that

$$\mathbf{H}_{(q_2 \mid q_1)} = \mathbf{H} - \mathbf{H}_{(q_1)} = \begin{pmatrix} 2.9776 & 4.0670 \\ 4.0670 & 5.5563 \end{pmatrix}.$$

Hence, finally,

$$(\hat{\mathbf{E}}'\hat{\mathbf{E}})^{-1}\mathbf{H}_{(q_2 \mid q_1)} = \begin{pmatrix} 0.4236 & 0.0525 \\ 0.0525 & 0.3548 \end{pmatrix} \begin{pmatrix} 2.9776 & 4.0670 \\ 4.0670 & 5.5563 \end{pmatrix}$$
$$= \begin{pmatrix} 1.4748 & 2.0145 \\ 1.5993 & 2.1849 \end{pmatrix}.$$

Eigenvalues of this matrix are 3.65955 and 0.00015, so that Wilks' Λ is given by $\Lambda = (4.65955 \times 1.00015)^{-1} = 0.21458$. In the present case $q_1 = 1$ and $q_2 = 1$, so Bartlett's statistic is $-\{6 - \frac{1}{2}(2 - 1 + 1)\} \log_e(0.21458) = 7.70$. If the second explanatory variable (age) carries no extra predictive value once height is included in the relationship, then 7.70 should be commensurate with values from the chi-squared distribution on $2 \times 1 = 2$ degrees of freedom. The 95 per cent point of this distribution is 5.99, and hence we reject the null hypothesis and conclude that age *is* an important adjunct to height in the predictive relationship for CC and MUAC.

Of course, any calculations for multivariate regression are usually done on a computer using one of the existing standard software packages. The above step-by-step calculations have been shown in full, however, to enable the reader to work through them and gain some insight into the workings of the technique. Other hypotheses that specify values for some or all elements of the **B** matrix can be carried out in exactly analogous fashion to the ones above.

A final inferential aspect of the multivariate regression model is the construction of confidence regions for some or all elements of **B**. However, the expressions that can be derived for these regions are complicated mathematically and difficult to visualize geometrically. They are therefore not given here, but the interested reader is referred to T. W. Anderson (1984, Section 8.7.2) or Mardia *et al.* (1979, Section 6.3).

15.2.4 Assessing the adequacy of a fitted model

Any model fitted to a set of data involves a number of assumptions. These assumptions can relate either to the systematic or to the random part of the model. In the former case the assumptions involve the *form* of the relationship between explanatory and dependent variables (e.g. which variables should be included in the model; is the relationship linear or quadratic), while in the latter case they relate to the *distribution* of the dependent variable(s) (e.g. assumptions of independence, normality, variance homogeneity of a dependent variable). If all such assumptions are not checked when a model is fitted to data, there is a danger than an inappropriate model may be accepted, and may subsequently assume an important role in all decision-making associated with the given subject area. Consequently, any proposed model should be subjected to critical scrutiny before being accepted and used, e.g. for prediction of future values or explanation of a system.

In the univariate case, with a single dependent variable y for which the multiple regression model (15.5) is proposed, all of the sample information on adequacy of the model is contained in the residuals $\hat{e}_i$ ($i = 1, \ldots, n$). Each residual $\hat{e}_i$ is an estimate of the error e_i. If the model is correct and all distributional assumptions are valid, the $\hat{e}_i$ will look *approximately* like independent observations from a normal distribution with a zero mean, and a constant variance. (Actually, the residuals will have non-zero correlations and variance depending on values of $\mathbf{X}$, but usually the correlations are small and the variance is nearly constant.) Hence departures from assumptions, and inadequacy of the fitted model, can be detected graphically by plotting the residuals in various ways. The most informative plots are usually:

1. A plot of the residuals $\hat{e}_i$ against the fitted values $\hat{y}_i$.
2. A plot of the residuals $\hat{e}_i$ against the values of an explanatory variable, x_{ij}.
3. A plot of the residuals $\hat{e}_i$ against time (or position in space, or any other factor determining the order in which the observations were taken).
4. A probability plot or histogram.

Case 4 will test the assumption of normality. The presence of unusual observations or severe departures from normality may require special attention; if n is large, then minor departures from normality will not greatly affect inferences about $\boldsymbol{\beta}$. In all other plots, consonance of the data with the model is indicated by a random scatter of points about the line $\hat{e} = 0$, between two lines parallel to, and approximately equidistant from, $\hat{e} = 0$. Such a pattern is illustrated in Fig. 15.1(a). Departures from model assumptions are indicated by systematic trends in the residual plots. For example, a pattern such as Fig. 15.1(b) for case 1 shows a dependence of the residuals on the fitted values caused by omission of an important explanatory variable from the fitted model; a pattern such as Fig. 15.1(c) for case 1 shows that the variance of the errors e_i is not constant, but depends on the means of the observations; a pattern such as Fig. 15.1(d) for case 2 shows that a quadratic term in x_j should be included in the model; while a pattern such as Fig. 15.1(e) indicates correlations between successive observations in time.

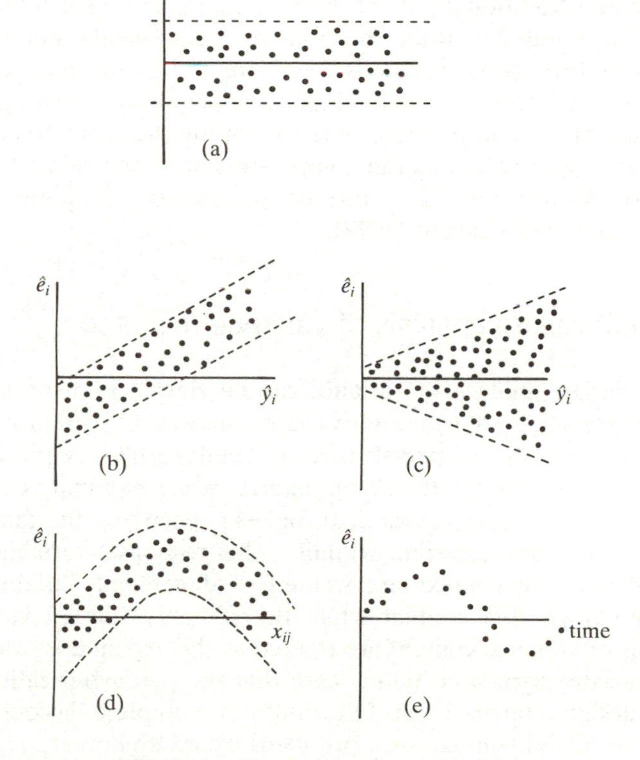

Fig. 15.1 Illustrative plots of residuals

Turning to the multivariate regression model (15.4), it is to be expected that the matrix of residuals $\hat{\mathbf{E}}$ will contain analogous information from which adequacy of the model can be assessed. Each *column* of $\hat{\mathbf{E}}$ contains the residuals from the fitted model for one particular dependent variable y_i. Hence the residuals in each column can be subjected separately to the plots described above, and this will help to detect major departures in the systematic part of the model. However, separate treatment of each column of $\hat{\mathbf{E}}$ ignores information passed through the correlations among dependent variables, and to make full use of these correlations we should consider each *row* of $\hat{\mathbf{E}}$ as a single entity. In the same spirit of approximation as used above, if the model is correct and all distributional assumptions are valid, then the rows of $\hat{\mathbf{E}}$ will look approximately like independent observations from a multivariate normal distribution with mean $\boldsymbol{0}$ and variance–covariance matrix $\boldsymbol{\Sigma}$ (Seber, 1984, Section 8.5). Thus the multivariate normality assumption of the model can be investigated graphically by the techniques described in Section 7.5, applied to the rows of $\hat{\mathbf{E}}$. Furthermore, principal component analysis of $\hat{\mathbf{E}}$ (viewed as a data matrix) will provide useful information. For example, a traditional plot of the principal component scores should help to identify outliers, while the principal components themselves can be viewed as *orthogonal residuals* to be treated in the manner of Fig. 15.1. Plotting the scores on individual principal components against any relevant variable should be preferable to plotting the separate columns of $\hat{\mathbf{E}}$ against such a variable, as the components are uncorrelated while the columns of $\hat{\mathbf{E}}$ are not. For further discussion of checking model adequacy, see Gnanadesikan (1977).

15.3 Multivariate analysis of variance revisited

Just as univariate analysis of variance can be viewed as a special case of multiple regression, so can multivariate analysis of variance (Section 13.3) be viewed as a special case of multivariate regression. The connection comes through the *design matrix,* which is a matrix of dummy variables (usually taking values 1, 0, or -1) specifying the factor levels appropriate to each experiment unit. The response variables in the analysis of variance context are treated as dependent variables in the multivariate regression context while the (dummy) design variables in the analysis of variance context are treated as the explanatory variables in the multivariate regression, taking care that the parameterization is such that the design matrix is of full rank. A complete account of the treatment of analysis-of-variance problems by multiple regression can be found in most textbooks on multiple regression analysis (e.g. Draper and

Smith (1981), Chapter 9), where appropriate parameterizations to ensure full-rank design matrices are discussed, and the standard situations are set out in detail. No new principles are involved in the multivariate case, the only difference being that sums-of-squares of response variables are replaced by sum-of-squares and -products matrices throughout. This formulation as multivariate regression is of greatest benefit in analysing non-standard designs. As far as specification of the design matrix is concerned, the fact that the multivariate model (15.5) is just a side-by-side assembly of the univariate models (15.4) means that the same design matrix is used for both univariate and multivariate versions of any given situation. Hence we need do no more here than point out the applicability of the methods of this chapter to such problems, leaving the interested reader to consult a suitable multiple regression text for appropriate details.

Note also that the smoothed continuous variable parameter estimates in the location model for mixed discrete/continuous data (Section 12.2.4) are obtained by multivariate regression. A linear model for the continuous variable means given the values of the discrete variables is precisely model (15.4), where the explanatory variables are dummy variables specifying cell membership of the multinomial table constructed from the discrete variables.

16 Explaining observed associations: latent-variable models

16.1 Background ideas and principles

The models considered in the previous chapter are ones linking the observed variables of a multivariate system. Their objective is explanation and prediction: by creating a model that adequately mimics the overall behaviour of the system, a better understanding can be gained into the mechanism of the system, and improved prediction of future behaviour is a natural consequence of such enhanced understanding. In order to make the model as accurate as possible, as much information as possible should be built into it. Given the observed associations between the variables of the system, therefore, it is natural these associations should be used in the model. In essence, the models of the previous chapter start from the observed associations and build outwards. It might be argued, however, that if we have a true understanding of the system then we should be able to predict reasonably accurately all aspects of the system, *including* the associations themselves. Thus a more fundamental model is one that *explains* the observed associations between variables, and considerable interest has been shown in such models in many fields of application. The impetus for this interest stemmed from the work of Psychologists involved in mental test theory at the turn of the century, and the pioneering ideas of Galton and Spearman provided fertile ground for the myriad developments (and no little controversy!) that have taken place in the succeeding eighty or ninety years.

Suppose that p variables $\mathbf{x}' = (x_1, \ldots, x_p)$ have been observed on each of n sample individuals. These variables can be of any type, but for the present introduction of ideas it is convenient to suppose that they are continuous, and that associations between them are therefore measured by correlation coefficients. (This restriction will be relaxed when considering the actual models later.) Let the sum-of-squares and -products, covariance, and correlation matrices obtained from the sample values be denoted as usual by $\mathbf{C}$, $\mathbf{S}$, and $\mathbf{R}$. Our objective is an adequate 'explanation' of all the entries in **R**. It is evident that any satisfactory explanation must draw on information from *outside* this set of correlation values, as otherwise we will become enmeshed in circular arguments. Let us therefore suppose that the relevant extra information resides in a further q variables $\boldsymbol{z}' = (z_1, z_2, \ldots, z_q)$ that could be measured on each sample member.

To see how these extra variables can explain the observed correlations, recollect from Section 14.4 that a correlation r_{ij} between two variables x_i and x_j could arise as a result of their mutual association with an extraneous variable z_k. The variation in z_k values over the sample, coupled with the associations between z_k and each of x_i and x_j, induces an association between x_i and x_j. The partial correlation $r_{ij.k}$ measures the association between x_i and x_j when the value of z_k is held fixed, and is therefore the 'residual' correlation between x_i and x_j after removal of the (linear) effect of z_k on each. If this partial correlation is close to zero, or more precisely if the null hypothesis of zero population partial correlation is tenable, then we can say that z_k has 'explained' the correlation r_{ij} between x_i and x_j. If $r_{ij.k}$ is not close to zero, then there still exists some association between x_i and x_j that is unexplained, and we need to consider a further variable z_l to account for this remaining correlation. If the partial correlation $r_{ij.kl}$ is not significantly different from zero, then z_k and z_l jointly explain the correlation r_{ij}. Otherwise, we consider a third variable z_m, and so on. A satisfactory explanation of all the entries in **R** will thus be obtained when we find a set of variables $\boldsymbol{z}' = (z_1, \ldots, z_q)$ such that the partial correlation between any two variables x_i and x_j, on fixing the values of all elements of $\boldsymbol{z}$, is not significantly different from zero. The most *parsimonious* explanation is achieved when q is as small as possible.

The converse of the above argument is that if the variables $z_1, \ldots, z_q$ are to provide a complete explanation of the entries of **R**, then the partial correlation between any two elements of $\boldsymbol{x}$ for a fixed value of $\boldsymbol{z}$ must be compatible with a population value of zero. This means that if only those individuals with specified values of the z_i were to be sampled, then no association would be detectable between any two elements of $\boldsymbol{x}$. In other words, x_i and x_j are conditionally independent given the values of $z_1, \ldots, z_q$, for all i and j. This property of *local independence* is a necessary requirement for a set of variables $\boldsymbol{z}$ to provide an explanation for the entries of **R**.

As it stands, the above discussion is an academic one because the z_i are not available: the data consist of n observations on each of the p variables x_i, and telling the investigator to search for q unspecified extra variables z_j with which the inter-correlations among the x_i might be explained is not a very positive contribution to the analysis! The great innovatory step made by the pioneers such as Galton and Spearman, however, was to say that the z_i were actually *unobservable* and so need not be sought physically; instead, they must be identified by means of statistical analysis. Since the z_i are unobservable, they are termed *latent* variables.

This idea seems a perfectly natural one in Psychology, where in dealing with human behaviour one often encounters traits that are not directly measurable. The classical example is in mental test theory, where the whole subject matter of the present chapter in fact originated. It is accepted that all humans possess characteristics such as 'intelligence', 'verbal ability', 'numerical ability', 'memory', and so on, but none of these characteristics can be measured exactly. The best that can be done is to offer an individual a battery of tests and measure their scores on each. Whereas each test might be designed to measure just one of the traits, in fact *all* of the traits will generally contribute to the score obtained by an individual on any test. Thus, while a mental arithmetic test might be intended to measure 'numerical ability', clearly 'intelligence' and 'memory' will also play a large part in determining the score on the test. Consequently, the observed inter-correlations among the tests for a set of individuals will be explainable by the presence of each trait in each test. Identifying the test scores with the x_i and the traits with the z_j above will therefore lead to a possible means of analysis.

The classical approach is to represent the associations among the x_i by a suitable model which embodies the z_j (in such a way that the property of local independence is satisfied), and then to estimate and test hypotheses about the parameters of this model. The resulting analysis is known as *factor analysis* if the x_i are continuous, and *latent-structure analysis* if the x_i are discrete, the z_j generally being assumed continuous in both of these techniques. The special case of discrete x_i and discrete z_j is termed *latent-class analysis*. These techniques will all be considered below.

Before proceeding to details of the analyses, however, a brief word about the underlying philosophy is appropriate. Since their original introduction in Psychology, the techniques have found many advocates in ever-wider fields of application. In particular, they are currently much in demand in the Social Sciences. Given that most of such work is concerned with human behaviour, use of the techniques can generally be justified in these circumstances. For example, in Sociology the concept of 'social class' may not be directly measurable and hence has to be used as a latent variable underlying a number of related manifest variables. Similarly, in Linguistics, one can believe in the existence of various latent variables associated with the measurement, e.g. of different speech patterns. However, as one moves further from the Social Sciences towards the Natural Sciences, it becomes more difficult to justify the use of the techniques; in the latter disciplines most entities are precisely measurable, and the existence of sensible latent variables becomes less defensible. At the extremes of, say, Physics or Chemistry, the models become totally unbelievable. Although the techniques may provide a formal 'explanation' of the observed inter-correlations, the context within which they do so is incompatible with the underlying model. We shall see later that many difficulties and arbitrary features exist in the methodology even if the techniques are used in appropriate circumstances; they should certainly not even be contemplated unless the models are fully justifiable.

We will consider the standard models for continuous x_i in Section 16.2, and those for discrete x_i in Section 16.3. No satisfactory means exists within this framework for encompassing mixtures of discrete and continuous x_i, but a unified approach for all situations has recently been attempted in a series of papers by Bartholomew. This approach will be outlined briefly in Section 16.4.

16.2 Continuous manifest variables: factor analysis

16.2.1 Basic model

We suppose that p continuous variables $x_1, \ldots, x_p$ have been observed on each of n sample individuals, and we wish to explain the resulting associations among these p manifest variables by means of q latent variables $z_1, \ldots, z_q$. The first step in the analysis is the formulation of an appropriate model, which in this case is any model that satisfies the requirement of local independence as outlined earlier. Thus the model must be one in which the x_i are conditionally uncorrelated given the values of all z_j. There are various possible models that satisfy this requirement, but the simplest is the following *factor analysis* model:

$$\begin{aligned} x_1 &= \mu_1 + \gamma_{11}z_1 + \gamma_{12}z_2 + \ldots + \gamma_{1q}z_q + e_1 \\ x_2 &= \mu_2 + \gamma_{21}z_1 + \gamma_{22}z_2 + \ldots + \gamma_{2q}z_q + e_2 \\ &\vdots \\ x_p &= \mu_p + \gamma_{p1}z_1 + \gamma_{p2}z_2 + \ldots + \gamma_{pq}z_q + e_p \end{aligned} \tag{16.1}$$

where the μ_i and γ_{ij} are constants, while the z_i and e_i are random variables ($i = 1, \ldots, p$; $j = 1, \ldots, q$). The minimal set of assumptions about these random variables (to ensure that the local independence property is satisfied) is that the e_i are uncorrelated with each other and with the z_i. It is then evident that if the values of the z_i are specified we can write $x_i = v_i + e_i$ ($i = 1, \ldots, p$), where the v_i are constants, so that $\text{corr}(x_i, x_j) = 0$ for all i, j.

We may collect the various entities in (16.1) together into vectors and matrices: $\boldsymbol{x}' = (x_1, \ldots, x_p)$; $\boldsymbol{\mu}' = (\mu_1, \ldots, \mu_p)$; $\boldsymbol{z}' = (z_1, \ldots, z_q)$; $\boldsymbol{e}' = (e_1, \ldots, e_p)$, and $\boldsymbol{\Gamma} = (\gamma_{ij})$, the $p \times q$ matrix with (i, j)th element γ_{ij}; whence (16.1) can be written

$$\boldsymbol{x} = \boldsymbol{\mu} + \boldsymbol{\Gamma z} + \boldsymbol{e}. \tag{16.2}$$

Furthermore, the assumption about the e_i given above is equivalent to the assumption that the dispersion matrix of $\boldsymbol{e}$ is diagonal. We can therefore write $\text{cov}(\boldsymbol{e}) = \boldsymbol{\Psi} = \text{diag}(\psi_1^2, \psi_2^2, \ldots, \psi_p^2)$. Since the constants $\boldsymbol{\mu}$ and $\boldsymbol{\Gamma}$ are unspecified at the outset, we can also assume that the means of $\boldsymbol{z}$ and $\boldsymbol{e}$ are both zero (and any subsequent adjustment necessary for the mean of $\boldsymbol{x}$ can be made within $\boldsymbol{\mu}$). Finally, what about the dispersion matrix of $\boldsymbol{z}$? In principle this could be *any* positive definite symmetric matrix $\boldsymbol{\Sigma}$. However, since $\boldsymbol{\Sigma}$ is symmetric and positive definite, we can find a lower triangular matrix $\mathbf{L}$ of full rank such that $\boldsymbol{\Sigma} = \mathbf{LL}'$ (the Cholesky decomposition: see also Appendix, Section A5). Consider therefore the transformed vector $\boldsymbol{z}^* = \mathbf{L}^{-1}\boldsymbol{z}$. This vector clearly also has mean zero, while its dispersion matrix is $\mathbf{L}^{-1}\boldsymbol{\Sigma}(\mathbf{L}^{-1})' = \mathbf{L}^{-1}\mathbf{LL}'(\mathbf{L}')^{-1} = \mathbf{I}$. If we therefore write $\boldsymbol{\Gamma}^* = \boldsymbol{\Gamma}\mathbf{L}$, we see that $\boldsymbol{\Gamma}^*\boldsymbol{z}^* = \boldsymbol{\Gamma}\mathbf{LL}^{-1}\boldsymbol{z} = \boldsymbol{\Gamma z}$. Hence the model $\boldsymbol{x} = \boldsymbol{\mu} + \boldsymbol{\Gamma}^*\boldsymbol{z}^* + \boldsymbol{e}$ is indistinguishable from (16.2), but the random vector $\boldsymbol{z}^*$ has standardized and uncorrelated elements. Without loss of generality, therefore, we can make this assumption directly about the elements of $\boldsymbol{z}$.

Hence the factor-analysis model can be formally stated as:

$$\boldsymbol{x} - \boldsymbol{\mu} = \boldsymbol{\Gamma z} + \boldsymbol{e} \tag{16.3}$$

where $\boldsymbol{\Gamma}$ is a $(p \times q)$ matrix of constants γ_{ij}, $\boldsymbol{z}$ is a random vector with mean zero and dispersion matrix $\mathbf{I}$, $\boldsymbol{e}$ is a random vector with mean zero and dispersion matrix $\boldsymbol{\Psi} = \text{diag}(\psi_1^2, \ldots, \psi_p^2)$, and $\boldsymbol{\mu}$ is a constant (the mean of the vector $\boldsymbol{x}$). In many applications, the assumption of normality is also made for the vectors $\boldsymbol{z}$ and $\boldsymbol{e}$, and this assumption will be used in some of the estimation and hypothesis-testing procedures below. However, just as in regression analysis, considerable progress can be made without this distributional assumption.

Let us first examine model (16.3) a bit more closely, in particular reviewing its relevance to the objectives set out in Section 16.1 above, identifying its connection with other familiar statistical models, and considering how it might be used in practice. It will help to base the discussion round a typical application, and a suitable choice is the intelligence testing situation already introduced earlier. Let $x_1, x_2, \ldots, x_p$ be the scores obtainable in a battery of tests (say: arithmetic, algebra, history, reading, and comprehension), so that the observations for the ith individual in a sample would be denoted by $\boldsymbol{x}_i' =$

$(x_{i1}, x_{i2}, \ldots, x_{ip})$ where $i = 1, \ldots, n$. The different tests exhibit associations, as given either by the sample covariance matrix $\mathbf{S}$ or by the sample correlation matrix $\mathbf{R}$. To explain these associations, model (16.3) postulates that each test score is made up of contributions from a number q of *common factors* (the z_i), together with a 'residual' *specific* to that test (the e_i). In the example under consideration, common factors relevant to the given tests might be 'intelligence' (z_1), 'numerical ability' (z_2), 'verbal ability' (z_3), and 'memory' (z_4). Each test requires a combination of these skills, but clearly each skill will be more important for some tests than for others. For example, we would expect arithmetic to have heavier contributions from z_1 and z_2 than from z_3 and z_4, while history would not require z_2 at all but would rely about equally on the other three qualities. The constant γ_{ij} in (16.3) expresses the importance of factor z_j in test x_i ($i = 1, \ldots, p$; $j = 1, \ldots, q$), and is usually known as the *loading* of factor j on test i. Each individual in the sample is assumed to possess a value for each of the factors (the set of *factor scores* for that individual), but these values are unobservable. If these values $\mathbf{z}_i' = (z_{i1}, z_{i2}, \ldots, z_{iq})$ for $i = 1, \ldots, n$ *were* known, then the individuals in the sample could be ranked with respect to each of these qualities (i.e. we could identify the 'most intelligent' individual, or the one with the worst numerical ability, the best memory, and so on). However, it is this spread of sample values along each of the common factors that, together with model (16.3), accounts for the observed associations between the tests. A factor such as z_1 contributes substantially to each test. Hence an individual with a high score on this factor will tend to have above-average scores on all tests, while an individual with a low score on this factor will tend to have below-average scores on all tests. An individual with high scores on all factors will have high scores on all tests, while an individual with low scores on all factors will have low scores on all tests, and so on.

While the reasonableness of such arguments may not be in doubt, it nevertheless remains true that model (16.3) seems a very flimsy basis from which to attempt an explanation of the system, because of the lack of *observable* quantities in it. In fact, the entire right-hand side of eqn (16.3) is composed of conceptual entities, and to obtain a complete description of the system within this framework we need to estimate the following quantities from the available data (i.e. from the n vectors $\mathbf{x}_i$):

1. The number q of common factors. Clearly we are interested in the smallest value of q that yields an adequate fit to the data, as this provides the most parsimonious explanation of the associations among the x_i. Also, re-expressing the data in terms of a small number of common factors is yet another method of dimensionality reduction for multivariate data (cf. the methods of Chapter 2).

2. The factor loadings γ_{ij}. Since the factors are all unobservable, the factor loadings provide the only means of 'labelling' each factor. By identifying which factors have high loading for each variable x_i, we can perhaps attach meanings to the factors (in similar fashion to the interpretation of principal components as described in Section 2.2.5). Thus high loadings on factor 1 for all tests in the example above would identify z_1 as 'general intelligence'; high loadings on factor 2 for x_1 and x_2 (arithmetic and algebra), but low loadings for the other x_i would suggest that z_2 was 'numerical ability'; high loadings on factor 3 for x_3, x_4, and x_5 (history, reading, and comprehension), but low loadings for x_1 and x_2 might identify z_3 as 'verbal ability' (although this labelling might be debatable), and so on.

3. The specific variances $\psi_1^2, \ldots, \psi_p^2$. These quantities determine how much of the variability of each variable is *not* attributable to the common factors, and hence how reliably each variable can be modelled by these factors.

4. The factor scores $z_i' = (z_{i1}, \ldots, z_{iq})$, which provide a ranking or scaling of the sample individuals with respect to each identified factor (trait, characteristic).

Even without any detailed analysis, it is clear that estimation of all these quantities from a single $(n \times p)$ data matrix is a very tall order, and some of the estimates may not be at all reliable. We shall consider estimation strategy below.

Before proceeding on to such details, however, let us briefly consider other techniques that are brought to mind by model (16.3). One obvious similarity is with the multivariate regression model of Chapter 15. If we write $\boldsymbol{x}_i'$ as the ith row of a matrix $\mathbf{X}$, z_i' as the ith row of a matrix $\mathbf{Z}$, $\boldsymbol{e}_i'$ as the ith row of a matrix $\mathbf{E}$ $(i = 1, \ldots, n)$, and assume columns of $\mathbf{X}$ to be mean-centered, then the form of eqn (16.3) applied to the whole set of sample data is $\mathbf{X} = \mathbf{Z}\boldsymbol{\Gamma}' + \mathbf{E}$. This is exactly the multivariate regression model (15.4), but with $\mathbf{X}$ in place of $\mathbf{Y}$, $\mathbf{Z}$ in place of $\mathbf{X}$, and $\boldsymbol{\Gamma}'$ in place of $\mathbf{B}$. However, the main distinguishing feature is the fact that $\mathbf{Z}$ is *unknown* in the present case, whereas it would have to be a matrix of *known constants* if it were part of a multivariate regression model. It is nevertheless still possible to estimate the model parameters, by embedding the multivariate regression within an E–M iterative scheme (Dempster *et al.*, 1977) as follows. An initial guess at the factor scores ($\mathbf{Z}$) is followed by a multivariate regression of $\mathbf{X}$ on $\mathbf{Z}$ to estimate $\boldsymbol{\Gamma}'$ (the M-step). With this estimate of $\boldsymbol{\Gamma}'$, the factor scores are then estimated as conditional expectations from a multivariate normal model (the E-step, see Section 16.2.7 below). This gives a new $\mathbf{Z}$ matrix, and the M-step can be carried out again. The E- and M-steps are repeated in turn in this way until the process converges. However, this procedure can be a very lengthy one, and it does not have the computational advantages of the maximum likelihood scheme described below. Consequently, multivariate regression has not played any significant role in the context of factor analysis (FA).

On the other hand, a technique that *has* played a significant role but has also caused a great deal of confusion in the same context is principal component analysis (PCA). Recall from Chapter 2 that in PCA we seek a new set of variables $y_1, \ldots, y_p$ as linear combinations of the observed (mean-centered) variables $x_1, \ldots, x_p$, in such a way as to maximize successively the variance of the y_i. If λ_i is the ith largest eigenvalue of the dispersion matrix of $\boldsymbol{x}' = (x_1, \ldots, x_p)$, and $\boldsymbol{\alpha}_i' = (\alpha_{i1}, \ldots, \alpha_{ip})$ is its corresponding eigenvector then the principal components are given by

$$\begin{aligned} y_1 &= \alpha_{11}x_1 + \ldots + \alpha_{1p}x_p \\ y_2 &= \alpha_{21}x_1 + \ldots + \alpha_{2p}x_p \\ &\vdots \\ y_p &= \alpha_{p1}x_1 + \ldots + \alpha_{pp}x_p, \end{aligned}$$

and $\operatorname{var}(y_i) = \lambda_i$ $(i = 1, \ldots, p)$. Now, since the matrix (α_{ij}) is orthogonal, we can invert this transformation to give

$$\begin{aligned} x_1 &= \alpha_{11}y_1 + \alpha_{21}y_2 + \ldots + \alpha_{p1}y_p \\ x_2 &= \alpha_{12}y_1 + \alpha_{22}y_2 + \ldots + \alpha_{p2}y_p \\ &\vdots \\ x_p &= \alpha_{1p}y_1 + \alpha_{2p}y_2 + \ldots + \alpha_{pp}y_p. \end{aligned}$$

Consequently, if the $(p-q)$ components with smallest variance are treated as 'noise' and set equal to a 'residual' η_i for the ith manifest variable $(i = 1, \ldots, p)$, then we obtain

$$\begin{aligned} x_1 &= \alpha_{11}y_1 + \alpha_{21}y_2 + \ldots + \alpha_{q1}y_q + \eta_1 \\ x_2 &= \alpha_{12}y_1 + \alpha_{22}y_2 + \ldots + \alpha_{q2}y_q + \eta_2 \\ &\vdots \\ x_p &= \alpha_{1p}y_1 + \alpha_{2p}y_2 + \ldots + \alpha_{qp}y_q + \eta_p, \end{aligned} \tag{16.4}$$

or equivalently

$$\begin{aligned} x_1 &= (\alpha_{11}\sqrt{\lambda_1})\frac{y_1}{\sqrt{\lambda_1}} + (\alpha_{21}\sqrt{\lambda_2})\frac{y_2}{\sqrt{\lambda_2}} + \ldots + (\alpha_{q1}\sqrt{\lambda_q})\frac{y_q}{\sqrt{\lambda_q}} + \eta_1 \\ x_2 &= (\alpha_{12}\sqrt{\lambda_1})\frac{y_1}{\sqrt{\lambda_1}} + (\alpha_{22}\sqrt{\lambda_2})\frac{y_2}{\sqrt{\lambda_2}} + \ldots + (\alpha_{q2}\sqrt{\lambda_q})\frac{y_q}{\sqrt{\lambda_q}} + \eta_2 \\ &\vdots \\ x_p &= (\alpha_{1p}\sqrt{\lambda_1})\frac{y_1}{\sqrt{\lambda_1}} + (\alpha_{2p}\sqrt{\lambda_2})\frac{y_2}{\sqrt{\lambda_2}} + \ldots + (\alpha_{qp}\sqrt{\lambda_q})\frac{y_q}{\sqrt{\lambda_q}} + \eta_p. \end{aligned}$$

Thus if we write $\gamma_{ij} = (\alpha_{ji}\sqrt{\lambda_j})$ and $z_i = (y_i/\sqrt{\lambda_i})$ for $i = 1, \ldots, p$ and $j = 1, \ldots, q$, then we have

$$\begin{aligned} x_1 &= \gamma_{11}z_1 + \gamma_{12}z_2 + \ldots + \gamma_{1q}z_q + \eta_1 \\ x_2 &= \gamma_{21}z_1 + \gamma_{22}z_2 + \ldots + \gamma_{2q}z_q + \eta_2 \\ &\vdots \\ x_p &= \alpha_{p1}z_1 + \gamma_{p2}z_2 + \ldots + \gamma_{pq}z_q + \eta_p. \end{aligned} \tag{16.5}$$

Putting back the means μ_i of the x_i into the right-hand side of these equations, we recover a set of equations exactly of the form (16.1). Moreover, since the y_i are uncorrelated and have variances λ_i ($i =$ 1, . . . , p) from the theory of Chapter 2, the z_i are also uncorrelated and each has variance 1 ($i = 1, \ldots, q$). Hence the standardized components obey the same assumptions as do the factors in (16.1). For this reason, PCA and FA have been inextricably linked and much confused as techniques over the years, and one technique is often applied erroneously when the other is the more appropriate. It is therefore important to appreciate precisely what the aims of the two techniques are. The vital distinction between them is that principal components are the optimal entities for describing or explaining the *variances* in a multivariate system, while factors are appropriate when trying to explain the *covariances* in the system. Although it appears from the above discussion that the factor-analysis model (16.1) can be reproduced exactly from a principal component standpoint, one important difference between models (16.1) and (16.5) has not yet been highlighted. This is that the e_i of (16.1) are *not* the same as the η_i of (16.5). The e_i are assumed to be uncorrelated with the z_i and with each other. Now consider the η_i of (16.5). From their definition we can write

$$\begin{aligned} \eta_1 &= \alpha_{q+1,1}y_{q+1} + \ldots + \alpha_{p1}y_p \\ \eta_2 &= \alpha_{q+1,2}y_{q+1} + \ldots + \alpha_{p2}y_p \\ &\vdots \\ \eta_p &= \alpha_{q+1,p}y_{q+1} + \ldots + \alpha_{pp}y_p. \end{aligned}$$

Since the y_i are all mutually uncorrelated, then the η_i are indeed uncorrelated with the z_i. However, since the same y_j occur in different η_i, the η_i are *not* mutually uncorrelated. Thus the z_i as derived from PCA in (16.5) *do not explain all the correlation structure in* $\boldsymbol{x}$. Hence PCA is not the same as FA. However, the close connections noted above often mean that the two techniques yield similar results on a given data set. Indeed, the close similarity between (16.1) and (16.5) has led to various FA estimation techniques being based on PCA. We shall return to this point and other related points below.

16.2.2 Implications of the model

There are two main drawbacks with the FA model as exemplified by eqns (16.1)–(16.3), and both arise from the fact that the right-hand side of each equation is composed entirely of conceptual entities. The first drawback is indeterminacy in defining the factors. We see from (16.3) that $\boldsymbol{x} - \boldsymbol{\mu} = \boldsymbol{\Gamma z} + \boldsymbol{e}$. Consider any orthogonal matrix $\mathbf{H}$ (satisfying $\mathbf{H'H} = \mathbf{HH'} = \mathbf{I}$). Then if we write $\boldsymbol{\Gamma}^* = \boldsymbol{\Gamma}\mathbf{H}'$ and $\boldsymbol{z}^* = \mathbf{H}\boldsymbol{z}$, we have $\boldsymbol{\Gamma}^*\boldsymbol{z}^* = \boldsymbol{\Gamma}\mathbf{H'H}\boldsymbol{z} = \boldsymbol{\Gamma z}$. Moreover $\mathrm{E}(\boldsymbol{z}^*) = \mathbf{H}\mathrm{E}(\boldsymbol{z}) = \boldsymbol{0}$ $(= \mathrm{E}(\boldsymbol{z}))$ and $\mathrm{cov}(\boldsymbol{z}^*) = \mathbf{H}\,\mathrm{cov}(\boldsymbol{z})\mathbf{H}' = \mathbf{HIH}' = \mathbf{HH}' = \mathbf{I}(= \mathrm{cov}(\boldsymbol{z}))$. Thus the model $\boldsymbol{x} - \boldsymbol{\mu} = \boldsymbol{\Gamma}^*\boldsymbol{z}^* + \boldsymbol{e}$ is completely indistinguishable from the model $\boldsymbol{x} - \boldsymbol{\mu} = \boldsymbol{\Gamma z} + \boldsymbol{e}$. In other words, if the factors $\boldsymbol{z}$ with loadings $\boldsymbol{\Gamma}$ provide an explanation for the observed covariances among elements of $\boldsymbol{x}$, then so do the factors $\boldsymbol{z}^* = \mathbf{H}\boldsymbol{z}$ with the loadings $\boldsymbol{\Gamma}^* = \boldsymbol{\Gamma}\mathbf{H}'$ for *any* orthogonal matrix $\mathbf{H}$.

This multiplicity of 'satisfactory' solutions is a potential source of embarrassment, particularly in situations where replicability of results is important. To ensure that a unique solution is obtained, it is necessary to impose a set of constraints on the parameters of the model, and the constraints that are generally imposed by computer algorithms require that all the off-diagonal elements of $\boldsymbol{\Gamma}'\boldsymbol{\Psi}^{-1}\boldsymbol{\Gamma}$ be zero. With these constraints, all computer programs employing a particular estimation method should yield the same results on the same data set (apart from possibly a reversal of signs in columns of $\boldsymbol{\Gamma}$). However, since $\boldsymbol{\Gamma}^* = \boldsymbol{\Gamma}\mathbf{H}'$ and $\boldsymbol{z}^* = \mathbf{H}\boldsymbol{z}$ are equally acceptable solutions to the problem, the investigator is subsequently at liberty to select an orthogonal matrix $\mathbf{H}$, and transform from $\boldsymbol{\Gamma}$, $\boldsymbol{z}$ to $\boldsymbol{\Gamma}^*$, $\boldsymbol{z}^*$ if the latter provide a more interpretable set of quantities. This procedure is known as *factor rotation*; we shall return to it again below.

The second drawback with model (16.3) is the number of unknown quantities relative to the number of items of data. In general, statistical estimation is only viable if we have more items of information than there are unknown parameters. For most statistical models, a large number of unknown parameters does not provide an insuperable obstacle, as we simply ensure that we have a large enough sample to enable estimation to be effected. In the present case, however, the number of unknown parameters *increases* with the sample size, because each new sample member provides q new, unknown, factor scores $\boldsymbol{z}$. Hence it is impracticable to expect simultaneous estimation of all unknown quantities $\boldsymbol{\Gamma}$, $\boldsymbol{\Psi}$, and $\boldsymbol{z}_i$ $(i = 1, \ldots, n)$. The usual strategy adopted is to focus initially on $\boldsymbol{\Gamma}$ and $\boldsymbol{\Psi}$ (which do not change with sample size). Once satisfactory estimates have been obtained for these quantities, then the factor scores $\boldsymbol{z}_i$ can be estimated as a subsidiary exercise. Note, however, that the precision of these latter estimates may not be very good.

To focus on estimation of $\boldsymbol{\Gamma}$ and $\boldsymbol{\Psi}$ alone, we must remove the $\boldsymbol{z}_i$ from the model. This is done by considering the dispersion matrix $\boldsymbol{\Sigma}$ of $\boldsymbol{x}$.

Now, by definition,

$$\begin{aligned}\boldsymbol{\Sigma} &= \mathrm{E}\{(\boldsymbol{x}-\boldsymbol{\mu})(\boldsymbol{x}-\boldsymbol{\mu})'\}\\ &= \mathrm{E}\{(\boldsymbol{\Gamma z}+\boldsymbol{e})(\boldsymbol{\Gamma z}+\boldsymbol{e})'\} \quad \text{from (16.3)}\\ &= \mathrm{E}\{\boldsymbol{\Gamma zz}'\boldsymbol{\Gamma}' + \boldsymbol{\Gamma ze}' + \boldsymbol{ez}'\boldsymbol{\Gamma}' + \boldsymbol{ee}'\} \quad \text{on expanding the inner brackets}\\ &= \boldsymbol{\Gamma}\mathrm{E}(\boldsymbol{zz}')\boldsymbol{\Gamma}' + \boldsymbol{\Gamma}\mathrm{E}(\boldsymbol{ze}') + \mathrm{E}(\boldsymbol{ez}')\boldsymbol{\Gamma}' + \mathrm{E}(\boldsymbol{ee}').\end{aligned}$$

But since $\boldsymbol{z}$ and $\boldsymbol{e}$ are uncorrelated random vectors with zero means, $E(\boldsymbol{ze}') = \mathrm{E}(\boldsymbol{ez}') = \mathbf{0}$. Also $\mathrm{E}(\boldsymbol{zz}')$ and $\mathrm{E}(\boldsymbol{ee}')$ are the dispersion matrices, $\mathbf{I}$ and $\boldsymbol{\Psi}$, of $\boldsymbol{z}$ and $\boldsymbol{e}$ respectively. Hence we obtain the fundamental relationship of factor analysis:

$$\boldsymbol{\Sigma} = \boldsymbol{\Gamma\Gamma}' + \boldsymbol{\Psi}. \tag{16.6}$$

We see from this relationship that if σ_{ij} is the (i, j)th element of $\boldsymbol{\Sigma}$ then $\sigma_{ii} = \sum_{j=1}^{q} \gamma_{ij}^2 + \psi_i^2$ for $i = 1, \ldots, p$. The quantity $\sigma_{ii} - \psi_i^2$ $\left(\text{which is equal to } \sum_{j=1}^{q} \gamma_{ij}^2\right)$ is termed the *communality* of the ith variable x_i. It is that portion of the variance of x_i accounted for exclusively (i.e. 'explained') by the q common factors. Also, we see from (16.3) that the covariances between elements of $\boldsymbol{x}$ and elements of $\boldsymbol{z}$ can be written in the $(p \times q)$ matrix

$$\begin{aligned}\operatorname{cov}(\boldsymbol{x}, \boldsymbol{z}) &= \mathrm{E}\{(\boldsymbol{\Gamma z}+\boldsymbol{e})\boldsymbol{z}'\}\\ &= \boldsymbol{\Gamma} \quad \text{since } \mathrm{E}(\boldsymbol{zz}') = \mathbf{I} \text{ and } \mathrm{E}(\boldsymbol{ez}') = \mathbf{0}.\end{aligned}$$

Thus the loading γ_{ij} is in fact the covariance between the ith original variable x_i and the jth factor z_j. This is comparable with the definition of principal component loadings as given in Section 2.2.5.

Now suppose that the manifest variables are standardized before being subjected to analysis. Thus instead of $x_i - \mu_i$ in model (16.1) or (16.3) we work with $(x_i - \mu_i)/\sqrt{\sigma_{ii}}$ for $i = 1, \ldots, p$, and $\boldsymbol{\Sigma} = (\sigma_{ij})$. Then the dispersion matrix of these standardized variables is just the correlation matrix $\mathbf{P}$ of the original variables. But $\mathbf{P} = \boldsymbol{\Delta}^{-1}\boldsymbol{\Sigma}\boldsymbol{\Delta}^{-1}$ where $\boldsymbol{\Delta} = \operatorname{diag}(\sqrt{\sigma_{11}}, \sqrt{\sigma_{22}}, \ldots, \sqrt{\sigma_{pp}})$ (cf. eqn (14.9)) so that, on substituting from (16.6),

$$\begin{aligned}\mathbf{P} &= \boldsymbol{\Delta}^{-1}(\boldsymbol{\Gamma}\boldsymbol{\Gamma}' + \boldsymbol{\Psi})\boldsymbol{\Delta}^{-1} \\ &= (\boldsymbol{\Delta}^{-1}\boldsymbol{\Gamma})(\boldsymbol{\Delta}^{-1}\boldsymbol{\Gamma})' + \boldsymbol{\Delta}^{-1}\boldsymbol{\Psi}\boldsymbol{\Delta}^{-1} \\ &= \boldsymbol{\Omega}\boldsymbol{\Omega}' + \boldsymbol{\Phi} \quad \text{with } \boldsymbol{\Omega} = \boldsymbol{\Delta}^{-1}\boldsymbol{\Gamma} \text{ and } \boldsymbol{\Phi} \text{ diagonal.}\end{aligned}$$

This is of exactly the same form as (16.6), so that the basic FA model can be applied either to dispersion or to correlation matrices. Although we shall concentrate mainly on dispersion matrices, this point should always be borne in mind; we shall raise it explicitly where necessary.

Finally, the FA model has a built-in restriction on the number of factors that can be used to explain the observed associations among the variables. The central equation expressing the connection between observable data and unknown parameters is (16.6). The data to be used in the estimation will come from the sample estimate $\mathbf{S}$ of $\boldsymbol{\Sigma}$, and hence will contain $\frac{1}{2}p(p+1)$ separate items of information. There are $pq+p$ unknown parameters (pq elements of $\boldsymbol{\Gamma}$ and p elements of $\boldsymbol{\Psi}$). However, the requirement that the off-diagonal elements of the symmetric matrix $\boldsymbol{\Gamma}'\boldsymbol{\Psi}^{-1}\boldsymbol{\Gamma}$ should all be zero introduces $\frac{1}{2}q(q-1)$ constraints on the parameters, reducing the number of free parameters to $pq+p-\frac{1}{2}q(q-1)$. Hence for estimability of parameters we require $\frac{1}{2}p(p+1) \geqslant pq+p-\frac{1}{2}q(q-1)$, or $(p-q)^2 \geqslant p+q$ on simplifying the algebra. There is thus a limit on the number q of factors that can be included in any given model. The small table below shows the maximum number of factors that can be postulated to explain inter-correlations among a given number of variables.

Number of variables	3	4	5	6	7	8	9	10	15	20
Maximum number of factors	1	1	2	3	3	4	5	6	10	14

In practice, of course, we would aim to explain the associations by as few factors as possible, and for large p in particular we would hope to use considerably fewer factors than the critical number given above.

16.2.3 A data set

Having established the basic model and its ramifications, it seems appropriate at this stage to present a set of data on which the various aspects of the model can be subsequently illustrated. These data originate from Yule *et al.* (1969), and concern the scores obtained by children on the following ten sub-tests of the Wechsler Pre-School and Primary Scale

Table 16.1 Covariance matrix (on and below the diagonal) and correlation matrix (above the diagonal) for the WPPSI data. (See text for identification of the tests)

Test No	1	2	3	4	5	6	7	8	9	10
1	9.078	0.755	0.592	0.532	0.627	0.460	0.407	0.387	0.461	0.459
2	7.049	9.597	0.644	0.528	0.617	0.497	0.511	0.417	0.406	0.583
3	5.191	5.808	8.471	0.388	0.529	0.449	0.436	0.428	0.412	0.602
4	3.796	3.872	2.670	5.604	0.475	0.442	0.280	0.214	0.361	0.424
5	4.912	4.969	4.005	2.922	6.756	0.398	0.373	0.372	0.355	0.433
6	3.694	4.102	3.483	2.787	2.753	7.102	0.545	0.446	0.366	0.575
7	3.500	4.512	3.618	1.890	2.767	4.142	8.138	0.542	0.308	0.590
8	3.300	3.655	3.528	1.436	2.740	3.367	4.376	8.026	0.375	0.654
9	4.245	3.844	3.668	2.613	2.822	2.984	2.691	3.253	9.360	0.502
10	3.745	4.896	4.745	2.719	3.048	4.153	4.558	5.021	4.160	7.340

of Intelligence (WPPSI):

1. Information
2. Vocabulary
3. Arithmetic
4. Similarities
5. Comprehension
6. Animal house
7. Picture completion
8. Mazes
9. Geometric design
10. Block designs.

The first five tests measure verbal and numerical skills, while the remaining five focus on visual perception, including 'everyday' pictures (tests 6 and 7) and geometric patterns (tests 8–10). Each child's score on each test was on the scale 0–20. The original sample consisted of seventy-six boys and seventy-four girls in the age range 4–6½ years; for the present illustration one half of each group has been chosen at random to yield a sample of seventy-five children. Boys will not be differentiated from girls in the analyses that follow.

The sample covariance and correlation matrices are given in Table 16.1, and show that considerable associations exist among the sub-tests. All correlations are positive and, with very few exceptions, lie in the range 0.4–0.6. A factor-analysis model seems eminently reasonable for this set of data, as the inter-test associations can plausibly be assumed to arise from the dependence of each test on a few unobservable factors such as 'verbal ability', 'geometric perception', 'general intelligence', etc.. We shall examine these aspects in more detail below.

16.2.4 Estimation of factor loadings and specific variances

Let us, for the present, suppose that we have fixed the value of q, the number of common factors to be fitted in the model. Then the fundamental relationship (16.6) links the $(p \times p)$ dispersion matrix $\boldsymbol{\Sigma}$ of the manifest variables to the $(p \times q)$ matrix $\boldsymbol{\Gamma}$ of factor loadings, and to the $(p \times p)$ diagonal matrix $\boldsymbol{\Psi}$ of specific variances. In practice we do not know $\boldsymbol{\Sigma}$, but can estimate it by the sample covariance matrix $\mathbf{S}$. Hence we seek corresponding estimates $\hat{\boldsymbol{\Gamma}}$ and $\hat{\boldsymbol{\Psi}}$ such that $\hat{\boldsymbol{\Gamma}}\hat{\boldsymbol{\Gamma}}' + \hat{\boldsymbol{\Psi}}$ is 'as close to $\mathbf{S}$ as possible'. ($\mathbf{S}$ can be replaced by the sample correlation matrix $\mathbf{R}$ if standardized variables are more appropriate.)

Estimation in factor analysis has something of a chequered history. A great number of different methods have been proposed over the years for obtaining estimates of $\boldsymbol{\Gamma}$ and $\boldsymbol{\Psi}$. However, many of these methods were founded on poor theoretical bases, so could be very unreliable in execution as well as of questionable accuracy in performance. Consequently the whole subject of factor analysis achieved a certain notoriety among statisticians, and it was rarely employed by them. That many social scientists, most notably psychologists, continued enthusiastically to use the unreliable methods did not help matters, as their results were thus treated with disdain by statisticians. Some improvement was achieved in these matters with the development of maximum likelihood methodology by Lawley and Bartlett in the 1930s and early 1940s, but even such a statistically 'respectable' approach proved embarrassingly prone to breakdown through lack of convergence. It was not until the computational breakthrough in the 1960s, due principally to Jöreskog, that sound and reliable algorithms for maximum likelihood estimation were developed. These algorithms form the basis of maximum likelihood solutions as implemented in modern computer programs. A legacy of this historical development is a continuing mistrust of factor analysis among some statisticians. Provided it is used in appropriate circumstances, however, it can be a very useful explanatory technique.

We shall content ourselves here merely with an outline of the general estimation principles. However, before turning to maximum likelihood, it is perhaps worth devoting just a brief paragraph to description of one of the older heuristic methods, as these methods are still very frequently used and are represented in all common general-purpose statistical computing 'packages'. The most popular of these methods is probably the iterated form of principal component analysis known as *principal factoring*. This method leans heavily on the close (if superficial) resemblance between FA and PCA elaborated in Section 16.2.1 above, and consists of the following simple iterative cycle:

1. Guess a set of specific variances, $\hat{\boldsymbol{\Psi}}$. (Most package implementations supply this guess automatically, usually by conducting a multiple regression of each manifest variable on all other manifest variables. The squared multiple correlation coefficient r_i^2 for the ith such regression is the proportion of variance in x_i explained by its connection with the other

variables (see Section 14.5), so $s_{ii}r_i^2$ can be used as an initial estimate of the ith communality. Hence an initial estimate of the ith specific variance is $s_{ii}(1 - r_i^2)$.)

2. Conduct a principal component analysis of $\mathbf{S} - \hat{\boldsymbol{\Psi}}$, and write the loadings of the first q components as columns of the matrix $\hat{\boldsymbol{\Gamma}}$.

3. Recalculate $\hat{\boldsymbol{\Psi}}$ as the diagonal of $\mathbf{S} - \hat{\boldsymbol{\Gamma}}\hat{\boldsymbol{\Gamma}}'$.

4. Return to step 2 with this new estimate of $\boldsymbol{\Psi}$.

The cycle through steps 2–4 is continued until two successive pairs $\hat{\boldsymbol{\Gamma}}$ and $\hat{\boldsymbol{\Psi}}$ are identical to within a pre-set level of accuracy. However, convergence may be very slow, and sometimes the method may altogether fail to converge. Moreover, since it leans so heavily on the connection between PCA and FA, it not only suffers from all the drawbacks given in Section 16.2.1, but it also shares the PCA property that different results will be obtained from a single set of data depending on whether the covariance or correlation matrix is used for the analysis. Indeed, this method of FA is probably at the root of the considerable confusion between FA and PCA which is exhibited in many disciplines. The one (and probably only) point in favour of the technique is that it does not require any distributional assumptions to be made about the data. Consequently it can provide a viable alternative to maximum likelihood if the multivariate normality assumption required by the latter is grossly violated.

It is to the method of maximum likelihood that we now turn. This method not only provides a firm theoretical basis for the estimation process but is also one of the few methods that permits an adequately objective determination of the number q of factors required to explain the data. We shall delay a discussion of this latter aspect to the next section, however, and for the present continue with the supposition that q is fixed. To employ maximum likelihood we require distributional assumptions, and as usual with continuous data we invoke the assumption of multivariate normality. The manifest variables $\boldsymbol{x}$ are therefore assumed to follow a p-variate normal distribution with a mean of $\boldsymbol{\mu}$, and a dispersion matrix $\boldsymbol{\Sigma}$, which in turn implies that the vectors $\boldsymbol{z}$ and $\boldsymbol{e}$ of model (16.3) must be assumed to come from q- and p-variate normal distributions respectively (with mean vectors and dispersion matrices as given previously). Now if the sample consists of n independent observations $\boldsymbol{x}_1, \boldsymbol{x}_2, \ldots, \boldsymbol{x}_n$ of this vector $\boldsymbol{x}$, the likelihood of the sample is obtained by multiplying together the probability density (7.9) for each observation. The exact algebraic expression is not of importance here (although its logarithm is given in eqn (8.10)); we only need to note that the likelihood is a function of the data $\boldsymbol{x}_i$ $(i = 1, \ldots, n)$ and the parameters $\boldsymbol{\mu}$ and $\boldsymbol{\Sigma}$, so may be written $L(\boldsymbol{x}; \boldsymbol{\mu}, \boldsymbol{\Sigma})$. Furthermore the factor model parameters of interest, $\boldsymbol{\Gamma}$ and $\boldsymbol{\Psi}$, may be substituted for $\boldsymbol{\Sigma}$ from (16.6), so the likelihood should in fact be written $L(\boldsymbol{x}; \boldsymbol{\mu}, \boldsymbol{\Gamma}, \boldsymbol{\Psi})$.

The maximum likelihood estimators $\hat{\boldsymbol{\Gamma}}$ and $\hat{\boldsymbol{\Psi}}$ are the values of $\boldsymbol{\Gamma}$ and $\boldsymbol{\Psi}$ that maximize this expression. A full maximization would also include the parameter $\boldsymbol{\mu}$, but $\boldsymbol{\mu}$ is irrelevant to these factor loadings and specific variances (i.e. it is a *nuisance parameter*). The general approach to nuisance parameters is to try and eliminate them from the likelihood, and hence to maximize a modified likelihood. This can be done in various ways; fortunately, in the present instance, they all lead to the same estimates of $\boldsymbol{\Gamma}$ and $\boldsymbol{\Psi}$. One approach is to replace $\boldsymbol{\mu}$ in the likelihood by its maximum likelihood estimator $\hat{\boldsymbol{\mu}} = \bar{\boldsymbol{x}}$ (thereby maximizing $\boldsymbol{\mu}$ out of the expression before estimating $\boldsymbol{\Gamma}$ and $\boldsymbol{\Psi}$). The resulting expression is known as the *profile likelihood* for $\boldsymbol{\Sigma} = \boldsymbol{\Gamma\Gamma}' + \boldsymbol{\Psi}$ and estimators derived from this likelihood have the same properties as those derived from the full likelihood. The second approach is to factorize the likelihood into the two components $L(\mathbf{S}; \bar{\boldsymbol{x}}, \boldsymbol{\Sigma})L(\bar{\boldsymbol{x}}; \boldsymbol{\mu}, \boldsymbol{\Sigma})$ where the sample mean $\bar{\boldsymbol{x}}$ and covariance matrix $\mathbf{S}$ are the joint sufficient statistics for $\boldsymbol{\mu}$ and $\boldsymbol{\Sigma}$ (Section 8.1). Since the parameters $\boldsymbol{\Gamma}$ and $\boldsymbol{\Psi}$ stem only from the dispersion matrix, we can consider the 'data' to be represented exclusively by $\mathbf{S}$ and hence obtain estimates by maximizing the first factor $L(\mathbf{S}; \bar{\boldsymbol{x}}, \boldsymbol{\Sigma})$ of the likelihood, i.e. the *conditional likelihood* (conditional on the given sample mean $\bar{\boldsymbol{x}}$). But for normal samples $\bar{\boldsymbol{x}}$ and $\mathbf{S}$ are independent (Section 8.1), so $L(\mathbf{S}; \bar{\boldsymbol{x}}, \boldsymbol{\Sigma})$ is also the *marginal likelihood.* Whichever likelihood is chosen, we find that the maximum likelihood estimators $\hat{\boldsymbol{\Gamma}}$, $\hat{\boldsymbol{\Psi}}$ are obtained as the values maximizing

$$\log_e L = c_1 - c_2\{\log_e |\boldsymbol{\Gamma\Gamma}' + \boldsymbol{\Psi}| + \text{trace}(\boldsymbol{\Gamma\Gamma}' + \boldsymbol{\Psi})^{-1}\mathbf{S}\} \qquad (16.7)$$

where c_1 and c_2 are constants.

Differentiating (16.7) in turn with respect to $\boldsymbol{\Gamma}$ and $\boldsymbol{\Psi}$, setting the resulting expressions to zero, and manipulating the algebra, leads to a pair of simultaneous equations. Taking these equations together with the constraint that $\boldsymbol{\Gamma}'\boldsymbol{\Psi}^{-1}\boldsymbol{\Gamma}$ be diagonal (Section 16.2.2) leads to an iterative scheme for estimation of $\boldsymbol{\Gamma}$ and $\boldsymbol{\Psi}$. This was the original method for the maximum likelihood procedure. The algebraic details are somewhat tortuous and may be found, for example, in Morrison (1976). They are largely irrelevant now, as this old iterative scheme is subject to severe difficulties and often fails completely. In fact, users tended to experience such problems with this scheme that it was rarely employed in practice, principal factoring being the much more common choice. The computational breakthrough came with the work of Jöreskog (1967), who approached the maximization of (16.7) in a two-stage process. For fixed $\boldsymbol{\Psi} > 0$ (i.e. such that diagonal elements of $\boldsymbol{\Psi}$ are all greater than zero) he showed that the likelihood equations are satisfied by

$$\hat{\boldsymbol{\Gamma}} = \boldsymbol{\Psi}^{\frac{1}{2}}\boldsymbol{\Omega}(\boldsymbol{\Theta} - \mathbf{I})^{\frac{1}{2}} \tag{16.8}$$

where $\boldsymbol{\Theta}$ is a $(q \times q)$ diagonal matrix containing the q largest eigenvalues of $\boldsymbol{\Psi}^{-\frac{1}{2}}\mathbf{S}\boldsymbol{\Psi}^{-\frac{1}{2}}$, and the columns of the $(p \times q)$ matrix $\boldsymbol{\Omega}$ contain the corresponding eigenvectors. For given $\boldsymbol{\Gamma}$, the likelihood can be maximized directly with respect to $\boldsymbol{\Psi}$ by using a numerical optimization procedure such as Newton–Raphson. These two stages provide the following iterative scheme:

(1) Guess $\hat{\boldsymbol{\Psi}}$ (as in principal factoring).
(2) Set $\boldsymbol{\Psi} = \hat{\boldsymbol{\Psi}}$ and use (16.8) to obtain $\hat{\boldsymbol{\Gamma}}$.
(3) Put $\hat{\boldsymbol{\Gamma}}$ into (16.7) and maximize $\log_e L$ numerically with respect to $\boldsymbol{\Psi}$ to obtain $\hat{\boldsymbol{\Psi}}$.
(4) Return to step (2) with this new estimate of $\boldsymbol{\Psi}$.

The cycle (2)–(4) is continued until two successive pairs of estimates $\hat{\boldsymbol{\Gamma}}$ and $\hat{\boldsymbol{\Psi}}$ are identical to within a pre-set level of accuracy.

Convergence of this scheme is far superior to that of the old one, and the only remaining problem is that one or more elements of $\hat{\boldsymbol{\Psi}}$ may become negative. In practice, therefore, the numerical maximization in step (3) is constrained to take place within the region of the parameter space in which $\psi_j^2 \geqslant \varepsilon$ for all j, where ε is a suitable small positive number. If any element ψ_j^2 attains the value ε in the course of the iterations, then it retains this value until the process terminates. The interpretation of this occurrence is that the corresponding manifest variable is entirely determined by the common factor structure, and it possesses no error variation. As this is conceptually an unlikely event, such situations are termed *improper solutions*. However, such solutions occur with disconcerting frequency in practice, and an adequate explanation has yet to be found for the phenomenon.

Apart from the interpretational aspect of improper solutions, the above iterative scheme is perfectly reliable in practice. Moreover, the maximum likelihood method is the only one that is independent of the scales of measurement. A simple proof of this property is given by Seber (1984, p. 218). This means that with maximum likelihood, the *same* result will be obtained whether the covariance or the correlation matrix is analysed. Similarly, it is irrelevant whether the maximum likelihood or unbiased estimator of $\boldsymbol{\Sigma}$ is used in the iterative cycle. For all these reasons, the maximum-likelihood method is the one that is to be recommended. For further details, both theoretical and computational, see Jöreskog (1967, 1977), Lawley and Maxwell (1971), and Clarke (1970). The modern algorithms are now implemented in most standard computer packages.

Table 16.2 Factor loading and communality estimates for a two-factor model fitted to the WPPSI data by principal factoring

Manifest variable	Factor loadings		Communalities	
	Factor 1	Factor 2	Initial	Final
1 (Information)	0.776	−0.333	0.659	0.713
2 (Vocabulary)	0.823	−0.224	0.689	0.728
3 (Arithmetic)	0.731	−0.055	0.527	0.537
4 (Similarities)	0.589	−0.248	0.403	0.419
5 (Comprehension)	0.678	−0.239	0.478	0.517
6 (Animal house)	0.668	0.148	0.451	0.468
7 (Picture completion)	0.647	0.293	0.469	0.505
8 (Mazes)	0.627	0.379	0.492	0.535
9 (Geometric design)	0.562	0.022	0.331	0.316
10 (Block designs)	0.789	0.321	0.664	0.726

To illustrate the above techniques, consider fitting a two-factor model (i.e. $q = 2$) to the WPPSI data described in Section 16.2.3. This was done on the computer package SAS (see *SAS User's Guide,* obtainable from SAS Institute, Box 8000, Cary, North Carolina 27511, USA), using both principal factoring and maximum likelihood for comparison. The sample correlation matrix formed the basis of each analysis. SAS uses the squared multiple correlations r_i^2 as the initial communality estimates for both methods. Results are given in Table 16.2 for principal factoring, and in Table 16.3 for maximum likelihood.

Table 16.3 Factor loading and communality estimates for a two-factor model fitted to the WPPSI data by maximum likelihood

Manifest variable	Factor loadings		Communalities	
	Factor 1	Factor 2	Initial	Final
1 (Information)	0.789	−0.403	0.659	0.786
2 (Vocabulary)	0.834	−0.234	0.689	0.751
3 (Arithmetic)	0.740	−0.034	0.527	0.548
4 (Similarities)	0.586	−0.185	0.403	0.378
5 (Comprehension)	0.676	−0.248	0.478	0.518
6 (Animal house)	0.654	0.140	0.451	0.447
7 (Picture completion)	0.641	0.234	0.469	0.466
8 (Mazes)	0.629	0.351	0.492	0.519
9 (Geometric design)	0.564	0.054	0.331	0.321
10 (Block designs)	0.808	0.414	0.664	0.824

Comparison of the two tables shows that, while the actual numerical values of the loadings may differ between the two methods, the differences are not substantial in the present case, and both methods give essentially the same qualitative result. (Of course, such agreement cannot be predicted in general, and for some data sets the two methods may give very different sets of results.) Interpretation of the factors may be made in the same way that interpretation of principal components (Section 2.2.5) or canonical variates (Section 11.1) was attempted previously, by picking out the variables with 'high' loadings on each factor and attempting a physical explanation of the factor in terms of these variables. Thus factor 1 in both tables contains high loadings for all variables and moreover these loadings are very roughly all of comparable size and sign; hence factor 1 might be termed 'general intelligence'. Factor 2, on the other hand, consists of five positive and five negative loadings, so it represents a contrast. Such contrasts are more difficult to provide simple labels for, and hence are generally disliked by factor analysts. Factor rotation will often get over this problem (see Section 16.2.6 below), but an interpretation can nevertheless be attempted even if rotation is not employed. In the present case the loadings for 'arithmetic', 'geometric design' and (probably) 'animal house' are close enough to zero to be ignorable, so the contrast is between verbal tests on the one hand and pictorial ones on the other (with the emphasis in the latter being on rectilinear shapes). The second factor thus picks out differences between sharply defined perceptions and those requiring diffuse or complicated descriptions. It will be seen later that factor rotation considerably simplifies the interpretation.

Broadly speaking, the communalities are also similar between the two methods. Those for 'information' and 'block designs' are noticeably greater for maximum likelihood than for principal factoring, while those for 'similarities' and 'picture completion' show the reverse relationship. However, the overall level of the communalities is the same in the two methods (5.73 against 5.56), and none of the individual discrepancies is very striking. A final point to note is that, in general, the communalities at the end of the iterative cycles are not very different from their initial estimates (with just one or two exceptions such as 'information' and 'block designs' for the maximum likelihood method). The final estimates are almost all larger than the initial ones for principal factoring, but some are smaller for maximum likelihood. The latter probably compensate for the two large communality improvements noted above. It is evident that the variances of 'information', 'vocabulary', and 'block designs' are well modelled by the two common factors, while the remaining variances are only moderately well modelled by them.

The main objective of the analysis is to explain the associations

Table 16.4 Residual correlation matrix computed from the two-factor maximum-likelihood fit to the WPPSI data

1	0.214									
2	0.003	0.249								
3	−0.005	0.019	0.452							
4	−0.005	−0.004	−0.052	0.622						
5	−0.006	−0.005	0.020	0.033	0.482					
6	0.000	−0.016	−0.030	0.085	−0.009	0.553				
7	−0.004	0.031	−0.030	−0.052	−0.002	0.093	0.534			
8	0.032	−0.025	−0.026	−0.089	0.034	−0.015	0.057	0.481		
9	0.038	−0.052	−0.004	0.040	−0.013	−0.010	−0.066	0.001	0.679	
10	−0.012	0.006	0.018	0.027	−0.011	−0.011	−0.025	0.000	0.024	0.176
	1	2	3	4	5	6	7	8	9	10

between the manifest variables, and one may see how successful the model is by computing the 'fitted' covariance or correlation matrix $\hat{\boldsymbol{\Gamma}}\hat{\boldsymbol{\Gamma}}'$ and comparing it with the input matrix (**S** or **R**). In fact the residual matrix ($\mathbf{S} - \hat{\boldsymbol{\Gamma}}\hat{\boldsymbol{\Gamma}}'$ or $\mathbf{R} - \hat{\boldsymbol{\Gamma}}\hat{\boldsymbol{\Gamma}}'$) will highlight the adequacy of the fitted model: if the model is a good one then all off-diagonal residuals will be close to zero, but if it is poor then some of the residuals will be noticeably non-zero. Any appreciable residuals pin-point areas of poor explanation by the model. The residual matrix from the two-factor maximum-likelihood model is given in Table 16.4.

It appears that the two-factor model provides a good fit. There are just three residual correlations of order 0.09 in magnitude, one of order 0.07, and one of order 0.06. The remainder are of order 0.05 at most, and generally considerably less than this. Hence we would feel reasonably satisfied with the fitted model. However, it would be advantageous to have a more objective and decisive test of goodness-of-fit of the model, as well as a systematic method for choosing q. We consider this aspect next.

16.2.5 Test of goodness-of-fit, and choice of q

With principal factoring, and other such *ad hoc* methods, the choice of q (i.e. number of factors to include in the model) must be based on arbitrary criteria such as the eigenvalues of the reduced covariance/correlation matrix, and the goodness-of-fit of the resulting model can only be judged by descriptive summaries such as the residual covariances/correlations discussed above. In the discussion below we use the sample covariance matrix **S**, but the arguments are unchanged if **S** is replaced by the correlation matrix **R**.

From (16.6), $\boldsymbol{\Sigma} - \boldsymbol{\Psi} = \boldsymbol{\Gamma}\boldsymbol{\Gamma}'$, which is a positive semi-definite matrix of rank q. Hence there are, at most, q positive eigenvalues of $\boldsymbol{\Sigma} - \boldsymbol{\Psi}$ (and at least $p - q$ zero eigenvalues). Because of sampling variation and estimation effects, $\mathbf{S} - \hat{\boldsymbol{\Psi}}$ need not be positive semi-definite, and we can expect it to have some negative eigenvalues. Most computer packages base the choice of q on the positive eigenvalues of $\mathbf{S} - \hat{\boldsymbol{\Psi}}_0$ (where $\hat{\boldsymbol{\Psi}}_0$ is the initial estimate of $\boldsymbol{\Psi}$ in the iterative cycle described in the previous section). Two popular choices are q as the number of positive eigenvalues, and q as the smallest value for which the sum of the q largest eigenvalues equals or exceeds the sum of all the eigenvalues (i.e. trace$(\mathbf{S} - \hat{\boldsymbol{\Psi}}_0)$, the total initial estimate of communality). The former method is likely to overestimate the number of factors that are necessary, so a multi-stage procedure can be adopted in which q is first chosen as the number of positive eigenvalues of $(\mathbf{S} - \hat{\boldsymbol{\Psi}}_0)$, the model is fitted to obtain final estimates $\hat{\boldsymbol{\Psi}}_1$ of $\boldsymbol{\Psi}$, and eigenvalues are extracted for $\mathbf{S} - \hat{\boldsymbol{\Psi}}_1$. Inspection of these eigenvalues then suggests a 'natural' value of q (by such devices as the scree plot described in Section 2.2.5 and illustrated in Fig. 2.13). The model is refitted using this modified value of q, and goodness-of-fit is assessed by inspecting the residual correlations or covariances. If some of these residuals are large then q can be increased and the model refitted. The process is continued until all the residuals are acceptably low.

The above is a highly subjective and arbitrary procedure. A much more systematic and theoretically sound procedure is available if multivariate normality is assumed and the maximum likelihood method is used, because a generalized likelihood ratio test of goodness-of-fit can be developed. This test is actually a test of $H_0 : \boldsymbol{\Sigma} = \boldsymbol{\Gamma}\boldsymbol{\Gamma}' + \boldsymbol{\Psi}$, where $\boldsymbol{\Gamma}$ is a $(p \times q)$ matrix of full rank and $\boldsymbol{\Psi}$ is diagonal, versus $H_1 : \boldsymbol{\Sigma}$ is unstructured. Under H_0, the maximum likelihood estimate of $\boldsymbol{\Sigma}$ is $\hat{\boldsymbol{\Sigma}} = \hat{\boldsymbol{\Gamma}}\hat{\boldsymbol{\Gamma}}' + \hat{\boldsymbol{\Psi}}$ (derived from (16.7)), while under H_1 the maximum likelihood estimate of $\boldsymbol{\Sigma}$ is just the usual sample variance–covariance matrix. Hence, evaluating the likelihood at each of these estimates in turn, we obtain the generalized likelihood ratio test statistic λ as

$$-2\log_e \lambda = n[\log_e |\hat{\boldsymbol{\Gamma}}\hat{\boldsymbol{\Gamma}}' + \hat{\boldsymbol{\Psi}}| + \text{trace}\{\mathbf{S}(\hat{\boldsymbol{\Gamma}}\hat{\boldsymbol{\Gamma}}' + \hat{\boldsymbol{\Psi}})^{-1}\} - \log_e |\mathbf{S}| - p],$$

and it can be shown that this reduces to

$$-2\log_e \lambda = n\{\log_e |\hat{\boldsymbol{\Gamma}}\hat{\boldsymbol{\Gamma}}' + \hat{\boldsymbol{\Psi}}| - \log_e |\mathbf{S}|\}. \tag{16.9}$$

Using the result of Bartlett (1951a),

$$W = \left\{n - \frac{2p + 11}{6} - \frac{2q}{3}\right\}\{\log_e |\hat{\boldsymbol{\Gamma}}\hat{\boldsymbol{\Gamma}}' + \hat{\boldsymbol{\Psi}}| - \log_e |\mathbf{S}|\} \tag{16.10}$$

approximately follows a χ^2 distribution on $\frac{1}{2}\{(p - q)^2 - (p + q)\}$ degrees of freedom when H_0 is true, and this statistic therefore provides a suitable

Table 16.5 Factor loadings and communality estimates for a one-factor model fitted to the WPPSI data by maximum likelihood

Manifest variable	Factor loadings	Communality estimates
1 (Information)	0.788	0.621
2 (Vocabulary)	0.841	0.707
3 (Arithmetic)	0.746	0.556
4 (Similaritics)	0.597	0.356
5 (Comprehension)	0.694	0.482
6 (Animal house)	0.652	0.425
7 (Picture completion)	0.630	0.397
8 (Mazes)	0.596	0.355
9 (Geometric design)	0.556	0.309
10 (Block designs)	0.750	0.563

test of H_0 against H_1. The usual procedure is to start with $q = 1$ and to fit models $q = 1$, $q = 2$, $q = 3$, etc. sequentially, testing the goodness-of-fit at each stage and stopping at the first value of q for which the calculated value of W is not significant (i.e. is less than the chosen percentage point of the appropriate χ^2 distribution).

Returning to the example of the WPPSI data, consider the above sequential maximum likelihood fitting procedure. For $q = 1$, the factor loadings and communalities given in Table 16.5 are obtained. (For the one-factor case, the communalities are just the squares of the corresponding loadings. Note also that the loadings for the one-factor model are *not* equal to the loadings of the first factor in the two-factor model (column one of Table 16.3), although in this case the two sets of loadings do not differ by much.) The value of W for the one-factor model is 58.83, and this value must be referred to the chi-squared distribution on $\frac{1}{2}\{(p-q)^2-(p+q)\} = \frac{1}{2}\{(10-1)^2-(10+1)\} = 35$ degrees of freedom. The probability of exceeding the value 58.83 for a χ^2_{35} distribution is 0.007, which is a significant result at the 1 per cent level. We therefore reject the null hypothesis, and conclude that the one-factor model does not provide an acceptable fit to the correlations among the manifest variables. Next we fit a two-factor model. Parameter estimates have already been given in Table 16.3. The value of W is 16.51, and the number of degrees of freedom is $\frac{1}{2}\{(10-2)^2-(10+2)\} = 26$. The probability of exceeding the value 16.51 for a χ^2_{26} distribution is 0.92, so we conclude that the two-factor model provides a highly acceptable fit to the correlations among the manifest variables.

Looking at the principal-factoring approach, the eigenvalues ϕ_i of the reduced correlation matrix $\mathbf{S} - \hat{\boldsymbol{\Psi}}_0$ are 4.82, 0.64, 0.14, 0.12, 0.02, −0.03,

-0.06, -0.12, -0.16, -0.18. Using the number of positive eigenvalues as the criterion for choice of q suggests fitting a five-factor model, while using the $\sum_{i=1}^{q} \phi_i \geqslant \sum_{i=1}^{p} \phi_i$ criterion suggests fitting a two-factor model. Having fitted the five-factor model, the positive eigenvalues of the recomputed matrix $\mathbf{S} - \hat{\boldsymbol{\Psi}}_1$ are very little changed from the original values, a scree diagram of which would suggest a two-factor model to be appropriate. Hence we would iterate towards agreement with the maximum-likelihood recommendation, but would not have the same objective support for our conclusions as we have with the latter method.

16.2.6 Factor rotation

We saw in Section 16.2.2 that any orthogonal transformation $\boldsymbol{\Gamma}^* = \boldsymbol{\Gamma}\mathbf{H}'$ of the factor loadings provides an equally valid formulation of the basic relationship (16.6). To ensure uniqueness, the estimated factor loadings are constrained to satisfy the requirement that $\hat{\boldsymbol{\Gamma}}'\hat{\boldsymbol{\Psi}}^{-1}\hat{\boldsymbol{\Gamma}}$ be diagonal. However, once this particular solution has been extracted, the investigator is at liberty to transform orthogonally the derived loadings matrix in any way that seems convenient. In particular, an orthogonal transformation may be sought so that the resultant factors are more readily interpretable. But if the p rows of $\boldsymbol{\Gamma}$ are plotted as points in a q-dimensional space (the *factor space*), then the orthogonal transformation $\boldsymbol{\Gamma}^* = \boldsymbol{\Gamma}\mathbf{H}'$ defines a rigid rotation of the axes in this space (Section 1.3). Hence a factor analysis is rarely complete without a search for a suitable factor rotation, in order to obtain the pattern of loadings $\boldsymbol{\Gamma}^*$ that is most easily interpreted or identified with the subject matter of the responses.

Consider the two-factor maximum likelihood solution of the WPPSI data. The raw factor loadings obtained directly from the iterative computational scheme are given in Table 16.3. Using the respective loadings on each factor as the coordinates of a point referred to orthogonal axes labelled γ_1 and γ_2, and hence plotting the ten responses as ten points in two-dimensional space, we obtain the configuration given in Fig. 16.1. The numbers on the plot correspond to the numbers of the variables as given in Section 16.2.3 and Table 16.3.

It was pointed out in the interpretation of the two factors in Section 16.2.4 that a bipolar factor such as factor 2, which contains some positive and some negative loadings, is generally disliked by factor analysts because it is more difficult to label succinctly and pertinently. It is shown up on the factor loadings plot by the presence of points on both sides of the dividing axis γ_1. We see from the plot that all negative loadings can be eliminated by a rotation of the factor axes through an angle of about $\theta = -27°$, the new axes γ_1^*, γ_2^* being shown by the dotted lines.

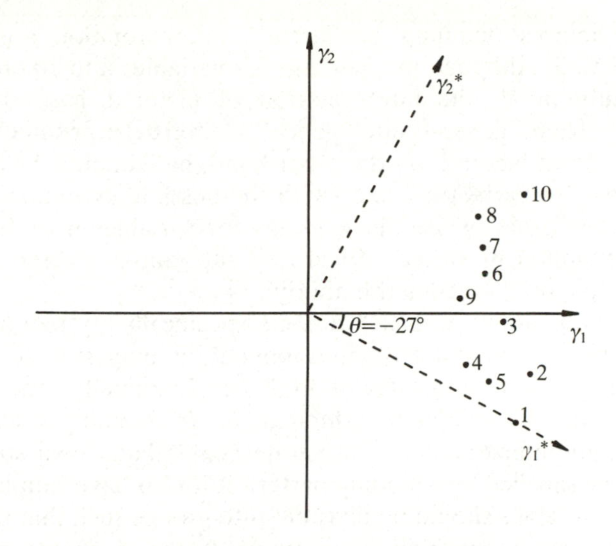

Fig. 16.1 Plot of factor loadings taken from Table 16.3, showing the minimum rotation of axes necessary to remove negative loadings

(Recollect from Section 1.3 that an anticlockwise rotation is positive while a clockwise one is negative.) The loadings of the rotated factors can be determined approximately from the figure, or directly by applying the transformation equations $\boldsymbol{\Gamma}^* = \boldsymbol{\Gamma}\mathbf{H}'$, where $\boldsymbol{\Gamma}$ is the (10×2) matrix of loadings in Table 16.3, and $\mathbf{H}$ is the appropriate orthogonal matrix for a rotation of axes through an angle $\theta = -27°$. From Section 1.3, in particular the discussion following eqn (1.5), we obtain

$$\mathbf{H} = \begin{pmatrix} \cos 27° & -\sin 27° \\ \sin 27° & \cos 27° \end{pmatrix} = \begin{pmatrix} 0.8910 & -0.4540 \\ 0.4540 & 0.8910 \end{pmatrix} \tag{16.11}$$

so that

$$\boldsymbol{\Gamma}^* = \begin{pmatrix} 0.789 & -0.403 \\ 0.834 & -0.234 \\ 0.740 & -0.034 \\ 0.586 & -0.185 \\ 0.676 & -0.248 \\ 0.654 & 0.140 \\ 0.641 & 0.234 \\ 0.629 & 0.351 \\ 0.564 & 0.054 \\ 0.808 & 0.414 \end{pmatrix} \begin{pmatrix} 0.891 & 0.454 \\ -0.454 & 0.891 \end{pmatrix} = \begin{pmatrix} 0.886 & -0.001 \\ 0.849 & 0.170 \\ 0.675 & 0.306 \\ 0.606 & 0.101 \\ 0.715 & 0.086 \\ 0.519 & 0.422 \\ 0.465 & 0.500 \\ 0.401 & 0.598 \\ 0.478 & 0.304 \\ 0.532 & 0.736 \end{pmatrix}.$$

Thus the heaviest loadings on factor 1 after rotation are those for variables 1 to 5. Although the loadings for variables 6 to 10 on this factor are still substantial, the interpretation of factor 1 has now changed perceptibly from 'general intelligence' to 'verbal + arithmetic ability'. Interpretation of factor 2, on the other hand, has simplified considerably. The loadings for variables 1 to 5 (with the possible exception of variable 3) are all negligible, while the loadings for variables 6 to 10 (with the possible exception of variable 9) are all substantial. Factor 2 therefore represents 'pictorial + geometric ability'.

The rotation above was determined specifically to remove negative loadings. While this is an obvious benefit, it may still not guarantee interpretability of resulting factor loadings. Essentially, one would like the rotated factor loadings to exhibit a *simple structure* which will avail itself of ready interpretation. Thurstone (1947) laid down some precise criteria to be satisfied by a loading pattern if it is to have simple structure: ideally the variables should be divisible into groups such that the loadings within each group are high on a single factor, perhaps moderate to low on a few factors, and negligible on the remaining factors. One could seek a rotation of axes by trial and error, until these criteria were satisfied. This is a very time-consuming process which is cumbersome when more than two factors are present (but see Cureton and D'Agostino (1983) for some details). Also, it is rare to find factors that meet Thurstone's rigid conditions for simple structure when extracted from real data. Consequently, various analytical procedures have been devised for obtaining an orthogonal matrix **H** which optimizes some less restrictive and single criterion of 'simple structure'.

The basic property that should be satisfied if a factor has a simple structure (for interpretation) is that some loadings should be 'reasonably large' while the remainder are 'reasonably small'. Neuhaus and Wrigley (1954) quantified this property by defining an index of 'simplicity' as the variance of the squares of all pq loadings. If $\boldsymbol{\Gamma} = (\gamma_{ij})$, then this index is

$$Q = \sum_{i=1}^{p} \sum_{j=1}^{q} \gamma_{ij}^4 - \frac{1}{pq} \left(\sum_{i=1}^{p} \sum_{j=1}^{q} \gamma_{ij}^2 \right)^2. \tag{16.12}$$

Kaiser (1958), on the other hand, adopted as his index the sum of the variances of the squared normalized loadings within each column of the factor loading matrix, i.e.

$$V = \frac{1}{p^2} \sum_{j=1}^{q} \left\{ p \sum_{i=1}^{p} \beta_{ij}^4 - \left(\sum_{i=1}^{p} \beta_{ij}^2 \right)^2 \right\}, \tag{16.13}$$

where $\beta_{ij} = \gamma_{ij} \Big/ \left(\sum_{j=1}^{q} \gamma_{ij}^2 \right)^{\frac{1}{2}}$. Other indices of 'simplicity' exist, but Q and V

are the two most commonly used; the larger the value of either index, the 'simpler' is the corresponding structure. The rotation matrix **H** is then chosen analytically to be the one that produces loadings that maximize the selected criterion. The rotation that maximizes Q of (16.12) is called the *quartimax* rotation; the one that maximizes V of (16.13) is called the *varimax* rotation; both of these rotations are included in most common statistical computer packages. We will not go into any of the computational details here; a sketch of the process may be found, for example, in Morrison (1976).

Returning to the WPPSI data, the varimax rotation yields the matrix $\mathbf{H} = \begin{pmatrix} 0.7278 & -0.6858 \\ 0.6858 & 0.7278 \end{pmatrix}$, corresponding to a rotation of axes through $-43°18'$, and a rotated matrix of factor loadings

$$\boldsymbol{\Gamma}^* = \begin{pmatrix} 0.851 & 0.248 \\ 0.768 & 0.402 \\ 0.561 & 0.483 \\ 0.554 & 0.268 \\ 0.662 & 0.283 \\ 0.379 & 0.550 \\ 0.306 & 0.610 \\ 0.218 & 0.687 \\ 0.373 & 0.426 \\ 0.304 & 0.855 \end{pmatrix}.$$

This set of loadings shows a similar interpretation to that obtained from the selected rotation (16.11); factor 1 loads most heavily on variables 1 to 5, while factor 2 loads most heavily on variables 6 to 10. However, whereas the contrast between variables 1–5 and variables 6–10 is now more pronounced in factor 1, it is less pronounced in factor 2. Figure 16.2 shows the varimax rotation relative to the original axes.

Sometimes there might exist a 'target' matrix **T** to which it is wished to match the estimated loadings matrix $\hat{\boldsymbol{\Gamma}}$ as closely as possible. For example, the investigator may have a hypothesized factor loading pattern, or may have conducted a previous analysis which yielded estimated factor loadings. In either case it is desired to rotate the new estimates $\hat{\boldsymbol{\Gamma}}$ (to $\hat{\boldsymbol{\Gamma}}^*$, say) so that they match as closely as possible the target matrix **T**. In a *confirmatory* analysis $\hat{\boldsymbol{\Gamma}}^*$ is then compared with **T** to see if the hypothesis is tenable, or if the new data set has similar factor structure to the previous set. An appropriate rotation and comparison

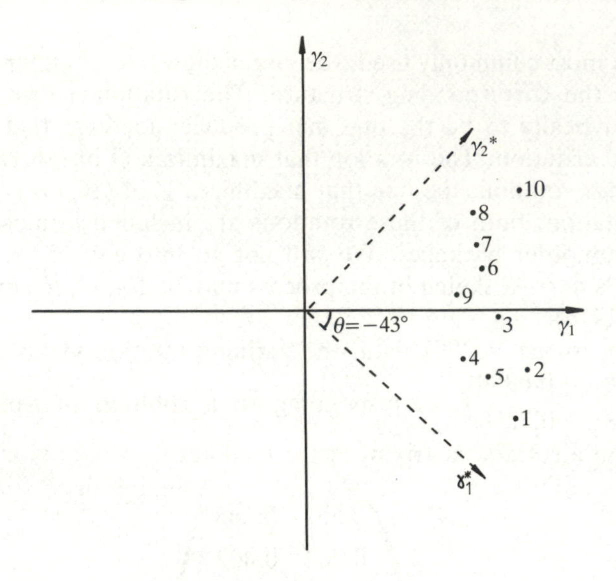

Fig. 16.2 Plot of factor loadings taken from Table 16.3, showing the varimax rotation of axes

technique to achieve this aim has already been described in Section 5.1 (procrustes analysis); for further discussion and alternative rotations, see Lawley and Maxwell (1971).

16.2.7 Estimation of factor scores

Return now to the model (16.3), namely $\boldsymbol{x} - \boldsymbol{\mu} = \boldsymbol{\Gamma z} + \boldsymbol{e}$, and suppose that $\boldsymbol{x}_i$ is the vector of manifest variable values for sample member i ($i = 1, \ldots, n$). If a factor analysis has established the presence of q factors, and each of these factors has been labelled, then it might be of interest to estimate the ith individual's vector $\boldsymbol{z}_i$ of *scores* on each of these factors ($i = 1, \ldots, n$). We can thereby rank the same individuals according to these factor scores. For example, if three factors in a particular data set were interpreted as 'general intelligence', 'verbal ability', and 'pictorial ability', then estimation of the factor scores would provide a ranking of the sample members in terms of their general intelligence, would identify those individuals who had greatest verbal or pictorial ability, and so on. For reasons already discussed in Section 16.2.2, estimation of factor scores can only be done after factor loadings and specific variances have been estimated, and even then the estimates may not be very precise or uniquely defined. Two methods exist in popular use for factor score estimation. Both have appealing intuitive rationale, one being reliant on multivariate normality, while the other uses least squares as its basis.

The method proposed by Thomson (1951) requires assumption of multivariate normality for both components $\boldsymbol{z}$ and $\boldsymbol{e}$ of model (16.3), and is therefore a natural method to take in conjunction with maximum likelihood estimation of loadings. Consider the $(q+p)$-component vector $\boldsymbol{y}' = (\boldsymbol{z}', \boldsymbol{x}')$ formed by compounding the factor variables and manifest variables. A basic assumption of model (16.3) is that $\boldsymbol{z}$ has dispersion matrix $\mathbf{I}$; in Section 16.2.2 we saw that the covariances between elements of $\boldsymbol{x}$ and elements of $\boldsymbol{z}$ are given by the elements of $\boldsymbol{\Gamma}$; and the dispersion matrix of $\boldsymbol{x}$ is $\boldsymbol{\Sigma}$, which is assumed equal to $\boldsymbol{\Gamma\Gamma}' + \boldsymbol{\Psi}$ from (16.6). Hence it follows that the dispersion matrix of $\boldsymbol{y}$ is $\begin{pmatrix} \mathbf{I} & \boldsymbol{\Gamma}' \\ \boldsymbol{\Gamma} & \boldsymbol{\Gamma\Gamma}' + \boldsymbol{\Psi} \end{pmatrix}$. Multivariate normality of $\boldsymbol{x}$ and $\boldsymbol{z}$ implies multivariate normality of $\boldsymbol{y}$, and the mean vector of $\boldsymbol{z}$ is zero by assumption. Hence, by Property 3 of the multivariate normal distribution (Section 7.2), the expected value of $\boldsymbol{z}$ *for given value* $\boldsymbol{x}_0$ *of the manifest variable vector* is

$$\mathrm{E}(\boldsymbol{z} \mid \boldsymbol{x}_0) = \boldsymbol{\Gamma}'(\boldsymbol{\Gamma\Gamma}' + \boldsymbol{\Psi})^{-1}(\boldsymbol{x}_0 - \boldsymbol{\mu}) \tag{16.14}$$

where $\boldsymbol{\mu}$ is the mean of the manifest vector $\boldsymbol{x}$. Thus if $\hat{\boldsymbol{\Gamma}}$, $\hat{\boldsymbol{\Psi}}$, and $\hat{\boldsymbol{\mu}}$ denote the estimates of the factor loadings, specific variances, and manifest vector mean respectively, then an obvious estimate of the factor score vector $\boldsymbol{z}_i$ for the ith sample member (having manifest vector value $\boldsymbol{x}_i$) is given by

$$\hat{\boldsymbol{z}}_i = \hat{\boldsymbol{\Gamma}}'(\hat{\boldsymbol{\Gamma}}\hat{\boldsymbol{\Gamma}}' + \hat{\boldsymbol{\Psi}})^{-1}(\boldsymbol{x}_i - \hat{\boldsymbol{\mu}}). \tag{16.15}$$

Note that here $\hat{\boldsymbol{\mu}} = \bar{\boldsymbol{x}}$ usually, and the $\hat{\boldsymbol{z}}_i$ $(i = 1, \ldots, n)$ would generally be standardized before use, to accord with the basic assumptions of model (16.3).

An alternative method of estimating factor scores is due to Bartlett (1937, 1938), who considered the factor model in the formulation (16.1). After the factor loadings and specific variances have been estimated this model can be written

$$\begin{aligned}
x_1 - \bar{x} &= \hat{\gamma}_{11}z_1 + \hat{\gamma}_{12}z_2 + \ldots + \hat{\gamma}_{1q}z_q + e_1 \\
x_2 - \bar{x} &= \hat{\gamma}_{21}z_1 + \hat{\gamma}_{22}z_2 + \ldots + \hat{\gamma}_{2q}z_q + e_2 \\
&\vdots \\
x_p - \bar{x} &= \hat{\gamma}_{p1}z_1 + \hat{\gamma}_{p2}z_2 + \ldots + \hat{\gamma}_{pq}z_q + e_p
\end{aligned}$$

where $\mathrm{var}(e_j) = \hat{\psi}_j^2$. Taking each sample member in turn, these equations can thus be considered as regression equations: $x_j - \bar{x}$ and $\hat{\gamma}_{jk}$ are known for $j = 1, \ldots, p$ and $k = 1, \ldots, q$, while z_j are to be estimated for $j = 1, \ldots, q$. Since the variances of the e_j are not all equal, weighted least squares is appropriate. Hence the vector of factor scores z_i for the ith sample member is estimated as the vector that minimizes $(\boldsymbol{x}_i - \bar{\boldsymbol{x}} - \hat{\boldsymbol{\Gamma}}\boldsymbol{z}_i)'\hat{\boldsymbol{\Psi}}^{-1}(\boldsymbol{x}_i - \bar{\boldsymbol{x}} - \hat{\boldsymbol{\Gamma}}\boldsymbol{z}_i)$. Differentiating this expression with respect to $\boldsymbol{z}_i$, equating the result to zero, and solving the consequent equation, yields the least squares estimates as

$$\hat{\boldsymbol{z}}_i = (\hat{\boldsymbol{\Gamma}}'\hat{\boldsymbol{\Psi}}^{-1}\hat{\boldsymbol{\Gamma}})^{-1}\hat{\boldsymbol{\Gamma}}'\hat{\boldsymbol{\Psi}}(\boldsymbol{x}_i - \bar{\boldsymbol{x}}). \quad (16.16)$$

Thus, either (16.15) or (16.16) provides estimated factor scores as linear combinations $\mathbf{A}(\boldsymbol{x}_i - \bar{\boldsymbol{x}})$ of the mean-centered manifest variable vectors. The complete $(p \times q)$ factor score matrix $\mathbf{Z}$ is thus given by $\mathbf{Z} = \mathbf{XA}'$, where the columns of $\mathbf{X}$ have first been mean-centered, and the columns of $\mathbf{Z}$ are also generally standardized after computation. The matrix $\mathbf{A}$ is often referred to as the *factor score estimation matrix,* and is provided from either (16.15) or (16.16) by general-purpose computing packages. The two methods (16.15) and (16.16) will give very similar estimates if all the elements of $\hat{\boldsymbol{\Gamma}}'\hat{\boldsymbol{\Psi}}^{-1}\hat{\boldsymbol{\Gamma}}$ are much larger than one.

16.2.8 Dimensionality reduction using factor analysis

Since factor analysis provides a means of reducing a large number (p) of manifest variables to a relatively small number (q) of uncorrelated common factors, together with interpretation of these factors and estimates of factor scores for each individual, it is frequently used as an alternative to principal component analysis for reducing the dimensionality of a large set of data. It therefore seems appropriate to conclude our discussion of factor analysis by comparing and contrasting it briefly with principal component analysis.

First and foremost, factor analysis is a model-based technique which has as a primary aim the explanation of the *associations* among the manifest variables of the system. The model embodies many assumptions, both structural and distributional, and as a consequence provides hypothesis-testing opportunities. By contrast, principal component analysis aims at explanation of the *variances* in the system, and has no underlying model as a basis. Consequently there are no distributional assumptions that have to be made; a benefit of this fact is that the technique can be applied validly to fairly broad data types, but a restriction is that there is no hypothesis testing associated with it and hence it is predominantly a descriptive technique. At the detailed level, there is one advantage and one disadvantage associated with each technique. Factor analysis (using maximum likelihood) is invariant under changes of scale of the variables, so the same results are obtained whether the correlation or the covariance matrix is analysed. In principal component analysis, on the other hand, different results are obtained from the correlation matrix than are obtained from the covariance matrix. Thus a careful decision (whether or not to standardize the data) must first be made before using the latter technique. However, principal component scores are obtained directly in the course of the analysis, whereas factor scores have to be estimated as a subsidiary (and not always satisfactory) part of the analysis.

In conclusion, therefore, there is room for both techniques in the context of general multivariate analysis, but due care must be paid to the objectives of the analysis when choosing between them. Principal component analysis simply provides a different, possibly more useful, way of looking at the data. Factor analysis is built on far more structured foundations, and should only be used if a positive answer is provided to the question, "Is the model valid?".

16.2.9 Binary data

If each x_i is dichotomous, it is possible to score the two categories 0 and 1 and treat the data as if they were numerical. An appropriate measure of association between any two binary variables is provided by their tetrachoric correlation (Section 14.1). A naive approach to the explanation of the associations in a binary data set is therefore to apply the standard factor analysis methodology to the matrix of tetrachoric correlations between pairs of variables. Often, such a procedure will yield perfectly acceptable results, despite the fact that the factor analysis model takes no account of the special data structure. Sometimes, however, the matrix of tetrachoric correlations is not positive definite, and complications arise with the analysis. Christoffersen (1975) and Muthen (1978) have proposed special models for the factor analysis of binary data; for a brief description see Everitt (1984). In general, however, it may be more satisfactory to view the dichotomous variables as qualitative and to apply the methods described in the next section.

16.3 Discrete manifest variables: latent-structure analysis

16.3.1 Background

We now suppose that the p variables $x_1, x_2, \ldots, x_p$ observed on each of the n sample individuals are discrete or qualitative; let s_i denote the number of distinct states or categories that x_i can assume $(i = 1, \ldots, p)$. Thus the sample space consists of $s = \prod_{i=1}^{p} s_i$ different patterns of response for the p-vector $(x_1, x_2, \ldots, x_p)$. The data therefore take the form of

Chapter 10, and the notation of that chapter will be appropriate here. For ease of illustration we will assume throughout that $p = 4$, but obviously the notation extends to any value of p in direct fashion.

Let the probability of the pattern $(x_1 = a,\ x_2 = b,\ x_3 = c,\ x_4 = d)$ be denoted by p_{abcd}, and let the number of sample members that exhibit this pattern be denoted by n_{abcd}. Then the sum of the p_{abcd} over all possible patterns is unity, and the corresponding sum of the four-way n_{abcd} is n. The full sample data can therefore be exhibited in the contingency table which has entries n_{ijkl} $(i = 1, \ldots, s_1;\ j = 1, \ldots, s_2;\ k = 1, \ldots, s_3;\ l = 1, \ldots, s_4)$. For illustration, consider the data of Example 10.4. Each of the four variables x_1, x_2, x_3, and x_4 has two states so that $s_1 = s_2 = s_3 = s_4 = 2$ and $s = 2^4 = 16$. Denoting by 1 the states 'small car' (of x_1), 'ejected' (of x_2), 'severe' (of x_3), and 'collision' (of x_4), and by 2 the other state of each variable, we have $n_{1111} = 23$, $n_{2111} = 161$, $n_{1211} = 150$, $n_{2211} = 1022$, $n_{1121} = 26$, and so on. We also require, in general, the entries and probabilities of the margins of such tables, on summing over the states of individual variables. We will again adopt the dot notation to indicate summation, so that

$$n_{.jkl} = \sum_{i=1}^{s_1} n_{ijkl},$$

$$n_{..kl} = \sum_{i=1}^{s_1} \sum_{j=1}^{s_2} n_{ijkl},$$

$$p_{.jkl} = \sum_{i=1}^{s_1} p_{ijkl}$$

$$p_{..kl} = \sum_{i=1}^{s_1} \sum_{j=1}^{s_2} p_{ijkl}, \quad \text{and so on.}$$

For Example 10.4, therefore,

$$n_{.111} = 23 + 161 = 184$$

$$n_{..11} = 23 + 161 + 150 + 1022 = 1356$$

$$n_{...1} = 23 + 161 + 150 + 1022 + 26 + 350 + 111 + 1878 = 3721$$

$$n_{....} = n = 4831$$

$$n_{12.1} = 150 + 350 = 500, \quad \text{and so on.}$$

The measurement of association between two discrete variables has been discussed in Chapter 14, and eqns (14.3), (14.4), and (14.5) provide relevant computational formulae. The associations that exist between the variables defining a multidimensional contingency table such as the one in Example 10.4, therefore, depend upon the patterns exhibited by the data in the possible two-way tables obtained by collapsing the four-way table over sets of variables. These associations are very different in nature from the correlations obtained when the variables are continuous. Hence we cannot expect factor analysis to be appropriate when attempting to explain associations among discrete variables, but we must look for alternative explanatory models.

The ideas of latent-structure analysis were first put forward by Lazarsfeld (1950, 1954), as an attempt to provide the analogue of factor analysis for discrete variables. The concept of an unobservable variable (or variables) is retained, together with the central idea of local independence as discussed in Section 16.1. However, instead of implying zero correlation for individuals that have a common value of the latent variable, local independence now means that such individuals exhibit no association in the contingency table formed from the manifest variables. The subsequent development of latent-structure analysis has, to some extent, mirrored that of factor analysis. The early work was almost exclusively concerned with the necessary *models*; identification, refinement, and interpretation were the chief concerns, whereas the statistical aspects were largely ignored. Methods of estimation and goodness-of-fit were initially rather primitive. More efficient estimation procedures were given by T. W. Anderson (1954), but twenty more years were to elapse before maximum likelihood techniques became popularized (Goodman 1974). There is now a plethora of different models, each one carrying associated problems of definition, estimation, and interpretation; a full discussion of all these aspects would require a book by itself. An early example of such a book is the one by Lazarsfeld and Henry (1968). More recent introductions and surveys have been given by Fielding (1977) and Andersen (1982). The latter in particular provides a concise summary, with clear discussion of the various models and a comprehensive bibliography. Some further discussion and practical examples are given by Everitt (1984). Here we shall present only a brief description of the main model, and sketch the estimation and goodness-of-fit procedures.

16.3.2 Model

The discrete manifest variables x_1, x_2, x_3, x_4 are observed on each of n sample individuals, which gives rise to the contingency table with entries n_{ijkl} and probabilities p_{ijkl} discussed above. Consider a (continuous) latent variable z which has an (unobservable) value z_i associated with the ith sample member ($i = 1, \ldots, n$). Thus z_i might denote the ith individual's 'intelligence', 'accident proneness', or 'degree of job satisfaction', for example. More than one latent variable might exist for each individual, and this situation can be catered for by making z a vector. In latent-structure analysis, however, a single latent variable is usually sufficient for all but the most complex models. This is a consequence of the extra 'looseness' provided by discrete manifest variables, and contrasts with the situation for continuous variables as exemplified by factor analysis. The models on which we shall focus are fairly simple ones, and a single variable z will prove adequate for them in practice. We shall also assume that the n observations are a sample from some population of interest, so that z will have a probability distribution over this population; let $\mathrm{f}(z)$ denote the corresponding probability density function of z.

Let us first focus attention only on those individuals that have a particular value of the latent variable. Denote by $p_{ijkl}(z)$ the probability that an individual *with value z of the latent variable* exhibits the manifest variable pattern $x_1 = i,\ x_2 = j,\ x_3 = k,\ x_4 = l$, and let the corresponding marginal probabilities for individuals with value z of the latent variable be denoted in the usual way by $p_{.jkl}(z)$, etc.. Then the property of local independence requires there to be no association between the manifest variables for all such individuals. In other words, for discrete manifest variables, the property of local independence is specified by the relationship

$$p_{ijkl}(z) = p_{i...}(z)p_{.j..}(z)p_{..k.}(z)p_{...l}(z). \tag{16.17}$$

But if individuals are sampled from some population, then the individuals obtained in the sample will in general all have *different* values of the latent variable, and p_{ijkl} is the marginal probability $\mathrm{pr}\{x_1 = i,\ x_2 = j,\ x_3 = k,\ x_4 = l\}$ for a chosen individual. By elementary probability theory, therefore,

$$\begin{aligned} p_{ijkl} &= \mathrm{pr}\{x_1 = i, x_2 = j, x_3 = k, x_4 = l\} \\ &= \int \mathrm{pr}\{x_1 = i, x_2 = j, x_3 = k, x_4 = l \mid \text{latent variable value} = z\}\mathrm{f}(z)\,\mathrm{d}z \end{aligned}$$

i.e.

$$p_{ijkl} = \int p_{ijkl}(z)\mathrm{f}(z)\,\mathrm{d}z. \tag{16.18}$$

If the latent variable is to explain all the associations among the manifest variables then it must be a variable with respect to which local independence holds good, so that from (16.17) and (16.18) we must have

$$p_{ijkl} = \int p_{i...}(z)p_{.j..}(z)p_{..k.}(z)p_{...l}(z)\mathrm{f}(z)\,\mathrm{d}z. \tag{16.19}$$

This is the fundamental relationship of latent structure analysis. Given the observed data, (i.e. the n_{ijkl} for all i, j, k, and l), we aim to use (16.19) in estimating the sample individuals' values of the latent variable

z (e.g. each individual's 'intelligence' or 'degree of job satisfaction'), and in making statements about the conditional probabilities $p_{i...}(z)$, $p_{.j..}(z)$, $p_{..k.}(z)$, and $p_{...l}(z)$ (e.g. the probability of obtaining a given state for each manifest variable for an individual with specified 'intelligence' or 'degree of job satisfaction'). To do this, we require to make assumptions about the density $f(z)$, and we usually also have to assume functional forms for the conditional probabilities. Some possible models are given in Lazarsfeld and Henry (1968).

One area in which such models have been used extensively is that of mental testing, where the discrete variables x_i are generally binary in nature (right/wrong responses to questions) and are termed *items*, the latent variable z relates either to general intelligence or to some specific ability, and conditional probabilities such as $p_{i...}(z)$ are termed *item characteristic curves*. Two established and popular models for such curves are the *Lawley–Lord model* and the *Birnbaum–Rasch model*. Much work, both technical and practical, has been done on these models and this specific application. For further details see Andersen (1980, Chapter 6), Andersen (1982, Section 6), and Lord and Novick (1968, Chapters 17–20).

The model that has found most widespread use for discrete manifest variables, however, assumes that the latent variable is also categorical; the resulting model is known as the *latent-class model*. Some latent variables, such as 'social class' for example, are naturally conceived in terms of classes, and hence are clearly discrete or categorical in form. Other latent variables may appear at first sight to be continuous, but little is lost if their range is grouped into classes and they are treated as categorical. For example, the three categories 'satisfied', 'neutral', and 'dissatisfied' may be perfectly adequate in practical terms to represent the variable 'degree of job satisfaction', and there will be minimal practical gain from using a complete continuous range of values. The appeal of a categorical latent variable is considerable. The probability density function $f(z)$ can be replaced by a few distinct probability values (e.g. the three probabilities of being 'satisfied', 'neutral', or 'dissatisfied' with one's job), while the continuous functions $p_{i...}(z)$, $p_{.j..}(z)$, $p_{..k.}(z)$, and $p_{...l}(z)$ can be replaced by a few sets of discrete values. Thus the associations between the manifest variables (as exhibited in the original contingency table formed from them) can be explained by estimating the proportion of individuals in each of a few latent classes (within which the manifest variables are conditionally independent), together with the probabilities of each manifest variable category in each latent class.

The latent-class model is a special case of the latent-structure model (16.19). Let us suppose that the latent variable z contains q categories or classes, denoted $z^{(r)}$ $(r = 1, \ldots, q)$. Let f_r denote $\text{pr}\{z = z^{(r)}\}$, the probability that an individual chosen at random from the population belongs to latent class r ($r = 1, \ldots, q$). Given that an individual belongs to latent class r, let the probability of obtaining the manifest variable pattern $x_1 = i$, $x_2 = j$, $x_3 = k$, $x_4 = l$ be denoted by $p_{ijkl}^{(r)}$. Thus

$$p_{i\ldots}^{(r)} = \text{pr}\{x_1 = i \text{ for individuals in latent class } r\},$$
$$p_{.j..}^{(r)} = \text{pr}\{x_2 = j \text{ for individuals in latent class } r\},$$

and so on. Hence, since integration is replaced by summation in the case of a discrete variable, we obtain directly from (16.19) the fundamental latent class relationship

$$p_{ijkl} = \sum_{r=1}^{q} p_{i\ldots}^{(r)} p_{.j..}^{(r)} p_{..k.}^{(r)} p_{\ldots l}^{(r)} f_r. \tag{16.20}$$

Given the observed values n_{ijkl} for a set of data, we estimate the p_{ijkl} by $\hat{p}_{ijkl} = n_{ijkl}/n$. Our objective in latent-class analysis is then to estimate all the quantities on the right-hand side of (16.20).

16.3.3 Statistical aspects of the latent-class model

There are many similarities between the latent-class model (16.20) and the factor-analysis model (16.1). The q latent classes of the former correspond to the q factors of the latter, and the p marginal probabilities of each manifest variable for the rth latent class correspond to the p loadings of each manifest variable on the rth factor, while the q class probabilities f_r in (16.20) replace the p specific variances in (16.1). In parallel fashion to the development of factor analysis, the first attempts at estimation schemes for these unknown parameters were *ad-hoc* ones. The most successful such scheme is based on the method of moments. Note first that the probabilities $p_{i\ldots}^{(r)}$, $p_{.j..}^{(r)}$, $p_{..k.}^{(r)}$, $p_{\ldots l}^{(r)}$ for a given latent class r form four separate probability distributions over the respective index variables i, j, k, l. Thus

$$\sum_{i=1}^{s_1} p_{i\ldots}^{(r)} = \sum_{j=1}^{s_2} p_{.j..}^{(r)} = \sum_{k=1}^{s_3} p_{..k.}^{(r)} = \sum_{l=1}^{s_4} p_{\ldots l}^{(r)} = 1.$$

Hence if we sum (16.20) successively over each suffix we obtain

$$\begin{aligned} p_{ijk.} &= \sum_{r=1}^{q} p_{i\ldots}^{(r)} p_{.j..}^{(r)} p_{..k.}^{(r)} f_r \\ p_{ij..} &= \sum_{r=1}^{q} p_{i\ldots}^{(r)} p_{.j..}^{(r)} f_r \\ p_{i\ldots} &= \sum_{r=1}^{q} p_{i\ldots}^{(r)} f_r \end{aligned} \tag{16.21}$$

and, clearly,

$$p_{....} = \sum_{r=1}^{q} f_r = 1.$$

Now, $E(n_{ijkl}) = np_{ijkl}$, $E(n_{ijk.}) = np_{ijk.}$, $E(n_{ij..}) = np_{ij..}$, and $E(n_{i...}) = np_{i...}$ (by standard results of multinomial distributions). Consequently, if we replace the probabilities on the left-hand side of the system (16.21) by their unbiased sample estimates, we obtain

$$\begin{aligned} n_{i...}/n &= \sum_{r=1}^{q} p_{i...}^{(r)} f_r \\ n_{ij..}/n &= \sum_{r=1}^{q} p_{i...}^{(r)} p_{.j..}^{(r)} f_r \\ n_{ijk.}/n &= \sum_{r=1}^{q} p_{i...}^{(r)} p_{.j..}^{(r)} p_{..k.}^{(r)} f_r. \end{aligned} \tag{16.22}$$

Of course, corresponding equations with $n_{.j..}/n$, $n_{..k.}/n$, and $n_{...l}/n$ on the left-hand side are supplementary to those in the first line of (16.22), ones with $n_{i.k.}/n$, $n_{i..l}/n$, etc. on the left-hand side are supplementary to those in the second line, and so on. These equations are called the *accounting equations*. The model has $5q-1$ parameters to be estimated if all variables are binary, and Lazarsfeld suggested selecting exactly $5q-1$ accounting equations and solving them directly for the unknown parameters. The best order of selection of equations seems to be the order given in (16.22), i.e. the four one-variable equations before the six two-variable equations before the four three-variable ones. Thus if $q=2$ then we need nine accounting equations, and we can do without the three-variable equations. Additionally, we only need five of the six two-variable equations, and the one to be omitted can be chosen arbitrarily. (Of course, for values of p other than four there will be different numbers of equations of each type, and possibly extra equations involving more than three variables.) A systematic approach to formulation and solution of these equations was provided by T. W. Anderson (1954). This approach is known as the determinantal method; for further details see Lazarsfeld and Henry (1968).

The accounting equations were used as the basis of parameter estimation for over twenty years. Although the maximum-likelihood equations were not difficult to derive, they were felt to be far too complex for a practical solution to be possible. However, Goodman (1974) finally produced an iterative method that demonstrated the feasibility of maximum-likelihood estimation. Imagine that the q separate p-way contingency tables formed from the p manifest variables (one table for each of the q classes of the latent variable) have been amalgamated into a single $(p+1)$-way contingency table, and write the cell probabilities of this expanded table as p_{ijklr}. Thus p_{ijklr} is the joint probability of obtaining an individual with values $x_1 = i$, $x_2 = j$, $x_3 = k$, $x_4 = l$, *and* class r of the latent variable. Note that this probability contrasts with the probabilities we had previously of the form $p_{ijkl}^{(r)}$, which were *conditional* probabilities. Here $p_{ijkl}^{(r)}$ denotes the probability of obtaining an individual with values $x_1 = i$, $x_2 = j$, $x_3 = k$, $x_4 = l$ given that the individual belongs to latent class r. (Hence p_{ijklr} is the probability of obtaining an individual with values $x_1 = i$, $x_2 = j$, $x_3 = k$, $x_4 = l$, and latent class r when sampling from the whole population, while $p_{ijkl}^{(r)}$ is the probability of obtaining an individual with values $x_1 = i$, $x_2 = j$, $x_3 = k$, $x_4 = l$ when sampling from individuals in latent class r only.)

Since the latent classes are unobservable, we only have data for the p-way contingency table with cell entries n_{ijkl} and corresponding probabilities p_{ijkl}. However, by elementary probability theory,

$$p_{ijkl} = \sum_{r=1}^{q} p_{ijklr} \tag{16.23}$$

and

$$p_{ijklr} = f_r p_{i\ldots}^{(r)} p_{.j..}^{(r)} p_{..k.}^{(r)} p_{\ldots l}^{(r)} \tag{16.24}$$

(using the local independence property in the latter case). One further conditional probability can be defined, namely the probability that an individual belongs to latent class r given that it has values $x_1 = i$, $x_2 = j$, $x_3 = k$, and $x_4 = l$. Denoting this conditional probability by $\mathrm{p}(r \mid ijkl)$, it follows again from elementary probability theory that

$$\mathrm{p}(r \mid ijkl) = p_{ijklr}/p_{ijkl}. \tag{16.25}$$

Equations (16.23), (16.24), (16.25) must also be satisfied by the maximum-likelihood estimates of the various parameters, so that we must have

$$\hat{p}_{ijkl} = \sum_{r=1}^{q} \hat{p}_{ijklr}, \tag{16.26}$$

$$\hat{p}_{ijklr} = \hat{f}_r \hat{p}_{i\ldots}^{(r)} \hat{p}_{.j..}^{(r)} \hat{p}_{..k.}^{(r)} \hat{p}_{\ldots l}^{(r)}, \tag{16.27}$$

and

$$\hat{\mathrm{p}}(r \mid ijkl) = \hat{p}_{ijklr}/\hat{p}_{ijkl}. \tag{16.28}$$

Standard calculus can be used on the observed contingency table to show that the maximum likelihood estimates $\hat{f}_r$, $\hat{p}_{i\ldots}^{(r)}$, $\hat{p}_{.j..}^{(r)}$, $\hat{p}_{..k.}^{(r)}$, $\hat{p}_{\ldots l}^{(r)}$ of the parameters in the latent class model satisfy the system of equations:

$$\hat{f}_r = \sum_i \sum_j \sum_k \sum_l n_{ijkl} \hat{p}(r \mid ijkl)/n, \tag{16.29}$$

$$\hat{p}^{(r)}_{i...} = \left\{ \sum_j \sum_k \sum_l n_{ijkl} \hat{p}(r \mid ijkl) \right\} \Big/ n\hat{f}_r, \tag{16.30}$$

$$\hat{p}^{(r)}_{.j..} = \left\{ \sum_i \sum_k \sum_l n_{ijkl} \hat{p}(r \mid ijkl) \right\} \Big/ n\hat{f}_r, \tag{16.31}$$

$$\hat{p}^{(r)}_{..k.} = \left\{ \sum_i \sum_j \sum_l n_{ijkl} \hat{p}(r \mid ijkl) \right\} \Big/ n\hat{f}_r, \tag{16.32}$$

$$\hat{p}^{(r)}_{...l} = \left\{ \sum_i \sum_j \sum_k n_{ijkl} \hat{p}(r \mid ijkl) \right\} \Big/ n\hat{f}_r. \tag{16.33}$$

Goodman (1974) suggested the following iterative scheme to solve these equations. Start with initial values for $\hat{f}_r$, $\hat{p}^{(r)}_{i...}$, $\hat{p}^{(r)}_{.j..}$, $\hat{p}^{(r)}_{..k.}$, $\hat{p}^{(r)}_{...l}$; use these values in (16.27) to obtain $\hat{p}_{ijklr}$; use the latter in (16.26) to obtain $\hat{p}_{ijkl}$; hence use (16.28) to obtain $\hat{p}(r \mid ijkl)$. Insert these last values into the right-hand sides of equations (16.29) to (16.33) to provide improved estimates $\hat{f}_r$, $\hat{p}^{(r)}_{i...}$, $\hat{p}^{(r)}_{.j..}$, $\hat{p}^{(r)}_{..k.}$ and $\hat{p}^{(r)}_{...l}$. Repeat the whole cycle until stability is reached, i.e. until successive estimates differ by less than a pre-set accuracy level. Convergence is guaranteed, provided that a latent class is deleted if the corresponding estimate $\hat{f}_r$ tends to zero. Goodman discusses suitable starting values for the process.

This iterative procedure is yet another special case of the E–M algorithm (Dempster *et al.* 1977) that we have met many times in the course of this book. The connection is made by treating each (unobservable) entry n_{ijklr} of the expanded $(p+1)$-way table as a *missing value,* and by using the fact that the entries of the various tables derivable from this one have multinomial distributions. In essence, the M-step comes from use of standard multinomial results, while the E-step uses the relationships among the parameters and the data given by the equations above. For technical details see Andersen (1982).

Since the latent-class model is a parametric multinomial model, the testing of hypotheses about parameters of the model and the testing of goodness-of-fit of the model itself is accomplished either by Pearson χ^2 tests, or by log-likelihood ratio tests (i.e. deviances in the terminology of Chapter 10). Using (16.26), expected frequencies in each cell are $n\hat{p}_{ijkl}$, so the test statistic for Pearson's goodness-of-fit test is $X^2 = \sum_i \sum_j \sum_k \sum_l (n_{ijkl} - n\hat{p}_{ijkl})^2 / n\hat{p}_{ijkl}$. The number of degrees of freedom is $f = s - q - q\left\{ \sum_{i=1}^{q} (s_i - 1) \right\}$. Hence if the computed value of X^2 is less than the $100(1-\alpha)$ per cent point of the χ_f^2 distribution, the q-class model provides a good fit to the data. This test has bearing on one other aspect of the latent class parameters not explicitly treated above, namely choice of q. All the above theory has been derived on the assumption that q is known and fixed. In some applications there may be an 'obvious' value of q, suggested by the substantive context. Where no such guidelines exist, a procedure analogous to that employed in the choice of number of factors for factor analysis may be adopted. Thus proceed sequentially: fit a latent class model with $q=2$ classes, and test the goodness-of-fit of the model; if the fit is satisfactory then stop there, otherwise fit a model with $q=3$ classes and test goodness-of-fit; continue in this way until the fit is satisfactory, or until no more classes can be fitted due to excess of parameters over data points (i.e. degrees of freedom for the χ^2 test become negative).

Everitt (1984) gives a numerical example of latent-class analysis. The data arose from four machine-design sub-tests given to 137 engineers; the data were dichotomized into positive (i.e. above the sub-test mean) and negative (i.e. below the sub-test mean). These two categories for each variable were scored 1 and 0 respectively. Thus $p=4$, $s_1=s_2=s_3=s_4=2$, $s=2^4=16$, and $n=137$. The number of degrees of freedom to test goodness-of-fit of a q-class model is therefore $16-5q$. The data had been analysed previously by McHugh (1956), who fitted a latent-class model with $q=2$ classes; Everitt fitted two latent-class models, for $q=2$ and $q=3$ respectively, using the E–M algorithm. Models with q greater than 3 cannot be fitted for these data. The observed and the expected frequencies (i.e. fitted values) are given in Table 16.6 for both the two-class and the three-class model. (These values are reproduced from Everitt (1984), with permission from Chapman and Hall.)

For the two-class model, Everitt found that $\hat{f}_1=0.48$ and hence $\hat{f}_2=0.52$, thereby showing two almost equal classes. The estimated probabilities $\hat{p}^{(1)}_{1\ldots}$, $\hat{p}^{(1)}_{.1..}$, $\hat{p}^{(1)}_{..1.}$, $\hat{p}^{(1)}_{\ldots 1}$ were all similar and large (approximately 0.75), thus making $\hat{p}^{(1)}_{0\ldots}$, $\hat{p}^{(1)}_{.0..}$, $\hat{p}^{(1)}_{..0.}$, $\hat{p}^{(1)}_{\ldots 0}$ all small (approximately 0.25). Conversely, $\hat{p}^{(2)}_{1\ldots}$, $\hat{p}^{(2)}_{.1..}$, $\hat{p}^{(2)}_{..1.}$, $\hat{p}^{(2)}_{\ldots 1}$ were all small (<0.20), thus making $\hat{p}^{(2)}_{0\ldots}$, $\hat{p}^{(2)}_{.0..}$, $\hat{p}^{(2)}_{..0.}$, $\hat{p}^{(2)}_{\ldots 0}$ all large (>0.80). Hence, in latent class 1, individuals have a high probability of getting each sub-test right, while in latent class 2, individuals have a low probability of getting each sub-test right. With regard to the nature of the sub-tests, therefore, latent class 1 could be interpreted as comprising 'creative' individuals, while latent class 2 comprised 'non-creative' individuals. Given an individual's response pattern for the x_i, the probability that the individual belongs to a specified latent class is given by (16.28). Everitt provides all these posterior probabilities. However, the goodness-of-fit of the two-class model is very poor, yielding a χ^2 value of 27.34 on 6 degrees of freedom. The fit of the

Table 16.6 Two-class and three-class latent-class models fitted to data from McHugh (1956); values reproduced from Everitt (1984), with permission from Chapman and Hall

Response pattern				Observed	Expected frequencies	
x_1	x_2	x_3	x_4	frequency	Two-class model	Three-class model
1	1	1	1	23	20.29	19.73
0	1	1	1	8	10.35	10.83
1	0	1	1	6	9.27	9.33
1	1	0	1	5	4.70	5.14
1	1	1	0	5	5.18	5.72
0	0	1	1	9	4.94	5.75
0	1	0	1	3	4.50	2.89
0	1	1	0	2	3.29	2.06
1	0	0	1	2	3.63	1.92
1	0	1	0	3	2.82	2.21
1	1	0	0	14	5.67	13.63
0	0	0	1	8	6.31	8.41
0	0	1	0	3	2.84	3.38
0	1	0	0	8	14.01	8.01
1	0	0	0	4	10.43	4.33
0	0	0	0	34	28.75	33.67

three-class model is much better (χ^2 value of 4.79 on 1 degree of freedom), the improvement being mainly attributable to the improvement in fit for individuals with $x_1 = x_2 = 1$, $x_3 = x_4 = 0$. The drawback here, though, is that the parameters of the three-class model are much more difficult to interpret.

For a further practical application of latent class analysis, the reader is referred to Aitkin *et al.* (1981).

16.4 A recent unified approach

In an interesting series of papers, Bartholomew (1980, 1981, 1983, 1984, 1985) has been exploring a new approach to factor analysis. This approach unifies the subject, in that all types of continuous (non-normal as well as normal) and categorical data are brought under the same umbrella. However, the factor model is provided with a new foundation, and the traditional analysis is viewed from a new angle, namely from the point of view of linear combinations of *observable* rather than of latent variables. This makes the approach somewhat akin to principal component analysis, but Bartholomew is careful to stress the differences between his 'components' and principal components. The benefits that he claims for the approach are that these new-style components are more 'real' for interpretation purposes than are the latent variables of the traditional model (thereby being more palatable to those practitioners who are uneasy with the concept of a latent variable), and that rotations in the factor space are linked with the firmly established statistical concept of a minimal set of sufficient statistics. The important paper as regards the mathematical and philosophical concepts is Bartholomew (1984). However, Bartholomew (1985) provides a more accessible introduction for the practitioner, and the following skeleton of the main ideas is extracted from this latter paper.

Return to the traditional factor analysis model (16.1) or (16.3). The central idea of the new approach is that once the manifest vector $\boldsymbol{x}$ has been observed, all that can be known about the unobservable (q-element) $\boldsymbol{z}$ is given by the conditional density

$$\mathrm{h}(\boldsymbol{z} \mid \boldsymbol{x}) = \mathrm{h}(\boldsymbol{z})\mathrm{g}(\boldsymbol{x} \mid \boldsymbol{z})/\mathrm{f}(\boldsymbol{x}) \tag{16.34}$$

where h, g, and f denote appropriate density or mass functions. Since

$$\mathrm{f}(\boldsymbol{x}) = \int \mathrm{h}(\boldsymbol{z})\mathrm{g}(\boldsymbol{x} \mid \boldsymbol{z})\,\mathrm{d}\boldsymbol{z}, \tag{16.35}$$

any model (such as (16.1) or (16.3)) is essentially a specification of $\mathrm{h}(\boldsymbol{z})$ and $\mathrm{g}(\boldsymbol{x} \mid \boldsymbol{z})$. If $\boldsymbol{z}$ is to explain $\boldsymbol{x}$ (in the sense of accounting for the dependencies between the x_i) then, as argued in Section 16.1, the axiom of local independence must be satisfied. In terms of the above quantities this axiom requires that

$$\mathrm{g}(\boldsymbol{x} \mid \boldsymbol{z}) = \prod_{i=1}^{p} \mathrm{g}_i(x_i \mid \boldsymbol{z}) \tag{16.36}$$

for some q (where g_i is the corresponding density or mass function for the single element x_i of $\boldsymbol{x}$).

Bartholomew (1984) asked whether it is possible to choose the conditional densities $\mathrm{g}_i(x_i \mid \boldsymbol{z})$ in such a way that $\mathrm{h}(\boldsymbol{z} \mid \boldsymbol{x})$ depended on $\boldsymbol{x}$ only through q specified functions $u_j(\boldsymbol{x})$. He showed that if $\mathrm{g}_i(x_i \mid \boldsymbol{z})$ belongs to an exponential family with natural parameter depending on the z_i in a particular way, then

$$u_j(\boldsymbol{x}) = \sum_{i=1}^{p} \alpha_{ij} v_i(x_i) \qquad (j = 1, \ldots, q) \tag{16.37}$$

where the functions v_i depend on the particular member of the exponential family that is chosen. The functions $u_j(\boldsymbol{x})$ thus contain all the information in the data about the distribution of z, and hence can be used in place of the z_i. Thus the unobservable 'factors' can be readily identified with these observable 'components'. Moreover, the exponential family is a broad one encompassing a variety of both normal and non-normal continuous models, as well as binary/categorical ones, so this simple summarization has wide applicability. Appropriate estimates of α_{ij} and $v_i(\boldsymbol{x})$ can be derived easily. In the normal case with assumptions as in (16.3), it can be shown that $u_j(\boldsymbol{x}) = \sum_{i=1}^{p} \alpha_{ij} x_i / \psi_i^{\frac{1}{2}}$ $(j = 1, \ldots, q)$. Bartholomew (1985) establishes the connections between the coefficients α_{ij} of these 'components' and the loadings γ_{ij} of the traditional 'factors' in (16.3). From these connections it follows that the coefficients α_{ij} can be estimated by the usual factor analysis routines described earlier. Bartholomew (1985) discusses a number of aspects of this new approach: the interpretation of the components; transformations of the components that correspond to rotations of the traditional model; and relationships between these components and the factor score estimates in the traditional model. He also provides an illustrative example.

At this stage it is not possible to say how far-reaching will be the consequences of this new approach. The interesting discussion on Bartholomew (1985), published in the November 1985 issue of the *British Journal of Mathematical and Statistical Psychology* (Volume **38**, pages 127–40), shows a division of opinion among some of the leading psychometric factor analysts. Undoubtedly, some users *are* more at home with the concept of 'components' than with that of 'factors', and unification of many disparate models within a single umbrella can only be welcomed. As to whether the ideas will gain wide adoption, however, only time will tell.

17 Conclusion: some general multivariate problems

The development of multivariate analysis in this book has been through a problem-orientated, rather than technique-orientated, approach, and the division of the material into five main parts has stemmed from this approach. However, a few general problems cut across such categorization, in the sense that they can arise irrespective of the form of data and in a variety of circumstances. We therefore conclude the book by discussing these problems briefly, and outlining the main ways in which they can be tackled. Detailed consideration of each would require a major portion of the book, so we do no more than convey their general flavour and present relevant references from which the reader can obtain full technical and practical details.

17.1 Variable selection

In many multivariate applications, it is evident that too many variables have been measured on each individual. This usually happens either because the researcher is unsure as to which responses are important for the study at hand (and hence he/she measures as many as possible to make sure that all important ones are included), or because the experiment or survey is very costly or difficult to organize (and once it is under way the researcher feels compelled to measure as many variables as possible in case the opportunity does not present itself again). There are three main problems with analysis of the whole data set in such circumstances:

(1) some of the variables may be totally irrelevant to the objectives of the study, and may mask any genuine effects that exist in the rest of the data;

(2) with a large number of variables present, the sample size necessary to obtain reliable parameter estimates may be unrealistically high;

(3) a large number of variables will induce a large number of unknown

parameters in the problem, and the larger the number of estimated parameters in any objective function, the more complications arise because of sampling variability.

For all these reasons, therefore, many problems require a prior selection of the 'best' variables to use in the analysis. Such selection of variables cuts across the categorization of multivariate problems that has been made in the earlier parts of the book: we may equally well want to select the best m out of p measured variables to highlight the main features of an unstructured data set, or to discriminate between two populations, or to provide a predictive relationship between two groups of variables.

The general approach to variable selection is similar in all cases: taking due account of the nature of the problem, define an objective function with the property that the maximum or minimum of this function over subsets (of given size) of the variables determines the 'best' such subset of variables, and then search through different subsets of variables to find the one that optimizes this function. Clearly, the best subset of size m can always be found by examining systematically all possible subsets of size m, but such a procedure may be computationally prohibitive if p is large. Consequently, various strategies have been suggested for restricting the search to a smaller number of 'good' subsets. The three most popular strategies are 'forward selection' (FS), 'backward elimination' (BE), and 'stepwise selection' (SS). Each strategy builds towards a subset of size m in stages, either omitting or adding a variable at each stage. In FS, all variables are first examined in turn to find the one that optimizes the chosen criterion. This single 'best' variable is retained, and then tried in conjunction with each of the remaining $(p-1)$ variables to find the pair that optimizes the criterion. This pair is retained, and then tried in conjunction with each of the remaining $(p-2)$ variables to find the triple that optimizes the criterion, and so on. The process is continued until m variables have been selected, which involves $p+(p-1)+\ldots+(p-m+1)=\frac{1}{2}m(2p-m+1)$ criterion function evaluations instead of the $p!/\{m!(p-m)!\}$ required with an 'all-subsets' scheme. BE works in the reverse direction. From the complete set of variables, each variable is omitted in turn, and the objective function is evaluated; hence the subset of $(p-1)$ variables that optimizes the criterion is found. From this set of $(p-1)$ variables, each variable is omitted in turn, and the objective function is evaluated; hence the subset of $(p-2)$ variables that optimizes the criterion is found. The process is continued until m variables are left; this again only involves $\frac{1}{2}m(2p-m+1)$ criterion function evaluations.

In each of these procedures, instead of specifying m, a critical value of the objective function can be specified, and selection of variables continued until this critical value is reached. Whatever the stopping rule, however, the FS procedure is such that, once a variable is entered into the 'retained' set it will always stay in this set, while the BE procedure is such that once a variable is removed from the 'retained' set it can never be re-admitted to it. These are sometimes undesirable characteristics. For example, suppose that the variables x_i, x_j are not 'important' separately, but because of the correlation structure the pair (x_i, x_j) *is* 'important'. Then the FS procedure will not lead to a subset including both x_i and x_j, because neither variable will be included singly. On the other hand, if the three pairs (x_i, x_j), (x_i, x_k), and (x_j, x_k) are equally 'important' but the single most 'important' variable is x_i, then the BE procedure may not find it because it may select arbitrarily (x_j, x_k) at the two-variable stage. These problems may be overcome by a judicious mixture of both FS and BE, because BE may obviate the FS problems, and vice-versa. This is essentially what the 'stepwise selection' procedure SS attempts to do. At each stage of the process, the variables *not* in the 'current subset' are examined in turn to see which is the best for inclusion, and then all the variables that *are* in the 'current subset' are examined in turn to see if any can be omitted without detriment.

Although the FS, BE, and SS procedures reduce the computations significantly and provide viable selection schemes in nearly all situations, they may not identify the globally optimum subset that would be uncovered by an 'all-subsets' search. For this reason, various authors have investigated computationally efficient methods of generating all subsets (Beale *et al.* (1967), Garside (1971)), and such methods will become increasingly attractive as computer power develops. In the main, however, the stepwise procedures remain the universally most popular tools for variable selection, and the user can justify appealing to them on the grounds that they should usually produce at least a satisfactory subset if not the globally optimum one. We will now briefly survey the application of these variable selection methods in each main categorization of multivariate problems.

With no a priori structure imposed on the data, one possible objective is to find the subset of m variables that captures as well as possible the main structural aspects of the complete set of p variables. By 'main structural aspects' we imply any distinguishing features of the data set, such as clusters of individuals, outliers, special patterns, etc. Suppose that a preliminary principal component analysis indicates that the essential dimensionality of the data is k (Section 2.2.5). Let the $(n \times k)$ matrix $\mathbf{X}$ contain the coordinates of the n sample individuals with respect to the first k principal components computed from the original p variables, and let the $(n \times k)$ matrix $\mathbf{Y}_{(m)}$ contain the corresponding coordinates on the first k principal components computed from a subset comprising m of the original p variables. Krzanowski (1987) suggests that the best subset is the one for which the configuration in $\mathbf{Y}_{(m)}$ is as close as possible to the configuration in $\mathbf{X}$. A natural measure of distance between the configurations is the residual sum of squares after a procrustes comparison between $\mathbf{X}$ and $\mathbf{Y}_{(m)}$ (Section 5.1), and this is therefore proposed as the objective function to be minimized. An 'all-subsets' search may be computationally prohibitive, but a BE procedure is relatively simple to implement, and this is described. Other authors have suggested different objective functions. McCabe (1984) defines 'principal variables' as being ones that satisfy certain optimality criteria related to principal component optimality criteria, and describes a stepwise selection procedure for choosing the subset of m variables that optimizes these criteria. For example, one might wish to select the m variables which maximize the total sample variance among all m-variable subsets of the data; this would be equivalent to choosing the $\mathbf{Y}_{(m)}$ for which a selected function of the eigenvalues of $\mathbf{Y}'_{(m)}\mathbf{Y}_{(m)}$ is maximized. McCabe lists various such criteria. Jolliffe (1972, 1973) discusses methods for discarding those variables in a principal component analysis that have little effect on the analysis, while further suggestions for variable selection in a principal component context are given by Hawkins and Eplett (1982), and by Robert and Escoufier (1976).

Perhaps the greatest amount of research effort into variable selection methods has been expended in those situations where there is a-priori grouping of individuals, and the aim is discrimination, classification, or testing differences between groups. Different objective functions are appropriate for each of these different aims; an objective function based on distances between groups is most appropriate for discrimination, one based on probabilities of misclassification is most appropriate for classification, and one based on a suitable test statistic is most appropriate for hypothesis testing. However, the three aims sometimes shade together in a given application, and any of the objective functions might be deemed to be suitable. A comprehensive review, with details of the relevant objective functions and a full set of references for the discrimination and classification applications, is provided by McKay and Campbell (1982a, 1982b). This reference should provide the best general source for the interested reader. Some specific methods are described by Habbema *et al.* (1974), McCabe (1975), and J. A. Anderson (1982), while comparison of different stepwise methods is provided by Lachenbruch (1975), Costanza and Afifi (1979), McLachlan (1980) and others. Methods specifically tailored to discrete data are given by Goldstein and Dillon (1978), and by Sturt (1981). McCabe's method extends to the hypothesis testing situation, and other references in this area are Rao (1970), Hawkins (1976), and McHenry (1978). In the main, the methodology in all these studies is of a technical nature, and no attempt will therefore be made here to summarize the results.

The third area in which variable selection can be useful is where a-priori structure exists among the variables, with some variables being classed as dependent, and some as predictor or explanatory. Thus we have the multivariate regression situation of Chapter 15, and it is frequently of interest to choose the *smallest* set of explanatory variables with which to predict the dependent variables. If there is just one dependent variable then we have the familiar multiple regression situation, in which stepwise variable selection methods have been popular for very many years. An excellent review of the methodology for this case is provided by Hocking (1976), and a critical examination focussing in particular on hypothesis testing aspects is given in Miller (1984). Extensions of some of these methods to the multivariate case are given by McKay (1979), and by Hintze (1980).

One thing that must always be remembered when conducting a study involving variable selection is the potential bias that can be incurred by the selection process itself. Since, inevitably, the variables selected will be the ones that best achieve the stated objectives on the data to hand, then any assessment of these variables *on the same data* is bound to be grossly over-optimistic in relation to the expected performance of the same variables on other (future) data. This phenomenon is similar to the assessment of an allocation rule by finding the error rates incurred by this rule on the same data from which the rule itself was derived (see Section 12.4). Thus some care must be exercised when assessing the performance of the selected variables. One possible approach is to split the available data into several portions, and then use one portion to select variables while reserving the other portion for a validation of the selection process. An excellent discussion of the problems encountered, and a description of such an approach to validation, is provided in the classic *Federalist* study by Mosteller and Wallace (1984). Further cautionary comments in the context of discriminant analysis are given by Murray (1977), while Miller (1984) also devotes considerable space to the whole topic. One possible way forward for assessment of a selection procedure is provided by the various re-sampling schemes that are discussed in the next section.

17.2 Non-parametric assessment of error: data re-sampling schemes

Most of the techniques that we have considered in this book have been based either explicitly or implicitly on an underlying model, and hence the quantities that have been derived as part of the analysis can be viewed as estimates of the corresponding model parameters. For example, the means of a set of groups on their canonical variate axes (Chapters 11 and 13) are the estimates of the corresponding population canonical variate means; the factor loadings in the factor analysis of a set of data are estimates of the factor model loadings (Chapter 16); the weights in an individual scaling analysis are estimates of the parameters of the INDSCAL model (Chapter 6); and so on. Given that these quantities are estimates, it is generally of interest to determine the extent of error present in them by calculating such measures as bias, standard error, or mean squared error. For some of the analyses, the models are specified sufficiently tightly to enable these measures to be derived analytically. For example, the normality assumption of the canonical variate model enables confidence regions for the canonical variate means to be constructed (Section 13.2). However, what happens if the data do not conform very accurately with the required assumptions? Are these confidence regions still appropriate, or must they be modified? In other analyses, such as the INDSCAL analysis of Chapter 6, the model is a much more loose one, and no direct means of obtaining standard errors or confidence regions is available. How reliable are the estimated weights? Finally, at the most extreme end of the spectrum, a technique may be used purely descriptively for a particular data set. For example, a single-link cluster analysis (Section 3.1) may be used to reduce a large number of micro-organisms, on each of which various microscopic measurements have been taken, to a simple dendrogram summarizing the relationships among the micro-organisms. How stable is this summary to perturbations in the data? Would addition or deletion of some data points change the shape of the dendrogram dramatically?

With the advent of high-speed computing power, a group of data-resampling procedures has been developed for obtaining non-parametric estimates of error of the kind described above. These procedures rely on heavy computation rather than sophisticated mathematical analysis, but are of interest because they use only the data available and no outside features or imposed models. Because of the heavy demand on computational power, use of these procedures is still in its infancy, particularly in the area of multivariate analysis. However, it seems certain that much more use will be made of them in the future. The two relevant methods as regards assessment of bias and standard error are the *bootstrap* and the *jack-knife*. Very closely related to the jack-knife is the idea of *cross-validation*. This latter procedure is used in the more specialized areas of model choice and assessment of performance of a prediction or allocation rule. We have already met it in connection with choice of number of components in principal component analysis (Section 2.2.5), and estimation of error rates in discriminant analysis (Section 12.4). We shall now briefly outline the ideas behind all three procedures and review some of their applications.

Suppose that a random sample $x_1, x_2, \ldots, x_n$ of size n has been taken from a population of interest, and an estimate $\hat{\theta}$ is obtained for some parameter θ of this population. To keep the description simple, we will assume for the present that the observations are univariate and the parameter is scalar. No conceptual changes are involved in extension to the multivariate/vector case, and this will be outlined later. Thus for the present we have just a single (univariate) estimator $\hat{\theta}$, from which we derive the estimate of θ for the data at hand. Interest centres on estimating the standard error of $\hat{\theta}$. Jack-knifing and bootstrapping are two methods of artificially introducing replication, so that $\hat{\theta}$ can be evaluated for each replicate, and the standard error obtained from the variation of these replicate values.

First, consider *jack-knifing*. Here, each sample member is omitted in turn from the data, thereby generating n separate samples each of size $n-1$. Let $\hat{\theta}_{(i)}$ denote the value of $\hat{\theta}$ obtained from the ith of these samples, i.e. when the ith sample member is omitted from the calculation, and let $\bar{\theta}$ denote the average of the n values $\hat{\theta}_{(i)}$. Then the jack-knife estimate of the standard error of $\hat{\theta}$ is given by

$$\hat{\sigma}_J = \left\{\left(\frac{n-1}{n}\right)\sum_{i=1}^{n}(\hat{\theta}_{(i)} - \bar{\theta})^2\right\}^{\frac{1}{2}}. \tag{17.1}$$

A subsidiary benefit of this jack-knife procedure is that an improved estimate of θ can be derived in those situations where the original estimator of θ is biased. The estimator $\tilde{\theta} = n\hat{\theta} - (n-1)\bar{\theta}$ is known as the *jack-knife estimator*. If $\hat{\theta}$ has bias of order $1/n$ then the bias of $\tilde{\theta}$ is substantially reduced, to order $1/n^2$.

Because of the explicit instructions for obtaining each of the sub-samples in the jack-knife procedure, it is possible to find an analytical expression for $\hat{\sigma}_J$ in those cases where $\hat{\theta}$ has a simple algebraic form. Also, the jack-knife estimate of standard error of $\hat{\theta}$ can be shown to be equal to the usual estimate in those circumstances where a simple analytical expression exists for the latter. The real power of the method, of course, lies in those circumstances where simple algebraic forms do *not* exist for either $\hat{\theta}$ or its standard error, and $\hat{\sigma}_J$ then comes from a purely data-based algorithm. We illustrate this aspect later, but to show the equivalence of jack-knife and 'usual' estimates of standard error in a simple case, let us first consider estimating a population mean μ by the sample mean $\bar{x}$. It is well known that the standard error of $\bar{x}$ is given by $s/\sqrt{n}$ where $s^2 = \frac{1}{n-1}\sum_{i=1}^{n}(x_i - \bar{x})^2$. The following simple algebra establishes that $\hat{\sigma}_J = s/\sqrt{n}$ also. However, readers who are not interested in algebraic details may pass over this proof without loss.

First, we have $$\hat{\mu} = \bar{x} = \frac{1}{n}\sum_{i=1}^{n} x_i, \quad \text{so that } \sum_{i=1}^{n} x_i = n\bar{x}.$$

On omitting x_j from the sample, we obtain a new sample of size $n-1$. The estimate of μ is the mean of this new sample. Hence

$$\hat{\mu}_{(j)} = \bar{x}_{(j)} = \frac{1}{n-1}\sum_{\substack{i=1\\ i\neq j}}^{n} x_i = \frac{n\bar{x}-x_j}{n-1}.$$

$\therefore$

$$\bar{\mu} = \frac{1}{n}\sum_{j=1}^{n}\hat{\mu}_{(j)} = \frac{1}{n}\sum_{j=1}^{n}\left(\frac{n\bar{x}-x_j}{n-1}\right)$$

$$= \frac{1}{n(n-1)}\sum_{j=1}^{n} n\bar{x} - \frac{1}{n(n-1)}\sum_{j=1}^{n} x_j$$

$$= \frac{n\bar{x}}{n-1} - \frac{\bar{x}}{n-1} = \left(\frac{n-1}{n-1}\right)\bar{x} = \bar{x}.$$

Thus

$$\hat{\sigma}_J^2 = \frac{n-1}{n}\sum_{j=1}^{n}(\bar{x}_{(j)}-\bar{x})^2.$$

Now

$$\bar{x}_{(j)} - \bar{x} = \frac{n\bar{x}-x_j}{n-1} - \bar{x} = \frac{n\bar{x}-x_j-n\bar{x}+\bar{x}}{n-1} = \frac{\bar{x}-x_j}{n-1}.$$

Hence

$$\hat{\sigma}_J^2 = \frac{n-1}{n}\sum_{j=1}^{n}\left(\frac{\bar{x}-x_j}{n-1}\right)^2 = \frac{1}{n(n-1)}\sum_{j=1}^{n}(x_j-\bar{x})^2 = \frac{s^2}{n},$$

i.e. $\hat{\sigma}_J = s/\sqrt{n}$ as required.

In the same spirit as above, it can be shown easily that if the variance σ^2 of a population is estimated by the biased estimator $\hat{\sigma}^2 = \frac{1}{n}\sum_{i=1}^{n}(x_i - \bar{x})^2$, then the jack-knife estimator derived from $\hat{\sigma}^2$ is the unbiased $\tilde{\sigma}^2 = \frac{1}{n-1}\sum_{i=1}^{n}(x_i-\bar{x})^2$. The reader is left to fill in the algebraic details.

Whereas jack-knifing generates subsamples in an entirely deterministic fashion, *bootstrapping* does so through random processes. There are several variants on the basic theme, but only the simple one will be described here. For a full discussion of the rationale and description of the different methods, see Efron (1982). In simple bootstrapping, k samples of size n are generated by means of random sampling *with replacement* from the n available data values. Thus each bootstrap sample is formed by successively choosing, with equal probability $1/n$, one of the original values $x_1, \ldots, x_n$ until n such values have been chosen. A typical bootstrap sample will thus contain repeats of some of the x_i, while other x_i appear either just once or not at all. Let $\hat{\theta}^{(i)}$ be the result of applying the estimator of θ to the ith bootstrap sample, and let $\bar{\hat{\theta}}$ be the mean of the k values $\hat{\theta}^{(i)}$. Then the bootstrap estimate of the standard error of $\hat{\theta}$ is given by

$$\hat{\sigma}_{\mathrm{B}} = \left\{ \frac{1}{k-1} \sum_{i=1}^{k} (\hat{\theta}^{(i)} - \bar{\bar{\theta}})^2 \right\}^{\frac{1}{2}} \tag{17.2}$$

For this to be an effective estimate of the standard error, the number k of bootstrap samples should be very large. Common values used in practice range between 100 and 1000, depending on the complexity of the estimator being studied.

To illustrate the two techniques, suppose that a random sample of ten observations is taken from a population of interest, yielding the values 1.22, 3.46, 1.56, 2.84, 7.03, 5.07, 1.18, 2.00, 9.33, and 1.95. The mean and median of this sample are 3.564 and $\frac{1}{2}(2.00+2.84)=2.42$ respectively, thereby providing estimates of the corresponding population quantities. Let us now estimate the standard errors of the two sample statistics. Table 17.1 shows the original sample, the ten jack-knife samples of size 9 obtained from the given data, and also twenty bootstrap samples of size 10 obtained from the same data. The mean and median are given for each of these samples. The bootstrap samples were obtained as follows. Since the sample size is 10, the probability of choosing any sample member for inclusion in a bootstrap sample is 1/10. This is also the probability of each of the digits 1, 2, 3, . . . , 9, 0 occurring next in a table of random digits. From such a table, therefore, random digits were read off in sets of ten and these gave the sample members that made up each bootstrap sample (associating the random digit 0 with unit 10). For example, the first ten random digits were 4, 4, 5, 9, 6, 2, 2, 6, 8, 2. Hence the first bootstrap sample contained three replicates of unit 2, two replicates of each of units 4 and 6, and one replicate of each of units 5, 8 and 9. The remaining bootstrap samples were obtained in similar fashion.

First consider estimation of the population mean. Using the jack-knife means $\hat{\theta}_{(i)}$, we find $\bar{\theta}=3.564$ $(=\hat{\theta})$ and hence $\hat{\sigma}_{\mathrm{J}}=0.871$ from (17.1). Using the bootstrap means $\hat{\theta}^{(i)}$ we find $\bar{\bar{\theta}}=3.859$ and hence $\hat{\sigma}_{\mathrm{B}}=0.878$ from (17.2). The variance of the original ten sample values is $s^2=7.59$, so the usual standard error of the mean is $(7.59 \div 10)^{\frac{1}{2}}=0.871$. Thus we confirm the general result $\hat{\sigma}_{\mathrm{J}}=s/\sqrt{n}$ that was established earlier for estimation of the population mean. Slightly more unexpected is the remarkably similar estimate of 0.878 that was obtained from the bootstrapping, given that as few as 20 bootstrap samples were used in this simple example. Note also that the sample mean is an unbiased estimator of the corresponding population quantity, so the jack-knife estimate $\tilde{\theta}$ cannot improve upon the original $\hat{\theta}$ in respect of bias. In fact, since $\bar{\theta}=\hat{\theta}$ as shown above, we have $\tilde{\theta}=10\hat{\theta}-9\bar{\theta}=10\hat{\theta}-9\hat{\theta}=\hat{\theta}$, so the jack-knife estimate of θ is the same as the original estimate of θ.

Table 17.1 Simple illustration of jack-knifing and bootstrapping

Type of sample	Sample members										Mean	Median
Original	1.22	3.46	1.56	2.84	7.03	5.07	1.18	2.00	9.33	1.95	3.564	2.42
Jack-knife (1)	3.46	1.56	2.84	7.03	5.07	1.18	2.00	9.33	1.95		3.824	2.84
(2)	1.22	1.56	2.84	7.03	5.07	1.18	2.00	9.33	1.95		3.576	2.00
(3)	1.22	3.46	2.84	7.03	5.07	1.18	2.00	9.33	1.95		3.787	2.84
(4)	1.22	3.46	1.56	7.03	5.07	1.18	2.00	9.33	1.95		3.644	2.00
(5)	1.22	3.46	1.56	2.84	5.07	1.18	2.00	9.33	1.95		3.179	2.00
(6)	1.22	3.46	1.56	2.84	7.03	1.18	2.00	9.33	1.95		3.397	2.00
(7)	1.22	3.46	1.56	2.84	7.03	5.07	2.00	9.33	1.95		3.829	2.84
(8)	1.22	3.46	1.56	2.84	7.03	5.07	1.18	9.33	1.95		3.738	2.84
(9)	1.22	3.46	1.56	2.84	7.03	5.07	1.18	2.00	1.95		2.923	2.00
(10)	1.22	3.46	1.56	2.84	7.03	5.07	1.18	2.00	9.33		3.743	2.84
Bootstrap (1)	3.46	3.46	3.46	2.84	2.84	7.03	5.07	5.07	2.00	9.33	4.456	3.46
(2)	1.22	1.22	2.84	2.84	7.03	7.03	2.00	9.33	9.33	1.95	4.479	2.84
(3)	1.22	1.22	3.46	1.56	2.84	7.03	7.03	2.00	1.95	1.95	3.026	1.98
(4)	1.22	3.46	3.46	3.46	1.18	2.00	2.00	2.00	9.33	1.95	3.006	2.00
(5)	3.46	3.46	3.46	7.03	7.03	7.03	2.00	2.00	9.33	1.95	4.675	3.46
(6)	3.46	3.46	7.03	5.07	1.18	1.18	9.33	9.33	1.95	1.95	4.394	3.46
(7)	3.46	2.84	2.84	5.07	5.07	1.18	2.00	9.33	1.95	1.95	3.569	2.84
(8)	1.22	1.56	7.03	5.07	5.07	2.00	2.00	9.33	1.95	1.95	3.718	2.00
(9)	3.46	1.56	1.56	2.84	2.84	2.00	2.00	9.33	9.33	1.95	3.687	2.42
(10)	2.84	7.03	7.03	7.03	5.07	1.18	9.33	9.33	1.95	1.95	5.274	6.05
(11)	1.22	3.46	3.46	7.03	7.03	1.18	1.18	9.33	9.33	1.95	4.517	3.46
(12)	3.46	3.46	3.46	1.56	2.84	7.03	7.03	5.07	1.18	1.95	3.704	3.46
(13)	3.46	3.46	3.46	2.84	7.03	7.03	7.03	1.18	1.18	1.18	3.785	3.46
(14)	1.22	1.22	1.22	1.22	1.56	2.84	7.03	5.07	1.18	1.18	2.374	1.22
(15)	1.22	1.22	1.22	1.22	3.46	3.46	1.56	2.84	2.84	5.07	2.411	2.20
(16)	1.22	1.22	2.84	7.03	5.07	1.18	1.18	2.00	9.33	9.33	4.040	2.42
(17)	3.46	3.46	1.56	2.84	7.03	1.18	1.95	1.95	1.95	1.95	2.733	1.95
(18)	7.03	7.03	7.03	5.07	5.07	1.18	1.18	9.33	9.33	1.95	5.420	6.05
(19)	1.22	1.22	1.22	3.46	2.84	7.03	1.18	9.33	9.33	9.33	4.616	3.15
(20)	1.22	3.46	1.56	1.56	2.84	2.84	7.03	1.18	9.33	1.95	3.297	2.84

Now consider estimation of the population median. Using the jack-knife medians $\hat{\theta}_{(i)}$, we find $\bar{\theta} = 2.42$ ($= \hat{\theta}$ again), and hence $\hat{\sigma}_J = 1.26$ from (17.1). Using the bootstrap medians $\hat{\theta}^{(i)}$ we find $\bar{\bar{\theta}} = 3.036$, and $\hat{\sigma}_B = 1.23$ from (17.2). In general, to find the standard error of a sample median we need to know the precise distribution from which the sample was taken (Kendall and Stuart (1969) p. 236). This distribution will rarely be known with certainty, and, if the wrong distribution is assumed, then a very poor estimate may result. For example, the ten observations used in the present illustration were taken from an exponential distribution; if this fact is not known and a normal distribution is assumed instead, an inaccurate standard error will result. In such circumstances, therefore, techniques such as bootstrapping and jack-knifing are extremely useful, and it is also reassuring to note the close agreement between the estimates $\hat{\sigma}_J$ and $\hat{\sigma}_B$ above.

Extension of the foregoing ideas to the multivariate case is very straightforward. The sample data $\boldsymbol{x}_1, \boldsymbol{x}_2, \ldots, \boldsymbol{x}_n$ are now p-variate vectors, while the unknown parameters and hence estimates, $\boldsymbol{\theta}$ and $\hat{\boldsymbol{\theta}}$ respectively, are m-element vectors. This does not affect the physical process of jack-knifing or bootstrapping, however, and the subsamples are obtained in exactly the same manner as before. For each subsample we now obtain the *vectors* $\hat{\boldsymbol{\theta}}_{(i)}$ or $\hat{\boldsymbol{\theta}}^{(i)}$, and hence $\bar{\boldsymbol{\theta}}$ or $\bar{\bar{\boldsymbol{\theta}}}$, depending on which technique is being employed. Instead of the sums of squares in (17.1) and (17.2) we therefore compute the corresponding sum-of-squares and -products matrices:

$$\boldsymbol{S}_J = \left(\frac{n-1}{n}\right) \sum_{i=1}^{n} (\hat{\boldsymbol{\theta}}_{(i)} - \bar{\boldsymbol{\theta}})(\hat{\boldsymbol{\theta}}_{(i)} - \bar{\boldsymbol{\theta}})' \tag{17.3}$$

and

$$\boldsymbol{S}_B = \frac{1}{k-1} \sum_{i=1}^{k} (\hat{\boldsymbol{\theta}}^{(i)} - \bar{\bar{\boldsymbol{\theta}}})(\hat{\boldsymbol{\theta}}^{(i)} - \bar{\bar{\boldsymbol{\theta}}})'. \tag{17.4}$$

These matrices summarize the variances of individual estimates $\hat{\theta}_i$, and the covariances between pairs of estimates $\hat{\theta}_i$, $\hat{\theta}_j$. They can be made the basis of confidence regions for the parameter vector $\boldsymbol{\theta}$ in m-dimensional space. Contours of ellipsoidal regions are given by the boundaries $(\boldsymbol{\theta} - \bar{\boldsymbol{\theta}})'\mathbf{S}_J^{-1}(\boldsymbol{\theta} - \bar{\boldsymbol{\theta}}) = c_1$ or $(\boldsymbol{\theta} - \bar{\bar{\boldsymbol{\theta}}})'\mathbf{S}_B^{-1}(\boldsymbol{\theta} - \bar{\bar{\boldsymbol{\theta}}}) = c_2$, depending on which resampling technique is employed. An example of such an application is provided by Weinberg *et al.* (1984). Further aspects of bootstrapping and jack-knifing are discussed in Efron (1982) and Efron and Gong (1983), the latter in particular being a useful introduction for the non-mathematical reader.

The final resampling technique we consider is *cross-validation.* Superficially this technique bears very close resemblance to jack-knifing, and is often erroneously given the latter name. The resemblance arises because cross-validation also involves omission in turn of each sample value. However, cross-validation is appropriate for assessment of the error inherent in a predictive rule or model, rather than for calculation of the standard error of an estimator. Each time an individual is omitted from the sample, the parameters of the rule or model are estimated afresh from the remaining data, and the estimated rule or model is then applied to the omitted unit. If the model is a strictly predictive one then the discrepancy between actual and predicted value is measured for each omitted unit, and these discrepancies are combined to form some overall assessment measure (such as predictive residual sum-of-squares). If the rule is an allocation rule (such as those in Chapter 12), then the number of omitted units that are wrongly classified usually forms the overall assessment measure.

The idea behind cross-validation is to mimic the behaviour, on future data, of the model or rule under consideration, when only a limited amount of data is available for both estimation of model parameters and assessment of model performance. Using the same data for both tasks leads to grossly biased assessment of performance, as has already been discussed in Chapter 12. By removing each unit when it is to be used for validation *before* estimation of model parameters, we mitigate this effect and obtain a much more accurate assessment of performance. The crucial difference between jack-knifing and cross-validation is that in the former technique the omitted unit takes no further part in each derived subsample, whereas in the latter technique it is used for the validation of the fitted model. The name 'cross-validation' thus derives from this process of alternating estimation and validation across all sample members. Note that cross-validation is also an ideal framework for *choosing* among competing models, since the overall assessment measure can be found for each competing model, and the model with optimum value of the measure can then be selected for future use. This aspect of cross-validation is fully explored by Stone (1974).

The area of intersection of all three resampling procedures as regards practical applications is in the assessment of performance of an allocation rule. Given training sets from each of two populations, a discrimination rule can be constructed by formulating a parametric model and estimating parameters from the data in the training sets (Chapter 12). As has been stressed above and also in Chapter 12, use of the same training sets to assess the performance of this rule will be grossly over-optimistic, i.e. the 'resubstitution' method leads to an estimate e_R of the error rate e for the rule that is heavily biased downwards. A cross-validation estimate e_C was described in Section 12.4, under the alternative name of the 'leave-one-out' error rate, and this error rate has a greatly reduced bias. Unfortunately, it also has much greater sampling variance than e_R, so alternative estimators have been sought that improve on the bias of e_R but do not inflate the variance as much as e_C. Jack-knifing and bootstrapping provide viable procedures. A bootstrap adjustment to e_R has been described in Section 12.4. A jack-knife estimator is obtained by leaving each individual out of the training sets in turn, finding the resubstitution error rate for the remaining individuals, and adjusting e_R in the way outlined earlier for jack-knife estimates. Full technical details of these, and related, schemes are given by McLachlan (1986). Other specifically multivariate applications of cross-validation are in the choice of number of components to use in factor and principal component models (Wold (1978), Golub *et al.* (1979), Eastman and Krzanowski (1982)). Bootstrapping and jack-knifing studies of variability in multivariate derived configurations are still relatively scarce, although the work of Weinberg *et al.* (1984) on variability in INDSCAL analysis is perhaps a pointer to the future in this general area.

Appendix A
Some basic matrix theory

The following introduction does no more than outline the fundamental aspects of matrix theory, to enable the reader unfamiliar with the concepts to follow some of the more algebraic arguments in the preceding text. For a fuller treatment, without unnecessary mathematical embellishment, the texts by Healy (1986) and Graybill (1969) are highly recommended.

A1 Definitions

An $r \times c$ *matrix* $\mathbf{A}$ is defined to be a rectangular array of elements arranged into r rows and c columns, where the elements come from some specified field F. For practical applications (and hence for the purposes of this book) the field that we require is the field of real numbers, and so we shall assume throughout that the elements are real numbers. Symbolically, it is customary to denote the elements of $\mathbf{A}$ by a_{ij} where i refers to the row and j to the column in which the element is located. Thus, in full,

$$\mathbf{A} = \begin{pmatrix} a_{11} & a_{12} & a_{13} & \dots & a_{1c} \\ a_{21} & a_{22} & a_{23} & \dots & a_{2c} \\ \vdots & \vdots & \vdots & & \vdots \\ a_{r1} & a_{r2} & a_{r3} & \dots & a_{rc} \end{pmatrix}$$

which is often written in shorthand as $\mathbf{A} = (a_{ij})$. The element in the ith row and jth column of the matrix is usually referred to as the (i, j)th element of the matrix.

An r-element *vector* is a one-dimensional set of r elements belonging to a field F, which again will be taken to be the field of real numbers. A one-dimensional array can be written either in row form or in column form; the latter is the usual convention. Thus the vector $\boldsymbol{x}$ which has elements $x_1, x_2, \ldots, x_r$ is written as

$$\boldsymbol{x} = \begin{pmatrix} x_1 \\ x_2 \\ \vdots \\ x_r \end{pmatrix}$$

and hence can be treated either as a column of a matrix or indeed as an $r \times 1$ matrix in its own right.

A *scalar* is a single element of the field F, i.e. a single real number for present purposes. Hence a scalar can be treated as a one-element vector, or as a 1×1 matrix. The convention we shall follow, as indicated by the notation above, is to write boldface upper-case letters to denote matrices, boldface italic lower-case letters to denote vectors, and italic letters to denote scalars.

The *transpose* of an $r \times c$ matrix is the $c \times r$ matrix obtained by interchanging the roles of rows and columns. A prime will denote the transpose. Thus for the matrix $\mathbf{A}$ with elements a_{ij} above, the transpose is

$$\mathbf{A}' = \begin{pmatrix} a_{11} & a_{21} & \ldots & a_{r1} \\ a_{12} & a_{22} & \ldots & a_{r2} \\ a_{13} & a_{23} & \ldots & a_{r3} \\ \vdots & \vdots & & \vdots \\ a_{1c} & a_{2c} & \ldots & a_{rc} \end{pmatrix}.$$

Hence if $\mathbf{A}$ is written as (a_{ij}), $\mathbf{A}'$ is written as (a_{ji}). Since a vector is an $r \times 1$ matrix, its transpose is a $1 \times r$ matrix. Thus the transpose of the column vector $\boldsymbol{x}$ above is the row vector $\boldsymbol{x}' = (x_1, x_2, \ldots, x_r)$. Scalars of course remain unaffected by transposition.

A matrix is *square* if it has the same number of rows and columns. It is *symmetric* if its elements are such that $a_{ij} = a_{ji}$ for all pairs of i and j. Thus, for example, the matrix $\mathbf{A} = \begin{pmatrix} 3 & 8 & 12 \\ 8 & 5 & 9 \\ 12 & 9 & 14 \end{pmatrix}$ is symmetric ($a_{12} = a_{21} = 8$, $a_{13} = a_{31} = 12$, $a_{23} = a_{32} = 9$) while $\mathbf{B} = \begin{pmatrix} 3 & 8 & 12 \\ 6 & 5 & 9 \\ 12 & 9 & 14 \end{pmatrix}$ is not (because $a_{12} \neq a_{21}$). Note that if a matrix is symmetric, it equals its transpose (check by writing down $\mathbf{A}'$ and $\mathbf{B}'$ above). For brevity, only the triangular part of a symmetric matrix need be written down (one of the duplicated upper or lower portions being omitted).

The *main diagonal positions* are occupied by the elements a_{ii} of a square matrix (the elements 3, 5, and 14 in **A** and **B** above). The sum of these diagonal elements is called the *trace* of the matrix, denoted by tr. Thus $\text{tr}(\mathbf{A}) = \sum_{i=1}^{n} a_{ii}$ for an $n \times n$ matrix. ($\text{Tr}(\mathbf{A}) = \text{tr}(\mathbf{B}) = 22$ above.)

A *diagonal matrix* has non-zero elements only in the main diagonal positions, and zeros everywhere else (but some of the main diagonal elements can be zero also, of course). The $p \times p$ diagonal matrix

$$\begin{pmatrix} a_1 & 0 & 0 & \dots & 0 \\ 0 & a_2 & 0 & \dots & 0 \\ 0 & 0 & a_3 & \dots & 0 \\ \vdots & \vdots & \vdots & & \vdots \\ 0 & 0 & 0 & \dots & a_p \end{pmatrix}$$

is often written $\text{diag}(a_1, a_2, \dots, a_p)$ for short.

The diagonal matrix with ones in all main diagonal positions is particularly important. It is known as the *identity matrix* and is denoted by $\mathbf{I}$, or by $\mathbf{I}_p$ if its dimension $(p \times p)$ is being emphasized. Thus

$$\mathbf{I} = \begin{pmatrix} 1 & 0 & \dots & 0 \\ 0 & 1 & \dots & 0 \\ \vdots & \vdots & & \vdots \\ 0 & 0 & \dots & 1 \end{pmatrix}.$$

A2 Elementary arithmetic operations

Two matrices can be *added or subtracted* if, and only if, they have the same number of rows and the same number of columns. The sum (difference) of two matrices of like dimensions is the matrix of sums (differences) of their corresponding elements. Thus for the matrices **A** and **B** given above,

$$\mathbf{A} + \mathbf{B} = \begin{pmatrix} 3+3 & 8+8 & 12+12 \\ 8+6 & 5+5 & 9+9 \\ 12+12 & 9+9 & 14+14 \end{pmatrix} = \begin{pmatrix} 6 & 16 & 24 \\ 14 & 10 & 18 \\ 24 & 18 & 28 \end{pmatrix}$$

while

$$\mathbf{A} - \mathbf{B} = \begin{pmatrix} 3-3 & 8-8 & 12-12 \\ 8-6 & 5-5 & 9-9 \\ 12-12 & 9-9 & 14-14 \end{pmatrix} = \begin{pmatrix} 0 & 0 & 0 \\ 2 & 0 & 0 \\ 0 & 0 & 0 \end{pmatrix}.$$

The *scalar multiple* $c\mathbf{A}$ of $\mathbf{A}$ is formed by multiplying each element of $\mathbf{A}$ by the scalar c. Thus for the matrix $\mathbf{A}$ above,

$$3\mathbf{A} = \begin{pmatrix} 3\times 3 & 3\times 8 & 3\times 12 \\ 3\times 8 & 3\times 5 & 3\times 9 \\ 3\times 12 & 3\times 9 & 3\times 14 \end{pmatrix} = \begin{pmatrix} 9 & 24 & 36 \\ 24 & 15 & 27 \\ 36 & 27 & 42 \end{pmatrix}.$$

Addition and subtraction obey the commutative and associative laws, and the usual sign rules apply to brackets. Thus:

$$\mathbf{A} + \mathbf{B} = \mathbf{B} + \mathbf{A},$$
$$\mathbf{A} + (\mathbf{B} + \mathbf{C}) = (\mathbf{A} + \mathbf{B}) + \mathbf{C}$$
$$\mathbf{A} - (\mathbf{B} - \mathbf{C}) = \mathbf{A} - \mathbf{B} + \mathbf{C}.$$

Scalar multiplication obeys the distributive law: $c(\mathbf{A} + \mathbf{B}) = c\mathbf{A} + c\mathbf{B}$. It is easily verified that $\text{trace}(\mathbf{A} \pm \mathbf{B}) = \text{trace}(\mathbf{A}) \pm \text{trace}(\mathbf{B})$, and $\text{trace}(c\mathbf{A}) = c\,\text{trace}(\mathbf{A})$.

For the *product* $\mathbf{AB}$ of the two matrices $\mathbf{A}$ and $\mathbf{B}$ to be defined, it is necessary for the number of columns of $\mathbf{A}$ to be equal to the number of rows of $\mathbf{B}$. If $\mathbf{A}$ is a $p \times r$ matrix, while $\mathbf{B}$ is an $r \times q$ matrix then $\mathbf{C} = \mathbf{AB}$ is a $p \times q$ matrix whose (i, j)th element is given by $c_{ij} = \sum_{k=1}^{r} a_{ik} b_{kj}$. This is the sum of products of the corresponding elements in the ith row of $\mathbf{A}$ and the jth column of $\mathbf{B}$. Thus for the matrices $\mathbf{A}$ and $\mathbf{B}$ given earlier,

$$\mathbf{AB} = \begin{pmatrix} 3\times 3 + 8\times 6 + 12\times 12 & 3\times 8 + 8\times 5 + 12\times 9 & 3\times 12 + 8\times 9 + 12\times 14 \\ 8\times 3 + 5\times 6 + 9\times 12 & 8\times 8 + 5\times 5 + 9\times 9 & 8\times 12 + 5\times 9 + 9\times 14 \\ 12\times 3 + 9\times 6 + 14\times 12 & 12\times 8 + 9\times 5 + 14\times 9 & 12\times 12 + 9\times 9 + 14\times 14 \end{pmatrix} = \begin{pmatrix} 201 & 172 & 276 \\ 162 & 170 & 267 \\ 258 & 267 & 421 \end{pmatrix}.$$

It is clear from the above definition that the order of multiplication is important. Thus if $\mathbf{A}$ is $p \times r$, and $\mathbf{B}$ is a $r \times q$, then $\mathbf{AB}$ can be formed but $\mathbf{BA}$ cannot, as the number of columns (q) of $\mathbf{B}$ does not equal the number of rows (p) of $\mathbf{A}$. For both $\mathbf{AB}$ and $\mathbf{BA}$ to exist, the number of rows of $\mathbf{A}$ must equal the number of columns of $\mathbf{B}$ *and vice-versa.* $\mathbf{AB}$ and $\mathbf{BA}$ are then both square matrices, but not necessarily of the same size. Since the matrices $\mathbf{A}$ and $\mathbf{B}$ of the example above are both 3×3, we can form **BA** as well as **AB**. Here

$$\mathbf{BA} = \begin{pmatrix} 3\times 3+8\times 8+12\times 12 & 3\times 8+8\times 5+12\times 9 & 3\times 12+8\times 9+12\times 14 \\ 6\times 3+5\times 8+9\times 12 & 6\times 8+5\times 5+9\times 9 & 6\times 12+5\times 9+9\times 14 \\ 12\times 3+9\times 8+14\times 12 & 12\times 8+9\times 5+14\times 9 & 12\times 12+9\times 9+14\times 14 \end{pmatrix} = \begin{pmatrix} 217 & 172 & 276 \\ 166 & 154 & 243 \\ 276 & 267 & 421 \end{pmatrix}.$$

From these two examples it is therefore clear that, in general, **AB** need not equal **BA**, i.e. the commutative law does not hold for matrix multiplication. However, the distributive and associative laws do hold:

$$\mathbf{A}(\mathbf{B}+\mathbf{C}) = \mathbf{AB}+\mathbf{AC}$$
$$(\mathbf{B}+\mathbf{C})\mathbf{A} = \mathbf{BA}+\mathbf{CA}$$
$$\mathbf{A}(\mathbf{BC}) = (\mathbf{AB})\mathbf{C}.$$

Also, it is always true that $\text{trace}(\mathbf{AB}) = \text{trace}(\mathbf{BA})$ $(= 792$ for **A** and **B** above).

Since the order of multiplication is important, any multiplication must be carefully specified. The product **AB** is said to be derived either by *pre-multiplication* of **B** by **A** or by *post-multiplication* of **A** by **B**. It can be verified easily that either pre- or post-multiplication of any matrix by a conformable identity matrix leaves the matrix unchanged, i.e. $\mathbf{IA} = \mathbf{AI} = \mathbf{A}$. Pre-multiplication by the diagonal matrix with diagonal elements $d_1, \ldots, d_p$ has the effect of multiplying each element in the ith *row* by d_i, while post-multiplication by such a diagonal matrix multiplies each element in the jth *column* by d_j.

The transpose of a sum or difference of matrices is the sum or difference of their transposes, i.e. $(\mathbf{A} \pm \mathbf{B})' = \mathbf{A}' \pm \mathbf{B}'$. However, the transpose of a product is the *reverse* product of the transposes, i.e. $(\mathbf{AB})' = \mathbf{B}'\mathbf{A}'$. Since transposition of a scalar leaves it unaffected, it is clear that $(c\mathbf{A})' = c\mathbf{A}'$.

Since an r-element vector can be treated as an $(r \times 1)$ matrix, it is evident that all the above definitions can be applied to vectors also. In the following, let the elements of $\boldsymbol{x}$ and $\boldsymbol{y}$ be denoted by x_i and y_i respectively.

1. Two vectors can be added or subtracted if, and only if, they have the same number of elements; the sum (difference) of two r-element vectors is the vector of sums (differences) of their elements. Thus if $\boldsymbol{x}$ and $\boldsymbol{y}$ both have r elements then $(\boldsymbol{x} \pm \boldsymbol{y})' = (x_1 \pm y_1, x_2 \pm y_2, \ldots, x_r \pm y_r)$.

2. The scalar multiple $c\boldsymbol{x}$ of $\boldsymbol{x}$ is formed by multiplying each element of $\boldsymbol{x}$ by the scalar c. Thus $c\boldsymbol{x}' = (cx_1, cx_2, \ldots, cx_r)$.

3. The product of two vectors can only be formed if one is a row vector and the other is a column vector. The *inner product* $\boldsymbol{x}'\boldsymbol{y}$ is formed when the row vector pre-multiplies the column vector. For this to be possible, the two vectors must have the same number of elements. Since it is equivalent to the multiplication of a $(1 \times r)$ matrix by an $(r \times 1)$ matrix, the result is a (1×1) matrix, i.e. a scalar. Following the usual rules of matrix multiplication we have

$$\boldsymbol{x}'\boldsymbol{y} = (x_1, x_2, \ldots, x_r)\begin{pmatrix} y_1 \\ y_2 \\ \vdots \\ y_r \end{pmatrix}$$

$$= x_1y_1 + x_2y_2 + \ldots + x_ry_r = \sum_{i=1}^{r} x_iy_i.$$

The *outer product* $\boldsymbol{xy}'$ is formed when the column vector pre-multiplies the row vector, and this can *always* be done. If $\boldsymbol{x}$ is an r-element vector, while $\boldsymbol{y}$ is an s-element vector, $\boldsymbol{xy}'$ is an $(r \times s)$ matrix with (i, j)th element x_iy_j. For example, let $\boldsymbol{x} = \begin{pmatrix} 2 \\ 3 \\ 4 \end{pmatrix}$, $\boldsymbol{y} = \begin{pmatrix} 6 \\ 7 \\ 8 \end{pmatrix}$, and $\boldsymbol{z} = \begin{pmatrix} 11 \\ 12 \end{pmatrix}$. Then

$$\boldsymbol{x} + \boldsymbol{y} = \begin{pmatrix} 8 \\ 10 \\ 12 \end{pmatrix}, \quad \boldsymbol{x} - \boldsymbol{y} = \begin{pmatrix} -4 \\ -4 \\ -4 \end{pmatrix}, \quad 4\boldsymbol{x} = \begin{pmatrix} 8 \\ 12 \\ 16 \end{pmatrix},$$

$$\boldsymbol{x}'\boldsymbol{y} = 2 \times 6 + 3 \times 7 + 4 \times 8 = 65,$$

$$\boldsymbol{xy}' = \begin{pmatrix} 2 \\ 3 \\ 4 \end{pmatrix}(6 \quad 7 \quad 8) = \begin{pmatrix} 2 \times 6 & 2 \times 7 & 2 \times 8 \\ 3 \times 6 & 3 \times 7 & 3 \times 8 \\ 4 \times 6 & 4 \times 7 & 4 \times 8 \end{pmatrix} = \begin{pmatrix} 12 & 14 & 16 \\ 18 & 21 & 24 \\ 24 & 28 & 32 \end{pmatrix},$$

$$\boldsymbol{zx}' = \begin{pmatrix} 11 \\ 12 \end{pmatrix}(2 \quad 3 \quad 4) = \begin{pmatrix} 11 \times 2 & 11 \times 3 & 11 \times 4 \\ 12 \times 2 & 12 \times 3 & 12 \times 4 \end{pmatrix} = \begin{pmatrix} 22 & 33 & 44 \\ 24 & 36 & 48 \end{pmatrix},$$

but $\boldsymbol{x}'\boldsymbol{z}$, $\boldsymbol{x} + \boldsymbol{z}$, and $\boldsymbol{y}'\boldsymbol{z}$ are all undefined.

Finally, if $\mathbf{A}$ is an $r \times c$ matrix, $\boldsymbol{x}$ an r-element vector, and $\boldsymbol{y}$ a c-element vector, then $\boldsymbol{x}'\mathbf{A}$ is a c-element row vector, $\mathbf{A}\boldsymbol{y}$ is an r-element column vector, and $\boldsymbol{x}'\mathbf{A}\boldsymbol{y}$ is a scalar. Because the last named is scalar, it follows that $\boldsymbol{x}'\mathbf{A}\boldsymbol{y} = (\boldsymbol{x}'\mathbf{A}\boldsymbol{y})' = \boldsymbol{y}'\mathbf{A}'\boldsymbol{x}$. The elements of these vectors and scalars are formed by the usual multiplication rules. Thus if

$$\mathbf{A} = \begin{pmatrix} 5 & 10 & 20 \\ 10 & 20 & 30 \end{pmatrix}, \qquad \boldsymbol{x} = \begin{pmatrix} 2 \\ 4 \end{pmatrix}, \quad \text{and} \quad \boldsymbol{y} = \begin{pmatrix} 1 \\ 2 \\ 3 \end{pmatrix},$$

then

$$\boldsymbol{x}'\mathbf{A} = (2 \quad 4)\begin{pmatrix} 5 & 10 & 20 \\ 10 & 20 & 30 \end{pmatrix}$$

$$= (2 \times 5 + 4 \times 10, 2 \times 10 + 4 \times 20, 2 \times 20 + 4 \times 30) = (50 \quad 100 \quad 160),$$

$$\mathbf{A}\boldsymbol{y} = \begin{pmatrix} 5 & 10 & 20 \\ 10 & 20 & 30 \end{pmatrix}\begin{pmatrix} 1 \\ 2 \\ 3 \end{pmatrix} = \begin{pmatrix} 5 \times 1 + 10 \times 2 + 20 \times 3 \\ 10 \times 1 + 20 \times 2 + 30 \times 3 \end{pmatrix} = \begin{pmatrix} 85 \\ 140 \end{pmatrix},$$

and $\boldsymbol{x}'\mathbf{A}\boldsymbol{y} = (2 \quad 4)\begin{pmatrix} 85 \\ 140 \end{pmatrix} = (50 \quad 100 \quad 160)\begin{pmatrix} 1 \\ 2 \\ 3 \end{pmatrix} = 730$. The reader can check that $\boldsymbol{y}'\mathbf{A}'\boldsymbol{x} = 730$, also.

A3 Determinants and inverses

The *determinant* $|\mathbf{A}|$ of a square $p \times p$ matrix $\mathbf{A}$ is the sum $\sum (-1)^N a_{1i_1}a_{2i_2} \ldots a_{pi_p}$, where the sum is taken over all possible permutations $i_1, i_2, \ldots, i_p$ of the numbers $1, 2, \ldots, p$, and N is the total number of inversions of adjacent pairs of indices necessary to reduce the given permutation $i_1, i_2, \ldots, i_p$ to the standard order $1, 2, \ldots, p$. (An inversion is required whenever a larger index precedes a smaller one in the given permutation.) For example, suppose that $\mathbf{A}$ is the 3×3 matrix $\begin{pmatrix} a_{11} & a_{12} & a_{13} \\ a_{21} & a_{22} & a_{23} \\ a_{31} & a_{32} & a_{33} \end{pmatrix}$. The possible permutations of $1, 2, 3$ are

(i) $1, 2, 3$
(ii) $1, 3, 2$
(iii) $2, 1, 3$
(iv) $2, 3, 1$
(v) $3, 1, 2$
(vi) $3, 2, 1$.

To recover the order $1, 2, 3$ we need the following number of inversions for each of these permutations:

(i) 0;
(ii) 1 (132→123);
(iii) 1 (213→123);
(iv) 2 (231→213→123);
(v) 2 (312→132→123);
(vi) 3 (321→231→213→123).

Hence the determinant is

$$\begin{aligned}|\mathbf{A}| &= (-1)^0 a_{11}a_{22}a_{33} + (-1)^1 a_{11}a_{23}a_{32} + (-1)^1 a_{12}a_{21}a_{33} \\ &\quad + (-1)^2 a_{12}a_{23}a_{31} + (-1)^2 a_{13}a_{21}a_{32} + (-1)^3 a_{13}a_{22}a_{31} \\ &= a_{11}a_{22}a_{33} - a_{11}a_{23}a_{32} - a_{12}a_{21}a_{33} + a_{12}a_{23}a_{31} \\ &\quad + a_{13}a_{21}a_{32} - a_{13}a_{22}a_{31}.\end{aligned}$$

Various alternative computational forms are available, which make systematic calculation of determinants of large matrices easier. Here, however, we are merely concerned with some basic properties of determinants, so we will not discuss these computational forms. Using the above formula, therefore, we can verify that the determinant of the matrix **A** used previously is $\begin{vmatrix} 3 & 8 & 12 \\ 8 & 5 & 9 \\ 12 & 9 & 14 \end{vmatrix} = 79$. If the determinant of a matrix is zero, that matrix is said to be *singular*; otherwise it is *non-singular*. It is therefore clear from the above formula for calculating a determinant that a matrix is singular if all elements in one row or column are zero. (Other conditions for singularity are deduced below.) The following are the most useful properties of determinants.

1. The determinant of *either* a diagonal matrix *or* of a triangular matrix (i.e. one having zeros in all positions to one side of the main diagonal) is given by multiplying together all the diagonal elements.
2. $|c\mathbf{A}| = c^p\,|\mathbf{A}|$ if $\mathbf{A}$ is a $p \times p$ matrix. Multiplying all elements in just one row or column of **A** by the scalar c multiplies the determinant by c.
3. Interchanging any two rows (columns) of the matrix reverses the sign of the determinant. (Hence if any two rows (columns) are equal, the determinant must be zero, i.e. the matrix is singular. Furthermore, use of (2) shows that proportionality of two rows (columns) is a sufficient condition for singularity of the matrix.)
4. Adding the multiple of any row (column) of the matrix to any other row (column) leaves the determinant unchanged.
5. If **A** and **B** are each $p \times p$ matrices, $|\mathbf{AB}| = |\mathbf{A}|\,|\mathbf{B}| = |\mathbf{BA}|$.

The *inverse* of the square $p \times p$ matrix $\mathbf{A}$ is that unique $p \times p$ matrix whose elements are such that $\mathbf{AA}^{-1} = \mathbf{A}^{-1}\mathbf{A} = \mathbf{I}$; it exists if, and only if, $|\mathbf{A}|$ is non-zero, i.e. $\mathbf{A}$ is non-singular. The inverse plays the corresponding role in matrix theory to the reciprocal in ordinary arithmetic or algebra. Thus multiplication by an inverse $\mathbf{A}^{-1}$ can be thought of as 'division' by the matrix $\mathbf{A}$. Clearly $(\mathbf{A}^{-1})^{-1} = \mathbf{A}$. The following properties of inverses are also frequently used:

1. The inverse of a symmetric matrix is symmetric.
2. The inverse of the diagonal matrix $\text{diag}(a_1, a_2, \ldots, a_p)$ is the diagonal matrix $\text{diag}(1/a_1, 1/a_2, \ldots, 1/a_p)$.
3. $(\mathbf{A}')^{-1} = (\mathbf{A}^{-1})'$.
4. $(c\mathbf{A})^{-1} = (1/c)\mathbf{A}^{-1}$.
5. $(\mathbf{AB})^{-1} = \mathbf{B}^{-1}\mathbf{A}^{-1}$ if $\mathbf{A}$ and $\mathbf{B}$ are both square and non-singular.

If $\mathbf{A}^{-1} = \mathbf{A}'$ (so that $\mathbf{A}'\mathbf{A} = \mathbf{AA}' = \mathbf{I}$), then $\mathbf{A}$ is said to be an *orthogonal* matrix. Inversion of a matrix is best done on a computer using a standard program, so methods of calculating inverses will not be discussed here. However, a fundamental idea underlying inversion of a matrix is the *rank* of the matrix, and this must therefore be covered briefly.

A set of p-component vectors is said to be *linearly independent* if it is impossible to write any one of them as some linear combination of the remaining vectors. Thus $\boldsymbol{x} = \begin{pmatrix} 1 \\ 2 \\ 3 \end{pmatrix}$ and $\boldsymbol{y} = \begin{pmatrix} 3 \\ 2 \\ 1 \end{pmatrix}$ are linearly independent, but $\boldsymbol{x} = \begin{pmatrix} 1 \\ 2 \\ 3 \end{pmatrix}$, $\boldsymbol{y} = \begin{pmatrix} 3 \\ 2 \\ 1 \end{pmatrix}$, and $\boldsymbol{z} = \begin{pmatrix} 8 \\ 8 \\ 8 \end{pmatrix}$ are *linearly dependent* since $\boldsymbol{z} = 2\boldsymbol{x} + 2\boldsymbol{y}$. Given a set of linearly dependent vectors, we can always find the largest subset of vectors that are linearly independent. For example, with the set $\boldsymbol{x}, \boldsymbol{y}, \boldsymbol{z}$ above, the largest linearly independent subset contains two vectors (any of the subsets $\boldsymbol{x}, \boldsymbol{y}$; $\boldsymbol{x}, \boldsymbol{z}$; or $\boldsymbol{y}, \boldsymbol{z}$ will do). The *rank* of a matrix is the largest number of linearly independent rows (or, equivalently, columns) of the matrix. It can never be greater than the *smaller* of the two dimensions of the matrix. Of particular interest is the rank of a $p \times p$ square matrix, which is a unique number between 0 and p. If the rank is p, the matrix is said to be *of full rank*; this is a necessary and sufficient condition for the matrix to be non-singular, and hence for its inverse to exist.

For example, the matrix $\mathbf{A} = \begin{pmatrix} 3 & 8 & 12 \\ 8 & 5 & 9 \\ 12 & 9 & 14 \end{pmatrix}$ has full rank 3, because none of its rows (or columns) can be written as a linear combination of the remaining rows (or columns). It can be verified (by checking that $\mathbf{A}^{-1}\mathbf{A} = \mathbf{I}$, and $\mathbf{A}\mathbf{A}^{-1} = \mathbf{I}$) that its inverse is given to 4 decimal places by

$$\mathbf{A}^{-1} = \begin{pmatrix} -0.1392 & -0.0506 & 0.1519 \\ -0.0506 & -1.2911 & 0.8734 \\ 0.1519 & 0.8734 & -0.6203 \end{pmatrix}.$$

The following properties are frequently useful:

(1) $\text{rank}(\mathbf{A}) = \text{rank}(\mathbf{A}')$;
(2) $\text{rank}(\mathbf{A}'\mathbf{A}) = \text{rank}(\mathbf{A}\mathbf{A}') = \text{rank}(\mathbf{A})$;
(3) $\text{rank}(\mathbf{AB}) = \text{rank}(\mathbf{CA}) = \text{rank}(\mathbf{A})$ if $\mathbf{B}$ and $\mathbf{C}$ are non-singular.

A4 Quadratic forms

If $\boldsymbol{x}$ is a p-component vector, and $\mathbf{A}$ is a $p \times p$ symmetric matrix, then the scalar $\boldsymbol{x}'\mathbf{A}\boldsymbol{x}$ has value $\sum_{i=1}^{p}\sum_{j=1}^{p} a_{ij}x_ix_j$ which is a *quadratic form* in the variables $x_1, \ldots, x_p$.

The symmetric matrix $\mathbf{A}$ (and its associated quadratic form) are said to be

(1) *positive definite* if $\boldsymbol{x}'\mathbf{A}\boldsymbol{x} > 0$ for all non-null $\boldsymbol{x}$;
(2) *negative definite* if $\boldsymbol{x}'\mathbf{A}\boldsymbol{x} < 0$ for all non-null $\boldsymbol{x}$;
(3) *positive semi-definite* if $\boldsymbol{x}'\mathbf{A}\boldsymbol{x} \geqslant 0$ for all $\boldsymbol{x}$;
(4) *negative semi-definite* if $\boldsymbol{x}'\mathbf{A}\boldsymbol{x} \leqslant 0$ for all $\boldsymbol{x}$;
(5) *indefinite* if $\boldsymbol{x}'\mathbf{A}\boldsymbol{x}$ can assume positive, negative, or zero values.

A positive-definite symmetric matrix always has full rank and is therefore non-singular.

A5 Latent roots and vectors

The *latent roots* (also known as *eigenvalues* or *characteristic roots*) of the $p \times p$ matrix $\mathbf{A}$ are solutions of the determinantal equation $|\mathbf{A} - \lambda\mathbf{I}| = 0$. Expanding the determinant in the manner outlined earlier yields a polynomial of degree p in λ. Hence, by the fundamental theorem of algebra, the determinantal equation has p solutions and therefore $\mathbf{A}$ possesses p latent roots $\lambda_1, \lambda_2, \ldots, \lambda_p$. In general, for a square matrix $\mathbf{A}$ with real elements, these roots λ_i will be complex numbers. However, if $\mathbf{A}$ is a square *symmetric* matrix then the roots are always real numbers. Since nearly all applications in this book require the roots to be found of either a symmetric variance–covariance matrix or a symmetric function of variance–covariance matrices, we shall concentrate mainly on the symmetric case here. Associated with every latent root λ_i of $\mathbf{A}$ is the *latent vector* (also known as the *eigenvector* or the *characteristic vector*) $\boldsymbol{x}_i$ whose elements satisfy $(\mathbf{A} - \lambda_i\mathbf{I})\boldsymbol{x}_i = \mathbf{0}$ or $\mathbf{A}\boldsymbol{x}_i = \lambda_i\boldsymbol{x}_i$ $(i = 1, \ldots, p)$. Since this equation is indeterminate up to scalar multiplication of $\boldsymbol{x}_i$, it is customary to normalize the vectors so that $\boldsymbol{x}_i'\boldsymbol{x}_i = 1$. As with the inverse of a matrix, the latent roots and vectors will normally be found on a computer using one of the many efficient standard packages that are available (particularly for symmetric matrices), so computational aspects will not be discussed here. For a brief outline, see Krzanowski (1971a). A geometric interpretation of latent roots and vectors has been given in Chapter 1; the following are the more important algebraic properties.

1. $|\mathbf{A}| = \lambda_1\lambda_2 \ldots \lambda_p$, the product of the latent roots.
2. $\text{Trace}(\mathbf{A}) = \lambda_1 + \lambda_2 + \ldots + \lambda_p$, the sum of the latent roots.
3. If $\mathbf{A} = \text{diag}(a_1, a_2, \ldots, a_p)$ then $\lambda_i = a_i$ $(i = 1, \ldots, p)$.
4. If $\mathbf{A}$ is positive definite, then $\lambda_i > 0$ $(i = 1, \ldots, p)$.
5. If $\mathbf{A}$ is positive semi-definite of rank r, then exactly r of the λ_i are positive while the remaining $(p - r)$ λ_i are zero.
6. If $\lambda_i \neq \lambda_j$ then $\boldsymbol{x}_i'\boldsymbol{x}_j = 0$, i.e. the latent vectors corresponding to distinct latent roots are orthogonal. Thus if $\mathbf{L}$ is the matrix whose columns are the $\boldsymbol{x}_i$ and the corresponding λ_i are all distinct, then $\mathbf{L}$ is an orthogonal matrix (i.e. $\mathbf{L}'\mathbf{L} = \mathbf{L}\mathbf{L}' = \mathbf{I}$).

The main use made of latent roots and vectors is in the *spectral decomposition* of a symmetric matrix $\mathbf{A}$, which states that every symmetric matrix $\mathbf{A}$ having distinct latent roots can be written as

$$\mathbf{A} = \lambda_1\boldsymbol{x}_1\boldsymbol{x}_1' + \lambda_2\boldsymbol{x}_2\boldsymbol{x}_2' + \ldots + \lambda_p\boldsymbol{x}_p\boldsymbol{x}_p' = \sum_{i=1}^{p} \lambda_i\boldsymbol{x}_i\boldsymbol{x}_i',$$

where the $\boldsymbol{x}_i$ are normalized to have unit sum of squared elements (i.e. $\boldsymbol{x}_i'\boldsymbol{x}_i = 1$ for all i). In matrix form we can write this decomposition as $\mathbf{A} = \mathbf{L}\mathbf{D}\mathbf{L}'$, where $\mathbf{L}$ is the orthogonal matrix whose ith column is given by $\boldsymbol{x}_i$, while $\mathbf{D}$ is the diagonal matrix whose ith diagonal element is λ_i $(i = 1, \ldots, p)$. Even if the λ_i are not distinct, the elements of the $\boldsymbol{x}_i$ can be chosen to ensure orthogonality of $\mathbf{L}$. It is worth contrasting this decomposition with another frequently used decomposition of a symmetric matrix, the *Cholesky decomposition*. The latter decomposition states that any positive-definite symmetric matrix $\mathbf{A}$ can be written in the form $\mathbf{A} = \mathbf{U}'\mathbf{U}$, where $\mathbf{U}$ is a non-singular *upper triangular matrix* (i.e. zeros everywhere below the main diagonal), and hence $\mathbf{U}'$ is a non-singular lower triangular matrix (i.e. zeros everywhere above the main diagonal). Standard computer programs are also readily available to find the matrix $\mathbf{U}$ for any given matrix $\mathbf{A}$.

Returning to the spectral decomposition, we see that if we pre-multiply $\mathbf{A}$ by $\mathbf{L}'$ and post-multiply by $\mathbf{L}$, we obtain $\mathbf{L}'\mathbf{A}\mathbf{L} = \mathbf{L}'\mathbf{L}\mathbf{D}\mathbf{L}'\mathbf{L} = \mathbf{D}$ (by orthogonality of $\mathbf{L}$). Thus we have *reduced* $\mathbf{A}$ *to diagonal form* using $\mathbf{L}$. If $\mathbf{A}$ is a symmetric matrix while $\mathbf{B}$ is a symmetric positive definite matrix, it is possible to *reduce simultaneously* $\mathbf{A}$ *and* $\mathbf{B}$ *to diagonal form* by finding a matrix $\mathbf{L}$ such that $\mathbf{L}'\mathbf{A}\mathbf{L} = \mathbf{D}$ (diagonal) and $\mathbf{L}'\mathbf{B}\mathbf{L} = \mathbf{I}$. The columns of $\mathbf{L}$ are given by the vectors $\boldsymbol{x}_i$ corresponding to the p solutions λ_i of the *generalized characteristic equation* $(\mathbf{A} - \lambda\mathbf{B})\boldsymbol{x} = \boldsymbol{0}$, while the diagonal elements of $\mathbf{D}$ are again given by the λ_i. The generalized characteristic equation can be solved by reducing it to the form of the ordinary characteristic equation in various ways. The simplest approach is to note that $\mathbf{B}$ must have an inverse $\mathbf{B}^{-1}$ (since it is symmetric and positive definite), to pre-multiply the generalized equation by $\mathbf{B}^{-1}$ thus yielding $(\mathbf{B}^{-1}\mathbf{A} - \lambda\mathbf{I})\boldsymbol{x} = \boldsymbol{0}$, and hence to obtain the λ_i and $\boldsymbol{x}_i$ as the latent roots and vectors of $\mathbf{B}^{-1}\mathbf{A}$. However, $\mathbf{B}^{-1}\mathbf{A}$ is not symmetric, so obtaining the solution in this way means that the efficient latent root and vector computer programs designed for symmetric matrices cannot be used. An alternative approach is to use the Cholesky decomposition on the symmetric positive definite matrix $\mathbf{B}$, and hence obtain a non-singular upper-triangular matrix $\mathbf{U}$ satisfying $\mathbf{B} = \mathbf{U}'\mathbf{U}$, $\mathbf{U}^{-1}\mathbf{U} = \mathbf{I}$ and $(\mathbf{U}')(\mathbf{U}')^{-1} = \mathbf{I}$. Then $(\mathbf{A} - \lambda\mathbf{B})\boldsymbol{x} = \mathbf{0}$ can be equivalently written as $\{\mathbf{U}'(\mathbf{U}')^{-1}\mathbf{A}\mathbf{U}^{-1}\mathbf{U} - \lambda\mathbf{U}'\mathbf{U}\}\boldsymbol{x} = \boldsymbol{0}$ (since $\mathbf{IAI} = \mathbf{A}$), i.e. $\mathbf{U}'\{(\mathbf{U}')^{-1}\mathbf{A}\mathbf{U}^{-1} - \lambda\mathbf{I}\}\mathbf{U}\boldsymbol{x} = \boldsymbol{0}$ (taking $\mathbf{U}'$ and $\mathbf{U}$ outside brackets) or $(\mathbf{C} - \lambda\mathbf{I})\boldsymbol{y} = \boldsymbol{0}$ where $\boldsymbol{y} = \mathbf{U}\boldsymbol{x}$, and $\mathbf{C} = (\mathbf{U}')^{-1}\mathbf{A}\mathbf{U}^{-1}$. Thus λ_i, $\boldsymbol{y}_i$ may be found as the latent roots and vectors of the *symmetric* matrix $\mathbf{C}$, and the $\boldsymbol{x}_i$ then found from $\boldsymbol{x}_i = \mathbf{U}^{-1}\boldsymbol{y}_i$ $(i = 1, \ldots, p)$.

A6 Matrix square root

The spectral decomposition of Section A5 can be used to define the square root of a positive (semi-)definite symmetric matrix $\mathbf{A}$. For such a matrix we can write $\mathbf{A} = \mathbf{LDL}'$, where the latent roots λ_i of $\mathbf{A}$ form the elements of the diagonal matrix $\mathbf{D}$, and the latent vectors $\boldsymbol{x}_i$ form the columns of the orthogonal matrix $\mathbf{L}$. Furthermore, since $\mathbf{A}$ is positive (semi-)definite, the λ_i must all be greater than or equal to zero. Hence we can obtain square roots of all latent roots, and write $\mathbf{D}^{\frac{1}{2}} = \text{diag}(\sqrt{\lambda_1}, \sqrt{\lambda_2}, \ldots, \sqrt{\lambda_p})$. Since $\mathbf{D}$ is diagonal, $\mathbf{D} = \mathbf{D}^{\frac{1}{2}}\mathbf{D}^{\frac{1}{2}}$. Thus $\mathbf{A} = \mathbf{L}\mathbf{D}^{\frac{1}{2}}\mathbf{D}^{\frac{1}{2}}\mathbf{L}'$. But since $\mathbf{L}$ is $p \times p$ orthogonal, then $\mathbf{L}'\mathbf{L} = \mathbf{I}$, and hence we can write

$\mathbf{A} = \mathbf{LD}^{\frac{1}{2}}\mathbf{L}'\mathbf{LD}^{\frac{1}{2}}\mathbf{L}'$. Setting $\mathbf{B} = \mathbf{LD}^{\frac{1}{2}}\mathbf{L}'$ we thus see that $\mathbf{A} = \mathbf{BB} = \mathbf{B}^2$, i.e. $\mathbf{B}$ is the matrix square root of $\mathbf{A}$. We can therefore write $\mathbf{B} = \mathbf{A}^{\frac{1}{2}}$.

A7 Partitioned matrices

Sometimes it is convenient to group certain rows and columns of a matrix together, and to treat the elements contained therein as submatrices of the whole matrix. For example, the 4×4 matrix

$$\mathbf{A} = \begin{pmatrix} 1 & 3 & 5 & 9 \\ 2 & 4 & 5 & 10 \\ 3 & 8 & 10 & 12 \\ 4 & 14 & 16 & 5 \end{pmatrix}$$

may be *partitioned* by grouping together the first two and last two rows and columns:

$$\left(\begin{array}{cc:cc} 1 & 3 & 5 & 9 \\ 2 & 4 & 5 & 10 \\ \hdashline 3 & 8 & 10 & 12 \\ 4 & 14 & 16 & 5 \end{array}\right).$$

If we now write $\mathbf{A}_{11} = \begin{pmatrix} 1 & 3 \\ 2 & 4 \end{pmatrix}$, $\mathbf{A}_{12} = \begin{pmatrix} 5 & 9 \\ 5 & 10 \end{pmatrix}$, $\mathbf{A}_{21} = \begin{pmatrix} 3 & 8 \\ 4 & 14 \end{pmatrix}$, and $\mathbf{A}_{22} = \begin{pmatrix} 10 & 12 \\ 16 & 5 \end{pmatrix}$ then $\mathbf{A}$ can be written as $\mathbf{A} = \begin{pmatrix} \mathbf{A}_{11} & \mathbf{A}_{12} \\ \mathbf{A}_{21} & \mathbf{A}_{22} \end{pmatrix}$. In general, any $r \times c$ matrix $\mathbf{A}$ may be partitioned as

$$\mathbf{A} = \begin{pmatrix} \mathbf{A}_{11} & \mathbf{A}_{12} & \dots & \mathbf{A}_{1k} \\ \mathbf{A}_{21} & \mathbf{A}_{22} & \dots & \mathbf{A}_{2k} \\ \vdots & & & \\ \mathbf{A}_{p1} & \mathbf{A}_{p2} & \dots & \mathbf{A}_{pk} \end{pmatrix}$$

where the only restriction is that all submatrices in a given row (column) of this partition must have the same number of rows (columns). Operations with partitioned matrices follow the same rules as with those that are not partitioned, except that it must be remembered that matrix addition, subtraction, and multiplication is being used in place of the scalar analogues for the unpartitioned case. (Thus the dimensions within each sub-matrix sum, difference, or product must conform.) For example, if $\mathbf{A} = \begin{pmatrix} \mathbf{A}_{11} & \mathbf{A}_{12} \\ \mathbf{A}_{21} & \mathbf{A}_{22} \end{pmatrix}$ with the $\mathbf{A}_{ij}$ as above, and $\mathbf{B} = \begin{pmatrix} \mathbf{B}_{11} & \mathbf{B}_{12} \\ \mathbf{B}_{21} & \mathbf{B}_{22} \end{pmatrix}$

with $\mathbf{B}_{11} = \begin{pmatrix} 1 & 0 \\ 0 & 1 \end{pmatrix}$, $\mathbf{B}_{12} = \begin{pmatrix} 1 & 1 \\ 1 & 1 \end{pmatrix}$, $\mathbf{B}_{21} = \begin{pmatrix} 1 & 1 \\ 1 & 1 \end{pmatrix}$, $\mathbf{B}_{22} = \begin{pmatrix} 1 & 0 \\ 0 & 1 \end{pmatrix}$, then

$$\mathbf{A} + \mathbf{B} = \begin{pmatrix} \mathbf{A}_{11} + \mathbf{B}_{11} & \mathbf{A}_{12} + \mathbf{B}_{12} \\ \mathbf{A}_{21} + \mathbf{B}_{21} & \mathbf{A}_{22} + \mathbf{B}_{22} \end{pmatrix} = \begin{pmatrix} 2 & 3 & 6 & 10 \\ 2 & 5 & 6 & 11 \\ 4 & 9 & 11 & 12 \\ 5 & 15 & 16 & 6 \end{pmatrix}$$

while

$$\mathbf{AB} = \begin{pmatrix} \mathbf{A}_{11}\mathbf{B}_{11} + \mathbf{A}_{12}\mathbf{B}_{21} & \mathbf{A}_{11}\mathbf{B}_{12} + \mathbf{A}_{12}\mathbf{B}_{22} \\ \mathbf{A}_{21}\mathbf{B}_{11} + \mathbf{A}_{22}\mathbf{B}_{21} & \mathbf{A}_{21}\mathbf{B}_{12} + \mathbf{A}_{22}\mathbf{B}_{22} \end{pmatrix}$$
$$= \begin{pmatrix} 15 & 17 & 9 & 13 \\ 17 & 19 & 11 & 16 \\ 25 & 30 & 21 & 23 \\ 25 & 35 & 34 & 23 \end{pmatrix}.$$

In general,

$$\mathbf{A} \pm \mathbf{B} = \begin{pmatrix} \mathbf{A}_{11} \pm \mathbf{B}_{11} & \dots & \mathbf{A}_{1k} \pm \mathbf{B}_{1k} \\ \vdots & & \vdots \\ \mathbf{A}_{p1} \pm \mathbf{B}_{p1} & \dots & \mathbf{A}_{pk} \pm \mathbf{B}_{pk} \end{pmatrix}$$

for

$$\mathbf{A} = \begin{pmatrix} \mathbf{A}_{11} & \mathbf{A}_{12} & \dots & \mathbf{A}_{1k} \\ \mathbf{A}_{21} & \mathbf{A}_{22} & \dots & \mathbf{A}_{2k} \\ \vdots & & & \\ \mathbf{A}_{p1} & \mathbf{A}_{p2} & \dots & \mathbf{A}_{pk} \end{pmatrix}$$

and

$$\mathbf{B} = \begin{pmatrix} \mathbf{B}_{11} & \mathbf{B}_{12} & \dots & \mathbf{B}_{1k} \\ \mathbf{B}_{21} & \mathbf{B}_{22} & \dots & \mathbf{B}_{2k} \\ \vdots & & & \\ \mathbf{B}_{p1} & \mathbf{B}_{p2} & \dots & \mathbf{B}_{pk} \end{pmatrix}$$

while

$$\mathbf{AB} = \begin{pmatrix} \sum_{j=1}^{p} \mathbf{A}_{1j}\mathbf{B}_{j1} & \dots & \sum_{j=1}^{p} \mathbf{A}_{1j}\mathbf{B}_{jk} \\ \vdots & & \vdots \\ \sum_{j=1}^{p} \mathbf{A}_{qj}\mathbf{B}_{j1} & \dots & \sum_{j=1}^{p} \mathbf{A}_{qj}\mathbf{B}_{jk} \end{pmatrix}$$

for

$$\mathbf{A} = \begin{pmatrix} \mathbf{A}_{11} & \mathbf{A}_{12} & \dots & \mathbf{A}_{1p} \\ \mathbf{A}_{21} & \mathbf{A}_{22} & \dots & \mathbf{A}_{2p} \\ \vdots & & & \\ \mathbf{A}_{q1} & \mathbf{A}_{q2} & \dots & \mathbf{A}_{qp} \end{pmatrix}$$

and

$$\mathbf{B} = \begin{pmatrix} \mathbf{B}_{11} & \mathbf{B}_{12} & \dots & \mathbf{B}_{1k} \\ \mathbf{B}_{21} & \mathbf{B}_{22} & \dots & \mathbf{B}_{2k} \\ \vdots & & & \\ \mathbf{B}_{p1} & \mathbf{B}_{p2} & \dots & \mathbf{B}_{pk} \end{pmatrix}.$$

A8 Vector differentiation

Suppose that f is a function of the p variables $x_1, x_2, \dots, x_p$. In many statistical procedures it is necessary to find the derivatives $\frac{\partial f}{\partial x_i}$ for each x_i $(i = 1, \dots, p)$, and this can conveniently be arranged in vector form. Let

$$\boldsymbol{x} = \begin{pmatrix} x_1 \\ x_2 \\ \vdots \\ x_p \end{pmatrix}$$

and define the derivative of f with respect to the vector $\boldsymbol{x}$ by

$$\frac{\partial f}{\partial \boldsymbol{x}} = \begin{pmatrix} \frac{\partial f}{\partial x_1} \\ \frac{\partial f}{\partial x_2} \\ \vdots \\ \frac{\partial f}{\partial x_p} \end{pmatrix}.$$

Then using term-by-term differentiation, it is easy to derive the following results for certain common functions.

(1) If $\boldsymbol{a}' = (a_1, \dots, a_p)$ and $f = \boldsymbol{a}'\boldsymbol{x} = \sum_{i=1}^{p} a_i x_i$, then $\frac{\partial f}{\partial \boldsymbol{x}} = \boldsymbol{a}$.

(2) If $\mathbf{A} = (a_{ij})$ is a $p \times p$ symmetric matrix and $f = \boldsymbol{x}'\mathbf{A}\boldsymbol{x} = \sum_{i=1}^{p} \sum_{j=1}^{p} a_{ij} x_i x_j$, then $\frac{\partial f}{\partial \boldsymbol{x}} = 2\mathbf{A}\boldsymbol{x}$. $\left(\text{Hence, on setting } \mathbf{A} = \mathbf{I} \text{ we see that } \frac{\partial f}{\partial \boldsymbol{x}} = 2\boldsymbol{x} \text{ for } f = \boldsymbol{x}'\boldsymbol{x}.\right)$

Appendix B
Postscript: further developments

In the dozen or so years that have elapsed since the first appearance of this book, computing power has increased dramatically and multivariate analysis has proved to be a fertile area for further research. There have been many new developments, some of which are potentially far-reaching and merit an update to the original text. However, it would not be possible to give full details of each new development without a huge increase in the size of the book, which would detract from its original purpose and intended audience. The aim of this appendix is therefore to present just a broad overview, tracing the main lines of development that have taken place, focusing on those aspects that are likely to have the most enduring impact, and picking out the most useful features from the point of view of the practical user. At the same time as providing such an update, the opportunity has been taken to revisit any aspects of the original treatment that may need some amplification or clarification.

To preserve the flavour of the original, this appendix is presented under the same headings that make up the main parts of the book. All previous notation and conventions are of course retained.

B1 Looking at multivariate data

Direct visualizations

Modern personal computers with windows systems offer many more possibilities for direct plotting of multidimensional data than was the case in the mid 1980s. The idea of producing a scatterplot matrix, for example, in which the scatterplots for all $p(p - 1)$ ordered pairs of variables are obtained and arranged in matrix fashion, had already been suggested but was deemed to be unduly confusing (see p. 35). Subsequent developments, allied with improved computing systems and monitor displays, now make this a much more useful concept. These developments include the linking of the separate plots by *tagging* individual points so that they are identified in each of the plots, by *brushing* the plots so that sets of points can be either highlighted or deleted from all plots, or by *painting* different sets of points in different colours to highlight grouping or overplotting.

The development of hardware also quickly opened up the possibility of three-dimensional scattergram depiction, in which the coordinate axes are identified with three chosen variables and in which the right sort of perspective can be achieved by depth-cueing devices such as *kinematic* or *stereoscopic* displays. This led on to the idea of rotating the display, and modern visualization packages now routinely include such facilities as dynamic rotations of three-dimensional scatterplots, tourplots, and rocking rotations. Furthermore, windowing systems allow several such displays to be presented in different windows and any observations highlighted or deleted in one window can be treated similarly in the other windows. A good overview of the whole area is provided by Cleveland (1993); technical details underlying the methodology, plus a summary of currently available systems, are given by Wegman and Carr (1993) and Young *et al.* (1993); while a detailed look at one particular system (XGobi) is provided by Swayne *et al.* (1997).

Projection methods

The idea of seeking low-dimensional projections of a multivariate data set, in such a way as to highlight specific features or reveal structure in the data, has been briefly discussed in Section 2.2.7 under the name 'projection pursuit'. This idea dates back to Friedman and Tukey (1974), and the projections that were available at the time of publication are also briefly reviewed in the same section. The main drawback that was seen at that time with this approach was the amount of computing necessary to find a particular projection using iterative numerical techniques. With the great strides made in the intervening years as regards computing power, this objection now has little force and much work has been done on defining various new 'projection indices' that can be optimized numerically to highlight specific features of the data. For example, Cook *et al.* (1993) discuss some possible indices, including the 'holes' index (to find projections displaying large areas of empty space at the centre of the data), the 'central mass' index (to find projections in which the data are concentrated centrally) and the 'skewness' index (to find projections highlighting any asymmetry in the data) while Eslava and Marriott (1994) discuss indices aimed at finding projections that emphasize any group structure in the data. Other more general aspects of projection pursuit are discussed by Nason (1995) and Posse (1995).

Principal component analysis is of course a particular form of projection pursuit, in which the projection index is the sample variance and for which analytical rather than numerical optimization is possible. A comprehensive account of principal component analysis from a practical perspective has been given by Jackson (1991). Some further connections between principal component analysis and projection pursuit are drawn by Bolton and Krzanowski (1999), who show that principal components result from minimising a projection index based on log-likelihood of the sample assuming multivariate normality—thereby relating the analysis back to the original motivation that "interesting" projections are ones showing departures from normality.

Non-linear generalizations

Projection pursuit in general, and principal component analysis in particular, are *linear* techniques in that they work by deriving new quantities $Y_1, Y_2, \ldots, Y_s$ which are linear combinations of the original variables $X_1, X_2, \ldots, X_p$. A natural avenue of investigation is thus to generalize a technique such as principal component analysis by allowing some form of non-linearity, and a number of ways have been explored for doing this.

The earliest attempt was by Gnanadesikan (1977), who suggested simply adding squares of the variables, pairwise cross-products between them, and perhaps higher-order powers and products, to the list of input variables, and then performing an ordinary principal component analysis on this expanded set of variables. To illustrate that the idea had merit, he applied the technique to an artificial set of data that lay exactly on a circle and used the reasoning set out on p. 451 to recover the equation of this circle from the elements of the eigenvector corresponding to the zero eigenvalue of the covariance matrix calculated on the expanded set of variables. He argued that by establishing relationships among the variables in this set, we are able to identify non-linear relationships among the original variables.

However, Flury (1995) criticized this approach on the grounds that it did not constitute a true generalization of principal component analysis, because the fundamental property of minimizing orthogonal distance from points to principal components (see p. 56) is no longer necessarily upheld when powers and products of variables are added to the data set. Flury suggested that a valid generalization of a technique requires any particular properties of that technique to remain true under the generalization, and hence considered the idea of principal curves put forward by Hastie and Stuetzle (1989) to be a more successful definition of non-linear principal components. The principal curve of a set of data points is a curve in the space containing those points, such that each point $\boldsymbol{x}$ on the curve is the mean of all those data points whose closest point on the curve is $\boldsymbol{x}$. For data from a multivariate normal distribution, or more generally from any member of the elliptic family (see p. 210), principal curves coincide with principal components, but data arising from other distributions have more general, non-linear, principal curves. Hastie and Stuetzle (1989) discuss these ideas much more fully, derive the principal curves of various distributions, describe iterative smoothing algorithms for fitting such curves to data, and provide illustrations of their use.

An alternative framework for non-linear multivariate data analysis was being simultaneously developed during the 1980s by a group of Dutch statisticians who worked under the collective pseudonym Gifi, and this framework provides a second non-linear generalization of principal component analysis. The central concept in this approach is that of homogeneity analysis, which requires an index $\phi(X_1, X_2, \ldots, X_p) = \frac{1}{np}\sum_{j=1}^{p}\sum_{i=1}^{n}(x_{ij} - z_i)^2$ measuring the loss of homogeneity among the variables $X_1, X_2, \ldots, X_p$ to be defined in terms of an unknown variable Z. Ordinary principal component analysis is fitted into this framework by defining Z to be the linear combination of the X_i that minimizes the loss of homogeneity, while extension to non-linear principal component analysis is obtained by allowing non-linear transformations of each variable in the function ϕ. Categorical and mixed variables can easily be incorporated into the system by requiring the non-linear transformations to return real values for each category of any categorical variable, so the system is very general. A set of components is derived in successive fashion, subjecting each new component to the restriction that it is orthogonal to all previous components. Solutions are obtained by a numerical alternating least squares procedure in which the process iterates between finding optimum weighting of variables for a given Z and optimum Z for a given set of weightings until convergence is achieved. The ideas of homogeneity analysis are set out by Heiser and Meulman (1995), while a full description of the non-linear approach including non-linear generalizations of many other techniques as well as principal component analysis is given by Gifi (1990).

Displaying proximity matrices

The classical algebraic solution to the problem of finding a low-dimensional representation of n objects by points $P_1, P_2, \ldots, P_n$ such that inter-point distances d_{ij} approximate the corresponding inter-object proximities δ_{ij}, contained in an $n \times n$ matrix $\boldsymbol{\delta}$, has been given in Section 3.2 under the name principal coordinate analysis. However, it should be noted that problems may be encountered with this method more frequently than has perhaps been implied by the account in that section. The main difficulty is that the scaling theorem on p. 105 only presents a *necessary* condition for an exact representation to be possible, namely that the input proximities satisfy the metric inequality. That this condition is not also a sufficient one means that proximities satisfying it may not yield an exact representation. To illustrate this fact, Gower and Legendre (1986) presented the following simple example of a 4×4 proximity matrix:

$$\delta = \begin{pmatrix} 0 & & & \\ 2 & 0 & & \\ 2 & 2 & 0 & \\ 1.1 & 1.1 & 1.1 & 0 \end{pmatrix}.$$

Consideration of all triples in this matrix easily establishes that the metric inequality holds. Suppose now that the rows of the matrix are to be represented by the four points P_1, P_2, P_3, and P_4. Then P_1, P_2, P_3 clearly form an equilateral triangle of side 2 units, and P_4 is equidistant at 1.1 units from each vertex of the triangle. But the smallest distance that P_4 can be from these vertices is when it is coplanar with them and at their centroid, giving a minimal distance of 1.15 units. Thus a valid Euclidean representation of objects by points does not exist, despite the metric inequality holding among the proximities.

In practice, the test of whether or not a valid representation exists will come at step 3 of the numerical procedure given on p. 107. If any of the calculated eigenvalues are negative, then a true Euclidean representation does not exist, and the discussion on pp. 108 and 109 becomes pertinent in deciding on the degree of approximation that can be achieved. Of course, treating the proximities in non-metric fashion is always an option for achieving a satisfactory representation, and any of the other approaches described in Section 3.4 may be used to this end. With the recent increases in routinely available computing power, such iterative numerical techniques no longer present any major disadvantages and indeed many variations on this theme have been proposed for scaling proximity data. The variations are too numerous to summarize here, but a good account of the whole area may be found in the monograph by Borg and Groenen (1997).

Two-way graphical representations

The biplot (Section 4.2) is a technique that has proved to be very useful in descriptive multivariate analysis. It is essentially a means of superimposing projections of the original coordinate axes (representing the measured variables as vectors) on principal component plots of scores (representing the sample units as points), thereby obtaining simultaneous representation of variables and units on a single plot.

An alternative way of viewing the biplot vectors is to imagine p pseudo-units, the ith of which has value r on the ith variable and zero on all other variables. Adding these p pseudo-units on to the principal component scores plot by the technique of Gower (1968), using the fact that principal component analysis operates on Euclidean distances among the units, produces an extra p points on the plot. Varying the value of r from zero to some desired upper limit l will trace out a path from the origin generated by each of these pseudo-units, and because principal component plots are obtained from (linear) orthogonal projections, these paths are all straight lines. The appropriate upper limit l is given by the length of the corresponding vector $\boldsymbol{g}_i$, which in turn depends on whether (4.11), (4.12), or (4.13) is used for the basic factorization of $\boldsymbol{Y}$. In particular, simply taking $l = 1$ will generate unit-length biplot vectors.

Taking this view opens up the area of biplots to possible generalization. The first generalization was by Gower and Harding (1988), who allowed any distance function embeddable in a Euclidean space to be used in place of Euclidean distance as specified above. This leads to the superimposition of biplot vectors on arbitrary scaling diagrams, but the paths traced out by these vectors are now in general non-linear ones. Some further developments of this idea then led to incorporation of qualitative variables into the framework, and also to the possibility of non-metric scaling being used at the first stage of the process. An overview of all these developments has been given by Gower (1995). Meulman and Heiser (1993) also considered the generalization of biplots to the non-linear case, but did so by embedding the biplot within a non-linear mapping (i.e. on minimizing one of the loss functions L or L^* of Section 2.2.7) rather than within principal component analysis (i.e. on minimizing loss function V of Section 2.2.7).

Useful monographs published recently on the material covered in Chapter 4 include the ones by Gower and Hand (1996) on biplots and by Greenacre (1993) on the practical aspects of correspondence analysis. All the two-way displays in Chapter 4 emanate from singular value decompositions (Section 4.1) of appropriate matrices, thereby defining *bilinear* models for two-way data in matrix form. Recent interest in types of data such as those obtained in spectroscopic work (see Section B4 below), where a third mode (e.g. wavelength, time, distance) is added to the basic two-way structure, has led to the study of trilinear or more general multilinear models. A good account of developments in this area can be found in the survey paper by Leurgans and Ross (1992).

B2 Samples, populations, and models

The multivariate normal distribution remains the essential model used as a basis for most analytical techniques of inference for multivariate data.

Some efforts have been made to widen the scope by allowing other members of the elliptic family of distributions (Section 7.4) to provide the random component, but resulting methodology has not yet found its way into general usage. Fang and Zhang (1990) provide a comprehensive summary of available results.

A more popular direction of development has been a much more pragmatically numerical one, in which joint probability density functions are either estimated nonparametrically from the data or are simulated in a variety of ways for subsequent incorporation into inferential techniques. The latter idea is deferred to the next section; here we just give a brief mention of nonparametric density estimation.

Various different approaches have been suggested for obtaining estimates of probability densities, but probably the universally most popular method is that of *kernel* density estimation. This method is essentially a smoothed generalization of the ordinary histogram. Suppose that the p-variate observations $\boldsymbol{x}_1, \boldsymbol{x}_2, \ldots, \boldsymbol{x}_n$ constitute a random sample of size n from a population with probability density function $f(\boldsymbol{x})$. Then the kernel estimate of this density is given by

$$\hat{f}(\boldsymbol{x}) = n^{-1}h^{-p}\sum_{i=1}^{n} K\left(\frac{\boldsymbol{x} - \boldsymbol{x}_i}{h}\right),$$

where $K(\cdot)$ is a non-negative and symmetric kernel function that integrates to one and h is a smoothing parameter generally termed the *bandwidth*.

The consensus of research done on this approach is that choice of appropriate bandwidth is critical, but many different kernel functions may be used with little effect on the results. A product of p Gaussian kernels $K(y) = \frac{1}{\sqrt{2\pi}}\exp\left(-\frac{1}{2}y^2\right)$ is generally the default choice, but the optimal bandwidth has to be estimated from the data. One popular way of doing this is by cross-validation: taking a particular value of h we omit each data point $\boldsymbol{x}_i$ in turn and compute the density $\hat{f}(\boldsymbol{x}_i)$ at that point from the other values, multiply the n resulting densities together to give a (pseudo-)likelihood of the sample, and then find the value of h that optimizes this likelihood by numerical search.

The most common applications of density estimation occur in discriminant analysis and cluster analysis, as well as in those applications that require estimation of quantities depending on the density. Notable among the latter are some forms of projection pursuit, and estimation of distances or dissimilarities between populations.

Full details of these and related ideas are given in the monographs by Silverman (1986) and Scott (1992). A very recent approach to density estimation uses the orthogonal set of basis functions known as wavelets (Tribouley 1995; Pinheiro and Vidakovic 1997), but it is too early yet to judge whether this will be a more useful approach than kernel density estimation.

B3 Analysing ungrouped data

The results of Chapter 8 remain the standard ones as regards parametric inference, but computer-intensive (nonparametric) methods are now gaining increasing prominence. Bootstrapping and jackknifing have already been introduced in Chapter 17, mainly as mechanisms for obtaining nonparametric estimates of sampling error and hence for constructing confidence regions for unknown population parameters. These applications continue to be among the most useful aspects of these techniques, and various refinements and improvements have been made to them in the past decade; good sources of reference for the reader are the review article by Young (1994) with its published discussion by various prominent researches, and the monographs by Efron and Tibshirani (1993) and Davison and Hinkley (1997).

In addition to facilitating such techniques as bootstrapping and jackknifing, the great improvements in computing power have led to increasing emphasis being placed on simulation and data-based inference for those situations in which analytical results are difficult to obtain. The most obvious place for such an approach is in hypothesis testing, where the requisite critical value may not be derivable analytically for a test statistic in a complex situation. However, if we can generate repeated samples from the underlying probability distribution $g(\boldsymbol{\theta})$ of the data under the null hypothesis, then by constructing the empirical distribution function of the consequent test statistic values we can find an appropriate cut-off point to satisfy any pre-specified requirements on size of test. Another common area of application of such an approach is in Bayesian inference, where some statistic obtained from the posterior distribution of the parameters given the data is generally required. Again, analytical expressions may not be derivable easily in complex situations but estimates could be obtained readily using samples generated from the appropriate distribution.

Computational issues arising from random variable generation are discussed by Boswell *et al.* (1993). Often, however, it may not be possible to generate samples directly from an arbitrary high-dimensional joint distribution, so indirect approaches must be sought. One of the most prolific areas of research in the 1990s has been into the use of the Markov Chain Monte Carlo method for obtaining such indirect samples. The basic idea behind the method is as follows. Suppose that we wish to generate a sample from $g(\boldsymbol{\theta})$, and we can construct a Markov Chain which is straightforward to sample from and whose equilibrium distribution is $g(\boldsymbol{\theta})$. We then run the chain (i.e. simulate successive values from it) for a suffficiently long time, termed the 'burn-in' period, so that at the end of this period the chain should have stabilized and values sampled from it must be ones from the equilibrium distribution $g(\boldsymbol{\theta})$. These values can then be used as a basis for summarizing those features of $g(\boldsymbol{\theta})$ that are of interest.

Various ways of constructing an appropriate chain have been explored; the most popular method to have emerged is that of *Gibbs sampling*. We suppose that we want a random sample of n realizations of the random vector $\boldsymbol{X} = (X_1, X_2, \ldots, X_p)'$. The Gibbs sampler assumes that the conditional distribution of each X_i given values for all the other Xs, i.e. $f(X_i|X_1 = x_1, X_2 = x_2, \ldots, X_{i-1} = x_{i-1}, X_{i+1}, = x_{i+1}, \ldots, X_p = x_p)$, is fully specified for all i and that samples from it can be straightforwardly and efficiently generated. Then we start with an arbitrary set of values $x_1^{(0)}, x_2^{(0)}, \ldots, x_p^{(0)}$ and generate successively

$$x_1^{(1)} \text{ from } f\left(X_1|x_2^{(0)}, \ldots, x_p^{(0)}\right),$$

$$x_2^{(1)} \text{ from } f\left(X_2|x_1^{(1)}, x_3^{(0)}, \ldots, x_p^{(0)}\right),$$

$$x_3^{(1)} \text{ from } f\left(X_3|x_1^{(1)}, x_2^{(1)}, x_4^{(0)}, \ldots, x_p^{(0)}\right),$$

and so on, up to $x_p^{(1)}$ from $f(X_p|x_1^{(1)}, x_2^{(1)}, \ldots, x_{p-1}^{(1)})$. These p simulations constitute one iteration of the process, at the end of which we have a new 'value' $\boldsymbol{x}^{(1)} = (x_1^{(1)}, x_2^{(1)}, \ldots, x_p^{(1)})'$ of the random vector. After the burn-in of i such iterations we arrive at $\boldsymbol{x}^{(i)} = (x_1^{(i)}, x_2^{(i)}, \ldots, x_p^{(i)})'$, which constitutes one observation from the required distribution $g(\boldsymbol{\theta})$. Repeating the whole process n times will thus yield a sample of size n from this distribution, for use as a basis for data-based inferences.

The above is clearly only a broad impression of the method; for further details see Gilks *et al.* (1996) or Gamerman (1997). Some typical applications described in the former include Bayesian mapping of disease, image analysis, and radiocarbon dating, while the method is more generally useful in making either probability or inferential calculations under very complex models where analytical calculations are not possible.

B4 Analysing grouped data

Descriptive methods

The standard descriptive method for grouped data, namely canonical variate analysis, has been described in Section 11.1. However, the method does have some shortcomings in particular circumstances.

First, it does not provide an orthogonal projection of the usual multidimensional model of the data (see p. 49) but, as explained on pp. 299 and 300, the canonical variate space is derived by deforming the axes in the original data space. In some applications it may be preferable to seek an *orthogonal* projection of the original data space that maximally separates existing groups, which means that we need to combine principal component analysis and canonical variate analysis in some way.

A second shortcoming arises when there are more variables than units in the data matrix. This type of situation has become much more prevalent since the development of high-precision instrumentation and automatic data recording devices which allow many readings to be taken on each individual in a sample. Many such situations are now encountered, particularly in the area of chemometrics; a typical example arises in infrared spectroscopy, where relatively few chemical samples (the units, often less than 100) are available but readings are taken on each sample at a very large number of wavelengths (the variables, often in excess of 1000). The problem with such data matrices is that the within-groups covariance matrix $\boldsymbol{W}$ is singular, and this may lead to numerical difficulties in extracting the necessary eigenvalues and eigenvectors for canonical variate analysis.

A third shortcoming is that canonical variate analysis is only applicable in the case of continuous (i.e. numerical) data, but mixed data sets containing both continuous and categorical variables are frequently encountered in practice.

Various approaches have been proposed for coping with these shortcomings. The first two can be handled by an adaptation of projection pursuit, already metioned in Section B1 above for description of unstructured data. Here we restrict ourselves to orthogonal projections only, to overcome the first shortcoming, but retain the canonical variate criterion (11.5) as the projection index for optimization. An iterative numerical approach now has to be adopted in place of an analytical solution, but the second shortcoming when there are more variables than units can be easily overcome within the same overall framework. General computational and algorithmic issues associated with such numerical procedures are discussed by Kiers (1995), while the specific 'orthogonal canonical variate analysis' is addressed by Krzanowski (1995) and Kiers

(1997). Once the general scheme is in place, we can also change the projection index to suit particular objectives. For example, Krzanowski (1998) considers some nonparametric indices that measure 'extremeness' of one group relative to all the others. Optimizing these indices numerically provides useful data displays in problems involving group selection (such as agricultural variety trials, for example).

To cope with the third shortcoming, that of handling general types of data, we can exploit the connection between canonical variate analysis and metric scaling of the matrix of Mahalanobis D^2 values between every pair of groups (p. 302). In the case of general rather than specifically continuous data, we simply need to replace the D^2 values by some more appropriate measures of distance between pairs of populations. Now distance between two *individuals* is a very familiar concept, and many distance functions are available to the user: a particularly useful one is Gower's general function d_{ij} defined on p. 29. All we need is a suitable distance between groups to be defined from such distances between individuals, and such a definition has been given by a number of authors including Rao (1982). If d_{ij} as above is the distance between individuals i and j, and if the data matrix is partitioned a priori into g groups $\pi_1, \ldots, \pi_g$ with n_r individuals in group π_r $(r = 1, \ldots, g)$, then the squared distance between π_r and π_s is given by

$$\delta_{rs}^2 = \frac{1}{n_r n_s} \sum_{i \epsilon \pi_r} \sum_{j \epsilon \pi_s} d_{ij}^2 - \frac{1}{2n_r^2} \sum_{i \epsilon \pi_r} \sum_{j \epsilon \pi_r} d_{ij}^2 - \frac{1}{2n_s^2} \sum_{i \epsilon \pi_s} \sum_{j \epsilon \pi_s} d_{ij}^2.$$

Replacing the matrix of Mahalanobis distances by the matrix of these distances and conducting metric scaling will give a representation of the differences between groups that generalizes canonical variate analysis to all types of data. Krzanowski (1994) gives further details, including how to superimpose points representing individuals on the metric scaling diagram.

Discrimination and classification

This area has traditionally been very heavily researched, and a plethora of methods exists for obtaining classification rules and assessing their performance under a wide variety of external conditions. The very comprehensive text by McLachlan (1992) gives full details of many of these methods, and provides references for those not covered in detail.

The treatment of discrimination and classification in Chapters 12 and 13 is pragmatically orientated to the relatively few general-purpose methods that find the most use in practice. In fact, through a mixture of availability in computer packages, ease of understanding, and robustness to variety of data types, the linear discriminant function is overwhelmingly the most used classification rule in practical applications, and the quadratic discriminant function is the one many people turn to when the linear function seems to be inappropriate. For these reasons, in this brief update on discrimination and classification we will highlight just three recent ideas; all of them involve either some mixture of, or variation on, linear and quadratic functions.

Friedman (1989) proposed the idea of 'regularized discriminant analysis', which is essentially a compromise between linear and quadratic discrimination. The classification rule is the quadratic rule (12.17), but the estimates $\boldsymbol{S}_1$ and $\boldsymbol{S}_2$ of $\boldsymbol{\Sigma}_1$ and $\boldsymbol{\Sigma}_2$ respectively are first 'shrunk' towards the pooled estimate $\boldsymbol{S}$ and the result is then further shrunk towards the identity matrix before being used. Specifically, we first form

$$\hat{\boldsymbol{\Sigma}}_i(\lambda) = \{(1-\lambda)(n_i-1)\boldsymbol{S}_i + \lambda(n-2)\boldsymbol{S}\}/\{(1-\lambda)(n_i-1)+\lambda(n-2)\}$$

where $i = 1, 2$ and $n = n_1 + n_2$, and then

$$\hat{\boldsymbol{\Sigma}}_i(\lambda, \gamma) = (1-\gamma)\hat{\boldsymbol{\Sigma}}_i(\lambda) + \gamma c \boldsymbol{I}_p$$

where $\boldsymbol{I}_p$ is the $p \times p$ identity matrix and $c = \{\text{tr}\,\hat{\boldsymbol{\Sigma}}_i(\lambda)\}/p$.

The two shrinkage parameters λ and γ are determined from the training data by minimization of the cross-validated overall error rate. The first shrinkage of the individual $\boldsymbol{S}_i$ towards the common $\boldsymbol{S}$ helps in those cases where p is large relative to n, in which case the individual $\boldsymbol{S}_i$ are either very unstable or singular, while the second shrinkage has the effect of decreasing the larger eigenvalues of $\hat{\boldsymbol{\Sigma}}_i(\lambda)$ and increasing the smaller ones, thus correcting for the bias inherent in the estimates provided by these eigenvalues. McLachlan (1992) gives further details, and summarizes the results obtained in various comparisons of this method with others.

Hastie *et al.* (1994) describe a data-based method for 'flexible' discriminant analysis. They take as their starting point the sample-space partition approach to multiple discriminant analysis described in Section 13.6.1, which leads to classifications derived from diagrams such as Fig. 13.2. They note that the coordinate axes in these diagrams are formed from linear discriminant functions, that such functions are equivalent to linear regression functions using optimal scorings to represent the groups, and that the resulting separating boundaries between groups are linear. They consider this linearity to be restrictive, so propose replacing the linear regression implicit in the discriminant analysis by some form of nonparametric regression (see Section B5 below). This produces a more flexible classification procedure with possibly non-linear boundaries, which they illustrate on various data sets.

The third recent modification of linear discriminant analysis is the method of 'high-breakdown' linear discriminant analysis proposed by Hawkins and McLachlan (1997). These authors note that the sample means and pooled covariance matrix used as estimates of parameters in the usual linear discriminant function are notoriously susceptible to outliers, and that the problem is compounded by the fact that these outliers may be invisible to conventional diagnostics. High-breakdown estimation is a procedure for finding estimates that are resistant to serious distortion by a minority of outliers, regardless of their severity. The authors develop a high-breakdown procedure for linear discriminant analysis and give an algorithm for its implementation.

A final study concerns high-dimensional data sets such as the ones in the chemometric applications described in the previous subsection. Discrimination and classification are frequently the objectives of analysis, but singularity of co-variance matrices precludes standard application of linear or quadratic discriminant functions. Krzanowski *et al.* (1995) discuss a range of possible approaches in such cases, all aimed at producing conditions in which the pooled covariance matrix is non-singular so that the linear discriminant function can be used. One line of approach is to reduce the number of variables used to a number that is less than the number of units in the samples, and here the possibilities include the application of the discriminant function to a small number of components selected in a prior analysis such as principal component analysis, partial least squares (see Section B5 below), or modified canonical variate analysis. A second line of approach is to modify the estimation of the pooled dispersion matrix in such a way that the resulting estimate is non-singular, and some options here include modelling the covariances appropriately or using methods analogous to ridge regression. For further details, and applications, of all these methods see the cited paper.

B5 Analysing associations

Manifest-variable models

The ideas of multivariate regression have been set out in Chapter 15. Essentially, the fitting of the regression model by least squares is equivalent to the separate fitting of p regression models, one for each of the dependent variables, and multivariate theory only needs to be specifically brought into play when testing hypotheses about, or constructing confidence regions for, the regression coefficients. The recent increases in computing power have brought with them many new ideas for nonparametric and computationally-intensive fitting of functions to data, and any of these could be used in place of least-squares regression. The area is again a large one, so we merely mention some of the possibilities and provide suitable references for the interested reader to follow up.

A purely data-based approach removes the need for pre-specifying any parametric functional forms (such as regression equations), and simply looks for the best 'smooth function' of the explanatory (X_i) variables to put through the values of the dependent variable Y. The simplest approach is to take the mean of Y replicates at each distinct combination of X_i values and join these means up in a smooth curve. If there is only one Y value at each distinct combination of the X_i then the next best thing is to average Y values that are 'close' in terms of their concomitant X_i. This is the basis of 'locally smoothed' non-parametric regression, and it bears close resemblance to methods of density estimation. We can use either kernel estimators, in which all points within a constant bandwidth of $\boldsymbol{X}$ are averaged to give the predicted Y at $\boldsymbol{X}$, or nearest neighbour estimates, in which the bandwidths vary in such a way as to always average the same number of points to obtain the predicted Y. Cross-validation can be used to determine the optimum bandwidth in the former case, or the optimum number of points to average in the latter. A good discussion of this approach is provided by Altman (1992).

A second general strategy is to estimate the predictive function by separate functions in different portions of the data space, a *piecewise parametric* fittitig procedure. The most flexible approach is to use different low-order polynomials (splines) in the subregions of space that are constructed as combinations of intervals (knots), and allied to this is the imposition of a *roughness penalty* to trade off the smoothness of the fitted function and its faithfulness to the data. Typically, the function parameters are estimated by locally-weighted least squares and the roughness parameters by cross-validation. Green and Silverman (1993) give a comprehensive account of this approach.

A variation on the above procedure is the idea of recursive partitioning regression. Here different functions are again fitted in different regions of the data space, but these regions are now determined in a systematic fashion by a recursive hierarchical splitting algorithm. The splitting is performed sequentially on individual explanatory variables, with a search being made through the data for the optimal sequence of variables and choice of critical values so as to partition the data as successfully as possible. Probably the best known method in this general category is CART ('classification and regression trees', Breiman *et al.* 1984) where the optimal splits and fitting functions are determined simultaneously by cross-validating an index that measures goodness of prediction.

A final variation on standard least-squares regression is necessary either when there is multicollinearity among the explanatory variables (i.e. near-singularity caused by near-linear relationships among them) or when there are more explanatory variables than there are individuals in the sample (such as occurs in the chemometric examples quoted earlier in B4 above). In these situations the matrix $\boldsymbol{X}'\boldsymbol{X}$ that is central to least-squares regression is either singular or near-singular, so that regression estimates are at best highly unstable and at worst are unobtainable. One approach previously adopted for such situations was to conduct a prior principal component analysis on the explanatory variables, and then to regress the dependent variable(s) on as many of the highest-variance principal components as possible whilst retaining non-singularity of regressors. In principle this idea is sound, but in practice there is no guarantee that the highest-variance components are the ones that afford best predictive power (since the principal component analysis involves only the explanatory variables). To overcome this problem, chemometricians have developed the technique of *partial least squares*, which attempts to blend regression and principal component analysis. The idea is that again a relatively few 'components' are selected as predictors, but instead of using variance explained among the X_i as the criterion of choice of components we use a combination of variance explained among the X_i and covariance between the X_i and Y. Much of the early work on this technique was algorithmic and difficult to disentangle, but a good recent account showing the connections with regression has been provided by Garthwaite (1994).

Latent-variable models

This has been an area of particularly vigorous research over the years and it has spawned many specialized texts and monographs, two of the more recent ones being the books by Bollen (1989) and Basilevsky (1994). Inevitably, therefore, the account of this topic in Chapter 16 is a very limited one, focusing just on the basic ideas and techniques and avoiding any of the more complex models. It would again be impossible to cover all developments in a brief update such as this one, so we focus just on three simple objectives: first, to amplify on some of the points raised in Chapter 16 in the light of recent research; second, to indicate some of the strands of development that have occurred in the past decade; and finally to point to some very general models and associated software that have came to play an important role in research and applications in the social sciences.

Two points that have been raised by reviewers of the book concern the calculation of tetrachoric and polychoric correlation coefficients (p. 409) and the occurrence of Heywood cases (p. 490). In the former case, much interest has been shown in the past twenty years in developing methods of calculation and estimation, and maximum likelihood methods are now available for all varieties of these coefficients. Details are provided by

Bollen (1989, pp. 441–5), who also shows the results of applying standard factor analysis methodology to a matrix of polychoric correlations (cf. remarks in Section 16.2.9). As to Heywood cases, many simulation studies have been conducted to investigate their causes and to establish their effects. Bollen (1989, pp. 282–5) summarizes some of these studies and their conclusions, which are essentially that small sample sizes and too small a variable-to-factor ratio frequently lead to negative error variances. One particular study (Anderson and Gerbing 1984) specifically recommends sample sizes in excess of 150 and three or more variables per factor as being necessary to lessen the chances of improper solutions. Indeed, large samples are essential in general for satisfactory results with factor analysis, and this point is often overlooked by users.

The only two latent variable situations which have been considered in Chapter 16 are factor analysis, in which both manifest and latent variables are continuous, and latent-class analysis, in which both manifest and latent variables are categorical. Of course, there is a full 2×2 set of possibilities when both manifest and latent variables can be either continuous or categorical, so we should mention the existence of methods for handling categorical manifest but continuous latent variables ('latent trait analysis') and ones for handling continuous manifest but categorical latent variables ('latent profile analysis'). There is now an extensive literature on all four areas so we cannot hope to go into details here; however, the previously referenced text by Everitt (1984) and also the one by Bartholomew and Knott (1999) will provide excellent introductions to this literature. More recent developments have included the treatment of ordered latent classes (Croon 1990), and latent variable models to deal with specialized forms of data including latent budget analysis (van der Heijden *et al.* 1992) and archetypal analysis (Cutler and Breiman 1994).

Situations often arise in practice in which the manifest variables can be partitioned into two logical groups, with those in one group being considered as explanatory for the values observed in those of the other group. We can try to explain the covariances among the variables in each of these groups by a latent variable model of some sort; if we now wish to build into the overall model an explanation of the variables of one group by the variables of the other, the most logical way of doing it is to postulate a linear model that links the *latent* variables of one group with those of the other. Thus if we denote the two sets of manifest variables by $\boldsymbol{X}$ (the explanatory set) and $\boldsymbol{Y}$ (the dependent set), then the full specification of the above model would have the components

$$\boldsymbol{Y} = \boldsymbol{\Lambda}_y \boldsymbol{\eta} + \boldsymbol{\epsilon},$$
$$\boldsymbol{X} = \boldsymbol{\Lambda}_x \boldsymbol{\xi} + \boldsymbol{\delta},$$

and

$$\boldsymbol{\eta} = \boldsymbol{\Gamma}\boldsymbol{\xi} + \boldsymbol{\zeta},$$

where $\boldsymbol{\eta}$ and $\boldsymbol{\xi}$ are latent vectors, $\boldsymbol{\Lambda}_g$, $\boldsymbol{\Lambda}_x$ and $\boldsymbol{\Gamma}$ are loading/coefficient matrices, and $\boldsymbol{\epsilon}$, $\boldsymbol{\delta}$ and $\boldsymbol{\zeta}$ are residual vectors.

This is a particular example of a *linear structural equation* model. Such models have become very important in social science applications, where they form the basis of what is known as *covariance structure modelling*. The point is that the model has specific implications for the covariances between all pairs of variables, and imposing constraints on the parameters of the model affects the resultant 'fitted' covariances. This in turn means that if distributional assumptions such as normality are made for the data, we can formulate the likelihood of the data in terms of the model parameters and hence derive their maximum likelihood estimates. Moreover, by imposing the relevant constraints we can obtain restricted likelihoods and hence likelihood ratio tests of hypotheses about these parameters. Even if suitable distributional assumptions cannot be made parameters can still be estimated by least-squares fitting of the postulated to the observed covariance matrix, but hypothesis testing is then more problematical.

As can be imagined, much of this area is highly technical. A good general introduction may be found in Everitt and Dunn (1991), while a comprehensive treatment is given by Bollen (1989). Various formulations of model are possible; the two most commonly used are LISREL (Jöreskog and Sorbom 1988) and EQS (Bentler 1989), which are also the acronyms for the computer software that fits the models, provides standard errors of parameter estimates, tests hypotheses about the model parameters, and assesses the goodness of fit of the models.

B6 New directions

Inevitably, given the impetus provided to multivariate research in the past dozen years by the computing revolution, new avenues have opened up and much work has proliferated in areas that scarcely existed at the time of the first edition of the book. To conclude our brief overview, therefore, it seems appropriate to point to a few of these developments as a taster for potential directions in the future.

One consequence of the rapid development of computing power has been the increased emphasis placed on computer-intensive methods for numerical optimisation of an objective function, as this process lies at the heart of many statistical (and more generally mathematical) techniques. Interest in such numerical optimization has led to a number of innovative ideas, including the introduction of random elements into the optimization process. This has resulted in the development of various *stochastic* optimization techniques, and a particular subset of these techniques are ones based on either biological or physical processes. *Neural networks* have found major uses in discrimination and classification. They are based on simulating the way in which the biological nervous system recognizes objects, with messages passing between successive layers of neurons. Good accounts of these techniques are given by Bishop (1995) and Ripley (1996). *Genetic algorithms*, on the other hand, view each candidate solution as an item of a biological population, the variables over which the optimization is being performed as alleles at p ordered sites, and the objective function as the overall 'fitness'. Optimization then proceeds by analogy with populations breeding to maximize fitness. A breeding sample is selected as a random sample with replacement (cf. bootstrapping), and individuals in the breeding sample mate to produce n offspring with each new member inheriting characteristics from its parents according to recognized genetic rules. The process is continued until equilibrium has been reached, at which point the fitness should be at its optimum. Michaelewicz and Janikov (1991) give a good account of the method. Finally, *simulated annealing* is based on an analogy with the annealing process of toughening metals, glass or ceramics by slow cooling, in which atoms make smaller or larger jumps to new positions depending on the temperature of the system. The way that this is built into the optimization process is to start with a standard search method (e.g. steepest ascent/descent) but to allow the process to jump to a *less* optimal value of the objective function at any stage, the probability of such a jump decreasing with the length of time that the optimization has been in operation. For further details of this method see Bertsimas and Tsitsiklis (1993).

The above methods are general numerical ones that can be applied in a variety of problem areas. There have also been developments in various specific areas. One such development has linked statistics with mathematical functional analysis. We have seen in Section B4 above that there is now increasing prevalence of high-dimensional data in which the variables are ordered in terms of some underlying continuum (e.g. wavelength in infrared data, time in repeated measurements, etc). The limiting case of such data is where the interval between each observation shrinks to zero, and each 'individual' becomes a continuous function plotted over some fixed interval of the underlying continuum. We are thus led to analysis of random samples of *functions* from an underlying population, as in the case of EEG or ECG traces in a medical context for example. Analysis will then involve the definition of mean and covariance *functions* over the range of the argument variable, the decomposition of the covariance function into a set of orthogonal eigen*functions*, and the development of techniques analogous to principal component analysis, canonical correlation analysis, and so on for such functions. Estimation of necessary quantities inevitably involves many of the smoothing and nonparametric methods already discussed earlier in this appendix. A full account of this topic may be found in Ramsay and Silverman (1997).

A second very specific focus of development has been into the analysis of the *shape* of individuals from multivariate observations made on them. This topic has a previous history in Biology, with multivariate morphometry (Reyment *et al.* 1984) being concerned with shape itself while allometry has focused on differences in shape associated with size (Sprent 1972). Distance-measures (such as length and height of a crustacean's shell) have typically provided the quantification, and traditional multivariate techniques such as principal component analysis the methodology, in such situations. More recent interest in Medicine, Geology, Geography, and, particularly, Archaeology has opened up a very rich vein of work in which *landmarks* located on each specimen have been used for the quantification, and differences in shape of specimens are then assessed by comparing their corrresponding landmarks. A landmark is basically some natural feature that can be identified easily an each specimen (such as the location of the eye tubercle on a crustacean's shell), although many pseudo-landmarks (such as site of maximum dorsal convexity on a crustacean's shell) also have good operational definition. Since shape is unchanged under translation, rotation or reflection of the specimen, Procrustes methods (pp. 153–67) play a central role in the theory. Indeed, the Procrustes residual sum of squares M^2 between corresponding landmarks is a natural basis for defining distance between two specimens, and this definition then leads on to the formulation of a variety of (non-Euclidean) shape spaces in which the relevant statistical analysis can be conducted. Recourse must be made to sophisticated mathematical techniques such as complex analysis and differential geometry in order to work in these non-Euclidean spaces, but a large body of analytical techniques now exists for handling such data. A comprehensive account of the whole area is provided by Dryden and Mardia (1998).

These few examples should serve to indicate the current level of interest in multivariate research as well as giving a sketch of state of the art in the area. There is no doubt that vigorous developments will continue to be made in the future.

References

Afifi, A. A. and Elashoff, R. M. (1966). Missing observations in multivariate statistics. I: Review of the literature. *J. Amer. Statist. Assoc.*, **61**, 595–604.

Afifi, A. A. and Elashoff, R. M. (1969). Multivariate two-sample tests with dichotomous and continuous variables. I: The location model. *Ann. Math. Statist.*, **40**, 290–8.

Aitchison, J., Habbema, J. D. F., and Kay, J. W. (1977). A critical comparison of two methods of statistical discrimination. *Appl. Statist.*, **26**, 15–25.

Aitkin, M., Anderson, D., and Hinde, J. (1981). Statistical modelling of data on teaching styles (with discussion). *J. Roy. Statist. Soc. A*, **144**, 419–48.

Ali, M. and Silvey, S. D. (1966). A general class of coefficients of divergence of one distribution from another. *J. Roy. Statist. Soc. B*, **28**, 131–42.

Alldredge, J. R. and Gilb, N. S. (1976). Ridge regression: an annotated bibliography. *I.S.I. Rev.*, **44**, 355–60.

Andersen, E. B. (1980). *Discrete statistical models with social science applications*. North-Holland, Amsterdam.

Andersen, E. B. (1982). Latent structure analysis. *Scand. J. Statist.*, **9**, 1–12.

Anderson, A. J. B. (1971). Numeric examination of soil samples. *Math. Geol.*, **3**, 1–14.

Anderson, E. (1960). A semi-graphical method for the analysis of complex problems. *Technometrics*, **2**, 387–92.

Anderson, J. A. (1972). Separate sample logistic discrimination. *Biometrika*, **59**, 19–35.

Anderson, J. A. (1982). Logistic discrimination. In *Handbook of statistics, Volume II: Classification, pattern recognition and reduction of dimensionality*. (eds P. R. Krishnaiah and L. Kanal), pp. 169–91. North-Holland, Amsterdam.

Anderson, T. W. (1951). Classification by multivariate analysis. *Psychometrika*, **16**, 31–50.

Anderson, T. W. (1954). On estimation of parameters in latent structure analysis. *Psychometrika*, **19**, 1–10.

Anderson, T. W. (1963). Asymptotic theory for principal component analysis. *Ann. Math. Statist.*, **34**, 122–48.

Anderson, T. W. (1984). *An introduction to multivariate statistical analysis*. John Wiley, New York.

Andrews, D. F. (1972). Plots of high-dimensional data. *Biometrics*, **28**, 125–36.

Andrews, D. F., Gnanadesikan, R., and Warner, J. L. (1971). Transformations of multivariate data. *Biometrics*, **27**, 825–40.

Andrews, D. F., Gnanadesikan, R., and Warner, J. L. (1973). Methods for assessing multivariate normality. In *Multivariate analysis III* (ed. P. R. Krishnaiah), pp. 95–116. Academic Press, New York.

Arabie, P. (1978). Random versus rational strategies for initial configurations in non-metric multidimensional scaling. *Psychometrika*, **43**, 111–13.

Ashton, E. H., Healy, M. J. R., and Lipton, S. (1957). The descriptive use of discriminant functions in physical anthropology. *Proc. Roy. Soc. B,* **146,** 552–72.

Baker, R. J. and Nelder, J. A. (1978). *The GLIM manual.* Numerical Algorithms Group, Oxford.

Banfield, C. F. and Bassill, S. (1977). A transfer algorithm for non-hierarchical classification. Algorithm AS 113, *Appl. Statist.,* **26,** 206–10.

Barlow, R. E., Bartholomew, D. J., Bremner, J. M., and Brunk, H. M. (1972). *Statistical inference under order restrictions.* John Wiley, London.

Barnett, V. (1974). *Elements of sampling theory.* English Universities Press, London.

Bartholomew, D. J. (1980). Factor analysis for categorical data. *J. Roy. Statist. Soc. B,* **42,** 293–321.

Bartholomew, D. J. (1981). Posterior analysis of the factor model. *Br. J. Math. Statist. Psychol.,* **34,** 93–9.

Bartholomew, D. J. (1983). Latent variable models for ordered categorical data. *J. Econometrics,* **22,** 229–43.

Bartholomew, D. J. (1984). The foundations of factor analysis. *Biometrika,* **71,** 221–33.

Bartholomew, D. J. (1985). Foundations of factor analysis: some practical implications. *Br. J. Math. Statist. Psychol.,* **38,** 1–10. (Discussion: *Br. J. Math. Statist. Psychol.,* **38,** 127–40.)

Bartlett, M. S. (1935). Contingency table interactions. *J. Roy. Statist. Soc. Suppl.,* **2,** 248–52.

Bartlett, M. S. (1937). The statistical conception of mental factors. *Br. J. Psychol.,* **28,** 97–104.

Bartlett, M. S. (1938). Methods of estimating mental factors. *Nature,* **141,** 609–10.

Bartlett, M. S. (1939). The standard errors of discriminant function coefficients. *J. Roy. Statist. Soc. Suppl.,* **6,** 169–73.

Bartlett, M. S. (1947). Multivariate analysis. *J. Roy. Statist. Soc. B,* **9,** 176–97.

Bartlett, M. S. (1951a). The effect of standardisation on an approximation in factor analysis. *Biometrika,* **38,** 337–44.

Bartlett, M. S. (1951b). An inverse matrix adjustment arising in discriminant analysis. *Ann. Math. Statist.,* **22,** 107–11.

Beale, E. M. L., Kendall, M. G., and Mann, D. W. (1967). The discarding of variables in multivariate analysis. *Biometrika,* **54,** 357–66.

Beale, E. M. L. and Little, R. J. A. (1975). Missing values in multivariate analysis. *J. Roy. Statist. Soc. B,* **37,** 129–45.

Belyavin, A. J. (1981). On skewness and kurtosis in multivariate analysis. Unpublished Ph.D. dissertation, University of Reading.

Bennett, J. F. and Hays, W. L. (1960). Multidimensional unfolding: determining the dimensionality of ranked preference data. *Psychometrika,* **25,** 27–43.

Benzecri, J.. P. (1969). Statistical analysis as a tool to make patterns emerge from data. In *Methodologies of pattern recognition* (ed. S. Watanabe), pp. 35–60. Academic Press, New York.

Birch, M. W. (1963). Maximum likelihood in 3-way contingency tables. *J. Roy. Statist. Soc. B,* **25,** 220–33.

Bishop, Y. M. M., Fienberg, S. E., and Holland, P. W. (1975). *Discrete multivariate analysis: theory and practice.* MIT Press, Cambridge, Massachusetts.

Bock, H. H. (1985). On some significance tests in cluster analysis. *J. Classification,* **2,** 77–108.

Bock, R. D. (1975). *Multivariate statistical methods in behavioural research.* McGraw-Hill, New York.

Box, G. E. P. (1949). A general distribution theory for a class of likelihood criteria. *Biometrika,* **36,** 317–46.

Box, G. E. P. and Cox, D. R. (1964). An analysis of transformations (with discussion). *J. Roy. Statist. Soc. B,* **26,** 211–52.

Box, G. E. P., Hunter, W. G. and Hunter, J. S. (1978). *Statistics for experimenters.* John Wiley, New York.

Bradu, D. and Gabriel, K. R. (1978). The biplot as a diagnostic tool for models of two-way tables. *Technometrics,* **20,** 47–68.

Brothwell, D. R. and Krzanowski, W. J. (1974). Evidence of biological differences between early British populations from Neolithic to Medieval times, as revealed by eleven commonly available cranial vault measurements. *J. Archaeolog. Sci.,* **1,** 249–60.

Buck, S. F. (1960). A method of estimation of missing values in multivariate data suitable for use with an electronic computer. *J. Roy. Statist. Soc. B,* **22,** 302–6.

Byth, K. and McLachlan, G. (1980). Logistic regression compared to normal discrimination for non-normal populations. *Aust. J. Statist.,* **22,** 188–96.

Campbell, N. A. (1980a). Shrunken estimators in discriminant and canonical variate analysis. *Appl. Statist.,* **29,** 5–14.

Campbell, N. A. (1980b). Robust procedures in multivariate analysis. I: Robust covariance estimation. *Appl. Statist.,* **29,** 231–37.

Campbell, N. A. (1982). Robust procedures in multivariate analysis. II: Robust canonical variate analysis. *Appl. Statist.,* **31,** 1–8.

Campbell, N. A. (1985). Updating formulae for allocation of individuals. *Appl. Statist.,* **34,** 235–6.

Campbell, N. A. and Atchley, W. R. (1981). The geometry of canonical variate analysis. *Syst. Zool.,* **30,** 268–80.

Carlstein, E., Richards, D., and Ruppert, D. (1985). Letter to the editor. *Amer. Statist.,* **39,** 326–7.

Carroll, J. D. and Chang, J.-J. (1970). Analysis of individual differences in multidimensional scaling via an N-way generalization of the 'Eckart–Young' decomposition. *Psychometrika,* **35,** 283–319.

Chang, W.-C. (1983). On using principal components before separating a mixture of multivariate normal distributions. *Appl. Statist.,* **32,** 276–86.

Chatfield, C. (1985). The initial examination of data (with discussion). *J. Roy. Statist. Soc. A,* **148,** 214–53.

Chatfield, C. and Collins, A. J. (1980). *Introduction to multivariate analysis.* Chapman and Hall, London.

Chauhan, J., Harper, R., and Krzanowski, W. J. (1983). Comparison between direct similarity assessments and descriptive profiles of certain soft drinks. In *Sensory quality in foods and beverages; its definition, measurement and control.* (eds A. A. Williams and R. K. Aitken), pp. 297–309. Ellis Horwood, Chichester.

Chernoff, H. (1973). The use of faces to represent points in k-dimensional space graphically. *J. Amer. Statist. Assoc.*, **68,** 361–8.

Chernoff, H. and Rizvi, M. H. (1975). Effect on classification error of random permutations of features in representing multivariate data by faces. *J. Amer. Statist. Assoc.*, **70,** 548–54.

Christofferson, A. (1975). Factor analysis of dichotomized variables. *Psychometrika,* **40,** 5–32.

Claringbold, P. J. (1958). Multivariate quantal analysis. *J. Roy. Statist. Soc. B,* **20,** 398–405.

Clarke, M. R. B. (1970). A rapidly convergent method for maximum likelihood factor analysis. *Br. J. Math. Statist. Psychol.*, **23,** 43–52.

Cochran, W. G. and Cox, G. M. (1957). *Experimental designs* (2nd edn). John Wiley, New York.

Constantine, A. G. and Gower, J. C. (1978). Graphical representation of asymmetric matrices. *Appl. Statist.*, **27,** 297–304.

Coombs, C. H. (1950). Psychological scaling without a unit of measurement. *Psychological Review,* **57,** 148–58.

Coombs, C. H. (1964). *A theory of data.* John Wiley, New York.

Coombs, C. H. and Kao, R. C. (1960). On a connection between factor analysis and multidimensional unfolding. *Psychometrika,* **25,** 219–31.

Corbet, G. B., Cummins, J., Hedges, S. R., and Krzanowski, W. J. (1970). The taxonomic status of British water voles, genus *Arvicola. J. Zool.*, **161,** 301–16.

Corsten, L. C. A. and Gabriel, K. R. (1976). Graphical exploration in comparing variance matrices. *Biometrics,* **32,** 851–63.

Costanza, M. C. and Afifi, A. A. (1979). Comparison of stopping rules in forward stepwise discriminant analysis. *J. Amer. Statist. Assoc.*, **74,** 777–85.

Courant, R. (1936). *Differential and integral calculus* (*Vol. II*). Blackie and Sons, Glasgow.

Cox, D. R. (1966). Some procedures associated with the logistic qualitative response curve. In *Research papers in statistics*: *Festschrift for J. Neyman.* (ed. F. N. David), pp. 55–71. John Wiley, London.

Cox, D. R. (1970). *Analysis of binary data.* Methuen, London.

Cox, D. R. and Small, N. J. H. (1978). Testing multivariate normality. *Biometrika,* **65,** 263–76.

Coxon, A. P. M. (1982). *The user's guide to multidimensional scaling.* Heinemann, London.

Crawley, D. R. (1979). Logistic discrimination as an alternative to Fisher's linear discriminant function. *N. Z. Statist.*, **14**(2), 21–5.

Cureton, E. E. and D'Agostino, R. B. (1983). *Factor analysis: an applied approach.* Lawrence Erblaum, New Jersey.

Das Gupta, S. (1972). Theories and methods in classification: a review. In *Discriminant analysis and applications* (ed. T. Cacoullos), pp. 77–137. Academic Press, New York.

Day, N. E. and Kerridge, D. F. (1967). A general maximum likelihood discriminant. *Biometrics,* **23,** 313–23.

Dempster, A. P., Laird, N. M., and Rubin, D. B. (1977). Maximum likelihood from incomplete data via the EM algorithm (with discussion). *J. Roy. Statist. Soc. B,* **39,** 1–38.

Devlin, S. J., Gnanadesikan, R., and Kettenring, J. R. (1975). Robust estimation and outlier detection with correlation coefficients. *Biometrika*, **62**, 531–45.

Devlin, S. J., Gnanadesikan, R., and Kettenring, J. R. (1981). Robust estimation of dispersion matrices and principal components. *J. Amer. Statist. Assoc.*, **76**, 354–62.

Dillon, W. R. and Goldstein, M. (1978). On the performance of some multinomial classification rules. *J. Amer. Statist. Assoc.*, **73**, 305–13.

Draper, N. and Smith, H. (1981). *Applied regression analysis* (2nd edn). John Wiley, New York.

Eastment, H. T. and Krzanowski, W. J. (1982). Cross-validatory choice of the number of components from a principal component analysis. *Technometrics*, **24**, 73–7.

Eckart, C. and Young, G. (1936). The approximation of one matrix by another of lower rank. *Psychometrika*, **1**, 211–18.

Efron, B. (1979). Bootstrap methods: another look at the jackknife. *Ann. Statist.*, **7**, 1–26.

Efron, B. (1981). Nonparametric standard errors and confidence intervals. *Canadian J. Statist.*, **9**, 139–72.

Efron, B. (1982). The jackknife, the bootstrap, and other resampling plans. *S.I.A.M.*, Monograph No. 38, Philadelphia.

Efron, B. and Gong, G. (1983). A leisurely look at the bootstrap, the jackknife, and cross-validation. *Amer. Statist.*, **37**, 36–48.

Escofier-Cordier, B. (1969). L'analyse factorielle des correspondances. *Cah. Bur. Univ. Rech. Oper. Univ. Paris*, **13**, 25–59.

Everitt, B. (1977). *The analysis of contingency tables*. Chapman and Hall, London.

Everitt, B. (1978). *Graphical techniques for multivariate data*. Heinemann, London.

Everitt, B. (1980). *Cluster analysis* (2nd edn). Heinemann, London.

Everitt, B. (1984). *An introduction to latent variable models*. Chapman and Hall, London.

Fielding, A. (1977). Latent structure models. In *The analysis of survey data* (*Vol. 1: exploring data structures*), (eds C. A. O'Muircheartaigh and C. Payne), pp. 125–57. John Wiley, Chichester.

Fienberg, S. E. (1981). *The analysis of cross-classified categorical data*. (2nd edn) MIT Press, Cambridge, Massachusetts.

Finden, C. R. and Gordon, A. D. (1985). Obtaining common pruned trees. *J. Classification*, **2**, 255–76.

Fisher, R. A. (1936). The use of multiple measurements in taxonomic problems. *Ann. Eugen.*, **7**, 179–84.

Fisher, R. A. (1938). The statistical utilisation of multiple measurements. *Ann. Eugen.*, **8**, 376–86.

Fisher, R. A. (1940). The precision of discriminant functions. *Ann. Eugen.*, **10**, 422–9.

Flury, B. N. (1983). Some relations between comparison of covariance matrices and principal component analysis. *Computat. Statist. Data Anal.*, **1**, 97–109.

Flury, B. N. (1984). Common principal components in *k* groups. *J. Amer. Statist. Assoc.*, **79**, 892–8.

Flury, B. N. (1985). Analysis of linear combinations with extreme ratios of variance. *J. Amer. Statist. Assoc.*, **80,** 915–22.

Flury, B. N. (1987). Two generalizations of the common principal component model. *Biometrika,* **74,** 59–69.

Fransella, F. (1977). *A manual of repertory grid techniques.* Academic Press, London.

Frets, G. P. (1921). Heredity of head form in man. *Genetica,* **3,** 193–384.

Friedman, H. P. and Rubin, J. (1967). On some invariant criteria for grouping data. *J. Amer. Statist. Assoc.*, **62,** 1159–78.

Friedman, J. H. and Tukey, J. W. (1974). A projection pursuit algorithm for exploratory data analysis. *IEEE Trans. Comput.*, **C-23,** 887–90.

Gabriel, K. R. (1971). The biplot graphical display of matrices with application to principal component analysis. *Biometrika,* **58,** 453–67.

Gabriel, K. R. and Odoroff, C. L. (1986). Use of three-dimensional biplots for diagnosis of models. In *Classification as a tool of research* (eds. W. Gaul and M. Schader), pp. 153–9. Elsevier Science Publishers, North Holland.

Garside, M. J. (1971). Some computational procedures for the best subset problem. *Appl. Statist.*, **20,** 8–15.

Gnanadesikan, R. (1977). *Methods for statistical data analysis of multivariate observations.* John Wiley, New York.

Gnanadesikan, R. and Wilk, M. B. (1969). Data analytic methods in multivariate statistical analysis. In *Multivariate analysis II* (ed. P. R. Krishnaiah), pp. 593–638. Academic Press, New York.

Goldstein, M. and Dillon, W. R. (1978). *Discrete discriminant analysis.* John Wiley, New York.

Golub, G. H., Heath, M., and Wahba, G. (1979). Generalised cross-validation as a method for choosing a good ridge parameter. *Technometrics,* **21,** 215–22.

Golub, G. H. and Reinsch, C. (1970). Singular value decomposition and least squares solutions. *Numerische Mathematik,* **14,** 403–20.

Good, I. J. (1969). Some applications of the singular value decomposition of a matrix. *Technometrics,* **11,** 823–31.

Goodchild, N. A. and Vijayan, K. (1974). Significance tests in plots of multidimensional data in two dimensions. *Biometrics,* **30,** 209–10.

Goodman, L. A. (1974). Exploratory latent structure analysis using both identifiable and unidentifiable models. *Biometrika,* **61,** 215–31.

Gordon, A. D. (1981). *Classification.* Chapman and Hall, London.

Gower, J. C. (1966a). Some distance properties of latent root and vector methods used in multivariate analysis. *Biometrika,* **53,** 325–38.

Gower, J. C. (1966b). A Q-technique for the calculation of canonical variates. *Biometrika,* **53,** 588–9.

Gower, J. C. (1971a). A general coefficient of similarity and some of its properties. *Biometrics,* **27,** 857–72.

Gower, J. C. (1971b). Statistical methods of comparing different multivariate analyses of the same data. In *Mathematics in the archaeological and historical sciences* (eds F. R. Hodson, D. G. Kendall, and P. Tautu), pp. 138–49. University Press, Edinburgh.

Gower, J. C. (1975). Generalized Procrustes analysis. *Psychometrika,* **40,** 33–51.

Gower, J. C. and Banfield, C. F. (1975). Goodness-of-fit criteria for hierarchical classification, and their empirical distributions. In *Proceedings of the 8th*

International Biometric Conference (eds L. C. A. Corsten and T. Postelnicu), pp. 347–61. Editura Academici Republicii Socialiste Romania.

Gower, J. C. and Ross, G. J. S. (1969). Minimum spanning trees and single linkage cluster analysis. *Appl. Statist.*, **18,** 54–64.

Graybill, F. A. (1969). *Introduction to matrices with applications in statistics.* Wadsworth, Belmont, California.

Greenacre, M. (1984). *Theory and applications of correspondence analysis.* Academic Press, Orlando, Florida.

Grime, J. P., Hunt, R., and Krzanowski, W. J. (1987). Evolutionary physiological ecology of plants. In *Evolutionary physiological ecology* (ed. P. Calow), pp. 105–26. Cambridge University Press.

Habbema, J. D. F., Hermans, J., and Remme, J. (1974). A stepwise discriminant analysis program using density estimation. In *Compstat 1974* (eds G. Bruckman, F. Ferschl, and L. Schmetterer), pp. 101–10. Physica-Verlag, Vienna.

Haberman, S. J. (1972). Log-linear fit for contingency tables. Algorithm AS51. *Appl. Statist.*, **21,** 218–25.

Haitovsky, Y. (1968). Missing data in regression analysis. *J. Roy. Statist. Soc. B,* **30,** 67–82.

Halperin, M., Blackwelder, W. C., and Verter, J. I. (1971). Estimation of the multivariate logistic risk function: a comparison of the discriminant function and maximum likelihood approaches. *J. Chron. Dis.*, **24,** 125–8.

Hamilton, M. (1948). Nomogram for the tetrachoric correlation coefficient. *Psychometrika,* **13,** 259–67.

Han, C.-P. (1979). Alternative methods of estimating the likelihood ratio in classification of multivariate normal observations. *Amer. Statist.*, **33,** 204–6.

Hand, D. J. (1981). *Discrimination and classification.* John Wiley, Chichester.

Hand, D. J. (1982). *Kernel discriminant analysis.* Research Studies Press, Letchworth, England.

Hartigan, J. A. (1975). Printer graphics for clustering, *J. Statist. Comput. Simul.*, **4,** 187–213.

Hartigan, J. A. (1985). Statistical theory in clustering. *J. Classification,* **2,** 63–76.

Hawkins, D. M. (1976). The subset problem in multivariate analysis of variance. *J. Roy. Statist. Soc. B,* **38,** 132–9.

Hawkins, D. M. and Eplett, W. R. (1982). The Cholesky factorization of the inverse correlation or covariance matrix in multiple regression. *Technometrics,* **24,** 191–8.

Hayes, S. P. jun. (1946). Diagrams for computing tetrachoric correlation coefficients from percentage differences. *Psychometrika,* **11,** 163–72.

Healy, M. J. R. (1986). *Matrices for statistics.* Oxford University Press.

Hill, M. O. (1973). Reciprocal averaging: an eigenvector method of ordination. *J. Ecol.*, **61,** 237–51.

Hill, M. O. (1974). Correspondence analysis: a neglected multivariate method. *Appl. Statist.*, **23,** 340–54.

Hills, M. (1966). Allocation rules and their error rates (with discussion). *J. Roy. Statist. Soc. B,* **28,** 1–31.

Hills, M. (1967). Discrimination and allocation with discrete data. *Appl. Statist.*, **16,** 237–50.

Hills, M. (1969). On looking at large correlation matrices. *Biometrika,* **56,** 249–53.

Hintze, J. L. (1980). On the use of elemental analysis in multivariate variable selection. *Technometrics,* **22,** 609–12.

Hirschfeld, H. O. (1935). A connection between correlation and contingency. *Proc. Camb. Phil. Soc.,* **31,** 520–4.

Hocking, R. R. (1976). The analysis and selection of variables in linear regression. *Biometrics,* **32,** 1–49.

Hotelling, H. (1933). Analysis of a complex of statistical variables into principal components. *J. Educ. Psychol.,* **24,** 417–41 + 498–520.

Huber, P. J. (1977). *Robust statistical procedures.* SIAM, Philadelphia.

Huber, P. J. (1985). Projection pursuit. *Ann. Statist.,* **13,** 435–75 (with invited discussion, 475–525).

Huynh, H. and Feldt, L. S. (1970). Conditions under which mean squares ratios in repeated measurements designs have exact *F* distributions. *J. Amer. Statist. Assoc.,* **65,** 1582–9.

Jacob, R. J. K. (1983). Investigating the space of Chernoff faces. In *Recent advances in statistics* (eds M. H. Rizvi, J. Rustagi, and D. Siegmund), pp. 449–68. Academic Press, New York.

James, A. T. (1973). The variance information manifold and the functions on it. In *Multivariate analysis III* (ed. P. R. Krishnaiah), pp. 157–69. Academic Press, New York.

Jardine, N. and Sibson, R. (1971). *Mathematical taxonomy.* John Wiley, London.

Jeffers, J. N. R. (1967). Two case studies in the application of principal component analysis. *Appl. Statist.,* **16,** 225–36.

Jewell, N. P. (1986). On the bias of commonly used measures of association for 2 × 2 tables. *Biometrics,* **42,** 351–8.

John, S. (1961). Errors in discrimination. *Ann. Math. Statist.,* **32,** 1125–44.

John, S. (1963). On classification by the statistics R and Z. *Ann. Inst. Statist. Math.,* **14,** 237–46.

Johnson, N. L. and Kotz, S. (1969). *Distributions in statistics: discrete distributions.* John Wiley, New York.

Johnson, N. L. and Kotz, S. (1972). *Distributions in statistics: continuous multivariate distributions.* John Wiley, New York.

Johnson, R. A. and Wichern, D. W. (1982). *Applied multivariate statistical analysis.* Prentice-Hall, New Jersey.

Jolicoeur, P. (1963). The degree of generality of robustness in *Martes americana. Growth,* **27,** 1–27.

Jolliffe, I. T. (1972). Discarding variables in a principal component analysis. I: Artificial data. *Appl. Statist.,* **21,** 160–73.

Jolliffe, I.. T. (1973). Discarding variables in a principal component analysis. II: Real data. *Appl. Statist.,* **22,** 21–31.

Jones, M. C. and Sibson, R. (1987). What is projection pursuit? (with discussion). *J. Roy. Statist. Soc. A,* **150,** 1–36.

Jöreskog, K. G. (1967). Some contributions to maximum likelihood factor analysis. *Psychometrika,* **32,** 443–82.

Jöreskog, K. G. (1977). Factor analysis by least-squares and maximum likelihood methods. In *Statistical methods for digital computers, Vol. 3* (eds K. Enslein, A. Ralston, and H. S. Wilf), pp. 125–53. John Wiley, New York.

Kaiser, H. F. (1958). The varimax criterion for analytical rotation in factor analysis. *Psychometrika,* **23,** 187–200.

Kelker, D. (1970). Distribution theory of spherical distributions and a location-scale parameter generalisation. *Sankhya A,* **32,** 419–30.

Kendall, D. G. (1971). Seriation from abundance matrices. In *Mathematics in the archaeological and historical sciences* (eds F. R. Hodson, D. G. Kendall, and P. Tautu), pp. 215–52. University Press, Edinburgh.

Kendall, M. G. and Stuart, A. (1967). *The advanced theory of statistics, Vol. 2* (2nd edn). Griffin, London.

Kendall, M. G. and Stuart, A. (1969). *The advanced theory of statistics, Vol. 1* (3rd edn). Griffin, London.

Kenward, M. G. (1979). An intuitive approach to the MANOVA test criteria. *The Statistician,* **28,** 193–8.

Kettenring, J. (1971). Canonical analysis of several sets of variables. *Biometrika,* **58,** 433–51.

Khatri, C. G. (1966). A note on a MANOVA model applied to problems in growth curves. *Ann. Inst. Statist. Math.,* **18,** 75–86.

Kihleberg, J. K., Narragon, E. A., and Campbell, B. J. (1964). Automobile crash injury in relation to car size. Cornell Aero. Lab. Report No. VJ-1823-R11. Cornell Aeronautical Laboratory.

Kleiner, B. and Hartigan, J. A. (1981). Representing points in many dimensions by trees and castles. *J. Amer. Statist. Assoc.,* **76,** 260–9.

Knoke, J. D. (1982). Discriminant analysis with discrete and continuous variables. *Biometrics,* **38,** 191–200.

Köster, E. P. (1971). Adaptation and cross-adaptation in olfaction. Thesis, University of Utrecht.

Kraemer, H. C. (1986). A measure of 2×2 association with stable variance and approximately normal small-sample distribution: Planning cost-effective studies. *Biometrics,* **42,** 359–70.

Kruskal, J. B. (1964a). Multidimensional scaling by optimising goodness-of-fit to a nonmetric hypothesis. *Psychometrika,* **29,** 1–27.

Kruskal, J. B. (1964b). Nonmetric multidimensional scaling: a numerical method. *Psychometrika,* **29,** 115–29.

Krzanowski, W. J. (1971a). The algebraic basis of classical multivariate methods. *The Statistician,* **20,** 51–61.

Krzanowski, W. J. (1971b). A comparison of some distance measures applicable to multinomial data, using a rotational fit technique. *Biometrics,* **27,** 1062–8.

Krzanowski, W. J. (1975). Discrimination and classification using both binary and continuous variables. *J. Amer. Statist. Assoc.,* **70,** 782–90.

Krzanowski, W. J. (1977). The performance of Fisher's linear discriminant function under non-optimal conditions. *Technometrics,* **19,** 191–200.

Krzanowski, W. J. (1979). Between groups comparison of principal components. *J. Amer. Statist. Assoc.,* **74,** 703–7 (correction in **76,** 1022).

Krzanowski, W. J. (1980). Mixtures of continuous and categorical variables in discriminant analysis. *Biometrics,* **36,** 493–9.

Krzanowski, W. J. (1982a). Between-group comparison of principal components—some sampling results. *J. Statist. Comput. Simul.,* **15,** 141–54.

Krzanowski, W. J. (1982b). Mixtures of continuous and categorical variables in discriminant analysis: a hypothesis-testing approach. *Biometrics,* **38,** 991–1002.

Krzanowski, W. J. (1983a). Cross-validatory choice in principal component analysis, some sampling results. *J. Statist. Comput. Simul.,* **18,** 299–314.

Krzanowski, W. J. (1983b). Distance between populations using mixed continuous and categorical variables. *Biometrika,* **70,** 235–43.

Krzanowski, W. J. (1984a). Sensitivity of principal components. *J. Roy. Statist. Soc. B,* **46,** 558–63.

Krzanowski, W. J. (1984b). Principal component analysis in the presence of group structure. *Appl. Statist.,* **33,** 164–8.

Krzanowski, W. J. (1986). Multiple discriminant analysis in the presence of mixed continuous and categorical data. *Comp. and Maths. with Appls.,* **12A(2),** 179–85.

Krzanowski, W. J. (1987). Selection of variables to preserve data structure, using principal component analysis. *Appl. Statist.,* **35,** 22–33.

Krzanowski, W. J. and Lai, Y. T. (1988). A criterion for determining the number of groups in a data set using sum-of-squares clustering. *Biometrics* **44,** 23–34.

Kurczynski, T. W. (1970). Generalized distance and discrete variables. *Biometrics,* **26,** 525–34.

Lachenbruch, P. A. (1975). *Discriminant analysis.* Hafner, New York.

Lachenbruch, P. A. and Mickey, M. R. (1968). Estimation of error rates in discriminant analysis. *Technometrics,* **10,** 1–11.

Lachenbruch, P. A., Sneeringer, C., and Revo, L. T. (1973). Robustness of the linear and quadratic discriminant functions to certain types of non-normality. *Commun. Statist.,* **1,** 39–56.

Lancaster, H. O. and Hamdan, M. A. (1964). Estimation of the correlation coefficient in contingency tables with possibly nonmetrical characters. *Psychometrika,* **29,** 383–91.

Lance, G. N. and Williams, W. T. (1966). A generalised sorting strategy for computerised classifications. *Nature,* **212,** 218.

Lawley, D. N. and Maxwell, A. E. (1971). *Factor analysis as a statistical method* (2nd edn). Butterworths, London.

Lazarsfeld, P. F. (1950). The logical and mathematical foundation of latent structure analysis. In *Measurement and prediction* (eds S. A. Stouffer, L. Guttman, E. A. Suchman, P. F. Lazarsfeld, S. A. Star, and J. A. Clausen), pp. 362–412. Princeton University Press.

Lazarsfeld, P. F. (1954). A conceptual introduction to latent structure analysis. In *Mathematical thinking in the social sciences* (ed. P. F. Lazarsfeld), pp. 349–87. The Free Press, Glencoe.

Lazarsfeld, P. F. and Henry, N. W. (1968). *Latent structure analysis.* Houghton-Mifflin, Boston.

Levine, D. (1978). A Monte Carlo study of Kruskal's variance based measure on stress. *Psychometrika,* **43,** 307–15.

Little, R. J. A. and Schluchter, M. D. (1985). Maximum likelihood estimation for mixed continuous and categorical data with missing values. *Biometrika,* **72,** 497–512.

Lord, F. M. and Novick, M. R. (1968). *Statistical theories of mental test scores.* Addison-Wesley, Reading, Mass.

Lubischew, A. A. (1962). On the use of discriminant functions in taxonomy. *Biometrics,* **18,** 455–77.

Lyons, R. (1980). A review of multidimensional scaling. Unpublished M.Sc. dissertation, University of Reading.

McCabe, G. P. (1975). Computations for variable selection in discriminant analysis. *Technometrics,* **17,** 103–9.

McCabe, G. P. (1984). Principal variables. *Technometrics,* **26,** 137–44.

McHenry, C. E. (1978). Computation of a best subset in multivariate analysis. *Appl. Statist.,* **27,** 291–6.

McHugh, R. B. (1956). Efficient estimation and local identification in latent class analysis. *Psychometrika,* **21,** 331–47.

McKay, R. J. (1979). The adequacy of variable subsets in multivariate regression. *Technometrics,* **21,** 475–9.

McKay, R. J. and Campbell, N. A. (1982a). Variable selection techniques in discriminant analysis. I: Description. *Br. J. Math. Statist. Psychol.,* **35,** 1–29.

McKay, R. J. and Campbell, N. A. (1982b). Variable selection techniques in discriminant analysis. II: Allocation. *Br. J. Math. Statist. Psychol.,* **35,** 30–41.

McLachlan, G. J. (1980). On the relationship between the F-test and the overall error rate for variable selection in two-group discriminant analysis. *Biometrics,* **36,** 501–10.

McLachlan, G. J. (1986). Assessing the performance of an allocation rule. *Comp. and Math. with Appls.,* **12A(2),** 261–72.

Mahalanobis, P. C. (1936). On the generalized distance in statistics. *Proc. Nat. Inst. Sci. India,* **2,** 49–55.

Malkovich, J. F. and Afifi, A. A. (1973). On tests for multivariate normality. *J. Amer. Statist. Assoc.,* **68,** 176–9.

Mardia, K. V. (1978). Some properties of classical multidimensional scaling. *Commun. Statist-Theor. Meth.,* **A7,** 1233–41.

Mardia, K. V., Kent, J. T., and Bibby, J. M. (1979). *Multivariate analysis.* Academic Press, London.

Marriott, F. H. C. (1971). Practical problems in a method of cluster analysis. *Biometrics,* **27,** 501–14.

Marriott, F. H. C. (1974). *The interpretation of multiple observations.* Academic Press, London.

Marriott, F. H. C. (1982). Optimisation methods of cluster analysis. *Biometrika,* **69,** 417–22.

Matusita, K. (1964). Distance and decision rules. *Ann. Inst. Statist. Math.,* **16,** 305–15.

Miller, A. J. (1984). Selection of subsets of regression variables (with discussion). *J. Roy. Statist. Soc. A,* **147,** 389–425.

Milligan, G. W. (1981a). A Monte Carlo study of 30 internal criterion measures for cluster analysis. *Psychometrika,* **46,** 187–99.

Milligan, G. W. (1981b). A review of Monte Carlo tests of cluster analysis. *Multiv. Behavioral Res.,* **16,** 379–407.

Milligan, G. W. and Cooper, M. C. (1985). An examination of procedures for determining the number of clusters in a data set. *Psychometrika,* **50,** 159–79.

Mitchell, A. F. S. and Krzanowski, W. J. (1985). The Mahalanobis distance and elliptic distributions. *Biometrika,* **72,** 464–7.

Montgomery, D. C. and Peck, E. A. (1982). *Introduction to linear regression analysis.* John Wiley, New York.

Mood, A. M., Graybill, F. A., and Boes, D. C. (1974). *Introduction to the theory of statistics* (3rd edn). McGraw-Hill Kogakusha, Tokyo.

Moore, D. H. (1973). Evaluation of five discrimination procedures for binary variables. *J. Amer. Statist. Assoc.*, **68,** 399–404.

Moran, M. A. (1975). On the expectation of errors of allocation associated with a linear discriminant function. *Biometrika,* **62,** 141–8.

Moran, M. A. and Murphy, B. J. (1979). A closer look at two alternative methods of statistical discrimination. *Appl. Statist.*, **28,** 223–32.

Morrison, D. F. (1976). *Multivariate statistical methods* (2nd edn). McGraw-Hill, New York.

Mosteller, F. and Wallace, D. L. (1984). *Applied Bayesian and classical inference: the case of the federalist papers.* Springer-Verlag, New York.

Mourant, A. E., Kopec, A. C., and Domaniewska-Sobczak, K. (1958). *The ABO blood groups.* Blackwells, Oxford.

Muirhead, R. J. (1982). *Aspects of multivariate statistical theory.* John Wiley, New York.

Murray, G. D. (1977). A cautionary note on selection of variables in discriminant analysis. *Appl. Statist.*, **26,** 246–50.

Muthen, B. (1978). Contributions to factor analysis of dichotomized variables. *Psychometrika,* **43,** 551–60.

Nelder, J. A. and Wedderburn, R. W. M. (1972). Generalized linear models. *J. Roy. Statist. Soc. A,* **135,** 370-84.

Neuhaus, J. and Wrigley, C. (1954). The quartimax method: an analytical approach to orthogonal simple structure. *Br. J. Statist. Psychol.*, **7,** 81–91.

Okamoto, M. (1963). An asymptotic expansion for the distribution of the linear discriminant function. *Ann. Math. Statist.*, **34,** 1286–301 (correction in **39,** 1358–9).

Olkin, I. and Tate, R. F. (1961). Multivariate correlation models with mixed discrete and continuous variables. *Ann. Math. Statist.*, **32,** 448–65 (correction in **36,** 343–4).

Pearson, K. (1901). On lines and planes of closest fit to a system of points in space. *Phil. Mag.*, **2** (series 6), 559–72.

Pearson, K. (1926). On the coefficient of racial likeness. *Biometrika,* **18,** 105–17.

Potthof, R. F. and Roy, S. N. (1964). A generalized multivariate analysis of variance model useful especially for growth curve problems. *Biometrika,* **51,** 313–26.

Press, S. J. and Wilson, S. (1978). Choosing between logistic regression and discriminant analysis. *J. Amer. Statist. Assoc.*, **73,** 699–705.

Ramsay, J. O. (1982). Some statistical approaches to multidimensional scaling data (with discussion). *J. Roy. Statist. Soc. A,* **145,** 285–312.

Rao, C. R. (1948). Tests of significance in multivariate analysis. *Biometrika,* **35,** 58–79.

Rao, C. R. (1952). *Advanced statistical methods in biometric research.* John Wiley, New York.

Rao, C. R. (1966). Covariance adjustment and related topics in multivariate analysis. In *Multivariate analysis* (ed. P. R. Krishnaiah), pp. 87–103. Academic Press, New York.

Rao, C. R. (1967). Least squares theory using an estimated dispersion matrix and its application to the measurement of signals. *Proc. Fifth Berk. Symp. Math. Statist. Prob.*, **I**, 355–72.

Rao, C. R. (1970). Inference on discriminant function coefficients. In *Essays in probability and statistics* (eds R. C. Bose, I. M. Chakravarti, P. C. Mahalanobis, C. R. Rao, and K. J. C. Smith), pp. 587–602. University of North Carolina Press, Chapel Hill, N.C.

Rao, C. R. (1973). *Linear statistical inference and its applications* (2nd edn). John Wiley, New York.

Rayner, J. C. W. (1985a). Maximum likelihood estimates of $\boldsymbol{\mu}$ and $\boldsymbol{\Sigma}$ for a Normal population. *Amer. Statist.*, **39,** 123–4.

Rayner, J. C. W. (1985b). Letter to the editor. *Amer Statist.*, **39,** 327.

Rayner, J. H. (1966). Classification of soils by numerical methods. *J. Soil Sci.*, **17,** 79–92.

Robert, P. and Escoufier, Y. (1976). A unifying tool for linear multivariate methods: the RV coefficient. *Appl. Statist.*, **25,** 257–65.

Roger, J. H. and Carpenter, R. G. (1971). The cumulative construction of minimum spanning trees. *Appl. Statist.*, **20,** 192–4.

Ross, J. and Cliff, N. (1964). A generalisation of the interpoint distance model. *Psychometrika,* **29,** 167–76.

Rothkopf, E. Z. (1957). A measure of stimulus similarity and errors in some paired-associate learning tasks. *J. Exp. Psychol.*, **53,** 94–101.

Sammon, J. W. (1969). A non-linear mapping for data structure analysis. *IEEE Trans. Comput.*, **C-18,** 401–9.

Schönemann, P. H. (1970). On metric multidimensional unfolding. *Psychometrika,* **35,** 349–65.

Schönemann, P. H. and Carroll, R. M. (1970). Fitting one matrix to another under choice of a central dilation and rigid motion. *Psychometrika,* **35,** 245–55.

Schweder, T. and Spjøtvoll, E. (1982). Plots of p-values to evaluate many tests simultaneously. *Biometrika,* **69,** 493–502.

Scott, A. J. and Symons, M. J. (1971). On the Edwards and Cavalli-Sforza method of cluster analysis. *Biometrics,* **27,** 217–9.

Seal, H. (1964). *Multivariate statistical analysis for biologists.* Methuen, London.

Seber, G. A. F. (1984). *Multivariate observations.* John Wiley, New York.

Shepard, R. N. (1962a). The analysis of proximities: multidimensional scaling with an unknown distance function I. *Psychometrika,* **27,** 125–40.

Shepard, R. N. (1962b). The analysis of proximities: multidimensional scaling with an unknown distance function II. *Psychometrika,* **27,** 219–46.

Shepard, R. N., Romney, A. K. and Nerlove, S. B. (eds) (1972). *Multidimensional scaling. Theory and applications in the behavioural sciences* (Volumes I and II). Seminar Press, New York.

Sibson, R. (1978). Studies in the robustness of multidimensional scaling: Procrustes statistics. *J. Roy. Statist. Soc. B,* **40,** 234–8.

Sibson, R. (1979). Studies in the robustness of multidimensional scaling: perturbational analysis of classical scaling. *J. Roy. Statist. Soc. B,* **41,** 217–29.

Sibson, R. (1984). Present position and potential development: some personal views. Multivariate Analysis. *J. Roy. Statist. Soc. A,* **147,** 198–207.

Slater, P. (ed.) (1976). *The measurement of intrapersonal space by grid technique.* John Wiley, London.

Sneath, P. H. A. and Sokal, R. R. (1973). *Numerical taxonomy.* Freeman, San Francisco.

Spence, I. and Young, F. W. (1978). Monte Carlo studies in nonmetric scaling. *Psychometrika,* **43,** 115–7.

Stone, M. (1974). Cross-validatory choice and assessment of statistical predictions (with discussion). *J. Roy. Statist. Soc. B,* **36,** 111–48.

Sturt, E. (1981). Computerised construction in FORTRAN of a discriminant function for categorical data. *Appl. Statist.,* **30,** 213–22.

Takane, Y., Young, F. W., and de Leeuw, J. (1977). Nonmetric individual differences multidimensional scaling: an alternating least squares method with optimal scaling features. *Psychometrika,* **42,** 7–67.

Thomson, G. H. (1951). *The factorial analysis of human ability.* University Press, London.

Thurstone, L. L. (1947). *Multiple factor analysis.* University Press, Chicago.

Torgerson, W. S. (1958). *Theory and methods of scaling,* John Wiley, London.

Vlachonikolis, I. G. and Marriott, F. H. C. (1982). Discrimination with mixed binary and continuous data. *Appl. Statist.,* **31,** 23–31.

Waddington, P. A. J. (1983). *The training of prison governors.* Croom Helm, London.

Webster, R. and McBratney, A. B. (1981). Soil segment overlap in character space and its implication for soil classification. *J. Soil. Sci.,* **32,** 133–47.

Weinberg, S. L., Carroll, J. D. and Cohen, H. S. (1984). Confidence intervals for INDSCAL using the jackknife and bootstrap techniques. *Psychometrika,* **49,** 475–91.

Welch, B. L. (1939). Note on discriminant functions. *Biometrika,* **31,** 218–20.

Wertz, W. and Schneider, B. (1979). Statistical density estimation: a bibliography. *ISI Rev.,* **47,** 155–75.

Wish, M. and Carroll, J. D. (1971). Multidimensional scaling with differential weighting of dimensions. In *Mathematics in the archaeological and historical sciences* (eds. F. R. Hodson, D. G. Kendall, and P. Tautu), pp. 150–67. University Press, Edinburgh.

Wish, M. and Carroll, J. D. (1982). Multidimensional scaling and its applications. In *Handbook of statistics, Vol. 2* (eds P. R. Krishnaiah and L. Kanal), pp. 317–45. North-Holland, Amsterdam.

Wish, M., Deutsch, M., and Biener, L. (1970). Differences in conceptual structures of nations: an exploratory study. *J. Personality, Social Psychology,* **16,** 361–73.

Wold, S. (1978). Cross-validatory estimation of the number of components in factor and principal component models. *Technometrics,* **20,** 397–405.

Wright, S. (1954). The interpretation of multivariate systems. In *Statistics and mathematics in biology* (eds O. Kempthorne, T. A. Bancroft, J. W. Gowen, and J. L. Lush), pp. 11–33. State University Press, Iowa.

Yule, W., Berger, M., Butler, S., Newham, V., and Tizard, J. (1969). The WPPSI: an empirical evaluation with a British sample. *Br. J. Educ. Psychol.,* **39,** 1–13.

Additional references for Appendix B

Altman, N. S. (1992). An introduction to kernel and nearest-neighbor nonparametric regression. *Amer. Statist.*, **46**, 175–85.

Anderson J. and Gerbing, D. W. (1984). The effects of sampling error on convergence, improper solutions and goodness-of-fit indices for maximum likelihood confirmatory factor analysis. *Psychometrika*, **49**, 155–73.

Bartholomew, D. J. and Knott, M. (1999). *Latent variable models and factor analysis* (2nd edn). Edward Arnold, London.

Basilevsky, A. (1994). *Statistical factor analysis and related methods*. Wiley, New York.

Bentler, P. M. (1989), *EQS. Structural equations manual.* BMDP Statistical Software, Los Angeles, California, USA.

Bertsimas, T. and Tsitsiklis, J. (1993). Simulated annealing. *Statist. Sci.*, **8**, 10–15.

Bishop, C. (1995). *Neural networks for pattern recognition*. Clarendon Press, Oxford.

Bollen, K. A. (1989). *Structural equations with latent variables*. Wiley, New York.

Bolton, R. J. and Krzanowski, W. J. (1999). A characterization of principal components for projection pursuit. *Amer. Statist.*, **53**, 108–9.

Borg, I. and Groenen, P. (1997). *Modern multidimensional scaling, theory and applications*. Springer, New York, USA.

Boswell, M. T., Gore, S. D., Patil. G. P., and Taille, C. (1993). The art of computer generation of random variables. In *Computational statistics* (ed. C. R. Rao), pp. 661–721, Handbook of statistics 9, North Holland, Amsterdam.

Breiman, L., Friedman, J. H., Olshen, R. A., and Stone, C. J. (1984). *Classification and regression trees*. Wadsworth, Belmont, California.

Cleveland, W. S. (1993). *Visualizing data*. Hobart Press, New Jersey.

Cook, D., Buja, A., and Cabrera, J. (1993). Projection pursuit indices based on orthonormal function expansions. *J. Comput. Graph. Statist.*, **2**, 225–50.

Croon, M. (1990). Latent class analysis with ordered classes. *Br. J. Math. Statist. Psychol.*, **43**, 171–92.

Cutler, A. and Breiman, L. (1994). Archetypal analysis. *Technometrics*, **36**, 338–47.

Davison, A. C. and Hinkley. D. V. (1997). *Bootstrap methods and their applications*. Cambridge University Press, Cambridge.

Dryden I. L. and Mardia K. V. (1998). *Statistical shape analysis*. Wiley, Chichester.

Efron, B. and Tibshirani, R. J. (1993). *An introduction to the bootstrap*. Chapman and Hall, London.

Eslava, G. and Marriott, F. H. C. (1994). Some criteria for projection pursuit. *Statist. Comput.*, **4**, 13–20.

Everrit, B. S. and Dunn, G. (1991). *Applied multivariate data anaylsis*. Edward Arnold, London.

Fang, K.-T. and Zhang, Y.-T. (1990). *Generalized multivariate analysis*. Science Press, Beijing, and Springer-Verlag, Berlin.

Flury, B. (1995). Developments in principal component analysis: a review. In *Recent advances in descriptive multivariate analysis* (ed. W. J. Krzanowski), pp. 14–33. Clarendon Press, Oxford.

Friedman, J. H. (1989). Regularized discriminant analysis. *J. Amer. Statist. Assoc.* **84**, 165–75.

Gamerman, D. (1997). *Markov Chain Monte Carlo: stochastic simulation for Bayesian inference*. Chapman and Hall, London.

Garthwaite, P. H. (1994). An interpretation of partial least squares. *J. Amer. Statist.*, **89**, 122–7.

Gifi, A. (1990). *Nonlinear multivariate analysis*. Wiley, New York.

Gilks, W. R., Richardson, S., and Spiegelhalter, D. J. (1996). *Markov Chain Monte Carlo in practice*. Chapman and Hall, London.

Gower, J. C. (1968). Adding a point to vector diagrams in multivariate analysis. *Biometrika*, **55**, 582–5.

Gower, J. C. (1995). A general theory of biplots. In *Recent advances in descriptive multivariate analysis* (ed. W. J. Krzanowski), pp. 283–303. Clarendon Press, Oxford.

Gower, J. C. and Hand, D. J. (1996). *Biplots*. Chapman and Hall, London.

Gower. J. C. and Harding, S. A. (1988). Non-linear biplots. *Biometrika*, **73**, 445–55.

Gower, J. C. and Legendre, P. (1986). Metric and Euclidean properties of dissimilarity coefficients. *J. Classific.*, **3**, 5–48.

Green, P. J. and Silverman, B. W. (1993). *Nonparametric regression and generalized linear models: a roughness penalty approach*. Chapman and Hall, London.

Greenacre, M. J. (1993). *Correspondence analysis in practice*. Academic Press, London.

Hastie, T. and Stuetzle, W. (1989). Principal curves. *J. Amer. Statist. Assoc.*, **84**, 502–16.

Hastie, T., Tibshirani, R., and Buja, A. (1994). Flexible discriminant analysis by optimal scoring. *J. Amer. Statist. Assoc.*, **89**, 1255–70.

Hawkins, D. and McLachlan, G. J. (1997). High-breakdown linear discriminant analysis. *J. Amer. Statist. Assoc.*, **92**, 136–43.

Heiser, W. J. and Meulman J. J. (1995). Nonlinear methods for the analysis of homogeneity and heterogeneity. In *Recent advances in descriptive multivariate analysis* (ed. W. J. Krzanowski), pp. 51–89. Clarendon Press, Oxford.

Jackson, J. E. (1991). *A user's guide to principal components*. Wiley, New York.

Jöreskog, K. G. and Sorbom, D. (1988). *LISREL 7: a guide to the program and applications*. SPSS, Inc., Chicago, Illinois.

Kiers, H. A. L. (1995). Maximisation of sums of quotients of quadratic forms and some generalisations. *Psychometrika*, **60**, 221–45.

Kiers, H. A. L. (1997). Discrimination by means of components that are orthogonal in the data space. *J. Chemomet.*, **11**, 533–45.

Krzanowski, W. J. (1994). Ordination in the presence of group structure, for general multivariate data. *J. Classific.*, **11**, 195–207 (with corrigenda in **12**, p. 12).

Krzanowski, W. J. (1995). Orthogonal canonical variates for discrimination and classification. *J. Chemomet.*, **9**, 509–20.

Krzanowski, W. J. (1998). Subspace projection for multivariate selection problems. *J. Classific.*, **15**, 81–92.

Krzanowski, W. J., Jonathan, P., McCarthy W. V., and Thomas, M. R. (1995). Discriminant analysis with singular covariance matrices: methods and applications to spectroscopic data. *Appl. Statist.*, **44**, 101–15.

Leurgans, S. and Ross, R. T. (1992). Multilinear models: application in spectroscopy (with discussion). *Statist. Sci.*, **7**, 289–319.

McLachlan, G. J. (1992). *Discriminant analysis and statistical pattern recognition.* Wiley, New York.

Meulman, J. J. and Heiser, W. J. (1993). Nonlinear biplots for nonlinear mappings. In *Studies in classification, data analysis and knowledge organization* (ed. O. Opitz, B. Lansen, and R. Klar). Springer-Verlag, Heidelberg.

Michaelewicz, Z. and Janikow, C. Z. (1991). Genetic algorithms for numerical optimization. *Statist. Comput.*, **1**, 75–91.

Nason, G. P. (1995). Three-dimensional projection pursuit. *Appl. Statist.*, **44**, 411–30.

Pinheiro, A. and Vidakovic, B. (1997). Estimating the square root of a density via compactly supported wavelets. *Comput. Statist. Data Anal.*, **25**, 399–415.

Posse, C. (1995). Projection pursuit exploratory data analysis. *Comput. Statist. Data Anal.*, **20**, 669–87.

Ramsay, J. O. and Silverman, B. W. (1997). *Functional data analysis.* Springer, New York.

Rao, C. R. (1982). Diversity and dissimilarity coefficients: a unified approach. *Th. Pop. Biol.*, **21**, 24–43.

Reyment, R. A., Blackith, R. E., and Campbell, N. A. (1984). *Multivariate morphometrics*, 2nd edn. Academic Press, New York.

Ripley, B. D. (1996). *Pattern recognition and neural networks.* Cambridge University Press, Cambridge.

Scott. D. W. (1992). *Multivariate density estimation*. Wiley, New York.

Silverman, B. W. (1986). *Density estimation for statistics and data analysis.* Chapman and Hall, London.

Sprent, P. (1972). The mathematics of size and shape. *Biometrics*, **28**, 23–37.

Swayne, D., Cook, D., and Buja, A. (1997). XGobi: interactive dynamic data vizualization in the X Window system. *J. Comput. Graph. Statist.*, **7**, 113–30.

Tribouley, K. (1995). Practical estimation of multivariate densities using wavelet methods. *Statist. Neerland.*, **49**, 41–62.

van der Heijden, P. G. M., Mooijaart, A., and de Leeuw, J. (1992). Constrained latent budget analysis. In *Sociological methodology 22* (ed. C. C. Clogg), pp. 279–320. Blackwell, Oxford.

Wegman, E. J. and Carr, D. B. (1993). Statistical graphics and visualization. In *Computational statistics* (ed. C. R. Rao), pp. 857–958. Handbook of statistics 9, North Holland, Amsterdam.

Young, F. W., Faldowski, R. A., and McFarlane, M. M. (1993). Multivariate statistical visualization. In *Computational statistics* (ed. C. R. Rao), pp. 959–98. Handbook of statistics 9, North Holland, Amsterdam.

Young, G. A. (1994). Bootstrap: more than a stab in the dark? (with discussion) *Statist. Sci.*, **9**, 382–415.

Index